BROADBAND COMMUNICATIONS
Convergence of Network Technologies

IFIP - The International Federation for Information Processing

IFIP was founded in 1960 under the auspices of UNESCO, following the First World Computer Congress held in Paris the previous year. An umbrella organization for societies working in information processing, IFIP's aim is two-fold: to support information processing within its member countries and to encourage technology transfer to developing nations. As its mission statement clearly states,

IFIP's mission is to be the leading, truly international, apolitical organization which encourages and assists in the development, exploitation and application of information technology for the benefit of all people.

IFIP is a non-profitmaking organization, run almost solely by 2500 volunteers. It operates through a number of technical committees, which organize events and publications. IFIP's events range from an international congress to local seminars, but the most important are:

- The IFIP World Computer Congress, held every second year;
- open conferences;
- working conferences.

The flagship event is the IFIP World Computer Congress, at which both invited and contributed papers are presented. Contributed papers are rigorously refereed and the rejection rate is high.

As with the Congress, participation in the open conferences is open to all and papers may be invited or submitted. Again, submitted papers are stringently refereed.

The working conferences are structured differently. They are usually run by a working group and attendance is small and by invitation only. Their purpose is to create an atmosphere conducive to innovation and development. Refereeing is less rigorous and papers are subjected to extensive group discussion.

Publications arising from IFIP events vary. The papers presented at the IFIP World Computer Congress and at open conferences are published as conference proceedings, while the results of the working conferences are often published as collections of selected and edited papers.

Any national society whose primary activity is in information may apply to become a full member of IFIP, although full membership is restricted to one society per country. Full members are entitled to vote at the annual General Assembly, National societies preferring a less committed involvement may apply for associate or corresponding membership. Associate members enjoy the same benefits as full members, but without voting rights. Corresponding members are not represented in IFIP bodies. Affiliated membership is open to non-national societies, and individual and honorary membership schemes are also offered.

BROADBAND COMMUNICATIONS

Convergence of Network Technologies

IFIP TC6 WG6.2 Fifth International Conference on Broadband Communications (BC '99)
November 10-12, 1999, Hong Kong

Edited by

Danny H. K. Tsang
Hong Kong University of Science and Technology
Hong Kong

Paul J. Kühn
University of Stuttgart - IND
Germany

KLUWER ACADEMIC PUBLISHERS
BOSTON / DORDRECHT / LONDON

Distributors for North, Central and South America:
Kluwer Academic Publishers
101 Philip Drive
Assinippi Park
Norwell, Massachusetts 02061 USA
Telephone (781) 871-6600
Fax (781) 871-6528
E-Mail <kluwer@wkap.com>

Distributors for all other countries:
Kluwer Academic Publishers Group
Distribution Centre
Post Office Box 322
3300 AH Dordrecht, THE NETHERLANDS
Telephone 31 78 6392 392
Fax 31 78 6546 474
E-Mail <services@wkap.nl>

Electronic Services <http://www.wkap.nl>

Library of Congress Cataloging-in-Publication Data has been applied for. The Library of Congress number for this volume is 99-47427.

Printed on acid-free paper.

Printed in the United States of America.

CONTENTS

PREFACE

Today, networking technologies evolve towards a quite heterogeneous scenario where many different technologies co-exist as, e.g., LAN/MAN, PSTN/ISDN, PSDN/Internet, Fixed/Mobile/Satellite, Cable/Fibre/HFC/XDSL, WDM/SDH/ATM etc. This diversity reflects various driving forces resulting out of new services and applications, technological developments, the tremendous growths of the internet and mobile communications and, last but not least, from the competitive environment. While diverging technologies basically characterize the physical and link layers, in the network layer a unifying trend towards IP based (IP: Internet Protocol) protocols is clearly visible. These trends open up a new dimension of questions how these different technologies interoperate or how they complement each other. Is there a convergence towards a unifying concept and what final architecture will result out of it? The outcome of these developments depends on a number of criteria, such as quality of service (QoS), flexibility with respect to new services and applications, scalability, security, manageability and so forth.

Broadband Communications '99 reflects the current state of the art precisely; its scope spans from switch technologies, protocols, performance modeling, traffic control to convergence questions, quality of service, pricing and management. BC '99, the fifth Conference on Broadband Communications supported by Working Group 6.2 of the Technical Committee 6 of IFIP, continues the topics of the previous conferences which were held in Estoril/Portugal in 1992, Paris/France in 1994, Montreal/Canada in 1996, and Stuttgart/Germany in 1998.

The conference theme of BC '99 "Convergence of Network Technologies" has been chosen to reflect exactly the transient phase of the current development. The organizers of BC '99 are thankful to all authors who have contributed by their submitted papers. From a total of 106 submissions about 50 % of papers have been chosen for publication and presentation. The work of the Scientific and Organizing Committees and of a large number of reviewers is greatly appreciated. Special thanks are due to IFIP WG 6.2 for the support, to the Invited and Tutorial Speakers and to the Sponsors of BC '99; without their support and co-operation BC '99 would not have been made possible.

Danny H. K. Tsang — Conference Chairman, HKUST, Hong Kong
Paul J. Kuehn — Conference Co-Chair, University of Stuttgart, Germany

BROADBAND COMMUNICATIONS '99

General Chair:

Danny H. K. Tsang,
Hong Kong University of Science & Technology, Hong Kong

Co-Chair:

Paul J. Kühn,
University of Stuttgart/IND, Stuttgart, Germany

Organizing Committee

Cheng, K. H.	Cable & Wireless HKT, Hong Kong
Kühn, P. J.	University of Stuttgart, Germany
Liu, E. Y. S.	Cable and Wireless plc, UK
Tsang, D. H. K.	Hong Kong University of Science & Technology, Hong Kong

Scientific Programme Committee

Ajmone-Marsan, M.	Politecnico di Torino, Italy
Albanese, A.	International Computer Science Inst., U.S.A.
Bensaou, B.	Centre for Wireless Commun., Singapore
Blondia, B.	University of Antwerp, Belgium
Butscher, B.	DeTeBerkom/GMD, Germany
Casaca, A.	IST/INESC, Portugal
Casals, O.	UPC, Spain
Chandran, S.	Ericsson, Malaysia
Chao, J.	Polytechnic University, U.S.A.
Chen, W. T.	National Tsing Hua, University, Taiwan
Cheng, S.	Southeast University, China
Chu, W.	Open University of Hong Kong, Hong Kong
Costa, B.	CSELT, Italy
Cuthbert, L.	QMW College London, UK
Denzel, W.	IBM Rueschlikon, Switzerland
Drobnik, O.	University of Frankfurt, Germany
Eberspaecher, J.	Technical University of Munich, Germany
El-Zarki, M.	University of Pennsylvania, U.S.A.
Fdida, S.	LIP6 Paris, France
Gallassi, G.	Italtel, Italy
Guerin, R.	University of Pennsylvania, U.S.A.

Gupta, S.	Motorola, U.S.A.
Hébuterne, G.	INT, France
Hubaux, J.-P.	EPFL, Switzerland
Hui, J.	Chinese University of Hong Kong, Hong Kong
Iversen, V. B.	Technical University of Denmark, Denmark
Kawashima, K.	NTT, Japan
Killat, U.	Technical University Hamburg-H., Germany
Koerner, U.	University of Lund, Sweden
Kofman, D.	Télécom Paris, France
Kühn, P. J.	University of Stuttgart, Germany
Leon-Garcia, A.	University of Toronto, Canada
Leslie, I.	University of Cambridge, UK
Li, V. O. K.	Hong Kong University, Hong Kong
Li, X.	Tsinghua University, China
Liang, X. J.	BUPT, China
Lin, X.	Tsinghua University, China
Liu, E. Y. S.	Cable and Wireless plc, UK
Low, S.	University of Melbourne, Australia
Mark, J.	University of Waterloo, Canada
Mason, L.	INRS-Telecommunications, Canada
Miyaho, N.	NTT, Japan
Niu, Z.	Tsinghua University, China
Nunes, M. S.	IST/INESC, Portugal
Pettersen, H.	Telenor R&D, Norway
Rathgeb, E.	Siemens AG/University of Essen, Germany
Roberts, J. W.	CNET, France
Rosenberg, C.	Nortel Imperial College, UK
Ross, K.	EURECOM, France
Saito, H.	NTT, Japan
Stuettgen, H.	NEC Europe Ltd., Germany
Spaniol, O.	RWTH Aachen, Germany
Takahshi, Y.	Nara Institute of Science and Tech., Japan
Tohmé, S.	Télécom Paris, France
Tran-Gia, P.	University of Wuerzburg, Germany
Tsang, D. H. K.	Hong Kong University of Sc. & Tech., Hong Kong
van As, H.	Vienna University of Technology, Austria
Van Landegem, T.	Alcatel, Belgium
Walke, B.	RWTH Aachen, Germany
Wolisz, A.	Technical Univ. of Berlin/GMD Fokus, Germany
Wong, J.	University of Waterloo, Canada
Yang, T.	Ascend Communications, U.S.A.
Zitterbart, M.	Technical Univ. of Braunschweig, Germany

Supporting Organizations

Gold Sponsors:

Cable & Wireless HKT
Sun Microsystems

Sponsors:

Cable & Wireless
Cisco Systems
EDTTC, Vocational Training Centre
Fore Systems
Global Technology Integrator
Hewlett Packard
Hongkong Telecom Institute of Information Technology
IEEE CAS/COM Hong Kong
IT Division, Hong Kong Institution of Engineers
Lucent Technologies – INS
Motorola

REVIEWERS

Ajmone-Marsan, M.
Anelli, P.
Asaka, T.

Banchs, A.
Barelo, J. M.
Bensao, B.
Bettsletter, C.

Blondia, C.
Butscher, B.

Casaca, A.
Casals, O.
Castelli, P.
Cerda, L.
Chandran, S.
Chao, J.
Cheng, S.
Cheung, M.
Chu, W.

Constantinescu, G.
Costa, B.
Cuthbert, L.

Delenze, C.
Denzel, W.
Dracinschi, A.
Drobnik, O.
Dümmler, M.

Eberspächer, J.
Einsiedler, H.
El-Zarki, M.

Fdida, S.
Frings, J.

Gallassi, G.
Glasmann, J.
Goerg, C.
Gupta, S.

Hamdi, M.

Hebuterne, G.
Heier, S.
Heiss, H.
Hettich, A.
Hubaux, J.-P.
Hui, J.

Ikeda, C.
Ishibashi, Y.
Iversen, V. B.

Jocher, P.
Jonas, K.

Kangasharju, J.
Kasahara, S.
Kawashima, H.
Killat, U.
Kind, A.
Koerner, U.
Köhler, S.
Kofman, D.
Koraitim, D.
Kühn, P. J.

Leon-Garcia, A.
Leslie, I.
Leung, Ka-Cheong
Liang, X. J.
Li, Bo
Li, V. O. K.
Li, Xing
Lin, X.
Liu, E. Y. S.
Liu, Xiaoming
Lo Cigno, R.
Lott, M.
Low, S.

Mason, L.
Mark, J.
Misic, J.
Miyaho, N.
Moret, Y.
Muppala, J.

Nakamura, H.
Nicola, V.F.
Ning, Chen Xiang
Niu, Z.
Nunes, M. S.

Peeters, S.
Petit, G.H.
Pettersen, H.

Quan Long Ding

Rathgeb, E.
Roberts, J. W.
Roca, V.
Rong, W.
Rose, O.
Rosenberg, C.
Saito, H.
Shen, Xuemin
Spaniol, O.
Spaey, K.
Stüttgen, H.

Takahashi, Y.
T'Joens, Y.
Tohme, S.
Tran-Gia, P.
Tsang, D. H. K.
Tutschka, K.

Van As, H.
Van Houdt, B.
Van Landegem, T.
Voegel, H.-J.

Walke, B.
Wallmeier, E.
Wen, Zong

Wilcox, P.
Wolisz, A.
Wong, J.
Wu, Yuanqing

Xu, Bangnan

Yang, T.

Zhang, Jianzhu
Zhang, Li
Zheng, Wentao
Zhu, Shihua
Zhuge, Lei
Zitterbart, M.

PLENARY SESSION

Opening and Invited Keynote

DEVELOPMENT AND REGULATORY SYSTEM REFORM OF TELECOMMUNICATION INDUSTRY IN CHINA

LIANG Xiongjian, ZHANG Wei, and ZHANG Xueyuan
College of Management & Humanities
Beijing University of Posts & Telecommunications, P.O.Box 164
10 Xi Tucheng Road, Beijing, China, 100876
E-mail: liangxj@bupt.edu.cn, b94000013@bupt.edu.cn

Abstract The paper reviews achievements and history of telecommunications development and regulatory system reform in China, discusses the evolution of reform, analyses influence of regulatory system reform to China's telecommunication industry, and finally makes some suggestions on the development of China Telecom.

Keywords: Telecommunication, Development, Regulatory System, China

1. DEVELOPMENT OF TELECOMMUNICATION IN CHINA

1.1 Achievements

China has got significant achievement in telecommunications over last about 50 years. In 1998, the switching capacity of the central office exchanges reached 135 million main lines. Some 17.04 million new telephone subscribers got access to the PSTN network, so that telephone subscribers in the whole country totaled 87.35 million. Mobile subscribers got to some 23.56 million, increased by 10.33 million over the year 1997. The telephone density of the whole country was raised to 10.53%, and that for the urban areas to 27.7%. The growth curve of the telephone density of the whole country, shown in Fig.1, can be used to indicate the development of telecommunications in China. From Fig.1, it is found that telecommunications in China has been developing very quickly since 1990.

The developing curve of total turnover of telecommunications services is shown in Fig.2. The total turnover of telecommunications reached 188.89 billion RMB in 1998, increasing by 17 times more than 10.95 billion RMB in 1990. The total turnover of telecommunications has been developing in an exponential curve.

Table 1 shows the development of some telecommunications services and capacity in China. It is also found that subscribers of telecommunications services and telecommunications capacity have been increasing with a high rate since 1990. At the same time, with the development of traditional services and the enlarging of the network size, many new value-added services, such as E-mail, facsimile store and forward, and Internet service, are introduced to subscribers. These new services are welcomed by subscribers and also get a high growth rate. Table 2 shows the development of new services during these years. Since 1994, most of new data services have been developing at the growth rate of over 100%. Especially, Internet subscribers have been increasing at the annual rate of 377.8% since 1995. It makes the telecommunications service market developing in a diversified way.

1.2 Review

In the early days after the People's Republic of China was just set up in 1949, it only held the capacity of switch system with over 310 thousand lines, and 208,750 telephones subscribers. The telephone density is only 0.05%. So Telecommunications, as a foundational industry, was very unstable at that time. But from 1949 to 1977, telecommunications developed very slowly in China because the role of telecommunications in national economy couldn't be recognized correctly. By the end of 1977, the telephone density was only 0.36%, increasing less 0.02% every year.

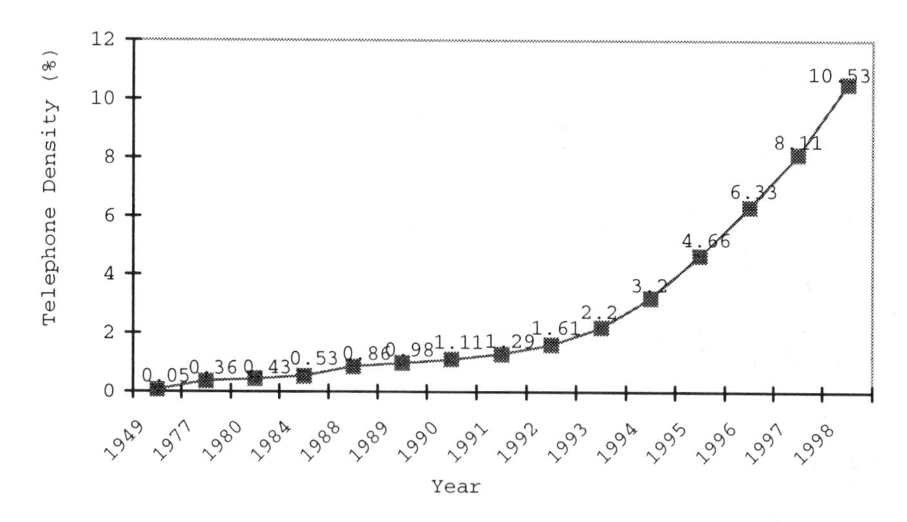

Fig.1. The Development of Telephone Density in China

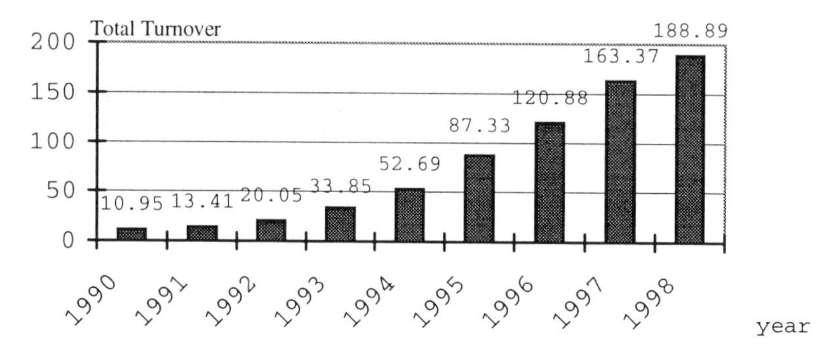

Fig.2. Developing Trend of Total Turnover of Telecommunications in China

Table 1. The Development of Telecommunications Services and Capacity

	unit	1978	1984	1989	1990	1995	1996	1997	1998
Long-distance Telephone Circuits	million	0.019	0.032	0.087	0.1124	0.865	1.04	1.24	1.74
Automatic Toll Switching Capacity	million	0.0019	0.0065	0.103	0.161	3.408	4.26	4.45	4.82
Central Office Exchanges	million	4.06	5.54	10.35	12.01	70.96	93.18	110.9	134.9
Telephone Subscribers	million	1.93	2.77	5.68	6.85	40.71	54.95	70.27	87.35
Mobile Subscribes	million	-	-	-	0.0183	3.63	6.85	13.23	23.56
Paging Subscriber	million	-	-	0.237	0.437	17.43	25.41	34.19	37.83

Since the beginning of reform and opening in China, the significance of telecommunications was recognized gradually. The state government brought forward some policies, such as first-installation fee of telephone, in order to support the development of telecommunications. It was very important to the development of telecommunications afterwards. As a whole, since the reform and opening, the development of telecommunications has undergone three periods in China.

Table 2. The Development of New Value-added Services

	Unit	1994	1995	1996	1997	1998
Packet switching subscribers	thousand	8.5	28	56.4	84.6	106
DDN subscribers	thousand	-	17	51.4	111.7	189
Videotext subscribers	thousand	1.083	2.495	1.987	2.247	1.7
EDI subscribers	thousand	-	0.132	0.113	0.216	0.403
Facsimile storage transfer subscribers	thousand	-	0.495	1.217	2.914	5.258
Internet subscribers (ChinaNet)	thousand	-	7.000	34.000	159.803	680
Frame relay subscribers		-	-	-	3084	8221
N-ISDN subscribers	thousand	-	-	-	0.319	25.533

The first period was from 1978 to 1984. In order to meet the needs of economic development, the State Council offered some special policies on telecommunications field, such as low tax, etc. The Ministry of Posts and Telecommunications (MPT) decided to transfer its first priority to the modernization of telecommunications so that telecommunications better served the modernization of Chinese economy. It ended the long-period slow development of telecommunications in China. During this period, the development of telecommunications insisted on some principles, the unitive control of MPT and specialization development and so on. The first priority problem to deal with was to release the pressure of telephone capacity shortage in big cities. Since then, MPT has begun to charge the first-installation fee of telephone. At the same time, MPT began to actively introduce into foreign fund, modern technologies and management methods to drive the development of telecommunications. In 1982, the first SPC in China was introduced into and installed in Fuzhou, which indicated the beginning of new period of telecommunications.

The second period was from 1984 to 1989. In this period, Chinese economy was rapidly growing. With deepening of reform and opening process and development of national economy, the demands for telecommunications was increasing speedily, which posed great strike on telecommunications of China. The significance of telecommunications was becoming obvious. Telecommunications was recognized as strategic pivot of national economy, and developed preferentially under some favorable policies approved by the State Council. In addition, local governments also supported its development greatly. All of these accelerated the development of telecommunications. It signed the beginning of high-speed takeoff of telecommunications in China.

The third period was from 1989 to now. In this period, telecommunications in China has been developing with great speed. The size of telecommunications network has been quickly enlarged year by year, with advanced technologies. The capacity of telecommunications network has been increasing. Since 1989, MPT has emphasized researches on management, which became urgent to the modernized telecommunications network. MPT began to adjust the network structure, and organized training program to administrative personnel in order to make them prepare in mind and work method for the transition from the manual network to the automated network. Up to 1998, telecommunications network of China has developed into the second largest telecommunications network in size in the world. With the development of network, the management of network began to step into a new period.

2. EVOLUTION OF REGULATORY SYSTEM REFORM OF CHINA'S TELECOMMUNICATION INDUSTRY

Telecommunication industry has been keeping high-speed development for the recent years in China. In this process, customer's demand is increasing, telecommunication technologies are in progress, and telecommunication market is becoming more and more open. All of these changes require adaptive regulatory system. Under the old system, the Ministry of Posts and Telecommunications (MPT) has both government and corporate functions, which results in nonstandard market behaviors and unfair competition. All of this malpractice is hampering the development of telecommunication industry in China. Therefore, it is very urgent to reform the old system.

In 1998, after a long period of argumentation and gestation, China took a key step and made great achievement in the regulatory system reform of telecommunications industry.

Firstly, with the reform of government organization, the Ministry of Information Industry (MII), newly founded on the base of the MPT and the Ministry of Electronic Industry, replaced the former MPT to manage the information industry. The functions of new MII are set as follows: to rebound manufacture of electric information production, communications and software industry to boost the informatization of national economy and social services; to set industry plan, policies and rules; to make overall plans of national communications backbone networks (including local and long-distance telecommunication network), broadcasting and television networks and other private communications networks; to rationally collocate all resource to avoid the reduplicate construction; to guarantee the information safety.

In this definition, we can observe that the new system has some features, different from the old one, as follows:

- The means of management is different. The new MII will control the information industry mainly by industry plans, policies and rules.
- The range of management is different. The range of management of MII will cover manufacture of electric information production, communications and software industry, exceeding the management range of former MPT.

However, the above changes will be realized only on the base of the divergence between government and enterprise functions.

Secondly, the divergence between government and enterprise functions will be really realized under the new regulatory system. Comparing with the functions of MPT ratified by the state council in 1994, we can find that the MII no longer has the function of managing and operating the national public communications network. It shows that the government organization reform will be helpful to realize the divergence of government and enterprise functions. Consequently, a variety of malpractice, such as indefinite property right relation, low interior efficiency and unresponsiveness to market needs, as a result of integration of government and enterprise functions, will be solved gradually.

China Telecom, after the divergence of government and enterprise functions and the separation from posts operation, will become a real corporate with self-operating and self-assuming sole responsibility for its profits or loss. It then can engage in market competition independently. In China, some value-added services have been opened to domestic operators since 1993. Especially in 1994, China Unicom entered into the telecommunications service market and obtained the license of operating the basic telecommunications services. Competition has been spreading in all areas of telecommunications service market in China, and will be more intense.

Thirdly, telecommunication operation will gradually separate from posts operation. In China, telecommunications and posts have been co-operated under MPT for a long period, so telecommunications always provided cross-compensation for posts in order to maintain the development of posts. Telecommunications is a technology-, capital- and brain-dense industry. With the development of telecommunication technologies, there are more and more differences in marketing, operating and management between telecommunications and posts. The old co-operation system begins to tie down the development of telecommunications. What's more, telecommunications enterprises have no their independent economic counting under the old system, which goes against the establishment of a fair, open, normative and ordered market environment. It means that the old co-operation system must be changed.

Currently, the divergence of telecommunications and posts operation is going on pressingly. It is expected to finish by the end of the year 1999.

3. A HISTORICAL REVIEW OF CHINA'S TELECOMMUNICATION REGULATORY SYSTEM REFORM

As an important sector of the national economy, the telecommunication industry was controlled by the government in the last 45 years. On the other hand, due to its unique characteristics of networking and cooperating, the telecommunication industry has to be planned, constructed and operated as a whole to realize large-scale benefits. These reasons caused the monopoly of China's telecommunication industry. The former Ministry of Posts and Telecommunications represented the government authorities for regulating the telecommunication industry and, at the same time, operated a national public telecommunication network and provided telecommunication services just as an enterprise did.

After China began the reform and openness process, the functions of MPT had changed and adjusted twice, which occurred in 1988 and 1994 respectively.

In 1988, the main functions of MPT were as following:

- to set the policies, rules, systems, plan and reform approach of the telecommunication industry and to supervise the performance;
- to regulate the domestic and international postal and telecommunication, to join the international communication organizations representing China and to deal with other issues concerning China's communication industry; to manage the national telecommunication standards and sequence, to coordinate private networks of different departments;
- to manage directly the telecommunication enterprises of the whole nation and to provide telecommunication services.

At that time the regulatory system combining the government and enterprises functions was a result of the country's planning economy. The advantages of this system was that it could collect resources from the whole country so as to be able to plan and build some huge projects and get the large-scale benefits. It had contributed greatly for the development of China's telecommunication industry.

In 1990's, China's economic reform entered a new phase and the telecommunication industry also saw some new changes. From 1993 on, with the approval of the State Council, some selected telecommunication services had been opened to the market. In 1994, China Unicom was founded as the second telecommunication provider in China. The monopoly of China Telecom was broken and competition was firstly introduced into the telecommunication industry.

Under this new circumstance, the State Council approved a new plan in 1994, which defined the functions of MPT. This plan summarized the functions of MPT into three categories:

- ♦ to macro-control the national telecommunication industry and carry on the planning, coordinating, serving and surprising responsibilities.
- ♦ to manage the national public telecommunication network, to ensure the unity, integrity and advance of the network and to regulate the national telecommunication market.
- ♦ to safeguard state benefits and to protect the customer's rights.

The 1994 plan made it clear that MPT was a government department responsible for the telecommunication regulation. This helped to set up the basis for the future divergence of government and enterprise functions. However, due to reasons that the national economic reform is not thorough, the market regulation is not perfect, the legal system is not strengthened and the overall social reform is not ready, the regulatory system reform of China's telecommunication industry had not progressed substantially.

4. INFLUENCE OF REGULATORY SYSTEM REFORM TO CHINA'S TELECOMMUNICATION INDUSTRY

It is no doubt that the system reform of China's telecommunication regulation in 1998 will have a significant effect. As a market with the biggest potential in the world, China is expecting to establish an united and efficient telecommunication regulatory body with coordinated operation, to ensure a fair, standardized and orderly market environment for competition and to set up telecommunication enterprises according with the demand of modern market economy. By these means, the telecommunication industry will become a new growth point of the national economy and find its position in the international market.

Generally speaking, the regulatory system reform has such influences as follow:

- ♦ The divergence between government and enterprise functions will guarantee that the fair competition of telecommunication market is standardized.

Under the conventional system, enterprises competing in the telecommunication market belonged to various government departments respectively. Because the benefits of these various departments were different, the enterprises tended to compete unstandardizedly. That is to say, some government departments would help their own enterprises by misusing their authorities. Therefore, the competition was difficult to be unified and fair. One of the purposes of the reform is to set up a market place of fair competition. With this market place, enterprises have to focus their attention on products, quality, price and services to satisfy their customers in order to survive and develop themselves, hence the competition of these enterprises will be promoted. On the other hand, the fierce competition will eliminate enterprises that did not perform well and then optimize the distribution of resources. This will improve the usage of resources and bring China's telecommunication industry with vitality and vigor in the forth coming 21-century.

♦ MII will regulate the manufacturing industry of electronics information equipment, telecommunication industry, and software industry as well as manage national public telecommunication networks, broadcasting and television networks, and various private telecommunication networks. It complies with the international trend of convergence of information services and gives much more opportunities for China's information industry in its future development.

It is widely known that, since early 1990's, due to the rapid development of technologies within the fields of computer, telecommunication and multimedia, the trend of convergence of various information services has been more and more obvious. In some countries with highly developed information technologies, such as the USA, the government has gradually removed the barriers for enterprises to enter difference services market and set up legal framework to guarantee it with the purpose of promoting the rapid development of information industry.

From the enterprises' point of view, because enterprises which traditionally operated in different industries, for example, telecommunication industry and television industry, are now becoming potential rivals, so the competition will be more fierce both in depth and in extension. However this will also broaden the enterprises vision while they work on their strategies of development.

♦ As a result of the regulatory system reform, it will be much easier to get into the telecommunication market. There will be more enterprises which are interested in the provision of information services joining the competition in telecommunication market. These new comers will challenge the conventional enterprises and the customers will hopefully get more convenient telecommunication services with lower price and better quality.

- The preparation for the opening of China's telecommunication market.

Resulted from the worldwide globalization trend of economy and the effort which China has made to join the WTO, the opening of China's telecommunication market will be inevitable. But China has not proposed a schedule for its telecommunication market opening, mainly because of some specific problems caused by the characteristics of China's telecommunication development, such as subsidiaries between posts and telecommunication enterprises, the unbalanced development among regional areas, the lack of laws and regulations and the preferential policies given by government to telecommunication enterprises, etc. Necessary actives to solve these problems have not been taken.

However, the regulatory system reform is now providing China an opportunity to solve these problems. By means of establishing new regulatory body, separating government functions from enterprises, dividing posts and telecommunications operation and improving laws and regulations, China will gradually find its way out of the difficulties which have blocked the openness of telecommunication market for a long time and merge the Chinese telecommunication market into the international one.

- Influence of the old system will still exist in a period.

Because China has not issued the telecommunication act, the newly established MII finds itself short of a legal foundation to perform the regulation functions of the industry. In this case, it becomes a critical task for MII to draft and promulgate the telecommunication act. But before the establishment of legal framework, the implementation of industry regulation has to use some administrative commands, just as the way it worked in the past. Meanwhile, the old system combining post and telecommunication will encounter lots of problems during the process of reform, which need the coordination of government authorities. Considering all these matters above, we will not expect the influence of old system can be eliminating in a short time. In fact, it will take a relative long time for the new system to replace the old one radically.

5. SUGGESTIONS ON THE DEVELOPMENT OF CHINA TELECOM

Being China's largest telecommunication services provider, China Telecom has enjoyed a growth rate of more than 40% for the last decade. By the end of 1997, China Telecom has built up the world's No.2 wireline telephone network and No.3 mobile telecommunication network and the number of customers has reached 83.5 million. Of course, the regulatory system reform of China's telecommunication industry will bring a significant impact to China Telecom. In short term, it seems China Telecom might suffer from the reform. But if we look at it in the long run, the divergence from government functions and the establishment of modern corporation system will definitely grant China Telecom with remarkable benefits. Here we would like to propose some suggestions for the future development of China Telecom.

- View and take the regulatory system reform of telecommunication industry actively instead of passively. As we just mentioned above, the reform will eventually benefit China Telecom. Think about the opportunities brought by the reform and try to make good use of them.
- China Telecom should convert itself from a product-oriented company to a customer-oriented one. That means China Telecom should pay more attention to the customers' needs, shift its strategies to satisfy the customer and maintain its market share in a competitive environment.
- Taking the opportunity of system reform, China Telecom could restructure its organization. By cutting down some nonproductive departments, adjusting services related departments and setting up some new marketing and customer care departments, China Telecom could optimize its management and get high efficiency.
- While setting up a modern corporation system, China Telecom will take some effective forms of capital constitution, such as share holding system. This will not only give China Telecom a better chance to get finance support from the capital market, but also help it begin its way of capital operation towards the international trend.

We are quite confident that with the system reform and the efforts that China Telecom will make to improve its service quality and customer satisfaction, China Telecom will definitely perform itself successfully and have a bright future.

References

[1] Zhu Younong and Li Guoliang. Economics of Posts & Telecommunications. Beijing: China Economy Publishing House, 1994.
[2] China Posts and Telecommunications 1996 Annual Report. Beijing: MPT, 1997.
[3] China Posts and Telecommunications 1997 Annual Report. Beijing: MII, 1998.
[4] Yu Hui. Regulation System of China's Government. Reform. No.3, pp93-103, 1998.
[5] Zeng Hongjian. Study on the operational mechanism of China's telecommunication enterprises. Proceedings of the Seminar of the Developing Strategy and Management of Communications, Cheng Du, 1995.

SESSION 1

Internet Services

BUFFER AND BANDWIDTH ALLOCATION FOR DIFFSERV CLASSES

Seyyed M-R Mahdavian and Alberto Leon-Garcia
University of Toronto
Dept. of Electrical and Computer Engineering
Toronto, Ontario, Canada M5S 3G4
{seyyed,alg}@nal.utoronto.ca

Abstract This paper proposes an optimal method for allocating buffer and bandwidth to different classes of a forwarding engine within a Differentiated Services domain in the Internet. The optimality criterion is based on the cost of the buffer and bandwidth. Based on this criterion, the best class that matches a certain traffic with a certain statistical characteristics and the maximum packet delay corresponding to this class is found. The results are general and can be applied to other networks such as ATM.

Keywords: Diffserv classes, Resource allocation, Pareto distribution

1. INTRODUCTION

The current Internet 'best effort' service does not provide any guarantee for packet loss or delay and is therefore not suitable for demanding applications such as high quality voice and video. The 'Differentiated Services' model [1], or diffserv, is a recent proposal to solve the above problem and with its simple architecture is particularly appealing for high speed routers carrying a large number of connections.

In diffserv, packets are differentiated based on the contents of the 'DS field' [2]. The DS field of a packet indicates the treatment or the so-called 'per hop behavior' (PHB) that this packet will receive at each diffserv node. Like ATM networks, it is desirable to define different classes of traffic and different loss priorities within each class. Each class and priority can then be associated with a certain PHB. A possible classification of PHB's is described in [3].

This paper focuses on matching different traffic to different classes. Using a simple mathematical model, we obtain a relationship between the characteristics of the traffic and the diffserv class that is optimum for serving this traffic.

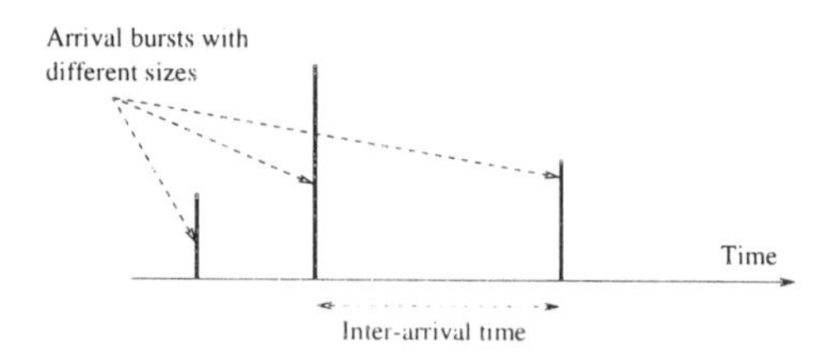

Figure 1 Illustration of the model

The optimality criterion is based on the cost of the buffer and the bandwidth available at a diffserv node. To model the traffic, we use both exponential and Pareto distributions for the burst size and the inter-arrival time. The Pareto distribution is a heavy-tail distribution which can simulate the fractal behavior present in the Internet traffic [4].

The rest of the paper is organized as follows: In section 2 we describe our mathematical model and assumptions. The method used for the analysis of the model is then explained in section 3. In section 4 we discuss and interpret numerical results. Finally in section 5 we summarize our findings.

2. DESCRIPTION OF THE MODEL

We consider a diffserv node allocating a certain amount of bandwidth and buffer to a diffserv class C. It is assumed that each diffserv class has a separate queue for incoming packets. The incoming traffic of class C is modelled as a sequence of bursts which arrive at the node as shown in figure 1. It is assumed that when the first bit of a burst arrives at the node, the whole burst is instantly loaded into a buffer with capacity X, if there is enough space available in the buffer, otherwise the fitting part of the burst is buffered and the excess part is lost. In practice, bursts are not transferred into the buffer instantly; the transfer time is equal to the burst size divided by the arrival rate of the burst. However, given that the average incoming rate of a single class is a small percentage of the whole traffic rate, the peak rate of a burst can be very high as compared to the average incoming rate. Therefore bursts can be considered to arrive almost instantly. This assumption is especially true when the input port of the node is connected to a high speed LAN.

It is furthermore assumed that the size of the burst is a continuous random variable with probability density function $u(x)$, independent of other bursts and the inter-arrival time between bursts. The inter-arrival time is also assumed to be a continuous random variable independent of other random variables. The buffer is depleted at constant rate during the inter-arrival time according to the bandwidth allocated to the class C. One can imagine an equivalent model in which there is an infinite sequence of 'pump up' and 'pump down'

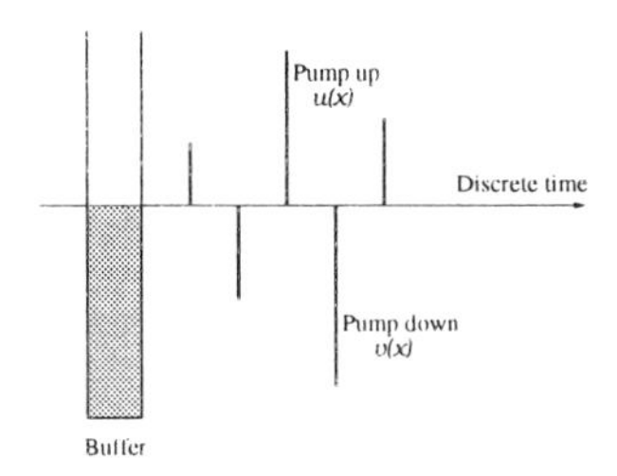

Figure 2 An equivalent representation of the model

cycles taking place at the buffer, as shown in Figure 2. The buffer content is increased or pumped up by a random number with probability density function $u(x)$ (to not more than X, the buffer size,) immediately followed by a pump down or decrease by a random number with probability density function $v(x)$ (to not less than zero) associated with the inter-arrival time and the allocated bandwidth for the class. Our model is therefore similar to the fluid flow model in [5] in that it does not capture the discrete nature of packets. This however has been shown to have negligible effect when the input buffer is large.

Two different distributions for $u(x)$ and $v(x)$ have been studied: The exponential and the Pareto distributions. It has been shown that using the Pareto distribution for the burst size [6] or the packet inter-arrival time [7][8] effectively simulates a self-similar behaviour which has been observed in many practical networks.

3. ANALYSIS METHOD

Given $u(x)$ and $v(x)$ in the model described in section 2, we wish to find the loss rate of incoming bursts. The loss rate is defined as the average of that part of a burst which does not fit into the buffer, divided by the average of a burst size:

$$\text{Loss rate} = \frac{\text{E(loss)}}{\text{E(burst)}} \tag{1}$$

E(loss) and E(burst) are given by:

$$\text{E(burst)} = \int_0^\infty x u(x) dx \tag{2}$$

$$\text{E(loss)} = \int_0^X f(x) E_c(X - x) dx + q E_c(X) \tag{3}$$

where:

$$\begin{aligned} f(x) &= \text{Equilibrium buffer state pdf} \\ E_c(x) &\stackrel{\text{def}}{=} \int_x^{\infty} (y-x)u(y)dy \\ q &= \text{Equilibrium probability of buffer being empty} \\ X &= \text{Buffer size} \end{aligned}$$

with $f(x)$ and q being observed at the time just prior to an incoming burst. Finding $f(x)$ and q is not trivial. They satisfy the following integral equation:

$$f(x) = \int_0^X K(x,y)f(y)dy + qK(x,0) \tag{4}$$

where $K(x,y)$ is the equilibrium pdf of the buffer state just prior to an incoming burst, given that the state of buffer just prior to the previous burst is y. $K(x,y)$ is given by the following formula:

$$K(x,y) = \int_{\max(x,y)}^X u(z-y)v(z-x)dz + v(X-x)\int_{X-y}^{\infty} u(z)dz \tag{5}$$

Equation 4 is a Fredholm integral equation of the second kind [9]. As an approximation, one can assume that the buffer is infinite and that loss happens when the buffer state is greater than X. In this case, Equation 4 becomes similar to the Wiener-Hopf equation [10] for which a closed form solution exits when $u(x)$ and $v(x)$ are both exponential. However, since we are interested in the Pareto distribution as well, we shall focus on a numerical solution to Equation 4.

By sampling $f(x)$ at N discrete points, Equation 4 can be transformed into a set of linear equations:

$$f_i = \sum_{j=1}^{N} A_{ij} f_j + qD_i \tag{6}$$

$$1 = q + \frac{X}{N}\sum_{i=1}^{N} f_i \tag{7}$$

Equations 6 and 7 can then be solved using any linear algebra software package. The coefficients A_{ij} and D_i can be computed using a numerical integration package such as [11]. E(loss) can therefore be approximated using discrete samples of $f(x)$ obtained from Equations 6 and 7. Simulation results showed that for $N = 50$, a fair approximation can be obtained.

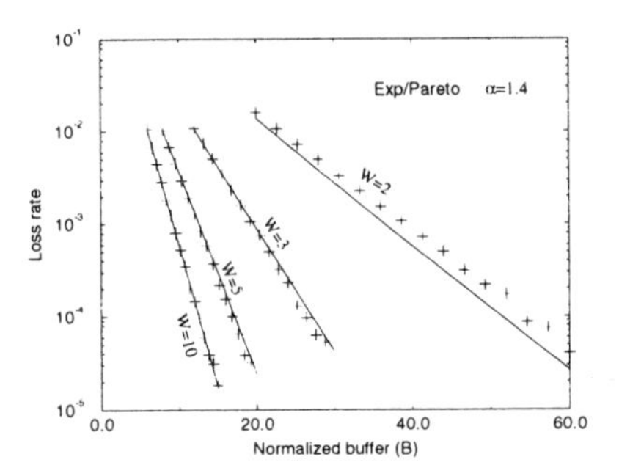

Figure 3 Comparison of the analysis with simulation, Exp/Pareto case

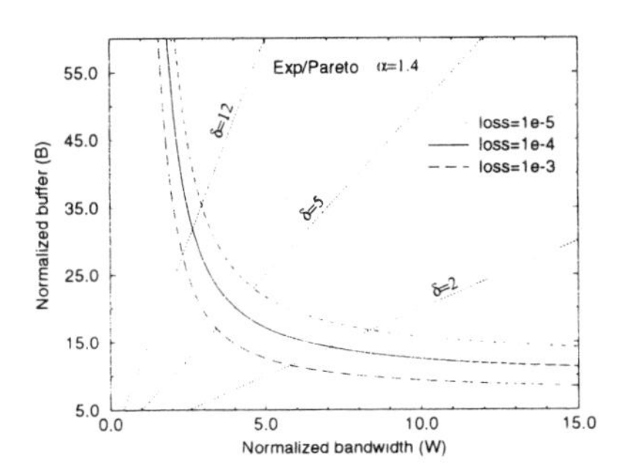

Figure 4 Buffer/Bandwidth tradeoff for different loss rates

4. RESULTS AND DISCUSSION

Consider a class C with a certain allocated bandwidth and buffer. Let B be the *normalized* allocated buffer, i.e. the ratio of the buffer size to the average input burst. Let W be the *normalized* allocated bandwidth, i.e. the ratio of the bandwidth to the average input rate. Given B and W, $u(x)$ and $v(x)$ can be determined for the exponential or Pareto distributions and the loss rate can be computed using the analysis of Section 3. Figure 3 shows the result of this analysis for $u(x)/v(x)$ being Exponential/Pareto and $N = 50$. Marks correspond to simulation results and lines correspond to the analysis.

Given a certain required loss rate, there is a tradeoff between buffer and bandwidth. The tradeoff between buffer and bandwidth has been previously examined in other contexts in the literature [12] [13] [14] [15]. In this paper, we are interested in the effect of this tradeoff on the optimal configuration of a diffserv node.

Figure 4 explicitly shows the tradeoff between buffer and bandwidth for the Exp/Pareto distribution case and for different loss rates. It is not economical

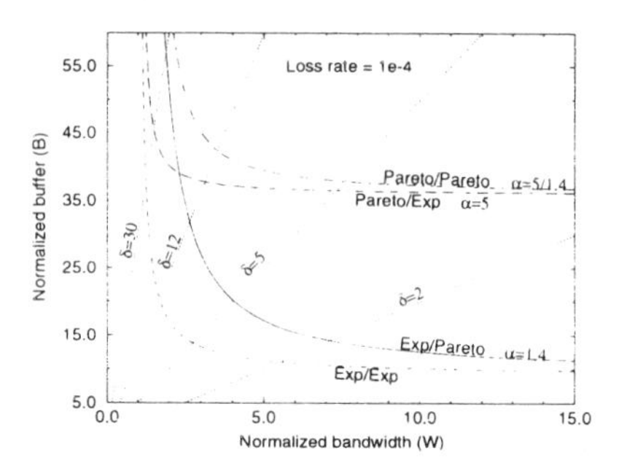

Figure 5 Buffer/Bandwidth tradeoff for different distributions

to push the (W, B) operating point toward the limits. If the buffer is too small, the cost of the bandwidth becomes very high without much gain in the buffer cost. Conversely, if the bandwidth is too low, the cost of the buffer will be too high without gaining much on the cost of the bandwidth. Therefore, the optimal operating point must be around the knee of the $B(W)$ tradeoff curve.

Let d_c be the maximum queueing delay of a diffserv class defined as:

$$d_c \stackrel{\text{def}}{=} \frac{\text{buffer size}}{\text{output rate}}$$

Let d_i be the 'input delay' defined as:

$$d_i \stackrel{\text{def}}{=} \frac{\text{average input burst}}{\text{average input rate}}$$

Then the 'normalized maximum delay' δ is defined as:

$$\delta \stackrel{\text{def}}{=} \frac{d_c}{d_i} = \frac{B}{W} \tag{8}$$

The dotted lines in Figure 4 correspond to different values of the normalized maximum delay δ. This figure shows that for the Exp/Pareto case, the knee of the curve corresponds to the value of $\delta \approx 5$, regardless of the desired loss rate value. Figure 5 compares different distributions with the same loss rate.

The Buffer/Bandwidth optimization can be made more precise. Let $\mathcal{C}$ be the normalized cost of allocating a normalized bandwidth W and a normalized buffer B. i.e.

$$\mathcal{C} = \kappa W + B \tag{9}$$

where κ is the factor by which bandwidth costs more than buffer, i.e.:

$$\kappa = \frac{\text{cost of unit } W}{\text{cost of unit } B} = \frac{\text{cost of unit bandwidth}}{\text{cost of unit buffer}} \times \frac{1}{d_i} \tag{10}$$

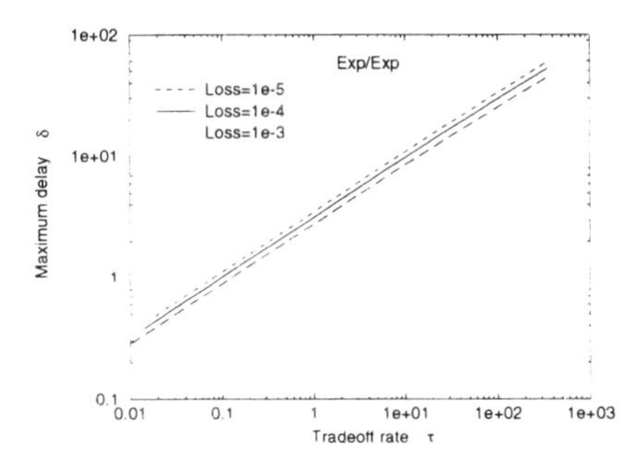

Figure 6 Maximum delay for different loss rates

To find the optimal point, we set the derivative of the cost function to zero:

$$\frac{dC}{dW} = 0 \Rightarrow \kappa + \frac{\partial B}{\partial W} = 0 \Rightarrow \left| \frac{\partial B}{\partial W} \right| = \kappa \tag{11}$$

If the absolute value of the slope of the $B(W)$ tradeoff curve in Figure 4 or 5 is called the 'tradeoff rate' τ, i.e.

$$\tau \stackrel{\text{def}}{=} \left| \frac{\partial B}{\partial W} \right| \tag{12}$$

then Equation 11 shows that the optimal point on the $B(W)$ tradeoff curve is the point where $\tau = \kappa$.

Figure 6 shows τ vs. δ for the Exp/Exp case with different loss rates. It can be seen that different loss rates will cause almost the same delay at a diffserv node. It can also be seen that the $\delta(\tau)$ curve is a straight line in the log/log scale. This has some important implications which we will now investigate.

Suppose that the slope of the line in Figure 6 is equal to $1/(1+a)$ (the slope in this figure is actually 0.5, corresponding to $a = 1$.) Also suppose that the line passes through the point $(\tau = b, \delta = 1)$. Then one can write the equation for this line as follows:

$$\log(\tau) = (1 + a)\log(\delta) + \log(b)$$

Taking out the log and replacing δ and τ from Equations 8 and 12 yields the following differential equation for B as a function of W:

$$\frac{dB}{dW} = -b\left(\frac{B}{W}\right)^{1+a}$$

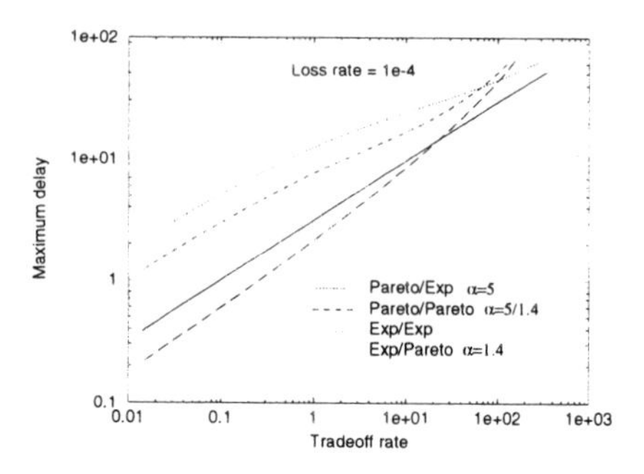

Figure 7 Maximum delay for different distributions

Solving the above differential equation, one obtains the following equation for the $B(W)$ tradeoff curve:

$$B = \frac{\mu W}{(W^a - \lambda^a)^{\frac{1}{a}}} \quad \text{with} \quad \left(\frac{\lambda}{\mu}\right)^a = b \tag{13}$$

λ and μ correspond to the asymptotes of the curves shown in Figure 5. Since the slope of the lines in Figure 6 is 0.5, Equation 13 in this case reduces to the well know bilinear function. λ is always equal to 1 and μ corresponds to the buffer size that gives the desired loss rate assuming that the buffer is completely empty. The $B(W)$ tradeoff curve is therefore readily available for the Exp/Exp case.

The linearity of the $\delta(\tau)$ curve has another important implication with respect to the delay of bursty traffic. Supposed that the burstiness of the traffic is increased by a factor of σ. Then given that the slope of the $\delta(\tau)$ line is $1/(1 + a)$, it turns out that the optimal class delay d_c becomes $\sigma^{\frac{a}{1+a}}$ times larger. For the Exp/Exp case where $a = 1$, the optimal class delay grows as the square root of the traffic burstiness.

The function $\delta(\tau)$ is more complicated in cases other than the Exp/Exp case. Figure 7 compares different distributions for the same loss rate of 10^{-4}. According to this figure, all curves corresponding to different distributions are almost linear with the same slope at low tradeoff rates (when there is plenty of bandwidth available at low cost.) These lines however have a considerable difference in offset. This means that the Buffer/Bandwidth tradeoff curve equation is the same bilinear equation for all distributions near the $B = \mu$ asymptote, with the value of μ itself being considerably different for different distributions.

Different diffserv classes are expected to have different maximum delays and loss probabilities in order to cover a wide range of traffic characteristics and quality of service requirements. Given a certain traffic burstiness, one can use the above analysis to match the diffserv class (and the corresponding set of PHB's) which is 'best' for that traffic from an economical point of view. Assuming that d_i is know for the aggregate traffic as a measure of its burstiness, κ can be determined from Equation 10. Therefore the optimal normalized delay δ can be obtained from a $\delta(\tau)$ curve such as in Figure 7 and the optimal class delay d_c can be computed from Equation 8. Then a class having a maximum delay close to this number can be assigned to that traffic.

The converse problem is a bit more complicated. Here the class delay d_c is known and we are interested in finding the traffic burstiness that best matches this class from an economical point of view. The easiest way to solve this problem is probably by trial and error. Make an initial guess for d_i and insert it in Equation 8. Then use the obtained δ to read the corresponding tradeoff rate τ which can be compared against the optimal value κ obtained from Equation 10 to make an adjustment for the initial guess.

5. CONCLUSION

A simple model for the aggregated traffic of a diffserv class was used to match the traffic with the best buffer/bandwidth configuration allocated to the class. The following results were obtained:

- The value of the desired loss rate has significant impact on the amount of the required buffer and bandwidth, but has little effect on the resulting maximum delay if buffer and bandwidth are configured cost effectively.

- A simple bilinear formula for the buffer/bandwidth tradeoff curve can be derived in the Exp/Exp distribution case. For other cases simple intuitive comparison with the bilinear formula can be made.

- The maximum delay corresponding to the optimal buffer/bandwidth configuration grows as the square root of the burstiness of the traffic if both the burst and the inter-arrival time are assumed to have exponential distribution. For other distributions, the exponent of the growth can be obtained from the slope of the $\delta(\tau)$ curve.

- Matching the traffic with the best class can be done using the $\delta(\tau)$ curve.

References

[1] S. Blake *et al.*, "An Architecture for Differentiated Services," RFC 2475

[2] K. Nichols *et al.*, "Definition of the Differentiated Services Field (DS Field) in the IPv4 and IPv6 Headers," RFC 2474

[3] Juha Heinanen *et al.*, "Assured Forwarding PHB Group," RFC 2597

[4] W. Willinger and V. Paxson, "Where mathematics meets the Internet," *Notices of the American Mathematical Society*, Vol. 45, No. 8, pp. 961-970, Sept. 1998.

[5] D. Anick *et al.*, "Stochastic theory of a data-handling system with multiple sources," *The Bell System Technical Journal*, Vol. 61, No. 8, Oct. 1982, pp. 1871-1894

[6] B. Tsybakov and N. Georganas, "On self-similar traffic in ATM queues: definitions, overflow probability bound, and cell delay distribution," *IEEE ACM Transactions on Networking*, Vol. 5 June '97, pp. 397-409

[7] J. Gordon, "Pareto process as a model of self-similar packet traffic," *IEEE GLOBECOM '95 Proceedings*, Vol. 3, pp. 2232-2236, Singapore 1995.

[8] R. Alexander *et al.*, "Modelling self-similar network traffic," Technical report, Department of Statistics, University of Auckland, http://www.stat.auckland.ac.nz/reports/report95/stat9514.html

[9] G.B. Arfken and H.J. Weber, "Mathematical methods for physicists," San Diego: Academic Press, ©1995

[10] A. Leon-Garcia, "Probability and random processes for electrical engineering," Reading, Mass.: Addison-Wesley, ©1989

[11] Clenshaw-Curtis-Quadrature automatic numerical integration package, http://momonga.t.u-tokyo.ac.jp/~ooura/intcc.html

[12] L. Kleinrock, "The latency/bandwidth tradeoff in gigabit networks," *IEEE Communications Magazine*, Vol. 30, Apr. 1992, pp. 36-40.

[13] S.H. Low, "Equilibrium allocation of variable resources for elastic traffics," *IEEE INFOCOM '98 Proceedings*, Vol. 2, pp. 858-864, San Francisco, USA, 1998.

[14] A. Elwalid *et al.*, "A new approach for allocating buffers and bandwidth to heterogeneous, regulated traffic in an ATM node," *IEEE Journal on Selected Areas in Communications*, Vol. 13, No. 6, August 1995, pp. 1115-1127.

[15] F. Lo Presti *et al.*, "Source time scale and optimal buffer/bandwidth tradeoff for regulated traffic in an ATM node," *IEEE INFOCOM '97 Proceedings*, Vol. 2, pp. 675-682, Kobe, Japan, 1997.

RED+ GATEWAYS FOR IDENTIFICATION AND DISCRIMINATION OF UNFRIENDLY BEST-EFFORT FLOWS IN THE INTERNET

Thomas Ziegler[†‡], Serge Fdida[†], Ulrich Hofmann[‡] [1]

[†] *Université Pierre et Marie Curie, Laboratoire Paris 6, Paris, France*

[‡] *Polytechnic University Salzburg, School for Telecommunications, Salzburg, Austria*

{Thomas.Ziegler, Serge.Fdida}@lip6.fr, {Thomas.Ziegler, Ulrich.Hofmann}@fh-sbg.ac.at

Abstract This paper proposes an add-on to the well known RED (Random Early Detection) algorithm called RED+. RED+ adds the functionality of identifying and discriminating high-bandwidth, unfriendly best-effort flows to RED gateways. It is based on the observation that unfriendly flows have higher arrival rates in times of congestion than friendly flows. Hence unfriendly flows can be discriminated if packets arriving at a router output-port are dropped as a function of RED's average queue size and the arrival rate of the packet's flow. RED+ is scalable regarding the amount per flow information stored in routers as it only allocates per-flow state for the n highest bandwidth flows, where n is a configurable parameter. As shown by simulation, RED+ is able to identify and discriminate unresponsive flows avoiding problems of unfairness and congestion collapse.

Keywords: Congestion Control, unfriendly Flows, RED Gateways, max-min Fairness

1. INTRODUCTION

So called "unfriendly flows" reduce their sending-rate into the net less conservatively in response to congestion-indications (packet-loss) than friendly flows[2]. Consequently, unfriendly flows tend to grab an unfairly high portion of the bottleneck-bandwidth. "Unresponsive flows" exhibit an extreme kind of unfriendly behavior as they do not back-off at all in response to congestion indications. Simulations in [6] show that for a friendly and an unresponsive flow sharing a link, throughput of the friendly flow and arrival-rate of the unresponsive flow are inversely proportional. In other words, throughput of the friendly flow converges to zero if a non responsive flow's arrival-rate approaches the link-bandwidth. This behavior strongly violates the goal of fair distribution of bandwidth among best-effort flows. Additionally, unresponsive flows can cause congestion collapse due congested links transmitting

---

[1] This work is partly sponsored by Telecom Austria and FFF Austria.

[2] A flow is defined by IP address pair and port-numbers, respectively flow-ids.

packets that are only dropped later in the network [6].

For the Internet, TCP flows can be considered friendly as TCP congestion control in its different derivates exhibits roughly homogeneous behavior. Due to the dominance of TCP in today's Internet it is reasonable to define "friendly" as "TCP-friendly" [6]. However, this definition of friendliness may not hold for the future in case of widespread deployment of alternative congestion-control mechanisms.

The objective of this paper is to propose a queue-management mechanism called RED+ adding the functionality of identifying unfriendly flows to RED gateways [7]. Using RED as a basis, RED+ inherits desirable properties of RED like controllable queueing-delay, avoidance of global synchronization and avoidance of a bias against traffic-bursts. Note, however, that this add-on for identification of unfriendly flows is rather orthogonal from a specific queue-management algorithm as it solely requires a packet-drop function which is monotonically increasing with the queue-size and a facility for preferential packet-dropping. Hence the mechanism proposed in this paper could be used in combination with other queue-management algorithms than RED.

Due to the current lack of mechanisms like RED+ users have an incentive to be misbehaving and to generate unfriendly flows obtaining a higher share of the bottleneck bandwidth. Contrary, the deployment of mechanisms identifying and discriminating unfriendly flows would encourage users to utilize conforming end-to-end congestion control.

2. RELATED RESEARCH

The Random-Early-Detection (RED) algorithm [7] employs the parameter-set {*minth, maxth, maxp*} in order to probabilistically drop packets arriving at a router output-port. If the average queue-size (*avg*) is smaller than *minth* no packet is dropped. If $minth < avg < maxth$, RED's packet-drop-probability varies between zero and *maxp*. If $avg > maxth$, each arriving packet is dropped. In order to take into account flows with different packet sizes, RED can be operated in "byte-mode" weighting the drop-probability by the incoming packet's size. WRED [3] and RIO [4], both enhancements of RED intended for service-differentiation in the Internet [1], relate arriving packets to the parameter-set $\{minth_{in}, maxth_{in}, maxp_{in}\}$, respectively $\{minth_{out}, maxth_{out}, maxp_{out}\}$ if the packet has been marked as *in*, respectively *out* according to its flow's service-profile at a network boundary. Assuming $minth_{in} \geq maxth_{out}$, in-profile packets are accommodated while out-of-profile packets have a drop-probability of one if $avg > maxth_{out}$. Hence out-of-profile packets are discriminated against in-profile packets. As opposed to WRED, which uses one average queue size for all packets in the queue, RIO computes an extra average queue-size only for in-profile packets.

In [6] routers execute a low-priority background task in periodic time-intervals. Unfriendly flows are identified as "non TCP friendly", "unresponsive" or "high bandwidth in times of congestion". [18] shows that the TCP-friendly test, as proposed in [6], is inaccurate as routers generally do not have knowledge of the end-to-end RTT. Hence unfriendly flows are unlikely to be detected by this test. Flows identified as unfriendly are discriminated by reclassification into a lower priority queue. Note that reclassifying flows from one queue into another implies the caveat of packet misordering possibly causing TCP fast retransmits and reduction of the congestion window in case more than three packets arrive out-of-order [11] [12].

The FRED algorithm [13] uses per-active-flow accounting and preferentially drops packets of flows having either more packets than a fair-share of the buffer-size stored or an outstanding-high number of packet drops. It has been shown in [18] that FRED is not able to restrict unresponsive flows to the fair-share in rather general scenarios. Additionally, per-active flow accounting means using very small time-scales enabling a bias against bursty traffic.

[20] proposes an architecture in the context of the diff-serv [1] for identification and discrimination of non-TCP-friendly flows. The principle is to detect non-TCP-friendly flows at the ingress-router by comparing arrival rates to equivalent TCP-friendly rates. If a flow is identified as non-TCP-friendly, its packets are marked as "unfriendly". Core routers discriminate packets marked as unfriendly with RED-based drop-preference mechanisms. The RTT is measured by means of a protocol between ingress- and egress routers hence the TCP-friendly test is significantly more accurate than in [6]. Additionally, [20] only requires storage of per-flow state and a flow-lookup in ingress- and egress routers and not in core-routers.

Another idea to identify unfriendly flows has been proposed in [15]. A alternative approach to identification of unfriendly flows is to allocate a fair-share to each flow (see [5], [18] and others). However, merely restricting unfriendly flows to their fair share does not necessarily create an incentive for users to implement end-to-end congestion control. Additionally, many unresponsive best-effort flows restricted to their fair share may still cause congestion collapse as shown in [6].

3. RED+ ALGORITHM

3.1 Principle of RED+

At packet-arrival a flow-lookup in a hash-table storing per-flow state-information is performed. Three counters are updated, measuring the number of bytes arrived per flow, of flows in state established or penalized and of all flows. Subsequently, RED+ determines if the packet's flow is in state "penalized" (i.e. the flow is considered unfriendly by RED+). If this happens to be the case, WRED is executed with a lower parameter-set $(min_l, max_l, maxp_l)$

else WRED is executed with the higher-parameter set $(min_h, max_h, maxp_h)$. Assuming $max_h > min_h > max_l > min_l$ and existence of sufficient demand from non-penalized flows the steady-state average queue-size converges between min_h and max_h, hence the penalized flows have a drop-probability of one. If the average queue size is below max_l in case of moderate congestion RED+ additionally accommodates flows in state penalized.

In order to determine which flows should be penalized RED+ computes an approximation of the max-min fair-share of the link-bandwidth in a periodic background task. We define "one period" as the constant interval of time between subsequent calls of the background task.

The background task additionally computes a flow's state as a function of its arrival rate in comparison to the fair-share. If a flow's arrival rate in times of congestion is higher than the fair share RED+ sets the flow into state "penalized", else the flow is set to state "non-existent, new or established". The transitions between these states are explained in section 3.2.

Figure 1 summarizes the operations of RED+ at packet-arrival and gives a rough overview of the background-task:

Packet arrival:
  Perform flow-lookup
  Update counters for measurement of arrival rates
  If the flow is in state penalized
    execute WRED with lower parameter-set $(min_l, max_l, maxp_l)$
  else
    execute WRED with higher parameter-set $(min_h, max_h, maxp_h)$
Periodic background task:
  Approximate max-min fair share of the link capacity
  Compute the state of flows

Figure 1 *RED+ pseudo-code*

3.2 RED+ State Transitions

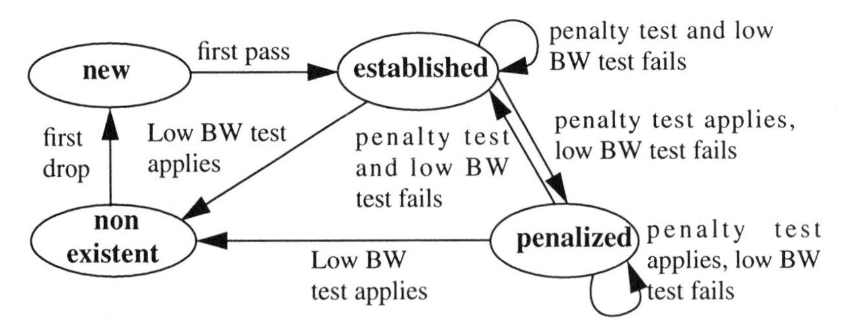

Figure 2 *RED+ state machine*

We examine the states a flow can pass during its lifetime as illustrated in figure 2:

1.) non existent: a flow is in state "non existent" if it has no per-flow-information stored. This state is virtual as RED+ is not aware of this flow.

2.) new: if a flow had its first packet drop and the hash table is not full, per flow state-information (IP Addresses, Port numbers, a counter for measurement of the per-flow arrival rate and 2 bits storing the flow's state) is allocated. The flow's state changes from "non-existent" to "new". The significance of the state new is to avoid false measurement of the per-flow arrival rate which is required for the "low-bandwidth test" and the "penalty test" (see point 3). Both tests compare the number of bytes received during the last period with the fair-share. The storage of flows in the hash-table happens asynchronously to the background task. Hence, if the penalty and the low-bandwidth test were performed when the background task recognizes the flow for the first time, the arrival rate of newly stored flows would be underestimated as their byte counters had been updated for a shorter interval of time than the counters of flows already stored in the hash-table for more than one period. The meaning of the state new is to avoid the problem of underestimation of the per-flow arrival rate for flows stored during the last period. Consequently, the time between a flow's first packet drop and storage in the hash-table and the first appliance of the penalty and low-bandwidth test is greater than one period and smaller than two periods.

3.) established: if the background task scans a flow in state new for the first time the flow's state is changed to "established". For established flows the arrival rate is measured and two test are performed: first, the penalty test determines if the flow's arrival rate since the last call of the background task has been higher than the fair-share. If the penalty test applies the flow's state is changed to "penalized". Second, the low-bandwidth test determines if the arrival rate of a flow is below the fair-share divided by the "*dealloc_param*"[3] or if this flow has the minimum arrival rate of all penalized and established flows (see section 3.3). If the low bandwidth test applies a flow's state changes to non-existent (i.e. its per flow information in the hash-table is deallocated).

4.) penalized: For penalized flows the arrival rate is measured and two test are performed: the penalty test determines if the flow's arrival rate since the last call of the background task has been lower than the fair-share and the flow's state should change to established; the low-bandwidth test is applied as explained above in point 3.

---

[3] The dealloc_param is a constant set to 16 in our simulations. Setting this parameter higher decreases the probability that flow information of unfriendly flows with variable demand is falsely deallocated but increases the probability that unfriendly flows can not be stored immediately if the hash-table is filled to a high degree.

3.3 Details on the Background Task

In order to allow router vendors to restrict the amount of stored per-flow information to fit into the processor-cache, RED+ assumes a fixed, rather small size of the hash-table. This implies, that the maximum number of flows stored in the hash-table will generally be significantly smaller than the total number of flows traversing the output port. Hence we need a mechanism that only keeps the highest bandwidth flows in the hash table and accommodates new flows even if the hash-table is filled to a high degree. RED+ allocates flow-state if a flow experiences a packet-drop and the hash-table is not full. Note that allocating flow state for flows having packet-drops implies that flow state is likely to be stored for the high-bandwidth flows as a flow's drop-probability with RED is directly proportional to its arrival rate [20]. In order to enable storage of new flows when the hash-table is filled to a high degree, the background task deletes at least one flow having the minimum arrival rate of all flows stored in the hash-table in each period (see low-bandwidth test described in section 3.2). Allocating flow-state to flows likely having a high arrival rate (i.e. flows experiencing drops) and deleting flow-state of the lowest-bandwidth flows converges to a "maximum arrival rate allocation of the hash-table" (i.e. only the highest bandwidth-flows are permanently stored in the hash-table). Deleting the flows with the lowest arrival-rate additionally solves the task of deallocating flows which have stopped transmitting packets.

For computation of the max-min fair-share we use a derivate of the iterative mechanism explained in [16]. On the contrary to the mechanism in [16], RED+ only has partial knowledge of the per-flow arrival rates as flows in state non-existent and new are not taken into account. However, this lack of information can be compensated, the max-min fair-share can be computed in all relevant cases and the maximum possible number of unfriendly flows given a certain capacity of the hash-table can be penalized if the algorithm in [16] is initialized differently (see appendix and [19] for details). After initialization, RED+ proceeds with the computation of the max-min fair share like the original mechanism.

In each iteration of the algorithm computing the fair share the whole hash-table has to be scanned. Although only comparisons and additions are required, scanning the hash-table arbitrary times would mean too much overhead for a real implementation. Hence we stop after n iterations (formally speaking, we perform an *n'th* order approximation to the fair-share in the max-min sense). In all our simulations, the max-min fair-share was approximated sufficiently accurate after two iterations (see [19]).

The penalty test seems straight forward: if the arrival-rate of a flow during the last period is greater than the fair share, the flow's state is set to penalized, else the flow's state is set to established. However, by simulation (see [19]) we figured out that this policy would cause global synchronization in scenarios with TCP flows because several TCP conversations would be penalized at one instance in time. Penalizing TCP flows means enforcing a high packet drop rate. The TCP-senders in turn significantly reduce their congestion window (and thereby their sending rate) at the same point in time causing global synchronization. To avoid this problem we only penalize one flow - the established flow with the maximum arrival-rate, and un-penalize another flow - the penalized flow with the minimum arrival rate. Obviously, this implies the drawback of longer convergence times. Consider a scenario with a period of k seconds. At one point in time l unresponsive CBR flows start to send with a rate above the fair share. It will take $k * l$ seconds until all CBR flows are in state penalized.

The periodic background task is invoked in constant time intervals. We have performed simulations with periods between three and eight seconds [19]. A period of 5 seconds seems appropriate in most cases. Obviously, longer time-intervals between calls of the background task cause longer convergence times for detection of multiple unfriendly flows. Shorter time-intervals between calls of the background task cause more load for the router. The background task can be made work-conserving by elongating the time-interval between calls of the background task directly proportional to the total arrival rate at the router output-port.

For detailed explanations of RED+ pseudo-code and discussion of further properties we have to refer to [19] due to space-limitations in this paper.

4. SIMULATION OF RED+

We have implemented the RED+ algorithm in the ns simulator [9], version 2. Simulations in [19] include scenarios investigating different RED+ parameter settings, topologies, multiple congested gateways and different mixes of TCP, CBR and ON/OFF flows. Additionally, scenarios showing RED+ with ECN packet-marking instead of dropping [10][17] for flows in state non-existent, new and established, different bottleneck-link capacities, different RTTs and number of TCP flows are investigated. RED+ succeeds in detection of unfriendly flows in all scenarios. Further simulations show that the max-min fair share is sufficiently accurately approximated if the maximum number of iterations of the algorithm computing the fair-share ("n" parameter) is set to 2.

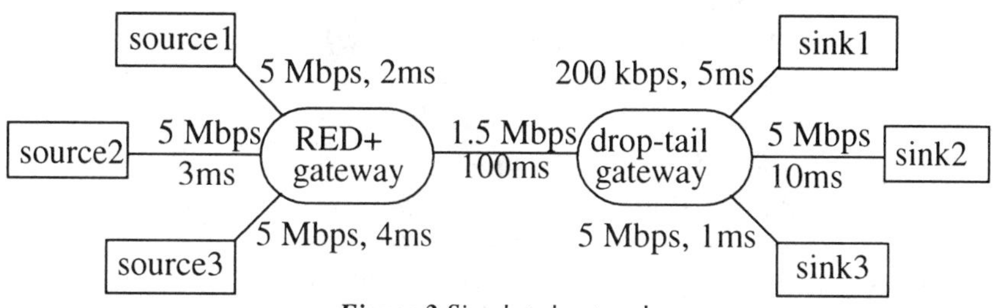

Figure 3 Simulated network

RED+ parameters (see [7] for an explanation of RED-specific parameters): *byte-mode* = false, w_q = 0.002, min_l = 20, max_l = 40, $maxp_l$ = 1, min_h = 50, max_h = 150 packets, $maxp_h$ = 0.1, ECN is disabled, *mean-pktsize* = 500 bytes, *period-length* = 4 sec, *dealloc-param* = 16, n = 2, the hash-table stores state-information of 15 flows at maximum.

Other parameters:Simulation duration: 50 seconds, buffers at router output ports store 200 packets. All output-ports except the output-port at the RED+ gateway served by the 1.5 Mbps link use drop-tail queue management.

Traffic: 3 CBR flows (flows 1,2,3) with rates of 400, 200 and 100 kbps start at 0, 5 and 10 seconds of simulation time. Flow 1 is routed from source 1 to sink1 hence it uses the 200 kbps link. 97 TCP-Reno and TCP-SACK flows transmit packets from sources to sinks and start randomly between zero and 10 seconds. Packet sizes of TCP flows are uniformly distributed with a mean of 500 bytes. None of the flows terminates prior to the simulation.

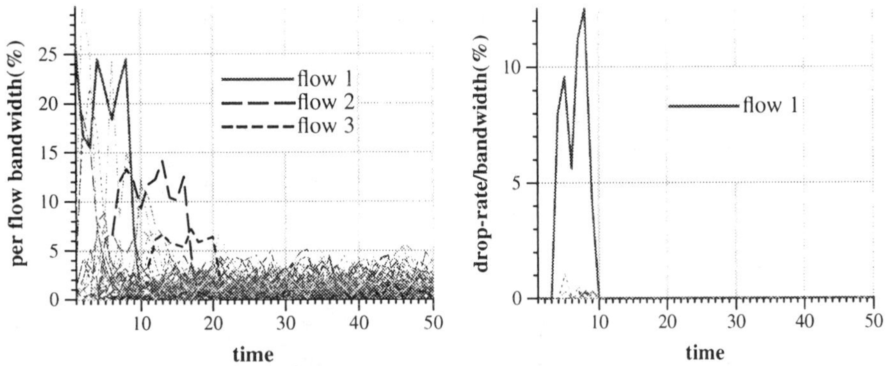

Figure 4 Left part: per-flow bandwidth allocation as a percentage of the bottleneck-capacity between the RED+ and the drop-tail gateway; right part: per-flow droprate at the 200 kbps link as a percentage of the link-capacity between the RED+ and the drop-tail gateway.

As shown in the left part of figure 4, RED+ sets the unresponsive CBR flows into state penalized between 10 and 20 seconds of simulation time, starting with the highest bandwidth flow. CBR flows are detected as unfriendly within 2 periods since their first packet-drop and allocation of flow-state. Due to their unresponsiveness they stay in state penalized and are shut out for the rest of the simulation.

During the first seconds of simulation time flow1 (the CBR flow traversing the 200 kbps link) is not penalized, hence it consumes a significant portion of the link-capacity between the RED+ and the drop-tail gateway and experiences vast packet drops at the second congested link. This behavior causes wastage of bandwidth and may cause congestion-collapse in the extreme case[4] [6]. As soon as flow1 is in state penalized its share of the link-capacity between the RED+ and the drop-tail gateway - and thereby its droprate at the 200 kbps link - equals zero (see figure 4, right part).

The simulation shows that RED+ is able to penalize unfriendly flows in case the total number of flows is significantly larger than the capacity of the hash-table. The hash-table is capable of storing state-information of 15 flows while the total number of flows traversing the RED+ output-port equals 100.

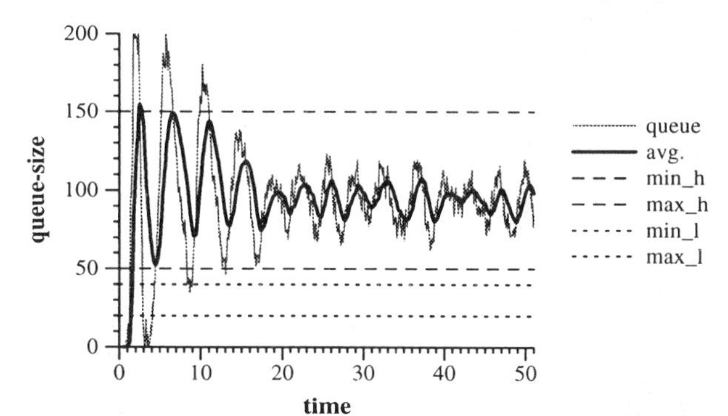

Figure 5 *Average and instantaneous queue-size over time at the RED+ gateway*

Figure 5 shows the reason why flows in state penalized are completely shut out: the average queue size converges between min_h and max_h as there is sufficient demand from CBR and TCP flows. Max_l is smaller than min_h, hence flows in state penalized experience a drop-probability of one.

5. CONCLUSIONS

As shown by simulation, RED+ is able to solve the problem of unfairness and congestion-collapse due to unresponsive best-effort flows. Unresponsive flows are completely shut out once they are detected. Due to this severe form of discrimination the current incentive for users to be misbehaving and to create unresponsive flows would disappear in case RED+ was deployed in the Internet. On the contrary, users would get the desirable incentive to implement conservative end-to-end congestion control.

While providing these functionalities RED+ only stores state-information of a small portion of the flows traversing the router output-port. Additionally, RED+ only requires a few additions (counter updates at packet arrival) besides the flow-lookup and the execution of a drop-preference mechanism (like WRED) in the data-forwarding-path. More complex operations are performed in a periodic background task which can be made work-conserving.

As opposed to [19] where the behavior of RED+ has been extensively investigated (see section 4. for a brief overview), the limited scope of this paper only allows showing a few simulations. [19] additionally shows detailed pseudo-code of the algorithm lacking in this paper.

---

[4] If the arrival rate of flow1 was greater than the link-capacity between the RED+ and the drop-tail gateway, flow1 had a throughput of 200kbps. All other flows were shut out completely.

Discrimination between friendly and unfriendly flows can not be perfect with mechanisms like RED+. Friendly TCP flows may be penalized falsely and unfriendly flows may not be penalized at all if their arrival rate in times of congestion is only marginally higher than the fair share. If a TCP flow is penalized most of its packets are dropped causing significant reduction of its arrival rate at the RED+ gateway. Consequently, the penalization is removed from the flow in the next period. Figuring out the operational bounds of RED+ with regards to falsely penalizing TCP flows and failing to identify unfriendly flows will be the task of a future paper. Additionally, the behavior of RED+ in the presence of unfriendly but responsive flows will be investigated.

References

[1] S. Blake et al., "An Architecture for Differentiated Services", RFC 2475, December 1998
[2] B. Braden, V. Jacobson et al., "Recommendations on Queue Management in the Internet", Internet draft, March 1997
[3] Cisco pages, http://www.cisco.com/warp/public/732/netflow/qos_ds.html
[4] D. Clark, "Explicit Allocation of Best Effort Packet Delivery Service", http://www.ietf.org/html.charters/diffserv-charter.html
[5] A. Demers, S. Keshav, S. Shenker, "Analysis and Simulation of a Fair Queueing Algorithm", Proc. of ACM SIGCOMM, 1989
[6] S. Floyd, K. Fall, "Promoting the Use of End-to-End Congestion Control", Submitted to IEEE/ACM Transactions on Networking, February 1998, http://www.aciri.org/floyd
[7] S. Floyd, V. Jacobson, "Random Early Detection Gateways for Congestion Avoidance", IEEE/ACM Transaction on Networking, August 1993
[8] S. Floyd, V. Jacobson, "On Traffic Phase Effects in Packet Switched Gateways", Computer Communications Review, 1991
[9] NS simulator homepage, http://www-mash.cs.berkeley.edu/ns/
[10]S. Floyd, "TCP and Explicit Congestion Notification", http://www-nrg.ee.lbl.gov/floyd/ecn.html
[11]V. Jacobson, "Congestion Avoidance and Control", Proc. of ACM SIGCOMM, Aug.1988
[12]V. Jacobson, "Modified TCP Congestion Avoidance Algorithm", Message to end2end -interest mailing list, April 1990
[13]D. Lin, R. Morris, "Dynamics of Random Early Detection", Proc. of ACM SIGCOMM, 1997
[14]M. Mathis et al, "The Macroscopic Behavior of the TCP Congestion Avoidance Algorithm", Computer Communications Review, July 1997
[15]T.J. Ott, T.V. Lakshman, L.H. Wong, "SRED: Stabilized RED", Proc. of IEEE INFOCOM, 1999
[16]K. K. Ramakrishnan, D. Chiu, R. Jain, "Congestion Avoidance in Computer Networks with a connectionless Network Layer; Part 4, A selective binary feedback scheme for general topologies", DEC-TR-510, 1987
[17]K.K. Ramakrishnan, S. Floyd, "A Proposal to add Explicit Congestion Notification (ECN) to IP", RFC2491, January 1999

[18]I. Stoica, S. Shenker, H. Zhang, "Core-stateless Fair Queueing: achieving approximately fair Bandwidth-Allocations in High-Speed Networks", Proc. of ACM SIGCOMM, 1998
[19]T. Ziegler, U. Hofmann, S. Fdida, "RED+ Gateways for detection and discrimination of unresponsive flows", Technical Report, December 1998, http://www-rp.lip6.fr/InfosTheme/Anglais/publicationsan.htm
[20]T. Ziegler, S. Fdida, U. Hofmann, "A distributed Mechanism for detection and discrimination of non-TCP-friendly Flows in the Internet", February 1999, http://www-rp.lip6.fr/InfosTheme/Anglais/publicationsan.htm

APPENDIX

[16] proposes an iterative mechanism computing the max-min fair-share in iteration-step i (A^i_{fair}) as follows:

$$A^i_{fair} = \frac{B - d^i_{low}}{h^i}$$

B denotes the link-capacity, d^i_{low} denotes the sum of the arrivalrates of the flows having arrival rates below or equal A^i_{fair}, h^i denotes the number of flows having arrival rates higher than A^i_{fair}. The d_{low} quantity is initialized to zero, h is initialized to the total number of flows. If $h^i = h^{i-1}$, A^i_{fair} equals the max-min fair share (A_{fair}) and the iteration can be terminated. A max-min fair allocation of B fully satisfies the demand of flows having arrival rates below A_{fair} and restricts flows having arrivalrates above A_{fair} to A_{fair}.

As mentioned in section 3.1, RED+ measures the per-flow arrival rate of flows in state penalized or established, the total arrival rate of flows in state penalized or established (λ_{ep}) and the total arrival rate of all flows (λ_{all}). For computation of the fair-share, RED+ initializes the d_{low} quantity to the total arrival rate of flows in state new or non existent, λ_{all} - λ_{ep}; h is initialized to the number of flows in state penalized or established. After the initialization RED+ continues with the computation of the max-min fair share as explained above for the original mechanism.

For the following considerations we assume that the process of maximum arrival rate allocation (see section 3.3, first paragraph) is in steady state, i.e. only the highest bandwidth flows are stored in the hash-table. The n parameter limiting the numbers of iterations of the algorithm computing the fair share, is assumed to be infinity.

Theorem: RED+ penalizes the maximum possible number of flows having arrival rates above the max-min fair share, given a certain size of the hash-table.

Explanation: Let m denote the maximum number of flows which can be permanently stored in the hash-table. We distinguish between two cases and show that in the first case exactly the flows having arrival rates above the max-min fair share are penalized, in the second case the maximum possible number of flows above the max-min fair share (i.e. m flows) is penalized.

First case: the number of flows with arrival rates higher than the max-min fair-share is smaller than or equal m. Under the assumption of maximum arrival rate allocation, the arrival rate of each flow in state non-existent or new has to be below the max-min fair-share in this case. The arrival rate of flows in state non-existent or new would contribute to d_{low} in the original mechanism, hence we may initialize d_{low} to the total arrival rate of flows in state new or non existent, λ_{all} - λ_{ep}. As only flows in state established or penalized remain to be considered h can be initialized to the number of flows in state penalized or established. After the algorithm has terminated, the max-min fair share computed by RED+ equals the value computed by the original mechanism with knowledge of all per-flow arrivalrates.

Second, inverse case: there are flows in state non-existent or new having an arrival rate above the max-min fair share. As convergence to a maximally arrival rate allocation of the hash table has been achieved, we know that any flow in state established or penalized has a higher arrival-rate than any flow in state non-existent or new. Hence all flows in state penalized or established have to have arrivalrates above the max-min fair share either.

The total portion of the link bandwidth RED+'s derivate for computation of the max-min fair share allocates to the established and penalized flows is given by *maximum(0, B -* d_{low}*)*, where d_{low} equals the total arrival rate of the flows in state non-existent or new. From the assumption of existence of flows in state non-existent or new having an arrival rate above the max-min fair share follows that RED+'s d_{low} is greater than the d_{low} value of the original mechanism. Consequently, RED+ allocates a smaller portion of the link-bandwidth to the established and penalized flows than the original mechanism; therefore the fair-share computed by RED+ is smaller than the max-min fair-share computed by the original algorithm. As flows in state penalized or established have arrivalrates above the max-min fair share (see last paragraph) and RED+'s fair-share is smaller than the max-min fair share all flows permanently stored in the hash-table are penalized.

Although the algorithm fails in computing the max-min fair share in the second case, it is correct to penalize the flows stored in the hash-table as these flows have arrival rates above the max-min fair-share, as shown above. Obviously, not all flows with arrival rates above the fair share are penalized as there are flows with an arrival rate above the max-min fair-share in state non-existent or new which are not taken into account by RED+. However, RED+ penalizes the maximum possible number of the highest bandwidth flows traversing the output-port, given a certain size of the hash-table.

ON THE PERFORMANCE OF DIFFERENTIATED SERVICES FOR IP INTERNETWORKS

Chie Dou*, Tian-Shiuh Jeng*
Shu-Wei Wang** and Kuo-Cheng Leu**
**Department of Electrical Engineering*
National Yunlin University of Science and Technology
***Computer & Communications Research Laboratories*
Industrial Technology Research Institute, Taiwan, R.O.C.

Abstract Traffic prioritization is an effective, yet relatively simple, tool for providing differentiated services. This paper investigates the performance of different traffic types for IP internetworks with traffic prioritization under various traffic patterns via numerical analysis. The following seven traffic types were considered: Network Control, 'Voice', 'Video', Controlled Load, Excellent Effort, Best Effort and Background, as suggested in IEEE P802.1D. Traffic patterns were appropriately selected to cover a wide range of traffic variations. This paper demonstrated how the performance of individual traffic types is affected by the distribution of server utilization of all types. The results obtained in this paper could also be helpful to network administrators in configuring the "bandwidth reservation" for IP internetworks.

Keywords: Traffic Prioritization, Differentiated Services, IP Internetworks

1. INTRODUCTION

Traditional IP internetworks were originally designed to provide only a "best-effort" delivery service for each application on a first-come, first-serve basis. This has been, and should continue to be, sufficient for the vast majority of network traffic. E-mail, file transfer and even network-based Fax services can be handled in this manner without suffering from its unpredictable (but finite) delay characteristics. However, several classes of application demand a higher level of service. For example [1]:

- New applications such as Voice over IP and Video Webcasting (distance learning, corporate broadcast, etc.) will not tolerate highly variable delays in the delivery of packets, even though the bandwidth they require is modest.

- Existing business-critical applications shifted to the corporate internetwork - for example, mainframe-based on-line transaction processing - may need to be prioritized above other tasks; in many cases, the applications themselves require highly predictable delivery service from the network.

The business and technical issues that network managers will need to understand in order to successfully deliver differentiated service in their converged networks were thoroughly discussed in [2]. Traffic prioritization is an effective, yet relatively simple, tool for providing differentiated services. Traffic from multimedia and business-critical applications should be assigned to higher-priority queues that are given preferential treatment over lower-priority queues and thus receive lower delay and better performance. However, a fundamental problem is arisen with this approach. The problem is that some applications seek to use the entire or a large portion of the available resource; if they use one or more higher-priority queues, they can completely lock out lower-priority queues for a period of time. This can cause lower-priority sessions to be dropped or to be so slow that they are unusable [3]. To ensure that no class achieves more than a given proportion when the line is under stress. Bandwidth can be reserved so that different streams of data are guaranteed a minimum quantity of bandwidth. This feature allows multimedia and business-critical traffic to have low latency while still permitting other network applications to run effectively. Class-based queuing (CBQ) [4] is one of such techniques which deliver differentiated services by reserving network bandwidth on a static (permanent until manually reconfigured) basis. While no class consumes more than a certain bandwidth under any circumstance, another fundamental problem is arisen with this approach. That is the guaranteed percentage of bandwidth for each traffic type needs to be determined appropriately in advance.

To deal with the second problem mentioned above, a prior knowledge about the performance of individual traffic classes under various traffic patterns could be helpful to network administrators in configuring the bandwidth reservation. For example, the network administrators may want to know what is the performance of best-effort and/or business-critical data traffic when the link utilization of multimedia traffic is high. If the multimedia traffic is further divided into voice and video traffic, what is the performance of other data streams when voice or video traffic becomes heavy? For IP internetworks, which adopt traffic prioritization, this paper provides an effective way through numerical analysis to obtain the useful information needed in configuring the bandwidth reservation.

2. TRAFFIC TYPES

As suggested in the Annex of [5], the following list of traffic types were considered: Network Control (NC), 'Voice' (VO), 'Video' (VI), Controlled Load (CL), Excellent Effort (EE), Best Effort (BE), and Background (BK). To fully understand the performance of individual traffic types under various traffic patterns, we assume that each traffic type has its own service queue. Thus, the server utilization of a particular queue can be regarded as the fraction of the bandwidth occupied by the corresponding traffic type. The bandwidth occupancy of the type is equal to the amount of bandwidth reserved for it if the data of the corresponding traffic type fully utilizes its reserved bandwidth and does not exceed it. To have a prior knowledge about whether a configuration of bandwidth reservation is effective, this paper provides a simple while systematic approach to get the answer. Our approach adopts strict priority queuing algorithm and assumes that all traffic types fully utilize their reserved bandwidths. That is, the server utilization of each priority queue can be used to represent the fraction of bandwidth reserved for the corresponding traffic type. It had already been shown that the average waiting time for each priority queue depends solely on the server utilization of every individual queue in the system. Hence, the numerical results of the waiting time analysis for strict priority queuing systems can be used to evaluate the performance of individual traffic types under various configurations of bandwidth reservation.

3. STRICT PRIORITY QUEUING SYSTEMS

Strict priority queuing algorithm is a simple while effective means to provide useful service differentiation in the IP internetworks. IEEE P802.1D (incorporating IEEE P802.1P) recommended this algorithm as the default algorithm for selecting frames for transmission on each Bridge port. This system is known also by the name of head-of-the-line (HOL) priority queuing or fixed priority queuing, as shown in Fig. 1. In this section we summarize some important results of the queuing analysis derived for the system from [6].

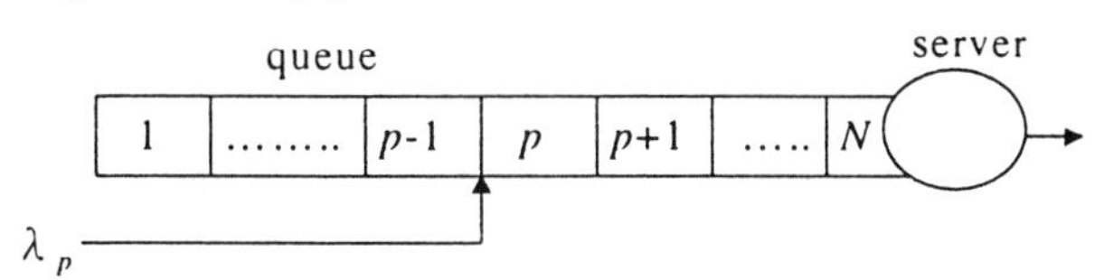

Figure 1 Head-of-the-line priority queue.

We assume that arriving packets belong to one of a set of N different priority classes, indexed by the subscript p (p = 1, 2, ..., N). We adopt the convention that the larger the value of the index associated with the priority group, the higher is the so-called priority associated with that group. We consider a fairly general model based on the system *M/G/1*. Thus we assume that packets from priority group p arrive in a Poisson stream at rate λ_p packets per second; each packet from this group has its service time selected independently from the distribution $B_p(x)$ with mean $\overline{x_p}$ sec. We define the following:

$$\lambda = \sum_{p=1}^{N} \lambda_p,\ \bar{x} = \sum_{p=1}^{N} \frac{\lambda_p}{\lambda} \overline{x_p},\ \rho_p = \lambda_p \overline{x_p},\ \text{and}\ \rho = \lambda \bar{x} = \sum_{p=1}^{N} \rho_p .$$

The interpretation of ρ here is, as usual, the fraction of time the server is busy (so long as $\rho < 1$). Moreover, ρ_p is the fraction of time the server is busy with packets from group p (again for $\rho < 1$).

3.1 Calculating Average Waiting Times

We consider the case of nonpreemptive systems. We study the system from the point of view of a newly arriving packet from priority group p (say); we shall refer to this packet as the "tagged" packet. Let us denote by W_0 the average delay to our tagged packet due to the packet found in service. Since we have a Poisson process, then ρ_i is the probability that our tagged packet finds a type-i packet in service. With Poisson arrivals, we have

$$W_0 = \sum_{i=1}^{N} \rho_i \frac{\overline{x_i^2}}{2\overline{x}_i} \quad (1)$$

where $\overline{x_i^2}$ is the second moment of service time for a packet from group i.
Let W_p be the average waiting time for packets from group p. From [6], we have

$$W_p = W_0 + \sum_{i=p}^{N} \overline{x}_i \lambda_i W_i + \sum_{i=p+1}^{N} \overline{x}_i \lambda_i W_p \qquad p=1, 2, \ldots, N \quad (2)$$

Solving (2) for W_p, we have

$$W_p = \frac{W_0 + \sum_{i=p+1}^{N} \rho_i W_i}{1 - \sum_{i=p}^{N} \rho_i} \qquad p=1, 2, \ldots, N \quad (3)$$

Solving recursively, we obtain the solution

$$W_p = \frac{W_0}{(1-\sigma_p)(1-\sigma_{p+1})} \qquad p=1, 2, \ldots, N \quad (4)$$

where $\sigma_p = \sum_{i=p}^{N} \rho_i$ and $\sigma_{N+1} = 0$.

From (4), we see the effect of those packets of equal or higher priority present in the queue when our packet arrives as given by the denominator term $1-\sigma_{p+1}$.

4. NUMERICAL RESULTS

This section evaluates the performance of different traffic types under various traffic patterns in an IP internetwork. Since we assume each traffic type has a corresponding priority queue, the server utilization of each queue can be used to represent the mean offered load of the corresponding traffic type. Thus, a specific traffic pattern can be denoted by a 7-tuple vector $(\rho_1, \rho_2, \rho_3, \rho_4, \rho_5, \rho_6, \rho_7)$, where ρ_i is the server utilization of queue i. Table 1 describes the characteristics of data packets belonging to each individual traffic type. We assume the packet length for each type is uniformly distributed between a minimum packet size (MIN) and a maximum packet size (MAX) of that type. To calculate the first and the second moment of the service time, i.e. $\overline{x}$ and $\overline{x^2}$, for each traffic type, we assume the link capacity C=1.54 Mbps. For convenience we further use ρ_l to denote the aggregate server utilization of the lower three traffic types, including the BK, BE and EE. That is, $\rho_{\mathrm{I}} = \rho_1 + \rho_2 + \rho_3$. Also, we use ρ_{II} to denote the aggregate server utilization of the higher four traffic types, including the CL, VI, VO and NC. That is, $\rho_{\mathrm{II}} = \rho_4 + \rho_5 + \rho_6 + \rho_7$. It is evident that larger ρ_{I} means that the more bandwidth is occupied by the lower priority traffic types. On the contrary, larger ρ_{II} means that the more bandwidth is occupied by the higher priority traffic types. This section investigates the performance of different traffic types by making comparisons between six appropriately selected different traffic patterns.

Table 1

Priority	Traffic type	Packet length (in byte) MIN MAX	$\bar{x}$	$\overline{x^2}$
7	NC	40 – 60	$2.60*10^{-4}$	$5.38*10^{-8}$
6	VO	64	$3.32*10^{-4}$	$1.10*10^{-7}$
5	VI	188	$9.75*10^{-4}$	$9.50*10^{-7}$
4	CL	150 – 250	$1.04*10^{-3}$	$1.10*10^{-6}$
3	EE	200 – 400	$1.56*10^{-3}$	$2.50*10^{-6}$
2	BE	400 – 800	$3.11*10^{-3}$	$1.00*10^{-5}$
1	BK	500 - 1500	$5.19*10^{-3}$	$2.91*10^{-5}$

Note: define

a= 8*MIN/C, b= 8*MAX/C;

$\bar{x}$ = (a+b)/2; $\overline{x^2} = \int_a^b \frac{1}{b-a} x^2 dx = (a^2+ab+b^2)/3$.

Figures 2 - 4 display the numerical results for three different traffic patterns considered in the first group, respectively. The first pattern assumes the server utilization of individual priority queues shares the same percentage of the total server utilization, ρ. That is $\rho_i = \rho/7$ for all *i*. Fig. 2 shows the average waiting time versus the total server utilization ρ for this case. Two scales are shown for the same case in Fig. 2. The second pattern assumes $\rho_{\mathrm{I}} = \rho/7$ and $\rho_1 = \rho_2 = \rho_3 = \rho_{\mathrm{I}}/3$; $\rho_{\mathrm{II}} = 6\rho/7$ and $\rho_4 = \rho_5 = \rho_6 = \rho_7 = \rho_{\mathrm{II}}/4$. The third pattern assumes $\rho_{\mathrm{I}} = 6\rho/7$ and $\rho_1 = \rho_2 = \rho_3 = \rho_{\mathrm{I}}/3$; $\rho_{\mathrm{II}} = \rho/7$ and $\rho_4 = \rho_5 = \rho_6 = \rho_7 = \rho_{\mathrm{II}}/4$. Evidently, the second pattern represents the case in which the server utilization is heavily dominated by the data of higher priority traffic types. The third pattern represents the case in which the server utilization is heavily dominated by the data of lower priority traffic types. Figures 3 and 4 show the numerical results of these two cases respectively. The bolded curve shown in each figure represents the corresponding *M/G/1* result of the system without traffic prioritization. Obviously, the *M/G/1* curve shown in Fig. 3 performs better than the corresponding curves shown in Figs. 2 and 4. This is because the packet lengths of higher priority types are smaller than that of lower priority types. In Fig. 3, the server utilization is heavily dominated by the data of higher priority types. But this does not means that all curves shown in Fig. 3 perform better than the corresponding curves shown in Figs. 2 and 4. From figures

2-4 we observe that the curves of W_1, W_6 and W_7 shown in Fig. 3 perform better than the corresponding curves shown in Figs. 2 and 4. On the contrary, the curves of W_2, W_3, W_4 and W_5 shown in Fig. 3 perform worse than the corresponding curves shown in Figs. 2 and 4 if the value of ρ is larger than some threshold for each type of curves. This observation clearly indicates that if the server is heavily loaded with the data of higher priority types as shown in Fig. 3, the corresponding performance of types including BE, EE, CL and VI becomes worse than the other two cases. But the performance of types BK, VO and NC still performs better than the other two cases. From Fig. 4 we notice that if the server is heavily loaded with the data of lower priority types, the performance of W_4, W_5, W_6 and W_7 is very close, and the individual performance of W_2, W_3, W_4 and W_5 is the best among three cases.

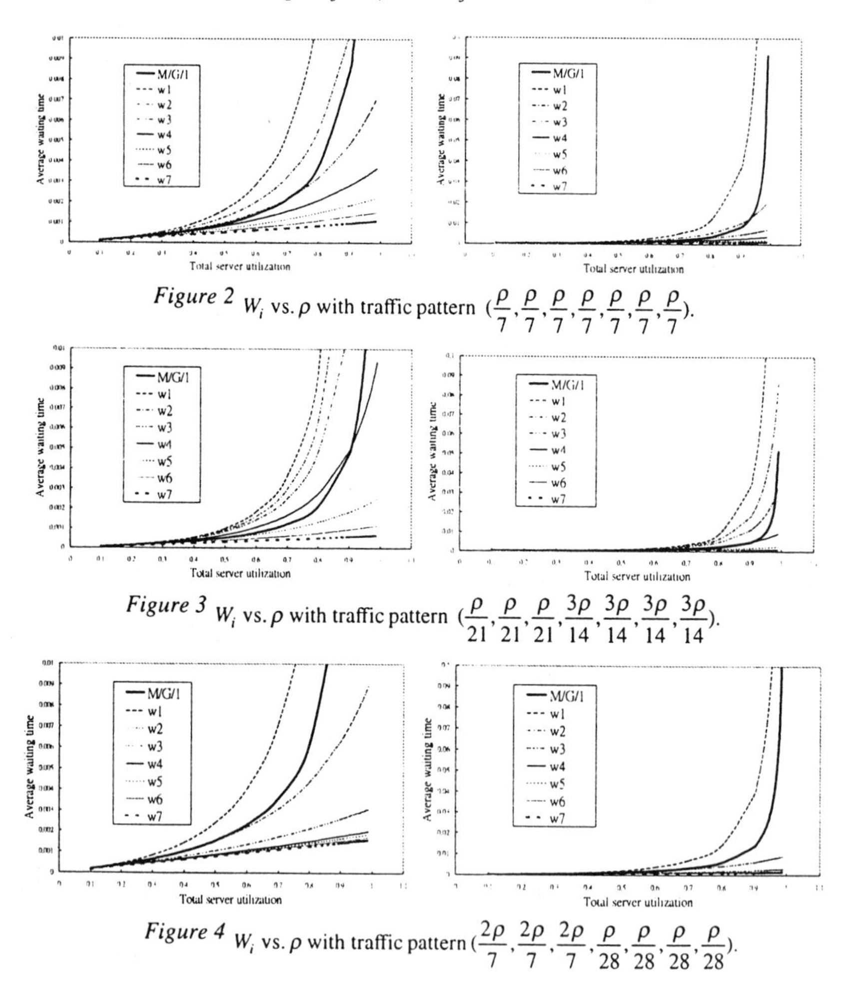

Figure 2 W_i vs. ρ with traffic pattern $(\frac{\rho}{7},\frac{\rho}{7},\frac{\rho}{7},\frac{\rho}{7},\frac{\rho}{7},\frac{\rho}{7},\frac{\rho}{7})$.

Figure 3 W_i vs. ρ with traffic pattern $(\frac{\rho}{21},\frac{\rho}{21},\frac{\rho}{21},\frac{3\rho}{14},\frac{3\rho}{14},\frac{3\rho}{14},\frac{3\rho}{14})$.

Figure 4 W_i vs. ρ with traffic pattern $(\frac{2\rho}{7},\frac{2\rho}{7},\frac{2\rho}{7},\frac{\rho}{28},\frac{\rho}{28},\frac{\rho}{28},\frac{\rho}{28})$.

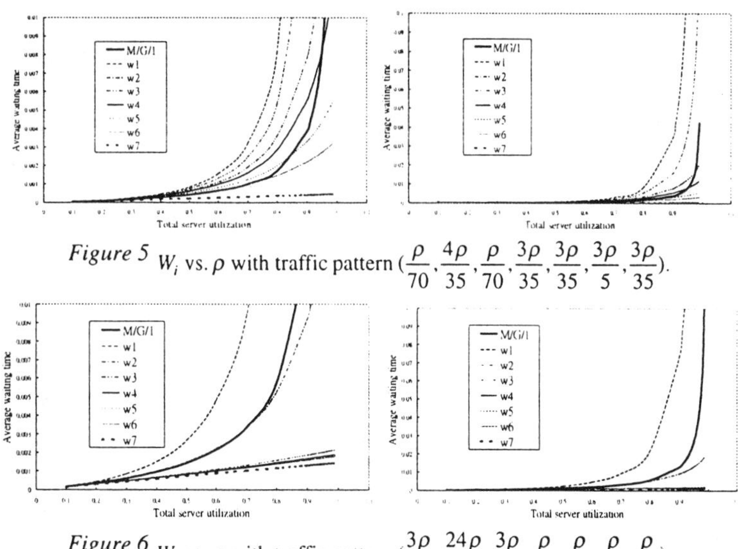

Figure 5 W_i vs. ρ with traffic pattern $(\frac{\rho}{70},\frac{4\rho}{35},\frac{\rho}{70},\frac{3\rho}{35},\frac{3\rho}{35},\frac{3\rho}{5},\frac{3\rho}{35})$.

Figure 6 W_i vs. ρ with traffic pattern $(\frac{3\rho}{35},\frac{24\rho}{35},\frac{3\rho}{35},\frac{\rho}{70},\frac{\rho}{70},\frac{\rho}{10},\frac{\rho}{70})$.

Figures 5 and 6 display the numerical results for two different traffic patterns considered in the second group, respectively. The first pattern assumes $\rho_{\mathrm{I}}=\rho/7$ and $\rho_{\mathrm{II}}=6\rho/7$. The second pattern assumes $\rho_{\mathrm{I}}=6\rho/7$ and $\rho_{\mathrm{II}}=\rho/7$. The server utilization of each traffic type for both patterns follows the same distribution: $\rho_1=\rho/10_{\mathrm{I}}$, $\rho_2=8\rho_{\mathrm{I}}/10$, $\rho_3=\rho_{\mathrm{I}}/10$, $\rho_4=\rho_{\mathrm{II}}/10$, $\rho_5=\rho_{\mathrm{II}}/10$, $\rho_6=7\rho_{\mathrm{II}}/10$ and $\rho_7=\rho_{\mathrm{II}}/10$. Similar to the results we have observed from figures 3 and 4, we also notice that the curves of W_2, W_3, W_4 and W_5 shown in Fig. 5 perform worse than the corresponding curves shown in Fig. 6 as ρ is getting larger. Since the total server utilization is dominated by the voice (VO) traffic only, we further observe that the curves of W_4, W_5 and W_6 shown in Fig. 5 move upwards more rapidly than the corresponding curves shown in Fig. 3 as ρ is getting larger. In Fig. 6 we assume the total server utilization is dominated by the best effort traffic. Hence, the curves of W_3, W_4, W_5, W_6 and W_7 are very close, and the performance of W_1 and W_2 is much worse than the corresponding performance shown in Fig. 4. From Fig. 6 we notice that if the total server utilization is dominated by lower priority traffic, the performance of higher priority types is almost not affected by lower priority traffic.

All the three different traffic patterns considered in the third group assume that $\rho_{\mathrm{I}}=\rho/7$ and $\rho_{\mathrm{II}}=6\rho/7$. We further assume all three patterns have $\rho_1=\rho_{\mathrm{I}}/10$, $\rho_2=8\rho_{\mathrm{I}}/10$, and $\rho_3=\rho_{\mathrm{I}}/10$, the same fraction as used in the previous two patterns. The difference between these three patterns occurred in the distribution of server utilization of higher priority types. For each pattern, ρ_{II} is dominated by a selected higher priority type with fraction 7/10. The other three higher priority types share ρ_{II} with equal fraction 1/10. The dominant types of three traffic patterns are the controlled load (CL), video (VI) and voice (VO) respectively. Table 2 listed the results of performance comparisons among these three traffic patterns for all priority types. From Table 2 we observe an important property of strict priority queuing systems. This property is that if the total server utilization is dominated by priority type *i*, then the performance of priority type *i*-1 will be the worst among three patterns.

Table 2

Pattern no.	Traffic type						
	BK(W_1)	BE(W_2)	EE(W_3)	CL(W_4)	VI(W_5)	VO(W_6)	NC(W_7)
1.CL dominant	W	W	(W)	B	B	B	W
2. VI dominant	M	M	M	(W)	M	M	M
3. VO dominant	B	B	B	M	(W)	W	B

(B:Best, M:Middle, W:Worst)

All the three different traffic patterns considered in the fourth group also assume $\rho_I=\rho/7$ and $\rho_{II}=6\rho/7$. The difference between these three patterns occurred in the distribution of server utilization of lower three priority types. For each pattern, ρ_I is dominated by a selected lower priority type with fraction 8/10. The other two lower priority types share ρ_I with equal fraction 1/10. The dominant types of ρ_I in three different traffic conditions are the BK, BE and EE respectively. We further assume all three patterns have $\rho_4=\rho_{II}/10$, $\rho_5=\rho_{II}/10$, $\rho_6=7\rho_{II}/10$ and $\rho_7=\rho_{II}/10$. Table 3 listed the results of performance comparisons among these three traffic conditions for all priority types. It is evident that for higher four priority types the third pattern (EE dominant) performs the best, and the first pattern (BK dominant) performs the worst. This is because the mean packet sizes of priority types are getting smaller as priority level increases. Smaller packet sizes always result in lower average waiting times. Table 3 clearly shows that the worst cases of BE and BK were EE dominant and BE dominant respectively. This matches the important property of strict priority queuing systems we have described above.

Table 3

Pattern no.	Traffic type						
	BK(W_1)	BE(W_2)	EE(W_3)	CL(W_4)	VI(W_5)	VO(W_6)	NC(W_7)
1. BK dominant	B	B	X	W	W	W	W
2. BE dominant	(W)	M	B	M	M	M	M
3. EE dominant	M	(W)	Y	B	B	B	B

Note: for $\rho>0.8$, X=M and Y=W;
$\rho<0.8$, X=W and Y=M.

Figures 7, 8 and 9 display the numerical results for three different traffic patterns considered in the fifth group, respectively. We assume $\rho_I=6\rho/7$ and $\rho_{II}=\rho/7$ for these three patterns. The distributions of server utilization of all traffic types for these three patterns are the same as the previous three patterns considered in the fourth group. We found that the comparison result for this group of traffic patterns is almost the same as Table 3 with some minor exceptions in BK and EE types. Although the respective curves belonging to two different groups are unlike in appearance.

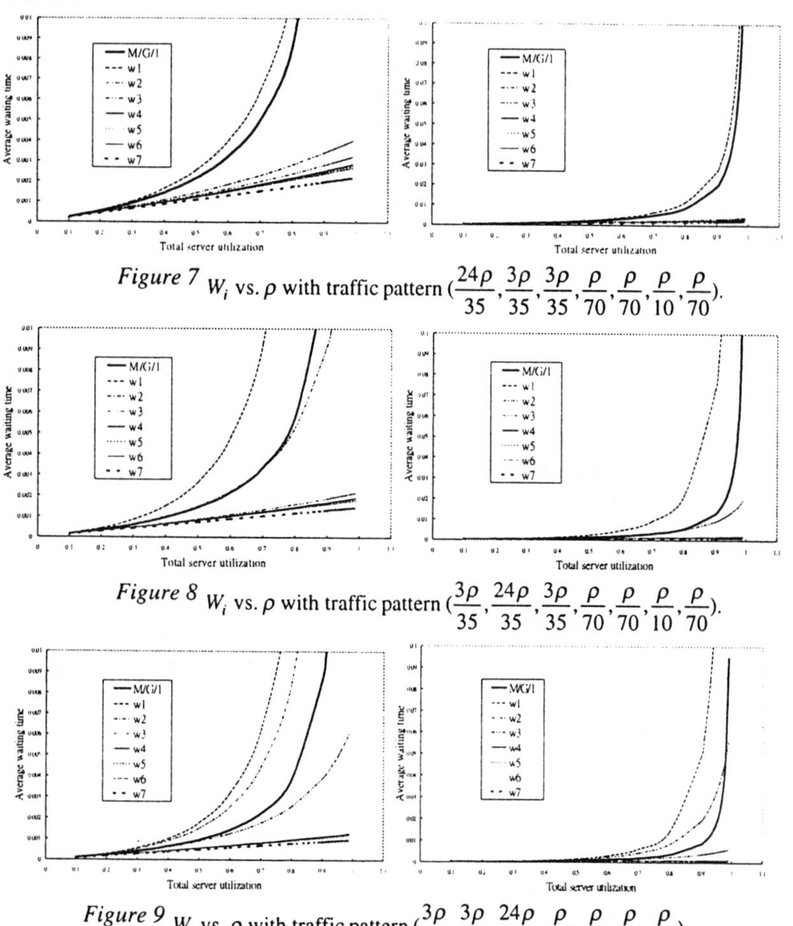

Figure 7 W_i vs. ρ with traffic pattern $(\frac{24\rho}{35},\frac{3\rho}{35},\frac{3\rho}{35},\frac{\rho}{70},\frac{\rho}{70},\frac{\rho}{10},\frac{\rho}{70})$.

Figure 8 W_i vs. ρ with traffic pattern $(\frac{3\rho}{35},\frac{24\rho}{35},\frac{3\rho}{35},\frac{\rho}{70},\frac{\rho}{70},\frac{\rho}{10},\frac{\rho}{70})$.

Figure 9 W_i vs. ρ with traffic pattern $(\frac{3\rho}{35},\frac{3\rho}{35},\frac{24\rho}{35},\frac{\rho}{70},\frac{\rho}{70},\frac{\rho}{10},\frac{\rho}{70})$.

All the three different traffic patterns considered in the sixth group also assume $\rho_{\text{I}}=6\rho/7$ and $\rho_{\text{II}}=\rho/7$. The distributions of server utilization of all traffic types for these three patterns are the same as those considered in the third group. The distribution of server utilization for lower three priority types follows $\rho_1=\rho_{\text{I}}/10$, $\rho_2=8\rho_{\text{I}}/10$ and $\rho_3=\rho_{\text{I}}/10$. For each pattern, ρ_{II} is dominated by a selected higher priority type with fraction 7/10. The other three higher priority types share ρ_{II} with equal fraction 1/10. Since the packet sizes of higher priority types are small and their aggregate utilization ρ_{II} only occupies a small portion of the total server utilization, the corresponding performance of W_1, W_2 and W_3 in all three traffic conditions is almost the same. Different distributions of server utilization for higher priority types affect the performance of W_4, W_5, W_6 and W_7 very slightly.

5. CONCLUSIONS

Numerical analysis for the strict priority queuing systems was used to investigate the performance of individual traffic types under various traffic patterns in IP internetworks. Traffic patterns were appropriately selected to cover a wide range of traffic variations, and were classified into six groups. Performance comparisons and discussions were made for each group of patterns. Some remarkable observations are listed as follows:

(1) If the server is heavily loaded with the data of traffic type *i*, then the performance of traffic type *i*-1 will become worse dramatically. This will also result in the degradation of the performance of traffic types with priority lower than *i*-1 to some degree.

(2) If the server is heavily loaded with the data of voice (VO) type, the performance of all traffic types will become worse obviously except NC.

(3) Since the mean packet sizes of traffic types are getting larger as priority level decreases, the performance of higher priority types is affected by the distribution of server utilization of lower priority types. Traffic patterns with EE dominant perform better than those with BE and/or BK dominant.

(4) If the server is heavily loaded with the data of lower priority types, the performance of higher priority types can be affected by the traffic of lower priority types only slightly.

(5) Since the packet sizes of higher priority types are rather small, if the aggregate utilization ρ_{II} only occupies a small portion of the total server utilization, the performance of individual traffic types will be affected by the distribution of server utilization of higher priority types very slightly.

References

[1] --"Implementing Guaranteed Service in IP Internetworks," *A technical paper from Torrent Networking Technologies.* http://www.torrentnet.com

[2] Chuck Semeria and Frank Fuller, "3Com's Strategy for Delivering Differentiated Service Levels," 3Com Systems Marketing, Feb. 6, 1998.

[3] *Cisco White Paper*, "Interface Queue Management," Posted: Dec. 5, 1997. http://www.cisco.com/warp/public/614/quemg_wp.htm

[4] Ashley Stephenson, "Class-based Queuing: Managing Broadband Access to the Internet," Network World, Vol. 14, Issue 21, May 26, 1997.

[5] ISO/IEC Final DIS 15802-3, Information Technology - local and metropolitan area networks - Common specifications -- Part 3: Media Access Control (MAC) Bridges, May 25,1998.

[6] Leonnard Kleinrock, *Queueing Systems, Volume II: Computer Applications*, John Wiley & Sons, Inc., 1977.

SESSION 2

Traffic Modeling

MAXIMUM LIKELIHOOD ESTIMATION OF THE PARAMETERS OF FRACTIONAL BROWNIAN TRAFFIC WITH GEOMETRICAL SAMPLING

Attila Vidács
High Speed Networks Laboratories
Dept. of Telecommunications and Telematics
Technical University of Budapest
Pázmány Péter sétány 1/D,
H-1117 Budapest, Hungary
vidacs@ttt-atm.ttt.bme.hu

Jorma T. Virtamo
Lab. of Telecommunications Technology
Helsinki University of Technology
P.O.B. 3000, FIN-02015 HUT, Finland
jorma.virtamo@hut.fi

Abstract Traffic model based on the fractional Brownian motion (fBm) contains three parameters: the mean rate m, variance parameter a and the Hurst parameter H. The estimation of these parameters by the maximum likelihood (ML) method is studied. Explicit expressions for the ML estimates $\hat{m}$ and $\hat{a}$ in terms of H are given, as well as the expression for the log-likelihood function from which the estimate $\hat{H}$ is obtained as the maximizing argument. A geometric sequence of sampling points, $t_i = \alpha^i$, is introduced in order to see the scaling behaviour of the traffic with fewer samples. It is shown that by a proper 'descaling' the traffic process is stationary on this grid leading to a Toeplitz-type covariance matrix. Approximations for the inverted covariance matrix and its determinant are introduced. The accuracy of the estimation algorithm is studied by simulations. Comparisons with corresponding estimates obtained with linear grid show that the geometrical sampling indeed improves the accuracy of the estimate $\hat{H}$ with a given number of samples.

Keywords: Traffic modeling, fractional Brownian motion, Maximum Likelihood Estimation.

1. INTRODUCTION

One of the simplest and most studied models for aggregated data traffic is the fractional Brownian motion (fBm) model [7], which is a model for truly self-similar Gaussian traffic. An important feature of the fBm model is its parsimonity [3]: in its basic form the model contains only three parameters, the mean rate m, the variance parameter a and the Hurst parameter H describing the scaling behaviour of the traffic. The estimation of even a small number of parameters poses a problem for long range dependent traffic. Some early work [7] suggested that to obtain a reasonable accuracy a very large number of sample points may be required. As H describes the scaling behaviour of the traffic variability, the sample points have to cover several time scales, i.e., the total time range must be several orders of magnitude greater than the finest time resolution in the measurement.

In this paper we show that by an appropriate choice of the sampling instants, the number of sampling points can be considerably reduced. In particular, we will introduce a grid of geometrically distributed sampling points $t_i = \alpha^{i-1}$, $i = 1, \dots, n$ where α is some constant less than one. The geometrical grid, being 'self-similar' fits well with the traffic process and gives rise to a simple structure in the covariance matrix.

Throughout this work we apply the maximum likelihood estimation (MLE) method [1]. MLE method has previously been applied to this problem by Deriche and Tewfik [2] and Ninness [5] using ordinary linear sampling. Explicit formulas for the estimators of m and a are given along with the log-likelihood function for determining the estimator for H. A major difficulty in this method is the calculation of the inverse and determinant of the covariance matrix. For the original fBm process the increment process is stationary. We show that another stationary process is obtained from the fBm process by 'descaling' and changing the process index to logarithmic time, i.e., on the geometrical sampling grid the descaled process is stationary. It turns out that the elements of the inverse covariance matrix far from the diagonal are small, enabling us to derive a simple approximation for the inverse matrix directly without using e.g. Whittle's method [1] based on the spectral analysis.

We compare the effectiveness of the MLE estimator based on ordinary evenly spaced sampling grid with that obtained with a geometrical grid by simulations.

The rest of this paper is organized as follows. In Section 2 we review the fractional Brownian motion traffic model with its three parameters. The general problem of the estimation of these parameters by the maximum likelihood method is considered in Section 3. The idea of geometrical sampling and the descaled process, along with an approximate form of the MLE, are introduced in Section 4. For comparison, in Section 5 we present the MLE method for the case of ordinary linear sampling. In Section 6, we present results for estimating the fBm parameters with the described methods from simulated realizations of the process. Section 7 concludes the paper.

2. FRACTIONAL BROWNIAN TRAFFIC

A normalized *fractional Brownian motion* with Hurst-parameter $H \in [0.5, 1)$, denoted by $Z(t)$, $(t \in \mathbb{R})$, is characterized by the following properties [6]:

1. $Z(t)$ has stationary increments;

2. $Z(0) = 0$, and $\mathrm{E}[Z(t)] = 0$ for all t;
3. $\mathrm{Var}[Z(t)] = \mathrm{E}[Z(t)^2] = |t|^{2H}$ for all t;
4. $Z(t)$ has continuous paths;
5. $Z(t)$ is a Gaussian process, i.e., all its finite-dimensional marginal distributions are Gaussian.

In the special case $H = 0.5$, $Z(t)$ is the standard Brownian motion. It follows from the above properties that $Z(t)$ is a self-similar process whose scaling behaviour is defined by the Hurst-parameter H as follows:

$$Z(\alpha t) \sim \alpha^H Z(t). \tag{1}$$

The covariance structure of the process is given by

$$\mathrm{Cov}[Z(t_1), Z(t_2)] = \frac{1}{2}\left\{t_1^{2H} + t_2^{2H} - |t_2 - t_1|^{2H}\right\}. \tag{2}$$

Fractional Brownian motion is a popular model for long-range dependent traffic. Norros [6] has suggested the following model

$$X(t) = mt + \sqrt{a}Z(t), \tag{3}$$

where $X(t)$ represents the amount of traffic arrived in $(0, t)$. The model has three parameters, m, a and H with the following interpretations and intervals for allowed values: $m > 0$ is the mean input rate, $a > 0$ is a variance parameter, and $H \in [0.5, 1)$ is the self-similarity parameter of $Z(t)$.

3. EXACT GAUSSIAN MLE

We use the notation of Beran [1]. Assume the traffic has been observed at n time instants forming the vector $\mathbf{t} = (t_1, \ldots, t_n)^{\mathrm{t}}$ where $(\cdot)^{\mathrm{t}}$ denotes the transpose. And let $\mathbf{X} = (X(t_1), \ldots, X(t_n))^{\mathrm{t}}$ be the vector of observed traffic values at these instants. Since $X(t)$ is Gaussian, the joint probability density function of $\mathbf{X}$ is

$$h(\mathbf{x}) = (2\pi)^{-\frac{n}{2}} |\mathbf{\Gamma}|^{-\frac{1}{2}} e^{-\frac{1}{2}(\mathbf{x}-\mathbf{m})^{\mathrm{t}}\mathbf{\Gamma}^{-1}(\mathbf{x}-\mathbf{m})}, \tag{4}$$

where $\mathbf{x} = (x_1, \ldots, x_n)^{\mathrm{t}} \in \mathrm{I\!R}^n$, $\mathbf{m} = m\mathbf{t}$, and $|\mathbf{\Gamma}|$ is the determinant of the covariance matrix

$$\mathbf{\Gamma} = \mathrm{Cov}\left[\mathbf{X}, \mathbf{X}^{\mathrm{t}}\right] = \mathrm{E}\left[\mathbf{X}\mathbf{X}^{\mathrm{t}}\right] - \mathrm{E}[\mathbf{X}]\,\mathrm{E}\left[\mathbf{X}^{\mathrm{t}}\right]. \tag{5}$$

The MLE for m is obtained by maximizing $\log h(\mathbf{X}; m)$ with respect to m, resulting in the estimator

$$\hat{m} = \hat{m}(H) = \frac{\mathbf{t}^{\mathrm{t}}\,\mathbf{\Gamma}^{-1}\mathbf{X}}{\mathbf{t}^{\mathrm{t}}\,\mathbf{\Gamma}^{-1}\,\mathbf{t}}. \tag{6}$$

Note, that the estimate is unbiased, irrespective whether our estimate for H is correct or not. The variance of $\hat{m}$ can also be calculated with the assumption that H is known exactly, $\hat{H} = H$. With straightforward calculations we get

$$\mathrm{Var}\left[\hat{m}\right] = \frac{a}{\mathbf{t}^{\mathrm{t}}\,\mathbf{\Gamma}^{-1}\,\mathbf{t}}. \tag{7}$$

The variance of our estimator is smaller than for the estimator based on the sample mean, by the factor in the denominator (which is close to one).

Next, consider the estimator for a. $\mathbf{\Gamma}$ is a simple linear function of a, $\mathbf{\Gamma} = a\,\mathbf{\Gamma}_H$, where $\mathbf{\Gamma}_H$ is independent of a and is given by

$$\mathbf{\Gamma}_H = \mathrm{E}\left[\mathbf{Z}\mathbf{Z}^{\mathrm{t}}\right] = \left[\mathrm{Cov}\left[Z(t_i), Z(t_j)\right]\right]_{i,j=1,\ldots,n}. \tag{8}$$

The MLE of a is obtained by maximizing the log-likelihood function $\log h(\mathbf{X}; a)$ with respect to a. If we do not know the mean input rate m in advance, $\mathbf{m}$ should be replaced by $\hat{m}\mathbf{t}$, and we get:

$$\hat{a}(H) = \frac{1}{n}\frac{(\mathbf{X}^{\mathrm{t}}\,\mathbf{\Gamma}_H^{-1}\,\mathbf{X})(\mathbf{t}^{\mathrm{t}}\,\mathbf{\Gamma}_H^{-1}\,\mathbf{t}) - (\mathbf{t}^{\mathrm{t}}\,\mathbf{\Gamma}_H^{-1}\,\mathbf{X})^2}{\mathbf{t}^{\mathrm{t}}\,\mathbf{\Gamma}_H^{-1}\,\mathbf{t}}. \tag{9}$$

Again, assuming for the time being that H is known correctly the expectation and variance of $\hat{a}$ can be calculated and finally we have

$$\mathrm{E}\left[\hat{a}\right] = \frac{n-1}{n}a, \qquad \mathrm{Var}\left[\frac{n}{n-1}\hat{a}\right] = \frac{2a^2(n-1)}{n^2}. \tag{10}$$

Thus $\hat{a}$ has the "normal" $(n-1)/n$ bias.

Finally, we are left with the maximization of the H-dependent part of the log-likelihood function, i.e., essentially we have to minimize

$$\tilde{L}(\mathbf{X}; H) = \log|\mathbf{\Gamma}_H| + n\log\frac{(\mathbf{X}^{\mathrm{t}}\,\mathbf{\Gamma}_H^{-1}\,\mathbf{X})(\mathbf{t}^{\mathrm{t}}\,\mathbf{\Gamma}_H^{-1}\,\mathbf{t}) - (\mathbf{t}^{\mathrm{t}}\,\mathbf{\Gamma}_H^{-1}\,\mathbf{X})^2}{\mathbf{t}^{\mathrm{t}}\,\mathbf{\Gamma}_H^{-1}\,\mathbf{t}}. \tag{11}$$

The minimum is obtained for some value $\hat{H}$ which is the MLE estimate; the corresponding MLE estimates for m and a are $\hat{m} = m(\hat{H})$ and $\hat{a} = a(\hat{H})$.

4. GEOMETRICAL SAMPLING

The Hurst parameter H describes the scaling behaviour of the traffic. Therefore, in order to determine its value from measured traffic, the sample points have to cover several time scales, i.e., the total time range of the measurements has to be many orders of magnitude greater than the finest resolution (smallest interval between the sampling points). With the ordinary linear sampling, i.e., sampling points at constant intervals, this leads to the requirement of very large number of sampling points. In order to use the measurements more efficiently we introduce a geometric sequence of sampling points, $t_i = \alpha^i$, $i = 1, \ldots, n$, with some $0 < \alpha < 1$.

In addition to distributing the sampling points in a better way on different time scales, geometric sampling fits neatly with the self-similar behaviour of the fBm traffic. We show first that by a simple transformation we can obtain from the fBm process another process which is a stationary process of logarithmic time. As a geometric sequence corresponds to equidistant points in logarithmic time, it follows that the samples of the modified process constitute a stationary sequence. This leads to a simple Toeplitz-type structure of the covariance matrix and allows us to develop approximations to the inverse and determinant of the covariance matrix.

4.1 DESCALED PROCESS

$Z(t)$ has the self-similar property $Z(\alpha t) \sim \alpha^H Z(t)$. Now consider the 'descaled' process $\breve{Z}(t) \stackrel{d}{=} t^{-H} Z(t)$ which has the scaling property

$$\breve{Z}(\alpha t) \sim (\alpha t)^{-H} Z(\alpha t) = t^{-H} Z(t) = \breve{Z}(t). \tag{12}$$

Further let us take a new time variable $u = -\log t$ and denote $\tilde{Z}(u) \stackrel{d}{=} \breve{Z}(e^{-u}) = \breve{Z}(t)$. Now we have

$$\tilde{Z}(u - \log\alpha) = \breve{Z}(e^{-u+\log\alpha}) = \breve{Z}(\alpha e^{-u}) = \breve{Z}(\alpha t) \sim \breve{Z}(t) = \tilde{Z}(u). \tag{13}$$

Thus the process $\tilde{Z}(u)$ is stationary and has the following covariance structure:

$$\mathrm{Cov}\left[\tilde{Z}(u_1), \tilde{Z}(u_2)\right] = \frac{1}{2} e^{H(u_2-u_1)} \left\{1 + e^{-2H(u_2-u_1)} - \left(1 - e^{-(u_2-u_1)}\right)^{2H}\right\}, \tag{14}$$

so the descaled process $\tilde{Z}(u)$ is short range dependent.

If we 'descale' the process $X(t)$ by the factor t^{-H} and use u as the process index, the covariance matrix $\tilde{\mathbf{\Gamma}}$ of the descaled samples $\tilde{\mathbf{X}} = (\tilde{X}(u_1), \tilde{X}(u_2), \ldots, \tilde{X}(u_n))^{\mathrm{t}}$ with $u_i = -\log t_i = (1-i)\log\alpha$ can be written as

$$\tilde{\mathbf{\Gamma}} = \mathrm{E}\left[\tilde{\mathbf{X}}\tilde{\mathbf{X}}^{\mathrm{t}}\right] = a \cdot \mathrm{E}\left[\tilde{\mathbf{Z}}\tilde{\mathbf{Z}}^{\mathrm{t}}\right]. \tag{15}$$

Note, that our geometrical grid is now equally spaced with regard to u. Thus, if we use the notation $\tilde{Z}_i = \tilde{Z}(u_i)$ the process $\tilde{\mathbf{Z}} = (\tilde{Z}_1, \tilde{Z}_2, \ldots, \tilde{Z}_n)$ is a stationary process in discrete time with zero mean and unit variance and its auto-correlation function $\rho(k)$ can be defined as

$$\rho(i-j) = \frac{1}{2}\alpha^{-H|i-j|}\left\{1 + \alpha^{2H|i-j|} - \left(1 - \alpha^{|i-j|}\right)^{2H}\right\}, \tag{16}$$

and thus

$$\tilde{\mathbf{\Gamma}}_{ij} = a\rho(i-j), \quad i, j = 1, 2, \ldots, n. \tag{17}$$

4.2 APPROXIMATE MLE

In practice, the exact MLE poses computational problems. And this is not just because of the computation time needed in case of large data sets, but because of the evaluation of the inverse and the determinant of the covariance matrix may be numerically unstable. To avoid these problems, one can use approximate methods for the calculations. In [1], several possible approaches are discussed, among them the well known Whittle's approximate MLE.

In our case we focus on the properties of the covariance matrix $\mathbf{\Gamma}_H$, trying to take advantage of the stationarity and short range dependent properties of the descaled process. Using the 'descaling matrix' $\mathbf{D} = \mathrm{diag}(t_1^{-H}, \ldots, t_n^{-H})$ we can easily derive $\tilde{\mathbf{\Gamma}} = \mathbf{D}\,\mathbf{\Gamma}\,\mathbf{D}$, and from this we get

$$\mathbf{\Gamma}_H^{-1} = \mathbf{D}\tilde{\mathbf{\Gamma}}_H^{-1}\mathbf{D}; \qquad |\mathbf{\Gamma}_H| = \alpha^{Hn(n-1)}|\tilde{\mathbf{\Gamma}}_H|. \tag{18}$$

The elements of the autocorrelation matrix $\tilde{\mathbf{\Gamma}}_H$ can be written as

$$(\tilde{\mathbf{\Gamma}}_H)_{i,j} = \alpha^{-H|i-j|}g(\alpha^{|i-j|}), \quad i,j = 1,2,\ldots,n \tag{19}$$

with

$$g(x) = \frac{1}{2}\left(1 + x^{2H} - (1-x)^{2H}\right). \tag{20}$$

It is interesting to note, that $g(x)$ is nearly completely linear for $x \in (0,1)$. Figure 1 shows the difference of $g(x) - x$ for different values of H. It can be seen from the plot that the largest absolute difference is less than 0.02 for each value of H. This observation gives us the idea to use the approximation $g(x) \approx x$. So $\tilde{\mathbf{\Gamma}}_H$ can be approximated as $\tilde{\mathbf{\Gamma}}_H \approx \mathbf{R}$, where $\mathbf{R}$ is a Toeplitz-type matrix of the form $[\mathbf{R}]_{ij} = \gamma^{|i-j|}$, $i,j = 1,2,\ldots n$, with $\gamma = \alpha^{1-H}$.

The inverse of $\mathbf{R}$ can be easily calculated as [8]

$$\mathbf{R}^{-1} = \frac{1}{\frac{1}{\gamma} - \gamma}\begin{pmatrix} \gamma^{-1} & -1 & 0 & \cdots & 0 \\ -1 & \gamma+\gamma^{-1} & -1 & \ddots & \vdots \\ 0 & -1 & \gamma+\gamma^{-1} & \ddots & 0 \\ \vdots & \ddots & \ddots & \ddots & -1 \\ 0 & \cdots & 0 & -1 & \gamma^{-1} \end{pmatrix}, \tag{21}$$

and the determinant of $\mathbf{R}$ is given as $|\mathbf{R}| = (1-\gamma^2)^{n-1}$ [8].

Using the fact $\mathbf{t}^{\mathrm{t}}\,\mathbf{D}\mathbf{R}^{-1}\mathbf{D}\,\mathbf{t} = 1$ and $\mathbf{t}^{\mathrm{t}}\,\mathbf{D}\mathbf{R}^{-1}\mathbf{D} = (1,0,\ldots,0)$, we get $\hat{m} = X(1)$ so using the above approximation the MLE estimate for m reduces simply to the sample mean. As for the estimate for a we get

$$\hat{a}(H) \;=\; \frac{1}{n}\left(\mathbf{X}^{\mathrm{t}}\,\mathbf{D}\,\mathbf{R}^{-1}\mathbf{D}\,\mathbf{X} - X_1^2\right). \tag{22}$$

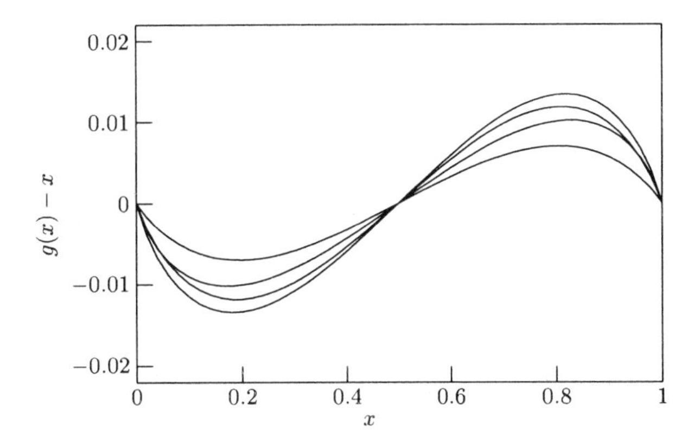

Figure 1 Error of approximation $g(x) \approx x$ for $H = 0.6, 0.7, 0.8$ and 0.9.

Finally, to get an estimate for H we have to minimize the function

$$L(\mathbf{X}; H) = \frac{n-1}{n} \log\left(\alpha^{nH}(1-\alpha^{2-2H})\right) + \log\left(\mathbf{X}^{\mathrm{t}}\,\mathbf{D}\,\mathbf{R}^{-1}\mathbf{D}\,\mathbf{X} - X_1^2\right). \tag{23}$$

It should be noted that though the linear approximation to $g(x)$ is rather accurate, the resulting inverse matrix $\mathbf{R}^{-1}$ of Eq.(21) is rather poor an approximation to $\tilde{\mathbf{\Gamma}}^{-1}$ for large n. Nevertheless, the use of $\mathbf{R}^{-1}$ in the log-likelihood function (23), as we will see, yields a good estimate for H, while the accuracy of the estimate $\hat{a}$ suffers more from this approximation.

4.3 IMPROVED APPROXIMATION

Since the matrix $\tilde{\mathbf{\Gamma}}$ is a Toeplitz-type matrix with decreasing elements as we go farther from the diagonal, we expect that its inverse can be well approximated with a band matrix of the form:

$$\mathbf{C} = \begin{pmatrix} c_1 & \cdots & c_p & 0 & \cdots & 0 \\ \vdots & c_1 & \vdots & c_p & \ddots & \vdots \\ c_p & \cdots & c_1 & \vdots & \ddots & 0 \\ 0 & c_p & \cdots & c_1 & \cdots & c_p \\ \vdots & \ddots & \ddots & \vdots & \ddots & \vdots \\ 0 & \cdots & 0 & c_p & \cdots & c_1 \end{pmatrix} \tag{24}$$

so that $\tilde{\mathbf{\Gamma}}_H^{-1} \approx \mathbf{C}$. Our aim is to set the p parameters $c_1, \ldots, c_p$ to get $\mathbf{C}\tilde{\mathbf{\Gamma}}_H \approx \mathbf{E}$. For example, this can be achieved by solving the equation

$$(c_p, \ldots, c_2, c_1, c_2, \ldots, c_p) \cdot \mathbf{G} = (0, \ldots, 0, 1, 0, \ldots, 0), \tag{25}$$

where $\mathbf{G} = (\tilde{\mathbf{\Gamma}}_H)_{(2p-1)\times(2p-1)}$ and from this we have

$$c_i = \mathbf{G}^{-1}_{p(p+i-1)}, \quad i = 1, 2, \ldots, p. \tag{26}$$

With this approximation we only need to calculate the inverse of a $(2p-1)$-by-$(2p-1)$ matrix.[1] (To improve the approximate inverse, its elements in the upper-left and lower-right corners can be further corrected.)

5. LINEAR SAMPLING

Let $\mathbf{X} = (X(t_1), X(t_2), \ldots, X(t_n))^{\mathrm{t}}$ be the vector of observed traffic values at instances

$$t_i = \frac{i}{n}, \quad i = 1, 2, \ldots, n. \tag{27}$$

The increment sequence $(Y_1, Y_2, \ldots)$ with $Y_i = X(t_i) - X(t_{i-1})$ (substituting $X(t_0) \equiv X(0) = 0$) is a strongly correlated stationary sequence with

$$\mathrm{Cov}\,[Y_i, Y_j] = \frac{1}{2} a n^{-2H} \left(|i-j+1|^{2H} + |i-j-1|^{2H} - 2|i-j|^{2H} \right) \tag{28}$$

for $i, j = 1, 2, \ldots, n$. The formulas for the exact Gaussian MLE for this increment process are nearly the same as in Section 3, we only need to replace the covariance matrix $\mathbf{\Gamma}$ with $\mathbf{\Sigma} = [\mathrm{Cov}\,[Y_i, Y_j]]_{i,j=1,2,\ldots,n}$, and the vector $\mathbf{t}$ with the vector $(1/n, 1/n, \ldots, 1/n)^{\mathrm{t}}$. After some minor simplifications we get an estimate for m

$$\hat{m} = \hat{m}(H) = \frac{\mathbf{1}^{\mathrm{t}}\, \mathbf{\Sigma}^{-1}\, \mathbf{Y}}{\mathbf{1}^{\mathrm{t}}\, \mathbf{\Sigma}^{-1}\, \mathbf{1}} \cdot n \tag{29}$$

where $\mathbf{1}$ is a vector of ones, and $\mathbf{\Sigma} = a\mathbf{\Sigma}_H$. For a we have the estimator

$$\hat{a}(H) = \frac{1}{n} \left(\mathbf{Y}^{\mathrm{t}}\, \mathbf{\Sigma}_H^{-1}\, \mathbf{Y} - \frac{(\mathbf{1}^{\mathrm{t}}\, \mathbf{\Sigma}^{-1}\, \mathbf{Y})^2}{\mathbf{1}^{\mathrm{t}}\, \mathbf{\Sigma}^{-1}\, \mathbf{1}} \right). \tag{30}$$

Again, finally we have to minimize

$$\tilde{L}(\mathbf{Y}; H) = \log |\mathbf{\Sigma}_H| + n \log \left(\mathbf{Y}^{\mathrm{t}}\, \mathbf{\Sigma}_H^{-1}\, \mathbf{Y} - \frac{(\mathbf{1}^{\mathrm{t}}\, \mathbf{\Sigma}^{-1}\, \mathbf{Y})^2}{\mathbf{1}^{\mathrm{t}}\, \mathbf{\Sigma}^{-1}\, \mathbf{1}} \right). \tag{31}$$

The minimum is obtained for some value $\hat{H}$ which is the MLE estimate.

However, to calculate the inverse and the determinant of $\mathbf{\Sigma}_H$ the same problems arise as in the case of geometrical sampling with the covariant matrix $\tilde{\mathbf{\Gamma}}_H$. Since $\mathbf{\Sigma}_H$ is also a Toeplitz type matrix, the same method as described in Section 4.3 can be used to approximate $\mathbf{\Sigma}_H^{-1}$ with $\mathbf{C}$ of Eq.(24).

[1] To be more exact, because of the symmetric structure we only need to calculate the inverse of a p-by-p matrix using slightly more complicated formulas.

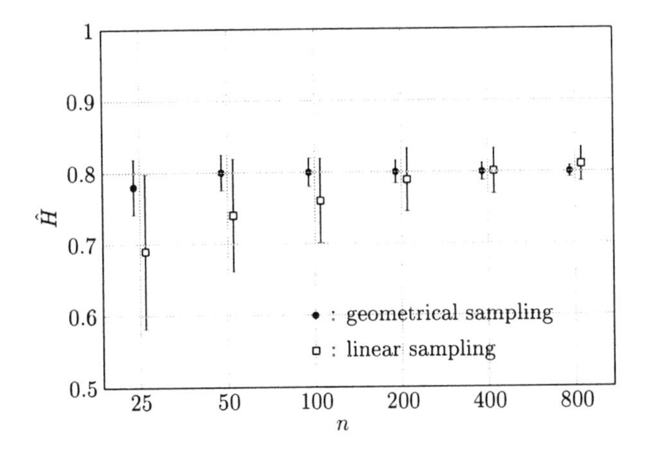

Figure 2 Estimates of H using geometrical and linear sampling.

6. SIMULATION RESULTS

The fBm samples were generated using the fact $\mathbf{Z} \sim \mathbf{\Gamma}_H^{1/2}\mathbf{N}$ where $\mathbf{N}$ is a vector of independent standard Gaussian variables. The model parameters were set as $m = 1$, $a = 1$ and $H = 0.8$ as an example, but similar results were obtained using different values of the parameters. The parameter α for the geometrical grid was chosen so that the difference between the nearest two measurement time instants (the 'resolution' of the measurement) was 10^{-6}. Figure 2 shows the results of H estimates as a function of the number of sample points using both geometrical and linear sampling. In the geometrical case Eq.(23) was minimized while for the linear sampling we used the formula Eq.(31) where the inverse of $\mathbf{\Sigma}_H$ was approximated with a band matrix of Eq.(24) with $p = 2$. The 95% confidence interval was obtained by repeating the simulations 100 times and calculating the sample variance of the estimates. The results show that the estimates using geometrical sampling have much smaller variance and are unbiased for sample sizes larger than 25. However, the variance of the estimates is always higher than in the geometrical case. For example, the variance for 800 samples using linear sampling is nearly the same as for only 50 geometrically sampled points.

The next question was how the two different sampling methods affect the estimates for the variance parameter a. Figure 3 displays the results, assuming that H is known. These simulations were useful to test whether our approximations in calculating the inverse and determinant of the covariance matrices are adequate or not. Figure 3 presents two different approximations for the geometrical sampling. First, we used the simple approximate inverse covariance matrix of Eq.(21) in Eq.(9) using Eq.(18) (denoted by light gray dots and

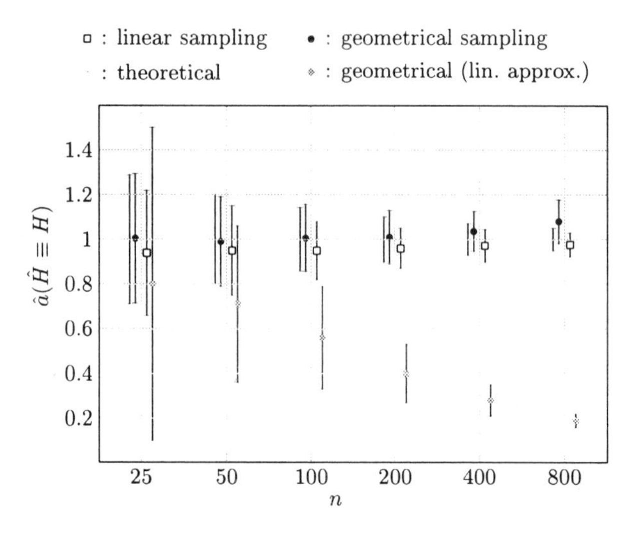

Figure 3 Estimates of a using geometrical and linear sampling and different approximations, assuming H is known.

labeled 'linear approximation' in the figure). As can be seen, the estimates of a are strongly biased and the bias is getting larger as the number of samples increases. So this estimate is clearly inadequate, the approximation of Eq.(21) had to be refined. Next, we used the approximation of Eq.(24) for $\tilde{\mathbf{\Gamma}}_H^{-1}$ with five parameters ($p = 5$). As we see from Figure 3, the strong bias from the $\hat{a}$ estimates disappeared and the variance of the estimates is only slightly higher than the theoretical value that can be calculated using Eq.(10). (Note, however, that the bias for sample sizes of 400 and 800 seems to be slightly increased.) Finally, the linear sampling method was used. Its estimates are asymptotically unbiased and have approximately the same variance as expected. The approximate inverse matrix used was as in Eq.(24) with only two parameters ($p = 2$). Figure 4 shows the MLE $\hat{a}$ estimates without any *a priori* knowledge about the model parameters. All the approximations used here were the same as in the previous cases. Since H is not known and can only be estimated with a given variance, the estimates of a have larger variances than in the previous simulations. The question is how robust those estimates are when $\hat{H}$ can have a slight bias (see Figure 2). As for the geometrical sampling, the bias of $\hat{a}$ gets smaller and its variance is also decreasing rapidly as the sample size increases. On the other hand, for the linear sampling case the estimates seem to be biased for larger sample sizes and their variance does not seem to decrease. The reason for this behaviour lies in the fact that the linear sampling for estimating H is less accurate than the geometrical sampling. The bias in $\hat{H}$ together with its higher variance is responsible for the bias and variance of $\hat{a}$, even if the linear

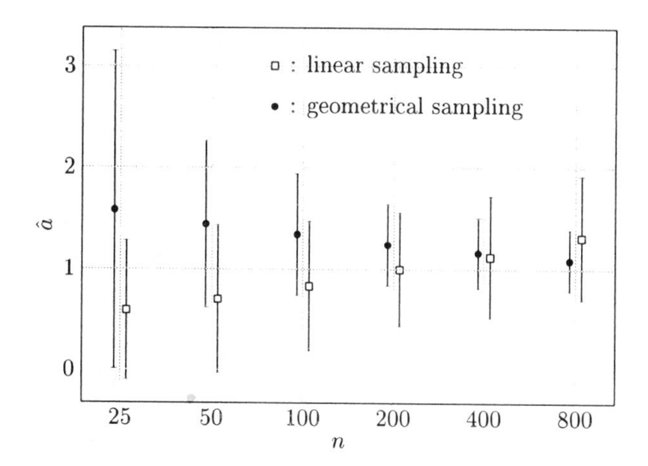

Figure 4 Estimates of a (when H is also estimated) using geometrical and linear sampling and different approximations.

sampling seems to be a better choice to estimate a than the geometrical one for known H (see Figure 3).

As for the MLE estimates for m the geometrical sampling does not give any extra advantage or disadvantage compared to the linear sampling. In fact, the MLE estimate gives almost negligible reduction in the variance of $\hat{m}$ when compared to the sample mean as an estimate for m.

7. CONCLUSION

In this paper we have introduced the idea of using geometrical sampling for the ML estimation of the parameters of fractional Brownian traffic. The intention with this sampling is to reduce the number of sampling points required for a given predefined confidence level. Intuitively, the geometrical sampling distributes the sampling points advantageously at different time scales, whereas linear sampling stresses the finest time scale.

We have derived expressions for the estimators of m and a and the log-likelihood function from which the estimator of H can be derived. Approximations were developed for the inverse and the determinant of the covariance matrix, needed for the calculation of the estimates. With these approximations the evaluation of the log-likelihood function is fast and the maximization with respect to H can easily be made.

The experiments with simulated traffic showed that the geometrical sampling does indeed give a better estimate for H leading to a reduction of sample points. In one example the number of required points was reduced from 800 to 50. For the estimation of a the geometrical sampling does not give any direct advantage, but as the estimator $\hat{a}$ actually depends on the estimator $\hat{H}$, the overall accuracy obtained is better. For the estimation of m, different sampling schemes give essentially the same result, the estimate is basically the observed average rate.

Acknowledgments

This work was done within the Com^2 project funded by the Academy of Finland. Useful discussions with Prof. V. Sharma are gratefully acknowledged.

References

[1] Beran, J. *Statistics for Long-Memory Processes.* Chapman & Hall, One Penn Plaza, New York, NY 10119, 1994.

[2] Deriche, M. and Tewfik, A.H. Maximum Likelihood Estimation of the Parameters of Discrete Fractionally Differenced Gaussian Noise Process. *IEEE Transactions on Signal Processing*, 41(10):2977–2989, October, 1993.

[3] Erramilli, A., Pruthi, P. and Willinger, W. Self-Similarity in High-Speed Network Traffic Measurements: Fact or Artifact? *Proc. in 12th Nordic Teletraffic Seminar—NTS12*, (Eds, Norros, I. and Virtamo, J.), pages 299–310, Espoo, Finland, 22-24 August, 1995.

[4] Krishnan, K.R., Neidhardt, A.L. and Erramilli, A. Scaling Analysis in Traffic Management of Self-Similar Processes. *Proc. of 15th International Teletraffic Congress—ITC15*, (Eds, Ramaswami, V. and Wirth, P.E.), pages 1087–1096, Washington, DC, USA, 22-27 June, 1997.

[5] Ninness, B. Maximum likelihood estimation of the parameters of fractional Brownian motions. *Proc. of the 34th IEEE Conference on Decision and Control*, 4:4018–4023, 1995.

[6] Norros, I. A storage model with self-similar input. *Queueing Systems*, 16:387–396, 1994.

[7] Roberts, J., Mocci, U. and Virtamo, J. eds. Broadband Network Teletraffic—Performance Evaluation and Design of Broadband Multiservice Networks. *Final Report of Action COST 242*, pages 354–369, Springer, 1996.

[8] Rózsa, P. *Linear Algebra and its Applications.* University Press, Budapest, 1974. (in Hungarian)

MPEG VIDEO TRAFFIC MODELS: SEQUENTIALLY MODULATED SELF-SIMILAR PROCESSES

Hai Liu, Nirwan Ansari, and Yun Q. Shi
New Jersey Center for Wireless Telecommunications
Department of Electrical and Computer Engineering
New Jersey Institute of Technology
University Heights, Newark, NJ 07102, USA

Abstract [1] New traffic models are called for to facilitate the design of effective admission and flow control algorithms, and network performance evaluation in accommodating video traffic. We propose a new approach, sequentially modulated self-similar processes (SMSSPs), to model MPEG coded video traffic. SMSSPs are shown to be able to capture both short range dependency (SRD) and long range dependency (LRD) of the video traffic accurately. Traffic data are decomposed according to the MPEG data structure, into several parts, each modeled as a self-similar process. These processes are then modulated sequentially in a manner similar to how the frames are grouped into the GOP (Group of Pictures) pattern.

Keywords: MPEG, modulated self-similar processes, video traffic modeling, long range dependency, short range dependency.

1. INTRODUCTION

The trend to transmit video over network, especially over ATM, is emerging. Traffic models are important to network design, performance evaluation, bandwidth allocation, and bit-rate control. It was, however, observed that traditional models fall short in describing video traffic because video traffic is strongly autocorrelated and bursty [1]. To accurately model video traffic, autocorrelations among data should be taken into consideration. A considerable

[1]This work was done in part while N. Ansari was on leave at the Department of Information Engineering, Chinese University of Hong Kong, Hong Kong, and Y.Q. Shi was on leave at the Information Engineering Div., Nanyang Technological University, Singapore.

amount of effort on video modeling has been reported. These models are used to capture two statistical factors: marginal distribution (first-order statistics) and autocorrelation function (second-order statistics) of traffic arrival times. The importance of long range dependency is among the most arguable issues in video modeling. Some of the results support the view that LRD has drastic impact on queuing performance [2],[3],[4],[5], while other results support the view that LRD has little impact on queuing performance because of the fact that the buffer capacity is limited in practice [6].

While the importance of long range dependency is arguable, the impact of short term autocorrelation in traffic processes on queuing performance with a finite buffer can be very drastic (see [7] and references in it). Simulation results show that the network queuing performance with strong and weak autocorrelation traffic may be quite different. Thus, a model should capture not only the first-order statistics, but also the second-order statistics. SRD models can capture short-term autocorrelation, but fail to capture long-term dependency. LRD models, on the other hand, can capture long-term dependency, but underestimate the short term dependency.

Markov-Renewal-Modulated TES (transform expand sample) models were used to model JPEG encoded motion pictures. One of the drawbacks of TES is that the ACF of a TES process for lags beyond one cannot be derived analytically. It can only be obtained by searching in the parameter space, and thus good results can hardly be guaranteed [8].

The $M/G/\infty$ input process model is a compromise between LRD and SRD models [8]. Simulation results were found to be better than those of a self-similar process when the switch buffer is relatively small. Better results than those of DAR(1) [9] model were found when the buffer size is large. The results were derived for JPEG and the I frames of MPEG sequences. As will be shown below, ACF of MPEG sequences is quite different from that of JPEG sequences and that of I sequences. In our opinion, it is almost impossible to accurately capture the ACF of MPEG compressed data by a simple function such as the exponential function, and thus this method fails to capture the second-order statistics of MPEG sequences.

In [10], I, P, and B frames were modeled separately. I frames were described by three parts: scene length, average I frame size during a scene, and variations from average frame size during the scene. P and B frames were modeled as $i.i.d$ processes. As we will show, I, P, and B frames are all LRD processes. Another characteristic of B frames is that its ACF exhibits a repeated pattern (see Fig. 1, 2, and 3) (The repeated pattern appears in B frames only!). The fact that B frames occupy a very large part of the whole sequence and B frames are also rather large (actually in *star wars*, the largest frame is a B frame) suggests that the impact of B frames cannot be ignored, and it is inappropriate to model P and B frames as $i.i.d$ processes.

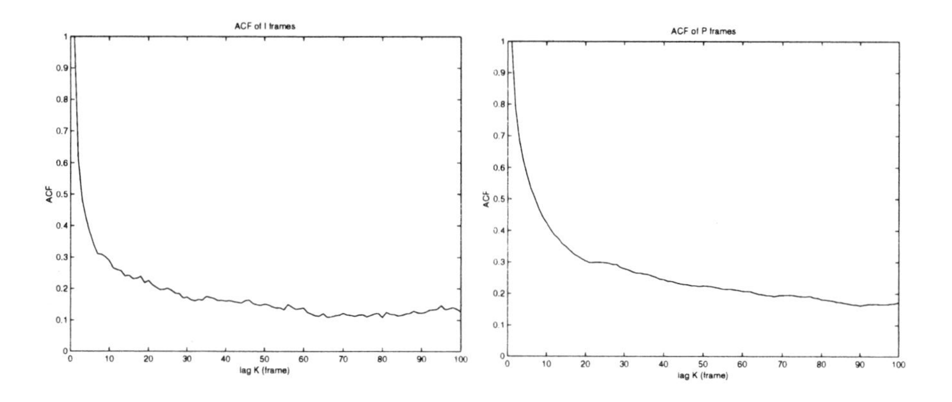

Figure 1 ACF of star wars' I frames.

Figure 2 ACF of star wars' P frames.

In this paper, we propose to model MPEG compressed video sequence by sequentially modulated self-similar processes, in which the original sequences are decomposed into several parts that can be modeled by self-similar processes. It has been found that video traffic possesses self-similarity, and thus it is natural to model video traffic by self-similar processes. Self-similar processes have very simple ACF forms, and therefore, are easier to analyze than other kinds of processes. The rest of the paper is organized as follows. Section 2 describes empirical data and ACF. Concepts of SRD, LRD and self-similar processes are presented in Section 3. Modeling of decomposed data and the whole data set along with simulation results are discussed and presented in Section 4.

2. EMPIRICAL DATA AND ACF

The empirical data used here was MPEG coded data of *Star Wars*[2]. The source contains materials ranging from low complexity/motion scenes to those with high and very high actions. The data file consists of 174,136 frames, each having a different frame size (bytes per frame). The movie length is approximately 2 hours at 24 frames per second. The original video was captured as 408 lines by 508 pixels, and then interpolated to 240×352 (Luminance - Y), and 120×176 (Chrominance - U and V). Every frame was partitioned into blocks of 8×8 pixels. These data blocks were transformed using DCT. After DCT transformation, coefficients were quantized and Huffman coded. Run length coding was further used to reduce bit rate. Motion estimation techniques were used to compress data volume. The frames were organized as follows: IBBPBBPBBPBB IBBPBB . . . , i.e., 12 frames in a Group of Pictures

[2]The MPEG coded data were the courtesy of M.W. Garrett of Bellcore and M. Vetterli of UC Berkeley.

(GOP). I frames are those which use intra frame coding method (without motion estimation), P frames are those which use inter frame coding technique (with motion estimation), and B frames can be predicted using forward and backward prediction.

The ACF of the MPEG coded *Star War* is shown in Fig 4. It fluctuates around three envelopes, reflecting the fact that, after the use of motion estimation and forward/backward prediction techniques, the dependency between frames is reduced. This characteristic should be taken into consideration in modeling MPEG coded video sequences. We propose to decompose the sequence into I, P, B_1, $B_2, \cdots$, B_8 according to the GOP pattern, and model every part by a different self-similar process.

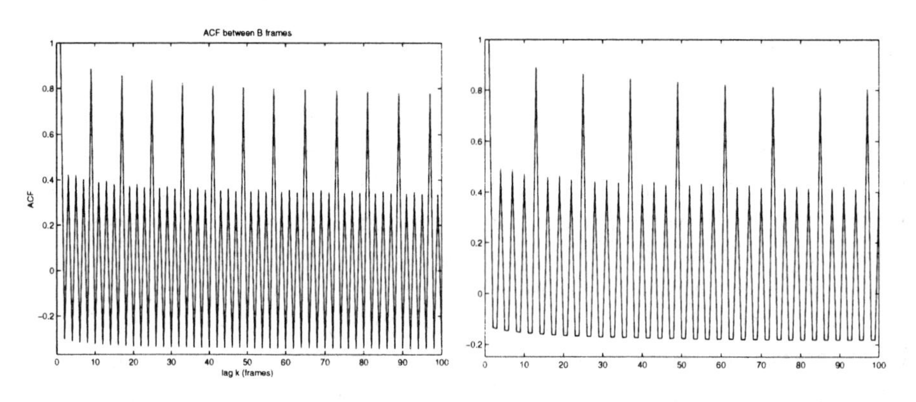

Figure 3 ACF of star wars' B frames.

Figure 4 ACF of the MPEG video.

3. SRD, LRD, AND SELF SIMILARITY

Consider a stationary process $X = \{X_n : n = 1, 2, \dots\}$ with mean μ and variance σ^2. The autocorrelation function and the variance of X are denoted as:

$$r(k) = \frac{E[(X_n - \mu)(X_{n+k} - \mu)]}{\sigma^2} \tag{1}$$

and

$$\sigma^2 = E[(X_n - \mu)^2]. \tag{2}$$

X is said to be SRD if $\sum_{k=0}^{\infty} r(k)$ is finite; otherwise, the process is said to be LRD [11].

Let X defined above have the following autocorrelation function:

$$r(k) \rightsquigarrow k^{-\beta} L(k), k \to \infty \tag{3}$$

where $0 < \beta < 1$, and L is a slowly varying function as $k \to \infty$, i.e., $\lim_{t\to\infty} L(tx)/L(t) = 1$ for all $x > 0$. Consider the aggregated process

$$X^{(m)} = \{X_t^{(m)}\} = \{X_1^{(m)}, X_2^{(m)}, \dots\},$$

where

$$X_t^{(m)} = \frac{1}{m}(X_{tm-m+1} + \cdots + X_{tm}), t \in Q, m \in Q, \tag{4}$$

and Q is a positive integer set. X is said to be exactly second-order self-similar [11] if

$$varX^{(m)} = \sigma^2 m^{-\beta} \tag{5}$$

and

$$r^{(m)}(k) = r(k) \tag{6}$$

for all $m \in \{1, 2, 3, \cdots\}$ and $k \in \{0, 1, 2, \cdots\}$. Here $r^{(m)}(k)$ is the autocorrelation function of $X^{(m)}$. In fact, Eq. (5) is sufficient to define a self-similar process since Eq. (3) and (6) can be derived from Eq. (5) [11].

Since empirical video traffic exhibits self-similarity and long range dependency, it is intuitive to use self-similar processes to model video traffic. It is one of the most often used processes to capture LRD of video traffic. Often times, a self-similar process is simply referred to as a LRD process.

Hurst parameter $H = 1 - \beta/2 (0 < \beta < 1)$ is used to measure the similarity of a process. It is the only parameter needed to describe a second-order self-similar process. For a process with self-similarity, $1/2 < H < 1$.

4. MODELING MPEG TRAFFIC

In order to model MPEG coded data, we decompose the MPEG traffic into 10 sub-sequences $X_I, X_P, X_{B_1}, X_{B_2}, \cdots$, and X_{B_8}. X_I consists of all I frames, X_P consists of all P frames, the first B frames in all GOPs constitute X_{B_1} , the second B frames in all GOPs constitute X_{B_2}, and so on. We have used $k^{-\beta}$, $e^{-\beta k}$, and $e^{-\beta\sqrt{k}}$, corresponding to the ACFs of a self-similar process, a Markov process, and an $M/G/\infty$ input process, respectively, to approximate ACFs of these processes. For illustrative purposes, some of these approximations are shown in Fig. 5 to 12. The sums of squares of errors obtained by the three kinds of methods are tabulated in Table 1. It is quite obvious that self-similar processes are better choices. We therefore use self-similar processes to model these data.

Using the least squares method, β = 0.4663, 0.3546, 0.4468, 0.4779, 0.4294, 0.4656, 0.4380, 0.4682, 0.4465, and 0.4606 are derived for X_I, X_P,

Table 1 Least square errors obtained by self-similar process, markov and $M/G/\infty$ method

	I	P	B_1	B_2	B_3	B_4	B_5	B_6	B_7	B_8
LRD	0.46	0.35	0.45	0.48	0.43	0.47	0.44	0.47	0.45	0.46
$M/G/\infty$	5.13	12.3	5.40	4.80	6.23	5.19	5.83	5.05	5.03	5.34
Markov	7.98	21.0	9.45	7.30	11.1	8.12	10.2	7.95	7.22	8.75

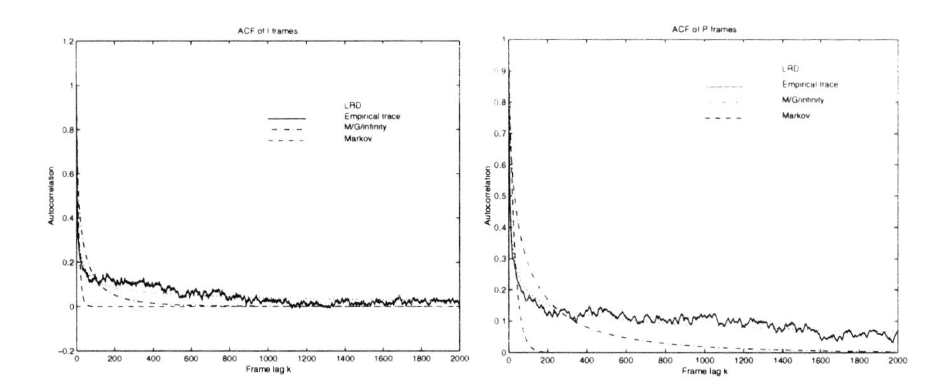

Figure 5 Approximation for ACF of I frames by : LRD, M/G/∞, and Markov processes.

Figure 6 Approximation for ACF of P frames by : LRD, M/G/∞, and Markov processes.

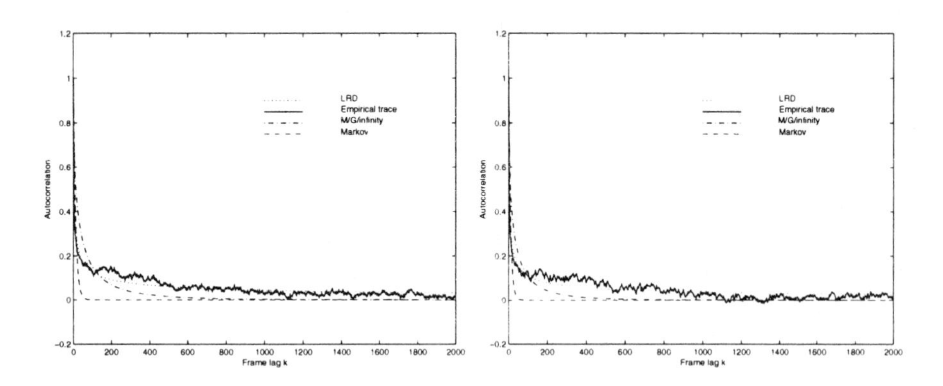

Figure 7 Approximation for ACF of B_1 frames by : LRD, M/G/∞, and Markov processes.

Figure 8 Approximation for ACF of B_2 frames by : LRD, M/G/∞, and Markov processes.

$X_{B_1}, X_{B_2}, \cdots$, and X_{B_8}, respectively. The corresponding Hurst parameters

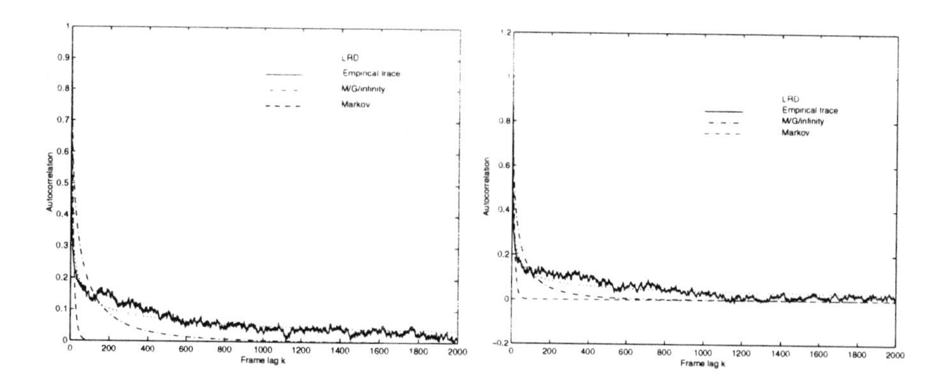

Figure 9 Approximation for ACF of B_3 frames by : LRD, M/G/∞, and Markov processes.

Figure 10 Approximation for ACF of B_4 frames by : LRD, M/G/∞, and Markov processes.

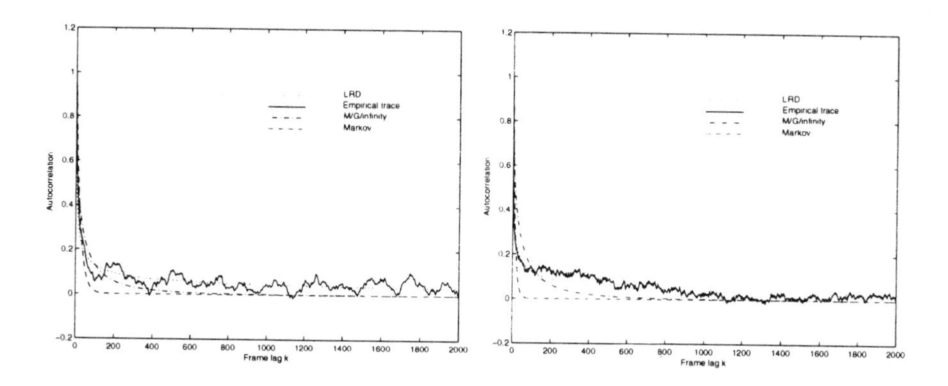

Figure 11 Approximation for ACF of B_7 frames by : LRD, M/G/∞, and Markov processes

Figure 12 Approximation for ACF of B_8 frames by : LRD, M/G/∞, and Markov processes

for these processes are $H = 0.7668, 0.8227, 0.7766, 0.7610, 0.7853, 0.7672, 0.7810, 0.7659, 0.7768, 0.7697$, respectively.

To model marginal distributions of these processes, we use Beta distributions which have the following form of probability density function:

$$f(x;\gamma,\eta,\mu_0,\mu_1) = \begin{cases} \frac{1}{\mu_1-\mu_0}\frac{\Gamma(\gamma+\eta)}{\Gamma(\gamma)\Gamma(\eta)}\left(\frac{x-\mu_0}{\mu_1-\mu_0}\right)^{\gamma-1}\left(1-\frac{x-\mu_0}{\mu_1-\mu_0}\right)^{\eta-1} & \\ \qquad \mu_0 \le x \le \mu_1, 0 < \gamma, 0 < \eta & \\ 0 \qquad \text{otherwise}, & \end{cases} \quad (7)$$

where γ and η are shape parameters, and $[\mu_0, \mu_1]$ is the domain where the distribution is defined. They can be estimated by the following formulae [12]:

$$\hat{\eta} = \frac{1-\bar{x}}{s^2}[\bar{x}(1-\bar{x}) - s^2] \tag{8}$$

$$\hat{\gamma} = \frac{\bar{x}\hat{\eta}}{1-\bar{x}} \tag{9}$$

where

$$\bar{x} = \frac{1}{N}\sum_{i=1}^{N} x_i, \tag{10}$$

$$s^2 = \frac{N\sum_{i=1}^{N} x_i^2 - \left(\sum_{i=1}^{N} x_i\right)^2}{N(N-1)}, \tag{11}$$

and N is the number of data in the data set. Using these formulae, $\hat{\eta} = 1.5237$, 1.5699, 1.4172, 1.3016, 1.6858, 1.6329, 1.7276, 1.4218, 4.0585, 1.5402, and $\hat{\gamma} = 12.7263$, 11.1939, 8.1089, 8.1604, 11.8499, 13.9278, 12.2180, 8.6536, 10.4233, 11.1768 are derived for X_I, X_P, X_{B_1}, X_{B_2}, $\cdots$, and X_{B_8}, respectively.

By combining X_I, X_P, X_{B_1}, X_{B_2},$\cdots$, and X_{B_8} in a manner similar to the GOP pattern, a model for MPEG coded traffic is obtained. This model can be used to generate traffic data.

Fig. 13 shows a trace of the empirical video traffic, and the trace generated by our model is shown in Fig. 14. Note the similarity between these two figures. Since traffic is random, the appropriateness of a traffic model should be judged by its statistical properties rather than the mere similarity between these two figures. This can be demonstrated by the ACF of the generated traffic shown in Fig. 15 (compare to Fig. 4), and the ACF of B frames shown in Fig. 16 (compare to Fig. 3), implying that the proposed model can capture both the LRD and SRD of B frames[3]. The ACF of P and I frames can also be captured very well (not shown here owing to the limited space.)

5. CONCLUSIONS

We have proposed a new traffic model, sequentially modulated self-similar processes, to model MPEG compressed video sequence. The model can match

[3]That is, the model can match the ACF of B frames for both large and small lag, k. Owing to the limited space, the ACF of B frames for large k is not shown.

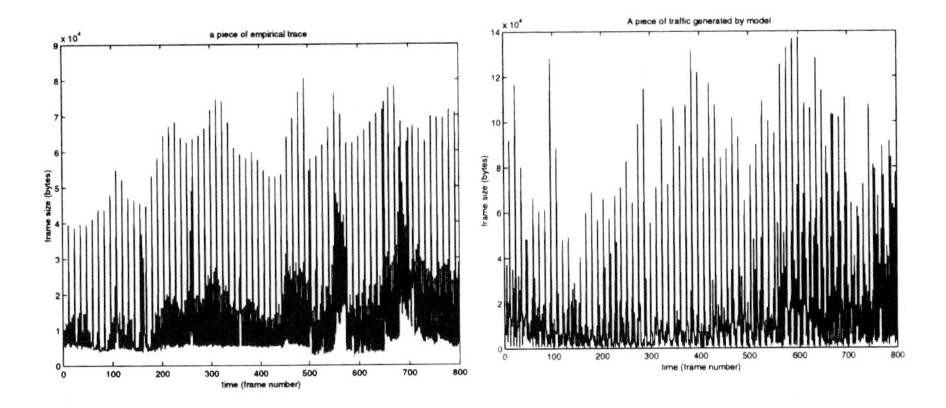

Figure 13 A trace of the empirical traffic data.

Figure 14 Traffic data generated by our model

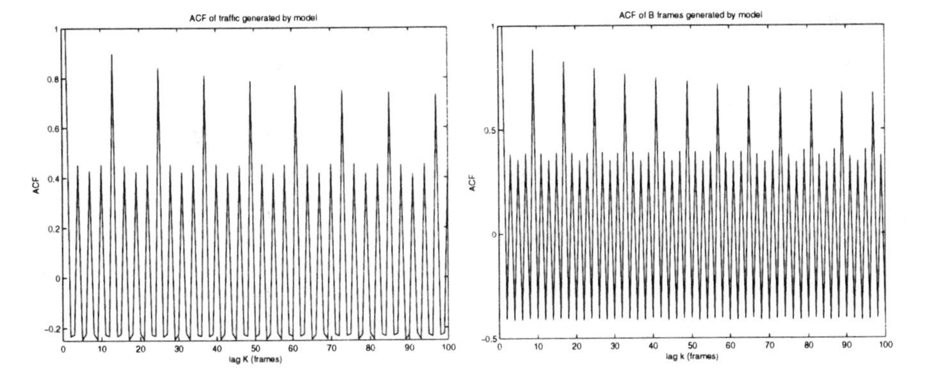

Figure 15 ACF of traffic data generated by our model

Figure 16 ACF of B frames generated by our model

the ACFs of the P, I, and B frame sequences very well. It can capture both the LRD and SRD. This model will play an important role in future network design and performance evaluation.

6. ACKNOWLEDGEMENT

The authors would like to further acknowledge M.W. Garrett for explaining their results.

References

[1] V. Paxson and S. Floyd, "Wide-area traffic: the failure of Poisson modeling," *Proc. SIGCOMM'94*, pp. 257–268, London, U.K., Aug. 1994.

[2] R. Addie, M. Zukerman and T. Neame, "Performance of a single server queue with self-similar input," *Proc. IEEE ICC'95*, volume 3, pp. 461–465, Seattle, WA, 1995.

[3] N. G. Duffield and N. O'Connell, "Large deviations and overflow probabilities for the general single server queue, with applications," DIAS-STP-93-30, Dublin Institute for Advanced Studies, 1993.

[4] C. Huang, M. Devetsikiotis, I. Lambadaris, and A. R. Kaye, "Fast simulation for self-similar traffic in ATM networks," *Proc. IEEE ICC' 95*, pp. 11–22, Seattle, June 1995.

[5] I. Norros, "A storage model with self-similar input," *Queueing Systems*, vol. 14, pp. 387–396, 1994.

[6] B. K. Ryu and A. Elwalid, "The importance of long-range dependence of VBR video traffic in ATM traffic engineering: myths and realities," *Proc. ACM Sigcomm'96*, pp. 3–14, Stanford University, CA, Aug. 1996.

[7] B. Melamed and D. E. Pendarakis, "Modeling full-length VBR video using markov-renewal-modulated TES models," *IEEE/ACM Trans. Networking*, vol. 5, pp. 600–612, Feb. 1998.

[8] M. Krunz and A. M. Makowski, "Modeling video traffic using $M/G/\infty$ input processes: a compromise between Markovian and LRD models," *IEEE J. Selected Areas in Comm.*, vol. 16, pp. 733–749, June 1998.

[9] D. P. Heyman and T. V. Lakshman, "Source models for VBR broadcast-video traffic," *IEEE/ACM Trans. Networking*, vol. 4, pp. 40–48, Feb. 1996.

[10] M. Krunz and S. K. Tripathi, "On the characterization of VBR MPEG streams," *Proc. SIGMETRICS'97*, pp. 192–202, Cambridge, MA, June 1997.

[11] B. Tsybakov and N. Georganas, "On self-similar traffic in ATM queues: definitions, overflow probability bound, and cell delay distribution," *IEEE/ACM Trans. Networking*, vol. 5, pp. 397–408, 1997.

[12] G. J. Hahn and S. S. Shapiro. *Statistical Models in Engineering*. John Wiley & Sons Inc., New York, 1967.

A PRACTICAL APPROACH FOR MULTIMEDIA TRAFFIC MODELING

Timothy D. Neame,[1] Moshe Zukerman[1] and Ronald G. Addie[2]

[1] *Department of Electrical and Electronic Engineering, The University of Melbourne, Parkville, Vic. 3052, Australia.*
{t.neame, m.zukerman}@ee.mu.oz.au

[2] *Department of Mathematics and Computer Science, University of Southern Queensland, Toowoomba, Qld. 4350, Australia.*
addie@usq.edu.au

Abstract This paper presents the M/Pareto process as a practical model for multimedia traffic. We explain the M/Pareto model, discuss its advantages and limitations, and demonstrate its ability to accurately predict the queueing performance of multimedia traffic.

Keywords: Multimedia traffic modeling, stochastic processes, queueing performance.

1. INTRODUCTION

Dimensioning and evaluation of traditional telephony networks is comparatively simple, with the well-known Erlang model accepted right around the world as the appropriate tool for the task. However traditional telephony is forming a smaller and smaller proportion of the total telecommunications load in most industrialised nations. Increasingly, packet-switched networks are replacing circuit-switched ones, and multimedia and data services are replacing simpler telephony services. While the Erlang model has served us well, it is simply not appropriate for this new environment.

We require a new model which is capable of consistently predicting the queueing behaviour of multimedia traffic. This model must be simple to use and give sufficiently accurate results. As yet, there is no widely accepted model which can be used in the modeling of multimedia and other modern traffic types. Many models have been proposed, but a consensus is needed. Meaningful comparison of results is impossible without such a consensus.

In this paper we propose the M/Pareto process as a suitable traffic model for this purpose. The M/Pareto process was presented in [8] as a model for long range dependent (LRD) traffic streams. By specifically taking LRD traffic into account, the M/Pareto process has a significant advantage over some of the other models which have been proposed, as there is now a large body of work which shows that multimedia and data traffic streams have this property of long range dependence [6, 7, 12]. Models which do not account for this property can still be used in certain specific situations, but are difficult to extend to general cases.

Recent work [3, 5, 10] has shown that the M/Pareto process can accurately predict the queueing performance of measured Ethernet traffic and VBR video in a single server queue. The Internet is becoming an increasingly significant transport mechanism for multmedia content, so in this paper we consider the suitability of the M/Pareto process as a model for an IP packet stream.

In Section 2 we describe the modeling of packet-switched networks in general terms. We explain a framework based on modeling each link as a single server queue (SSQ), within which we examine suitability of the M/Pareto process. Section 3 describes the M/Pareto process itself. In Section 4 we explain the method we use to assign values to the parameters of the M/Pareto process so as to fit the queueing results of a given realistic traffic stream. In Section 5 we show that this method of fitting the M/Pareto process has given accurate estimates of queueing performance for Ethernet traffic streams and VBR video traffic. We extend these results to include a comparison with IP traffic in Section 6. In Section 7 we briefly discuss some obstacles which must be overcome to make the M/Pareto process a more useful traffic model.

2. MODELING A LINK

Regardless of the higher layer protocols being used, the single server queue (SSQ) forms a fundamental building block in the modeling of packet based networks. A packet-switched network can be considered to be a network of SSQs, with each link represented by an SSQ. We evaluate a given model in terms of its ability to characterise the buffer overflow probability of measured traffic in a SSQ for a range of threshold values and service rates. To do this, we measure the statistics of a given traffic stream, then fit the parameters of the model process so as to match the measured statistics. If the matched process consistently produces similar buffer overflow probability values as the modeled stream for a range of thresholds and service rates, then we say it is a good model of the given traffic stream.

If we consider each link to have a fixed maximum capacity, then the service rate of the SSQ representing that link will have a constant service rate. We use a $G/D/1$ discrete time SSQ with an infinite buffer as our model for a link. We consider the queueing system to have a general arrival process $\{A_n\}$ which is assumed to be stationary and ergodic. Each value A_n represents the amount of work arriving at the buffer in time interval n. In this paper we discuss two specific arrival processes - the Gaussian process and the M/Pareto process - and compare them with cases in which the $\{A_n\}$ sequence is given by measured traffic values.

We define the statistics of the input process as follows. Let the mean arrival rate be $\mu = \mathrm{E}(A_n)$, and let A_n have variance σ^2. We assume that the process A_n is long range dependent (LRD), with Hurst parameter H. The Hurst parameter represents the level of correlation in the process. We use the techniques described in [1] to estimate the Hurst parameter in the measured traffic streams.

The arrival process is fed into an SSQ with a constant service rate of τ per interval. For the infinite buffer queue the queueing curve is given by considering the values of the buffer overflow probability, $\Pr(Q > t)$, for a range of values for the threshold t.

3. THE M/PARETO PROCESS

It is now widely accepted that LRD streams are an important part of the traffic carried on broadband networks. Since LRD traffic, regardless of its source, is characterised by significant long bursts (see [12] and references therein), it is appealing to model LRD traffic using a model involving long bursts. The M/Pareto model is such a process, generating an arrival process based on overlapping bursts. The M/Pareto model described below is closely related to that given in [8], and is one of a family of such processes which form a sub-group of the $M/G/\infty$ models explored in [11].

M/Pareto traffic is composed of overlapping bursts. The duration of each burst is a random variable chosen from a Pareto distribution. Using the Pareto distribution here allows the significant long bursts which characterise LRD traffic to appear in the model traffic. The complementary distribution function for a Pareto-distributed random variable is given by

$$\Pr(X > x) = \begin{cases} \left(\frac{x}{\delta}\right)^{-\gamma}, & x \geq \delta, \\ 1, & \text{otherwise}, \end{cases}$$

$1 < \gamma < 2$, $\delta > 0$. The mean of X is $\frac{\delta\gamma}{(\gamma-1)}$ and the variance of X is infinite. Note that the mean of the Pareto process in [5] was given incorrectly, and was a factor of γ too small.

Burst arrivals are given by a Poisson process with rate λ. The arrival rate for each burst is constant for the duration of that burst, and has rate r. All bursts generate work at the same rate r. Thus the mean amount of work within one burst is: $\frac{r\delta\gamma}{\gamma-1}$. The mean total amount of work arriving from all bursts within an interval of length t is

$$\mu = \frac{\lambda t r \delta \gamma}{(\gamma - 1)}. \tag{1}$$

Although the Pareto process has infinite variance, the variance of the M/Pareto process is finite. In [12] the term "Poisson burst process" was used to refer to processes like the M/Pareto process, where i.i.d. bursts of fixed rate start according to a Poisson process. For a Poisson burst process the variance function is given by repeatedly integrating the distribution function:

$$\sigma^2(t) = 2\lambda r^2 \int_0^t dt \int_0^u du \int_v^\infty dx \, \Pr\{X > x\}$$

Calculating for Pareto distributed burst durations gives

$$\sigma^2(t) = \begin{cases} 2r^2\lambda t^2 \left(\frac{\delta}{2}\left(1 - \frac{1}{1-\gamma}\right) - \frac{t}{6}\right), & 0 \leq t \leq \delta \\ 2r^2\lambda \left\{\delta^3 \left(\frac{1}{3} - \frac{1}{2-2\gamma} + \frac{1}{(1-\gamma)(2-\gamma)(3-\gamma)}\right) \right. & \\ \quad +\delta^2 \left(\frac{1}{2} - \frac{1}{1-\gamma} + \frac{1}{(1-\gamma)(2-\gamma)}\right)(t-\delta) & \\ \quad \left. - \frac{t^{3-\gamma}}{\delta^{-\gamma}(1-\gamma)(2-\gamma)(3-\gamma)}\right\}, & t > \delta \end{cases} \tag{2}$$

This corresponds with the variance function for processes of this type given in [2]. It represents a correction to the variance function quoted in [3, 5, 10].

Examining the expression for the variance given in Equation (2), we see that for large t, the dominant term is $2r^2\lambda\frac{\delta^\gamma t^{3-\gamma}}{(1-\gamma)(2-\gamma)(3-\gamma)}$. If we define $H = \frac{3-\gamma}{2}$, then we can observe that for increasing t the growth of this function is proportional to t^{2H}. This implies that this model is *asymptotically self similar* with Hurst parameter

$$H = \frac{3-\gamma}{2}. \tag{3}$$

The rate of the Poisson process, λ, controls the frequency with which new bursts commence. The superposition of two independent M/Pareto processes with identical burst length distributions will itself be an M/Pa-reto process with Poisson arrival rate equal to the sum of the arrival rates of the two constituent processes. Thus, increasing λ can represent an increase in the number of sources making up an M/Pareto stream, and λ can be thought of as representing the level of aggregation in the stream.

4. MATCHING THE M/PARETO

In the previous section we have seen that there are four parameters which define the behaviour of an M/Pareto process. These are: the Poisson arrival rate, λ; the arrival rate within a burst, r; the rate of decrease of the Pareto tail, γ; and the starting point of the Pareto tail, δ. Equations (1), (2) and (3) define relationships between these four controlling parameters and three measurable statistics – the mean (μ), variance (σ^2) and Hurst parameter (H).

Using these relationships, we can define M/Pareto processes which will produce given values of μ, σ^2 and H. That means we can fit the M/Pareto process to match the statistics of an arbitrary real traffic stream. When fitting the M/Pareto in this way we have a system with three known values and four unknowns, so we can define an infinite number of M/Pareto processes with different parameters but which will all produce the same values of μ, σ^2 and H. These processes will produce differing queueing performance.

In our analysis of the M/Pareto process we consider families of M/Pa-reto processes with different Poisson arrival rates, but identical values of μ, σ^2 and H. There are a number of ways we could achieve this. In this work we limit ourselves to trading off between the Poisson arrival rate for bursts, λ, and the cell arrival rate contributed by each burst, r. The parameters controlling the burst duration (δ and γ) are held fixed. We do this so as to model increasing aggregation in the stream without altering the nature of the sessions making up the stream.

The Hurst parameter, H, depends only on the parameter γ, so altering λ and r will have no impact on that value. Given that we have already chosen δ and γ, the mean and variance values will be given by:

$$\mu = C_\mu \lambda r \quad \text{and} \quad \sigma^2 = C_\sigma \lambda r^2$$

where C_μ and C_σ are constants given by the values of δ, γ and the interval length, t. Clearly these two relationships imply that if we arrange the trade off between λ and r such that the mean is unchanged, the variance will be altered, and vice versa.

We take an approach in which we give first priority to fitting the variance. For each value of λ we choose r such that the variance is unchanged. Thus if λ is multiplied by a factor N, then r will be divided by a factor of $\sqrt{N}$. Making this adjustment will cause the mean to increase by a factor of $\sqrt{N}$. This prevents us from matching the statistics of a given measured process with a pure M/Pareto process with the same statistics. However, we can introduce a constant bit rate (CBR) component, κ, (which may be negative), so as to a match the mean arrival rate of the M/Pareto process to that of the modeled stream. The addition of κ units of work per interval to every arrival interval will not affect the values of σ^2 or H, nor the queueing performance of the process.

5. PREVIOUS RESULTS

In previous work [3, 5, 10] we have shown that the choice of λ in the M/Pareto model has a significant impact on the queueing performance. Specifically, for identical values of μ, σ^2 and H, different M/Pareto processes chosen according to the method described above yield different queueing curves. In [5] it was shown that as the value of λ increases, the probability of buffer overflow

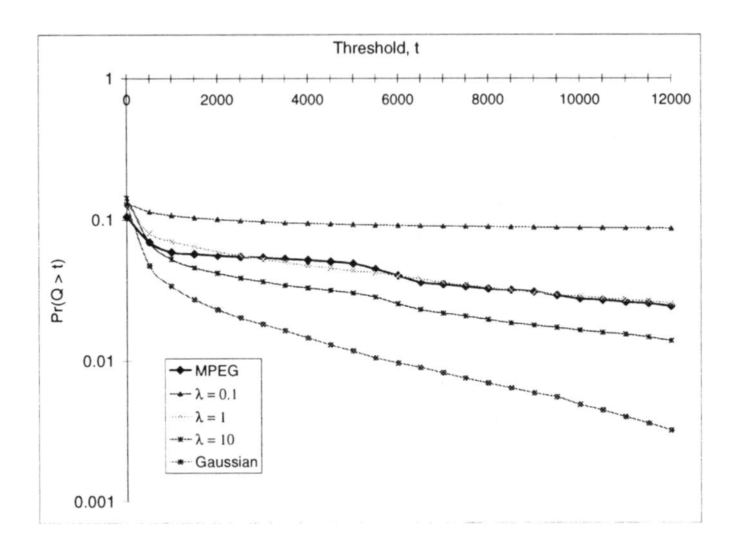

Figure 1 Fitting the M/Pareto process to a VBR video trace

converges to that given by a Gaussian process with the same mean, variance and Hurst parameter. Lower values of λ give higher overflow probabilities.

This convergence to Gaussian is significant as it implies that as levels of aggregation increase, traffic may become more Gaussian in nature. If and when traffic becomes Gaussian, the analytic results given in [2, 4, 5] will provide a simple basis for traffic modeling. Furthermore, the properties of the Gaussian process are such that when multiple Gaussian streams are aggregated together a multiplexing gain is possible.

However, studies of existing broadband networks, such as [9], have clearly shown that at present traffic is not Gaussian in nature. We suggest that, in data networks at least, this is because the level of aggregation in the stream is insufficient to reach Gaussian behaviour. We propose the M/Pareto process, with lower levels of aggregation (values of λ) as a model for this non-Gaussian traffic.

In [3] we showed that the M/Pareto model can accurately predict the queueing performance of an Ethernet trace when the right value for the parameter λ is selected. We found that for lower λ the M/Pareto process over-estimates the overflow probabilities of the Ethernet trace, but when λ is chosen correctly the M/Pareto matches well.

In [10] we showed that matching the queueing curves of VBR video streams is also possible using the M/Pareto process. In this case, even though the modeled traffic is not an aggregated stream, the "level of aggregation" parameter, λ, must still be correctly chosen for an accurate modeling of the VBR video

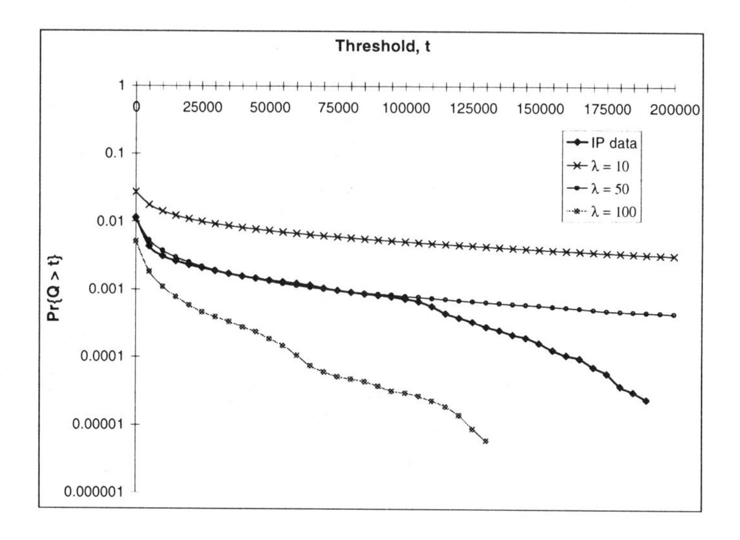

Figure 2 Matching the M/Pareto process to an IP trace

stream to be possible. Figure 1 shows the fitting of an M/Pareto process to one of the VBR MPEG sequences generated by Rose and analysed in [13].

6. IP RESULTS

In Figure 2 we show that the M/Pareto process sucessfully predicts the queueing performance of an IP traffic stream. The link traffic was recorded as a sequence of IP packet header summaries, which were reduced to a sequence of integers, where each value represents the number of bytes transmitted on the link in a 0.1 second interval. For this sequence, we measured a mean arrival rate of 5225 bytes per second and a variance of 21.233×10^6. We found $H \approx 0.90$. As for previous traffic streams, the correct value must be chosen for λ before a fitting is possible. In this case $\lambda = 50$ seems to give an accurate match.

The results given above show that the M/Pareto model can predict the queueing curves of both aggregated data traffic and VBR video streams. Accurate fitting is achieved only when the level of aggregation parameter, λ is correctly assigned. An arbitrary matching the mean, variance and Hurst parameter of the M/Pareto process to that of the original traffic stream is not sufficient to accurately predict the queueing curve.

7. LIMITATIONS OF THE M/PARETO

We have seen that choosing the right value for λ is vital in creating an M/Pareto process capable of matching the queueing curves of real traffic

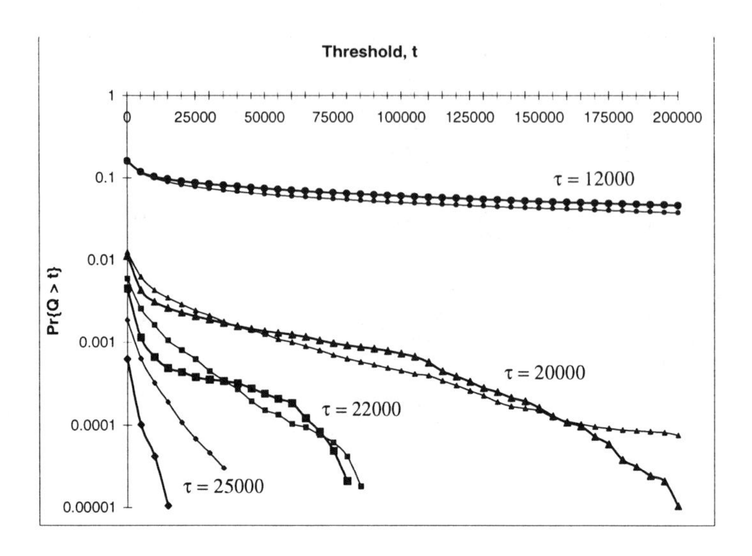

Figure 3 Comparing an M/Pareto with the IP trace for a range of service rates

streams. However the choice of λ is complicated by the fact that the correct value of λ differs depending on the service rate τ.

Figure 3 shows that a single value of λ may not be appropriate for all possible service rates. The figure shows four pairs of curves. In each pair the heavier line represents the queueing performance of the IP trace when fed into an SSQ with service rate τ. The lighter line represents the performance of an M/Pareto process matched to the properties of the IP trace in an identical SSQ. The M/Pareto process used has λ chosen so as to provide a good fit with the traffic when $\tau = 20,000$, i.e. $\lambda = 50$.

Choosing the correct value for λ is not a trivial exercise. As yet we have no systematic method for determining the value of λ to be used in modeling a given traffic stream. Trial and error must be used for each different modeled traffic stream, and for each different service rate. Until a heuristic for determining λ is developed, the practical usefulness of the M/Pareto process will be limited.

The bursty nature of the M/Pareto process makes consistently estimating the queueing performance a difficult task. As yet we have no analytic expressions for the queueing performance of an M/Pareto process, so we must rely on simulation results. The LRD nature of the process means that long simulations are required to give meaningful accuracy in queueing results. However this will be true of any bursty process, and a bursty process is required to model bursty traffic, so any proposed model for bursty traffic will be similarly limited.

8. CONCLUSIONS

In this paper we have shown that the M/Pareto process has much to recommend it as a model for multimedia traffic. We have seen that it can be used to accurately predict the queueing performance of traffic traces taken from both Ethernet and IP links. Between them, these protocols provide carriage for much of the multimedia content distributed today. We have also seen that the queueing performance of VBR video streams can be accurately predicted using the M/Pareto process.

We have seen that for the M/Pareto process to predict the queueing performance of these traffic types, the correct value must be chosen for the level of aggregation, λ, in the model. As yet, we do not have a simple formula or heuristic to determine the correct value for λ. If such a formula can be developed, the M/Pareto process will present a useful practical approach for the modeling of multimedia and data traffic.

Acknowledgment

The authors wish to thank Danielle Liu of AT&T for supplying the IP packet trace analysed in this paper.

References

[1] P. Abry and D. Veitch. Wavelet Analysis of Long-Range-Dependent Traffic. *IEEE Trans. on Information Theory*, Vol. 44, No. 1, January 1998, pp 2–15.

[2] R. Addie, P. Mannersalo and I. Norros. Performance Formulae for Queues with Gaussian Input. *Proceedings of ITC 16*, June 1999, pp 1169–1178.

[3] R. G. Addie, T. D. Neame and M. Zukerman. Modeling Superposition of Many Sources Generating Self Similar Traffic. *Proceedings of ICC '99*, June 1999.

[4] R. G. Addie and M. Zukerman. An Approximation for Performance Evaluation of Stationary Single Server Queues. *IEEE Trans. on Communications*, December 1994.

[5] R. G. Addie, M. Zukerman and T. D. Neame. Broadband Traffic Modeling: Simple Solutions to Hard Problems. *IEEE Communications Magazine*, Vol. 36, No. 8, August 1998.

[6] J. Beran, R. Sherman, M. S. Taqqu, and W. Willinger. Long-Range-Dependence in Variable-Bit-Rate Video Traffic *IEEE Trans. on Communications*, Vol. 43, No. 2/3/4, February/March/April 1995, pp 1566–1579.

[7] W. E. Leland, M. S. Taqqu, W. Willinger, and D. V. Wilson. On the Self-Similar Nature of Ethernet Traffic (Extended Version). *IEEE/ACM Trans. on Networking*, Vol. 2, No. 1, 1994, pp 1–15.

[8] N. Likhanov, B. Tsybakov and N. D. Georganas. Analysis of an ATM Buffer with Self-Similar ("Fractal") Input Traffic. *Proceedings of Infocom '95*, April 1995.

[9] T. D. Neame, R. G. Addie, M. Zukerman, and F. Huebner. Investigation of Traffic Models for High Speed Data Networks. *Proceedings of ATNAC '95*, December 1995.

[10] T. D. Neame, M. Zukerman and R. G. Addie. Applying Multiplexing Characterization to VBR Video Traffic. *Proceedings of ITC 16*, June 1999.

[11] M. Parulekar and A. M. Makowski. Tail Probabilities for a Multiplexer with Self-Similar Traffic. *Proceedings of Infocom '96*, 1996.

[12] J. Roberts, U. Mocci, and J. Virtamo. *Broadband Network Teletraffic, Final Report of Action COST 242*. Springer, 1996.

[13] O. Rose. Statistical Properties of MPEG Video Traffic and Their Impact on Traffic Modeling in ATM Systems. *Proceedings of the 20th Annual Conference on Local Computer Networks*, October 1995.

SESSION 3

Internet Traffic Control

INTEGRATION OF A TRAFFIC CONDITIONER FOR DIFFERENTIATED SERVICES IN END-SYSTEMS VIA FEEDBACK-LOOPS

M. Bechler, H. Ritter, J. Schiller
University of Karlsruhe (TH), Institute of Telematics
Zirkel 2, 76128 Karlsruhe, Germany
[mbechler, ritter, schiller]@telematik.informatik.uni-karlsruhe.de

Abstract More and more applications on the Internet would benefit from Quality of Service provisioning. Unfortunately, the IP-based Internet works according to a best-effort model and gives no QoS guarantees. Thus, the (compared to Integrated Services) simple Differentiated Services architecture has been developed to provide a service better than best-effort. This paper presents components of the Differentiated Services architecture and shows the benefits of local pre-marking and traffic conditioning. This traffic conditioning is combined with a feedback architecture and a simple user interface where a user can express his or her request for more QoS via a simple button. The traffic is formed in a way that QoS is supported and no packet is discarded or is downgraded at the next domain. First results demonstrate the benefits of the approach in a real testbed.

Keywords: Differentiated Services, Operating Systems, QoS, Feedback Mechanisms

1. INTRODUCTION

As more and more people do business on the Internet, many multimedia applications have been introduced, and people pay for network access, it was soon discovered that the traditional best effort packet switching is not sufficient any longer to provide satisfying Quality of Service (QoS). While in the ATM-world an elaborated scheme for QoS provisioning exists (at the expense of higher complexity), the IP-based Internet in its original design lacks QoS support. Thus, the Integrated Services architecture [1] (with the resource reservation protocol RSVP) has been developed to provide "soft" guarantees for traffic streams (so-called flows) across the Internet. However, it was soon discovered that this model has only a limited scalability as all intermediate systems (the IP routers) have to store and manage state per traffic flow. As the Internet community did not want to adopt ideas of signaling, connections, and resource reservation as done in ATM (which would provide QoS), a very simple architecture, Differentiated Services (DS), has been developed. DS does neither use signaling nor does it store per flow state in intermediate systems. The basic idea of DS is the aggregation of traffic streams with a common identifier and the per hop treatment of data packets. Intermediate systems receive a packet, check a single identifier (or few fields), and forward the packet. All packets carrying the same identifiers are treated identically, no matter what source or destination addresses they contain. This local view on packet forwarding makes the approach scalable compared to per flow treatment.

The following section gives a short introduction of the DS approach and focuses on its architectural elements [2, 3]. The third section proposes two different types of DS enabled end-systems, a simpler system that is able to mark packets according to a traffic contract between the sender and a service provider, and a more complex system that is able to shape traffic in a way that outgoing traffic never violates the traffic contract. Section 4 discusses our overall architecture of a DS capable end-system with easy-to-use user interaction via a Q(uality) button in the user interface, while the 5th section presents the elements of our system which condition traffic according to profiles. Finally, section 6 explains the implementation and test scenarios, and presents first results demonstrating the behavior of our implementation.

2. DIFFERENTIATED SERVICES

Primary goal of the DS approach is to provide a simple, efficient, and thus scalable mechanism that allows for better than best effort services in the Internet [2, 3]. DS does not perform any reservation of resources as needed in, e.g., Integrated Services or ATM, nor does it differentiate traffic based on single applications or flows in each networking element (typically a router in IP networks, a switch in ATM networks). In DS, a user has a service agreement with, e.g., an ISP (Internet Service Provider), the ISP gives service guarantees based on aggregates of traffic. For example, the ISP could offer 2 Mbit/s Premium Service [4, 5] and up to 5 Mbit/s total bandwidth to the customer. As long as the user stays within 2 Mbit/s with Premium Service traffic, the ISP guarantees special treatment. Premium Service traffic exceeding 2 Mbit/s will be dropped. The agreements between a user and an ISP or between ISPs are rather of long-term nature and reflect the typical traffic characteristics and not the current requirements of a single short-living connection. The special treatment of different types of traffic is based on the PHB (Per Hop Behavior) of all nodes within a DS domain. Each node only checks the DS field (which carries the DS code point) and applies simple rules (e.g., remarking the packet, inserting into special queues) before forwarding. Several PHBs have been proposed, e.g., "Expedited Forwarding" [4] and "Assured Forwarding" [6]. For the DS field the TOS field in IPv4 or the Traffic Class in IPv6 is used respectively. Using a single DS field allows for a scalable service discrimination without per-flow state and signaling at every hop. Nodes only perform a simple mapping of the DS code point to a local behavior (PHB), there are no rules how to achieve a certain end-to-end service for users by setting the DS field to a certain value.

The DS architecture defines several architectural elements [3]: A DS domain consists of DS interior nodes and DS boundary nodes. DS boundary nodes build the border of a DS domain and can be ingress nodes (handling traffic entering the DS domain) or egress nodes (handling traffic leaving the DS domain). DS boundary nodes connect a DS domain to a node in another DS domain or to a non-DS-compliant node. DS boundary nodes classify and possibly condition ingress traffic in addition to the application of the PHB. Traffic conditioning of ingress traffic is based on the TCA (Traffic Conditioning Agreement) and may comprise metering, shaping, policing, and remarking of the DS field. The DS boundary node may be an ingress router of an ISP or may be co-located with a host generating traffic. Only DS boundary nodes perform classification and conditioning, DS interior nodes only apply the appropriate PHB based on the DS field. Thus, scalability is achieved by shifting all complexity to some boundary nodes.

The classification of traffic in a DS boundary node can be based on two different classifiers: the BA (Behavior Aggregate) classifies traffic based on the DS field only. The MF (Multi Field) classifier may use several fields of the packet such as source address, destination address, DS field, protocol type, ports etc. Now assume the following scenario: a user has three applications running and decides to favor one of them (i.e., the user wants to have a better quality for this application, e.g., higher bandwidth, lower delay). How can this application on a host get a better forwarding behavior? After aggregation the DS domain cannot distinguish different applications or different hosts. But also classification at a boundary node (which is not the sending host) based only on packet fields is problematic. Users do not know in advance which application may be more important as this often depends on the content. Consider, for example, a file upload using ftp – one file could be time-critical business data, the other file just some updates of manuals. Although using the same program from the same host, both packet streams have a different importance for the user and should be treated differently. Thus, nodes besides the host cannot classify packets according to their importance and the use of MF for classification is useless in this case. Signaling the “importance” of traffic originating from single applications as done in RSVP or ATM (via resource reservation) is no option due to the well-known scalability and complexity problems.

RFC2475 states that traffic sources may perform traffic classification and conditioning. Traffic originating from the source domain across a boundary may be marked by the traffic sources directly or by intermediate nodes before leaving the domain which is called initial marking or pre-marking. The advantage of this initial marking is that the traffic source can more easily take an application’s preferences into account when deciding which packets should receive better forwarding treatment. This classification of packets is much simpler compared to classification of aggregated traffic (e.g., aggregated traffic from all hosts of a company) as less classification rules are needed. The host can now set the bits in the DS field according to the local importance of an application and according to the TCA with an ISP. The host can ensure the local conformance of its traffic to the TCA to avoid packet loss or remarking due to a TCA violation. So-called bandwidth brokers may support mid-term changes in QoS requirements or the distribution of resources within administrative domains.

3. END-SYSTEMS FOR DIFFERENTIATED SERVICES

A host performing initial marking is inside an own DS domain. Thus the boundary node of the DS domain is co-located with a host generating traffic. Two approaches of such hosts, realizing different functions of traffic conditioning, are presented in this paper.

3.1 The marking approach

A simple approach to integrate service differentiation into the end-system is to implement the function of packet marking without the knowledge of the TCA. This simple approach will be further called the marking approach. The end-system does not perform any traffic shaping or policing. Therefore, the DS node closest to the end-system acts as a DS boundary node [3]. This DS boundary node is responsible for ensuring that the traffic generated by this end-system conforms to the appropriate TCA. The complete set of traffic conditioning elements as described above (metering, shaping, policing and/or remarking) may be performed on the aggregated traffic coming from the end system.

The data packets of an application can be marked according to the service level currently needed. If an application is started, it will use a default service, in most cases the best effort service. If this service is not sufficient for the application to perform well, the packets can be marked as assured service or even premium service packets. In general, best effort packets are more likely to be discarded than packets marked as assured packets, respectively suffer more delay than premium service packets. Therefore, marking packets as assured service packets will in most cases result in a better throughput of the respective application.

Due to traffic conditioning in the ingress boundary node there is no guarantee for packets marked by the end-system to keep their DS field. If the end-system marks packets for example as premium service packets, the ingress boundary node only passes the packets not exceeding the data rates specified in the TCA. Out-of-TCA packets are discarded. This can be seen as a disadvantage, because traffic generated in the end-system may be discarded at the next hop. Therefore, the end-system should only mark packets with the premium service DS code point if an application really needs this service; an example is a telephony application requiring a minimum amount of guaranteed rate.

Considering assured service packets, the situation differs. If assured service packets exceed the TCA, they are not discarded, only their DS field is cleared and they will be handled further as best effort packets. Due to this behavior, the application generating assured service packets uses the maximum amount of assured service without suffering from packet loss at the ingress boundary node. The main advantage of this approach is its simplicity. There is no need for anything else than a simple packet marker, resulting in a small, CPU-saving code especially suited for small mobile devices.

3.2 The shaping approach

Nevertheless, if the end-system provides additional functions like shaping based on the TCA, this has major advantages: In the case of premium service, the end-system can inform the applications if the requested rate of premium service traffic is supported. Furthermore, the traffic generated by the end-system will most likely not be discarded at the next ingress boundary node, as the traffic already conforms to the TCA. The end-system can also adapt to the available, probably very small bandwidth on the link up to the next boundary node. This more sophisticated approach realizing TCA-aware shaping, is further on called shaping approach. The end-system acts as DS boundary node for traffic from applications running on that host and monitors the conformance to the TCA, i.e., it performs both marking and shaping.

Realizing the shaping approach, the end-system must be aware of the TCA and the rules of classification and conditioning defined by the TCA. This is the main difference to the marking approach described before. The traffic generated by the end-system is therefore more likely not to be discarded at the next hop, and thus the goodput of the end-system can be increased. However, the realization of the end-system as a DS boundary node requires additional functionality and thus complexity compared to the simple marking approach.

4. USERS, FEEDBACK, AND HIERARCHICAL SCHEDULERS

As mentioned above, the classification of traffic flows of different applications in the end-system by the boundary node is quite problematic if it is not the sending host, because the boundary node does not know the relative importance of certain applications. The proposed marking and shaping approaches avoid this problem by performing the traffic conditioning in the end system, where the application's preference can be more easily taken into account. Nevertheless, the user preferences regarding an application can change rapidly depending on the content or a change of the user's focus.

Therefore, the user needs an interface for expressing his or her current preferences. A simple button related to each application is sufficient to express the demand for a better performance of the application. On top of Figure 2 an example of this simple button is shown. A button labeled with the letter Q (denoting Quality of Service) is integrated into the window title bar of all applications running on the system. The user expresses the current preference for an application by clicking on the Q-Button. This approach might be implemented in any windowing system (e.g., X Windows as well as Windows CE) independent of the application. The operating system is provided via a system call with the basic information which comprises the user request of a somehow "better" performance of the application running in the respective window.

This information influences the sharing of resources in favor of the related application. In a first step, the resources "CPU time" and "network bandwidth" are taken into account. The resulting feedback architecture is shown in Figure 1. The upper feedback loop is constituted by the process scheduler. Clicking on the Q-Button influences the share of computation time in favor of the related application. In the simple case of a multimedia application this might be enough to improve its performance, e.g., given a MPEG player requiring more computation time for encoding. Feedback to the user is therefore provided via the visible improvement of the application's performance. This is described in more detail in [7].

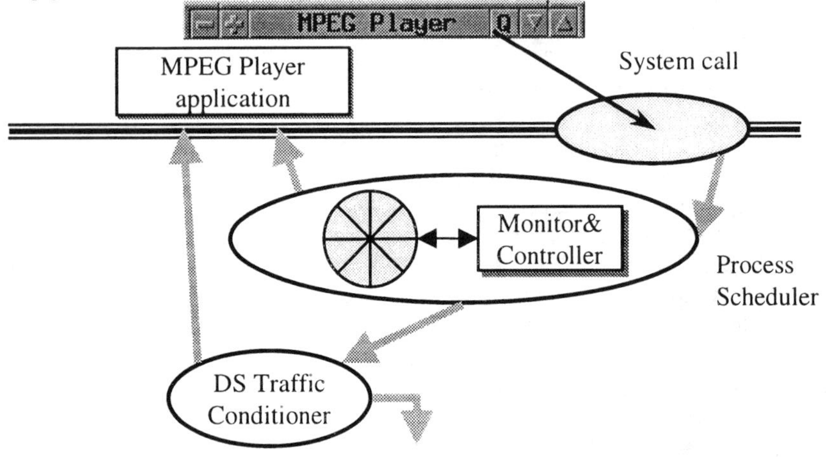

Figure 1 Integration of the DS Traffic Conditioner in the end system.

The focus of the research described in this paper is on the lower feedback loop constituted by the DS traffic conditioner. The DS traffic conditioner realizes the functions of packet marking and/or shaping. The information about the user's preference (the click on the Q-Button) is passed down from the monitor&control-entity of the process scheduler to the traffic conditioner. The elements of the DS traffic conditioner are described in the next section. Feedback to the user is provided by a better performance of applications requiring more bandwidth or a better service level.

5. TRAFFIC CONDITIONER IN END SYSTEMS

In order to integrate the functionality of service differentiation in an end system, the operating system has to be extended by a DS Traffic Conditioner. The main component is the (generic) traffic conditioner (adapted from [8]), that realizes an adaptive resource manager for scheduling the data packets. The consideration of user preferences is realized by a profile that contains the DS parameters that are needed if the application sends data. For every (running) application such a profile is implicitly specified. If a user requests a better transmission quality for an application, a profile manager is responsible for modifying the DS parameters in the profile of this application.

These three components are sufficient for realizing the marking approach described above. As an example, the profile manager modifies the profile of an application to send packets using premium service. Then, the traffic conditioner marks the DS field of the packets generated by this application with the corresponding values. Note that this approach does not require shaping functionality as it is the job of the boundary node to perform the complete set of traffic conditioning functions according to the TCA. The traffic conditioner in the end-system just sets the DS field in this marking approach, independent of the TCA.

The realization of the more interesting shaping approach requires a conditioner that takes the TCA into account. The traffic conditioner is additionally responsible for shaping the packet stream that leaves the network interface, conforming to the TCA, i.e., the functionality of a boundary node is realized in the end system. The traffic conditioner has to be adaptive, as the sharing of a service has to consider the dynamically varying user's preferences on the one hand, and the restrictions determined by the TCA on the other hand. Therefore, the traffic conditioner works in two steps:

1. First, the DS field of each packet is marked according to the requirements stored in the profile of each application. In contrast to the marking approach, the profile for a connection has to be more extensive as more information about the transmission characteristics is needed. Note that the content of the profile does not depend on the TCA.
2. In the second step, traffic conditioning is performed in order to fulfill the TCA. Therefore, the requirements of the applications are adapted to the available resources of the service.

The strategy for traffic conditioning depends on the algorithm that is used. Mainly, two algorithms are conceivable: Priority Based Sharing and Proportional Fair Sharing. Using the Priority Based Sharing, an order of the applications will be established that expresses the priority of each application. The requirements of the applications for a service are then satisfied in turn, depending on the priority.

Using Proportional Fair Sharing leads to an even degradation of the requirements of the applications competing for a service. Therefore, the resource of the service is proportionally shared between the competing applications. A short example shows this in more detail. Assume, there are two applications A1 and A2 that like to send data using premium service. The data rate specified in the profile for A1 is 8 Mbit/s, for A2 it is 4 MBit/s. The specified premium service (in the TCA) is only 9 MBit/s, and thus the requirements of A1 and A2 (12 MBit/s together) cannot be met. Therefore, a degradation of A1 and A2 will be performed by the traffic conditioner to conform to the TCA. The Proportional Fair Sharing reduces the rate of A1 to 6 MBit/s, the rate of A2 to 3 MBit/s, respectively, in order to distribute the resources proportionally.

However, it is quite difficult to decide what a request for a better transmission quality really means. An improvement can be reached, e.g., by extending the share of the current service, but also by using a better service with the original requirements. Thus, the decision for the best conditioning strategy mainly depends on the requirements of the applications. We implemented the Proportional Fair Sharing algorithm as it is more flexible compared to the Priority Based Sharing. It is intuitive for a user, that the sending rate of an application using a service will be degraded, if another application uses this service for transmission. In this case, the user has the possibility to click several times on the Q-Button until the wanted characteristics of the service is reached.

6. IMPLEMENTATION

The integration of DS functionality in an end-system requires changes in the kernel of the operating system. Especially for the transmission of data, modifications on the IP stack are necessary to set the DS field in the header of the outgoing IP packets. Therefore, we use Linux as the platform for implementation of the marking and the shaping approach.

The notification of the DS Traffic Conditioner to achieve a better characteristic for data sent by an application was discussed in chapter 4. This characteristic is stored in the profile that exists for every application running on the system. A profile comprises the following components:

- *service:* Specifies the service that should be used for sending data packets. This could be (currently) the mentioned premium service, assured service, or best effort service.
- *send_priority:* This parameter specifies the current priority of an application for sending data. Using this value, the share of the service for this application is calculated to define the sequence of the outgoing packets.
- *last_send_time:* The time- stamp of the last packet t7he application transmitted is stored in this parameter. This value is also needed to calculate the sequence of the outgoing packets.

The marking approach only uses the service entry of the profile as it just sets the DS field of all data packets generated by an application depending on this value. Therefore, the traffic conditioner identifies the application that sent the data, fetches the service of the profile, sets the DS field, and computes the checksum of the IP header to generate a valid IP packet.

The more sophisticated shaping approach also sets the DS field in the IP header as mentioned above. In a further step the time stamp for transmitting the packet is computed, based on the *send_priority* and the *last_send_time* parameters with the following formula:

$$next\_send\_time = last\_send\_time + length_{packet} \cdot \frac{rate_{CPU} \cdot \sum send\_priority}{rate_{service} \cdot send\_priority}$$

where $length_{packet}$ is the length of the current packet, $rate_{CPU}$ the speed of the CPU in the end system, and $rate_{service}$ the specified rate of the service. The sum of the *send_priorities* covers the *send_priority* of every application that currently sends data. The traffic conditioner is then responsible for sending the packets at the correct time.

In order to show the feasibility of the approach and to gain experience with the handling of a DS-capable end-system we set up measurements using the marking and the shaping approach. The tests and measurements were performed using the DS implementation of the UNIQuE environment [9].

Marking approach: Clicking the Q-Button results in both approaches in a better service for the related application. In the marking approach applications are thus shifted from best effort to assured service and finally up to premium service. Pressing the Q-Button in favor of an application already using premium service nevertheless does not change the service class, because there is currently no "better" service class defined. Using the marking approach the user can not further influence the bandwidth share of two applications using premium service. The fraction of totally available bandwidth an application gets is equal to its fraction of the total bandwidth demand. Given a link of 1 Mbit/s reserved for premium service and applications demanding 0.7 Mbit/s (application A) and 0.466 Mbit/s (application B) respectively, this results in an *average* share of 60% of the totally available bandwidth for application A and an *average* share of 40% for application B. This behavior is a result of the premium service queuing discipline realized in the boundary node closest to the end system.

Shaping approach: Using the shaping approach, the user can influence the bandwidth share of two applications using the same service. The fraction of totally available bandwidth an application gets is equal to the relation of the current user priorities. The user priorities are stored in the profile of each application as described above. Given again a link of 1 Mbit/s reserved for premium service and applications demanding 0.7 Mbit/s (application A) and 0.466 Mbit/s (application B) respectively, the share depends now on the relation of the user priorities for application A and B. Figure 2 shows a scenario with application A beginning to send 0.7 Mbit/s at 0 s and application B starting at 10 s. Given equal user priorities for both applications, the shaper limits the throughput of application A to 0.6 Mbit/s and that of application B to 0.4 Mbit/s as in the marking approach. After 20 s the user prioritizes application A by clicking the Q button and thus rises the throughput to 0.65 Mbit/s. The throughput of application B drops to 0.35 Mbit/s. Given limited bandwidth, this partitioning perfectly reflects the current user preference, expressed by clicking on a simple button.

7. CONCLUSION

The future Internet will need some kind of QoS support. As neither protocols nor applications currently integrate mechanisms for resource reservation and the Internet should remain very simple in its architecture, the Differentiated Services approach claims to introduce at least some better QoS compared to the current best-effort model into the Internet. While typically routers at the edge of administrative domains act as boundary nodes in the DS architecture to protect the domain, we believe in the benefits of hosts acting as additional boundary nodes which already pre-mark packets according to application and user demands, and perform traffic conditioning to avoid packet downgrading or discarding.

Future work comprises more detailed evaluation of the approach in our UNIQuE [9] testbed as well as the integration of the approach in mobile and wireless scenarios. Particularly these environments could benefit from preconditioning as it makes no sense at all to transmit packets via the bandwidth limited air interface only to drop them at a boundary router due to the violation of a traffic contract.

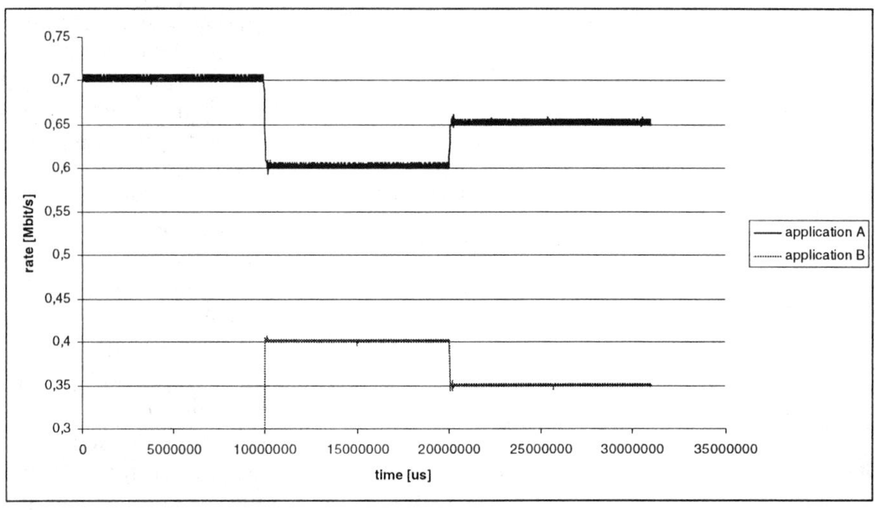

Figure 2 Example result of the shaping approach.

References

[1] R. Braden, D. Clark, S. Shenker: Integrated Services in the Internet Architecture: An Overview, RFC 1633, 1984.

[2] K. Nichols, S. Blake, F. Baker, D. Black: Definition of the Differentiated Services Field (DS Field) in the IPv4 and IPv6 Headers, RFC 2474, 1998.

[3] S. Blake, D. Black, M. Carlson, E. Davies, Z. Wang, W. Weiss: An Architecture for Differentiated Services, RFC 2475, December 1998.

[4] V. Jacobson, K. Nichols, K. Poduri: An Expedited Forwarding PHB, Internet Draft, RFC2598, 1999.

[5] K. Nichols, V. Jacobson, L. Zhang: A Two-bit Differentiated Services Architecture for the Internet, Internet Draft, <draft-nichols-diff-svc-arch-00.txt>, November 1997.

[6] J. Heinanen, F. Baker, W. Weiss, J. Wroclawski: Assured Forwarding PHB Group, Internet Draft, RFC2597, 1999.

[7] J. Schiller: Feedback controlled scheduling for QoS in communication systems, IFIP Conference on High Performance Networking (HPN'98), Vienna, Austria, 1998.

[8] J. Schiller, P. Gunningberg: Feasibility of a Software-Based ATM Cell-Level Scheduler with Advanced Shaping, Broadband Communications '98, Stuttgart, 1998.

[9] UNIQuE, Universal Networking Infrastructure with QoS Enhancements, http://www.telematik.informatik.uni-karlsruhe.de/forschung/UNIQuE/.

A SCHEDULER FOR DELAY-BASED SERVICE DIFFERENTIATION AMONG AF CLASSES

*Mudassir Tufail[1], Geoffroy Jennes[1], Guy Leduc[2]
University of Liege, Belgium.
[1] {mtufail,jennes}@run.montefiore.ulg.ac.be
[2]leduc@montefiore.ulg.ac.be

Abstract In the Differentiated Services (DS) framework, service differentiation is performed among the aggregates (collection of one or more microflows) rather than among the microflows (data stream pertaining to a single connection). We analyse three quality metrics namely bandwidth, loss and delay (that might be used for defining a service differentiation at a DS node) on two criteria 1) service differentiation should be **respected at all loads** and 2) service provision at aggregate level should scale down to microflow level **without being microflow aware**. We find that bandwidth requires microflow aware management, loss lacks in simplicity (though it satisfies the criterion # 2), and delay is the right candidate.

Ensuring better delays at an aggregate level also means ensuring better delays for all the included microflows, and additionally it is easier to define a scheduler that can adapt itself to the relative loads of the aggregates so that relative delays between aggregates are preserved at all loads. Our objective is to provide **relative quantification service** in DiffServ by a delay-based scheduler while satisfying both criteria.

Delay is also a meaningful QoS parameter for both interactive real-time applications and TCP applications, since the mean TCP throughput is roughly inversely proportional to the RTT.

We, therefore, develop a scheduler for Assured Forwarding (AF) PHB where service differentiation among aggregates is based on delays. We provide simulation results that prove that relative delays among aggregates are perfectly respected at all loads.

Keywords: Differentiated Services (DiffServ), application level QoS, relative quantification service, delay-based DiffServ, adaptable scheduling, Assured Forwarding

*This work was supported by the Flemish Institute for promotion of Scientific and Technical Research in the Industry under the IWT project for which University of Liege and Alcatel Alsthom CRC (Antwerp) are the two partners.

1. INTRODUCTION TO DIFFSERV

In Differentiated Services (DiffServ or DS) [1], service differentiation is performed at aggregate level rather than at microflow level. The motivation is to render the framework scalable. The service differentiation is ensured by employing appropriate packet discarding/forwarding mechanisms called Per Hop Behaviours (PHB) at core nodes along with traffic conditioning functions (metering, marking, shaping and discarding) at boundary nodes.
The DiffServ working group has defined three main classes: Expedited Forwarding (EF), Assured Forwarding (AF) and Best Effort (BE). The EF can be used to build a low loss, low latency, low jitter, assured bandwidth, end-to-end **quantitative** service through DS domains. The AF class is allocated a certain amount of forwarding resources (buffer and/or bandwidth) in each DS node. The level of forwarding assurance, for an AF class, however depends on 1) the allocated resources, 2) the current load of AF class and 3) the congestion level within the class. The AF class is further subdivided into four AF classes: AF1, AF2, AF3 and AF4 [2]. Each AF subclass may have packets belonging to three drop precedences which eventually makes 12 levels of service differentiation under AF PHB group. The AF encompasses **qualitative** to **relative quantification** services [6]. In qualitative service, the forwarding assurance of the aggregates is not mutual-dependent, i.e. an aggregate may get forwarded with low loss whereas other with low delay [4]. In relative quantification, the service given to an aggregate is quantified relatively with respect to the service given to other aggregate(s). For example, an aggregate A would get x time better service than an aggregate B.

Motivation:. Despite of fact that the DiffServ proposition is simple and scalable, there are some important issues, notably:

- how would the service differentiation, which is performed at aggregate level, be at microflow level?

- how would the network resource allocation among the aggregates adapt with load so that service differentiation is preserved at all load?

This work aims at resolving the above mentioned two issues and focuses on AF PHB group meant for providing relative quantification service in DiffServ. Our paper is structured as follows: section 2 evaluates three metrics (bandwidth, loss and delay) for service differentiation with respective pros and cons and finally select one on which the rest of the paper is based, section 3 describes a formal description of the selected metric, section 4 develops a VirtualClock-based scheduling algorithm and finally section 5 presents the simulation results and their analysis.

2. DIFFERENT METRICS FOR SERVICE DIFFERENTIATION

There are three quality metrics which might be used for defining a service differentiation among AF classes at a DS node. These are: bandwidth, loss and delay. In the following sections, we study each of these metrics individually.

2.1 BANDWIDTH

If service differentiation at aggregate level is bandwidth-based then one needs to know the number of included microflows (for each aggregate) in order to determine the service differentiation at microflow level. For example, an aggregate getting 50 Mbps would deliver 5 Mbps per microflow if they are 10 whereas it would be 25 Mbps if there are just two microflows inside. Consequently, a microflow in an aggregate (supposed to give highest quality) may get a worse service (than a microflow in any other aggregate) if the aggregate contains a big number of microflows. This can be avoided by PHBs which are microflow aware. It's typically that kind of complexity that we would like to avoid in Differentiated Services deployment.

2.2 LOSS

The loss is often determined in terms of percentage of total data transmitted. Therefore, defining a certain loss ratio for an aggregate can easily be scaled down to all its microflows. Although the loss-based service differentiation does not require microflow aware PHB, it is rendered tedious when combined with packet precedence levels within an aggregate as explained below.
There are three packet drop precedences in an aggregate. The precedence of a packet defines how much it is prone to be discarded in case of congestion. The precedence level of a packet may be selected by the application or by an edge router. Introducing two levels of services differentiation (aggregates & precedences within an aggregate) based on a same metric (i.e. loss) needs to implement extra control and intelligent discard mechanisms. This is to manage all the thresholds (for aggregates & precedences) not only to respect the relative quality of services, at all loads, among aggregates, but also to ensure the relative quality of services among packets of different drop precedences within an aggregate.

2.3 DELAY

The delay is a parameter which provides numerous advantages. The delay metric itself is microflow independent as ensuring better delays at an aggregate level also means ensuring better delays for all the included microflows. If a class X should have lesser delay (i.e. better service) than class Y then the

queue length of X should be kept smaller than that of Y. It can be done either by discarding packets at a higher rate or by serving the queue at higher rate. Discarding packets at higher rates, although keeps the delay shorter, does not offer a reliable service for loss-sensitive applications. On the other hand, servicing a class with a higher rate, so as to limit its queue length, offers a shorter delay as well as a better throughput to its applications. A delay-based service differentiation is thus required to have a self-regulation property, i.e. the service rate for an aggregate would then be calculated dynamically. Note that this dynamic calculation of the service rate does not need to be microflow aware and preserves the service differentiation at all loads. Self-regulation of an aggregate's service rate can be done by knowing just the current queue length of the aggregate and its relative quality index (refer to section 3).
Let us consider that the transport protocol is TCP as is the case with most of the Internet applications these days. The throughput of a TCP application depends on two factors, refer to relation 1 from [5]: RTT (delay [1]) and loss probability.

$$throughput = \frac{Constant}{RTT * \sqrt{loss\_probability}} \tag{1}$$

The self-regulation property of delay-based service differentiation ensures that an aggregate experiences packet loss, during congestion, in proportion to its current load. Consequently, the loss probability of all microflows of all aggregates tend to attain a same value at a DS node. Naturally, an application with lesser delay will end up getting a better goodput, an expected compensation for paying more.

2.4 CONCLUDING REMARKS

We presented three metrics for service differentiation among aggregates. Bandwidth is dropped as it requires the microflow aware PHB whereas the loss metric, when coupled with packet precedence level, is tedious to manage. The delay-based service differentiation, on the other hand, is easy to self-regulate and is microflow independent. We select the delay as a metric for service differentiation and present it formally in the next section.

3. FORMAL DESCRIPTION OF DELAY-BASED SERVICE DIFFERENTIATION

We consider a generic case of N backlogged classes. Let q_i represents the quality index associated with an aggregate i and service differentiation among aggregates follows the relation:

$$q_1 \leq q_2 \leq q_3 \ldots \leq q_N \tag{2}$$

The relation 2 declares the class N with the highest quality index and thus ought to be serviced better than all other aggregates. Let the quality index represents the delay-based (relative) service of an aggregate and d_i represents the delay for an aggregate i, then:

$$\frac{q_i}{q_j} = \frac{d_j}{d_i} \tag{3}$$

We introduce now a self-regulation property in the model. It means that the service rate r_i of an aggregate i is modified with respect to its current load, determined by its current buffer occupation, b_i. This self-regulation is *weighted* as it takes into account the aggregates quality index also:

$$\frac{r_i}{r_j} = \frac{b_i}{b_j} * \frac{q_i}{q_j} \tag{4}$$

In order to maintain the scheduling server work conserving, $\sum^N r_i = r$ where r is the speed of the scheduling server, we may rewrite the relation 4:

$$r_i = \frac{rb_iq_i}{\sum_{j=1}^N b_jq_j} \tag{5}$$

The above relation adjusts the service rate of an aggregate as its queue length changes but without violating the relative service differentiation among aggregates.

4. EX-VC ALGORITHM

This section presents an Extended VirtualClock (Ex-VC) scheduling algorithm and emulates the model presented in section 3. The Ex-VC algorithm has an additional instruction of self-regulation compared to the traditional VC algorithm [7], hence the term "extended". Note that the Ex-VC algorithm is not restricted to four aggregates (of DiffServ) only. It may be used with any number of aggregates (or queues). However, the cost of self-regulation increases with the number of aggregates[2]. Each packet is stamped at its arrival. The packets are then served in increasing order of the stamp values. $v(t)$ represents the system virtual time at time t and is defined equal to the stamp value of the packet receiving service at time t. $v(t)$ is initially set to zero. The stamp value of kth packet of the ith aggregate is written as $stamp_i^k$ whereas the packet itself is denoted as p_i^k. s_i^k and f_i^k represents the instants of service-start and service-finish of a packet p_i^k. Each aggregate i maintains two registers $flow_i$ and VS_i (Virtual Spacing). The $flow_i$ registers are initially set to zero and $VS_i = \frac{L_i^k}{r_i}$ where L_i^k is the size of packet p_i^k and r_i is the service rate of aggregate i. The Ex-VC algorithms works as follows:

At an arrival of a packet p_i^k at instant t

- increase b_i by L_i^k
- $r_i = \frac{r b_i q_i}{\sum_{j=1}^{4} b_j q_j}$ /*self-regulation*/
- $VS_i = \frac{L_i^k}{r_i}$
- $stamp_i^k = max(v(t), flow_i) + VS_i$
- $flow_i = stamp_i^k$

At selecting a packet $p_{i'}^{k'}$, having the minimum stamp value, for service at instant t

- $v(t) = stamp_{i'}^{k'}$ where $s_{i'}^{k'} < t \leq f_{i'}^{k'}$

At departure of the packet $p_{i'}^{k'}$

- decrease $b_{i'}$ by $L_{i'}^{k'}$

About existing algorithms: In [3], a similar delay-based approach for service differentiation is presented. It studies two schedulers, Backlog-Proportional Rate (BPR) and Waiting Time Priority (WTP). The BPR adjusts the service rate (self-regulation property) of an aggregate with its backlog whereas in the WTP the priority of a packet increases proportionally with its waiting time. Both schedulers require separate queues per aggregate (note that this constraint does not exist in the Ex-VC scheduler). The simulation results in [3; 9] show that the WTP is significantly better than the BPR. We envisage comparing the Ex-VC with the WTP in our future simulations.

5. SIMULATIONS

We simulate four AF classes: AF4, AF3, AF2 and AF1. These aggregates have relative quality indexes as: $q_4 = 4$, $q_3 = 3$, $q_2 = 2$ and $q_1 = 1$. We define a warm-up period during which the rate of packet arrival is higher than the packet service rate (i.e. the scheduler speed r). This makes the queue lengths grow. Once the warm-up period is over, the packet arrival rate becomes equal to packet service rate. Moreover, we define three scenarios of packets arrival: symmetrical, equal and asymmetrical (refer to figure 1).

- In the symmetrical packets arrival scenario, the queues are loaded proportionally to their delay guarantees. That is to say that aggregate AF4 will receive packets at rate half of that at which aggregate AF2 would receive the packets. Remember that aggregate AF4 experiences half the delay of AF2. This yields a buffer loading where Ex-VC algorithm self-regulates at a rather easy-going pace.

- In the equal packets arrival rate scenario, all aggregates receive packets at equal rates regardless of their quality indexes (i.e. delay guarantees).

- The third scenario, asymmetrical packets arrival, tests the Ex-VC algorithm in tending-to-worst buffer loading configuration and algorithm self-regulates at a hard-going pace. Here, queues are loaded inversely proportionally to their delay guarantees. In other words, the aggregate AF4 will receive packets at rate double of that at which aggregate AF2 would receive.

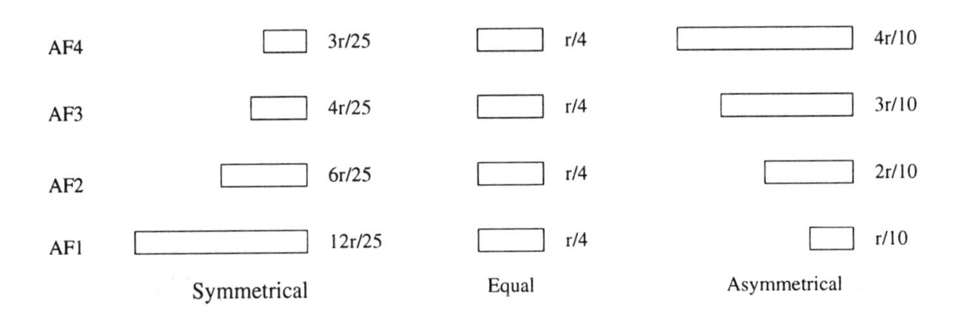

Figure 1 The rate of packet arrivals in three scenarios of buffer loadings

For each simulation type, we analyse two parameters: aggregate's service rates and aggregate's packets delay. All results are shown on the same respective scale. The simulation parameters are: warm-up period is 500 packet slots, simulation time is 10000 packet slots, buffer over-loading factor during warm-up is 2 and service scheduler speed r is 1 packet/time-slot. The first packet of AF1 arrives at $t = 0$, that of AF2 at $t = 1$, that of AF3 at $t = 2$ and that of AF4 at $t = 3$.

- Figure 2: We can notice that, by imposing symmetrical packet arrivals during the warm-up period, queue lengths reach an ideal relative load in order to offer the desired delay-based service differentiation. So it takes a relatively small time for the algorithm to stabilise itself once the arrival rates become equal (post warm-up phase). The aggregates experience relative packet delays in accordance to their respective quality indexes. The packets of the AF4 aggregate ($q_4 = 4$) have a delay four times shorter than those of aggregate AF1 ($q_1 = 1$).

- Figure 3: With equal rates of packet arrivals during the warm-up time, queues do not grow in required relative sizes and the algorithm self-regulates significantly (observe the crossing and fluctuating service rate curves) and takes longer than figure 2 to reach stabilisation. In post warm-up phase, packet arrivals are symmetrical and the algorithm works at easy-pace and ensures the relative service differentiation among aggregates.

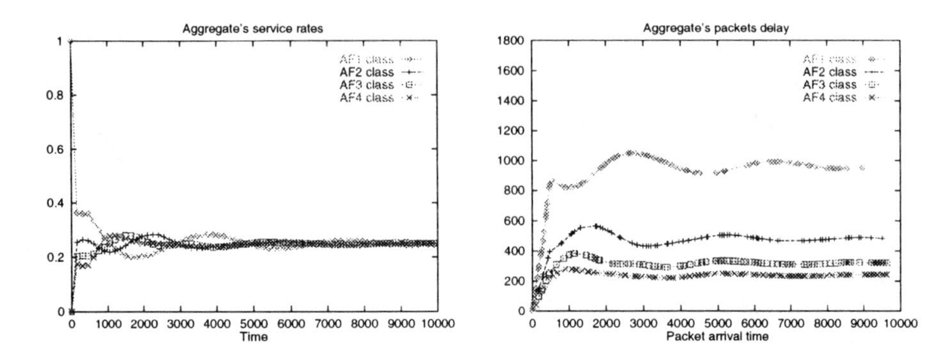

Figure 2 Symmetrical warm-up and equal post warm-up packet arrivals

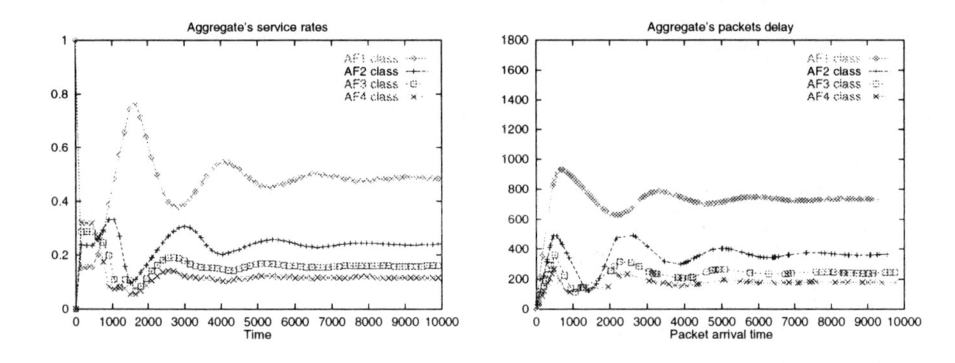

Figure 3 Equal warm-up and symmetrical post warm-up packet arrivals

- Figure 4: After an equal packet arrival warm-up period, the packet arrival rates are then changed to asymmetric rates. The algorithm self-regulates with changing queue lengths (due to changing packet arrival rates). We notice that aggregates suffer individually, though, more delays[3] (due to asymmetric packet arrivals), relative delay differentiation is still respected.

- Figure 5: The asymmetric packet arrivals during the warm-up period have a great influence on the algorithm behaviour even though post warm-up packet arrival is symmetrical. This is because of the fact that queues are over-loaded (arrival rate is twice the service rate) during warm-up. Hard self-regulation is needed to respect the desired relative quality of service between aggregates and a longer period to reach stability.

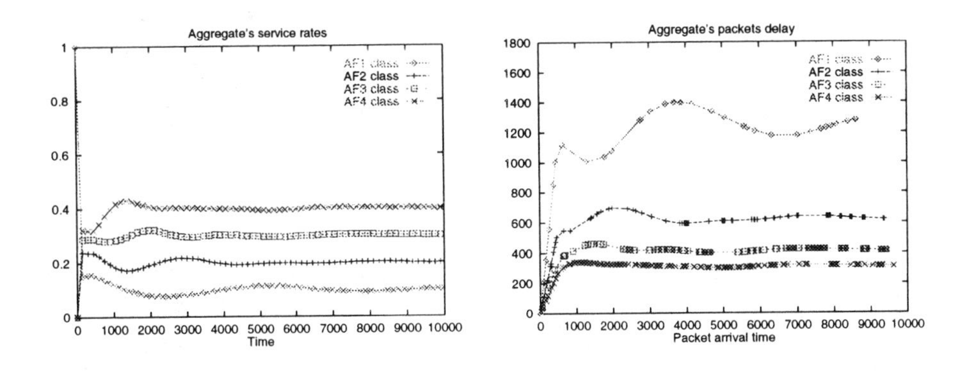

Figure 4 Equal warm-up and asymmetric post warm-up packet arrivals

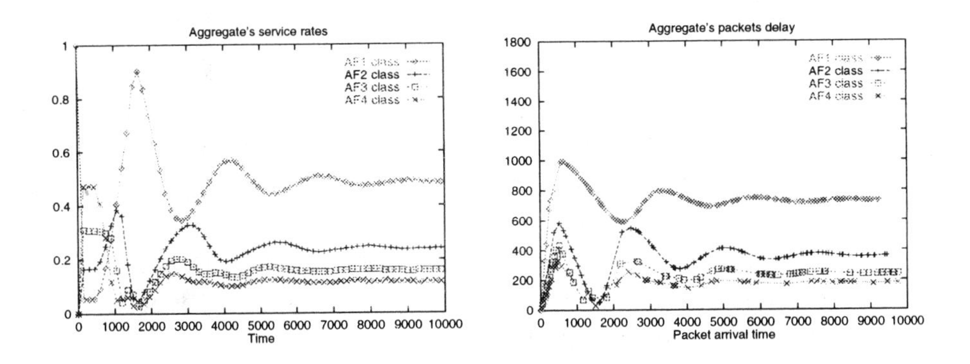

Figure 5 Asymmetric warm-up and symmetrical post warm-up packet arrivals

6. CONCLUSION

This paper evaluates three possible metrics (bandwidth, loss and delay) for service differentiation among four AF classes. Each definition is individually analysed: bandwidth-based service differentiation suffers from scalability problem, loss-based definition lacks in simplicity, delay comes out as the rational quality metric to maintain a relative service differentiation among all aggregates independently of number of microflows in the aggregates. A formal model of delay-based service differentiation is described and is tested via simulations. The simulation results prove that the proposed model provides differentiated service to all aggregates in proportion to their quality indexes. Moreover, this differentiation is respected at all queue loads.

For a PHB to be defined completely, we need, in addition to a scheduling algorithm like Ex-VC, an algorithm which decides whether to accept or discard a packet, at it's arrival, depending upon the factors like congestion level and packet drop precedence. The two algorithms (packet accept/discard and scheduler), when implemented together at a node, helps ensuring the SLAs (Service Level Agreements). Currently we are working on packet accept/discard algorithms and on their simulations for TCP flows when coupled with the Ex-VC scheduler [8].

Notes

1. The delay comprises queueing delay as well as transmission delay. The transmission delay, being insignificant (due to high speed links), does not contribute much in RTT. On the other hand queueing delay stays a non-negligible factor in RTT estimation.

2. One may not perform the self-regulation at each packet arrival (i.e. the instance of its stamp calculation). It has been noted that during the stable periods (i.e. fewer burst arrivals) reducing the frequency of self-regulation by 10 does not have a significant effect on algorithm performance.

3. Note that the average delay value (calculated on the whole buffer) per packet is same for all the configurations.

References

[1] S. Blake, D. Black, M. Carlson, E. Davis, Z. Wang and W. Weiss. *"An Architecture for Differentiated Services"*. Internet RFC 2475, 1998.

[2] J. Heinanen, F. Baker, W. Weiss, J. Wroclawski. *"Assured Forwarding PHB Group"* Internet RFC 2597, 1999.

[3] C. Dovrolis and D. Stiliadis. "Proportional Differentiated Services: Delay Differentiation and Packet Scheduling". To appear in *ACM SIGCOMM-99*, (http://www.cae.wisc.edu/ dovrolis/).

[4] P. Hurley, J. Y. Le Boudec, M. Hamdi, L. Blazevic, P. Thiran. "The Asymmetric Best-Effort Service". *SSC/1999/003*, (http://icawww.epfl.ch).

[5] M. Mathis, J. Smeke, J. Mahdavi and T. Ott. "The macroscopic behaviour of the TCP congestion avoidance algorithm". *Computer Communication Review*, July 3 1997.

[6] Y. Boram, J. Binder, S. Blake, M. Carlson, Brian E. Carpenter. "A Framework for Differentiated Services". *<draft-ietf-diffserv-framework-02.txt>*, Feb. 1999.

[7] L. Zhang. "VirtualClock: A new traffic control algorithm for packet switching". *ACM Transactions on Computer Systems, 9(2), May 1991.*

[8] M. Tufail, G. Jennes & G. Leduc. "Attaining per flow QoS with class-based differentiated services". To appear in *SPIE Symposium on Voice, Video and Data Communications, Conf: Broadband Network, Sept 1999.*

[9] S. De Cnodder and K. Pauwels. "Relative delay priorities in a differentiated services network architecture". *Alcatel Alsthom CRC (Antwerp) deliverable, May 1999.*

RSSP: RSVP SESSION SET-UP PROTOCOL FOR RSVP-INCAPABLE MULTIMEDIA APPLICATIONS

Xiaobao Chen
Lucent Technologies
xchen1@lucent.com

Andrea Paparella
Lucent Technologies
apaparella@lucent.com

Ioannis Kriaras
Lucent Technologies
ikriaras@lucent.com

Abstract RSVP is the QoS signalling protocol used with Integrated Service Framework for QoS provisioning in IP networks. The existing RSVP mechanism requires that an RSVP daemon be embedded with the applications so as to initiate RSVP sessions. This requirement has been envisaged to be a major limitation over the use of RSVP and the associated QoS control mechanisms for those applications or platforms which are either RSVP-incapable or impossible to be modified to introduce a built-in RSVP capability. This paper introduces a mechanism, RSVP Session Set-up Protocol, which overcomes the limitations of existing RSVP implementations by initiating RSVP sessions for RSVP-incapable applications running on different platforms. The proposed mechanism features application and platform independence and eliminates the need for built-in RSVP processing capability for those applications or terminals of short battery life and limited processing power. Moreover it facilitates dynamic QoS provision for wireless mobile devices during hand-off control.

Keywords: QoS, RSVP, Proxy Server, and Multimedia

1. INTRODUCTION

QoS provisioning and Differentiating Classes of Services have been a major issue in research, commercial, industrial, and standardisation activities in both wired and wireless telecommunications. Two major elements need to be provided for a successful provisioning of QoS control: the QoS signalling mechanism and the traffic control. As one of the many alternatives in providing QoS, IETF's Integrated Services architecture provides a framework for applications to choose between multiple controlled levels of delivery services for their traffic flows. Applications need to communicate their QoS requirements to the QoS-control capable nodes along the transit path, as well as for the network nodes to communicate between one another. RSVP, a receiver-oriented QoS signalling protocol, is proposed by IETF as an approach to provide QoS requests to all routers along the transit path of the traffic flows and to maintain the state necessary in the router required to provide different levels of delivery services for applications using RSVP to signal their QoS requests. All existing RSVP implementations have been platform dependent and require an RSVP functional module or a daemon to be embedded with applications so as to initiate a RSVP session. This has incurred serious limitations of using RSVP and IntServ for QoS provisioning for those applications and platforms that are RSVP-incapable or difficult or even impossible to be changed to build the RSVP functionality. Moreover it makes it difficult to support seamless hand-off with QoS control using RSVP for a mobile device or terminal which roams between different base stations and thus requires instant updating of an RSVP session between itself and its communication peer(s). It is, therefore, an major motivation for inventing a mechanism as proposed in this paper to overcome the limitations of existing RSVP implementation mechanisms and facilitate the use of RSVP in a wireless mobile environment. Section 2 introduces the proposed mechanism, RSVP Session Set-up Protocol (RSSP) and RSSP Agent and Client, with detailed discussions on its design and implementation. Section 3 provides a performance evaluation. In Section 4 a review of the related work is given. Finally Section 5 concludes the paper with some observations.

2. RSVP SESSION SET-UP PROTOCOL (RSSP) AND ITS DESIGN

RSSP is defined to specify the RSVP session invocation procedures between a RSSP agent and a RSSP client. A RSSP agent is set up which is able to process standard RSVP messages and, at the same time, is able to intercept the RSVP messages destined for RSSP clients that have registered with the RSSP agent.

2.1 RSSP Agent Discovery

(i) RSSP Agent Advertisement

The location of an RSSP agent is advertised by sending (multicasting or broadcasting) Client Request Messages (CRQM). The CRQM message can be an UDP packet that bears the information about the current location and the provided services by the RSSP agent. A RSSP client uses the information in these advertisements to register and issues an RSSP service request to a RSSP agent. The advertisement information as carried in a CRQM can include:

- The current location, specifically the IP address, of the RSSP Agent.
- The service access point, specifically a port number, on which the RSSP Agent is listening for any incoming service request and to which a RSSP client sends its RSSP service request.
- The lifetime of the agent. The default lifetime is eternal.
- The security control information such as security key or index used between the proxy agent and the client.
- The services provided by the RSSP agent. For example, it may only provide a subset of IntServ services in its RSVP session set-up service (Controlled Load or Guaranteed Service only). It may provide all or just one or two of the three reservation styles, FF, WF, SE.

(ii) RSSP Agent Soliciting

For a RSSP client which requires the set-up of RSVP session but does not have an agent registry in its own list of valid RSSP agents, it sends (multicasting or broadcasting) messages called Agent Soliciting Message (ASM). An RSSP agent in the vicinity running in its normal service state will respond with a unicast agent advertisement message Agent Response Message (ARM) sent directly to the soliciting client. After receiving the ARM, the soliciting client responds by sending a Client Registration Message (CRGM) to the agent as it does upon receiving an Agent Advertisement Message. The ASM can be an UDP message with a broadcast address as its destination address. In addition, it contains:

- The location, specifically the IP address, or other identity information of the client.
- The requested RSVP session services, including the reservation style, unicast or multicast service, Controlled-Load or Guaranteed service, etc.

- The requested lifetime of service. The lifetime has to be specifically indicated otherwise the RSSP agent rejects the RSVP service request.
- The security control information such as the encryption key.

The ARM message can be an UDP message with the soliciting client's address as the destination address. It contains:

- The confirmation of the validity of the solicitation from the client.
- The availability or the confirmation of the requested RSVP session service.
- The suggested lifetime of the service. It should be no longer than the required service time.
- The security control information such as the encryption key or the authentication information to be used during the interaction between the client and the agent.

A RSSP agent advertisement message, CRQM, uses a broadcast UDP message to advertise its presence, location and services.

2.2 RSSP Agent Registration

A RSSP client registers with its RSSP agent by sending a unicast Client Registration Message (CRGM) message to the Agent. The CRGM message can be an UDP message sent to the RSSP agent with specific QoS/RSVP session service requirements including:

- The explicit QoS requirements such as the specifications on average data rate, maximum delay, delay variation, peak rate and packet loss rate.

Or

- The implicit QoS requirement such as specific coding algorithm s and the codecs being used by the clients, e.g. H.263, MPEG-4, etc.
- The selected signalling protocol type, Type "1" is reserved for RSVP.

The selected QoS control type, Type "1" is reserved for Integrated Service.

- The requested service time during which the client expects the RSSP agent to set up and maintain its RSVP session(s).
- Security control information such as the authentication information and encryption key.

An CRGM message can be one of the following five types:

- RSSP_REQ: A RSSP service request message sent from a client to a RSSP agent. It is usually sent by a client serving as a data sender to invoke the transmission of RSVP *Path* messages through the RSSP agent.
- RSSP_IND: An RSSP service indication message issued by a RSSP agent and sent to a specific RSSP client serving as a data receiver. It indicates the arrival of *Path* messages at the RSSP agent for a particular flow to be received by the client.
- RSSP_REP: An RSSP session service response message issued by a RSSP client as an indication to the RSSP agent to start sending RSVP *Resv* messages.
- RSSP_CON: An RSSP session service confirmation message issued by a RSSP agent at the data sender's side to confirm the arrival of *Resv* messages at the RSSP agent and the successful set-up of a RSVP session.
- RSSP_REJ: An RSSP session service rejection message issued by a RSSP agent as an indication of the failure of setting up of a RSVP session, in particular, when the RSSP agent receives *PathErr/ResvErr* messages reporting errors during the set-up of *Path/Resv* states at certain routers.

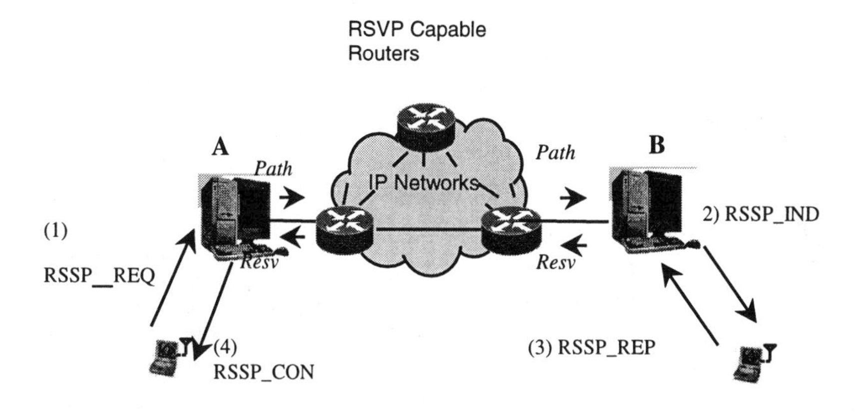

Figure 1. The RSSP Agent Registration Process

Figure 1 shows the control procedures and the control messages exchanged between a RSSP client and its RSSP agent and between the peer RSSP agents. It demonstrates an example for using RSVP as the QoS signalling protocol to inform the intermediate IP routers of the QoS specifications for a specific data transmission and set up and maintain the RSVP states (*Path/Resv* states). RSSP agent A serves a RSSP client to be a data sender while RSSP gent B serves a RSSP client to be a data receiver. A RSVP session is maintained by the RSSP agent(s) which sends *Path/Resv* message periodically to refresh the RSVP soft-states until an explicit agent de-registration request is received or the requested service time expires.

3. RSSP SERVERS AND PERFORMANCE EVALUATION

To test the control effectiveness and evaluate the feasibility of the proposed mechanism, two RSSP Servers are set up with the functions of RSSP agent's and connected to our multimedia QoS test-bed. The network configuration of the test-bed is shown in Figure 2. Two sub-networks including both wired and wireless multimedia servers and terminals are interconnected through RSVP-capable routers. Multimedia servers provide video/audio/web services to the multimedia access and display terminals as what is shown as the RSSP client. RSSP clients (the wireless terminal and the multimedia server) register with its local RSSP Server to invoke the set-up of a RSVP session across the inter-networking routers. A successful RSSP Server operation in response to a RSVP session set-up request from the multimedia server and the terminal results in the establishment of a RSVP session along the transit path of the multimedia flow(s) between the multimedia server and the multimedia terminal(s).

3.1 Qualitative Performance Evaluation

For the current configuration of the test-bed, Video-on-Demand (VoD) Servers and the multimedia terminals run on machines running Windows 95/NT4.0/SunSoloaris where no changes or modifications are introduced to make the VoD servers and our wired/wireless terminals RSVP-capable. To test the control effectiveness of the proposed RSSP scheme, two multimedia display terminals (laptops) running Windows95 are used to request the same video service from the VoD server. The VoD session and the RSVP session configuration parameters are shown in Table 1 and Table 2, respectively. Two scenarios with different network loading conditions are investigated, one without network overload and the other with network overloaded and thus with traffic congestion at the routers.

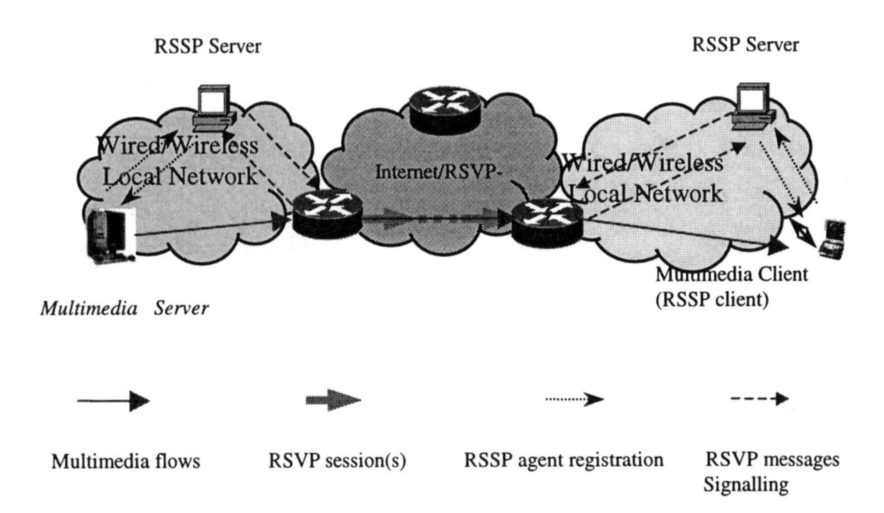

Figure 2. RSSP/QoS Multimedia Test-bed Network Configuration

For the non-overloading scenario, the routers are configured to provide a total transmission bandwidth of no less than 2Mbps. It is observed that the perceptive video and audio display quality at both terminals are satisfactory without observable delay or standstill or broken images. There is no perceivable loss of audio and video inter-media synchronisation.

Multimedia Terminals	Source/Port Number	Dest./ Port Number	Protocol	Codec	Frame Rate (Frames/sec)
Terminal I	135.86.224.86 /1555	192.128.71. 20/ 1122	UDP	H.261	33
Terminal II	135.86.224.86 /1556	192.128.71. 68/ 1234	UDP	H.261	33

Table 1. VoD Session Parameters

For the network-overloading scenario the routers are configured to provide a total transmission bandwidth of 1Mbps. It is clearly observed that the audio/video perceptive quality for terminal I is guaranteed in that no difference is observable in the perceptive qualities from that in the non-overloading network scenario. However, for terminal II without RSVP session service and thus no QoS control over its VoD session, a serious deterioration in perceptive audio/video quality is observed. A proximately two seconds delay is observed in the video display in comparison with that

RSVP Session	Source/ Port No.	Dest./ Port No.	Pro- tocol	Service	Avg./ Peak Rates (kbps)	Min/Max MTU (bytes)	Buf. (bytes)	Slack Term (us)
Session I	135.86.224.86/1555	192.128.71.20/1122	UDP	Guaranteed	650/ 750	64/9138	81250	100
Session II	X	X	X	X	X	X	X	X

Table 2. RSVP Sessions for Terminal I and II.

for terminal I. Broken and even standstill pictures happen from time to time during its traffic bursts. A total loss of audio and video inter-media synchronisation is also clearly observed.

3.2 Quantitative Performance Evaluation

In order to investigate detailed behaviours of QoS control of the RSVP-capable routers under network overloading conditions with RSSP Servers, further experiments are carried out with detailed statistics analysis on a series of packet transmission quality parameters, including effective received data rates, packet loss, maximum end-to-end transmission delay and delay variations (jitter). Traffic generators and receivers are set up running on separate machines in the two sub-networks as shown in Figure 2. At the transmission side, the traffic generator is configured to transmit five simultaneous traffic streams with transmission bursts. The detailed configuration parameters are shown in Table 3.

A RSVP session is set up between the traffic generator and the traffic receiver for Stream II via the RSSP Server in the local sub-network across the inter-networking routers. The RSVP Session State for Stream II is shown in Table 4.

At the receiver side the traffic is monitored and receiving statistics of the all five streams are collected. In order to test the control effectiveness of the RSVP session established through the RSSP Server under the network over-loading condition, the routers are configured to provide maximum 500 kpbs bandwidth in total so that the routers work under over-loaded conditions with five simultaneous streams, each of which generates an average date rate of 128 kbps.

Traffic Streams	Src./ Port Number	Dest/. Port Number	Traffic Pattern	Protocol	Avg. Data Rate (kbps)	Peak Data Rate (kbps)	Packet Size (bytes)	Tx. Time (packets)
Stream I	135.86.224.50/4000	192.128.71.20/5001	Poisson	UDP	128	256	1000	100000
Stream II	135.86.224.50/4000	192.128.71.20/5002	Poisson	UDP	128	256	1000	100000
Stream III	135.86.224.50/4000	192.128.71.20/5003	Poisson	UDP	128	256	1000	100000
Stream IV	135.86.224.50/4000	192.128.71.20/5004	Poisson	UDP	128	256	1000	100000
Stream V	135.86.224.50/4000	192.128.71.20/5005	Poisson	UDP	128	256	1000	100000

Table 3. Traffic Generation Statistics

RSVP Session	Src./ Port Number	Dest. / Port Number	Protocol	Service	Avg./ Peak Rates (kbps)	Min/Max MTU (bytes)	Buf. (bytes)	Slack Term (us)
Stream II	135.86.224.50/4000	192.128.71.20/5002	UDP	Guaranteed	128/230	64/9138	16000	100

Table 4. RSVP Session Configuration for Stream II.

Two separate experiments are carried out. In the first experiment, no RSVP session is set up for any of the streams and therefore the routers provide best-effort service to all streams. In the second experiment, a RSVP session is set up for Stream II as shown in Table 4. The measured performance statistics include the effective received data rate, packet drop, end-to-end maximum transmission delay and the delay variation. The results are shown in Figure 3, 4, 5 and 6, respectively.

Figure 3 shows that, for stream II with a RSVP-enabled QoS provision, it receives data at the average rate of 129.8 kbps. The minor difference from its requested average data rate 128 kbps is caused by the uneven arrival of packet transmission which uses Poisson distribution and thus the data rate varies in time. The measured received data rate is calculated on the average over 100000 packets. Therefore, there is a slight variation from the requested average transmission bandwidth. It is also observed that all other streams, III, IV, V and I, receive data at an average rate of equable share of the left bandwidth after the resource reservation for stream II. It is also interesting to point out that all five streams will share the total available bandwidth when no RSVP service is invoked as shown on the right side of

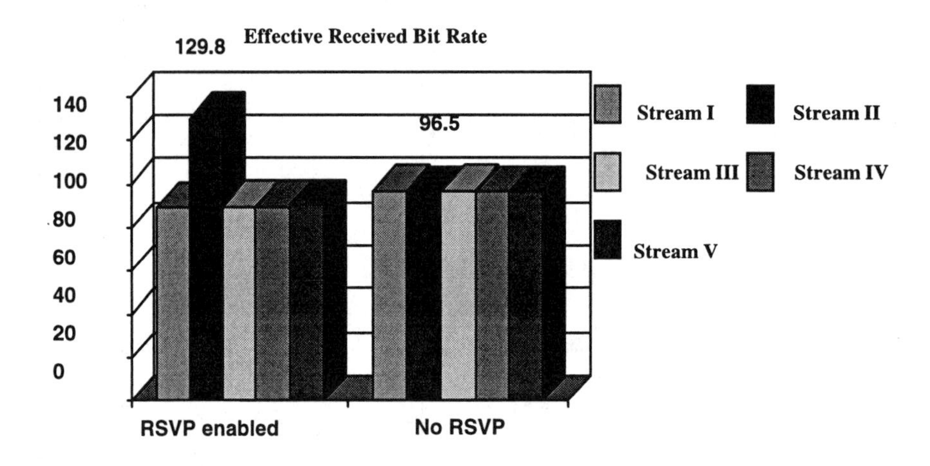

Figure 3. Per-Flow Effective Bandwidth with/without RSVP Sessions

Figure 3. This has clearly signified the best-effort service nature of IP routers when there is no resource reservations. The distribution of packet loss is shown in Figure 4. It clearly shows that stream II is subject to no packet loss while all other streams suffer from dramatic high level of packet loss due to the traffic congestion at each router and uncontrolled packet drop. It is also observed that the overall network over-load does not affect the packet loss level of stream II with RSVP-enabled resource reservation as long as the reserved resources (buffer length) can accommodate the burst of its traffic, while for all other traffic streams without resource reservations, more network overload leads to higher levels of packet loss. Again for the best-effort service nature of IP networks as featured by each IP router when RSVP or other QoS provisioning mechanism is not in operation, all streams will be subject to the same level of packet drop as shown by the histogram on the right-hand side in Figure 4

It can be seen from Figure 5 that stream II with RSVP enabled QoS control over its data packets at each router has its end-to-end delay minimised in comparison to all other streams without RSVP session service. Similar to the bandwidth and packet loss control as shown in Figure 3 and 4, the same level of maximum end-to-end transmission delays is incurred on all streams as the best-effort service provided by each router if there is no QoS control enabled. It needs to be noted that the values as shown in Figure 5 are measured by synchronising the system clocks across the networks, i.e.,

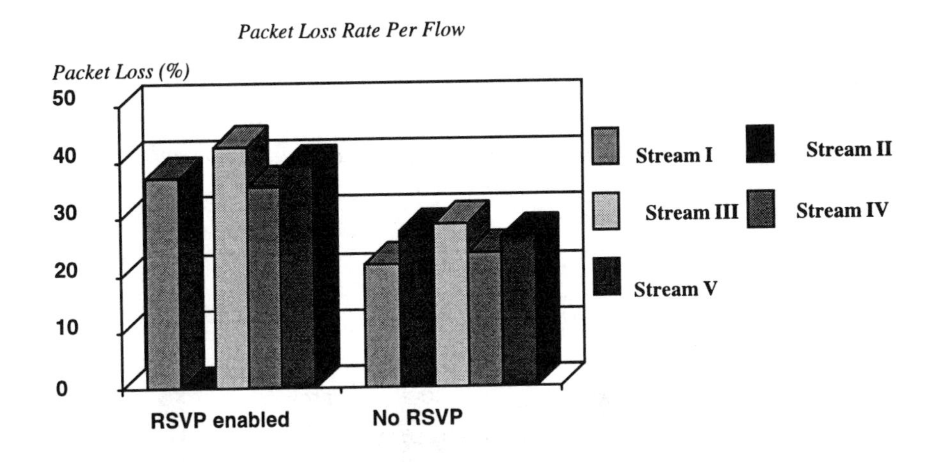

Figure 4. Per-Flow Packet Loss with/without RSVP Sessions

between the traffic generators and receivers as well as across the routers, via NTP running as a daemon in each machine. The estimated clock synchronisation skew achieved by the NTP is about half a second under the light-load network condition. Although the Integrated Service QoS Framework is deployed together with RSVP for QoS provisioning in the current test-bed, according to the IntServ RFC specifications, it does not provide control over the maximum achievable delay variations. But the actual delay variation as experienced by each packet for stream II is limited and dramatically reduced in comparison with those experienced by all other streams without RSVP enabled resource reservations. This is clearly shown by the histogram on the left-hand side in Figure 6.

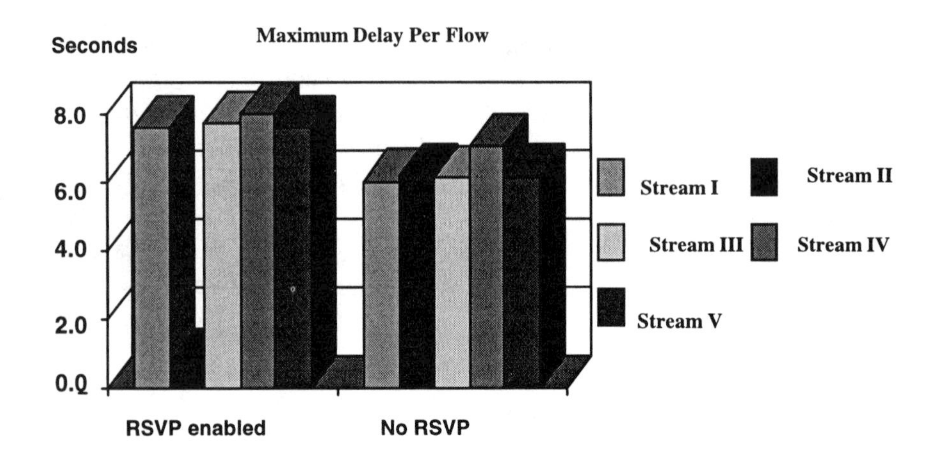

Figure 5. Per-Flow Maximum End-to-End Transmission Delay with/without RSVP Sessions

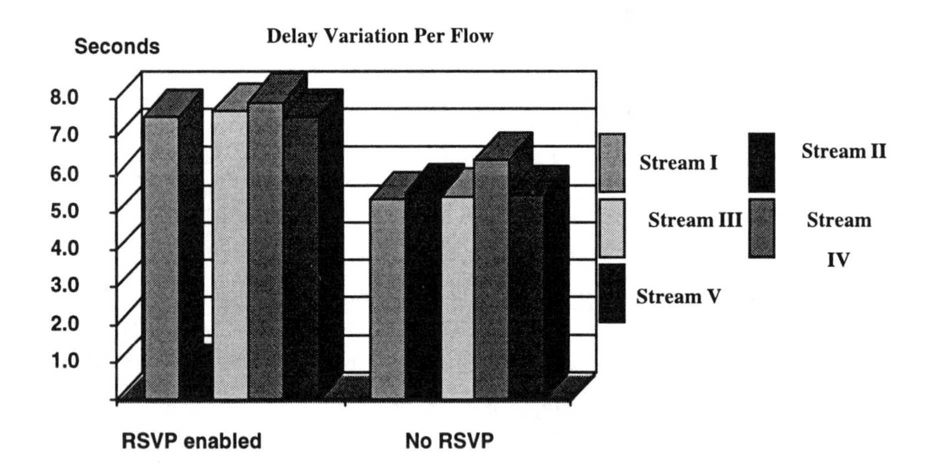

Figure 6. Per-Flow Delay Variation with/without RSVP Sessions

Similar to the measurement of end-to-end delay, the measured delay variation is subject to the clock synchronisation skew which fluctuates over time, in particular, during the network over-load condition which will cause long delay and loss of NTP messages used to refresh the system clocks across the networks.

4. RELATED WORK

3Com supports RSVP both on the router family NetBuilderII and on the switch family CoreBuilder. What is particularly interesting is the CoreBuilder 3500. It is a 10/100 Mbps Ethernet L2/L3 switch that provides differentiated levels of QoS for multimedia applications. As an alternative approach to the SBM mechanism, using RSVP for local resource reservations in CoreBuilder is obviously a complement to the proposed RSSP Server that provides resource reservations over inter-networking nodes. Ascend Communications (http://www.ascend.com) GRF Multi-gigabit Router supports both Integrated Service/RSVP, offering Controlled Load and Guaranteed services, and Differentiated Service models. At the time of writing the paper, it is still in development status. The Cisco Systems routers provide differentiated levels of QoS for multi-media applications. They support Integrated Service/RSVP Control Load service, which are RSVP and IntServ compliant. The proposed RSSP and the implementation of RSSP Server/Client has been tested with the Cisco Systems Routers for RSVP/IntServ QoS control over multimedia traffic with full inter-operability. There have been other RSVP implementations on MS-Windows (95/NT4.0/98/NT5.0), Sun Solaris, DEC Alpha machines and Line. From their descriptions they are not targeting at resolving the application and platform dependence problems of RSVP.

5. CONCLUSION AND FUTURE WORK

RSVP Session Set-up Protocol, RSSP, is proposed in this paper to overcome the limitations of current RSVP design and implementation such as application and system independence which can cause serious limitations to the deployment of RSVP services over simple machines and terminals with limited and expensive processing powers to build a full RSVP session set-up and maintenance functions. It also provides those applications that are difficult or even impossible to embed an RSVP daemon or its equivalent functions within the platforms. Both qualitative and quantitative tests have been carried out and the experimental results demonstrate the control effectiveness through the RSSP Server in QoS provisioning over real-time applications. The proposed scheme has been successfully deployed in providing QoS control over multimedia traffic to and between mobile nodes in our mobile multimedia QoS control test-bed. Through the design and implementation work the proposed mechanism has been proved to play a significant role in providing dynamic QoS provisioning during the seamless hand-off control of mobile multimedia terminals.

References

[1] S. Shenker, J. Wroclawski, RFC 2215 "General Characterisation Parameters for Integrated Service Network Elements", RFC 2215, September 1997

[2] S. Shenker, J. Wroclawski, "Network Element Service Specification Template", RFC 2216, September 1997.

[3] J. Wroclawski, "The Use of RSVP with IETF Integrated Services", RFC2210, September 1997.

[4] J. Wroclawski, " Specification of the Controlled-Load Network Element Service", RFC2211, September 1997.

[5] S. Shenker, C. Partridge, R. Guerin, "Specification of Guaranteed Quality of Service", RFC2212, September 1997.

[6] R. Braden, D. Clark, S. Shenker, "Integrated Services in the Internet Architecture: an Overview", Informational RFC1633R, June, 1994.

[7] R. Braden, L. Zhang, S. Berson, S. Herzog, S. Jamin, "Resource ReSerVation Protocol (RSVP) – Version 1 Functional Specifiction",, September 1997.

[8] A. Mankin, F. Baker, B. Braken, S. Bradner, M. O'Dell, A. Ramanow, A. Weinrib, L. Zhang, "Resource ReServation Protocol (RSVP) – Version 1 Applicability Statement", Informational RFC2208, September 1997.

[9] R. Braden, L. Zhang, "Resource ReSerVation Protocol (RSVP) -- Version 1 Message Processing Rules", RFC2209, September 1997.

[10] R. Yavatkar, D. Hoffman, Y. Bernet, F. Baker, M. Speer, "SBM (Subnet Bandwidth Manager): Protocol for RSVP-based Admission Control over IEEE 802-style Networks", IETF Internet-Draft <draft-ietf-issll-is802-sbm-06.txt>, March 1998.

[11] http://www.isi.edu/rsvp/DOCUMENTS/ietf_rsvp-qos_survey_02.txt

SESSION 4

Performance Evaluation I

A NEW SELF-SIMILAR TRAFFIC MODEL AND ITS APPLICATIONS IN TELECOMMUNICATION NETWORKS

David G. Daut[1] and Ming Yu[2]

[1] *Rutgers - The State University of New Jersey*
Dept. of Elec. & Comp. Eng.
Piscataway, NJ 08855
USA
daut@ece.rutgers.edu

[2] *AT&T Labs*
480 Red Hill Road
Room 1D-328
Middletown, NJ 07748
USA
mingyu@att.com

Abstract This paper presents a new self-similar traffic model derived from the arrival processes of $M/G/\infty$ queue. It has a structure similar to that of a fractional ARIMA, with a driven process of fBm (fractional Brownian motion). The coefficients of the fBm are derived from the Pareto distribution of the active periods of the arrival process. When applied to a single server with self-similar input, the model results in an explicit buffer level equation which matches Norros' storage model. So this method can be also served as a verification of Norros' assumptions. The effectiveness of the proposed model has been verified by some practical examples.

Keywords: Self-similar, long-range dependence, network traffic.

1. INTRODUCTION

Network traffic modeling is of primary importance for network design, performance prediction, and control. It has been observed recently that packet loss and delay are more serious than expected because network traffic is more bursty and exhibits greater variability than previously suspected. This phenomenon has led to the discovery of network traffic's self-similar, or fractal, characteristic [6]. A covariance-stationary process $X(t)$ is called self-similar if $X(t) - X(0)$ and $r^H(X(u) - X(0))$ are identical in distribution, where the time t is rescaled in the ratio r, i.e., $u = t/r$. Therefore the area of network traffic modeling, buffer design, and performance evaluation needs to be reexamined.

There are many investigations into self-similar models. The fractional Brownian motion (fBm) and fractional Auto-Regressive Integrated Moving Average (f-ARIMA) model are the two popular mathematical models that are used to describe self-similar processes. F-ARIMA is much more flexible than fBm. But both can be only used to simulate short time series. Cox and Isham proposed a model to describe an immigration-death process. Kosten applied this model to a multi-entry buffer in the general context of self-similarity. There are two other related models of asymptotically self-similar processes. One is a sum of short-range independent attenuated and weighted stationary processes. Another is taking the limit of the aggregated traffic of M individual On-Off sources, as $M \longrightarrow \infty$, while keeping the source rate and the distribution of active period remain unchanged, but the distribution of idle period went to zero. The $M/G/\infty$ and aggregated $AR(1)$ processes are the two models often used to generating asymptotically self-similar traffic although without analytical expressions. Other self-similar models are heavy tailed On-Off process, stable self-similar process, fractal shot-noise (point process approaches), fractal renewal process (point process approaches) [3], deterministic nonlinear chaotic map models, and stochastic difference equations.

The complexity of self-similarity has rendered existing traffic and performance models to be either analytically intractable in the evaluation of performance, or usually inaccurate with respect to dynamic queueing behavior. In order to investigate the impact of self-similar traffic modeling parameters on network performance, it is necessary to develop new analytic traffic models.

2. CONSTRUCTION OF SELF-SIMILAR TRAFFIC

Consider a stationary random process $Y = \{Y_t\}$, where $t \in I_{-\infty} = \{..., -1, 0, 1, ...\}$. We assume that there are ξ_t traffic sources which begin their active periods at time t, where ξ_t is independent and Poisson with mean λ_1. A source s is associated with its active period s, where $s \in I_{-\infty}$. The beginning instant of active period s is denoted by ω_s. The length of active period s is denoted by $\tau_s \in I_1 = \{1, 2, ...\}$. The number of cells generated by source s at time instant t is denoted by $\vartheta_t^{(s)}$, which is also called the source rate of source s. In this section, we will focus on the derivation of our self-similar traffic model instead of the procedure of traffic construction. Details can be found in [7]. The number of new cells Y_t that appeared in traffic Y at time t is the sum of numbers of cells generated by all active (at time t) sources, including those sources began at time t and those began before time t but still in active at time t,

$$Y_t = \sum_{s=-\infty}^{\infty} \vartheta_t^{(s)}. \tag{1}$$

If the source rate is constant, $\hat{\vartheta}_t^{(s)} = R \in I_1$, the distribution of Y_t for given t is Poisson. If the source rate is Poisson, $\hat{\vartheta}_t^{(s)}$ are independent (for all s and t) and Poisson with mean λ_2. The intensity of this traffic is $E\{Y_t\} = \lambda_1 \lambda_2 E\{\tau\}$. However, the distribution of Y_t is not Poisson (since the sum of a random number of Poisson random variables is not necessarily a Poisson variable). Moreover, Y_t are dependent variables. We will find a mathematical model for this process for the purpose of queueing analysis and prove its self-similarity.

The number of new cells Y_t at time t can be expressed by

$$Y_t = \sum_{n=-\infty}^{t} R\xi_n P_{rob}^{(n)}(\tau > t - n) \tag{2}$$

where $P_{rob}^{(n)}(\tau > l)$ is the distribution of the length of active period at time n. If $P_{rob}^{(n)} = P_{rob}$, for all n; let $t - n = l$, then we have

$$Y_t = \sum_{l=0}^{\infty} R\xi_{t-l} P_{rob}(\tau > l) \tag{3}$$

The above model can be rewritten as

$$Y_t = RP_{rob}(\tau > 0)\xi_t + RP_{rob}(\tau > 1)\xi_{t-1} + RP_{rob}(\tau > 2)\xi_{t-2} + ... \tag{4}$$

After some rearrangement, we have

$$\frac{1}{R}Y_{t+1} - \frac{1}{R}Y_t = \xi_{t+1} - a_1\xi_t - a_2\xi_{t-1} - ... - a_{-\infty}\xi_{-\infty} \tag{5}$$

where $a_j = P_{rob}(\tau > j-1) - P_{rob}(\tau > j) = P_{rob}(\tau = j)$, $j = 1, 2, 3, ...$, $\{\xi_j\}$ is Poisson. Suppose the distribution $P_{rob}(\tau \geq l+1)$ is Pareto-type,

$$P_{rob}(\tau \geq l+1) = \frac{E\tau}{2}\beta(1-\beta)(2-\beta)l^{-(\beta+1)}, \beta \to \infty. \tag{6}$$

the process Y_t will be proven to be self-similar with $H = 1 - \frac{\beta}{2}$, $0 < \beta < 1$ (see next section).

A Poisson process $N(t)$ with parameter m can be approximated by a diffusion process

$$N(t) \approx mt + \sqrt{am}W(t) \tag{7}$$

Substitute ξ in Eqn. (5) for $N(t)$ in Eqn. (7), and note that $\sum_{j=0}^{\infty} a_{t-j} = 1$, and $\sum_{j=1}^{\infty} a_{t-j} j = E\tau$, we have

$$Y_{t+1} - Y_t = mRE\tau + R\sqrt{am}z(t+1), \tag{8}$$

where m is the diffusion approximation parameter; a is the ratio of the variance to the mean of the traffic process; τ is the length of an active period; $z(t)$ is a fractional Brownian motion which can be expressed as

$$z(t+1) = W(t+1) - a_t W(t) - a_{t-1} W(t-1) - \ldots - a_{-\infty} W(-\infty), \tag{9}$$

where the coefficients a_t are the probabilities of the lengths of the active periods; and $W(t)$ is the standard Brownian motion.

3. SELF-SIMILARITY OF THE NEW MODEL

In this section, we attempt to give a rough proof of the self-similarity of the traffic process Y_t represented by the new model developed in the previous section. Before we present our proof, we need to introduce an LRD property:

A covariance-stationary process $X = (X; i = 1, 2, 3, \ldots)$ with mean μ, variance σ^2 and autocorrelation function $r(k)$, $k \geq 0$, is self-similar iff X is long-range dependence (LRD), i.e., for some $\frac{1}{2} < r < 1$,

$$r(k) \sim c k^{2H-2} \quad as \quad k \to \infty \tag{10}$$

where c is a finite positive constant. This property can be derived from the definition fo LRD and its relationship to self-similarity.

Define the auto-covariance function of Y_t,

$$\omega(k) = cov\{Y_t, Y_{t+k}\} = E\{(Y_t - E\{Y_t\})(Y_{t+k} - E\{Y_{t+k}\})\} \tag{11}$$

because of its stationary, we have $\omega(k) = r(k) - (E\{Y_t\})^2$. Therefore $\omega(k)$ and $r(k)$ will have the same asymptotic property for large k. Now the problem becomes to prove that $\omega(k)$ will have the form of Eqn. (10) for large k. What we know is that Y_t has the form of Eqn. (8) and $z(t)$ is fBm which can be expressed by $W(t)$, standard Bm.

From $Y_{t+1} = Y_t + mR\bar{\tau} + R\sqrt{am} z(t+1)$, we have

$$Y_{t+k} = Y_t + kmR\bar{\tau} + R\sqrt{am}[z(t+1) + z(t+2) + \ldots + z(t+k)] \tag{12}$$

$$Y_{t+k} - EY_{t+k} = Y_t - EY_t + \sqrt{am} R[z(t+1) + z(t+2) + \ldots$$

$$+ z(t+k) - \bar{z}(t+1) - \bar{z}(t+2) - \ldots - \bar{z}(t+k)] \tag{13}$$

$$\omega(k) = E\{(Y_t - \bar{Y}_t)^2\} + maR^2[\omega_z(0) + \omega_z(1) + \ldots + \omega_z(k)] \tag{14}$$

where $\omega_z(k) = E\{[z(t) - \bar{z}(t)][z(t+k) - \bar{z}(t+k)]\}$. Note that $\bar{z}(t+i) = 0, i = 0, 1, 2, \ldots, k$. We have

$$z(t+i) = W(t+i) - a_1 W(t+i-1) - a_2 W(t+i-2) - a_3 W(t+i-3) - ...$$

$$-a_{i-1} W(t+1) - a_i W(t) - a_{i+1} W(t+1) - ... - a_{-\infty} W(-\infty), \quad (15)$$

Also note $W(i)W(j) = \delta_{ij}\sigma^2$ Suppose $\sigma^2 = 1$, we have

$$\omega_z(i-1) = E\{z(t+1)z(t+i)\} = 1a_{i-1} + a_1 a_i + a_2 a_{i+1} + a_3 a_{i+2} + ... \quad (16)$$

Because of the Pareto-type distribution of the active periods: $a_i = ci^{-\alpha-1}$, where c is constant, which can be determined from $\bar{\tau}$ and β. Thus

$$\omega_z(i) = ci^{-\alpha-1} + \sum_{n=1}^{\infty} c^2 n^{-\alpha-1}(i+n)^{-\alpha-1}, \quad (17)$$

We know $E\{(Y_t - EY_t)^2\} = R^2(E\xi)(E\tau) = R^2 m\bar{\tau}$ therefore

$$\omega(k) = mR^2\bar{\tau} + amR^2 \sum_{j=0}^{k} \omega_z(j) \quad (18)$$

when n_0 is large, using

$$\sum_{n=n_0}^{\infty} n^{-\alpha} \approx \frac{n_0^{-\alpha+1}}{\alpha - 1}, \quad (19)$$

therefore when k is large we have

$$\sum_{n=1}^{\infty} (nk + n^2)^{-(\alpha+1)} \approx \frac{(k+1)^{-\alpha}}{\alpha}, \quad (20)$$

$$\sum_{j=0}^{k} \omega_z(j) \approx \sum_{j=0}^{k} [cj^{-\alpha-1} + c^2 \frac{(j+1)^{-\alpha}}{\alpha}] \quad (21)$$

$$\sum_{j=0}^{k} \omega_z(j) \approx c\frac{k^{-\alpha}}{\alpha} + \frac{c^2}{\alpha} \frac{(k+1)^{-\alpha+1}}{\alpha - 1} \quad (22)$$

$$\omega(k) \approx \frac{c^2}{\alpha(\alpha-1)}(k+1)^{1-\alpha} \quad (23)$$

That is, $\omega(k)$ has the form of Eqn. (10). Therefore Y_t is a self-similar process, with Hurst parameter $H = (3 - \alpha)/2$ according to the LRD property. So far, we've developed a new self-similar traffic model and proven that the traffic sequences of $\{Y_t\}$ is self-similar with parameter $H = (3 - \alpha)/2$, which will be used to analytically investigate the queueing behavior of networks with self-similar traffic input.

4. QUEUEING ANALYSIS

4.1 BUFFER LEVEL EQUATION

While there are several excellent papers dedicated to the queueing analysis for self-similar traffic, there is still a lack of exact results concerning queueing analysis such as queue length distributions. Interested readers may check the references [4; 1; 2; 5]. Norros' storage model has been widely used to estimate asymptotic performance bounds in the case of large buffers, although it has not been proven because of the complexity of the problem. In this Section, we will apply the new traffic model to develop a buffer level equation that can be used to determine the multiplexing behavior of self-similar traffic and the bounds of the queue length. This development can be also served as a proof of Norros' model.

The new self-similar traffic model can be expanded as

$$Y_t = RP_{rob}(\tau > 0)\xi_t + RP_{rob}(\tau > 1)\xi_{t-1} + RP_{rob}(\tau > 2)\xi_{t-2} + ... \quad (24)$$

Define the coefficients using the notation

$$b_j = P_{rob}(\tau > j) \quad (25)$$

then, we have

$$Y_t = Rb_0\xi_t + Rb_1\xi_{t-1} + Rb_2\xi_{t-2} + ... \quad (26)$$

Similar to Section 2, the Poisson process $\xi(t)$ with parameter m can be written in the form

$$\xi(t) = mt + W(t), \quad (27)$$

where $W(t)$ is the martingale $\xi(t) - mt$. It is well known that $\xi(\alpha t) - \alpha mt)/\sqrt{\alpha m}$ converges towards the standard Brownian motion $W(t)$ as $\alpha \to \infty$. This suggests the approximation of $\xi(t)$ by a diffusion process

$$\xi(t) \approx mt + \sqrt{am}W(t) \quad (28)$$

and upon substitution into our model, we have

$$Y_t = R\sum_{j=0}^{\infty} b_j mt - Rm\sum_{j=0}^{\infty} jb_j + R\sqrt{am}\sum_{j=0}^{\infty} b_j W(t-j). \quad (29)$$

Note that

$$\sum_{j=0}^{\infty} b_j = \bar{\tau} \quad (30)$$

and also

$$\sum_{j=0}^{\infty} jb_j = D(\bar{\tau}, \beta) \tag{31}$$

is a constant which is related the distributions of the length of active periods. Therefore the model can be simplified as

$$Y(t) = mt - D(\bar{\tau}, \beta)/\bar{\tau} + \sqrt{am}Z(t) \tag{32}$$

where $Y(t) = Y_t/R\bar{\tau}$ being defined as the normalized traffic process; $Z(t)$ being defined as

$$Z(t) = \sum_{j=0}^{\infty} \frac{b_j}{\bar{\tau}} W(t-j) \tag{33}$$

is a normalized version of the fractional Brownian motion $z(t)$ in the new model. Suppose the system has a constant leak rate C, the net input process is

$$X(t) = Y(t) - Ct. \tag{34}$$

Thus, the stationary storage model with fractional Brownian net input is the stochastic process $V(t)$, where

$$V(t) = sup_{s \leq t}(Y(t) - Y(s) - C(t-s)),\ t \in (-\infty, \infty), \tag{35}$$

That is equivalent to the process

$$V(t) = sup_{s \leq t}(A(t) - A(s) - C(t-s)),\ t \in (-\infty, \infty), \tag{36}$$

with an arrival process of

$$A(t) = mt + \sqrt{am}Z(t) \tag{37}$$

It is similar to the well-known expression for the amount of work (or virtual waiting time) in a queueing system with service rate C and cumulative work arrival process $A(t)$ [4].

4.2 PERFORMANCE EVALUATIONS

The performance requirement in telecommunication applications can be described by following relation

$$\epsilon = P_{rob}(V > x) \tag{38}$$

that the probability that the queue length exceeds a certain level x is required to be at most equal to a "Quality of Service" parameter ϵ. Norros' lower bound is asymptotically (in a logarithmic sense) exact for the Brownian model. Massoulie and Simonian have reported a tighten lower and upper bounds which have only a constant underdetermined. Recently, Narayan found an exact asymptotic queue lenght distribution for $1/2 < H < 1$. For case of $H = 1$, Norros' lower bound is the exact probability. In this paper we will apply these bounds to a practical example to verify our models.

5. EXAMPLES

Example 1 A set of practical traffic data we collected from a large-scale data network was plotted in Fig.1. The Pox diagram of the traffic data is shown in Fig.2. Fig.3 is the power spectrum.

The traffic Y_t can be modeled by

$$Y_{t+1} - Y_t = mRE\{\tau\} + R\sqrt{am}z(t+1), \tag{39}$$

where $m = 0.5709$ is a diffusion approximation parameter; $a = 1.4194$; $\tau = 1.9977$ is the length of an active period; the source rate $R = 1.0$; $z(t)$ is a fractional Brownian motion which can be expressed as

$$z(t+1) = W(t+1) - a_t W(t) - a_{t-1} W(t-1) - \ldots - a_{-\infty} W(-\infty), \tag{40}$$

where the coefficients $a_t = ct^{\alpha-1}, c = 0.7188, \alpha = 3 - 2H = 1.3800, H = 0.81$; and $W(t)$ is the standard Brownian motion. This model has been successfully used to design an alarm processor in a larg-scale network. Fig.4 is the Pox diagram of the traffic generated by this model.

Example 2 The performance bound has been calculated from the proposed buffer level equation when a single sever fed by the above traffic data. In order to compare the bound with other performance bounds, it is plotted in Fig.5 (lower line) against the bound derived from Reference [2]. The traffic would be modeled as Poisson arrival process if self-similarity was ignored. The resulted performance bound was shown in Fig.6 also against the bound derived from Reference [2]. It can be seen that the performance will be much worse than predicted by Poisson models. This has also demonstrated the need of self-similar traffic model.

6. CONCLUSION

In this paper, we developed a new tractable model for self-similar traffic processes. We also derived an explicit buffer level equation based on the proposed traffic model, which matches Norros' storage model. So this method can be also served as a verification of Norros' assumptions. The queueing behavior of a single server to self-similar input can be analytically investigated with the proposed model with respect to each of the model parameters. This

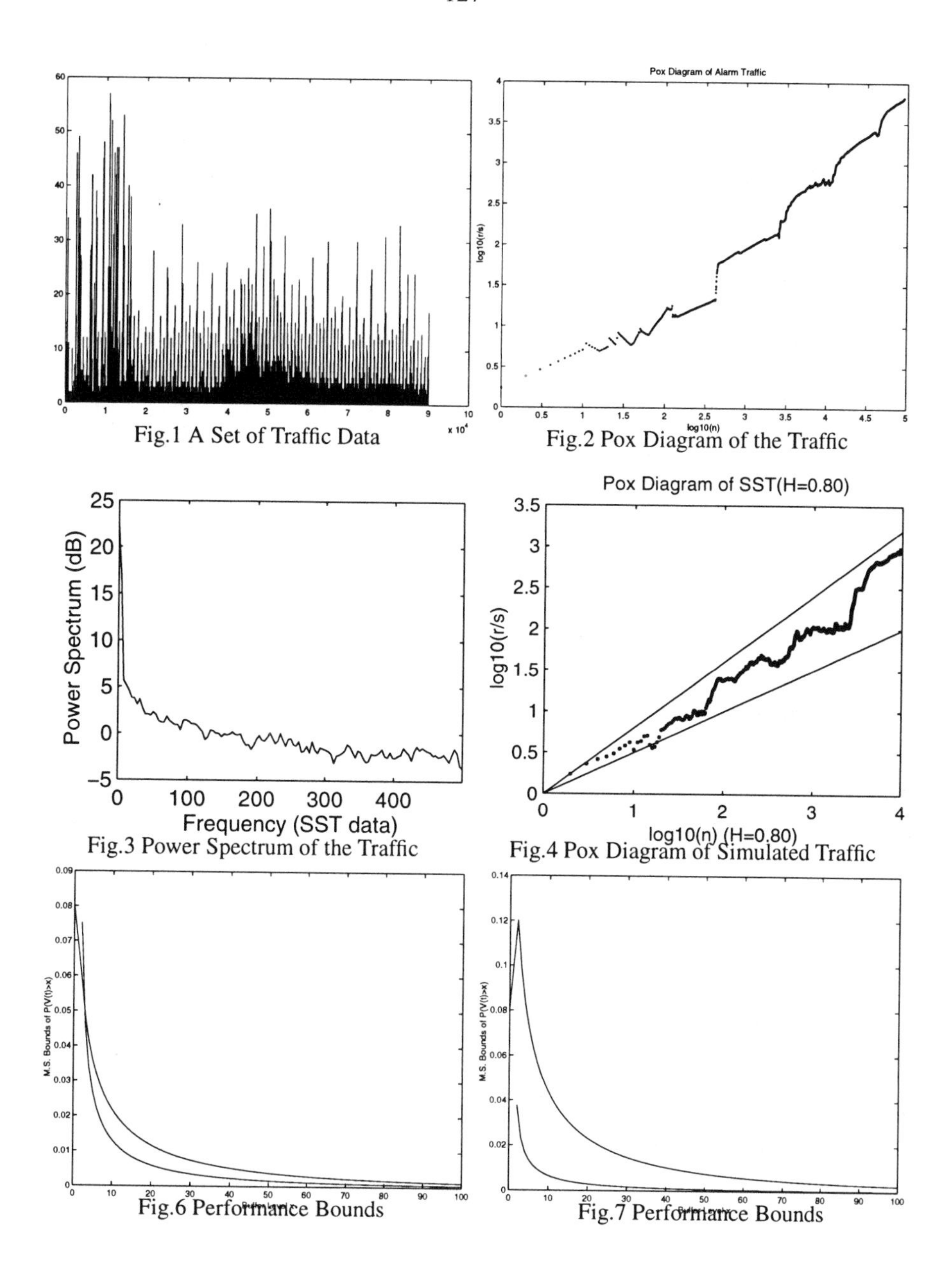

Fig.1 A Set of Traffic Data

Fig.2 Pox Diagram of the Traffic

Fig.3 Power Spectrum of the Traffic

Fig.4 Pox Diagram of Simulated Traffic

Fig.6 Performance Bounds

Fig.7 Performance Bounds

will be our future work. Finally, we presented some practical examples to demonstrate the effectiveness of the proposed methods.

Acknowledgments

The authors would like to acknowledge the reviewers for their many suggestions and comments. We also thank N.G.Duffield, W.Willinger, and V. Ramaswami for giving us very useful references.

References

[1] O. Narayan, "Exact Asymptotic Queue Length Distribution for Fractional Brownian Traffic,", Advance in Performance Analysis, vol.1, pp. 39-63, no. 1, Oct. 1998.

[2] L. Massoulie and A. Simonian, "Large Buffer Asymptotics for the Queue with FBM Input,", unpulished, 1998.

[3] W. E. Leland, M. S. Taqqu, W. Willinger, , and D. V. Wilson. On the self-similar nature of ethernet traffic. IEEE/ACM Transactions on Networking, vol. 2:1–14, March 1994.

[4] I. Norros. A storage model with self-similar input. Queueing Systems, vol. 16:387–396, 1994.

[5] N. G. Duffield. Large deviatioins and overflow probabilities for the general single-server queue, with applications. Math. Proc. Camb. Phil. Soc., vol. 118:363–374, 1995.

[6] W. Stallings. Self-similarity upsets data traffic assumptions. IEEE Spectrum, vol. 34:28–29, 1997.

[7] B. Tsybakov and N. D. Georganas. Overflow probability in an atm queue with self-similar traffic. In ICC'97, 1997.

BIASED INITIAL DISTRIBUTION FOR SIMULATION OF QUEUES WITH A SUPERPOSITION OF PERIODIC AND BURSTY SOURCES

Fumio Ishizaki

The University of Tokushima, 2-1 Minamijosanjima, Tokushima 770-8506, Japan

ishizaki@is.tokushima-u.ac.jp

Abstract In this paper, we focus on queues with a superposition of periodic and bursty sources where the initial states of the sources are randomly and independently selected. The dynamics of the queue are described as a *reducible* Markov chain. For such queues, we consider the tail probability of the queue length as the performance measure and try to estimate it by simulations. To do simulation efficiently, we develop a sampling technique which replaces the initial distribution of the reducible Markov chain with a new biased one. Simulation results show that the sampling technique is useful to reduce the variance of the estimates.

Keywords: Importance Sampling, Monte Carlo Simulation, Queues

1. INTRODUCTION

In ATM networks, the arrival process of cells is essentially a superposition of sources which are typically bursty and periodic either due to their origin or their periodic slot occupation after traffic shaping. For instance, VBR (variable bit rate) MPEG traffic, which will become a large volume of traffic in future high-speed networks, has both bursty and periodic natures because of its coding scheme. Thus, queues with a superposition of periodic and bursty sources have been studied by many researchers.

Queues with a superposition of periodic sources have the following special characteristic: The dynamics of the queues are *not* ergodic due to the periodicity of the sources, when the periodic sources are randomly superposed (i.e., the initial states of the sources are randomly chosen). As a result, depending on the combination of the initial states, the system performance such as cell loss ratio and queueing delay can drastically change [9]. This situation will be observed in Section 5.

Unfortunately, it is difficult to estimate the performance in such queues. For example, since the matrix analytical approaches [12] require to deal with prohibitively large size matrices, the analysis is extremely complicated. On the other hand, by standard Monte Carlo (MC) simulations, it is hard to get reliable estimates of the performance measure. The reason is that there are rare events whose contribution to the performance measure cannot be neglected. In other words, there exist some bad initial state combinations which lead to extreme degradation of the system performance, although they are rarely realized.

In this paper, we focus on queues with a superposition of periodic and bursty sources where the initial states of the sources are randomly and independently selected. The dynamics of the queue are described as a *reducible* bivariate Markov chain (i.e., they are not ergodic). We consider the tail probability of the queue length as the system performance, and try to estimate it by simulations. In order to overcome the difficulty in MC simulations, we develop a sampling technique which is a kind of importance sampling (IS) technique (see, e.g., [3; 6; 13; 16] and references therein). In most of the previous works (including IS based on large deviation theory [14], RESTART [15], DPR [7] and so on), ergodic queues have been considered, and replacing only the state transition probabilities with new biased ones has been investigated. Contrary to the previous works, the sampling technique developed in this paper replaces the initial distribution with a new biased one. This means that the realization probabilities of the combinations of the initial states are modified. Existing IS techniques and our sampling technique are not exclusive but complementary. We can expect that combined with existing IS techniques, our sampling technique will lead to more efficient MC simulations.

In order to get an appropriate biased initial distribution, we apply the asymptotic queueing analysis (see, e.g., [1; 2; 4; 5] and references therein). Unfortunately, for this purpose, most of the asymptotic queueing analyses are not suitable. The reason is that they do not take the effect of the initial state combinations on the system performance into account. In contrast, taking the initial state combinations into account, Ishizaki and Takine [8] have studied discrete-time queues where the arrival process is a superposition of the general periodic Markov sources. They have derived bound formulas for the asymptotic tail distribution of the queue length. We consider a sample path version of their upper bound as a rough estimate of the queue length and use it to get a biased initial distribution.

The remainder of the paper is organized as follows. In Section 2, we describe a queue with a superposition of periodic and bursty sources. The background material for the asymptotic queueing analysis [8] is provided in Section 3. In Section 4, we develop the sampling technique which replaces the initial distribution with a new biased one. In Section 5, we provide simulation results to examine the usefulness of the sampling technique. Conclusions are drawn in Section 6.

2. MODEL

We consider a discrete-time infinite-buffer queueing system. The arrival process at the system is a superposition of K general periodic Markov sources [8]. The kth ($k = 1, \ldots, K$) source is governed by an underlying discrete-time $M^{(k)}$-state Markov chain with period $R^{(k)}$. We assume that the number of cells from each source arriving in a slot depends on the state of the underlying

Markov chain in the slot. The service time of cell is assumed to be constant and equal to one slot.

We now describe our model in more detail. We first introduce notations for each source. Let $S_n^{(k)}$ $(n = 0, 1, \ldots)$ denote a random variable representing the state of the underlying Markov chain of the kth source in the nth slot, where $S_n^{(k)} \in \mathcal{S}^{(k)} = \{0, 1, \ldots, M^{(k)} - 1\}$. We assume that all the underlying Markov chains are irreducible and stationary. The state transition matrix of the underlying Markov chain is denoted by $\boldsymbol{U}^{(k)} = \{U_{i,j}^{(k)}\}$ $(i, j = 0, \ldots, M^{(k)} - 1)$. We denote the stationary state vector of the underlying Markov chain for the kth source by $\boldsymbol{\pi}^{(k)} = (\pi_0^{(k)}, \ldots, \pi_{M^{(k)}-1}^{(k)})$. Let $D_r^{(k)}$ denote the rth *moving class* of the kth source, i.e., if $i \in D_r^{(k)}$ $(r = 0, \ldots, R^{(k)} - 1)$, then $\Pr\{S_{n+1}^{(k)} = j | S_n^{(k)} = i\} = 0$ for all $j \notin D_{r \oplus_k 1}^{(k)}$, where $\oplus_k$ is defined as $r \oplus_k l = (r + l) \bmod R^{(k)}$ for $k = 1, \ldots, K$. Let $P^{(k)}$ $(k = 1, \ldots, K)$ denote a random variable representing the index of the moving class where the kth source is in the 0th slot. We then have $S_0^{(k)} \in D_{P^{(k)}}^{(k)}$. Hereafter, we call the random variable $P^{(k)}$ the phase of the kth source.

Let $A_n^{(k)}$ $(n = 1, 2, \ldots)$ denote a random variable representing the number of cells arriving from the kth source in the nth slot. $\{A_n^{(k)}\}$ is governed by the underlying Markov chain for the kth source. We assume that given $S_n^{(k)}$, $A_n^{(k)}$ is conditionally independent of all other random variables. Let $\hat{a}_{ij}^{(k)}(l)$ denote the conditional joint probability of the following events: l cells arrive from the kth source and the underlying Markov chain is in state j in the next slot given that the underlying Markov chain is in state i in the current slot. Namely, $\hat{a}_{ij}^{(k)}(l) = \Pr\{A_{n+1}^{(k)} = l, S_{n+1}^{(k)} = j \mid S_n^{(k)} = i\}$. Let $\boldsymbol{A}^{(k)}(l)$ denote an $M^{(k)} \times M^{(k)}$ matrix whose (i, j)th element is given by $\hat{a}_{ij}^{(k)}(l)$. Note that $\boldsymbol{A}^{(k)}(l)$ represents the transition matrix of the underlying Markov chain when l cells arrive from the kth source at the system. We define the probability matrix generating function for the arrival process: $\hat{\boldsymbol{A}}^{(k)}(z) = \sum_{l=0}^{\infty} \boldsymbol{A}^{(k)}(l) z^l$. Let $\rho^{(k)}$ denote the traffic intensity of the kth source which is given by $\rho^{(k)} = \boldsymbol{\pi}^{(k)} \sum_{l=0}^{\infty} l \boldsymbol{A}^{(k)}(l) \boldsymbol{e}$, where $\boldsymbol{e}$ is an $M^{(k)} \times 1$ vector whose elements are all equal to one.

Next we consider a superposition of the K general periodic Markov sources. We assume that the K periodic sources are independently superposed. This assumption implies that all the phases of sources are independent with each other. Note here that the assumption makes the underlying Markov chain for the superposed process *reducible*, because of the periodicity of the sources [8]. As a result, the arrival process is *not* ergodic.

We define a random variable $S_n \in \mathcal{S} = \{0, \ldots, M - 1\}$ $(n = 0, 1, \ldots)$ as $S_n = f(S_n^{(1)}, \cdots, S_n^{(K)})$, where $M = \prod_{k=1}^{K} M^{(k)}$ and the function f is defined as $f(j^{(1)}, \cdots, j^{(K)}) = \sum_{i=1}^{K} j^{(i)} \prod_{k=i+1}^{K} M^{(k)}$. S_n then represents the state of the underlying Markov chain for the superposed arrival process in the nth slot. Hereafter, we use the expressions $\boldsymbol{S}_n = (S_n^{(1)}, \ldots, S_n^{(K)})$ and S_n interchangeably to express the state of the underlying Markov chain for the superposed arrival process. Let $\boldsymbol{P}$ denote a random vector representing the phase vector of the superposed arrival process, which is defined as $\boldsymbol{P} = (P^{(1)}, \ldots, P^{(K)})$.

Let A_n denote a random variable representing the number of cells arriving in the nth slot. We then have $A_n = \sum_{k=1}^{K} A_n^{(k)}$. Note that $\{A_n\}$ is not ergodic, though it is stationary. The overall traffic intensity for the superposed arrival process is then given by $\rho = \sum_{k=1}^{K} \rho^{(k)}$. Hereafter we assume that $\rho < 1$. Let X_n $(n = 0, 1, \ldots)$ denote a random variable representing the queue length at the beginning of the nth slot. Its evolution is described by $X_{n+1} = (X_n - 1)^+ + A_{n+1}$ with $X_0 = 0$, where $(x)^+ = \max(x, 0)$.

Note that $\{(X_n, S_n)\}$ $(n = 0, 1, \ldots)$ becomes a *reducible* bivariate Markov chain. The one-step state transition matrix $\boldsymbol{Q}$ of the Markov chain is given by

$$\boldsymbol{Q} = \begin{bmatrix} \boldsymbol{A}_0 & \boldsymbol{A}_1 & \boldsymbol{A}_2 & \boldsymbol{A}_3 & \cdots \\ \boldsymbol{A}_0 & \boldsymbol{A}_1 & \boldsymbol{A}_2 & \boldsymbol{A}_3 & \cdots \\ \boldsymbol{O} & \boldsymbol{A}_0 & \boldsymbol{A}_1 & \boldsymbol{A}_2 & \cdots \\ \boldsymbol{O} & \boldsymbol{O} & \boldsymbol{A}_0 & \boldsymbol{A}_1 & \cdots \\ \vdots & \vdots & \vdots & \vdots & \ddots \end{bmatrix},$$

where $\boldsymbol{O}$ denotes the $M \times M$ zero matrix and $\boldsymbol{A}_l$ is given by

$$\boldsymbol{A}_l = \underbrace{\sum_{l_1} \sum_{l_2} \cdots \sum_{l_K}}_{l = l_1 + l_2 + \cdots + l_k} \boldsymbol{A}^{(1)}(l_1) \otimes \boldsymbol{A}^{(2)}(l_2) \otimes \cdots \otimes \boldsymbol{A}^{(K)}(l_K),$$

$\otimes$ denotes the Kronecker product. The assumption that all the sources are independent with each other uniquely determines the initial distribution of the reducible bivariate Markov chain. We will provide the form of the initial distribution. To do so, we let $j = f(j^{(1)}, \ldots, j^{(K)})$ and $j^{(k)} \in D_{r^{(k)}}^{(k)}$ $(k = 1, \ldots, K)$. Then, the initial distribution μ of the reducible bivariate Markov chain is given by

$$\mu(X_0 = i, S_0 = j) = \begin{cases} \dfrac{\prod_{k=1}^{K} q_{j^{(k)}, r^{(k)}}^{(k)}}{\prod_{k=1}^{K} R^{(k)}} & (i = 0, j = 0, \ldots, M-1), \\ 0 & (\text{otherwise}), \end{cases}$$

where $q_{j,r}^{(k)}$ is defined as

$$q_{j,r}^{(k)} = \Pr\{S_n^{(k)} = j \mid S_n^{(k)} \in D_r^{(k)}\}. \tag{1}$$

We define $g_{N_0,N}^m$ as $g_{N_0,N}^m(\omega) = \sum_{n=N_0}^{N_0+N-1} \mathbf{1}_{\{X_n > m\}}(\omega)/N$, where ω denotes a sample, m, N_0 and N are nonnegative integer-valued parameters.

In this paper, we consider the performance measure $J^m_{N_0,N} = E[g^m_{N_0,N}] = \sum_{n=N_0}^{N_0+N-1} \Pr\{X_n > m\}/N$.

3. ASYMPTOTIC QUEUEING ANALYSIS

In this section, we present the result of the asymptotic queueing analysis as a background material.

Let $\hat{\delta}^{(k)}(z)$, $\hat{\boldsymbol{u}}^{(k)}(z)$ and $\hat{\boldsymbol{v}}^{(k)}(z)$ denote the Perron-Frobenius (PF) eigenvalue for the kth source associated with $\hat{\boldsymbol{A}}^{(k)}(z)$, its associated left and right eigenvectors with normalizing conditions $\hat{\boldsymbol{u}}^{(k)}(z)\hat{\boldsymbol{v}}^{(k)}(z) = \hat{\boldsymbol{u}}^{(k)}(z)\boldsymbol{e} = 1$, respectively. The probability matrix generating function $\boldsymbol{A}(z)$ for the superposed arrival process is then given by $\boldsymbol{A}(z) = \hat{\boldsymbol{A}}^{(1)}(z) \otimes \cdots \otimes \hat{\boldsymbol{A}}^{(K)}(z)$. The PF eigenvalue $\delta(z)$ for the superposed arrival process is then given by $\delta(z) = \prod_{k=1}^{K} \hat{\delta}^{(k)}(z)$, and its associated left and right eigenvectors $\boldsymbol{u}(z)$ and $\boldsymbol{v}(z)$ are given by $\boldsymbol{u}(z) = \hat{\boldsymbol{u}}^{(1)}(z) \otimes \cdots \otimes \hat{\boldsymbol{u}}^{(K)}(z)$ and $\boldsymbol{v}(z) = \hat{\boldsymbol{v}}^{(1)}(z) \otimes \cdots \otimes \hat{\boldsymbol{v}}^{(K)}(z)$.

Note that we can easily construct an arrival process in such a way that the queue length is bounded and has no tail. The following assumption guarantees that the queue length (including a cell in service) has a simple asymptotic form [8].

Assumption 1:

- There exists at least one zero of $\det[z\boldsymbol{I} - \boldsymbol{A}(z)]$ outside the unit disk.
- Among those, there exists a real and positive zero z^*, and the absolute values of z^* is strictly smaller than those of other zeros.
- $\boldsymbol{O} < \boldsymbol{A}(z) \ll +\infty$, $1 \le z \le z^*$, $z \in \mathcal{R}$, where $\mathcal{R}$ denotes a set of all real numbers.

For $\boldsymbol{r} = (r^{(1)}, \ldots, r^{(K)})$ ($r^{(k)} \in \{0, \ldots, R^{(k)} - 1\}$ for $k = 1, \ldots, K$), we define $c(\boldsymbol{r}, z)$ as

$$c(\boldsymbol{r}, z) = \frac{1}{R} \sum_{l=0}^{R-1} \prod_{k=1}^{K} \theta^{(k)}_{r^{(k)} \oplus_k l}(z), \tag{2}$$

where

$$\theta^{(k)}_r(z) = \sum_{j \in D^{(k)}_r} q^{(k)}_{j,r} s^{(k)}_j \hat{v}^{(k)}_j(z),$$

$$s^{(k)}_j = \Pr(A^{(k)}_n = 0 \mid S^{(k)}_n = j), \quad \hat{v}^{(k)}_j(z) = [\hat{\boldsymbol{v}}^{(k)}(z)]_j,$$

$q^{(k)}_{j,r}$ is defined as (1), R is defined as $R = \mathrm{LCM}\{R^{(1)}, \ldots, R^{(K)}\}$, and $\hat{\boldsymbol{v}}^{(k)}(z)$ denotes the right eigenvectors which is associated with the PF eigenvalue $\hat{\delta}^{(k)}(z)$ and satisfies the normalizing condition.

The following proposition holds under Assumption 1. From the ergodic theorem [11] and Theorem 4.1 in [8], it immediately follows. The proof is shown in [10].

Proposition 1: Under Assumption 1, for a sufficiently large m, any nonnegative integer N_0 and any phase vector $\boldsymbol{P}$, we have

$$\lim_{N\to\infty} \frac{1}{N} \sum_{n=N_0}^{N_0+N-1} \mathbf{1}_{\{X_n>m\}} \leq \frac{(z^*)^{-m}}{\delta'(z^*)-1} c(\boldsymbol{P}, z^*) \quad \text{a.s.},$$

where $\delta(z)$ denotes the PF eigenvalue for the superposed arrival process and z^* is the minimum real solution of

$$z = \delta(z) \tag{3}$$

for $z \in (1, \infty)$.

4. SAMPLING TECHNIQUE

We consider estimating the performance measure $J^m_{N_0,N}$ by simulation. In order to do simulation efficiently, we replace the initial distribution μ of the bivariate Markov chain with a new biased one ν. For this purpose, we use Proposition 1 in the previous section. Recall here that the system is described as a reducible Markov chain.

We let $j^{(k)} \in D^{(k)}_{r^{(k)}}$ $(k = 1, \dots, K)$, $j = f(j^{(1)}, \dots, j^{(K)})$ and $\boldsymbol{r} = (r^{(1)}, \dots, r^{(K)})$. Also, let $\bar{c}(z)$ denote the sum of $c(\cdot, z)$ over all the phase combinations which is defined as

$$\bar{c}(z) = \sum_{r^{(1)}=0}^{R^{(1)}-1} \cdots \sum_{r^{(K)}=0}^{R^{(K)}-1} c(\boldsymbol{r}, z), \tag{4}$$

where $c(\cdot, \cdot)$ is defined in (2). We then define the biased initial distribution ν of the reducible bivariate Markov chain as

$$\nu(X_0 = i, S_0 = j) = \begin{cases} \frac{c(\boldsymbol{r}, z^*)}{\bar{c}(z^*)} \prod_{k=1}^{K} q^{(k)}_{j^{(k)}, r^{(k)}} & (i = 0, j = 0, \dots, M-1), \\ 0 & (\text{otherwise}), \end{cases} \tag{5}$$

where z^* is the minimal real solution of $z = \delta(z)$ for $z \in (1, \infty)$.

Roughly speaking, the heuristic idea to get the biased initial distribution ν is as follows. The optimal biased initial distribution ν^* satisfies [13]

$$\nu^*(\mathcal{B}_0) \Pr(\mathcal{B} \mid \mathcal{B}_0) = \frac{g^m_{N_0,N}(\omega)}{J^m_{N_0,N}} \mu(\mathcal{B}_0) \Pr(\mathcal{B} \mid \mathcal{B}_0),$$

where ω is any sample satisfying $\omega \in \mathcal{B} \cap \mathcal{B}_0$, $\mathcal{B}$ and $\mathcal{B}_0$ are events given by

$$\mathcal{B} = \{X_1 = i_1, S_1 = j_1, \dots, X_{N_0+N-1} = i_{N_0+N-1}, S_{N_0+N-1} = j_{N_0+N-1}\},$$

$$\mathcal{B}_0 = \{X_0 = i, S_0 = j\},$$

respectively. We consider the upper bound given in Proposition 1 as the approximation. If for any $\omega \in \mathcal{B} \cap \mathcal{B}_0$ the approximation

$$g_{N_0,N}^m(\omega) \approx \frac{(z^*)^{-m}}{\delta'(z^*) - 1} c(\boldsymbol{P}(\omega), z^*)$$

is a rough estimate of $g_{N_0,N}^m$, then $\nu^*(\mathcal{B}_0)$ may be expressed as $\nu^*(\mathcal{B}_0) \approx A'c(\boldsymbol{r}, z^*)$, where A' is an unknown constant. Noting that $\boldsymbol{P}(\omega) = \boldsymbol{r}$ for $\omega \in \mathcal{B}_0$, we can determine A' from the normalizing condition for ν and get (5).

We now describe MC simulation procedure with the sampling technique which uses the biased initial distribution ν below.

1. Numerically solve (3) to get z^* and calculate $\bar{c}(z^*)$.
2. Do the following steps (a)–(d) for $n = 0, \ldots, N_s - 1$ to get N_s samples.
   (a) Randomly choose a vector $\boldsymbol{r} = (r^{(1)}, \ldots, r^{(K)})$ with probability $c(\boldsymbol{r}, z^*)/\bar{c}(z^*)$ ($r^{(k)} \in \{0, \ldots, R^{(k)} - 1\}$ for $k = 1, \ldots, K$).
   (b) Then, for $k = 1, \ldots, K$, randomly select a initial state $j^{(k)}$ for the kth source with probability $q_{j^{(k)},r^{(k)}}^{(k)}$.
   (c) Do simulation with this selected initial state and $X_0 = 0$, and then get the nth sample performance $\hat{Y}_n$.
   (d) Get the unbiased estimator $\hat{Z}_n$ by multiplying $\hat{Y}_n$ by its likelihood ratio: $\hat{Z}_n = \bar{c}(z^*)/(c(\boldsymbol{r}, z^*)\prod_{k=1}^{K} R^{(k)})\hat{Y}_n$.
3. Finally, obtain an estimate as the sample mean $\overline{Z}$ of the unbiased estimators $\hat{Z}_n$ ($n = 0, \ldots, N_s - 1$) by $\overline{Z} = \sum_{n=0}^{N_s-1} \hat{Z}_n / N_s$.

Note here that if we set $c(\boldsymbol{r}, z^*)$ constant, the MC simulation procedure with the sampling technique reduces to the standard MC simulation procedure.

When we get the biased initial distribution ν, we need to calculate $\bar{c}(z^*)$. However, for this purpose, we do not need to compute $c(\boldsymbol{r}, z^*)$ for all $\boldsymbol{r}$. Instead, we can compute $\bar{c}(z^*)$ from the following equation:

$$\bar{c}(z^*) = \prod_{k=1}^{K} \sum_{l=0}^{R^{(k)}-1} \theta_l^{(k)}(z^*). \tag{6}$$

Recall here that the number of the phase combinations is prohibitively large. Thus, getting $\bar{c}(z^*)$ from (4) is expensive in terms of computational cost. Using (6), we can greatly reduce the computational cost to get $\bar{c}(z^*)$.

In Step 2 (a), we need to select a vector $\boldsymbol{r}$ with the probability $c(\boldsymbol{r}, z^*)/\bar{c}(z^*)$. Since the number of phase combinations is very large, it is not so easy to determine the selected phase combination in a short time. To overcome this difficulty, we use a sophisticated algorithm described as follows:

```
/*generate a random number uniformly distributed in [0, c̄(z*)]*/
u = uniform[0, c̄(z*)];
α = 1.0; β = c̄(z*); s = 0.0;
for(k = 1; k ≤ K; k++){
  β = β/ Σ_{r=0}^{R^(k)-1} θ_r^(k)(z*);
  /*determine the phase of the kth source*/
  r^(k) = min{m | s + αβ Σ_{r=0}^{m} θ_r^(k)(z*) ≥ u};
  s = s + αβ Σ_{n=0}^{m-1} θ_n^(k)(z*); α = αθ_{r^(k)}^(k)(z*);
}
```

5. SIMULATION RESULTS

In this section, we apply the sampling technique to queueing systems where the arrival process is a superposition of periodic Bernoulli sources [9], and examine the usefulness of the sampling technique. For this purpose, we provide some simulation results and compare the variance of the estimates obtained through the MC simulation using the sampling technique with that obtained through the standard MC simulation. Due to limit of space, we show only one simulation example in this paper. Various simulation examples are shown in [10].

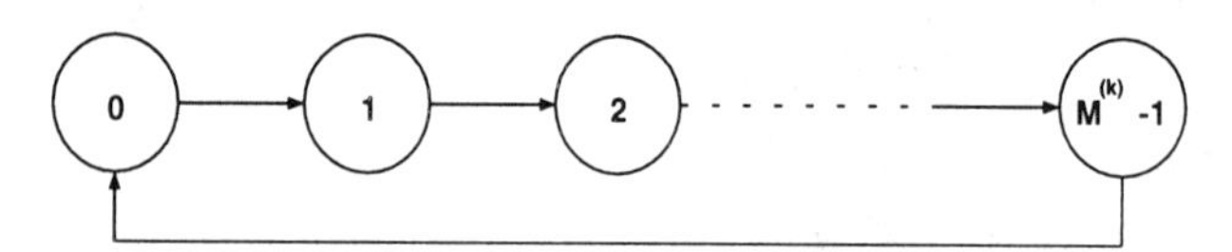

Figure 1 Periodic Bernoulli source

We first describe a periodic Bernoulli source, which is a special case of the general periodic Markov source described in Section 2 and used as a source model in the simulation. As shown in Fig. 1, the kth periodic Bernoulli source is governed by an underlying discrete-time $M^{(k)}$-state Markov chain with period $M^{(k)}$. If its underlying Markov chain is in the state $m^{(k)}$ ($m^{(k)} \in \{0, \dots, M^{(k)}-1\}$) in the nth slot, the source generates one cell with probability $\hat{\lambda}^{(k)}_{m^{(k)}}$ and it is in the state $m^{(k)} \oplus_k 1$ in the $(n+1)$st slot with probability one. For the periodic Bernoulli sources, the PF eigenvalue and its associated left and right eigenvectors with the normalizing conditions are derived in [9].

In particular, the periodic Bernoulli sources which is used in the simulation are set as follows. Each source is described by the three parameters $l^{(k)}$, $p^{(k)}$ and $\gamma^{(k)}$. From state 1 through state $l^{(k)}$, it generates exactly one cell every $p^{(k)}$ slots with probability one. From state $l^{(k)}+1$ through state $M^{(k)}$, it generates one cell with probability $\gamma^{(k)}$. We consider a queue with a superposition of 7 identical periodic Bernoulli sources where we set $M^{(k)} = 100$, $l^{(k)} = 40$, $p^{(k)} = 4$ and $\gamma^{(k)} = 0.05$ for $k = 1, \dots, 7$. Throughout the simulation, the performance measure $J^m_{N_0,N} = \sum_{n=N_0}^{N_0+N-1} \Pr\{X_n > m\}/N$ where $m = 20$, $N_0 = 4.0 \times 10^3$ and $N = 1.0 \times 10^6$ will be estimated.

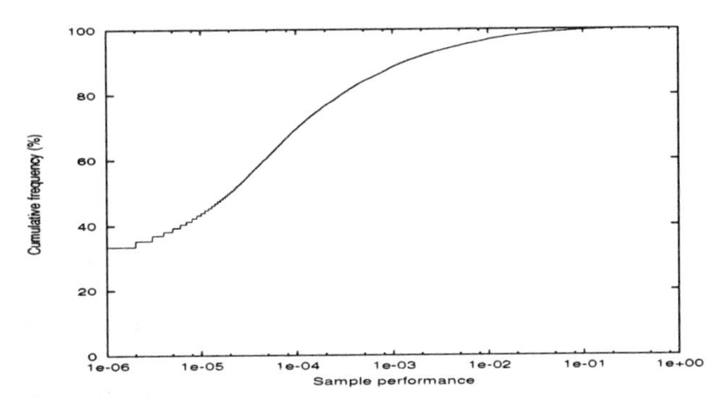

Figure 2 Cumulative frequency of the sample performance

Before examining the usefulness of the sampling technique, we will observe that the sample performance can drastically change depending on the combination of the initial states. Fig. 2 shows the cumulative frequency of the sample performance $\sum_{n=N_0}^{N_0+N-1} \mathbf{1}_{\{X_n > m\}}/N$ for randomly chosen 20000 initial states (with the initial distribution μ). In Fig. 2, we observe that the samples widely spread. While 50% of the samples are less than 2.00×10^{-5}, 0.01% of the samples are greater than 1.97×10^{-1}. We can see that as mentioned in Introduction, there exist some bad initial state combinations which lead to extreme degradation of the system performance, although they are rarely realized. This makes it hard to get estimates with low variance by standard MC simulation.

Table 1.1 Simulation results

Method	standard MC	MC using sampling technique
Estimation	1.85×10^{-3}	1.91×10^{-3}
Variance	9.10×10^{-7}	2.10×10^{-8}

We now investigate the usefulness of the sampling technique for reducing the variance of the estimates. For this purpose, we obtain 100 estimates and compute the sample variance of these estimates. Each estimate is the sample mean of 200 samples (i.e., $N_s = 200$). Table 1.1 displays the simulation results obtained through the standard MC simulation and the MC simulation using the sampling technique. In Table 1.1, we observe that the utilization of the sampling technique can reduce the variance of the estimates.

For both the standard MC simulation and the MC simulation using the sampling technique, the average total CPU time required to obtain an estimate was 523 sec. Thus, the overhead computational cost to get a biased initial distribution is negligible in this case. We define the time-reliability product [3] as (time-reliability product) = (the average total CPU time required to obtain an estimate) $\times$ (the variance of the estimates). We then consider the speed-up factor [3], which is defined as the ratio of the time-reliability product of the standard MC simulation to that of the simulation using the sampling technique. In this case, the speed-up factor was 43.

6. CONCLUSIONS

For MC simulation of queues with a superposition of periodic and bursty sources, we have developed a sampling technique which replaces the initial distribution with a new biased one. The simulation result shows that the sampling technique is useful to reduce the variance of the estimates. We are now investigating the usefulness of the sampling technique combined with existing IS techniques which replace the state transition probabilities with biased ones.

References

[1] Abate, J., Choudhury, G. L., and Whitt W. (1994). "Asymptotics for steady-state tail probabilities in structured Markov chains of $M/G/1$-type". *Stoch. Mod.*, 10:99–144.

[2] Chang, C.-S. (1995). "Sample path large deviations and intree networks". *Queueing Systems*, 20:7–36.

[3] Devetsikiotis, M., and Townsend, J. K. (1993). "Statistical optimization of dynamic importance sampling parameters for efficient simulation of communication networks". *IEEE/ACM Trans. Networking*, 1:293–305.

[4] Duffield, N. G. (1994). "Exponential bounds for queues with Markovian arrivals" . *Queueing Systems*, 17:413–430.

[5] Falkenberg, E. (1994) "On the asymptotic behavior of the stationary distribution of Markov chains of $M/G/1$-type". *Stoch. Mod.*, 10:75–98.

[6] Glynn, P. W., and Iglehart, D. L. (1989). "Importance sampling for stochastic simulations". *Management Sci.*, 35:1367–1392.

[7] Haraszti, Z., and Townsend, J. K. (1998). "The theory of direct probability redistribution and its application to rare event simulation". *Proc. of ICC '98*, 1443–1450.

[8] Ishizaki, F., and Takine, T. (1997). "Bounds for the tail distribution in a queue with the superposition of general periodic Markov sources". *Proc. of INFOCOM '97*, 1088–1095.

[9] Ishizaki, F., and Ohta, C. (1998). "Finding good combinations of initial phases for periodic sources in discrete-time queues". *Proc. of ICC '98*, 1510–1514.

[10] Ishizaki, F., (1999). "Sampling technique for queues with superposition of periodic and bursty sources". in preparation.

[11] Karlin, S., and Taylor, H. M. (1975). *"A first course in stochastic processes, second edition"*. Academic press, San Diego.

[12] Neuts, M. F. (1989). *"Structured stochastic matrices of M/G/1 type and their applications"*. Marcel Dekker, New York.

[13] Rubinstein, R. Y. (1981). *"Simulation and the Monte Carlo method"*. John Wiley & Sons, New York.

[14] Sadowsky, J. S. (1991). "Large deviations theory and efficient simulation of excessive backlogs in a GI/GI/m queue". *IEEE Trans. Autmat. Contr.*, 36:1383–1394.

[15] Villén-Altamirano, M., Martínez-Marrón, A., Gamo, J., Fernández-Cuesta, F. (1994). "Enhancement of the accelerated simulation method RESTART by considering multiple thresholds". *Proc. of ITC 14*, 797–810.

[16] Wang, Q., and Frost, V. S. (1993). "Efficient estimation of cell blocking probability for ATM systems". *IEEE/ACM Trans. Networking*, 1:230–235.

PERFORMANCE EVALUATION OF SCHEDULING ALGORITHMS FOR BLUETOOTH

Niklas Johansson, Ulf Körner
Department of Communication Systems
Lund University
Box 118, S-221 00 Lund, Sweden
niklasj@tts.lth.se, ulfk@tts.lth.se

Per Johansson
Ericsson Radio Systems AB
S-126 25 Stockholm, Sweden
Per.Johansson@ericsson.com

Abstract During the last couple of years, much attention has been brought to research and development of mobile ad hoc networks. In an ad hoc network, a collection of peer wireless mobile users within range of each other may dynamically form a temporary network without the use of any existing network infrastructure or centralized server, as opposed to cellular systems.

This paper analyses the performance of a wireless ad-hoc network concept called Bluetooth which was presented in February 1998 by its five promoters - Ericsson, Nokia, IBM, Toshiba and Intel. We discuss a modified exhaustive scheduler, proposed by the authors, and show its applicability under various operating conditions. A number of scenarios are analyzed and we also address the importance of multi-slot packets to increase throughput and to keep the delays low.

Keywords: Ad hoc networking, Bluetooth, performance, polling, wireless networks

This work was partially funded by the Foundation for Strategic Research - Personal Computing and Communication

1. INTRODUCTION

Battery powered, untethered computers, Personal Digital Assistants (PDAs), cameras, multimedia terminals, etc. are likely to become a pervasive part of our computing and communication infrastructure. An *ad hoc network* is the cooperative engagement of a collection of these mobile terminals. In such a network, each mobile node operates not only as a host but also as a router, forwarding packets for other mobile nodes in the network that may not be within direct wireless transmission range of each other. Traditionally, ad-hoc packet radio networks have mainly concerned military applications, where a decentralized network configuration is an advantage or even a necessity. For the commercial sector, equipment for wireless mobile computing and communication has not been available at a prize affordable for any larger market. However, as capacity of common mobile computers is steadily increasing, the need for untethered networking is expected to do likewise.

This paper analyses the performance of a wireless ad-hoc network concept called Bluetooth which was presented in February 1998 by its five promoters - Ericsson, Nokia, IBM, Toshiba and Intel. These five companies have formed a special interest group, the Bluetooth SIG. The purpose of the consortium is to establish a de facto standard for the air interface and the software that controls it, thereby ensuring interoperability between devices of different manufacturers. Today, more than 800 companies have joined the SIG. The name Bluetooth was taken from Harald Blåtand, a Danish Viking king from the early Middle Age.

Today the media access scheme used in Bluetooth is based on polling, i.e. one of the communicating devices within a Bluetooth picocell acts as a master and polls the other devices. The basic feature of a polling system is the action of centralized intelligence in polling each node on the network following some rules in order to provide access to the channel. As each node is polled, the connected entity may use the full data rate to transmit its backlog for a time depending on how the channels capacity has to be time-shared among the entities, i.e. the multiplexing technique. In between polls, the connected entities accumulate workload, but do not transmit until polled.

In this paper, the original Bluetooth media access principle, i.e. a strict round robin polling algorithm is challenged by a round robin exhaustive polling algorithm and also by a new polling algorithm proposed by the authors. The objective was to analyze some key performance metrics, like delays and throughput, for the above mentioned algorithms under various operating conditions. A number of scenarios are analyzed and we especially stress the importance of multi-slot transmissions to increase throughput and keeping the delays low. The first and second papers on Bluetooth performance was presented at the IFIP 4th Workshop on Personal Wireless Communications in Copenhagen, April 1999, [1] and at ICC'99 in Vancouver, June 1999, [2].

2. SYSTEM DESCRIPTION

A few years ago it was recognized within Ericsson Mobile Communications that a truly low cost, low power radio based cable replacement, or wireless link, was feasible. Hence, the original intension of Bluetooth was to eliminate the cables between cordless phones, laptops, wireless headsets etc. As the project progressed, it became clear that such an ubiquitous link could provide the basis for small portable devices to communicate together in an ad hoc fashion. Today, Bluetooth is a true ad-hoc wireless network intended for both synchronous traffic, i.e. voice, and asynchronous traffic, i.e. IP-based data traffic.

In Bluetooth all units are peer units with identical hard and software interface. Two or more units sharing the same channel form a piconet, where one unit acts as a master (any unit can become a master), controlling traffic on the piconet, and the other units act as slaves. The communication channel operates in the unlicensed ISM band (Industrial-Scientific-Medical band) at 2.45GHz (bandwidth 80 MHz). A frequency hop transceiver is applied to combat interference and fading. The system offers a gross bit rate of 1 Mbps per piconet and allows a mix of voice and data communication channels. The system uses a slotted Time-Division Duplex (TDD) scheme for full duplex transmission, where each slot is *0.625 ms* long (two slots form one frame). Time slots may carry both synchronous information (Synchronous Connection Oriented, SCO, voice links) and dynamically allocated asynchronous information (Asynchronous Connection-less, ACL, data links). A master unit controls the traffic on the channel by allocating capacity for SCO links, i.e. reserving slots for this link, and by using a polling scheme for ACL links. In this paper we especially focus on the efficiency of three different polling algorithms, the strict round robin polling algorithm, the exhaustive round robin polling algorithm and a polling algorithm proposed by the authors. Up to 8 active devices, but to that many inactive devices, may form a Bluetooth piconet. Several piconets can be linked together, thus forming a scatternet in which each piconet is identified by a unique hopping sequence, and as long as not too many, each provides a capacity of almost 1Mbps.

In each slot, a packet can be exchanged between the master and one of the slaves. Each packet begins with a 72 bit access code, which is unique for each piconet and is derived from the master identity. Recipients in the piconet compare the incoming packet access code with the access code of its own piconet. If the two do not match, they simply discard the received packet. The access code also contains information used for synchronization and compensation for offset. A header trails the access code. The header contains control information such as a three bit MAC-address, flow control, a type field, an ARQ field and a HEC field. Payload may or may not trail the header. The length of the payload may vary from 0 to 2745 bits.

In order to increase throughput under high data rates, so called multi-slot packets may be used. These packets may cover three or five slots. A three slot packet may contain 6 times as much information as a single slot packet and a five slot packet 12 times as much. In this paper we, among others, analyze how performance is increased by using multi-slot packets instead of single-slot packets, but also how the three mentioned polling algorithms behave under different loads.

Data packets are protected by an ARQ scheme in which lost data packets are automatically retransmitted in the next slot. This results in a fast ARQ scheme - delay due to errors is just one slot in duration. Voice traffic is protected by a robust voice encoding scheme and thus retransmission of voice packets never takes place. For a more detailed description of the system, the reader is referred to [3].

3. MODEL DESCRIPTION

We have built up a rather comprehensive model of an ad-hoc wireless network that in most respects maps Bluetooth. It is our main purpose to grasp especially the polling mechanism used for the asynchronous links (ACL links).

The simulation studies were made for one piconet comprising of seven units, i.e. one master and six slaves. The traffic is assumed to be symmetric, i.e. one slave sends the same amount of data to all the other slaves. The master does not send any traffic to the slaves but just forwards traffic from one slave to another.

When we allow multi-slot packets, these packets are formed and sent according to the following scheme: a queue with 0 to 2 1-slot packets sends a 1-slot packet, a queue with 3 to 7 1-slot packets sends a 3-slot packet and a queue with more than 7 one slot packets sends a 5-slot packet.

Transmitted packets, i.e. packets on OSI layer 2, may be lost due to bit errors and this is modeled with a constant packet loss probability. All lost packets are being retransmitted according to the Bluetooth ARQ scheme. In our simulations the packet loss probability is set to 10^{-4} for all types of packet.

3.1 THE QUEUING MODEL

The arrival process of IP-packets is assumed to be bursty. To capture the discrete events of packet arrivals as well as the burstiness, arrivals are modeled as a discrete time Interrupted Bernoulli Process (IBP) (see chapter 3.2, The Arrival Process). The arriving IP-packets are being fragmented in order to fit the length of Bluetooth layer 2 packets. The different queues are of infinite size, a valid assumption for this preliminary study.

3.2 THE ARRIVAL PROCESS

The arrival process used in this paper is based on an IP trace generated by a single user connected to a 10Mbps Ethernet LAN. Measurements were performed at the Department of Communication Systems, Lund University. A Linux station running the program *tcpdump* was used to observe the web traffic to the station. All IP packets to the station during a period of approximately 2 hours were captured and the packet sizes were logged to a file together with a time-stamp. The measurements have an accuracy of 1 μs. It is assumed that the behavior of the wireless IP-traffic, which we try to emulate in this paper, shows similar characteristics as the LAN IP-traffic.

IP-Packet Generator. The number of users in the modeled network is relatively small and they are assumed to transfer web-like traffic. This combination is expected to result in bursty traffic streams in the network, and hence, Poissonian assumptions are not valid. To be able to make a performance analysis we have to use some other arrival model that reflects these bursty arrivals.

In recent literature it is illustrated that e.g. LAN packet traffic has a packet arrival process that displays variability over a variety of time-scales and causes queuing problems at relatively low loads [5] [6]. This has spurred research in the area of traffic models inherently more suited for modeling second order self similar behavior, e.g. Fractional Brownian Motions (FBM) [7] and chaotic maps [8]. For these models, however, the tools for analyzing queuing behavior are still in their embryonic state. Additionally most of the novels are focused on capturing the first- and second order characteristics of counts that, in general, are known to be insufficient when attempting to predict queuing behavior [9].

Since the knowledge of traffic patterns in ad-hoc networks is limited and those traffic patterns also shift from application to application, we have chosen to model our packet generator as an Interrupted Bernoulli Process (IBP). That is, for a geometrically distributed period (ON state) arrivals occur according to a Bernoulli process. This period is followed by another period (OFF state), also geometrically distributed, during which no arrivals occur (The IBP is a special case of the Markovian Arrival Process (MAP) in discrete time). This is a very straightforward model that provides a bursty, but controlled, traffic characteristic, sufficient for this preliminary analysis.

In an IBP the probability for sending a packet during a slot (Note that the time scale of the IBP is almost 50 times shorter than that of the modeled piconet) is set to 1 in the ON state and to 0 in the OFF state. If p is the transition probability from the OFF to the ON state and q is the transition probability in the opposite direction, it can be shown that the probability generating function of the inter arrival times is given by,

$$P(z) = \frac{(1-q-z(1-p-q))z}{1-(1-p)z} \tag{1}$$

thus we get the load, ρ, on the piconet channel as,

$$\rho = \frac{N \cdot N_B \cdot \frac{S_B}{S_{IBP}}}{1+q/p} \tag{2}$$

where,

- N=Number of active units.
- N_B=E[Number of Bluetooth packets per IP packet].
- S_B=Bluetooth slot length.
- S_{IBP}=IBP slot length.

The above equation holds for the case when only single slot packets are used. If multi-slot packets are allowed, a term compensating for this has to be multiplied to the right hand side of equation (2). This term is hard to predict since the fraction of the different packets is not known, but with the aid of simulation one can measure the occurrence of the different packet types and calculate the actual load.

The squared coefficient of variation, C^2, for the IP-packet inter arrival times were used as a measure of the burstiness in the simulations. From equation (1) we get the squared coefficient of variation as,

$$C^2 = \frac{q(2-p-q)}{(p+q)^2} \tag{3}$$

Now, by varying p and q we can alter the load while keeping the squared coefficient of variation C^2 constant.

The IBP only generates the instances when IP packets are generated. When a packet is generated the length of this packet is drawn from essentially a two point distribution, retrieved from the LAN-trace. The IP packets are further fragmented into Bluetooth layer 2 packets.

4. SCHEDULING

To be able to provide Quality of Service (QoS) capabilities that support bandwidth allocation and latency control, some form of intelligent scheduling algorithm is needed. The decision process, based on polling, should be managed by the master alone and handled by a slave scheduler located in the master entity. The decision should be based on the aggregate state of the slave queue and the master output queue for that particular slave.

There are generally two ways to guarantee the negotiated QoS. One is to poll the slaves in a round robin fashion, the other is to associate a local time-stamp to each packet and serve the packet with the least time-stamp. One example of a scheduler using local time-stamps is Weighted Fair Queuing (WFQ) [10]. It has the advantage that it may guarantee a minimum rate of service with arbitrarily fine granularity, however, it is computationally complex and require storage of time-stamps for each packet. Round robin schedulers can guarantee a minimum rate of service and require minimal computation to service each packet, however, they have finite bandwidth granularity and it may therefore be more difficult to modify the bandwidth allotment. Since the schedulers are supposed to run in "thin clients", minimal computation complexity is very important and this is the reason why we decided to focus on round robin schedulers.

In this paper we have focused on three different scheduling algorithms, namely: round robin, exhaustive and a modified exhaustive scheduling algorithm herein called Fair Exhaustive Polling (FEP).

4.1 STRICT ROUND ROBIN POLLING

In the round robin service system [11], in some literature also known as limited service system [12], the master polls the slaves in consecutive order. Each slave is allowed to send just one packet per service cycle. With this round robin scheduler the master can assign different bandwidths to different units by allowing different packet types to the users (see chapter 2, System Description). This has the effect that users with higher demands on bandwidth get a fair share of the excess bandwidth, which is a desirable feature since we then can guarantee max-min fairness [10].

4.2 EXHAUSTIVE POLLING

In the exhaustive service system, the master continuous to poll the addressed slave until the slave queue and the master output queue for the specific slave is emptied. This means, that not only those packets present in the queue at the beginning of the service cycle but also those packets generated during the service cycle are served before the master moves on to the next slave. This is of course not an optimal scheduling algorithm since this procedure will favor slaves generating packets at maximum rate. Therefore, an additional parameter has been added, which limits the service time to some predetermined maximum time.

With this scheduler the master may assign different bandwidths to different units by allowing different packet types, i.e. 1, 3 or 5-slot packets, and different maximum poll time for each individual slave. Also for this scheduler one can guarantee the max-min fairness, at least over time scales larger than the maximum sojourn time [13].

4.3 FAIR EXHAUSTIVE POLLING (FEP)

The fair exhaustive polling service system can be viewed as a combination of the two earlier mentioned polling algorithms. The main idea behind FEP is to poll slaves that probably has nothing to send as seldom as possible.

The slaves belong to one of two complementary states, the *active state* and the *inactive state.* A polling cycle starts with the master moving all slaves to the active state, and then begins one of possibly several polling sub cycles. In a polling sub cycle all *active* slaves are polled once in a round robin fashion. As mentioned above, the master performs the task of packet scheduling for both the downlink and uplink flows, however, the master has only limited knowledge of the arrival processes at the slaves. This means that the scheduling, of the uplink flows, has to be based on the feedback it gets when polling the slaves. Based on this feedback and the current state of the master output queues, slaves are moved between the different states. A slaves is moved from the active state to the inactive state when both of the following conditions are fulfilled:

- The slave has no information to send

- The master has no information to send to the specific slave

A slave is moved to the active state when the master has information to send to it.

This is an iterative process that continues until the active state is emptied (the exhaustive part of the algorithm) and when it is, a new polling cycle starts.

Also this algorithm will favor slaves generating packets at maximum rate and therefore an additional parameter has been added, which limits the polling interval of any slave to some predetermined maximum time. For the algorithm this means that a slave, whose maximum polling interval timer has expired, is moved to the active state and therefore will be polled in the next polling sub cycle.

FEP behaves, asymptotically, as an exhaustive scheduler at low loads and as a round robin scheduler at high loads. This is a very attractive feature since the mean delay for the exhaustive scheduler is less than or equal to the mean delay for the round robin scheduler for low traffic, whereas the opposite holds for heavy traffic [12]. We have, through simulation, been able to show that the FEP scheduler experiences lower delays than both the exhaustive- and the round robin scheduler at all loads.

With this scheduler the master may assign different bandwidths to different units by allowing different packet types, i.e. 1, 3 or 5-slot packets, and different maximum poll intervals for each individual slave. Also for this scheduler one should be able to guarantee the max-min fairness, at least over time scales larger than the maximum sojourn time.

5. SIMULATION RESULTS

In [1] and [2] we have analyzed a number of different situations under the scenario where slaves send web-like traffic to other slaves. There we analyzed symmetric as well as non symmetric load situations for both bursty and non bursty traffic characteristics. We also analyzed the transient behavior of the Bluetooth network when one of the slaves send a very dense burst for a short period.

In this paper the aim of our simulations is twofold. On one hand a case using only single slot packets is compared with one that allows multi slot packets, i.e. 3 or 5 slot packets as well. A 3 slot packet may hold up to 6 times the payload of a 1 slot packet and a 5 slot packet up to 12 times. Furthermore, which is the main focus of this article, the three scheduling algorithms mentioned in the previous chapter are compared. The strict round robin was initially chosen as the media access protocol for Bluetooth. Here we compare performance metrics of the strict round robin with the classical exhaustive polling algorithm and what we refer to as the Fair Exhaustive Polling algorithm, (FEP).

It is well known that the round robin scheduler outperforms the exhaustive scheduler under high symmetric loads and vice versa under low loads. Our algorithm, the FEP, more or less works as the exhaustive under low loads and as the round robin under high loads and thus shows to be a good alternative for a Bluetooth network especially and for most polling based networks in general.

In order to evaluate these algorithms, we have formed a worst case scenario in that we feed the Bluetooth network with traffic that really stresses the network. The traffic is modeled first at an IP level with a rather high squared coefficient of variation equal to more than 60. (For the reader we again mention that the squared coefficient of variation for the arrival process is equal to the ratio of the variance of the inter arrivals between subsequent IP packets and that of the squared mean.). To that comes the fact that IP-packets normally are much longer in bits than that of a Bluetooth-packet. Thus, at most IP-packet arrivals, a number of Bluetooth packets are generated. We experience extremely long delays but this is so just for us to compare the algorithms under high loads.

In figure 1, where we allow multi slot packets, the average delays are plotted against the total arrival rate in bits to the entire Bluetooth piconet. We observe here that the exhaustive algorithm outperforms that of round robin, while our algorithm, the FEP, outperforms that of the others at high as well as low loads. We also observe from figure 2 that the same behavior also holds for single slot packets though the delays are even longer. It is also worth noting that the round robin policy performs better than the exhaustive policy at high loads and vice versa for low loads. Figure 3 and 4 show the squared coefficient of variations for the end to end delays for the two cases discussed above. Here also our FEP outperforms the two others by a large factor.

6. CONCLUSIONS AND FURTHER WORK

In this paper we have analyzed and compared the behavior of three different polling algorithms used as medium access schemes for an ad hoc wireless network called Bluetooth. When Bluetooth was announced, in late spring 1998, the polling scheme proposed was based on a strict round robin principle. This requirement has now been relaxed and we have in this paper been able to demonstrate the strength of a proposed scheduler, herein called the Fair Exhaustive Polling (FEP) scheduler.

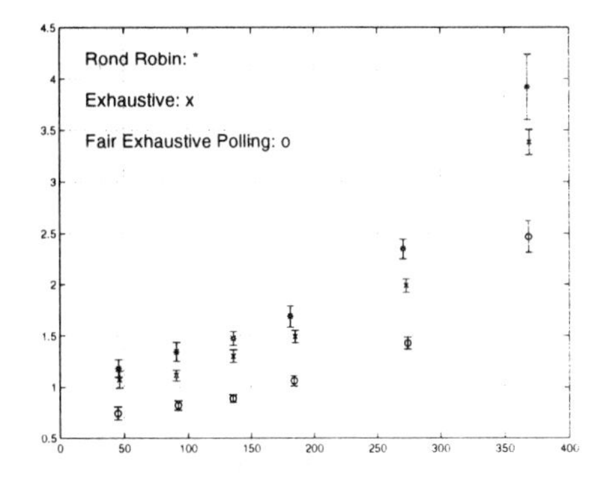

Figure 1 Average delay [s] against bitrate [kbps] for the three scheduling algorithms, multi-slot packets allowed.

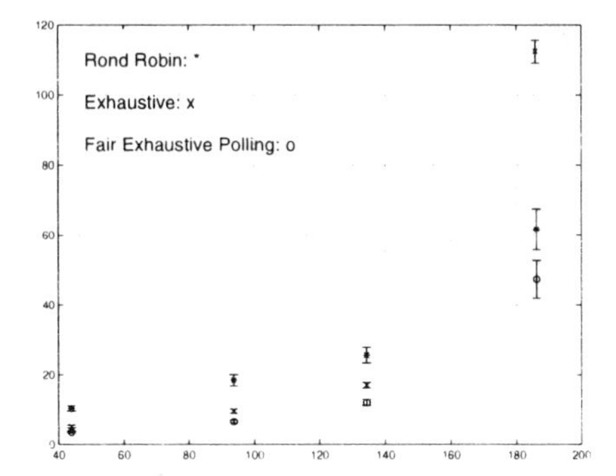

Figure 2 Average delay [s] against bitrate [kbps] for the three scheduling algorithms, only single-slot packets allowed.

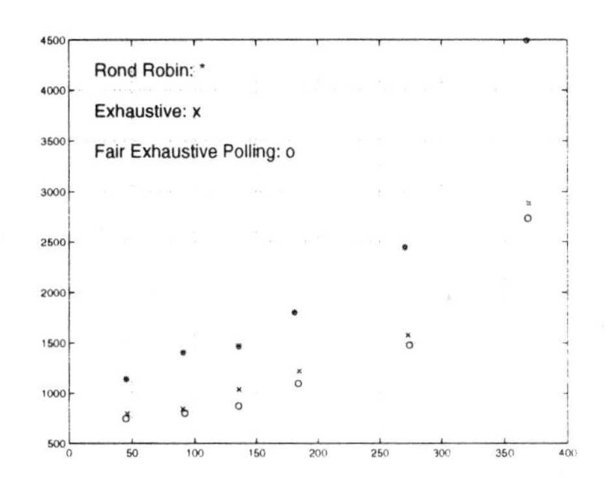

Figure 3 Squared coefficient of variation against bitrate [kbps] for the three scheduling algorithms, multi-slot packets allowed.

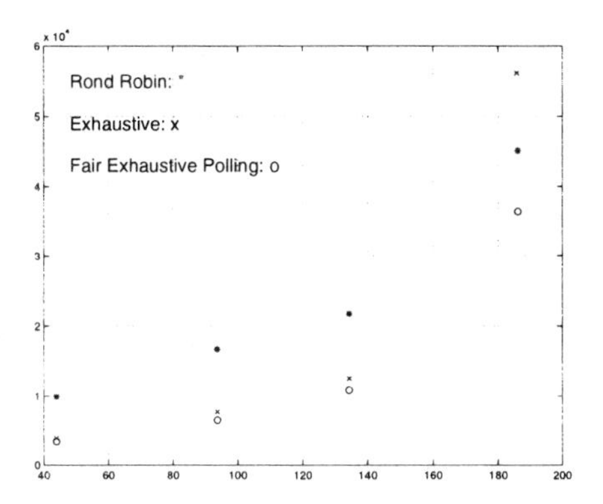

Figure 4 Squared coefficient of variation against bitrate [kbps] for the three scheduling algorithms, only single-slot packets allowed.

We have also clearly demonstrated the increase in performance when allowing packets to be sent in so called multi-slots. Not at least when traffic tends to be very bursty, the use of multi-slots has proven to be efficient. This is of course natural as the transmission overhead dramatically decreases.

We will in coming papers refine the FEP scheduler to also include inter-piconet capabilities. We will also address the question of under which circumstances new piconets should be established for devices in an existing piconet when traffic patterns change dynamically.

References

[1] Niklas Johansson, Ulf Körner and Per Johansson. Wireless Ad-hoc Networking with Bluetooth. *4th Workshop on Personal Wireless Communications*, Copenhagen, 1999.

[2] Per Johansson et al. Short Range Radio Based Ad-hoc Networking: Performance and Properties. *ICC'99*, Vancouver, 1999.

[3] Jaap Haartsen. Bluetooth - The universal radio interface for ad hoc, wireless connectivity. *Ericsson Review*, no 3, 1998

[4] The Bluetooth Special Interest Group, Documentation available at http://www.bluetooth.com/

[5] Ashok Erramilli, Onuttom Narayan and Walter Willinger. Experimental Queueing Analysis with Long-Range Dependent Packet traffic. *IEEE/ACM Transactions on Networking*, 4(2):209-223, April 1996.

[6] Will E. Leland, Murad S. Taqqu, Walter Willinger and Daniel V. Wilson. On the Self-Similar Nature of Ethernet Traffic (Extended Version). *IEEE/ACM Transactions on Networking*, 2:1-15, February 1994.

[7] I. Norros. A storage model with self similar input. *Queueing Syst.*, vol. 16, pp. 387-396, 1994.

[8] A.Erramilli, R.P. Singh and P. Pruthi. An application of deterministic chaotic maps to model packet traffic. *Queueing Syst.*, vol. 20, pp. 171-206, 1995.

[9] Kerry W. Fendick and Ward Whitt. Measurements and Approximations to Describe the Offered Traffic and Predict the Average Workload in a Single-Server Queue. *Proceedings of the IEEE*, 77(1):171-194, January 1989.

[10] S. Keshav. An Engineering Approach to Computer Networking. *part of the Addison-Wesley Professional Computing Series*, 1998. ISBN 0-201-63442-2.

[11] Hideaki Takagi. Exact analysis of round-robin scheduling of services. *IBM J. Res. Dev. 31*, 4 (July), 484-488.

[12] Hideaki Takagi. Queuing Analysis of Polling Models. *ACM Computing Surveys*, Vol. 20, No. 1, March 1988.

[13] Uri Yechiali. Optimal Dynamic Control of Polling Systems. *ITC-13*, Copenhagen, 1991.

SESSION 5

Billing, Pricing and Admission Policy

THE ECONOMICS AND COMPETITIVE PRICING OF CONNECTIVITY AT INTERNET EXCHANGES

Joseph Y. Hui, FIEEE, FHKIE
Department of Information Engineering
Chinese University of Hong Kong
Shatin, Hong Kong

Abstract With the emergence of Internet hubs, it is important to understand the history, the functions and the economics of the formation of Internet hubs and exchanges. These exchanges are often used for the multicasting of multimedia information, with quality of service control provided. This paper first delineates the history and types of Internet hubs. We then look at the economics of network interconnections with application to the development of Internet hubs. Such economics include aspects of increasing return and externalities, resource substitution, interconnection fee, price competition, and quality of service provisioning. We conclude by looking at policy and competition issues aimed at promoting Internet hubbing.

Keywords: Internet, network economics, bandwidth pricing, Internet exchanges

1. INTRODUCTION

Internet hubs have emerged as strategic facilities for economic developments. Traffic at these facilities typically grows at a compound rate which more than doubles every year [1]. It is anticipated that the importance of these facilities will increase as new applications such as Internet telephony, Intranets, and interactive Video on Demand services will employ Internet hubs for interconnection and facility location. These applications demand Quality of Service (QoS) guarantees for service delivery.

Foreseeing their emerging strategic importance, governments have indicated intention and issued policies to facilitate the emergence of Internet hubs within their jurisdiction [2]. These hubs not only serve the local economy and service industries, but also are set up competitively to attract traffic from the surrounding regions. Therefore, the emergence of Internet hubs and multimedia exchanges within cities reinforces their status as service and goods entrepot for their surrounding regions. These Internet hubs naturally compete for traffic within the surrounding region. Given the increasing return or externality provided by a hub with large size, regional dominance of a hub is predicated by a head start in traffic buildup and liberal telecommunication policies.

This paper is organized as follows. In section 2, we shall first examine the historical development of Internet hubs. The many functions of these hubs are explained. In section 3, we look at the externality and economic benefits of a hub as a function of its size and other factors. In section 4, we form simple models of price competition for various modes of interconnection. In section 5, we draw simple conclusions from these models concerning means of achieving prominence of an Internet hub.

2. HISTORY AND FUNCTIONS OF INTERNET HUBS

The Internet was developed as a protocol for the internetworking of independently administered networks. Initially the network was funded by the US federal government for educational and government usage. As the network grew and usage became popular, access and backbone transport services are commercialized and are offered by large number of Internet Service Providers (ISP).

This explosive growth is a result of the low entry cost for setting up an ISP for the following reasons. First, the Internet Protocol (IP) is standardized, simple, and highly distributed. The IP switching equipment is relatively inexpensive and easy to operate compared with switching for the telephone network. Second, interconnection cost, which was based on a settlement free model to start with, is low and simple to administer. This offers a significant competitive advantage compared to the archaic, costly, and hardly competitive settlement system devised for the internetworking of telephony. Third, the transport facility of an IP network is built on top of the transmission facility of the telephony network, and therefore there is no expensive network investment. In addition, the global trend of liberalizing Public Switched Telephone Networks (PSTN) tends to drive down cost of subscriber access to ISP, and ISP access to the local and global Internet.

Internet hubs are developed for exchanging traffic among ISPs and for access to the global Internet. Internet hubs are termed Internet eXchanges (IX) or Network Access Points (NAP) [3]. In essence, Internet hubs are points of interconnections for Internet Access Providers (IAP), Content Providers (CP), and Internet Backbone Providers (IBP). In a region where there is no Internet hub or exchange, an ISP would have to buy global access using long distance transmission facilities to connect to a remote exchange point, often in the US. This was the case for most countries until recent years.

The development of IXs and NAPs is closely related to the development of the Internet in the US. Initially, the Internet is a backbone network funded by the Defense Department (DARPA) and by the National Science Foundation (NSF) of the US. As the Internet is commercialized, NSF established NAPs, called MAE, were turned over to commercial companies for operation. Later, major regional telephone companies in the US set up their own NAPs, providing Internet exchange functions using their own transmission facilities and switching centers. Still later, computer hardware and software companies entered the field. These Internet hubs focus not only on exchange functions, but also on the hosting of bandwidth intensive content.

The development of Internet hubs around the world is a logical necessity. Non-US ISPs market subscriber access in terms of the amount of dedicated bandwidth they use to connect to the US. As multiple ISPs develop within the local region, it is only natural for these ISPs to negotiate a local and preferably neutral point for themselves to exchange local traffic. The rationale is simply that local transmission facilities are much cheaper and of better quality in terms of delay and throughput than international transmission facilities. Thus we witness the emergence of one or more Internet exchanges within many countries in the Asian Pacific region.

While much of the Internet web content remains in the US, there is a common trend that both the volume and importance of local content are increasing. Also, mirror sites for non-local content are established locally in order to reduce international bandwidth consumption and for improved speed of retrieval. Consequently, local traffic volume is expected to approach the volume of international traffic.

The economics of these Internet hubs involves the tradeoff in the use of local bandwidth, international bandwidth, and storage at the hub for caching both local and international content. Also, global access can be provided at the hub, rather than having each ISP acquiring international bandwidth by itself. The sharing of international bandwidth at the hub itself is termed a transit service, which is cost effective due to sharing of a large international bandwidth. A good price discount is often obtained for a high capacity international link.

Internet hub could be seen as an example of the Internet information marketplace, where different players such as IAP, CP, IBP, etc meet, creating a vibrant market of services for each other. Such a market place allows players to flexibly acquire services from anyone they so choose. The market place allows these players to choose alternative service suppliers without a physical reconnection, provided the alternative suppliers are at the same hub.

3. THE ECONOMICS OF INTERNET HUBBING

The economics of Internet hubbing bears a certain resemblance to the hubbing of airline routes. The size of the hub brings about an externality (economic benefit), which is the ability to reach a large number of Internet sites efficiently. This externality in turn attracts more ISPs to connect to the hub. Therefore, the success of a hub depends strongly on achieving dominance early on.

To achieve dominance, the cost of accessing the hub must be low compared to other aspiring hubs. This is a common strategy in Internet product and service marketing, namely the use of low or even below cost introductory prices in order to achieve dominance early on, so that an increasing return is gained by signing on a large number of subscribers.

Consider the economics of two competing hubs for interconnecting N ISPs. Let us assume that each ISP generate an equal amount of traffic A, and this traffic is evenly distributed to the other $N-1$ ISPs. Suppose a fraction f of the N ISPs is connected to the first hub, leave the rest connected to the second hub.

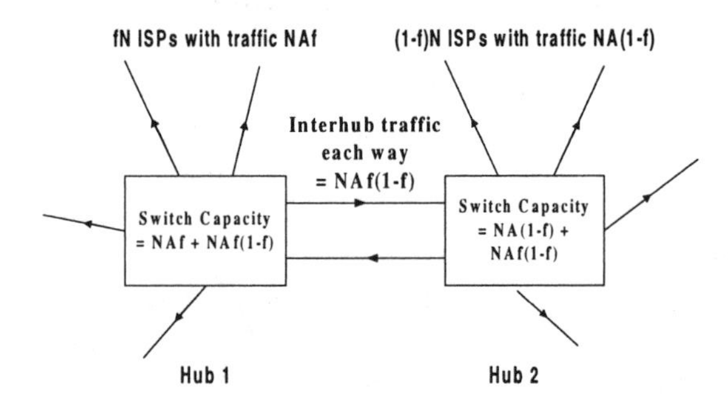

Figure 1. Traffic flow for two Internet Hubs

As seen in figure 1, a traffic volume of NAf is generated by ISPs connected to the first hub, out of which a traffic volume of $NAf(1-f)$ is destined for the second hub. This inter-hub traffic volume is symmetrical since similarly, a traffic volume of $NA(1-f)f$ is destined from the second hub to the first hub. Thus the use of two hubs requires an inter-hub transmission link which increases cost as shown in table 1.

Beyond increased transmission cost, switching cost is also increased. The first hub would require a switching capacity of $NAf + NAf(1-f) = NAf(2-f)$. Similarly, the second hub would require a switching capacity of $NA(1-f) + NA(1-f)f = NA(1-f- 2)$. The total switching capacity required is then $NA(1+2f-2f 2)$.

Table 1. Switching and Inter-Hub Link Cost per ISP

	Switch Capacity		Inter-hub capacity	
	Total	Per ISP	Total	Per ISP
Hub1	$NAf(2-f)$	$A(2-f)$	$Naf(1-f)$	$A(1-f)$
Hub2	$NA(1-f^2)$	$A(1+f)$	$Naf(1-f)$	Af

Compare the cases of $f=1$ (one hub only) and $f=.5$ (two equal size hubs). We see readily that the case of two equal size hubs requires a total switching capacity of $1.5NA$, and an inter-hub transmission link of capacity $0.5NA$. This is a significant increase in cost compared to the case of one hub, which requires a switching capacity of NA and no inter-hub link.

If $f>0.5$, the first hub acquires a superior externality compared to the second hub, if we consider the per subscriber cost of switching and inter-hub transmission at each hub as shown in table 1. For the extreme case of f close to 1, namely that the first hub have a close to 100% dominance, we see that the switching cost per ISP of the second hub is twice that of the first hub, and the inter-hub capacity required per ISP is 0 for the first hub versus A for the second hub.

The placement of content at an Internet hub promotes the same externality as that generated by size. An Internet hub is an ideal place for the placement and caching of information for ISP connected to that hub. Placing content at the hub removes the need to set up a link to the hub. Any ISP connected to the hub enjoys fast one-hop retrieval for content placed at the hub. Therefore, Internet hubs may derive significant income via content co-location and facility management at these hubs.

Therefore, Internet exchanges can provide other services than simply traffic exchange, as shown in figure 2. Facility management provided at the hub reduces the operation cost of enterprises since managing networks may not be the core competency of most, and particularly the small and medium enterprises.

Another advantage of Internet hubbing is the provisioning of international Internet transit services. Many Asian ISPs maintain two links separately for local and international Internet access. International transit service allows an ISP to access the local hub via a single local link. At the hub, international traffic is sorted out and allowed to make transit onto shared international links. This sharing allows multiple ISPs to enjoy two types of economy of scale. First, together they may subscribe a larger capacity link and thereby enjoy a volume discount for the price of the link. Second, they may enjoy the economy of statistical multiplexing, which reduces the size of the link they may have to subscribe since their traffic patterns are not likely to peak at the same time.

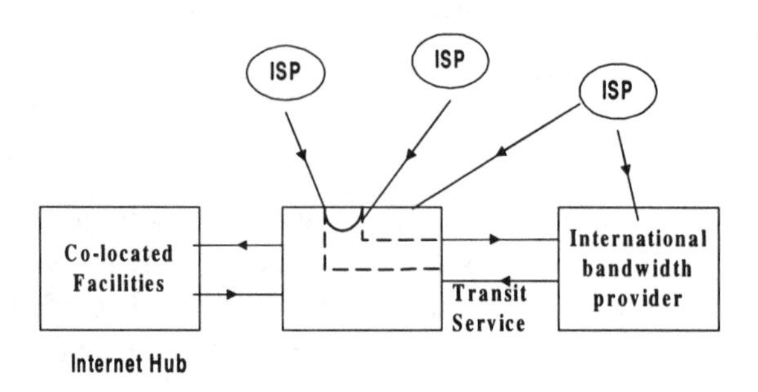

Figure 2. Services Provided at Internet Hubs

An Internet hub can achieve other intangible externalities besides the above externalities generated by the number of connected ISPs, the amount of co-located content, and the cost effectiveness of transit service. Value-added services provided at a hub make the hub more attractive for connection or service placement. It is often these services, rather than plan vanilla connectivity, which attract connectivity and service placement. Also, an open and flexible placement of services for the co-inhabitants of an Internet hub is by itself an attraction, since this kind of environment produces a vibrant information market place which safe-guards competition and consumer choice.

An important value added function for Internet exchanges is a multimedia exchange, which is a co-located storage of multimedia information at an Internet exchange for local Web based retrieval. This provides an open platform for Internet based VOD services. A multimedia exchange can also facilitate good quality of service provisioning, an externality for attracting connectivity to an Internet exchange.

4. PRICE COMPETITION FOR INTERCONNECTIONS AND QUALITY OF SERVICE

The competitive advantage of the larger hub depends strongly on the per unit cost of switching and transmission. If the inter-hub link is local, high bandwidth transmission can be obtained for inter-hub communication at very low per unit cost. Therefore, the competitive advantage of the larger hub is not particularly strong. Therefore, it could be argued that multiple hubs might be sustained within a locality, if smaller hubs can achieve substantial added values to their connectivity services.

If the two hubs are not in the same locality, inter-hub transmission is expensive. Also, the cost of an ISP connecting to a non-local hub can be very expensive. In choosing a hub to connect to for regional connectivity, an ISP is strongly influenced by the transmission cost to the hub. If a hub is situated in a region with an open and competitive market for international bandwidth, the region readily becomes a magnet for connectivity, and thereby its hub can readily become a dominant hub in the region.

The purpose of this section is to examine price competition for quality of service for various types of connectivity. A large literature concerning Internet economics can be found in [4]. We propose a new model concerning regional price competition of several network connectivity types as a non-cooperative game. The purpose of the analysis is to provide insights into collusive and competitive pricing practices for internetworking. This analysis can be applied to pricing access and network traffic exchange.

A network consists of a graph $G = (N,L)$ where N is the set of nodes n (where switches are placed for interconnections) and links l in the set of links L. To form a model of price competition, each link l is assigned ownership m_l , which is in the set M of service providers. Each service provider m owns a set of links L_m. Each service provider is interested in maximizing revenue, which is the product of price p_l and demand x_l , summed over all links l in L_m. A more general revenue model can also be assumed for non-linear pricing as a function of demand, as well as the incorporation of a demand dependent cost charged against revenue.

Naturally, traffic chooses the least costly connection for meeting an end-to-end demand. The cost per unit traffic for a link l depends both on the price p_l as well as the quality of service degradation cost $Cl(xl)$. This QoS cost may be a consequence of congestion induced by the traffic x_l on the link. To route traffic end-to-end, the total cost is given by the sum of the cost on the links in the end-to-end path. For an end-to-end traffic demand using more than one path, the cost of these paths must be the same, or else traffic could be shifted from the more costly path to the lest costly path.

We may assume two models concerning price-demand elasticity. The first model fixes demand for end-to-end service regardless of cost. The second model is cost elastic, which renders demand as a function of the lowest end-to-end cost.

This network of service providers with traffic seeking the least costly path is a non-cooperative game. Each service provider sets a price for each link it owns to attract traffic seeking the lowest cost paths. This non-cooperative game produces in a unique supply-demand equilibrium if the total cost suffers from a diseconomy of scale (i.e., the function $C_l(x_l)$ is a convex function).

We shall defer the solution of this competitive pricing model and the analysis of its properties for a later paper. In the remainder of this section, we analyze the simple networks shown in figure 3 to illustrate collusive and competitive pricing for internetworking.

In figure 3, we show three basic types of connectivity, namely the serial type, the parallel type, and the series-parallel type. Serial type connectivity consists of concatenation of two or more links. Parallel type connectivity consists of two or more links in parallel. Series-parallel type connectivity is formed by combining serial and parallel networks recursively.

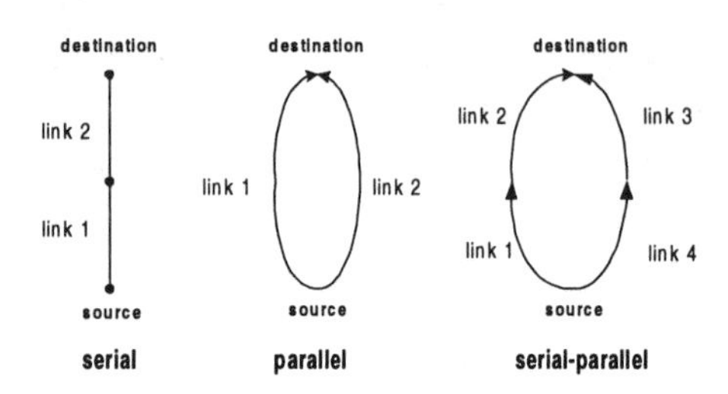

Figure 3. Types of Interconnection

We use these network types to illustrate collusive pricing for serial type network and competitive pricing for parallel type network. For a simple series-parallel network, we illustrate how a dominant network service provider can unfairly uncut its competitor.

For the sake of simplicity of illustration, we assume a linear congestion cost for each link. In other words, the total cost of using a link l is given by $y_l = p_l + c_l x_l$. We may assume further that cl is the same for all l.

We also assume a traffic demand of d between the source and the destination. This demand may be inelastic, in which case d is a constant. For elastic traffic demand as a function of the cost y, we may further assume a linear demand versus cost function $d(y) = f - ey$ for y between 0 and f/e, and $d(y)=0$ otherwise.

4.1 Competitive Pricing of Parallel Networks

Let us assume that the demand is inelastic, i.e. d is a constant. Let us assume traffic is split among the two links as x_1 and x_2 such that $x1 + x2 = d$. By symmetry, we have for optimal routing of traffic $x_1 = x_2 = d/2$. If the two links set prices competitively in order to maximize the revenue for each link, it can be readily shown that the optimal prices are $p_1 = p_2 = c$.

If there are k parallel competing links, it can be shown that each link would set a competitive price of $c/(k-1)$. In this case, revenue is driven to zero in an environment with many competing links .

If both links of the network belong to the same service provider, a monopoly exists and given an inelastic demand, the provider would set an exorbitant price with infinite revenue. If we assume a price elastic demand of $d(y) = f - ey$ for y between 0 and f/e, we can readily solve for the optimal monopolistic prices. Nonetheless, it can be readily shown that demand is suppressed and price is set high for the monopolistic case relative to the competitive case.

For the highly competitive Internet market where many service and route alternatives are available, significant allocative efficiency results from consumer choice and cost reduction. In that regard, the internetworking philosophy produces a highly efficient network economy.

4.2 Price Collusion for Serial Networks

Contrary to keen price competition for a parallel network of competing links, a serial network of competing links tends to produce collusive prices. For the case of inelastic demand, it can be readily seen that the two links of a serial network would raise prices indefinitely.

If we assume a price elastic demand of $d(y) = f - ey$, it can be readily shown that a serial network of competing links suppresses demand even more than the case of monopolistic ownership of both links.

An example of collusive monopolies is the provisioning of International Private Leased Circuit (IPLC). Two national and monopolistic PSTNs each owns a half-circuit for an IPLC. They tend to raise the price of their half-IPLC to levels which have no bearing with cost, thereby suppressing demand to an artificially low level.

The opening up of the IPLC market for International Simple Resale (ISR) of Internet and other value-added communication services is therefore an important step for development a vibrant information economy. Lowered IPLC cost will provide favorable conditions for the formation of a regional Internet hub.

4.3 Price Distortion for Series-Parallel Networks with Dominant Carriers

For series-parallel network, the existence of parallel alternative paths eases the collusive pricing effect of serial links. In the presence of a dominant carrier, significant price distortion and unfair competition may occur as shown by the following example.

For the series-parallel network shown in figure 3, suppose link 1 is owned by a non-dominant carrier, while links 2, 3, and 4 are owned by a dominant carrier. We may view link links 2 and 3 as local loops and the non-dominant carrier is purchasing local access from the dominant carrier.

The question then is how much should the non-dominant carrier pay the dominant carrier for local access. In both cases of elastic or inelastic demand, there is an incentive for the dominant carrier to charge excessive prices for link 2, so that demand would be redirected to links 3 and 4. The prices charged for links 2 and 3 would be different, even if the costs of these links are the same for the dominant carrier.

The regulation of the proper level of charges for access is difficult in this case. Improper determination may artificially suppress the demand for value-added services such as Internet access.

5. CONCLUSION

Many countries and cities aspire to become the Internet hub for the Asia-Pacific region. Similar to the creation of a hub for maritime, aviation, and ground transshipment of goods, a regional hub for Internet must achieve an economy of scale so that a significant sized induced externality is established.

There are a number of factors identified to create the conditions conducive to the establishment of a pre-eminent hub. First and foremost is an open and fair regime of interconnection. Second, local and international circuit cost must be low, as a result of a liberalized telecommunication market. Low cost local and international circuits attract connectivity and help build a large network infrastructure. A large traffic volume provides an attractive environment for the caching and further value-added processing of information. Third, promotion of value-added services further enhances the attractiveness of connecting to a hub.

In this paper, we modeled the economy of interconnection as a non-cooperative game among competing network service providers and service users. We identified network structures which produce competitive or collusive pricing practices. Examples of price distortion are given. Further work is being done to examine policies for promoting competition in the Internet.

References

[1] HKIX traffic statistics, http://www.hkix.net/hkix/stat/aggt/aggregate_new.html
[2] 1998 HK SAR Policy Address http://www.info.gov.hk/pa98/english/index.htm
[3] For a good list of NAPs around the world, see the link concerning exchange points in the page http://www.isi.edu:80/div7/ra
[4] The information economy home page organized by Hal Varian contains diverse discussion on pricing and policy, http://www.sims.berkeley.edu/resources/infoecon/

POLICY-BASED BILLING ARCHITECTURE FOR INTERNET DIFFERENTIATED SERVICES

Felix Hartanto and Georg Carle
GMD FOKUS, Kaiserin-Augusta-Allee 31, D-10589 Berlin, Germany
{hartanto, carle} @fokus.gmd.de

Abstract The Differentiated Services architecture allows a service provider to configure new services dynamically using a policy protocol. This benefit, however, may not be fully realized if the service provider need a high effort to update its billing system to charge for the services. Thus, there is a real need for a flexible billing architecture. To meet this need, a policy-based billing architecture is proposed in this paper. This architecture allows a service provider to define policies for configuring various processes of a billing system based on the charging and pricing schemes used for individual services. Definitions of policies for various charging and pricing schemes are discussed and the potential complexity of each of them is analyzed. Based on the complexity analysis four classes of services, which utilize the least complex charging schemes and require the minimum traffic metering effort, are recommended for meeting different application requirements.

Keywords: Policy-based Architecture, Charging, Pricing, Differentiated Services.

1. INTRODUCTION

The current Internet supports a single level of best-effort service. Every packet has the same probability of being delayed or discarded in case of congestion. Any lost packets can be recovered by using higher layer protocols (e.g. TCP), which incorporate an acknowledgement procedure. However, this mechanism degrades the achievable throughput and incurs additional delay, making it less suitable for emerging real-time applications, which have strict delay requirements.

In an attempt to enrich this service model, the Internet Engineering Task Force (IETF) is considering a number of architectural extensions that permit the allocation of different service levels to different users. One of the outcomes of this effort is the Integrated Services (IntServ) architecture that integrates guaranteed and predictive service quality with the best-effort service of the Internet. This service model provides service discrimination through explicit allocation and scheduling of resources in the network using RSVP (Resource Reservation Setup Protocol) [6]. The feasibility of maintaining the state of each reservation in all intervening routers near the core of the Internet is questioned [21]. Because of the IntServ scalability problems, the Differentiated Services (DiffServ) architecture [25,3] has been proposed to provide a means of offering a spectrum of services without having to maintain per-flow state in every router.

The DiffServ architecture is based on an interconnecting of administrative domains. Within each domain most of the resource management complexity is pushed to the domain edge. On domain ingress, incoming traffic is classified by the per-hop behavior (PHB) bits into aggregates. The aggregated traffic is forwarded and policed within the domain according to the aggregate profiles in place.

The definitions of the PHB within a domain define the different services that can be provided by the DiffServ architecture. Policy protocols, such as COPS (Common Open Policy Service) [5], have been suggested to provide dynamic and automatic configuration of various network elements in implementing the PHB. This offers high flexibility for a domain administrator or service provider to define a wide variety of services to meet market needs. This benefit, however, may not be fully realized if the service provider can not charge for the new services, or if an update of their billing system to charge for the services requires a high effort. Thus, a flexible billing architecture is needed to complement the flexibility offered by the differentiated service model. To meet this need, a policy-based billing architecture is proposed in this paper. Prior to describing the architecture, a structure review of existing charging and pricing schemes is presented in the next section.

2. CHARGING MODELS

2.1 Charging Structure

As charging for other telecommunication services, e.g. telephony service, the Internet charging is also structured into *subscription charge* and *session charge*. Each of these in turns has a setup component and a recurring or usage part.

The subscription-setup charge is sometimes termed the "joining fee", for setting up the user account and provision of software or hardware required for connection to the Internet. The subscription-recurring charge is often termed "access or rental". It is often a simple flat charge.

The session-setup charge is often termed "session-access" charge. It is often a simple flat charge, such as for setting up a multicast session. The session-usage charge usually varies according to the amount of resources reserved or consumed.

The time scale for the subscription charge is normally longer than or equal to the time scale of the session charge. Thus, the subscription charge can be used to reduce the need for a session charge. In view of this, we can differentiate the charging into two categories, i.e.

1. *Flat rate charging*, where the session charge is zero and the session-usage charge is absorbed by the subscription-recurring charge. This charge allows the user to receive unlimited network access, regardless of the amount of time connected to the network and the amount of traffic sent or received.
2. *Usage sensitive charging*, where the session-usage charge is non-zero and varies according to the level of resources used or reserved by the customers.

The usage sensitive charge can be defined by a formula which expresses the usage charge UC as a function of setup charge SC, pricing p and usage U parameters, i.e.

$$UC = SC + \Sigma_i p_i U_i \tag{1}$$

The usage parameters quantify the number of units of usage. The possible parameters used are duration (D) and volume (V). Based on the usage parameters, we can differentiate two categories of usage-sensitive charging, namely *duration-based* and *volume-based charging*. Duration-based charging is commonly used for charging reserved resources, while the volume-based charging is commonly used for charging consumed resources. A combination of both charging categories has also been suggested, for example, to charge out-profile traffic differently from in-profile traffic [28,8], or to charge consumed resources on top of the reserved resources [12]. The combined or two-tier charging can be expressed as:

$$UC = SC + p_1 D_1 + p_2 V_2 \tag{2}$$

The benefits of such a combined pricing is to allow the user to lower the 'per unit time' cost at the cost of raising the 'per unit volume' cost.

The pricing or tariffing parameters define the price per unit of reserved or consumed resources. The reserved resources can be expressed in terms of bandwidth and buffer (B), or token bucket filter (F) parameters (e.g. leaky rate and bucket size). The reserved bandwidth can be specified explicitly by the users, measured [9], or derived from the source parameters (e.g. mean rate, peak rate, and burstiness) and the required quality of service usage, for example, using the equivalent bandwidth concept [14]. On the other hand, the consumed resources can be given in terms of packets, octets or bits. In this paper, we assume the bandwidth based pricing for reserved resources and packet based pricing for consumed resources. The packet can be of different priority or precedence level (P), where different levels of resources are allocated implicitly to meet specific quality of service.

In general we can write the pricing parameters as a function of allocated resources (R), i.e. $p = f(R)$, where $R = B$ or P. For example, the function $f(B)$ can simply be a linear function, e.g. $f(B) = (B/B_u)\, p_{Bu}$, where Bu is unit of bandwidth and p_{Bu} is the price per unit of bandwidth.

The price per unit bandwidth p_{Bu} can be static or dynamically varied depending on the current demand on the network resources, which gives rise to the two pricing categories, namely *static pricing* and *dynamic pricing*.

2.1.1 Static Pricing

The price in this category is set in the contract between the user and the service provider. The time scale of this price change is much longer than the session duration. Prices normally do not change simply because of instantaneous congestion within the network, but rather due to long term observation of network usage and market conditions. For example, the service provider will give advance notice to lower prices to stay competitive, or raise prices to meet increasing cost.

The static pricing parameters can be modified by other session characteristics. Three price modifiers, which are commonly used in the telephony pricing, are *time-of-day* (T), *destination* or *end-points* (E), and *usage* (U).

With the *time-of-day* modifier, the price depends on the calendar time, such as the time of day (peak or off-peak), day of the week (weekday or weekend), or public holiday. It is a form of congestion pricing based on long-term observations. This factor has been suggested in [26] for Internet pricing.

With the *destination* modifier, the price depends on the distance between the sender and the receiver. It can be based on the number of hops traversed or simply the location of the receiver [10]. While hop count and domain name may reveal information on distance and location, no ubiquitous method is available for accurately determining distance or location. With the diminishing distance pricing in the telephony industry, it is also expected that the distance pricing in the Internet will play a less significant part. Thus, this argues for a distance independence pricing or other simple alternatives, such as a zoning approach. With this approach user traffic is categorized as in-zone if it is addressed to a receiver within the same service provider domain and out-zone otherwise. A surcharge is applied to out-zone traffic which accounts for the interconnecting charge incurred by the traffic.

With the *usage* modifier, the price depends on the amount of usage, i.e. the duration of a session or the level of traffic volume. This is to encourage long session in order to minimize the overhead of session setup process or to encourage high volume customers [7].

Taking into account these three factors as surcharges or discounts to the base charges $f(R)$, we can write the pricing formula as:

$$p = f(R,T,E,U) = f(R)\,f(T)\,f(E)\,f(U), \text{ where } U = D \text{ or } V \quad (3)$$

2.1.2 Dynamic Pricing

The price in this category varies depending on the demand on the network resources or congestion level within the network. The price changes instantaneously or on the spot (thus, the name spot-pricing). The intention is that the price should be zero when the network is uncongested, but when there is congestion the price should reflect the incremental social cost determined by the marginal delay cost to other users and the willingness of the user to pay for the cost. Price adjusting can be performed by auction pricing and by feedback pricing.

With *auction pricing* or a "smart market" approach [22], prices are determined based on consumers bids. Users include a bid in each packet [23]. At congested routers, packets are prioritized based on these bids. In case of congestion, packets containing the lowest bid are discarded first, and accepted packets are priced at a rate determined by the highest bid among the rejected packets. The cost of carrying each packet is thus related to the marginal value (represented by the bid) of the traffic which has been pushed out. At the equilibrium price the user's willingness to pay for additional data packets equals the marginal increase in delay cost generated by those packets [22]. Bandwidth auctioning, rather than per-packet auctioning, has been considered in [20,13]. In this approach bandwidth is split into small units and users bid for the required bandwidth at each auction period.

With *feedback pricing*, prices are calculated by the provider dynamically based on current network load. For example, in [24] prices are calculated based on the instantaneous filling level of the buffers at network nodes. Price feedback can be initiated by a customer query, e.g. by sending a request to convey the source demand followed by the network feeding back the price [27], or by a load threshold within the network [13]. The basic unit for pricing can be sent packet, or units of bandwidth reserved over a fixed time period [24,13]. The users decide whether or not to send packets or to reserve bandwidth based on the prices given.

3. POLICY-BASED BILLING ARCHITECTURE

3.1 Billing System Framework

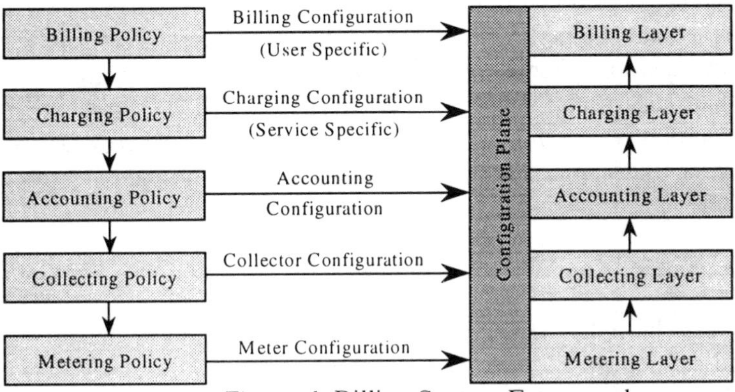

Figure 1. Billing System Framework

From capturing the usage to creating a bill to be sent to a customer, a billing system goes through processes, which can be modeled by a layering framework as shown in Figure 1.

The *metering layer* tracks and records usage of resources by observing the traffic flows. The metering policy, used for configuring the metering layer, specifies the attributes of the traffic flows to be observed. In a connectionless network, such as Internet, where it is difficult to locate the end-point of a flow, the metering policy can also be used to define the flow duration.

The *collecting layer* accesses data provided by metering entities as well as collecting charged related events and forward them for further processing to accounting layer. This layer can collect information from multiple meters, as for multicast and distribute to home domains, as for user roaming. For this reason, the efforts in standardizing data exchange format and protocol at this layer will be beneficial. The meters from where to collect the data, the type of data and the frequency in collecting them are defined by the accounting policy.

The *accounting layer* consolidates the collected information from the collecting layer either within the same provider domain or from other provider domains and creates *network accounting data sets or records* which are passed to the charging layer for the assignment of prices. For supporting multicast charging, the multicast topology including splitting points can be reconstructed by entities of this layer (see [16,7] for further information on multicast charging).

The *charging layer* derives session charges for the accounting records based on service specific charging and pricing schemes as per equation (1), which are specified by the charging policy.

The *billing layer* collects the charging information for a customer over a time period, e.g. one month, and include subscription charges and possible discounts into a bill. Billing policy can be used to specify the bill details.

Not all components of the reference model will appear in every billing system. For example, a service provider which only provides a single service and charges the customers flat-rate will only implement the functionality of the billing layer. On the other hand, a service provider offering multiple services may implement the policy-based architecture to allow different charging schemes to be used for different services or customers without having to hard-coded the charging formula into the billing system. For further discussion on the implementation issues of the proposed billing system framework as well as the ways for handling policy differences among multiple providers, interested readers are referred to [16].

3.2 Policy Setup

To illustrate the policy definitions and their relationship to the user subscription and session profile, we consider a subscriber with a default service and being allowed to use two other services. For each service, there are associated charging policies, accounting policies and metering policies which are set up based on the subscriber's service contract or service level agreement (SLA). We assume that a combined charging scheme is applied to one of the service, where duration-based charging is applied to the reserved resources and volume-based charging is applied to the consumed resources or out-profile traffic. A sample of policy setup for this customer is shown in Figure 2. Due to page limitations, we combine collecting and accounting policy and refer to them simply as accounting policy. Interested readers are referred to [16] for more descriptions and examples.

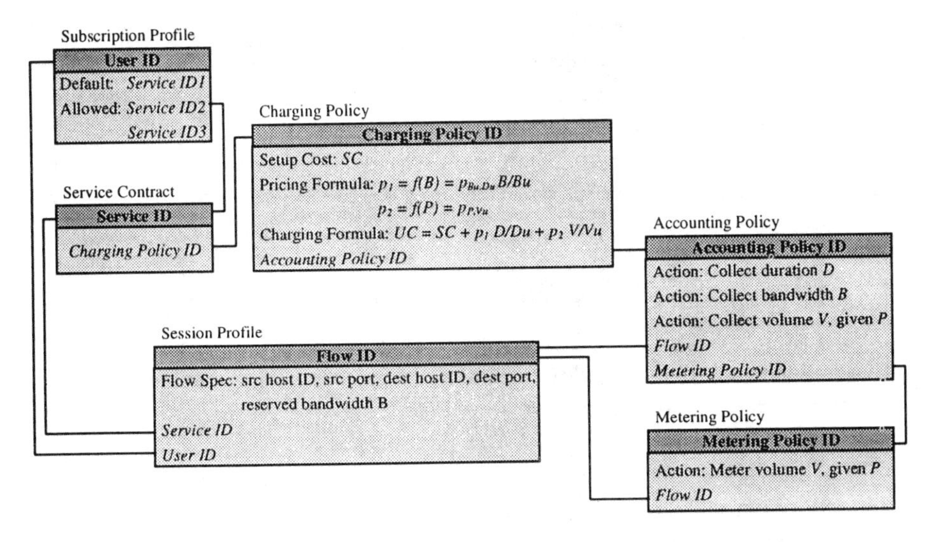

Figure 2. Relationship between Profiles and Policies

In the figure the identifications appearing in italic indicate the link between the records. For example, *Accounting Policy ID* links a charging policy to its corresponding accounting policy. In the charging policy a combined charging formula is used for Service ID2. Based on the charging formula, an accounting policy is defined to collect the required usage (duration and volume for packets with priority P) and pricing (reserved bandwidth) parameters for a specific Flow ID. A metering policy is defined to meter the traffic volume required by the accounting policy.

The policies are used by authentication server [1] in conjunction with admission control or bandwidth broker. The authentication process checks the subscriber profile based on User ID and Service ID from the user service request and decide whether the request is admissible or not. Once a request is admitted the process will create a session profile and retrieve the accounting and metering policy from policy database and distribute them to the accounting and collecting layer and the metering layer, respectively.

3.2.1 Applicability for Various Charging Schemes

In discussing the policy setup in the previous section a combined charging has been used. This charging scheme represents a general scheme as the three charging schemes, flat-rate, duration-based and volume-based, can be treated as special cases of this scheme. For the flat-rate charging no charging policy, and hence accounting policy and metering policy, need to be defined. For the duration-based charging no metering policy needs to be defined. Assuming that the bandwidth is static throughout the session, then the accounting policy just needs to obtain the bandwidth from the session profile and the duration from the flow start time and flow stop time. This means that the scheme can be considered simple. For the volume-based, all three types of policies are needed and the metering can be quite complex and resource consuming [4]. This means that the volume-based charging is the most complex among the three charging schemes. Moreover, as the accounting and metering policy definitions are based on the charging policy, which in turns based on the charging formula used, we can conclude that the complexity of a billing system depend on the charging formula.

3.2.2 Applicability for Various Pricing Schemes

The pricing strategies used in the above example is based on static resource pricing, where only bandwidth has been used for pricing the reserved resources and priority packet pricing has been used for pricing the consumed resources.

The use of price modifiers, such as time-of-day, destination and usage, in conjunction with the resource pricing may require additional policy definitions. For example, let us consider a flow, which starts at 17:55 and lasts until 18:30 and the peak rate to off-peak rate transition is at 18:00. The simplest alternative for the service provider is to charge the peak rate for the entire flow duration at 1 minute granularity, i.e. $UC = 35\, f(B, peak)$. In this case no additional policy is needed. However, in order to be competitive and fairer to the users, the service provider may use an alternative charging scheme where the peak rate is charged up to 18:00 and off-peak rate is charged for the rest of the flow duration, i.e. $UC = \Sigma_i\, U_i\, f_i(B,T) = 5\, f(B, peak) + 30\, f(B, off\text{-}peak)$. In this case additional accounting policy is needed to specify the creation of an accounting record at 18:00. This increases the complexity of the charging scheme.

Additional accounting policy will also be needed for dynamic pricing, which can be viewed as an extension of the time of day resource pricing where the price changes on the spot depending on the network condition. Unlike time-of-day, it is imperative for the dynamic pricing to generate accounting records when the price changes. These records form a dynamic contract between the customer and the service provider and thus, can be used as evidence in any disputes with the customers. The resulting number of accounting records can be large depending on the price update period. For example, let us consider a flow, which lasts for three minutes and generate an average of 250 packets. Using a bandwidth auction pricing with an auction period of 30 seconds (same as default refresh RSVP's soft-state period) six accounting records needs to be created. On the other hand, if the auction is based on the packet, then up to 250 accounting records need to be generated if all packets successfully gained access to the network. This number is obviously much larger than one or two accounting records created for static resource pricing. In addition to the potential large overhead, the dynamic pricing has been considered complex due to the need for modifying the bandwidth reservation protocol or packet transfer protocol to include the price information [26].

Table 1 summarizes the policies required by different charging and pricing schemes along with the complexity of the schemes.

Table 1. Complexity of Charging and Pricing Schemes

Charging and Pricing Schemes	**Charging Policy**	**Accounting Policy**	**Metering Policy**	**Complexity/ Overhead**
Flat rate charging	None	None	None	Simple
Duration charging, bandwidth pricing	$p = f(B) = p_{Bu}\ B/Bu$ $UC = SC + p\ D/Du$ (*Bu* is bandwidth unit, *Du* is duration unit)	Collect duration *D* and bandwidth *B*	None	Simple Require additional policies if price modifiers are used
Duration charging, dynamic pricing (auction and feedback)	$p_i = p_{ci}\ Bi/Bu$ $UC = SC + \Sigma_i\ p_i\ D_i/Du$ (*Bu* is unit of bid bandwidth)	Collect duration D_i, reserved bandwidth B_i and price per unit bandwidth p_{ci} within *i*-th auction/ feedback period	None	Complex Increasing session duration or auction/ feedback period increases number of accounting records
Volume charging, packet pricing	$p_{DS} = f(P) = p_{DS,Vu}$ $UC = SC + \Sigma_{DS}\ p_{DS}\ V_{DS}/Vu$ (*Vu* is volume unit, *P* indicated by DS-byte)	Collect number of packets V_{DS} per priority *P*	Count packets per priority *P*	Medium Large overhead in counting packets.
Volume charging, auction pricing	$p_i = p_{ci}$ $UC = SC + \Sigma_i\ p_i$	Collect current price per packet p_{ci}	None	Very complex Increasing number of packets generated within session duration increases number of accounting records
Volume charging, feedback pricing	$p_i = p_{ci}$ $UC = SC + \Sigma_i\ p_i\ V_i$	Collect volume V_i and current price per packet p_{ci} within *i*-th feedback period	Count packets within feedback period	Complex Increasing session or feedback period increases number of accounting records

4. CONCLUSIONS AND RECOMMENDATIONS FOR DIFFSERV CHARGING AND SERVICES

The proposed policy-based architecture has been demonstrated to support flat-rate, duration-based and volume-based charging, which are the three charging schemes identified through a structure review in Section 2. The architecture has also been shown to support static and dynamic pricing schemes.

The analysis in the previous section has shown that some schemes are more complex than others, especially the dynamic pricing schemes. Based on this complexity analysis, four classes of services as listed in Table 2, which use the least complex charging and pricing schemes, are recommended.

By providing various *default services* with the subscription charge varied according to the loss priority level, users can select the most suitable default service based on their long-term requirements. The users can use an adaptive traffic control, such as packet marking engine [11], to monitor their traffic and select to transmit at higher priority than their chosen default service (i.e. using the *priority service*) if the observed service rate falls below the minimum target rate. With the network dropping lower priority packets first, this approach approximates the dynamic pricing where the high priority indicates the willingness of the users to offer higher bids in order to ensure the delivery of their packets. It avoids the complexity of the dynamic pricing scheme as the charge for the priority service is simply volume-based with static priority packet pricing.

In regard to the marking of out-profile traffic in this service, out-profile packets can be marked down to the default service of the user, which is not necessary best-effort, since users value their packets as low as their default service. The default marking can be conveyed to the routers using AAA policy protocol [1] during authentication process.

The four service classes can be implemented using a multi-queue with protective buffer policies [17] in order to ensure that the traffic from one class does not affect the traffic from other classes during overload, while making full use of network resources during light load.

Table 2. Proposed Service Classes

Service	Description	Examples	Recommended Charging Scheme
Premium service [19]	Peak bandwidth reservation, strict delay, jitter and loss guarantee, highest delay and loss priority	Real-time applications that concern about jitter and cannot tolerate loss, e.g. CBR video transmission	Duration charging, bandwidth pricing
Assured service [18]	Expected bandwidth reservation, guarantee delay and jitter, but tolerate some loss Marking of out-profile traffic to user default service	Real-time applications that concern about jitter but tolerate loss, e.g. VBR video transmission	Duration charging, bandwidth pricing
Priority service [2]	Relative delay priority with loss priority option as per default services	Applications that require some delay or throughput guarantee, e.g. priority data transmission	Volume charging, priority packet pricing
Default services	Relative loss priority, best effort and above best effort	Non-real time applications, e.g. normal data transfer	Flat rate charging

5. ACKNOWLEDGEMENT

This research is carried out within the framework of European Commission ACTS SUSIE project. The authors would like to thank Mikhail Smirnov and

Tanja Zseby for their valuable discussions and anonymous reviewers for their useful comments.

References

[1] IETF Authentication, Authorization and Accounting (AAA) Workin Group. http://www.ietf.org/html.charters/aaa-charter.html.

[2] Y. Bernet, et al. A Framework for Differentiated Services. *IETF Internet Draft <draft-ietf-diffserv-framework-02.txt>*, February 1999.

[3] S. Blake, et al. An Architecture for Differentiated Services. *IETF Request for Comments, RFC2475*, December 1998.

[4] M. S. Borella, V. Upadhyay and I. Sidhu. Pricing Framework for a Differential Services Internet. *European Transactions on Telecommunications*, Vol. 10(2), March/April 1999.

[5] J. Boyle, et al. The COPS (Common Open Policy Service) Protocol. *IETF Internet draft <draft-ietf-rap-cops-06.txt>*, February 1999.

[6] R. Braden, et al. Resource ReSerVation Protocol (RSVP) - Version 1 Functional Specification. *IETF Request for Comments RFC2205*, September 1997.

[7] G. Carle, F. Hartanto, M. Smirnov and T. Zseby. Charging and Accounting for QoS-Enhanced IP Multicast. *Proc. of Protocols for High-Speed Networks (PfHSN)*, Salem, MA, August 1999.

[8] D. Clark. A Model for Cost Allocation and Pricing in the Internet. *Technical Report*, Laboratory for Computer Sciences, MIT, Cambridge, MA, August 1995.

[9] C. Courcoubetis, F. Kelly and R. Weber. Measurement-based Charging in Communication Networks. *Statistical Laboratory Research Report 1997-19*, Univ. of Cambridge, 1997.

[10] H. Einsiedler, P. Hurley, B. Stiller and T. Braun. Charging Multicast Communications Based on A Tree Metric. *Proc. of Multicast Workshop*, Braunschweig, Germany, 1999.

[11] W. Feng. D. Kandlur, D. Saha and K.G. Shin. Adaptive Packet Marking for Providing Differentiated Services in the Internet. *Proc. of ICNP*, Austin, TX, October 1998.

[12] D. Ferrari and L. Delgrossi. Charging for QoS. *Keynote Paper at the IWQoS'98*, Napa, CA, May 1998.

[13] G. Fankhauser, B. Stiller, C. Vögtli and B. Plattner. Reservation based Charging in an Integrated Services Network. *Proc. of INFORMS Telecommunications*, Bocca Raton, FL, 1998.

[14] R. Guerin, H. Ahmadi and M. Naghshineh. Equivalent Bandwidth and Its Application to Bandwidth Allocation in High-Speed Networks. *IEEE J. Select. Areas Commun.*, Vol. 9(7), September 1991, pp. 968-981.

[15] G. Guthrie and M. Carter. User Charges for Internet: the New Zealand Experience. *Telecommunication Systems*, Vol. 6, September 1996.

[16] F. Hartanto and G. Carle. Policy-Based Billing Architecture for Internet Differentiated Services (extended version). *Technical Report*, GMD FOKUS, August 1999. Available at htp://www.fokus.gmd.de/usr/hartanto.

[17] F. Hartanto, H. Sirisena and K. Pawlikowski. Protective buffer policies at ATM switches. *Proc. IEEE ICC*, Seattle, WA, June 1995, pp. 960-4.

[18] J. Heinanen, F. Baker, W. Weiss and J. Wroclawski. Assured Forwarding PHB Group. *IETF Request for Comments, RFC 2597*, June 1999

[19] V. Jacobson, K. Nichols and K. Poduri. An Expedited Forwarding PHB. *IETF Request for Comments, RFC2598*, June 1999.

[20] A. Lazar and N. Semret. Auctions for Network Resource Sharing. *CTR Technical Report*, Columbia University, February 1997.

[21] A. Mankin, et al. Resource ReSerVation Protocol (RSVP) Version 1 Applicability Statement Some Guidelines on Deployment. *IETF RFC2208*, September 1997.

[22] J. MacKie-Mason and H. Varian. Pricing the Internet. *In Public Access to the Internet (Kahin B. and Keller J., Eds.)*, Prentice Hall 1995.

[23] J. MacKie-Mason. A Smart Market for Resource Reservation in a Multiple Quality of Service Information Network. *Technical Report*, University of Michigan, September 1997.

[24] J. Murphy, L. Murphy and E. Posner. Distributed Pricing for Embedded ATM Networks. *Proc. of ITC-14*, Antibes, France, June 1994.

[25] K. Nichols, S. Blake, F. Baker and D. Black. Definition of the Differentiated Services Field (DS Field) in the IP v4 and IP v6 Headers. *IETF RFC2474*, December 1998.

[26] S. Shenker, D. Clark, D. Estrin and S. Herzog. Pricing in Computer Networks: Reshaping the Research Agenda. *Communications Policy*, Vol. 20(3), 1996, pp. 183-201.

[27] V.A. Siris, C. Courcoubetis and G.D. Stamoulis. Integration of Pricing and Flow Control for Available Bit Rate Services in ATM Networks. *Proc. of IEEE GLOBECOM*, London, U.K., November 1996.

[28] D. Songhurst and F. Kelly. Charging Schemes for Multiservice Networks. *Proc. of ITC-15 (V. Ramaswami and P.E. Wirth, editors)*, Elsevier, June 1997.

TELETRAFFIC ISSUES IN HIGH SPEED CIRCUIT SWITCHED DATA SERVICE OVER GSM

Dayong Zhou and Moshe Zukerman
Department of Electrical and Electronic Engineering
The University of Melbourne, Parkville, Victoria 3052, Australia
Email: m.zukerman@ee.mu.oz.au

Abstract This paper considers a range of channel allocation schemes for High Speed Circuit Switched Data (HSCSD) over GSM and reports performance (blocking probability) and efficiency results. The channel allocation schemes studied differ in the way channels are packed (First Fit, Best Fit and Repacking), and in the connection admission policy. An overall performance comparison of the schemes is provided in order to gain insight into simplicity/efficiency tradeoffs.

Keywords: GSM, mobile networks, teletraffic, HSCSD, performance evaluation

1. INTRODUCTION

High Speed Circuit Switched Data (HSCSD) is a new GSM service that provides multi-slot high speed data service through GSM [1]. GSM supports a TDMA [8] based digital cellular mobile network. The capacity allocated to a cell is a function of frequency carriers allocated to that cell. In particular, each frequency carrier (or *frame*) supports eight TDMA channels henceforth referred to as time-slots.

Under HSCSD, each data service can obtain from one to eight time-slots. In other words, a data service may occupy an entire frame. A service is not allowed to use time-slots from different frames and the time slots of a particular service must be consecutive.

Three channel allocation schemes are studied: First Fit, Best Fit, and Repacking. Focusing on the simplicity/efficiency tradeoffs, their performance (blocking probabilities and utilisation) will be compared by simulation. Analytic solution is beyond the scope of this paper but readers who are interested in related analyses may be referred to [3][9].

We shall distinguish between two cases: (1) inflexible customers and (2) flexible customers. Inflexible customers specify the required number of time slots for their connection and will not accept any number lower than that. The flexible customers specify a lower bound which allow the service provider to allocate less than the required level (upper bound) but not less than that lower bound. The flexible customers scenario leads to an interesting situation whereby the service provider may choose to allocate less than the required level even if it has the capacity to allocate the required level. Actually, if the lower bound is one, the service provider may choose to allocate always one channel whereby ignoring the main premise of HSCSD. All these alternatives will be studied in this paper.

The remainder of the paper is organised as follows: Section 2 describes the simulation model used in the paper. Section 3 gives a description of each of the three channel allocation schemes. Sections 4 and 5 present simulation results and provide insight into the peculiarities of the different schemes and their effects on network performance and Quality of Service (QoS) levels.

2. THE MODEL

As mentioned, each carrier can support several data services. In a GSM system, one channel within each cell must be reserved for broadcasting, meaning that only $8n$-1 time-slots are available for user traffic in an n carrier cell. We will consider the cases with n= 1, 2 and 3.

We assume call arrivals as a Poisson stream and have exponential holding times. Let λi and $1/\mu i$ be the Poisson arrival rate and holding time of calls that require i consecutive time-slots. The aim is to assign for each arrival the optimal available set of consecutive time-slots.

According to [9] and references therein, upon set up of an HSCSD connection, two values are specified by the user, they are denoted B and b, where B denotes the maximum acceptable capacity and b denotes the minimum acceptable capacity. In this paper, we consider three cases.

(1) $B = b$ This is the **inflexible customers scenario** – it is the worst for the network, and it is still a possibility that the network should consider.

(2) $B>b$ *and* $b=1$ This case represents the **flexible customers scenario**. We assume two extreme policies in this case:

i) The network provides the highest amount of bandwidth possible (not exceeding *B*). This will be designated as **Low Delay Policy (LDP)**.

ii) The network always allocate one channel (*b* value in this case) regardless what the *B* value is. This will be designated as **High Utilization Policy (HUP)**.

3. CHANNEL ALLOCATION SCHEMES

In this section, we describe the three channel allocation schemes as relevant to the inflexible customers scenario. Notice that it is straightforward to apply these descriptions to the inflexible customers scenario, and notice also that in the case of the network allocating always b=1 channel, all three schemes lead to the same performance.

3.1 First Fit [4]

In this scheme, the frames are ordered and designated as Frame 1, Frame 2 and Frame 3 etc. When a service which requires m time-slot arrives, we allocate it m consecutive time-slots based on availability first in frame 1, then in frame 2, etc. In a particular frame we allocate the first available m time-slots. This channel allocation scheme is simplest to implement [5].
The eight time-slots in the frame are permanently allocated ID numbers as a two-dimensional array. For example, the first eight time-slots in the first carrier will have ID number (1,1), (1,2), ..., (1,8), the 8 time-slots in the second carrier will have ID number (2,1), (2,2), ... , (2.8).

Each frequency carrier may be in any one of the following two states :

(i) no HSCSD data service in progress.

(ii) any feasible combination of HSCSD data services each of which may occupy between one and eight time-slots.

The allocation algorithm then functions as follows :

(a) Each incoming HSCSD call which needs one time slot channel is allocated one EMPTY time slot whose ID-number is the lexicographically smallest among all time-slots that are currently EMPTY. (Henceforth, we shall use the word smallest to mean lexicographically smallest ID numbers of time-slots.)

(b) Each incoming data service which requires n time-slots will be allocated n empty consecutive time-slots of which the first time slot has the smallest ID-number among all EMPTY time-slots.

(c) When there is no possibility to fit that call into one carrier, try the next carrier. The call must not be split across carriers.

(d) The HSCSD call will be blocked after channel allocation in all carriers in the cell have been tried and failed.

(e) No reordering of calls is performed at any time a set of times slots remained assigned for that service until it terminates.

3.2 Best Fit [4][5]

Let a *hole* be a consecutive set of empty time-slots. Under Best Fit for each incoming m time-slots service we try to find an m slot hole. If such search fails, we search for an $m+1$ slot hole. The aim is to keep the allocated time-slots close together. If more than one hole of the same size is available we select based on the smallest ID number. The HSCSD call will be blocked if no such area exists. No action is taken upon call departure and call arrival.

3.3 Repacking [3][5]

Here, for new calls we implement the Best Fit approach. Unlike Best Fit, under this approach if a new call arrives and cannot find a suitable hole, the time-slots allocated to the calls in progress are rearranged to find a suitable hole for the new call. This rearrangement is implemented by solving the bin packing problem [6] using Branch and Bound algorithm [7].

If such suitable hole cannot be found even with rearrangement of time-slots, the new call is blocked.

Implementation of the Repacking strategy makes use of intracell handover including Repacking across different radio frequency carriers within the same cell. A large number of intracell handovers during a call may have a negative effect on the QoS. It is therefore important to limit this number.

4. SIMULATION RESULTS: INFLEXIBLE CUSTOMERS

In this section, we present performance results for the inflexible customers scenario. In Section 5 we shall present results for the flexible customers scenario, and compare them with the results of this section. The following performance measures will be considered:

a. Blocking Probabilities of each scheme in the case of three carriers, two carriers and one carrier in a cell.
b. Maximal utilisation of each scheme subject to meeting blocking probability constraints in case of 1, 2, 3 carriers in a cell.

In the simulations presented in this paper, we assume that the arrival rate of the different services in any particular run are equal (i.e. $\lambda_1 = \lambda_2 =\ldots, =\lambda_8$). This is a worst case scenario from the service provider point of view. We shall demonstrate the significant wastage occurs with such traffic scenario.

4.1 Blocking Probability

It is easily noticed that in the case of only one carrier, because one channel is reserved for broadcasting and signalling, the services of eight time-slots is always blocked. Since we assumed equal arrival rate, the blocking probability in this scenario must be higher than 1/8. Actually, very high blocking probability is observed also for two and three carriers under this worst case scenario. Of course, the worst case scenarios we are presenting here are theoretical and may not occur in practice too often. They nevertheless signifies the worst case wastage and performance degradation. Based on above considerations, sufficiently long simulations were run such that 95% confidence intervals are as presented in the following tables :

Table 1 Blocking probabilities of three schemes in case of one carrier (7 time-slots for HSCSD service, 1 for signalling)

Load (calls/s)	Blocking_Prob of Repacking	Blocking_Prob of Best Fit	Blocking_Prob of First Fit
1	0.364±0.0069	0.403±0.0022	0.403±0.0022
2	0.470±0.0063	0.534±0.0029	0.534±0.0029
3	0.540±0.0060	0.621±0.0032	0.612±0.0030
4	0.588±0.0057	0.633±0.0032	0.671±0.0038
5	0.625±0.0054	0.681±0.0039	0.689±0.0040

Table 2 Blocking probabilities of three schemes in case of two carriers (15 time-slots for HSCSD service, 1 for signalling)

Load (calls/s)	Blocking_Prob of Repacking	Blocking_Prob of Best Fit	Blocking_Prob of First Fit
1	0.094±0.0041	0.116±0.0047	0.108±0.0044
2	0.211±0.0055	0.246±0.0052	0.248±0.0049
3	0.299±0.0072	0.333±0.0057	0.348±0.0053
4	0.377±0.0088	0.402±0.0066	0.407±0.0062
5	0.440±0.0100	0.468±0.0069	0.465±0.0066

Table 3 Blocking probabilities of three schemes in case of three carriers (23 time-slots for HSCSD service, 1 for signalling)

Load (calls/s)	Blocking_Prob of Repacking	Blocking_Prob of Best Fit	Blocking_Prob of First Fit
1	0.016±0.0010	0.019±0.0022	0.022±0.0018
2	0.074±0.0018	0.089±0.0023	0.097±0.0024
3	0.149±0.0043	0.162±0.0041	0.168±0.0047
4	0.220±0.0061	0.245±0.0062	0.244±0.0061
5	0.273±0.0082	0.290±0.0069	0.318±0.0087

Discussion on the Blocking Probability Results

Table 1 shows that, in the case of one carrier, for arrival rates between 1 and 5 calls/s there is no significant difference between the blocking probabilities of First Fit and Best Fit. This difference remains small (however somewhat more noticeable) in the case of 2 and 3 carriers in Tables 2 and 3. This is because these schemes do not exhibit good packing compression. There will be a group of free time-slots be created upon departures of different calls. This is even worse in First Fit because it puts calls into carriers without considering following arrivals. Best Fit performs better than First Fit as available carrier goes from 1 to 3. This is because Best Fit is trying to leave as many time-slots as it can for following arrivals though it does not repack time-slots either. As the number of carriers increases, Best Fit has greater selection to chose an optimal one. But in case of one carrier, the selection is limited.

Best Fit does not show the same advantage in this simulation as it in a simulation which only has full rate and half rate calls [2]. The reason is that HSCSD provides a large diversity of channel (1 to 8 time-slots) which adversely affects Best Fit performance. Best Fit does not give significant benefit in case of one carrier system. But we can get higher benefit when there are more carriers.

From Table 1, Table 2 and Table 3, it is evident that Repacking has the best performance in any case. This is because Repacking leaves free time-slots as many as possible and fully uses them.

4.2 Utilisation

As mentioned above, this wide diversity of traffic leads to low performance or alternatively high wastage. This will be demonstrated in this section. We will seek to find what the maximal utilisation is subject to meeting blocking probability requirements. We will start with blocking probability level of 20% and then we will consider a more realistic blocking probability of 2% and we shall demonstrate the enormous wastage required to provide such low blocking probability. As discussed above, the 2% will only apply for the two and three carriers cases because 1/8 of the traffic (namely the 8 time-slots service) is always blocked in one carrier case.

In this paper, utilisation is defined as the average number of occupied time-slots divided by the total number of time-slots. For every given blocking probability, we will find by using bisection the maximal utilisation subject to meeting required blocking probability level (in our case 20% or 2%).

We have obtained the following results:

Table 4 Utilisation of three schemes in case of 1, 2 and 3 carriers (given 20% blocking probability)

Number of Carriers in a cell	Utilisation of First Fit	Utilisation of Best Fit	Utilisation of Repacking
1	0.081±0.0053	0.082±0.0059	0.1619±0.0350
2	0.3574±0.0161	0.3729±0.0287	0.4280±0.0316
3	0.4538±0.0290	0.4800±0.0277	0.5030±0.0251

Table 5 Utilisation of three schemes in case of 2 and 3 carriers (given 2% blocking probability)

Number of Carriers in a cell	Utilisation of First Fit	Utilisation of Best Fit	Utilisation of Repacking
2	0.09338±0.0041	0.09486±0.038	0.11316±0.0051
3	0.17924±0.028	0.19064±0.037	0.19351±0.026

The results presented in Table 5 signify the enormous wastage caused by wide traffic diversity in HSCSD in the inflexible customers scenario with large diversity of services.

Discussion On Utilization Results

A clear observation of the results is that allowing large diversity of services, in other word, allowing multiple time-slots (from 1 to 8), assuming inflexible customers ($B=b$), may lead to unacceptable wastage with significant cost implications. This is also true for Repacking - the most efficient scheme. The problem can be resolved by either not allowing $b>4$ service or by charging heavily inflexible customers who insist on having more than four channels. Optimal charging scheme for HSCSD in GSM is a topic for further research. From tables 4 and 5 it has been observed that as carrier goes from 1 to 3 :

1. The utilization of First Fit, Best Fit and Repacking increases rapidly from around 10% to 60%. This is due to the fact that for a given blocking probability more carriers means the system can have higher arrival rate and thus provide more free time-slots for incoming calls. In another word, the efficiency of usage of time-slots increases while the number of carriers increase.
2. Utilisation of Repacking is better than that of the other two schemes.
3. In case of only one carrier in the system, Table 4 shows that utilisation of Repacking is much better than First Fit and Best Fit. We observe that when the system has very limited resources, the allocation scheme turns out to be very important. As the number of carriers goes up, Repacking

does not seem to be much better than Best Fit especially in case of three carriers in the system. Best Fit works well when we have enough carriers.

5. SIMULATION RESULTS: FLEXIBLE CUSTOMERS

We now focus on the flexible customers case. Figure 1 demonstrates that in this case, Repacking is the worst performer. This is because under Repacking, because of the creation of large holes, more customers get larger chunks of capacity leaving no space for others. First Fit and Best Fit, on the other hand, have smaller holes (but more holes) and are forcing the network to allocate less than the required capacity to customers leaving more holes for others. Figures 2, 3 and 4 present results on comparison between LDP and HUP. In all three Figures the First Fit scheme is implemented. (Recall that under HUP, all three schemes give the same performance results). Comparing the blocking probability results presented in Figure 2 between LDP versus HUP clearly show higher blocking in the case of LDP because HUP accepts more calls at lower rate and hence the lower blocking.

In Figure 3 we present the utilization of HUP and LDP. The utilization values are calculated by fixing blocking probability levels and finding by simulation the maximal load which gives the required (allowable) blocking probability. We clearly observe a significant increase in utilization in favor of HUP for all levels of blocking probability. This is intuitively clear and it is consistent with the objective of HUP; however, the significant gain demonstrated here of 50% is important to notice.

The results of Figure 4 which demonstrate the significantly lower delay of the Low Delay Policy (LDP) over HUP are also important to notice. The tradeoff between high utilization and low blocking and cheaper calls under HUP versus low delay and more expensive calls under LDP.

6. CONCLUSIONS

This paper has studied by simulation the average blocking probabilities and utilisation for a range of channel allocation schemes for HSCSD over GSM. Systems with 1, 2, and 3 carriers were considered. We have observed that allowing large diversity of multiple time-slots (1 to 8), and assuming inflexible customers ($B=b$), can lead to significant decrease in efficiency. Under the inflexible customers scenario, for HSCSD traffic, Repacking is found to incur the lowest blocking probability and highest utilisation, but it introduces higher complexity and excessive processing costs. Best Fit is better than First Fit and has a competitive performance in a system with three carriers, but not as efficient as Repacking when the system resource is limited (for example, one carrier in a cell).

First Fit has been the worst performer. On the other hand, under the flexible customers scenario with Low Delay Policy, the results have been reversed. Repacking became the worst performer. The tradeoff between high utilization and low blocking and cheaper calls under HUP versus low delay and more expensive calls under LDP.

References

[1] Nokia's Vision for a Service Platform Supporting Circuit Switched Applications, White Paper http://www.nokia.com/products/networks/cellular/gsm/person1.htm.

[2] M. Ivanovich, M.Zukerman, P. Fitzpatrick and M. Gitlits, "Channel allocation schemes for half and full rate connections in GSM", *Proceedings of IEEE International Conf. on Comm.*, Dallas, 1996.

[3] M. Zukerman, "Circuit allocation and overload control in a hybrid switching system", *Computer Networks and ISDN Systems*, vol. 16, pp. 281-298, 1988/1989.

[4] D. E. Knuth, *The Art of Computer Programming*, Volume 1: Fundamental Algorithms, NY: Addison Wesley, 1973.

[5] M. Zukerman, "Bandwidth allocation for bursty isochronous traffic in a hybrid switching system", *IEEE Trans. on Comm.*, December 1989.

[6] M. R. Garey and David S. Johnson, *Computers and Interactability: A Guide to the Theory of NP-Completeness*, W. H. Freeman and Company, San Fransico, 1979.

[7] R. S. Garfinkel, George L. Nemhauser, *Integer Programming*, USA, John Wiley & Sons, 1972.

[8] ETSI GSM Specifications, Series 01-12.

[9] D. Calin and D. Zeghlache, Performance Analysis of High Speed Circuit Switched Data (HSCSD) over GSM, *Proceedings of ICC'98*, Atlanta, Georgia, June 1998.

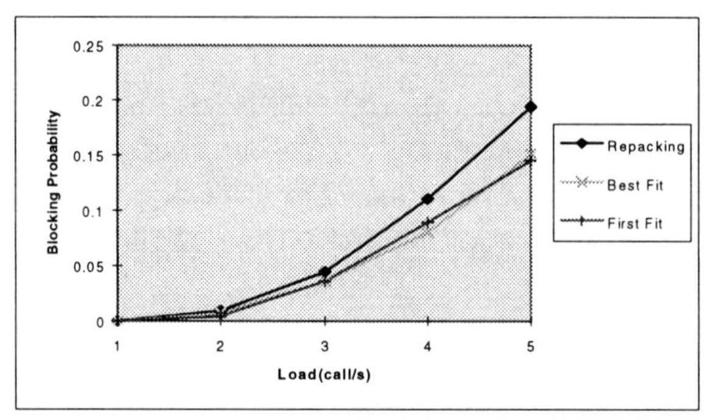

Figure. 1 Comparison of Blocking Probability of the three schemes under LDP with three frequency carriers

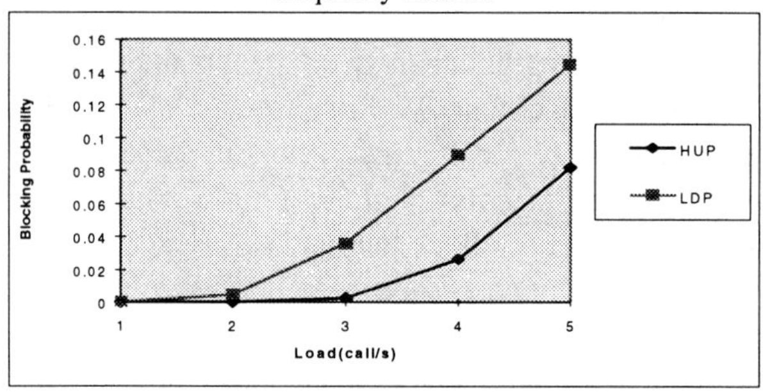

Figure. 2 Comparison of Blocking Probability between LDP versus HUP

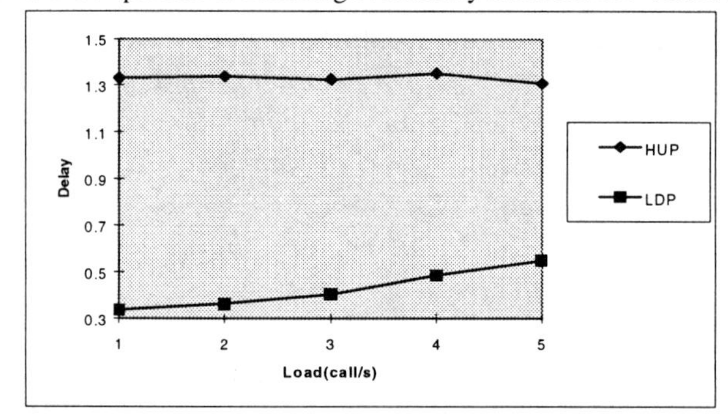

Figure. 3 Comparison of Utilisation achieved by LDP versus HUP

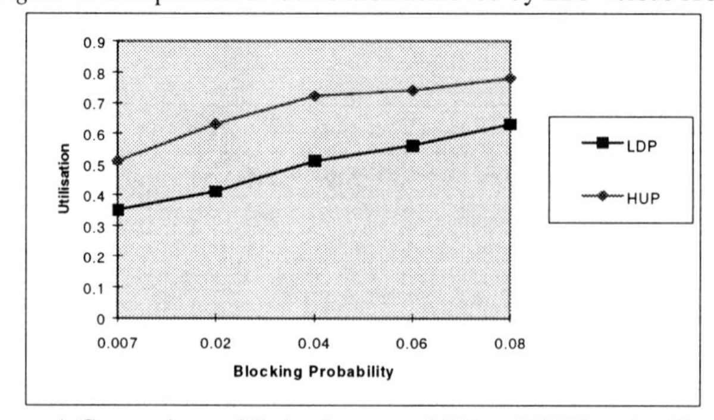

Figure 4 Comparison of Delay between LDP and HUP under First Fit

SESSION 6

Performance Evaluation II

STOCHASTIC FEATURES OF VBR VIDEO TRAFFIC AND QUEUEING WORKING CONDITIONS: A SIMULATION STUDY USING CHAOTIC MAP GENERATOR

Rosario G. Garroppo, Stefano Giordano, Michele Pagano *
University of Pisa
Department of Information Engineering
Via Diotisalvi 2, I-56126 Pisa, Italy
{garroppo,giordano,pagano}@iet.unipi.it

Abstract The paper presents a simulation study on the impact of LRD and SRD on a simple queueing system and shows that, while for short buffers an analytically tractable SRD model may suffice to capture the long–term loss probability, for systems with large buffers the effect of LRD can be significant. The main contribution of this paper consists in highlighting this evidence through simulations driven by synthetic traces (generated by traditional AR models and LRD Chaotic Map Generator), measured data and shuffled versions of them.

Keywords: Packet Video, Chaotic Map Generator, Short Range Dependence (SRD), Long Range Dependence (LRD).

1. INTRODUCTION

The discovery of long memory characteristics in the traffic generated by new broadband services, such as LAN-to-LAN interconnection, high speed data transport over geographical backbones and VBR video, indicates the self-similar models as promising mathematical tools able to take into account the LRD of real traffics in a parsimonious manner [1]. These long memory properties presented by self-similar processes are indicated like the new burstiness measure: the fractal behaviour of the acquired traces arises from a sort of independence of variations and fluctuations in the activity of sources over several

*This work was partially supported by the project "Techniques for QoS guarantee in multiservice telecommunication networks" of the Italian Ministry of University and Tech. Research (MURST Research Program - N. 9809321920)

time scales. For this reason a definition of burstiness linked to a specific time scale or to the burst length concept is inadequate.

Two kinds of problems come out from these experimental evidences: self-similar queueing performance analysis and discrete event simulation driven by synthetic self-similar traces. They are needed in the design of new traffic management and control schemes [2] which should take into account the actual statistical behaviour of multimedia sources.

The analytical works on queueing systems loaded by LRD arrival processes are restricted to the evaluation of asymptotic relations or bounds for the complementary probability of the queue occupancy with infinite buffer [3][4].

The generation of self-similar traces (see [1] for some generation algorithms), needed in the simulation studies, is much more complex than producing arrivals corresponding to renewal or SRD processes.

Teletraffic researchers are in agree with the strong influence of the long memory features of real traffic on performance of infinite buffer size queueing systems [5]. On the other hand the accuracy of LRD models in operating networks with relatively small buffers is becoming a key debate topic. In this paper we want to emphasise the relevance of the process memory vs. the queueing memory, i.e. the effects of short and long range correlation for different buffer sizes and in different loading conditions. The queueing performance comparison is carried out by means of discrete event simulation considering a single server queue with deterministic service time (simplified model of an ATM network element), fed by videoconference, videophone and entertainment video sequences, whose long memory nature was shown in [6], and by synthetic traces. Traditional self-similar generators, based on statistical methods, are computationally expensive; on the other hand, in this paper we use a fast traffic generator that follows a completely different approach. The generator has been defined developing the idea of producing LRD sequences by means of chaotic deterministic monodimensional systems that were introduced as suitable traffic models in [7]. The model is very simple and can be characterised by few parameters (parsimonious modelling).

In the paper, after a survey of the analysed traces and a description of the Chaotic Map Generator (CMG), we analyse the implication of SRD and LRD on queueing performance. Finally, the main results are collected in the Conclusions.

2. DESCRIPTION OF VBR PACKET VIDEO TRACES

The actual data used to drive the discrete event simulations are obtained by measuring sessions of three different video services: videoconference, videophone and entertainment video. The first one is a videoconference session corresponding to three people in front of a 3-CCD camera, while the second

Table 1.1 Analysed Trace Data

Trace Name	Number of frame	Average (Mbit/s)	Peak/ Mean Ratio	Frame rate (frame/s)	Mean Compr. Ratio
1-Vconf	48496	1.668	4.8	25	39.78
2-Vphone	182353	1.634	8.25	25	56.52
3-"Star Wars"	171000	5.336	2.82	24	8.7

is a long videophone session (a man carrying on a videophone call). These two sequences were produced by a prototype coder described in [8]. The last sequence is an example of entertainment video, namely the movie "Star Wars", corresponding to a frame by frame coding scheme similar to JPEG [9]. Some statistical characteristics of the traces are shown in Table 1.1.

Each sequence corresponding to the number of bytes per frame can be seen as a sample path of a discrete-time second order wide sense stationary process $\{X_n\}_{n\in\mathbb{N}}$ with autocorrelation function $r(k)$ and power spectral density

$$S(f) = \sum_{k=-\infty}^{\infty} r(k)e^{-i2\pi kf}$$

Considering the hyperbolic decay of the autocovariance function of LRD processes

$$r(k) \propto k^{-\beta} \text{ as } k \to \infty$$

and that in the self similar modelling $\beta = 2 - 2H$, we can use the Hurst parameter, H, as a measure of the long memory properties of actual traffics. Among the different methods employed for the estimation of H [1] the so called Variance Time statistic is one of the most widely used; the results of its application to the considered traces are reported in Table 1.2 (see [6] for further details).

Table 1.2 Estimation of the Hurst parameter by means of the v-t plot

Trace	Videoconf.	Videophone	"Star Wars"
$\hat{H}$	0.68	0.69	0.78

3. CHAOTIC MAP GENERATOR (CMG) FOR BROADBAND TRAFFIC

Recent measurements carried out in Bellcore have shown that the long memory of real traffic in a LAN interconnection scenario arises as a consequence of network element interactions that can be modelled as heavy tailed ON–OFF processes [10]. According to this physical approach and taking into account that VBR video traffics present equivalent statistical features, we consider a traffic generator obtained by the superposition of independent ON–OFF sources. Each elementary source is modelled by a discrete time monodimensional dynamic system, whose state evolution is described by a chaotic map. The synthetic data are obtained by a two-level quantization, with an adequately chosen threshold c, of the map state $x(n)$ at step n. Several deterministic maps (such as Bernoulli Shift, the Liebovitch Map and the Intermittency Map [7]) have been proposed for traffic traces generation. As far as broadband networks are concerned, a modified version of the Intermittency Map is particularly attractive since it generates sequences exhibiting the characteristics of asymptotic self-similar processes. The deterministic mapping that defines the Intermittency Map is the following:

$$x(n+1) = |x(n) + x^z(n)|_{\mathrm{mod}\,1} \tag{1}$$

while the two-state output of the relative ON–OFF model is:

$$y(n) = \begin{cases} 0 & \text{if } 0 < x(n) \leq c \\ k & \text{if } c < x(n) < 1 \end{cases} \tag{2}$$

The output data are generated producing, by means of the map, the number of packets transmitted during each time interval, which can either be empty or contain k packets. The relevant property of the proposed map is that, unlike classical ON–OFF models, the marginal distribution of the OFF period length is heavy tailed [7] for values of c close to zero; the z parameter permits to control the decay of this marginal and the long term decay of the autocorrelation function. This characteristic of the generated sequence makes the model particularly suitable as the basic component of a long memory traffic generator and real traffics can be reproduced by the aggregation of many independent heavy-tailed sources:

$$\begin{cases} x_i(n+1) & = & |x_i(n) + x_i^z(n)|_{\mathrm{mod}\,1} \\ y_i(n) & = & \begin{cases} 0 & \text{if } 0 < x_i(n) \leq c \\ k & \text{if } c < x_i(n) < 1 \end{cases} \end{cases} \tag{3}$$

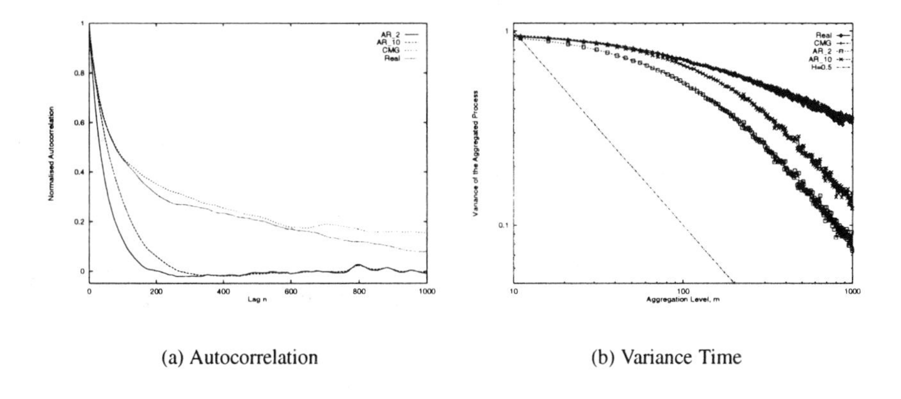

(a) Autocorrelation

(b) Variance Time

Figure 1 Comparison among the different "Star Wars"-related sequences

$$y(n) = \sum_{i=1}^{N} y_i(n) \tag{4}$$

where $y_i(n)$ represents the output of the i-th map (with $i = 1, \ldots, N$) at time n and the sum (4) gives the output sequence $y(n)$, representing the number of cells per frame. The fitting of traffic data, described in [11], involves the determination of the four parameters that define the CMG (z, c, k and N) considering the statistical features of the trace itself (mean, variance and autocorrelation function).

The results of the fitting procedure, carried out considering for instance the "Star Wars" trace, highlight that the CMG is unable to reproduce some very pronounced peaks characterising the real trace. However it permits to capture the persistence characteristics of data pattern, eyeballing evidence of the LRD property.

To compare the fitting properties of the proposed model with respect to traditional SRD ones, figure 1 presents the autocorrelation functions and the VT plots for the "Star Wars" sequence (similar results are obtained with the other traces), the one of the proposed CMG and those of two Autoregressive models (AR(2) and AR(10) respectively), which were used in the past as suitable models for VBR traffic [12]. It is remarkable that the growth of the AR model order is not particularly relevant since that approach is not able by itself to capture the LRD behaviour in a parsimonious way.

4. SRD & LRD IN VBR VIDEO TRAFFIC

To identify in which cases the LRD of VBR traffic is more critical than short lag correlation in queueing performance, we extended to a finite buffer scenario the approach originally proposed in [5]. In order to consider only the effect of the correlation structure without any influence from the marginal distribution of the model, we produced several traces, some of which maintain only the LRD of the original video sequences while others preserve only their SRD. This was possible by shuffling fixed size blocks of the considered traces. In particular, after subdividing the original data into non overlapping blocks of adequate size, two different procedures can be employed:

- external shuffling: the positions of the blocks are changed preserving the inside sample arrangement. This operation destroys LRD, while maintains SRD;
- internal shuffling: only the sample arrangement inside each block is modified, destroying in this way only SRD.

The choice of the block size determines the threshold lag of the correlation structure over which the statistical properties of the original trace are modified or preserved. In order to highlight the difference between LRD and SRD, we have chosen a block size B_s around 100 samples.

5. SIMULATION ANALYSIS OF QUEUEING PERFORMANCE

Our analysis is directed to evaluate, by means of trace driven simulations, the implications that traffic process memory has on performance of a single server queueing system with finite buffer size, N_w, and deterministic service time ($VBR/D/1/N_w$). To highlight the effects of correlation, we considered either two traditional autoregressive (AR) models of different orders (namely 2 and 10), able to fit the autocorrelation function only for relatively short lags, and CMG sequences, that preserve the LRD properties of real traffics. Moreover, as mentioned in the previous section, we shuffled the original data in order to test the effects of other statistics (shuffled sequences preserve the marginal and the chosen term – LRD or SRD – of the autocorrelation structure) on queueing performance.

The results are compared in terms of cell loss probability P_L, varying the normalized offered load A_0, i.e. the ratio between the mean arrival rate and the service rate.

Analysing the simulation results, we observe that for relatively short buffers (N_w=5000 cells) the performance of the shuffled data are very close to those obtained with the real trace (see figure 2(a)) in spite of the captured autocorrelation component. A slightly different behaviour can be observed only for low values of A_0 (around 0.5), where no loss is produced by the LRD-preserving trace, while the P_L determined with external shuffling is comparable to the actual value.

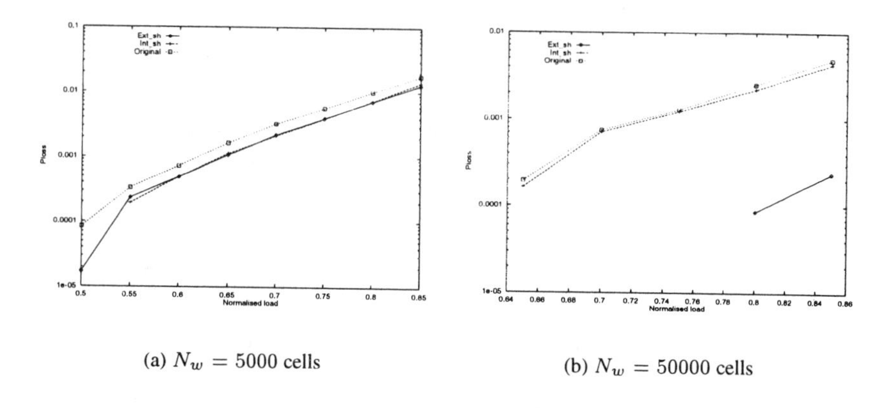

(a) $N_w = 5000$ cells

(b) $N_w = 50000$ cells

Figure 2 Effects of internal and external shuffling over P_L

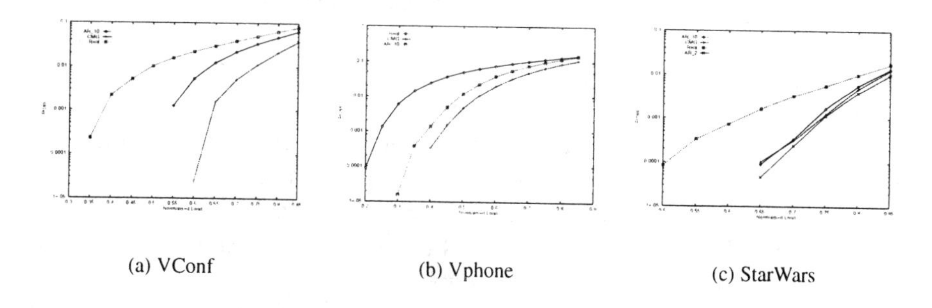

(a) VConf

(b) Vphone

(c) StarWars

Figure 3 P_L Comparison: LRD vs. SRD models ($N_w = 5000$ cells)

In figure 3 the comparison is extended to the mentioned synthetic traces for the three test sequences: the performance obtained with all the different models (characterised by gaussian marginals) is very similar, but quite different from the real curves, whose distributions present heavier tails than the empirical distribution of synthetic sequences (for "Star Wars", see [9]). This is confirmed by figure 4, which presents the Q-Q plots for the CMG and AR(10) synthetic data fitting the three considered sequences. The Q-Q plot is a goodness-of-fit test for the marginal distribution of a data set and it is obtained representing the quantile of the distribution model under examination vs. the quantile of the empirical distribution of considered data [13]. In our case, we substitute the distribution model with the empirical distribution of the synthetic traces in order to take into account the differences from the theoretical model introduced by the setting of all negative samples to zero.

The P_L analysis for $N_w = 5000$ points out the poor influence of the long-term autocorrelation structure for short buffers and highlights the relevance of other statistics, like marginal distribution of the cell rate.

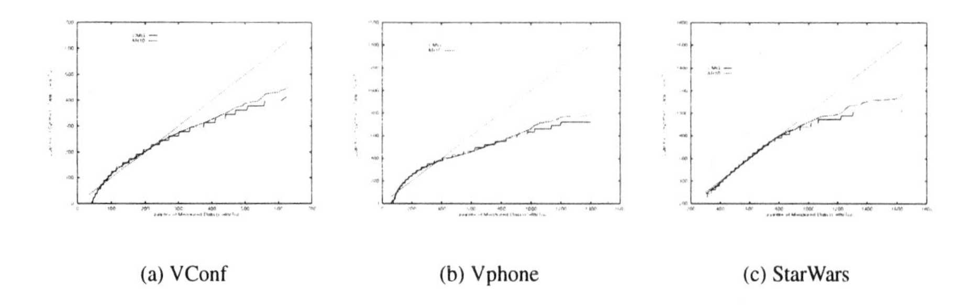

(a) VConf (b) Vphone (c) StarWars

Figure 4 Q-Q Plots for the CMG and AR(10) synthetic data for the considered sequences

Increasing the buffer size (N_w=50000 cells), the performance with SRD traces goes away from the real one, while that of LRD data draws near (see figure 2(b)). In particular we have observed that long term correlation provokes a dramatic decrease in queueing performance as soon as the offered load exceeds a given threshold (no losses are observed for $A_0 < 0.65$).

The comparison of P_L for the original trace, the CMG and the AR models (see figure 5) points out that, as buffer size grows, only the CMG generator permits to estimate reasonable P_L values (at least for traces b and c), although even the latter seems to be less critical than the original data (compare the curves in figure 2(b) and 5(c)), because of the marginal distributions that have not been considered in the fitting procedure.

For sake of simplicity, we have considered the AR(2) model only for the "Star Wars" trace, observing no loss over the whole range of normalised loads. The AR(10) traces provoke losses only for $A_0 \geq 0.85$ for trace c and no loss for all normalised load values for trace a.

The simulations confirm that in case of short buffers, also traditional models could be a correct approach in simple queueing performance evaluation, while LRD must be taken into account in case of longer buffers.

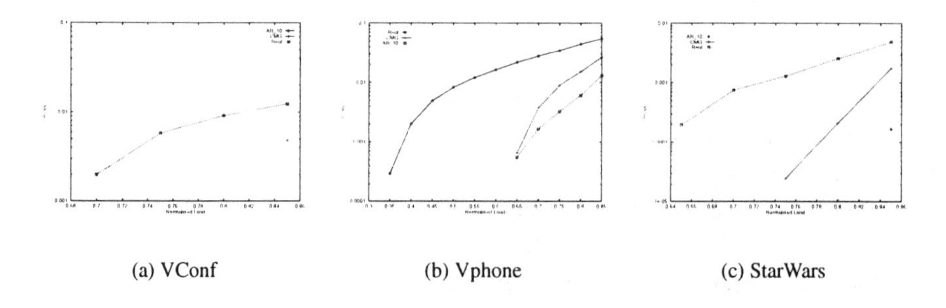

(a) VConf (b) Vphone (c) StarWars

Figure 5 P_L Comparison: LRD vs. SRD models ($N_w = 50000$ cells)

6. CONCLUSIONS

The goal of this paper is to show, by means of simulation studies, the implications of LRD and SRD on queueing performance considering finite buffer systems. The modelling of VBR video traces using self-similar processes implies a more complicated analytical evaluation of the queueing behaviour. Unlike the case of Markovian modelling approaches used to design networks with narrowband services, closed-form results have not been obtained yet. At the state of the art, few bound approximations, involving complicated mathematical tools, such as Large Deviations Theory [3], were presented only for infinite buffers.

From this study we deduce that in case of short buffer systems it is not needed to consider long term correlation because the queueing behaviour is mainly determined by the short term correlation structure of ingoing traffic, while in case of long memory queueing systems we have to take into account the LRD of the input traces.

This result is particularly relevent since in the real network environment the trend is toward the use of short buffers (around some thousands of cells), especially when real-time services, such as the transport of VBR video information (i.e. videoconference and videophone services) in a broadband network, are involved.

Moreover, irrespective of the buffer size, a relevant role is also played by the marginal distribution. This is confirmed by the different results obtained with shuffled traces, with the same marginals of the original data, and synthetically generated sequences, characterised by gaussian distribution: only the first case gives a queueing behaviour very close to the actual one (if the above rule for the correlation structure is matched).

Acknowledgments

The authors want to express special thanks to Helmut Heeke from Siemens AG Munich and Mark W. Garrett from Bellcore for having kindly provided some of the VBR traffic traces used in this analysis.

References

[1] W. Willinger, M.S. Taqqu, and A. Erramilli. A bibliographical guide to self-similar traffic and performance modelling for modern high speed networks. In F.P.Kelly, S.Zachary, and I.Ziedins, editors, *Stochastic networks : Theory and applications*. Oxford University Press, Oxford, 1996.

[2] S. Giordano, M. Pagano, R. Pannocchia, and F. Russo. A new call admission control scheme based on the self similar nature of multimedia traffic. In *Proceedings of IEEE ICC 96*, pages 1612–1618, Dallas, June 1996.

[3] N.G. Duffield and N. O'Connell. Large deviations and overflow probabilities for the general single server queue, with applications. DIAS Technical Report DIAS-STP-93-30, Dublin Institute for Advanced Studies, Dublin, Ireland, 1993.

[4] O. Narayan. Exact asymptotic queue length distribution for fractional brownian traffic. *Advances in performance Analysis*, 1(1):39–64, March 1998.

[5] A. Erramilli, O. Narayan, and W. Willinger. Experimental queueing analysis with long-range dependent packet traffic. *IEEE/ACM Transactions on Networking*, 4:209–223, 1996.

[6] J. Beran, R. Sherman, M.S. Taqqu, and W. Willinger. Long-range dependence in variable-bit-rate video traffic. *IEEE Transactions on Communications*, 43(2/3/4):1566–1579, february/march/april 1995.

[7] A. Erramilli, R.P. Sing, and P. Pruthi. Modeling packet traffic with chaotic maps. Technical Report 7, Royal Institute of Technology, Stockholm–Kista, Sweden, August 1994.

[8] H. Heeke. Statistical multiplexing gain for VBR video codecs in ATM networks. *Intern. Journal of Digital and Analog. Comm. Systems*, 4:261–268, 1991.

[9] M.W. Garrett and W. Willinger. Analysis, modeling and generation of self-similar VBR video traffic. In *Proc. of the ACM Sigcomm'94*, pages 269–280, London, UK, September 1994.

[10] W. Willinger, M.S. Taqqu, R. Sherman, and D.V. Wilson. Self-similarity through high-variability: Statistical analysis of ethernet LAN traffic at the source level. *IEEE/ACM Transactions on Networking*, 5(1):1–16, 1997.

[11] R.G. Garroppo, S. Giordano, M. Pagano, and F. Russo. Chaotic maps generation of broadband traffic. In *2nd IEEE MICC*, pages 7.7.1 – 7.7.8, Langkawi, Malaysia, November 1995.

[12] M. Nomura, T. Yasuda, and N. Otha. Basic characteristics of VBR coding in ATM environment. *IEEE JSAC*, 7(5):752–760, June 1989.

[13] A.M. Law and W.D. Kelton. *Simulation modeling & analysis.* McGraw-Hill, 1991.

EFFECTS OF VARIATIONS IN THE AVAILABLE BANDWIDTH ON THE PERFORMANCE OF THE GFR SERVICE

Norbert Vicari
University of Würzburg, Computer Science, Am Hubland, D - 97074 Würzburg, Germany
e-mail: vicari@informatik.uni-wuerzburg.de

Abstract The Guaranteed Frame Rate service category is currently under discussion for incorporation into the specification documents of the ATM-Forum. The concept of the GFR service is to provide a minimum service guarantee to classical best-effort services, taking into account the frame-based nature of todays data-traffic. In this paper we present a discrete-time analysis of the GFR service and discuss the effects of variations of the bandwidth available for the GFR service. The presented method can be applied to dimension the thresholds of the algorithms used to enforce the guaranteed service.

Keywords: GFR, ATM, Performance, Discrete Time Analysis, F-GCRA

1. INTRODUCTION

The ATM-Forum recently introduced the *Guaranteed Frame Rate* (GFR) service category [6], which is currently considered to be incorporated into the ATM-forum specification [3]. This new service category is motivated by several intentions. Today most applications are not equipped to select the proper traffic parameters required to establish ATM connections. Thus, choosing CBR or VBR service categories will fail either by causing inefficiency by overestimating required resources or not being able to give any QoS guarantees. The ABR service category is regarded to be too complex to be implemented in the majority of systems.

Transferring data traffic with the best-effort service category UBR would avoid the problem of estimating traffic descriptors, but will also give no QoS guarantees at all. Worse, the throughput seen at higher protocol layers is severely reduced. Most of todays applications utilize the *Transmission Control Protocol/Internet Protocol* (TCP/IP) for transferring data in frame based structures. When transmitting these frames over an ATM network the data is fragmented into cells. The loss of a single cell will cause an irrecoverable damage to the whole frame and induce retransmission. To cope with these problems, the GFR service category provides the user with a *Minimum Cell Rate* (MCR) guarantee under the assumption of a given *Maximum Frame Size* (MFS) and *Maximum Burst Size* (MBS). The user is allowed to send traffic in excess of the negotiated parameters, but this traffic will only be delivered within the limits of available resources.

The resources available for the GFR service alternate in different time-scales. Fast fluctuations are caused by higher priority VBR connections, while, the average share of bandwidth for the GFR service is mainly influenced by the establishment and release of other ATM connections. In [9] we considered constant available bandwidth reflecting the steady-state effects of long-term fluctuations of the availability of resources. In this paper we will also take short-term variations of the resource availability into account.

The paper is organized as follows: in Section 2 an overview of the GFR service category and its key components are given. Section 3 describes the modeling and analysis of the system. Numerical examples derived with the presented analysis method are provided in Section 4. The paper is concluded with a summary.

2. THE GFR SERVICE CATEGORY

Motivated by the needs of a guaranteed minimum bandwidth for best-effort ATM connections the introduction of a MCR for the UBR service category was suggested [5]. This considerations resulted in the definition of the so called UBR+ service category [4]. The main idea of this service category was to preserve the best-effort properties of the UBR service category while adding the guarantee of a minimum bandwidth. The newer specification of this service category [6] names the service GFR, which reflects the approach of taking frames into account for the minimum guaranteed bandwidth.

In comparison to ABR, which also provides a guaranteed best-effort service, GFR is easier to implement and does not add a new flow control scheme. Thus, implementation of GFR in adapter cards and network nodes is expected to be faster and cheaper. Further, the coupling of different flow control schemes – like TCP over ABR – may lead to unpredictable and unintended complications.

The VBR.3 service category [1] also allows the user to send traffic in excess to the traffic contract, but in comparison to the GFR service the traffic is not regarded as flow of frames. Since, most currently available applications use the TCP/IP protocol, data is organized in frames and has to be fragmented into cells for transport over an ATM net. Thus, random loss of a single cell leads to corruption of the whole frame and reduces the goodput of the transmission. Discarding whole frames that are not eligible for guaranteed transmission increases the goodput of the net.

The GFR service is intended to support non-real-time traffic and requires the data being organized in frames which can be delineated at the ATM layer. The user is provided with a MCR guarantee when transmitting frames that do not exceed the MFS in a burst that does not exceed the MBS. Frames sent in excess to this parameters will be delivered only within the limits of available resources.

The GFR service [2] provides a guarantee to deliver complete unmarked frames that are *conforming* and *eligible*. A frame is defined to be conforming if the CLP bit of all cells of a frame has the same value as the CLP bit of the first cell of the frame, the number of cells on the frame is less than MFS and the rate of the cells conforms to the parameter *Peak Cell Rate* (PCR), which is monitored with a conventional *Generic Cell Rate Algorithm* (GCRA). The eligibility of frames is defined with the *Frame-Based-GCRA* (F-GCRA), which controls if the rate of the cells of a frame is less than MCR and the length of burst is less than MBS. In order to stick to the frame-based approach all cells of a frame are valued identically to the first cell of a frame.In order to provide the above defined service guarantee the network has to discriminate between eligible and non-eligible frames when transmitting data with the GFR service. Therefor, a queuing discipline with two thresholds is applied to guarantee the transmission of eligible frames while providing best effort service for the remaining frames.

In the following we will review the queuing discipline applied for the GFR service after introducing the exact functionality of the F-GCRA.

2.1 Frame-Based Generic Cell Rate Algorithm: F-GCRA

Figure 1 shows how the *F-GCRA(I,L)* algorithm decides if an arriving frame is eligible or not according to a given increment parameter I and a limit parameter L. The only basic difference of *F-GCRA(I,L)* to a *GCRA(I,L)* – as formerly been defined by the ATM traffic management standard [1] – is that only the first cell of a frame is checked according to the GCRA while effectively tagging the whole frame as eligible or non-eligible. Frame cells arriving later only update the F-GCRA state if the first cell was eligible but they are never checked against the limit parameter L.

The simplified F-GCRA algorithm above assumes that the frame stream is already conforming to the *PCR* with frame sizes no bigger than *MFS*. For the

```
INPUT cell arrival on time time
temp = X - (time - last_eligible_time)
if (first cell in frame) then
          if  (CLP == 1) or (temp > L) then
                    frame_eligible = false
          else
                    frame_eligible = true
          endif
endif

if (frame_eligible) then
          X  = max(0, temp) + I
          last_eligible_time = time
endif

OUTPUT cell eligible according to F-GCRA(I,L) if (frame_eligible = true)
```

Figure 1: F-GCRA(I,L) implemented as virtual scheduling algorithm.

treatment of non-conforming frames no binding rule exists. However, it is very common practice to regard a non-conforming frame as a strict violation of the traffic contract with the immediate consequence of complete frame rejection. Hence, for the addressed GFR performance evaluation an inclusion of further details in the algorithm is neither necessary nor sensible. Figure 2 shows an snap-shot of the F-GCRA state for an example scenario with three consecutively arriving frames.

The first two frames are declared to be eligible according to *F-GCRA(I,L)* as the F-GCRA state upon arrival of the first cell does not exceed the limit parameter *L*. For each arriving cell of these two frames the F-GCRA state is increased by the increment parameter *I*. The third frame is declared non-eligible and the F-GCRA is never increased upon arrival of any frame cell.

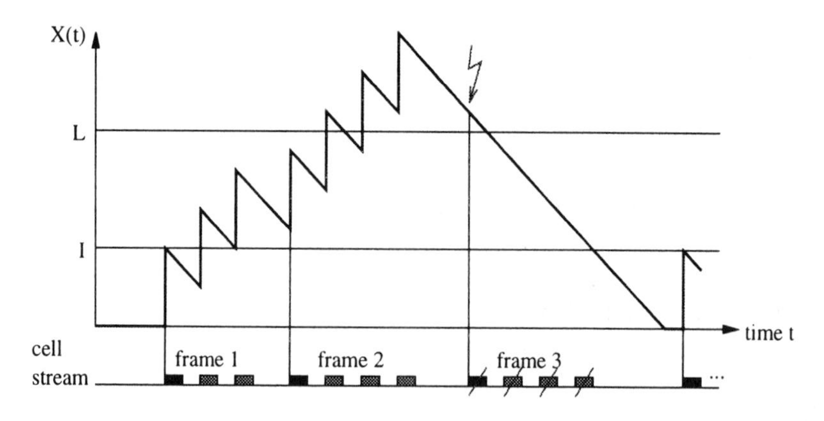

Figure 2: F-GCRA state diagram with three arriving frames.

2.2 Queuing Discipline

After classifying the frames with help of the F-GCRA the network node has to transmit the frames eligible for service guarantee with low loss probability. If additional resources are available on the transmission link, frames sent in excess to the traffic contract should be also transferred. Naturally these frames will suffer a higher loss rate than frames with guaranteed service. Cells which could not be transferred immediately are stored in a buffer of size *Q_MAX*. When a cell of a frame – eligible or non-eligible – could not be stored in the buffer, this cell and all subsequent cells of this frame are discarded since it is assumed that the loss of a single cell of a frame leads to the retransmission of the whole frame.

To discern eligible and non-eligible frames two threshold values are introduced. The Low Buffer Occupancy (*LBO*) value indicates the limit for the acceptance of non-eligible frames. That is, if at the time instant of the arrival of the first cell of a non-eligible frame at least *LBO* cells are waiting in the buffer the whole frame is discarded. Analogously the High Buffer Occupancy (*HBO*) value defines the limit for the acceptance of eligible frames. Once the first cell of a frame is accepted the subsequent cells of this frame could be only discarded due to buffer overflow. In order to investigate the influence of different values of *LBO* and *HBO* we model the correlated system of F-GCRA and queue.

3. MODELING AND ANALYSIS

In the following, a model and its corresponding discrete-time analysis of the GFR service are presented. The issue of conforming/non-conforming frames will be neglected, that is, all arriving frames are considered conforming. After a description of the frame arrival process we will evaluate the system state of the coupled model of F-GCRA and transmission line as shown in Figure 3.

For the analysis of the model we will describe the system state by a two dimensional random variable (X_f, X_q). The first dimension represents the

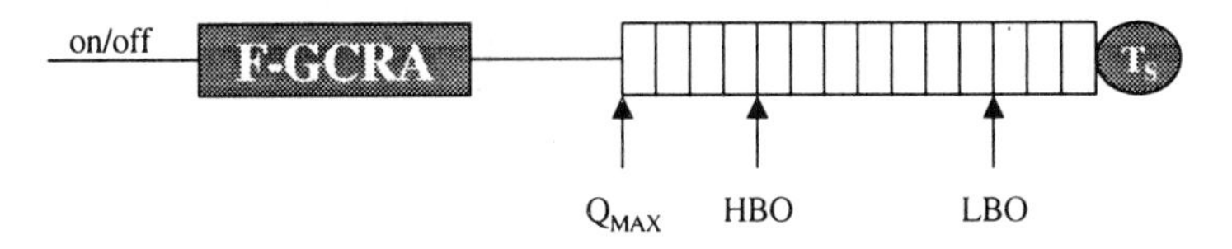

Figure 3: The basic GFR model

leaky bucket counter of the F-GCRA and the second dimension represents the time required to transmit the cells waiting in the buffer.

All random variables to describe the system state are measured in multiples of the duration of a cell transmission at PCR. The capacity of the transmission link is denoted by the RV T_S, that is, the duration of the transmission of a cell takes T_S time units. Then, the capacity QL of the buffer including the transmission unit can be approximated using the average cell transmission duration $\overline{T_S}$:

$$QL = (Q_{MAX} + 1)\overline{T_S} - 1 . \tag{1}$$

Analogously the limits for the acceptance of eligible and non-eligible frames HBL and LBL – expressed in time-slots – are defined:

$$\begin{aligned} LBL &= (LBO + 1)\overline{T_S} - 1, \\ HBL &= (HBO + 1)\overline{T_S} - 1. \end{aligned} \tag{2}$$

An analysis of the F-GCRA for determining the parameters MCR and MBS can be found in [8].

3.1 Modeling the Arrival Process

The functionality of the GFR service is based on the organization of data in frames. Most of the currently used protocols e.g. TCP/IP transport data in frames that have to be split up in several cells for transport over an ATM network. This sort of traffic can be modeled by the class of on/off-processes. An on-state represents the transmission of a frame, while the off-state represents the time between the frames. In our analysis we consider an on/off source as depicted in Figure 4.

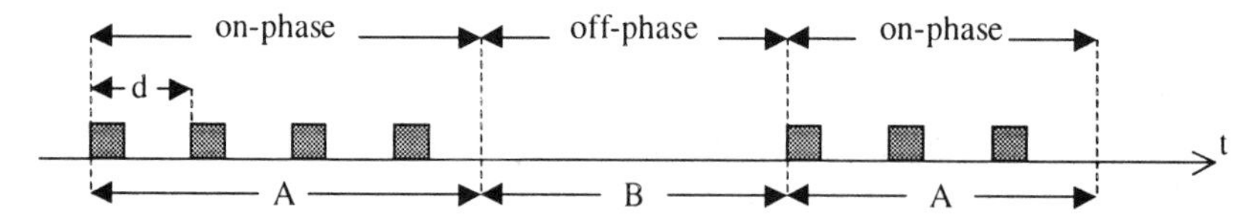

Figure 4: On/off source

The duration of the on- and off-phases are distributed according to discrete general and independent distributions. At the beginning of a on-phase a cells arrives immediately. During an on-phase cells arrive in intervals of d time units, which correspond to the transmission time of a cell *1/PCR*. The end of an on-phase is not generally synchronized to a cell arrival.The duration of the on-phase is denoted by the random variable A and the duration of the off-phase by B, respectively.

3.2 System State Evolution

The random variable (X_f^{on}, X_q^{on}) describes the system state at the beginning of an on-phase. Tracing every cell arrival the system state after the arrival of the last cell of a frame is iteratively derived. Taking into account the time remaining in the on-phase the system state at the beginning of the off-phase (X_f^{off}, X_q^{off}) is given. Since no cell arrivals occur during the off-phase the system state is decreased until the beginning of the next on-phase is reached.

Transition Off-Phase to On-Phase

The state of the F-GCRA is reduced during the off-phase by B units, since no cell arrival occurs. Before the arrival of the next frame the state of the F-GCRA computes as follows:

$$X_f^{on} = max(X_f^{off} - B, 0) . \tag{3}$$

Analogously at most B units of virtual work could be served at PCR during the off-phase, thus the state of the queue at the beginning of the on-phase is given by the following equation:

$$X_q^{on} = max(X_q^{off} - B, 0) . \tag{4}$$

Transition On-Phase to Off-Phase

During the on-phase cell arrivals occur starting with the first time slot and continuing every d time slots, c.f. Figure 4. The system state is denoted recursively, and the system is observed at the end of time slot k.

At the beginning of the on-phase the F-GCRA takes the following state:

$$X_f^{on, 0} = X_f^{on} . \tag{5}$$

The state of the F-GCRA at the end of an on-phase of A time units is computed recursively as follows:

$$X_f^{on, A} = \begin{cases} X_f^{on, A-1} - 1 + I, & \text{if } X_f^{on, 0} \le L \text{ and cell arrival} \\ max(0, X_f^{on, A-1} - 1), & \text{else} \end{cases} . \tag{6}$$

For the derivation of the buffer state we discern the arrival of the first and the subsequent cells of a frame. The state of the buffer before the arrival of the first cell is given by:

$$X_q^{on, 0} = X_q^{on} . \tag{7}$$

Upon the arrival of the first cell of a frame the frame acceptance is decided. If the state of the buffer at most *LBL* all frames are accepted. Eligible frames are accepted even if the state of the buffer is higher then *LBL* but at most *HBL*. If a cell could not be accepted all remaining cells of the frame are also discarded. We denote this changing the value of the RV *DF* from 0 to 1.

$$X_q^{on,1} = \begin{cases} X_q^{on,0} - 1 + T_{S,} & \text{if } X_f^{on,0} \le L \text{ and } (X_q^{on,0} \le HBL - T_S) \\ X_q^{on,0} - 1 + T_{S,} & \text{if } X_f^{on,0} > L \text{ and } (X_q^{on,0} \le LBL - T_S) \\ max(0, X_q^{on,0} - 1), \text{ else} & \end{cases} \tag{8}$$

The remaining cells of a frame are discarded only if either the buffer is occupied or some cells of the frame have already been discarded. Thus the state of the buffer is recursively computed as follows:

$$X_q^{on,A} = \begin{cases} X_q^{on,A-1} - 1 + T_{S,} & \text{if } X_q^{on,A-1} \le QL - T_S \\ & \text{and } DF = 0 \text{ and cell arrival} \\ max(0, X_q^{on,A-1} - 1), & \text{if } DF = 1 \text{ and cell arrival} \\ max(0, X_q^{on,A-1} - 1), & \text{if not cell arrival} \\ max(0, X_q^{on,A-1} - 1), & \text{else} \end{cases} \tag{9}$$

The 'else' branch of equ. (8) and (9) indicates the first discarding of a cell and thus initiates a change of the value of *DF* from 0 to 1.

3.3 Discrete-Time Analysis

In order to obtain the frameloss probability the probability mass function describing the system in equilibrium state are derived. Due to space limitations the probability mass function description corresponding to the RV description of equ. (3) to equ. (9) is omitted here, but can be found in an extended version of this report [10]. In difference to the common discrete time analysis approach the state space is divided in 4 semi-distributions to realize the memory of the system regarding frame loss.

4. NUMERICAL EXAMPLES

For the presentation and discussion of numerical examples we will refer to the following basic parameter set unless otherwise expressed.

The length of the on- and off-phase of the considered traffic stream is distributed geometrically. In order to represent a cell indicating the beginning and the end of a frame, the minimum length of an on-phase is set to 2. The system load is chosen to be 25% of the *PCR*, for example for $d=1$ – the cells of a frame are sent back to back – the on-phase is set to be 10 slots and the off-phase to 30 slots in average. For other values of d the duration of the on- and off-phase is adopted accordingly.

The *MCR* of the F-GCRA is set to 20% of the link cell rate, that is $I=5$. With a limit $L=100$ for cells sent back to back we obtain a ratio of 74% eligible frames. For varying burstiness of the traffic stream the limit of the F-GCRA is adopted accordingly. The required parameters could be easily computed with the analysis method presented in [8].

The transmission-time distribution t_s is chosen to adopt only two values – one below and one above the expectation – to reduce the computational complexity of the analysis. The values are adjusted accordingly to fit the selected expectation T_S and the coefficient of variation c.

4.1 GFR Functionality

First we look at the queuing behavior of a single link carrying GFR-traffic. In order to reflect the best-effort characteristics of the GFR service the bandwidth available for the connection is set to 1/3 of the *PCR*, which should fulfill the bandwidth requirement of the reference traffic stream. The coefficient of variation of the available bandwidth is $c=1.0$. Figure 5 shows the conditional frame loss probability for eligible and non-eligible frames in dependence of the *LBO*. Lower *LBO* values mean that even in the case of low buffer occupancy non-eligible frames are discarded. Thus, eligible frames are served with higher reliability and suffer lower loss. But on the other hand the transmission buffer for the whole traffic stream is reduced by preferential treat-

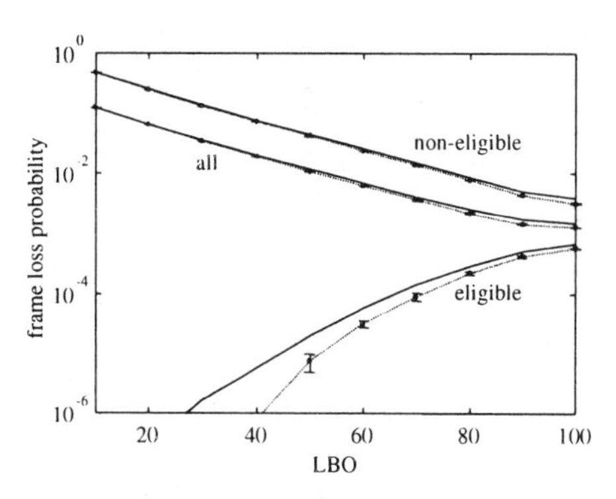

Figure 5: Functionality of the GFR service.

ment of the eligible frames, which leads to a reduction of the total throughput of the system. Thus, the functionality of the GFR service depends on the proper dimensioning of the *LBO*.

To demonstrate the approximation accuracy of the analysis some simulation results are included in the graphic. Generally the simulation results – depicted with dashed lines and 95% confidence interval – show good accordance with the results of the analysis. Since, the approximation in the analysis depends on the length of the queue the results are more accurate for higher share of the queue between eligible and non-eligible frames. Reducing the variance of the transmission time distribution also increases the accuracy of the analysis. For deterministic transmission time the analysis even is exactly.

4.2 Factors Influencing the Performance of GFR

To assess the effects of long-term and short-term fluctuations of the available bandwidth the blocking probability of eligible frames for different amounts of available bandwidth is depicted in Figure 6 (left side). In our example the traffic stream causes a load of 25%. The ratio of eligible frames is 74%. Thus, the system is capable to transfer eligible frames with a available bandwidth of 20% ($T_S = 5$) of the PCR if the *LBO* is dimensioned properly. If higher capacity is available for the GFR service, the *LBO* can be chosen higher to obtain the same maximum blocking probability for eligible frames and to increase the overall throughput, cf. Figure 5.

In most cases the short-term variation of the available bandwidth increases the blocking probability only by a negligible amount, since, the buffer available for eligible frames is sufficient to compensate short-term variations of the transmission time. If the average available bandwidth changes the effects are by far stronger and require a modification of the threshold *LBO* to guarantee the service for eligible frames.

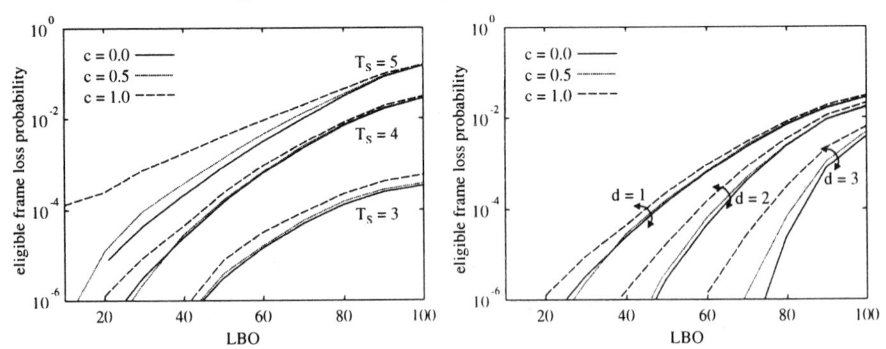

Figure 6: Influence of the available capacity and the traffic-streams burstiness.

To evaluate the impact of the burstiness of the traffic stream on the GFR service the following system configuration is used: the bandwidth available for the GFR service is set to 25% of the PCR, that is equal to the minimum bandwidth required by the traffic source. The cells in a frame are spaced by d time slots, cf. Section 3.1. The duration of the on- and off-phase are adjusted accordingly to ensure a traffic load of 25%. The parameters of the F-GCRA are adjusted to classify 74% of the frames as eligible, utilizing the analysis introduced in [8]. As shown in Figure 6 (right side) the blocking probability of eligible frames increases with increasing burstiness. Short-term variation of the available bandwidth compared with the traffic streams burstiness has minor impacts on the performance. Again, to obtain an identical maximum blocking probability the parameter *LBO* has to be adjusted accordingly.

The numerical examples show that the dimensioning of the parameters for the GFR service depends heavily on the traffic characteristics and average amount of available bandwidth. Short-term variations of the available bandwidth do not reduce the quality of the GFR service significantly. The Information about the traffic characteristics could be gained from the attributes of a GFR-connection [7] and taken into account. If sufficient bandwidth is available the – with regard to the service guarantee – highest possible *LBO* value should be chosen to obtain a high throughput of eligible- and also non-eligible frames. But if the only the minimum bandwidth MCR for the GFR-connection is available the *LBO* has to be chosen more restrictively. A restrictive selection of the *LBO* will fulfill the service requirements in any case, but not the expectations to a best-effort service class.

5. SUMMARY

In this paper we presented a discrete-time analysis of the GFR service category, which is currently defined by the ATM Forum. The both key components of the GFR service – the F-GCRA algorithm and the transmission queue – are described and modeled. While the F-GCRA discriminates eligible and non-eligible frames the buffer discipline ensures the service quality guaranteed by the GFR service category. To model the frame-based cell arrivals a on/off-process with generally distributed on- and of-phase was chosen. Since the state of the F-GCRA and the queue are correlated, a two-dimensional discrete-time analysis approach is applied.

The average available bandwidth and the traffic characteristic are identified in the numerical examples as the determining factors for the performance of the GFR service, while short-term variations of the available bandwidth can be neglected. The results indicate a discrepancy in the dimensioning of the parameters of the queuing discipline. A restrictive selection of the relevant parameter *LBO* ensures a transmission of eligible frames with the guaranteed service quality, but the best-effort spirit of the GFR service is given away. Thus, a dimensioning of the queuing discipline of the transmission buffer in dependence of the currently available bandwidth is an interesting approach to guarantee the service and to preserve the best-effort characteristic of the GFR service.

Acknowledgment

The author would like to thank Robert Schedel for his programming efforts during the course of this work. Further the financial support of the Deutsche Telekom AG (Technologiezentrum Darmstadt) is appreciated.

References

[1] *Traffic Management Specification Version 4.0.* The ATM Forum Technical Committee, April 1996.

[2] *Traffic Management Baseline Text Document, BTD-TM-01-02.* The ATM Forum Technical Committee, July 1998.

[3] N. Giroux and J. Kenney. *Proposed Additions to the Traffic Management 4.1 Draft Specification.* The ATM Forum Technical Committee 98-0626, October 1998.

[4] R. Guerin and J. Heinanen. *UBR+ Service Category Definition.* The ATM Forum Technical Committee 96-1598, December 1996.

[5] J. Heinanen. *MCR for UBR.* The ATM Forum Technical Committee 96-0362, April 1996.

[6] S. Jagannath, N. Yin, J. B. Kenney, J. Heinanen, J. Axell, K. K. Ramakrishnan. *Modified Text for Guaranteed Frame Rate Service Definition.* The ATM Forum Technical Committee 97-0833, December 1997.

[7] G. Koleyni. *Updating Table 2-1.* The ATM Forum Technical Committee 98-0450, July 1998.

[8] M. Ritter. *Discrete Time Modeling of the Frame-Based Generic Cell Rate Algorithm.* Univ. of Würzburg, Research Report, No. 190, January 1998.

[9] N. Vicari and R: Schedel: *Performance of the GFR-Service with Constant Available Bandwidth.* Proceedings of the IEEE Infocom'99, March 1999, New York, USA.

[10] N. Vicari: *Effects of Variations in the Available Bandwidth on the Performance of the GFR Service.* Univ. of Würzburg, Research Report, No. 214, October 1998.
ftp://www-info3.informatik.uni-wuerzburg.de/pub/TR/tr214.pdf.

PERFORMANCE EVALUATION OF THE CONFORMANCE DEFINITION FOR THE ABR SERVICE IN ATM NETWORKS

L. Cerdà[1], B. Van Houdt[2], O. Casals[1] and C. Blondia[2] *

[1] *Polytechnic University of Catalonia*
Computer Architecture Dept.
c/ Jordi Girona, 1-3, Modulo C6,
E-08034 Barcelona, Spain
{llorenc,olga}@ac.upc.es

[2] *University of Antwerp*
Dept. Math and Computer Science
Universiteitsplein, 1
B-2610 Antwerp, Belgium
{vanhoudt,blondia}@uia.ua.ac.be

Abstract The standardization bodies have defined the Dynamic Generic Cell Rate Algorithm (DGCRA) as the Conformance Definition for ABR. This algorithm may be implemented in the Usage Parameter Control (UPC) for policing. In this paper we describe an equivalent queuing model of the DGRCA and we solve this model using a matrix analytic approach. The analytical results are validated by simulation.

Keywords: ATM Networks, Available Bit Rate Service, Conformance Definition, CDV Tolerance, Queuing Model, Performance Evaluation, Matrix Analytic.

1. INTRODUCTION

The ABR Service has been defined to efficiently multiplex sources that can adapt their transmission rate to the congestion state of the network. In this scheme, a source maintains the Allowed Cell Rate (ACR) parameter that defines the maximum rate at which cells may be scheduled for transmission. This parameter is controlled by special control cells called Resource Management Cells (RM-Cells). RM-Cells are transmitted embedded in the Data-Cell flow from the Source End System (SES) to the Destination End System (DES). The DES "turns around" the RM-Cells, which are sent back to the SES along the same path carrying congestion information. Depending on the congestion

*The first and third authors of this work were supported by the Ministry of Education of Spain under grant TIC96-2042-CE. The second and fourth authors were supported by the *Vlaams Actieprogramma Informatietechnologie* under project ITA/950214/INTEC.

information received in the RM-Cell, the SES increases or decreases the ACR. A detailed description of the ABR Service can be found in [1].

The Conformance Definition for ABR is the Dynamic Generic Cell Rate Algorithm (DGCRA). It is based on the GCRA, which has been defined as the conformance definition for the PCR/SCR of CBR and VBR service categories. The decision of cell conformance in the DGCRA is made by measuring the inter-cell arrival time of a connection and checking whether it deviates from the inverse of the expected rate less than a tolerance called Cell Delay Variation Tolerance (CDVT). The CDVT is negotiated at the connection set up and is an upper bound of the unavoidable CDV introduced by the ATM layer functions and multiplexing stages up to the measuring point. A network operator may use a Usage Parameter Control (UPC) which considers as non-conforming the cells with a CDV higher than the CDVT. Non-conforming cells may be marked or discarded. Consequently, a correct dimensioning of the CDVT is needed to guarantee a low non-conforming cell probability at the UPC.

In order to compute the expected rate, the UPC keeps track of the rate changes conveyed by the RM-Cells. These rate changes are applied in the forward direction, after the round trip delay between the UPC and the source. Since this round trip delay is variable, the standard specifies that two time constants referred to as τ_2 and τ_3 have to be given with the CDVT at the connection set up. These are respectively an upper and a lower bound of the round trip delay between the UPC and the source. If these delay bounds are not set correctly, the UPC may not compute the expected rate properly, causing cell rejection.

The paper is organized as follows. We first give a definition of the DGCRA in Section 2. In Section 3 we describe an equivalent queuing model for the DGCRA. Section 4 presents a detailed description of the analytical model used in the evaluation. This model is solved using a matrix analytical approach as shown in Section 7. Section 8 gives numerical results obtained from the analytical model. The results are validated by simulation. Finally, in Section 9 some concluding remarks are formulated.

2. SPECIFICATION OF THE DGCRA

The DGCRA has been defined by the ATM Forum as the conformance definition for an ABR connection. At arrival instant of cell n, conformance is decided by measuring the CDV value $y_n = c_n - a_n$, where a_n is the arrival epoch and c_n is the theoretical arrival time. Cell n is non-conforming if y_n is greater than τ_1 and is conforming otherwise. The parameter τ_1 is the CDVT for the ABR connection. This notation is used in the ATM Forum standard and we will indistinctly use CDVT and τ_1 in the rest of the paper.

After the initializations $LVST_0 = a_0$, $I_0^{old} = I_0$, the set of theoretical arrival times $\{c_n\}_{n>0}$ is computed at the arrival epochs a_n as:

$$\begin{aligned} c_n &= LVST_{n-1} + \min(I_{n-1}^{old}, I_n) \\ LVST_n &= \begin{cases} \max(c_k, a_k) & \text{if} \quad y_k \leq \tau_1 \quad \text{(cell conforming)} \\ LVST_{n-1} & \text{if} \quad \tau_1 < y_n \quad \text{(cell non conforming)} \end{cases} \\ I_n^{old} &= \begin{cases} I_n & \text{if} \quad y_n \leq \tau_1 \quad \text{(cell conforming)} \\ I_{n-1}^{old} & \text{if} \quad \tau_1 < y_n \quad \text{(cell non conforming)} \end{cases} \end{aligned} \tag{1}$$

$LVST_{n-1}$ stands for the Last Virtual Scheduled Time at cell n arrival. The theoretical arrival time c_n is given by $LVST_{n-1}$ plus an increment $\min(I_{n-1}^{old}, I_n)$ equal to the inverse of the expected emission time that should be used by the source between cell $n-1$ and n. I_n is the inverse of the last rate change to be received by the source before cell n emission. This is computed by the DGCRA based on the feedback conveyed by the backward RM-cell flow up to cell n arrival time a_n. The algorithm takes $\min(I_{n-1}^{old}, I_n)$ because the first cell received after a new increase I_n is scheduled may be received at this increase or at the previous increment I_{n-1}^{old}. By taking the minimum the algorithm stays on the safe side.

The computation of the sequence I_n is not an easy task because a change of rate conveyed by a backward RM-cell received at the measuring point at a given time may be applied to the forward cell flow after a delay equal to the round trip delay between the measuring point and the source. To cope with this problem two time constants τ_2 and τ_3 have been introduced which are respectively an upper bound and a lower bound of this round trip delay. Furthermore, two algorithms "A" and "B" have been defined to determine I_n (see [1] for details).

3. QUEUING MODEL

The GCRA can be modeled as a single server queue with a workload (waiting time) limited to the CDVT (see e.g. [5]). This queuing model can be extended to the DGCRA conformance definition given by equations (1). In the model we make the following assumptions: (i) the same rate changes followed by the source are policed by the DGCRA, (ii) the DGCRA is able to properly schedule the increments I_n corresponding to these rate changes, (iii) the source schedules cell transmissions at a new rate at cell emission times. With these assumptions, when cell n arrives at the DGCRA the increment I_n is equal to the source emission interval between cell n and cell $n+1$. This will be our definition of I_n in the rest of the paper.

The behavior of the queuing system which we use to model the DGCRA is shown in the time diagram of Figure 1. In this queuing model the cells correspond to customers. The service time is the increment applied by the DGCRA and the theoretical arrival time of cell n is the departure time of the

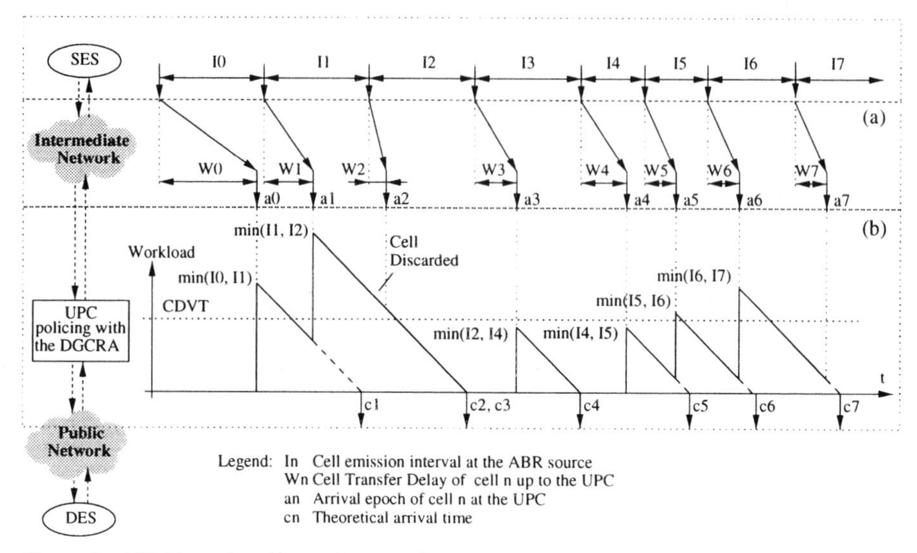

Figure 1 CDV introduced by an intermediate network (a) and the workload at the equivalent queuing model of the DGCRA (b).

previous accepted cell (c_n in the figure). The CDV value y_n of Equations (1) is given by the workload of the queue when $y_n > 0$. Therefore, when a cell arrival finds a workload higher than the CDVT, a non-conforming condition is given.

Note that the increment added at the cell n arrival epoch a_n in (1), is the service time to be added at the cell arrival epoch a_{n-1} in the equivalent queuing model. In other words, the workload increment added at the arrival epoch a_n in the queuing model is the increment that would be added by the DGCRA at the cell arrival epoch a_{n+1}, and is therefore given by $\min(I_n^{old}, I_{n+1})$ if cell $n+1$ is accepted, and 0 otherwise. For sake of simplicity, in Figure 1 we have approximated this by adding $\min(I_n^{old}, I_{n+1})$ to the workload if the cell n is accepted and 0 if cell n is not accepted.

4. MODEL USED IN THE EVALUATION

In order to evaluate the DGCRA we consider an ABR source multiplexed with a VBR source into a single switch . The VBR source has full priority over the ABR source. This multiplexing stage models the jitter introduced in the intermediate network shown in Figure 1.

The assumptions made in the equivalent queue of the DGCRA described in Section 3 apply to our model. Note that these assumptions imply that the ABR source generates as much traffic as possible without violating the allowed cell rate (ACR). We also assume that the ACR is not fixed by the intermediate network but by the switches located after the UPC (such that the UPC can keep track of these rate changes). This is a foreseeable situation, at least as long as the VBR and the ABR sources do not cause a heavy congestion condition at the switch in the intermediate network. As a likely consequence of these assumptions, we consider the VBR and the ABR rate changes to be independent.

Finally, we assume a propagation delay equal to zero between the UPC and the ABR source. This is a plausible assumption since it is a fixed delay which should have no influence on the policing function. We also consider that a backward RM cell experiences no delay between the UPC and the ABR source. With these assumptions the only component of the round trip delay between the ABR source and the UPC considered in our model is the delay introduced by the switch in the intermediate network.

In the following we give a detailed description of the analytical model considered for each device.

4.1 THE ABR SOURCE BEHAVIOR

The ABR traffic that is multiplexed together with the VBR background traffic is described here. A set of N cell rates $r_s,\ s = 1, ..., N$ is associated with the ABR source. The ABR source always transmits at one of these rates for which we have: MCR $\leq r_1 \leq \ldots \leq r_{N-1} \leq r_N \leq$ PCR

The parameters MCR and PCR represent the minimum and peak cell rate of the ABR connection. When the ABR source is transmitting at rate $r_i,\ i = 1, ..., N$, a cell is forwarded respectively every $1/r_i,\ i = 1, ..., N$, slots. In our analytical model we need $1/r_i,\ i = 1, ..., N$ to be an integer value.

The source behavior can be modeled by associating $1/r_i$ states to each possible rate r_i. We denote these states by $(i, j),\ i = 1, ..., N;\ j = 1, ..., 1/r_i$. We say that the ABR source is in state $(i, j),\ j = 1, ..., 1/r_i$, when it is transmitting at rate r_i. As long as the cell rate remains the same, e.g. r_i, the states $(i, 1)$ to $(i, 1/r_i)$ are traversed periodically. i.e. at each slot a transition occurs from (i, j) to $(i, j+1),\ j = 1, ..., 1/r_i - 1$ (note that in case of $1/r_i = 1$ there is only one of these states, the state $(i, 1)$, and the ABR source remains in it until a transition to a different rate state occurs). A cell is only transmitted in state $(i, 1/r_i)$ and no cell is generated in states $(i, 1)$ to $(i, 1/r_i - 1)$.

The ABR source considered here schedules cell transmissions at a new rate at cell emission times (the $(i, 1/r_i),\ i = 1, ..., N$ states in our model). Therefore, when a rate change occurs, e.g. from r_i to $r_{i'}$, there is a transition from state $(i, 1/r_i)$ to state $(i', 1)$. We denote by $P_{\text{ABR}}(i, i'),\ i, i' = 1, ..., N$ the probability that such state transition occurs and assume that they are Markovian.

4.2 THE BACKGROUND TRAFFIC

We consider as background traffic a VBR source which will be multiplexed together with the ABR source at the switch. This VBR source is modeled by a Markov Chain with M states, each with an associated rate. We say that the source is in state k when it is transmitting at rate v_k, $k = 1, ..., M$. These rates obey the following relation: $v_1 < v_2 < \ldots < v_M$

A cell is generated by the VBR source with probability v_k while being in state k, $k = 1, ..., M$. The VBR source can change its state at the end of each slot. We define $P_{\text{VBR}}(k, k')$ as the probability that a transition occurs from state k to state k', $(k, k' = 1, ..., M)$. Notice that these transitions are independent of the ABR rate changes.

4.3 THE SWITCH

The switch which precedes the UPC device is used to model the jitter caused by the intermediate network on the ABR traffic. The input of this switch consists of an ABR and a VBR traffic stream generated by the traffic sources described above. As the VBR traffic has full priority over the ABR source and the VBR source never generates more than one cell in a slot, the switch only needs a buffer to store delayed ABR cells. Delayed ABR cells are forwarded by the switch towards the UPC device when there is no VBR cell arrival. If a VBR cell arrives, this cell is forwarded and the ABR cells have to wait.

4.4 THE UPC DEVICE

In our model the current state of the UPC device is characterized by its workload U as described in Section 3. Recall from Section 3 that the workload U is increased by $\min(I_n^{old}, I_{n+1})$ upon the arrival of cell n, if the cell is accepted, and is decremented by one at the end of each slot. For sake of simplicity we have taken I_n instead of $\min(I_n^{old}, I_{n+1})$ in the analytical model. The validity of this approximation is checked by simulation.

As described in Section 2, the expected rate at the interface is computed taking into account the upper and lower bounds of the round trip delay between the UPC and the source, τ_2 and τ_3 respectively. In order to see the influence of these delay bounds we have considered two scenarios. In the first one we model a UPC which immediately applies a rate change when scheduled. This is equivalent to setting $\tau_2 = \tau_3$. Clearly, if there are delayed cells at the switch buffer when a rate decrease occurs, these will be likely considered as non conforming (because an increment higher than the emission interval used by the source will be applied at the UPC).

In the second scenario we consider a UPC with τ_2 properly set to an upper bound of the round trip delay between the UPC and the source. Such a UPC guarantees that no higher increment than the emission interval used by a "well behaving" source is applied at the UPC. We refer as "well behaving" a source that follows the rate changes conveyed by the backward RM-Cells. Note that this is the kind of source we use in our model. In the following the analytical model we use for these two scenarios is described.

5. SCENARIO WITH $\tau_2 = \tau_3$

In this case we consider a UPC which does not apply a time tolerance to the scheduled rate changes. Since in our model we consider a propagation delay equal to zero, we have $\tau_2 = \tau_3 = 0$. In our model this is equivalent to using the inverse of the rate associated with the ABR state at the time that a cell arrives at the UPC. Clearly this is not necessarily the rate of the ABR source when this cell was generated, as the rate of the ABR source might have changed if the cell was delayed in the switch.

We recall that as long as the ABR cell rate remains the same, e.g. r_i, it traverses the states $(i, 1), ..., (i, 1/r_i)$ emitting one cell and possibly changing the rate in state $(i, 1/r_i)$. Therefore, we use $1/r_i$, $i = 1, ..., N$ as the increment I_n associated with the states (i, j), $j = 1, ..., 1/r_i - 1$. The increment associated with the state $(i, 1/r_i)$ depends on whether or not a rate change occurs. If no rate change occurs the increment $1/r_i$ is used, otherwise we use $1/r_{i'}$ where $r_{i'}$ represents the new rate.

Note that we are making the following approximation. In a real situation with $\tau_2 = \tau_3 = 0$ the UPC would apply I_n immediately after the backward RM-Cell conveying the new rate traversed the UPC. In our model I_n is applied when the ABR source effectively performs the rate change, which happens at the cell emission epoch. This approximation is confronted with simulation results.

6. SCENARIO WITH τ_2 PROPERLY TUNED

In this case the UPC postpones the scheduled rate decreases until after a delay bound $\tau_2 > \tau_3$. Since in our model we consider a propagation delay equal to zero, this implies that during the first τ_2 slots after the scheduling of a rate reduction (from r_i to $r_{i'}$), the UPC will continue to use the smaller increment $1/r_i$.

We approximate this scenario by flushing the switch buffer each time that a rate reduction occurs. To assess the increments at the UPC we use the same rules as in the previous scenario. Note that by doing this we guarantee that no higher increment than the emission interval used by the source will be applied to a cell arriving at the UPC.

To be able to solve the analytical model, the flushing is only performed with probability $1 - \alpha$, where α is small, e.g. $\alpha < 10^{-12}$ (see Appendix A.2).

7. PERFORMANCE ANALYSIS

The system is observed at the end of each time slot. A Markov Chain is obtained by looking at the stochastic vector:

$$(Q, U, (i, j), k) \tag{2}$$

Where Q represents the queue length of the buffer inside the switch, U equals the remaining workload at the UPC, $(i, j), i = 1, ..., N; j = 1, ..., 1/r_i$ is the state of the ABR source and $k, k = 1, ..., M$ the state of the VBR source.

Denote by $P(S, S')$ the one slot transition probability from state $S = (Q, U, (i, j), k)$ to state $S' = (Q', U', (i', j'), k')$. By ordering the states $(Q, U, (i, j), k)$ lexicographically the probabilities $P(S, S')$ define a stochastic transition probability matrix $\mathbf{P}$ with the block structure $\mathbf{P} = (\mathbf{Q}_{m,n})$. The sub-matrices $\mathbf{Q}_{m,n}$ govern the state $(U, (i, j), k)$ transitions when a queue length change from m to n occurs. Therefore, $\mathbf{Q}_{m,n}$ are square matrices of order equal to $U_{max} + 1$ times the number of ABR states times the number of VBR states, where U_{max} is the maximum workload. Notice that $U_{max} = \tau_1 + 1/r_{min}$, where r_{min} is the minimum of the cell rates considered for the ABR source. In the Appendix we describe how to derive the matrix $\mathbf{P}$ and how to find the stationary probabilities in each scenario.

Having calculated the stationary probability vector of the process (2), we denote its components as $\pi(Q, U, (i, j), k)$, i.e. the probability that we are in state $(Q, U, (i, j), k)$. The rejection probability can then be found as:

$$\frac{\sum\limits_{Q\geq 1}\sum\limits_{\substack{U>\\ \tau_1+1}}\sum\limits_{i,j,k}\pi(Q, U, (i, j), k)\,(1 - v_k) + \sum\limits_{\substack{U>\\ \tau_1+1}}\sum\limits_{i,k}\pi(0, U, (i, \frac{1}{r_i}), k)\,(1 - v_k)}{\sum\limits_{Q\geq 1}\sum\limits_{U,i,j,k}\pi(Q, U, (i, j), k)\,(1 - v_k) + \sum\limits_{U,i,k}\pi(0, U, (i, \frac{1}{r_i}), k)\,(1 - v_k)}$$

8. NUMERICAL RESULTS

In this section we show the numerical results obtained with the analytical model described in Section 7. In order to validate the analytical model, all the results shown in this section have been verified by simulation.

We have used two different rates for the ABR source. In order to assess $P_{\text{ABR}}(i, i'),\ i, i' = 1, 2$ we have assumed that a Markovian process governs the rate changes. This process alternatively changes between two states, namely $E_i,\ i = 1, 2$. In this model a change into a certain state E_i represents a backward RM-Cell arrival conveying a new rate equal to r_i. Therefore, when the ABR source is in state $(i, 1/r_i),\ i = 1, 2$, a change to state $(i', 1)$ occurs if $E_{i'},\ i' = 1, 2$ is the current state. We have taken the sojourn time in each state $E_i,\ i = 1, 2$, to be the same and equal to p slots. With these assumptions we have:

$$P_{\text{ABR}}(i,i') = \sum_{n=0}^{1/r_i} \binom{1/r_i}{n} 1/p^n \, (1-1/p)^{1/r_i - n} \, .$$
$$1\big[(i = i' \text{ and } n \text{ is even}) \text{ or } (i \neq i' \text{ and } n \text{ is odd})\big], \; i,i' = 1,2$$

where we use the indicator function $1\big[condition\big]$ equal to 1 if *condition* is true and 0 otherwise. Note that as long as the inverse of the ABR rates is small compared to p, the mean time between two consecutive rate changes of the ABR source is equal to p. Thus, we will refer to $1/p$ as the ABR rate change frequency.

For the VBR source only one state is possible, and thus one rate v_1. Obviously $P_{\text{VBR}}(k,k') = 1, \; k,k' = 1$. Note that the switch load ρ in this model is approximately given by: $\rho = v_1 + (r_1 + r_2)/2$

In the following we first investigate the validity of the analytical results by comparing them with the simulation results. Then the figures are analyzed, in order to derive some engineering rules.

8.1 VALIDATION

In the simulation we have used the DGCRA given by Equations 1. Moreover, the approximations made in the analytical model to make it tractable have been removed.

Figures 2.A and 2.B correspond to the scenario where τ_2 is properly tuned. The figures show a good agreement between the analytical and simulation results. Only when there is a heavy load and the ABR rate change frequency is high, the model yields an underestimated rejection probability.

This can be explained by the following reasoning. If the queue length flushed when there is a rate reduction, e.g. from r_i to $r_{i'}$, is small compared to the cells emitted by the source while the rate was r_i, this approximation will clearly have a small influence on the rejection probability. Therefore, the approximation is worse for high loads and small sojourn times in the states with higher rates.

Figures 3.A and 3.B show the analytical and simulation results for the scenario where $\tau_2 = \tau_3$. In these figures we can see that the analytical model gives a good approximation for high loads, but it becomes worse when the load decreases. To explain these results we have to take into account the two main reasons that may lead to a cell rejection in this scenario: (i) the jitter of the ABR cell stream and (ii) the usage of an increment higher than the source cell emission interval.

The analytical model is able to capture very well the cells rejected due to condition (i). When the load is high, the condition (i) is predominant and thus the analytical model yields a good approximation. For lower loads, condition

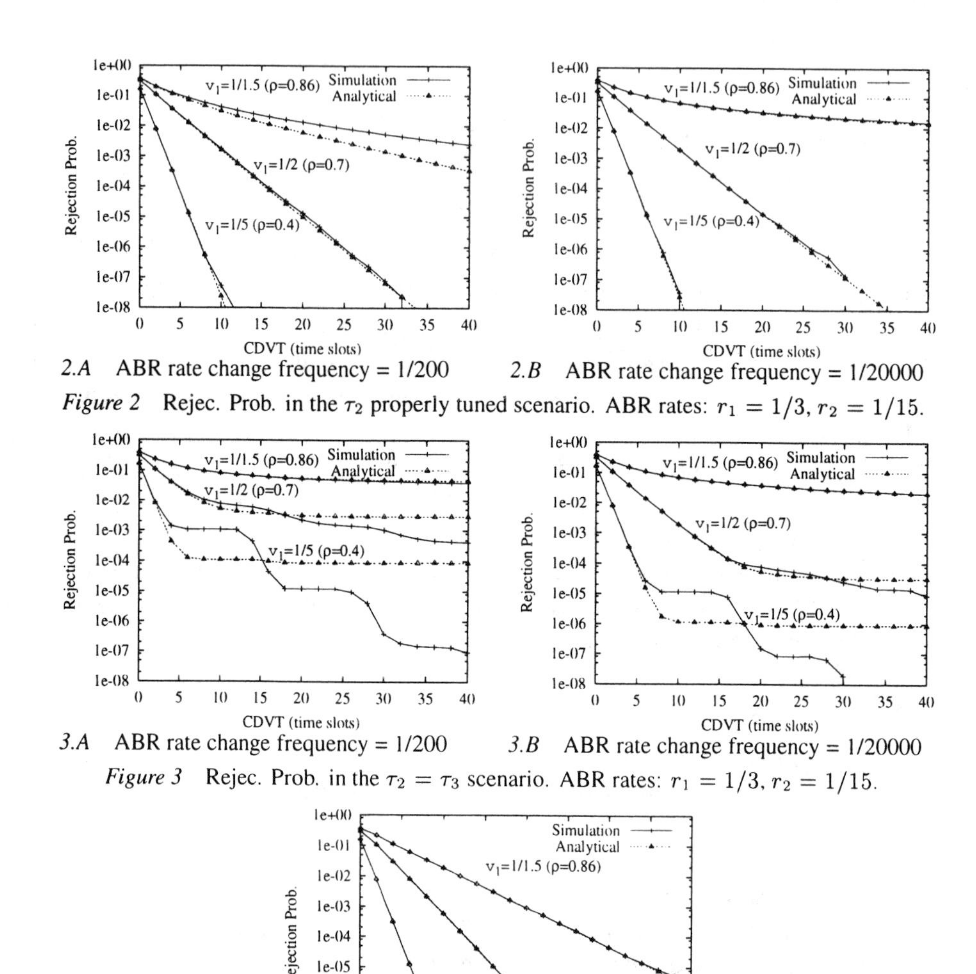

2.A ABR rate change frequency = 1/200 *2.B* ABR rate change frequency = 1/20000

Figure 2 Rejec. Prob. in the τ_2 properly tuned scenario. ABR rates: $r_1 = 1/3, r_2 = 1/15$.

3.A ABR rate change frequency = 1/200 *3.B* ABR rate change frequency = 1/20000

Figure 3 Rejec. Prob. in the $\tau_2 = \tau_3$ scenario. ABR rates: $r_1 = 1/3, r_2 = 1/15$.

Figure 4 Rejec. Prob. when there are no ABR rate changes. ABR rate: $r_1 = 1/5$.

(i) is only predominant when the CDVT is close to zero. When the CDVT increases, cell rejection is mainly due to (ii). When this happens the analytical model shows that the rejection probability remains nearly constant.

When condition (ii) is predominant, Figures 3.A and 3.B show that the analytical model is initially more optimistic (it yields lower rejection probability), and then the curve obtained by simulation falls down showing a step behavior. The optimistic results are explained because the rate changes are applied later in the analytical model than in the simulation (remember that rate changes occur at the cell emission epochs in the analytical model and at the backward RM-Cell arrivals in the simulation). The steps are caused because $\min(I_n^{old}, I_{n+1})$ is used in the simulation while I_n is used in the analytical model. This makes that each time a rate increase occurs in the simulator, e.g. from r_i to $r_{i'}$, the workload is increased by $1/r_{i'}$ instead of by the emission interval $1/r_i$, which causes an average reduction of $1/r_i - 1/r_{i'}$. Consequently, each time the CDVT added to these reductions compensates the wrong increments mentioned above, there is a reduction on the rejection probability.

The former differences between the simulator (which models the real DGCRA) and the analytical model could be removed. However, this would considerably increase the complexity of the analytical model. For dimensioning purposes the scenario with τ_2 properly tuned would be used, and in this scenario these differences do not have any influence.

Finally, Figure 4 shows the analytical and simulation results when there are no ABR rate changes. Without rate changes there are no differences between the two scenarios considered in this paper. Moreover, in this case there are no differences between the simulation and analytical models. This is confirmed by Figure 4 which shows a perfect agreement of the analytical and simulation results.

8.2 ANALYSIS

The Figures 2-4 show the influence of the following items on the rejection probability:

- Tuning of the τ_2 delay bound,
- jitter on the ABR stream (the higher the VBR rate, the higher the jitter),
- difference between the ABR rates,
- frequency of the ABR rate changes.

In Figure 2 we can see that when τ_2 is properly tuned, the rejection probability decreases exponentially when increasing the CDVT for low to moderated loads. In this case rejection probabilities of 10^{-9} can be reached with CDVT in the order of tens. Moreover, the frequency of the ABR rate changes has little influence on the results.

Figure 3 shows that when τ_2 is not properly set, the rejection probability has a major degradation. Moreover, the rejection probability does not decrease exponentially with the CDVT and is much more sensitive to the frequency of the ABR rate changes.

Figure 4 shows the rejection probability when the ABR source rate does not change. Note that the average load of the ABR source is maintained constant in all the figures (approximately equal to 0.2). By comparing Figure 4 with the other figures we can see that the higher the differences between the ABR rates (while maintaining the same load) the higher the rejection probability. The figures also show that this effect increases very rapidly with an increasing overall load.

9. CONCLUSIONS

In this paper we have investigated the parameter dimensioning in the conformance definition for the ABR Service, the Dynamic Generic Cell Rate Algorithm (DGCRA). We have proposed an equivalent queuing model of the DGCRA and we have solved this model using a matrix analytic approach. The analytical results have been validated by simulation.

In the DGCRA three parameters are negotiated at the connection setup: the Cell Delay Variation Tolerance (CDVT), and the upper and lower bounds of the round trip delay between the UPC and the source, τ_2 and τ_3 respectively.

In the model we have considered the jitter introduced by a VBR source on an ABR cell stream sharing a common multiplexing stage. The model shows the influence of the following parameters on the rejection probability at the UPC: (i) Tuning of the τ_2 delay bound, (ii) jitter on the ABR stream (the higher the VBR rate, the higher the jitter), (iii) the difference between the ABR rates, (iv) the frequency of the ABR rate changes. These are investigated in two scenarios which show the influence of τ_2:

1. Scenario with τ_2 properly tuned:

- For loads low to moderate, the rejection probability decreases exponentially when increasing the CDVT. In this case, rejection probabilities of 10^{-9} can be reached with CDVT in the order of tens.

- The higher the differences between the ABR rates (while maintaining the same load) the higher the rejection probability. This effect increases very rapidly with increasing overall loads.

- The frequency of the ABR rate changes has a minor influence on the rejection probability.

2. Scenario with $\tau_2 = \tau_3$ (the UPC does not apply a time tolerance to the scheduled rate changes):

- Compared with the former scenario, results show a major degradation of the rejection probability. This probability does not decrease exponentially when increasing the CDVT, but decreases at a slower rate.

- Frequency and amplitude of the ABR rate changes have a remarkable influence on the rejection probability.

Appendix: Solution of the Transition Probability Matrix

A.1 SCENARIO WITH $\tau_2 = \tau_3$

Since the ABR source can only emit one cell at each slot, $P(S, S')$ elements with queue length increments or decrements higher than one (i.e. $|Q' - Q| > 1$) are zero. Moreover, the transitions are exactly the same for all values with $Q \geq 1$. Therefore, we define: $\mathbf{B}_0 = \mathbf{Q}_{0,0}$, $\mathbf{A}_0 = \mathbf{Q}_{n+1,n}$, $n > 0$, $\mathbf{A}_1 = \mathbf{Q}_{n,n}$, $n > 0$ and $\mathbf{A}_2 = \mathbf{Q}_{n,n+1}$, $n \geq 0$. This yields the following structure for the matrix $\mathbf{P}$:

$$\mathbf{P} = \begin{pmatrix} \mathbf{B}_0 & \mathbf{A}_2 & \mathbf{0} & \mathbf{0} & \ldots \\ \mathbf{A}_0 & \mathbf{A}_1 & \mathbf{A}_2 & \mathbf{0} & \ldots \\ \mathbf{0} & \mathbf{A}_0 & \mathbf{A}_1 & \mathbf{A}_2 & \ldots \\ \ldots & \ldots & \ldots & \ldots & \ddots \end{pmatrix} \tag{A.1}$$

The following partitioned solution $(\pi_0, \pi_1, ...)$ of the stationary probabilities exists for this type of processes [4] (π_i, $i = 0, 1, ...$ are vectors of length equal to the order of the blocks of the matrix $\mathbf{P}$):

$$\begin{aligned} \pi_k &= \pi_0 \, \mathbf{R}^k \qquad , k \geq 1 \\ \pi_0 &= \pi_0 \left[\mathbf{B}_0 + \mathbf{R}\mathbf{A}_0\right] \\ \pi_0 &\left(\mathbf{I} - \mathbf{R}\right)^{-1} \mathbf{e} = 1 \end{aligned} \tag{A.2}$$

where $\mathbf{R}$ has the same order as $\mathbf{A}_i$. $\mathbf{I}$ is the unity matrix with the same order and $\mathbf{e}$ is an all ones vector with corresponding length. To find $\mathbf{R}$ we use the logarithmic reduction algorithm of Latouche and Ramaswami [2].

Now we show how $\mathbf{A}_2$ may be derived, similar reasoning can be applied to derive $\mathbf{A}_0$, $\mathbf{A}_1$ and $\mathbf{B}_0$. In this case the queue length at the switch is increased by one, i.e. $Q' = Q + 1$. This can only happen if a cell is emitted by the ABR source (and thus $j = 1/r_i$) and a cell is emitted by the VBR source. Remember from Section 4.1 that the ABR source changes to state $j' = 1$ after a cell emission. Since there is no cell arrival at the UPC the workload is decreased by 1, and thus $U' = \max(U - 1, 0)$. Therefore, we have:

$$\begin{aligned} \mathbf{A}_2(S, S') = P_{\text{ABR}}(i, i') \cdot P_{\text{VBR}}(k, k') \cdot v_k \cdot \\ 1\left[j = 1/r_i \text{ and } j' = 1 \text{ and } U' = \max(U - 1, 0)\right] \end{aligned}$$

where $\mathbf{A}(S, S')$ denotes the element (S, S') of the matrix $\mathbf{A}$.

A.2 SCENARIO WITH τ_2 PROPERLY TUNED

We approximate this scenario by flushing the switch buffer each time that a rate reduction occurs, i.e $r_i > r_{i'}$ (see Section 6). For the reasons explained below, the flushing is only performed with probability $1 - \alpha$. A transition from

$Q \geq 0$ to $Q' = 0$ occurs when the queue is flushed, therefore, we obtain the following structure for the transition probability matrix $\mathbf{P}^{(p)}$ (by the superscript $^{(p)}$ we shall distinguish the matrices derived in this scenario from those used in Appendix A.1):

$$\mathbf{P}^{(p)} = \begin{pmatrix} \mathbf{B}_0^{(p)} & \mathbf{A}_2^{(p)} & \mathbf{0} & \mathbf{0} & \mathbf{0} & \cdots \\ \mathbf{A}_0^{(p)} + \mathbf{A}_3^{(p)} & \mathbf{A}_1^{(p)} & \mathbf{A}_2^{(p)} & \mathbf{0} & \mathbf{0} & \cdots \\ \mathbf{A}_3^{(p)} & \mathbf{A}_0^{(p)} & \mathbf{A}_1^{(p)} & \mathbf{A}_2^{(p)} & \mathbf{0} & \cdots \\ \mathbf{A}_3^{(p)} & \mathbf{0} & \mathbf{A}_0^{(p)} & \mathbf{A}_1^{(p)} & \mathbf{A}_2^{(p)} & \cdots \\ \vdots & \vdots & \vdots & \ddots & \ddots & \ddots \end{pmatrix} \tag{A.3}$$

The submatrices are given by: $\mathbf{A}_0^{(p)} = \mathbf{A}_0$, $\mathbf{A}_1^{(p)} = \mathbf{A}_1'$, $\mathbf{A}_2^{(p)} = \mathbf{A}_2'$, $\mathbf{A}_3^{(p)} = \mathbf{A}_1'' + \mathbf{A}_2''$, $\mathbf{B}_0^{(p)} = \mathbf{B}_0 + \mathbf{A}_2''$. where the matrices $\mathbf{A}_0$ and $\mathbf{B}_0$ are obtained as in Appendix A.1 . The matrices $\mathbf{A}_1'$ and $\mathbf{A}_2'$ are respectively obtained replacing $P_{\text{ABR}}(i,i')$ by $P_{\text{ABR}}(i,i') \cdot \left(1\left[r_i \leq r_{i'}\right] + \alpha \cdot 1\left[r_i > r_{i'}\right]\right)$ in the relations given for $\mathbf{A}_1$ and $\mathbf{A}_2$ in Appendix A.1 . Similarly, the matrices $\mathbf{A}_1''$ and $\mathbf{A}_2''$ are obtained replacing $P_{\text{ABR}}(i,i')$ by $P_{\text{ABR}}(i,i') \cdot (1-\alpha) \cdot 1\left[r_i > r_{i'}\right]$ in the relations of $\mathbf{A}_1$ and $\mathbf{A}_2$.

To solve the former matrix $\mathbf{P}^{(p)}$, the matrix $\mathbf{A}_0^{(p)} + \mathbf{A}_1^{(p)} + \mathbf{A}_2^{(p)}$ must be irreducible [3]. By choosing $\alpha > 0$, this condition is fulfilled. $\mathbf{P}^{(p)}$ is then solved using equations (A.2) except that the second equation has to be replaced by $\pi_0 = \pi_0 \left[\mathbf{B}_0^{(p)} + \mathbf{R}\,\mathbf{A}_0^{(p)} + \left((\mathbf{I}-\mathbf{R})^{-1} - \mathbf{I}\right)\mathbf{A}_3^{(p)}\right]$. The matrix $\mathbf{R}$ still obeys a similar equation as before and therefore can be found using the L-R algorithm [2].

References

[1] ATM Forum Technical Committee Traffic Management Working Group. *"ATM Forum Traffic Management Specification Version 4.0"*, April 1996.

[2] G. Latouche and V. Ramaswami. "A logarithmic reduction algorithm for Quasi-Birth-Death processes". *Journal of Applied Prob.*, 30, 650–674, 1993.

[3] M.F. Neuts. "Markov Chains with Applications in Queueing Theory, which have a Matrix-Geometric Invariant Probability Vector". *Journal of Applied Prob.*, 10, 185–212, 1978.

[4] M.F. Neuts. *"Matrix-Geometric Solutions in Stochastic Models"*. The John Hopkins University Press, Baltimore, 1981.

[5] J. Roberts, U. Mocci, and J. Virtamo, editors. *"Broadband Network Teletraffic - Final Report of Action COST 242"*. Springer Verlag, 1996.

SESSION 7

Mobile Network Protocols

IMPACTS OF POWER CONTROL ON OUTAGE PROBABILITY IN CDMA WIRELESS SYSTEMS

Kenji Leibnitz
University of Würzburg, Dept. of Computer Science, Am Hubland, 97074 Würzburg, Germany
leibnitz@informatik.uni-wuerzburg.de

Abstract In this paper we analyze the power control loops on the reverse link of a CDMA system. We obtain an expression for the transmission power of a mobile station and its signal-to-interference ratio (SIR) received at the base station. This allows the derivation of outage probability, which is the probability for not fulfilling the SIR requirements. We will show that power control in the North-American IS-95 system is very robust and that the number of users in the cell does not have much influence on outage. Furthermore, we compare outage which is an event at the receiver with the mobile station exceeding its maximum transmission power.

Keywords: CDMA, power control, signal-to-interference ratio, outage probability.

1. INTRODUCTION

Wideband Code Division Multiple Access (W-CDMA) is the upcoming RF technology for future mobile communication systems, like *Universal Mobile Telecommunications System* (UMTS) or IMT-2000. This is mainly due to its superior capacity compared to second generation systems employing F/TDMA. In CDMA the transmitted signal is spread over the frequency bandwidth by modulating each user's data signal with a pseudo-noise carrier of much higher frequency. Since the signals appear like noise over the channel, all other users in the cell constitute to a certain level of interference.

This limitation leads to the necessity of controlling the interference induced by other users to a minimum. *Mobile stations* (MS) will be located in the cell at varying distances from the *base transceiver station* (BTS). It has to be avoided that a mobile near the BTS transmits at a too high level and causes too much interference for other MS farther away ("near-far" problem). Additionally, due to shadowing and multi-path fading as well as fluctuations in user traffic there will be variations in the received signal strength. This is overcome by a tight power control performed on the *reverse link* (mobile-to-base station path).

With this power control algorithm the BTS tries to perform a balancing of the received *signal-to-interference ratio* (SIR) of all users in the cell.

The performance of closed loop power control based on local SIR estimates has been studied by several authors. In [14] an analytical model was constructed containing an inner loop based on SIR processing and an outer loop based on frame error performance. Simulation studies of single and multi-cell systems were conducted in [1] and the dependence of signal and interference statistics on step size and processing delays was examined. A similar approach was taken in [2], where the impact of update rate, loop delay and vehicle speed on the bit error performance was examined in simulations. The effects of power control non-idealities on performance were investigated in [3] and [8].

The model proposed here is based on the one presented in [4] and is a Markov state space representation of the closed loop power control which allows direct computation of the statistics of the MS transmit power. We will investigate how the power control loops affect the SIR and derive an expression for outage probability. Furthermore, we will show the relationship between outage, which is an event occurring at the BTS and the event of the mobile station exceeding its maximum transmission power.

This paper is organized as follows. Section 2 gives an overview of the closed loop power control which is implemented in the IS-95 system [11]. In Section 3 we present the system parameters and Gaussian channel and derive our analytical Markov chain model. Based on this model, we will discuss the impacts of power control on outage probability in Section 4. This paper is concluded in Section 5 by giving an outlook on future work.

2. MODEL OF THE IS-95 CDMA SYSTEM

In CDMA systems it is essential that all users are received with nearly equal strength at the BTS in order to be demodulated and decoded correctly. To overcome the near-far problem, several mechanisms to control the MS transmit power take place in the IS-95 standard at the base and mobile stations.

In *open loop* power control the MS uses the received signal strength on the forward link as estimation for the path loss and sets its transmit power accordingly. Contrary to that, *closed loop* power control works in a tight cooperation between mobile and base station to overcome fluctuations on the traffic channel. The closed loop itself consists of an *inner loop* and *outer loop.* Within the inner loop, the BTS continually monitors the link quality of the reverse link in terms of received SIR and compares it with a certain threshold. If the received value is too high, then the MS is told to decrease its power. On the other hand, if it is too low, the link quality is not good enough and a "power-up" command is sent. This power update is performed every 1.25 *ms*. The power control command itself consists of one bit of information that is multiplexed on the traffic channel after the convolutional encoding and is therefore not error protected. The transmission of a single bit also results in the fact that there are only commands for increasing or decreasing, but none for maintaining a certain power level.

After every 16 such inner loop cycles, one frame has been transmitted and the power control algorithm enters the outer loop. Its main goal is to maintain an acceptable *frame error rate* (FER) by readjusting the SIR threshold of the inner loop after every frame. The interaction between inner and outer loop is illustrated in Fig. 1.

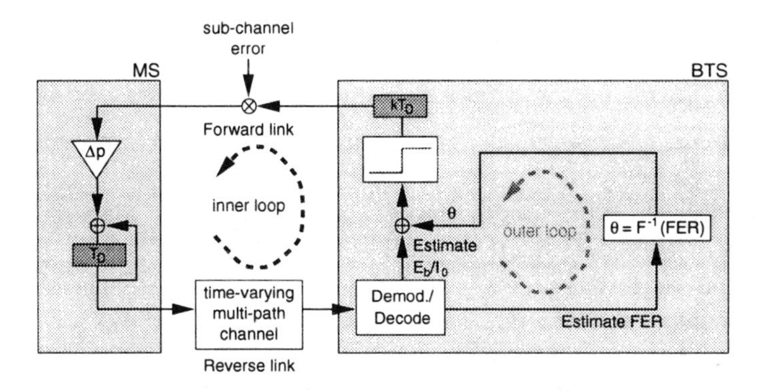

Figure 1 CDMA reverse link closed loop power control

3. ANALYSIS OF CDMA POWER CONTROL

In this section we derive an analytical model of the reverse link power control loops which will permit the computation of the transmit power of an arbitrary MS. Our model is based on a Markov chain which is discrete in time (1.25 *ms* steps) and discrete in state space. In the IS-95 system, the MS transmit power is limited to the range between -50 and 23 *dBm* and the update step size is 1 *dB*. In the following, we will denote the mobile's power with the variable $j = 0, \ldots, J$, where J is the total number of possible steps within the range given above.

3.1 System Parameters and Gaussian Channel

Let an MS be located at a distance x from the BTS and $\hat{S}$ be its transmission power at an observed time instant. On the path to the base station, the signal power is being attenuated by propagation loss $L(x)$ [9], shadow fading, and multi-path fading and is received with power T at the BTS. The received SIR is given by the bit-energy-to-noise ratio (E_b/N_0) denoted by the variable ϵ. Note that variables in linear space will be given as $\hat{\chi}$ which is $\chi = 10 \log \hat{\chi}$ in dB.

$$\hat{\epsilon} = \frac{\hat{G}_s \hat{T}}{\sum_{i=1}^{k-1} \hat{T}_i \nu_i + N_0 W} \tag{1}$$

The spreading gain of the system is $\hat{G}_s = W/R$ with a data bitrate of $R = 9.6$ *kbps* and frequency spectrum of $W = 1.25$ *Mhz*. The term N_0 is the thermal noise power density. By introducing $k_{\text{pole}} = W/(Rm_{\hat{\epsilon}}\rho) + 1$ as the *pole capacity* [12] of the system for a mean SIR $m_{\hat{\epsilon}}$ we can eliminate the dependency on the received power levels $\hat{T}$ and can rewrite Eqn. (1) as

$$\hat{\epsilon} = \hat{G}_s \frac{\hat{S}\hat{L}(x)}{\hat{N_0}} \hat{\varphi}(k) \qquad \Rightarrow \qquad \epsilon = G_s + S + L(x) + \varphi(k) - N_0. \tag{2}$$

The *multi-access interference* (MAI) induced by the other users in the same cell is represented by $\hat{\varphi}(k) = (k_{\text{pole}} - k)/(k_{\text{pole}} - 1)$ and is a function of the number of users k. If the system supports only a single user, the term $\hat{\varphi}$ will be one and there will be no interference from other users. However, if k is near k_{pole} then the MAI causes that the SIR approaches zero. Note that at this point we don't know the exact value of $m_{\hat{\epsilon}}$. However, for our purposes it is sufficient to assume a fixed k_{pole} and observe the loading percentage of the cell.

We will use an *Additive White Gaussian Noise* (AWGN) channel model where the total attenuation caused by fading is considered to be an i.i.d. Gaussian random variable C with mean μ_C and standard deviation σ_C. We can write $\mu_C = G_s + L(x) + \varphi(k) - N_0$ in our case as the sum of all non-random components of Eqn. (2). We can now express ϵ as the sum of the random variables S and C. In the following we will derive the distribution of S.

3.2 Closed Loop Power Control

As mentioned in Section 2, the closed loop power control in IS-95 consists of the interworking of inner and outer loop. The following section will describe how both loops are taken into account in our model.

3.2.1 Inner Loop

In the inner loop the received SIR level ϵ is compared with the target threshold θ of the outer loop whether to increase or decrease the signal strength. Thus, the probability for a power-up command is $p_u = \Pr(\epsilon < \theta)$.

Using the properties of the Gaussian distribution, the probability for a power up command p_u under the condition that the MS is transmitting at power level j and the outer loop threshold is i can be given as

$$p_u(i,j) = \Pr(C \leq i - j | S = j, \theta = i) = \frac{1}{2} + \frac{1}{2} \operatorname{erf}\left(\frac{i - j - \mu_C}{\sqrt{2}\sigma_C}\right). \tag{3}$$

Since it is only possible to increase or decrease the power, the probability for a power-down command is $p_d(i,j) = 1 - p_u(i,j)$.

As previously mentioned, the power control command is transmitted unprotected. We will therefore model the forward link channel as a *binary symmetrical channel* with a bit error probability p_b. It is well known [15] that the probability of bit error in a QPSK modulated channel can be approximated by $p_b = \frac{1}{2} Q(\sqrt{E_b/N_0})$.

3.2.2 Outer Loop

The outer loop is performed after every frame, i.e., 16 inner loop cycles. Its purpose is to update the SIR threshold to achieve an acceptable link quality in terms of frame error rate. Whereas the inner loop is specified in IS-95, the algorithm for the outer loop is up to the manufacturer. In the following we will use a simple algorithm similar to the one presented in [10].

Let θ be the threshold level at a certain frame. If the frame is in error, the threshold is increased by K *dB*, e.g. $K = 5$, otherwise it is decreased by 1 *dB*. The frame is considered to be in error if at least one bit in the frame is in error, which can occur independently. Therefore, for $N = 192$ bits in a frame and the bit error probability p_b given above, the probability for increasing the SIR threshold is $q_u(i) = 1 - (1 - p_b(i))^N$ and for decreasing is $q_d(i) = 1 - q_u(i)$. It is assumed that the threshold will be limited by a maximum value M.

3.2.3 Markov Chain Model

To model the power control loops we will use a Markov chain with two-dimensional states $s(i,j), i = 1, \ldots, M, j = 1, \ldots, J$. The first index i describes the SIR threshold value and j is the power level at which the MS is transmitting. The state transitions to and from a state $s(i,j)$ of this Markov Chain are illustrated in Fig. 2.

The power level will be increased and decreased with $p_u(i,j)$ and $p_d(i,j)$ in Eqn. (3), respectively. Since the inner loop is performed more frequently than the outer loop, where a threshold update is done every 16th cycle, the transitions with the same SIR threshold are weighted with $\frac{15}{16}$. The transitions from $s(i,j)$ to states with other thresholds must consider the probability for threshold updates $q_u(i)$ and $q_d(i)$. Therefore, the following transition probabilities are used.

$$\alpha(i,j) = \frac{15\, p_u(i,j)}{16} \qquad \beta(i,j) = \frac{15\, p_d(i,j)}{16} \qquad \gamma(i,j) = \frac{p_d(i,j)\, q_d(i)}{16}$$

$$\delta(i,j) = \frac{p_u(i,j)\, q_d(i)}{16} \qquad \xi(i,j) = \frac{p_d(i,j)\, q_u(i)}{16} \qquad \zeta(i,j) = \frac{p_u(i,j)\, q_u(i)}{16}$$

Note that special care has to be taken at the range boundaries for $i \in \{0, M\}$ and $j \in \{0, J\}$. We can now order all probabilities in a transition probability matrix $\mathcal{P}$, see Eqn. (4).

$$
\mathcal{P} = \begin{pmatrix}
A_0 + B_0 & \overbrace{0 \quad \cdots \quad 0}^{K-1} & C_0 & & & \\
B_1 & A_1 & & C_1 & & \\
& \ddots & \ddots & & \ddots & \\
& & B_{M-K} & A_{M-K} & \cdots & C_{M-K} \\
& & \ddots & \ddots & & \vdots \\
& & & B_{M-1} & A_{M-1} & C_{M-1} \\
& & & & B_M & A_M + C_M
\end{pmatrix} \tag{4}
$$

The matrices A_i, B_i, and C_i can be given in the following way. If we define a matrix $P_i, i = 1, \ldots, M$, by

$$
P_i = \begin{pmatrix}
p_d(i,0) & p_u(i,0) & & & \\
p_d(i,1) & & p_u(i,1) & & \\
& \ddots & & \ddots & \\
& & p_d(i,J-1) & & p_u(i,J-1) \\
& & & p_d(i,J) & p_u(i,J)
\end{pmatrix}
$$

then $A_i = \frac{15}{16} P_i$, $B_i = \frac{q_d(i)}{16} P_i$, and $C_i = \frac{q_u(i)}{16} P_i$. With Eqn. (4) it is now possible to compute the equilibrium state distribution $s(i,j)$. The stationary MS power distribution can be derived by the sum of all states with common second index and the distribution of the outer loop threshold by summation of the states with common first index, see Eqn. (5).

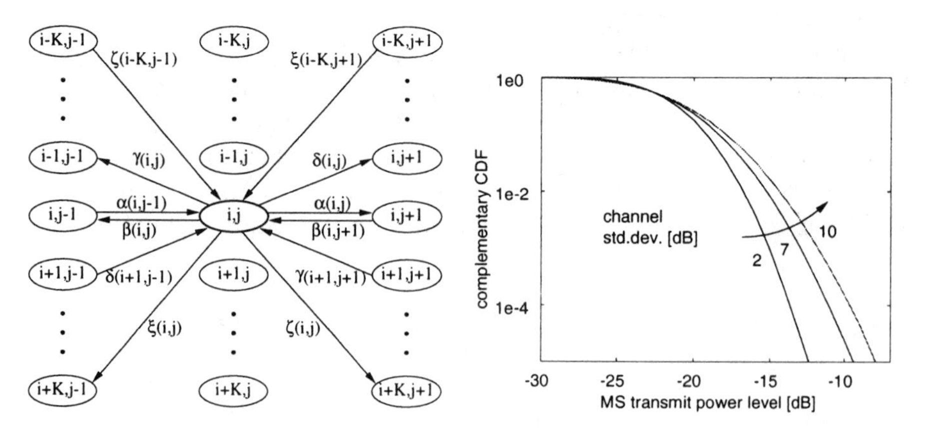

Figure 2 Markov Chain state transitions

Figure 3 MS transmit power distribution

$$\Pr(S = j) = s(j) = \sum_{i=0}^{M} s(i,j) \quad \text{and} \quad \Pr(\theta = i) = \sum_{j=0}^{J} s(i,j) \tag{5}$$

Figure 3 depicts the complementary cumulative distribution function (CDF) of the MS transmit power with $k = 15$ users in the cell and the observed user at a distance of $x = 2000\ m$ from the BTS. It can be seen that when the channel gets worse, a higher transmit power is required.

We have performed computations of the mean and standard deviation of the random variable S for varying parameters k and x. It could be seen that variations of the traffic load had almost no impact on the statistics of S. However, the distance plays an important role on the transmit power. Fig. 4 shows that there is a logarithmic relationship between the mean transmit power $E[S]$ and the distance due to path loss. The standard deviation decreases with growing distance, since the MS will be transmitting almost deterministically with maximum power at the cell boundaries.

4. IMPACTS OF POWER CONTROL ON OUTAGE

So far we have derived an expression for the transmit power of the MS at a distance x from the base station and k other users in the cell. We will now use this description to analyze the performance of the system by investigating the effects of these two parameters on the probability of outage.

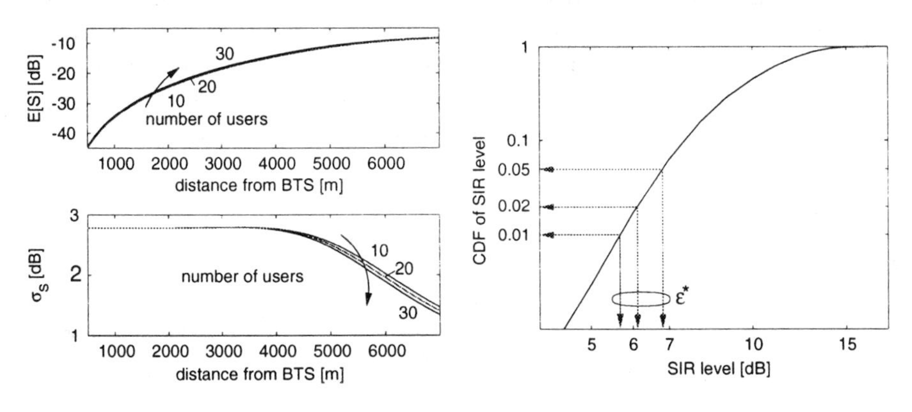

Figure 4 Mean and standard deviation of S

Figure 5 Cumulative SIR distribution

4.1 Outage Condition

The probability of outage is an important performance measure in a CDMA system [12, 13]. It is defined as the probability that the SIR of a user lies below a minimum requirement threshold $\epsilon^\star$ for a certain amount of time, usually a few frames. Since the time scale is such small (about 100 *ms*), it is assumed that the dynamics of the spatial user distribution can be neglected. The SIR is measured at the BTS, making outage an event occurring at the receiver, i.e., $P_{\text{out}} = \Pr(\epsilon < \epsilon^{\star})$.

In Section 3.2 we described the variable ϵ as the sum of S and C. With the MS transmit power in Eqn. (5) and C being Gaussian, both distributions are now known. The distribution of ϵ can be computed in this case by using the total probability law and conditioning the transmit power for a fixed channel gain.

The resulting SIR distribution has the shape of a Gaussian distribution, a fact which has also been confirmed by simulation and empirical results [14]. For our purpose of determining the outage probability, the CDF of the SIR is more useful. For a certain minimum required link quality given by $\epsilon^{\star}$, we can find the corresponding probabilities to be below this level. In the same way, we can give the minimal SIR requirement for a maximum allowable outage probability (Fig. 5).

Since the distribution of ϵ still has the two parameters x and k, we have investigated the dependence on these two values. Figure 6 shows the variations of the mean and standard deviation of ϵ as function of the distance. It can be seen that for distances less than 4 *km* the power control can keep the mean and standard deviation at nearly equal levels. With greater distance, however, the mean decreases due to path loss. The inability of the power control loops at this distance causes also that the standard deviation of the SIR increases at greater distances. Users at the cell boundaries will therefore experience a higher outage probability than users in the cell. Note also that up to about 4 *km* the load of the cell does not influence the distribution of SIR. Therefore, for an examination of outage probability the distribution of the users is only of minor importance.

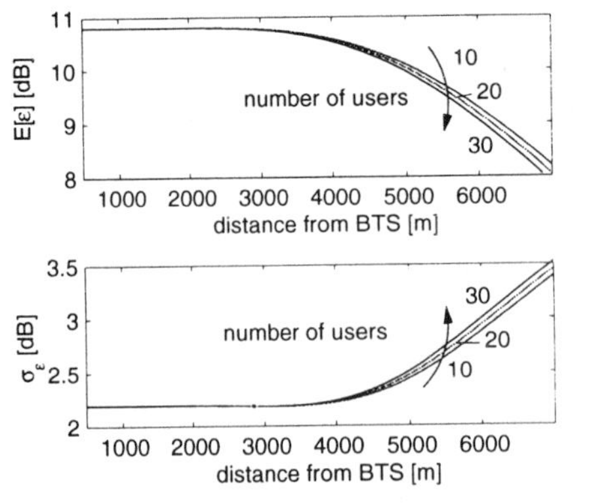

Figure 6 Mean and standard deviation of SIR

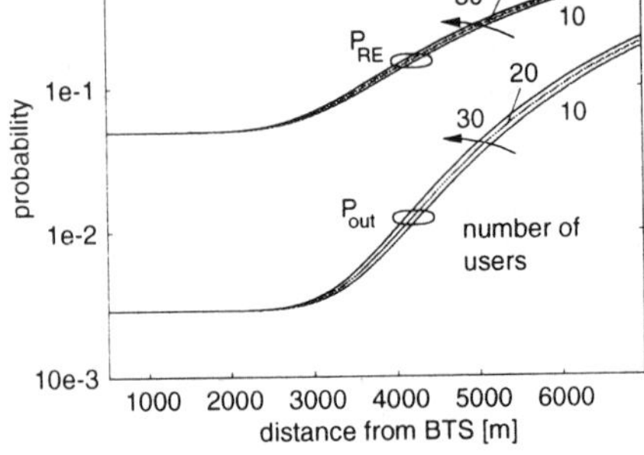

Figure 7 P_{out} versus P_{RE}

4.2 Probability of Exceeding the Transmit Power Range

Another term often used to describe outage probability is the probability of the MS exceeding its dynamic range since this will be the obvious cause for not being able to fulfill the SIR requirements [7, 12].

We obtain P_{RE} from our Markov chain model from the probability of power-up under the condition of transmitting at maximum power level J, i.e.,

$$P_{\mathrm{RE}} = \Pr\left(S > S_{\max}\right) = \sum_{i=0}^{M} \left(\alpha(i,J) + \delta(i,J) + \zeta(i,J)\right) \Pr\left(\theta = i\right).$$

Figure 7 shows both interpretations of outage probability. The curve computed for P_{out} is as expected almost independent of the distance x up to about 4 *km*. Then it increases due to the higher propagation loss. The threshold value in this case was selected as $\epsilon^{\star} = 6\ dB$ which corresponds to a maximum tolerable FER. Furthermore, we have also plotted the probabilities for exceeding the dynamic range of the mobile transmitter. It can be seen that for increasing cell load both curves have a similar shape. However, the range exceeding probability P_{RE} always overestimates P_{out} by a magnitude.

5. CONCLUSION AND OUTLOOK

In this paper we have presented a simple analytical model for the closed loop power control in a CDMA system. With our model, we obtain the distribution of the transmit power of the MS, as well as the SIR received at the base station from this mobile. We have shown that the power control algorithm currently implemented in the IS-95 system is robust enough to overcome the near-far problem and fluctuations in cell traffic. Furthermore, the often used relationship between the probability of exceeding the MS transmit power and outage probability cannot be confirmed. However, both probabilities are functions of the transmit power, which is dependent on the distance of the MS to the base station and only to some extent by the number of users in the cell.

We have also seen that the examination of outage is quite straightforward in the stationary case. However, when observing the temporal fluctuations of the channel by considering a correlated fading channel, the computation of the power control loops will become a challenging task. Some research has already been done on the outage characterization over time [5, 6]. In future work we will extend our model to include the temporal behavior of the channel, as well as examine power control in a multi-cell environment with soft-handoff. The performance analysis of CDMA systems becomes especially important, as future generation W-CDMA systems like UMTS will have a power control mechanism very similar to the one described in this paper.

Acknowledgments

The author would like to thank Prof. P. Tran-Gia and N. Gerlich (University of Würzburg) and J.E. Miller (NORTEL Wireless Networks, Richardson) for the valuable discussions during the course of this work. Part of this work has been supported by NORTEL External Research.

References

[1] Ariyavisitakul, S. and Chang, L.F. (1993) Signal and interference statistics of a CDMA System with feedback power control. *IEEE Trans. on Comm.*, vol. 41 nr. 11.

[2] Chockalingam, A., Dietrich, P., Milstein, L.B., and Rao, R.R. (1998) Performance of Closed-Loop Power Control in DS-CDMA Cellular Systems. *IEEE Trans. on Veh. Tech.*, vol. 47 nr. 3.

[3] D'Avella, R., Marizza, D., and Moreno, L. (1994) Power Control in CDMA Systems: Performance Evaluation and System Design Implications. *Proc. of IEEE ICUPC*, San Diego, CA.

[4] Leibnitz, K., Tran-Gia, P., and Miller, J.E. (1998) Analysis of the Dynamics of CDMA Reverse Link Power Control. *Proc. of IEEE GLOBECOM*, Sydney, Australia.

[5] Mandayam, N.B., Chen, P.-C., and Holtzman, J.M. (1998) Minimum Duration Outage for CDMA Cellular Systems: A Level Crossing Analysis. *Journal of Wireless Pers. Comm.*.

[6] Mukherjee, S. and Viswanathan, H. (1998) Minimum duration outages for diversity systems. *Proc. of IEEE GLOBECOM*, Sydney, Australia.

[7] Patel, P.R., Goni, U.S., Miller, E., and Carter, P.P.S. (1996) A Simple Analysis of CDMA Soft Handoff Gain and its Effect on the Cell's Coverage Area. In *Wireless Information Networks*, Kluwer Academic, Boston.

[8] Pichna, R., Wang, Q., and Bhargava, V.K. (1993) Non-Ideal Power Control in DS-CDMA Cellular. *Proc. of IEEE Pacific Rim Conf. on Comm., Computers and Signal Processing*.

[9] Rappaport, T.S. (1996) *Wireless Communications – Principles & Practice*. Prentice Hall, Upper Saddle River, NJ.

[10] Sampath, A., Kumar, P.S., and Holtzman, J.M. (1997) On setting reverse link target SIR in a CDMA system. *Proc. of IEEE Veh. Tech. Conf.*, Phoenix, AZ.

[11] TIA/EIA/IS–95 (1995) Mobile station – Base station compatability standard for dual-mode wideband spread spectrum cellular systems.

[12] Veeravalli, V.V., Sendonaris, A., and Jain, N. (1997) CDMA Coverage, Capacity and Pole Capacity. *Proc. of IEEE Veh. Tech. Conf.*, Phoenix, AZ.

[13] Viterbi, A.M. and Viterbi, A.J. (1993) Erlang capacity of a power controlled CDMA system. *IEEE Journal on Sel. Areas in Comm.*, vol. 11 nr. 6.

[14] Viterbi, A.J., Viterbi, A.M., and Zehavi, E. (1993) Performance of power–controlled wideband terrestrial digital communication. *IEEE Trans. on Comm.*, vol. 41 nr. 4.

[15] Yang, S.C. (1998) *CDMA RF System Engineering*. Artech House.

HOME AGENT REDUNDANCY AND LOAD BALANCING IN MOBILE IPV6

F. Heissenhuber, W. Fritsche, and A. Riedl
IABG, Einsteinstr. 20, 85521 Munich, Germany
{heissenhuber;fritsche}@iabg.de
Lehrstuhl für Kommunikationsnetze, TU München, Arcisstr. 21, 80290 Munich, Germany
Anton.Riedl@ei.tum.de

Abstract This work proposes an optional extension to Mobile IPv6 to enhance the performance and the reliability of home agent clusters. First, a load balancing mechanism is presented which allows a set of home agents to distribute the load equally among them. Furthermore, we specify a Home Agent Redundancy Extension which increases the resilience of home agent clusters. In case of home agent failures, this extension enables the remaining peers to take over the functions of the missing home agents and, thus, keep up reachability of all currently registered mobile nodes. The extension is completely transparent for mobile nodes. The two mechanisms work independently of each other and, therefore, can be implemented separately. For better performance, though, it is advisable to apply both mechanisms at the same time.

Keywords: Mobile IPv6, Redundancy, Load Sharing

1. INTRODUCTION

Mobile IP will be an integral part of the next generation Internet Protocol [1,2]. It can be expected that the importance of mobility in the Internet will increase. Commercial providers could offer some kind of "Mobility Service" which allows their customers to go online at any point across the Internet and at the same time be reachable through their permanent home address.

In Mobile IPv6 each node is always identified by its home address, regardless of its current point of attachment to the Internet. While a mobile node is away from home it sends information about its current location to a home agent on its home link. The home agent stores the information about the remote location in its Binding Cache and advertises its own link-layer address for the IP address of the mobile node. Packets which are addressed to the mobile node's home address, are then intercepted by the home agent and tunneled to the mobile node's present location. After a certain time, a Binding Cache entry is timed out unless the mobile node refreshes it by sending a new Binding Update.

Home agents are single points of failure. If the home agent of a mobile node crashes, all communication to this mobile node which is routed over the home agent is disrupted. In Mobile IPv6, a mobile node discovers the loss of its home agent by sending registration messages periodically. When the mobile node does not receive acknowledgements, it performs Dynamic Home Agent Address Discovery to find a different suitable home agent. However, the time interval between registration requests is often set to be fairly large to avoid frequent traffic across the Internet. Thus, the most obvious way of detecting home agent failures may be too slow.

Although Mobile IPv6 allows several home agents on one link, it does not specify a redundancy mechanism in case one home agent fails. Our paper addresses this problem and proposes a Home Agent Redundancy Extension to enhance resilience. Furthermore, we describe a mechanism which enables load balancing among several home agents on one link. This becomes necessary when the number of Mobile IP users increases and the load cannot be handled by a single home agent anymore.

2. HOME AGENT LOAD BALANCING MECHANISM

In this work, we define load as the number of mobile nodes which are registered with a home agent. This load value is simple to determine and gives at the same time implicit information about some other load factors like the number of packets forwarded per time unit.

The load balancing mechanism utilizes Dynamic Home Agent Address Discovery which is initiated by a mobile host to find out the address of its home agent. When a mobile host is away from home it sends a Binding Update to the Home-Agents anycast address [3] of its home link. In response, one home agent on this link returns a list of all available home agents which are listed in order of decreasing preference value. The mobile node selects one of the listed home agents and resends a Binding Update to this specific home agent. This completes the registration procedure.

Dynamic Home Agent Address Discovery requires every home agent to know the preference values of all other home agents. Therefore, the preference values are locally advertised by adding Home Agent Information options to Router Advertisements which are sent periodically. Preference values are specified as 16-bit signed, two-complement integers. As there is no standardized procedure to determine the Home Agent Preference value, we correlate it to the number of current registrations. Most preferable (32767) means 0 registrations. For every mobile host which is registered with a home agent, the preference value is decremented by one. Thus, the value -32768 (least preferable) represents 65535 bindings and more.

To realize load balancing, it is necessary that mobile nodes are not configured with a fixed IP address of their home agent. Instead, they must use Dynamic Home Agent Address Discovery to determine their home agent. The mobile node, upon receiving the Home Agents List, must try to register with each home agent in this list in the given order. Thus, it sends the first Binding Update to the home agent with currently the least number of registrations (= most preferable home agent). Normally, this home agent accepts the Binding Update, returns a Binding Acknowledgement, and decreases its preference value. Accordingly, when an entry in the Binding Cache of a home agent expires, a home agent increases its preference value.

Within one Router Advertisement period (0.5 to 1.5 seconds) the preference values in the Home Agents Lists of all home agents are updated, and a different home agent may become the most preferable one. Assuming a large number of mobile hosts, the load balancing mechanism should assure that all home agents have approximately the same number of registrations.

3. HOME AGENT REDUNDANCY EXTENSION

3.1 Definitions and New Messages

Rank: Every active home agent of a cluster is uniquely identified by its rank which is determined by the home agent's link-local address. The home agent with the lowest link-local address has rank 0. Higher link-local addresses result in higher ranks. When determining the rank at a certain time, only active home agents are considered.

All-home-agents multicast address: This is a multicast address which is assigned to all home agents using the Home Agent Redundancy Extension.

Home Agent Sequence Number: A home agent incrementally assigns a Home Agent (HA) Sequence Number to every accepted binding and forwards this number together with the binding information to all other home agents.

Extended Home Agents List: Every home agent maintains an Extended Home Agents List. It contains, among other information, a copy of the Binding Caches of all home agents which are compliant with the redundancy extension. The list is an extended version of the Home Agents List as described in [2].

Binding Update Forward Message: The Binding Update Forward message is used by a home agent to forward registration information about mobile nodes to the other home agents. It contains the Home Agent (HA) Sequence Number associated with the specific binding. The HA Sequence Number is used by home agents that receive a Binding Update Forward to detect missing Binding Update Forward messages.

Binding Cache Request Message: The Binding Cache Request message is used by a home agent to request the Binding Cache entries of another home agent.

Binding Cache Reply Message: The Binding Cache Reply message is sent in response to a Binding Cache Request. It contains the Binding Cache entries of home agents.

Home Agent Information Option: The Home Agent Information option is the same as specified in [2]. It is included in the Router Advertisements of a home agent. The Home Agent Redundancy Extension defines one additional bit, the Redundancy (R) bit. This bit is set by the sending home agent indicating that this home agent is compliant with the Home Agent Redundancy Extension.

3.2 Protocol Description

3.2.1 Redundancy Mechanism

When a home agent receives a Binding Update from a mobile node requesting home registration, the home agent processes it as usual. If it decides to reject the request, it sends a Binding Acknowledgement with the status field set accordingly. No further action takes place.

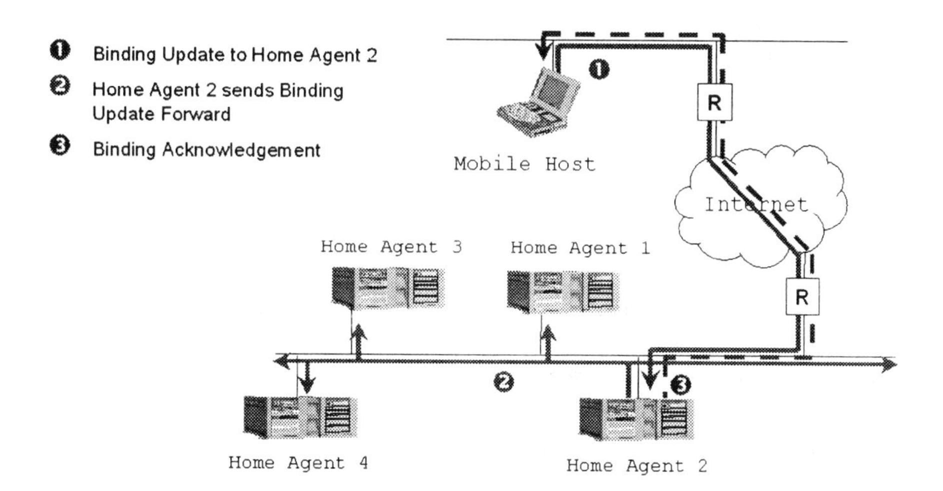

Figure 1. Forwarding Binding Updates

If, on the other hand, the home agent accepts the Binding Update, it stores information about the mobile node in its own Binding Cache and sends a Binding Update Forward message to the all-home-agents multicast address. This message contains the current address of the mobile host and a sequence number (HA Sequence Number) which is consecutively assigned to every accepted binding. Furthermore, the Update Forward Option carries a Lifetime and a Refresh field whose values are identical to the same fields in the respective Binding Acknowledgement. They indicate how long a binding will be valid until it has to be updated by the mobile host. All home registrations of a home agent are forwarded to the other home agents in the same way. Finally, to conclude the registration process the home agent sends a Binding Acknowledgement to the mobile node indicating successful registration (*Figure 1*). Sending a Binding Update Forward message for every mobile host registration enables all home agents to set up their Extended Home Agents Lists appropriately.

When a home agent receives a Binding Update Forward message, it first checks the HA Sequence Number whether it equals the expected value. If so, it searches all Binding Caches (also its own) in the Extended Home Agents List for an entry with the same home address as the new binding. If necessary, it deletes the old entry and adds the new binding to the Binding Cache of the home agent which has sent the Binding Update Forward message. It also stores the HA Sequence Number with this binding and starts a timer with the value that was given in the Lifetime field. If, however, an out-of-order HA Sequence Number indicates that one ore more Binding Update Forward messages have not been received, the home agent tries to synchronize itself with the owner of the Binding Cache. Synchronizing the Extended Home Agents Lists at this point is similar to the Recovery Mechanism described in 3.2.3.

To reduce the traffic load on the home link, Binding Update Forward messages as well as Binding Cache Requests and Replies are sent by multicast. Since inconsistencies usually arise when Binding Update Forward messages are lost or the home link is divided, it is very likely that other home agents notice the same inconsistency. Therefore, Binding Cache Requests are sent to the all-home-agents multicast address. Moreover, the sending of this request is delayed by each home agent depending on its current rank. If a home agent receives a multicast Binding Cache Request while delaying the same request, it does not have to send this request itself. It just has to wait for the incoming multicast Binding Cache Reply. Using multicasting in this situation will reduce the number of packets on a large scale.

It should be noted that the entries of all Binding Caches are timed out independently. The deletion of a registration is not synchronized among the home agents. However, Binding Cache entries should time out at nearly the same time because multicast Binding Update Forward messages are received by all home agents almost simultaneously. Hence, the timer for the Binding Update is also started at almost the same time. Considering that the lifetime of a binding is assumed to be greater than one hour, the differences between the timers of different home agents for the same bindings are negligible.

3.2.2 Monitoring and Proxy Mechanism

A home agent which implements the Home Agent Redundancy Extension does not have to be configured with the addresses or names of the other home agents on a link. Every home agent listens to the Router Advertisements of all other home agents and uses this information to maintain the Home Agents List. A Router Advertisement which contains a Home Agent Information Option with the Redundancy (R) bit set indicates that the sender is compliant with the Home Agent Redundancy Extension. Thus, every home agent is able to set up the Extended Home Agents List accordingly.

For monitoring the state of the peer home agents, no special messages are required. The monitoring mechanism is based on the periodic, unsolicited Router Advertisements. Every time a home agent receives a Router Advertisement of a peer home agent, it resets a timer. When a home agent fails to receive Router Advertisements from an already known peer for a certain amount of time (e.g., three times the Router Advertisement interval), it assumes that this home agent has been shut down or has crashed.

In case of a home agent failure, the remaining home agents act as proxies. They distribute the bindings of the respective Binding Cache among each other and take over the task of forwarding packets addressed to mobile hosts on behalf of the failed home agent. To determine which bindings are assigned to which home agent, every home agent performs a simple modulo computation for each entry in the Binding Cache. It is important to note that the home agents perform the calculation independently and that no additional messages are required. The calculation is based on the HA Sequence Number S which is stored with the binding and the number N of remaining active home agents. First, a home agent determines its current rank. Then, it computes $R = S \bmod N$ for every binding and assumes responsibility for this binding if R equals its rank.

The splitting of the bindings has been designed to be non-critical for unsynchronized expiration of bindings. Since the HA Sequence Numbers are unique within one Binding Cache, the modulo mechanism together with the rank of the home agents, guarantees an implicit and unequivocal distribution method. Furthermore, the monitoring mechanism ensures - even if several home agents fail - that every home agent is able to determine the correct rank of all home agents when it is necessary to do so.

3.2.3 Recovery Mechanism

When a home agent comes up, it immediately starts sending Router Advertisements without having set the Home Agent (H) bit. It starts a timer and listens for Router Advertisements which have the Redundancy (R) bit in the Home Agent Information Option set. If the timer expires and the home agent has not received such a Router Advertisement, it assumes that there is no other home agent (which is compliant with the Home Agent Redundancy Extension) on this link and starts without previous bindings.

However, if the home agent receives Router Advertisements which have the Redundancy (R) bit in the Home Agent Information Option set, it tries to synchronize its Binding Cache. To do so, it sends a Binding Cache Request message (*Figure* 2). Binding Cache synchronization can be done with any home agent since all active home agents know the Binding Caches of all other home agents. If a home agent does not reply or if transmission fails, the home agent may choose to try it again or synchronize with a different home agent.

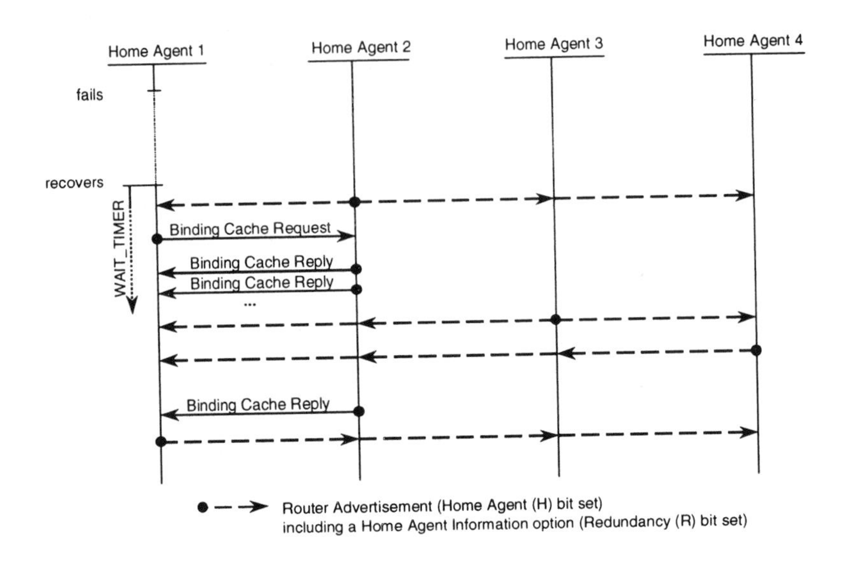

Figure 2. Data flow diagram: A home agent recovers

In case a home agent recovers after a crash, it also receives its own previous Binding Cache. For every entry of this Binding Cache, it immediately sends a Neighbor Advertisement to advertise its own link-layer address for the IP address of that mobile node.

When the home agent has completed the synchronization procedure, it sets the Home Agent (H) bit in the Router Advertisements and the Redundancy (R) bit in the Home Agent Information option. From now on, it is ready to accept registration requests and act as a proxy if another home agent fails.

3.3 Combination with Load Balancing

The Home Agent Redundancy Extension and the load balancing mechanism (Chapter 2) are independent of each other. It is very advisable, though, to use both mechanisms in parallel.

The load balancing mechanism allows the home agents to roughly share the load while doing regular operation. In case of failures, the Home Agent Redundancy Extension ensures that the remaining home agents take over the task of forwarding packets.

Assuming a large number of mobile hosts, the distribution mechanism of the Home Agent Redundancy Extension should split the additional load equally among the remaining home agents. However, an unfavorable constellation could result in an unequal distribution of the bindings. In this case, the load balancing mechanism would restore a well-balanced situation.

3.4 Mobile Host Considerations

One of the main goals of this work is to require as little changes to the Mobile IPv6 protocol as possible. The Home Agent Redundancy Extension affects only the home agent's behavior and is completely transparent for mobile nodes. There is no need to make any changes to mobile nodes. Any Mobile IPv6 implementation for mobile nodes that is compliant with [1] is able to use the Home Agent Redundancy Extension without changes.

3.5 Bandwidth and Storage Requirements

Table 1 shows the messages which are used for redundancy and monitoring purpose. For every Binding Update which is accepted by a home agent, a Binding Forward and a Binding Acknowledgement has to be sent. Compared to standard Mobile IPv6 (only Binding Update + Binding Ack), the redundancy extension generates an additional load of 96 bytes (Binding Forward) for every Binding Update process.

Table 1. IP Packet Size

Message Type	IP packet size in bytes
Binding Update	104
Binding Acknowledgement	80
Binding Forward	96
Router Advertisement	96

However, the redundancy mechanism allows reducing the Binding Update frequency significantly. Home agent failures are remedied within three times the Router Advertisement interval (between 0.5 and 1.5 seconds by standard) independently of the Binding Update interval. To achieve similar fault tolerance in standard Mobile IPv6, Binding Updates would have to be sent at least every 4.5 seconds. Assuming a scenario with 10000 mobile hosts, this would create a load of about 3 Mbps across the Internet and on the local link. Choosing a Binding Update frequency of 1 hour could reduce this load to about 4 kbps across the Internet and 6kbps on the local link.

In case of a home agent failure, no extra load is created. Only when a home agent comes up after a failure, the redundancy extension generates additional messages (Binding Cache Request + Binding Cache Replies). The respective load which only appears on the local link depends on the number of registered mobile hosts.

The storage requirement when using the redundancy extension is somewhat higher than without the extension. A home agent needs to store information about every mobile host which is registered with any of the other home agents. This is not necessary in regular Mobile IPv6. However, the higher storage demand does not affect a home agent since a binding entry requires only little space (52 bytes with and 42 bytes without redundancy extension).

4. CONCLUSION

As Mobile IPv6 is implemented in commercial products, load balancing and fault tolerance become essential features of home agents. The current Mobile IPv6 specification leaves the employment of adequate mechanisms to the producers of network equipment. Therefore, we have presented two optional extensions to Mobile IPv6 which achieve load balancing and home agent redundancy, thus, enhancing the resilience of home agent clusters to a great extent. The two mechanisms are mainly based on already existing features of Mobile IPv6 and can be implemented without great effort. Only a few additional functions and messages are required. Although independent of each other, the complementary nature of the two mechanisms suggests their concurrent employment.

References

[1] S. Deering and R. Hinden, *Internet Protocol, Version 6 (IPv6) Specification*, RFC 2460, 1998

[2] David B. Johnson and Charles Perkins, *Mobility Support in IPv6*, Internet Draft (work in progress), draft-ietf-mobileip-ipv6-08.txt ,1999

[3] D. Johnson and S. Deering, *Reserved IPv6 Subnet Anycast Addresses*, RFC 2526, 1999

VERIFICATION AND ANALYSIS OF AN IMPROVED AUTHENTICATION PROTOCOL FOR MOBILE IP

Qing Gao
Dept of Electrical Engineering
Faculty of Engineering
National University of Singapore
10 Kent Ridge Crescent
Singapore 119260
engp8682@nus.edu.sg

Winston Seah, Anthony Lo, Kin-Mun Lye
Centre for Wireless Communications[1]
National University of Singapore
20 Science Park Road
#02-34/37 TeleTech Park
Singapore Science Park II
Singapore 117674

Abstract In this paper, we proposed an authentication protocol for Mobile IP to improve its scalability and efficiency. The protocol was verified and analysed with BAN logic. We discovered that there is a tradeoff between security and efficiency. To improve the efficiency, we omitted the direct authentication between the Mobile Host (MH) and the Foreign Agent (FA) which resulted in weaker beliefs. However, through discussion, we were able to show that the result does not compromise the security of the protocol.

Keywords: Mobile IP, Scalable Tripartite Authentication, public-key cryptographic mechanism, secret-key cryptographic mechanism, BAN logic.

1. INTRODUCTION

In previous work [2], an improved authentication protocol for Mobile IP [5, 7] which is based on both the secret-key and the public-key cryptographic mechanism has been proposed. The protocol realizes a scalable tripartite authentication among the Mobile Host (MH), the Foreign Agent (FA) and the Home Agent (HA). Meanwhile, it improves the efficiency as compared to existing scalable protocols. In order to check the security of the protocol, we adopted the widely used BAN Logic [1] to verify the proposed authentication protocol. In the verification process, we found that the result was weaker than the other tripartite authentication schemes based on public key methods [4] or

[1]CWC is a national R & D Centre funded by the Singapore National Science and Technology Board.

shared key methods [6]. However, we were able to show that the result does not compromise the security of the protocol, and furthermore, our proposed protocol is definitely more computationally efficient and scalable.

This paper is organized as follows. Section 2 presents the improved Mobile IP registration authentication protocol. Section 3 briefly presents the authentication verification logic, BAN logic. Section 4 describes the verification procedure of the proposed authentication protocol. Section 5 analyses the plausible security problem incurred because of the lack of direct MH-FA authentication. Section 6 concludes the paper.

2. IMPROVED AUTHENTICATION PROTOCOL FOR MOBILE IP

In [2], we proposed an improved authentication protocol for Mobile IP, as shown in Figure 1, which realizes the tripartite authentication among the MH, the FA and the HA. It improves the security and scalability of Mobile IP but at the same time minimizes computational overheads and re-registration delay as much as possible. In the proposal, we assume that both the MH and the HA belong to the same administrative domain, and the number of MHs an HA handles is not too large. Therefore, the symmetric key mechanism as in the standard Mobile IP can be used in the MH-HA authentication in order to take advantage of its computational efficiency and simplicity. As the MH could travel across a considerably large area, neither the HA nor the MH would know how many FAs they have to authenticate. Hence, for scalability reason, we use the public-key cryptographic mechanism for the authentication between the FA and the HA. The MH-FA pair is implicitly authenticated without compromising the security when the authentication between the MH-HA and FA-HA is successfully performed. Also, this minimizes the re-registration delay and further improves the efficiency of the protocol.

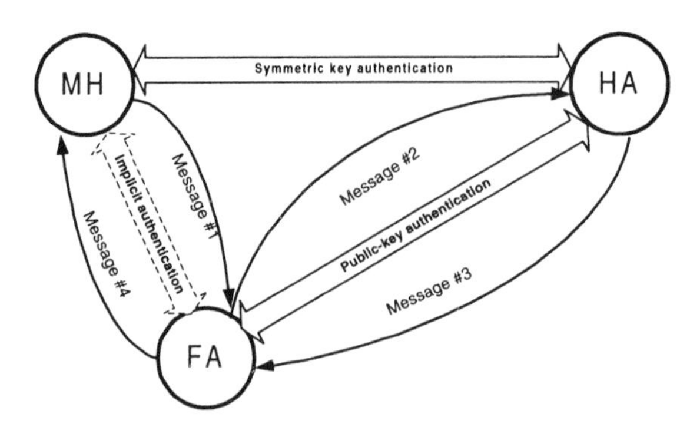

Figure 1 Tripartite two-way authentication protocol

The steps in the improved authentication protocol are explained below and Table 1 defines the terminologies used in the description.

Table 1: Definitions of the nomenclture used in the authentication protocol[a]

Nomenclature	Definition
E(X, Y)	the encryption of *Y* under key *X*
Auth(X)	the message digest function value on contents *X*
Priv_FA	the private key of foreign agent
Pub_FA	the public key of foreign agent
Priv_HA	the private key of home agent
Pub_HA	the public key of home agent
Sig(X, Y)	the signature of *Y* with key *X* where *Sig(X, Y) = E(X, MD(Y))*.
Cert_FA	certificate of foreign agent
Cert_HA	certificate of home agent
Pub_CA	public key of the certification authority

[a] A certificate is a data structure that binds an entity's identity to a public key with a CA's digital signature.

Step 1: When an MH detects that it is away from home, it initiates a registration procedure by sending Message #1 (Figure 2) to the FA.

Registration Request	ID	Auth(Registration Request, ID, K)

Figure 2 Message #1 MH → FA

K is the secret key shared between the MH and the HA. ID is set as the timestamp or a nonce as described in the basic Mobile IP protocol [5] to perform replay protection. Auth(Registration Request, ID, K) is the message digest function value on contents (Registration Request, ID, K).

Step 2: When the FA receives Message #1, it adds its certificate (Cert_FA) to the message, then signs it and creates Message #2 (Figure 3). Then the FA relays Message #2 to the HA. *(Note: The unshaded part is the standard Mobile IP registration message expunging the authentication extension. The shaded part contains the fields that differ from the standard protocol, including the new authentication extension.)*

Registration Request	ID	Auth(Registration Request, ID, K)	Cert_FA	Sig{Priv_FA, (Registration Request, ID, Auth(Registration Request, ID, K), Cert_FA)}

Figure 3 Message #2 FA → HA

Step 3: When the HA receives Message #2, its actions are shown in Figure 4.

Then the HA sends Message #3 (Figure 5) to the FA. It indicates whether the MH's registration is accepted or refused as well as the refusal reason. Note that, in Message #3, HA adds the FA identifier, FA_ID, into the message indicating that it trusts FA.

```
HAAction()
{
  Validate CA's signature on Cert_FA;
  if Cert_FA is invalid then return with a code indicating invalid Pub_FA;
    else Validate FA's signature with Pub_FA;
    if FA's signature is invalid then return with a code indicating invalid FA;
      else  Check ID;
      if ID is invalid then return with a code indicating timeout or replay attack;
        else Check the authenticator
        if Authenticator is invalid then return with a code indicating invalid MH;
          else accepts Registration Request;
}
```

Figure 4 Actions of the HA when receives Message #2

Registration Reply	ID	FA_ID	Auth(Registration Reply, ID, FA_ID, K)	Cert_HA	Sig{Priv_HA, (Registration Reply, ID, FA_ID Auth(Registration Reply, ID, FA_ID, K), Cert_HA)}

Figure 5 Message #3 HA → FA

Step 4: When the FA receives Message #3, its actions are shown in Figure 6.

```
FAAction()
{
  Validate CA's sigature on Cert_HA;
   if Cert_HA is invalid then return with a code indicating invalid Pub_HA;
    else Validate HA's signature with Pub_HA;
    if HA's signature is invalidthen return with a code indicating invalid HA;
      else Forward  Message #4 to the MH;
}
```

Figure 6 Actions of the FA when receives Message #3

If HA is valid, no matter what the contents in the Registration Reply is, FA will relay the first half of Message #3, i.e., Message #4 (Figure 7).

Registration Reply	ID	FA_ID	Auth(Registration Reply, ID, FA_ID, K)

Figure 7 Message #4 FA → MH

Step 5: When the MH receives Message #4, its actions are shown in Figure 8.

3. OUTLINE OF THE VERIFICATION METHODOLOGY BASED ON BAN LOGIC

3.1 BAN LOGIC

BAN logic [1] is a widely used logic specifically designed for analyzing the protocols for the authentication of principals in distributed computing systems.

MHAction()
{
Verify the authenticator;
if the authenticator is invalid **then** MH-HA authentication failed, return;
else Check if the FA is trusted by the HA;
if the FA is not trusted by the HA **then** invalid FA, return;
else successful authentication;
}

Figure 8 Actions of the MH when receives Message #4

We use this logic to verify the improved authentication protocol for Mobile IP [2].

To use BAN logic for verification, one has to formalize the protocol with the BAN notations. The reason is that the conventional notations for describing an authentication protocol is not convenient for manipulation in a logic since the contents of each message often means more than necessary in the security sense. With the BAN notations, one can transform each message into a logical formula which is an idealized version of the original message. The notations shown in Table 2 will be used throughout the paper.

BAN logic uses some logical postulates to annotate an idealized protocol with assertions. An assertion usually describes beliefs held by the principals at the point in the protocol where the assertion is inserted. An annotation consists of a sequence of assertions inserted before the first statement and after each statement. To express that the statement Z follows from a conjunction of statements X and Y, we write: $\frac{X,Y}{Z}$. The postulates are listed as follows:

- The *message-meaning* rules:

  For shared keys:

  Rule 1: $\frac{P|\equiv P \overset{K}{\leftrightarrow} Q,\, P \triangleleft \{X\}_K}{P|\equiv Q|\sim X}$ *(That is, **if** P believes that the key K is shared with Q and sees a message X encrypted under K, **then** P believes that Q once sent X.)*

  Rule 2: $\frac{P|\equiv Q \overset{K}{\rightleftharpoons} Q,\, P \triangleleft \langle X\rangle_Y}{P|\equiv Q|\sim X}$ *(That is, **if** P believes that the secret Y is shared with Q and sees $\langle X\rangle_Y$, **then** P believes that Q once sent X.)*

  For public keys:

  $\frac{P|\equiv \overset{K}{\mapsto} Q,\, P \triangleleft \{X\}_{K^{-1}}}{P|\equiv Q|\sim X}$ *(That is, **if** P believes that the key K is the public key of Q and sees a message X encrypted under Q 's private key K^{-1}, **then** P believes that Q once sent X.)*

- The *nonce-verification* rule:

  $\frac{P|\equiv \#(X),\, P|\equiv Q|\sim X}{P|\equiv Q|\equiv X}$ *(That is, **if** P believes that X could have been uttered only recently and that Q once sent X, **then** P believes that Q believes X.)*

- The *jurisdiction* rule:

Table 2: Notations in BAN logic

Notation	*Definition*
A, B, S	specific principals
K_{ab}, K_{as}, K_{bs}	specific shared keys
K_a, K_b, K_s	specific public keys
N_a, N_b	specific statements
P, Q	general principals
X, Y	general statements
K	general encryption keys
$P \mid\equiv X$	P is persuaded of the truth of X
$P \triangleleft X$	P receives a message containing X
$P \mid\sim X$	P is known to have sent a message containing X
$P \Rightarrow X$	*P has jurisdiction over X* ,or P is trusted as an authority on X
$\#(X)$	X has not been sent in a message belonging to a previous run of the protocol
$P \stackrel{K}{\leftrightarrow} Q$	P and Q have a *shared key* K
$\stackrel{K}{\mapsto} P$	P has K as a *public key* and K^{-1} as a *private key*
$P \stackrel{X}{\rightleftharpoons} Q$	X is a *secret* known only to P and Q
$\{X\}_K$	X is encrypted under the *key* K
$\langle X \rangle_Y$	X is *combined with* Y; Y is a *secret* whose presence proves the identity of whoever utters $\langle X \rangle_Y$

$\frac{P|\equiv Q \Rightarrow X,\, P|\equiv Q|\equiv X}{P|\equiv X}$ *(That is, **if** P believes that Q has jurisdiction over X and that P believes that Q believes X, **then** P trusts Q on the truth of X.)*

The assertions are expressed in the same notations used to write messages. Moreover, annotations can be concatenated and new assertions can be derived from established ones. For example, if the X is an assertion in a legal annotation A, if X' is provable from X, and if A' is the result of substituting X' for X in A, then A' is a legal annotation.

The desired goals of an authentication protocol can be specified in BAN logic. Usually, for secret key authentication protocols, we might deem that an authentication is complete between A and B if there is a K such that:

$A \mid\equiv A \stackrel{K}{\leftrightarrow} B$ *(A believes that A and B has a shared key K .)*

$B \mid\equiv A \stackrel{K}{\leftrightarrow} B$ *(B believes that A and B has a shared key K .)*

Or in addition:

$A \mid\equiv B \mid\equiv A \stackrel{K}{\leftrightarrow} B$ *(A believes that B believes that A and B has a shared key K .)*

$B \mid\equiv A \mid\equiv A \overset{K}{\leftrightarrow} B$ *(B believes that A believes that A and B has a shared key K .)*
Or even weaker goals.

For public-key authentication protocols, the goal may be:
$A \mid\equiv \overset{K}{\mapsto} B$ *(A believes that B has K as a public key.)*
Or in addition: $A \mid\equiv A \overset{N_a}{\rightleftharpoons} B$*(A believes that A and B has a secret N_a.)*

3.2 THE VERIFICATION METHODOLOGY

The verification methodology is outlined as follows:

1. Define the original authentication protocol with the conventional notations.

2. Formalize the authentication goals using BAN logic notations.

3. Derive the idealized protocol from the original one. This generally involves simplifying the protocol by omitting the parts of the message that do not contribute to the beliefs of the recipient. The idealized protocols do not include cleartext message parts because it can be easily forged.

4. Formalize the assumptions about the initial state with BAN logic notations.

5. Verify the protocol. This includes attaching the logical formulas to the statements of the protocol as assertions about the state of the system and applying the logical postulates to the assumptions and the assertions until the beliefs held by the parties in the protocol is deduced.

4. VERIFICATION OF THE IMPROVED AUTHENTICATION PROTOCOL USING BAN LOGIC

4.1 ORIGINAL PROTOCOL

The following notations are defined to be used in the verification procedure.
S: CA A: MH B: FA C: HA
X_a: Registration Request produced by A(MH)
X_c: Registration Reply produced by C(HA)
T_a: $ID\_RegRequest$ *(ID in Figure 2)* T_c: $ID\_RegReply$ *(ID in Figure 5)*
K_{ac}: the key shared between MH and HA *(K in Figure 2,3,5,7)*
K_b, K_c, K_s: the respective public keys of FA, HA and CA
K_b^{-1}, K_c^{-1}, K_s^{-1}: the respective private keys of FA, HA and CA

The messages in the improved authentication protocol as described in Section 2 are defined in the conventional notations as follows:

Message #1. $A \longrightarrow B$: $\{X_a, T_a, \{X_a, T_a, K_{ac}\}_{K_{ac}}\}$

(See also Figure 2, $\{X_a, T_a, K_{ac}\}_{K_{ac}}$ corresponds to Auth(Registration Request, ID, K).)

Message #2. $B \longrightarrow C$: $\left\{ \begin{array}{c} \{X_a, T_a, \{X_a, T_a, K_{ac}\}_{K_{ac}}\}, \{B, K_b\}_{K_s^{-1}}, \\ \left\{\{X_a, T_a, \{X_a, T_a, K_{ac}\}_{K_{ac}}\}, \{B, K_b\}_{K_s^{-1}},\right\}_{K_b^{-1}} \end{array} \right\}$

(See also Figure 3, $\{B, K_b\}_{K_s^{-1}}$ corresponds to Cert_FA, the last term corresponds to FA's signature.)

Message #3. $C \longrightarrow B$: $\left\{ \begin{array}{c} X_c, T_c, B, \{X_c, T_c, B, K_{ac}\}_{K_{ac}}, \{C, K_c\}_{K_s^{-1}}, \\ \left\{\{X_c, T_c, \{X_a, T_c, B, K_{ac}\}_{K_{ac}}\}, \{C, K_c\}_{K_s^{-1}},\right\}_{K_c^{-1}} \end{array} \right\}$

(See also Figure 5, $\{X_c, T_c, B, K_{ac}\}_{K_{ac}}$ corresponds to Auth(Registration Reply, ID, FA_ID, K), $\{C, K_c\}_{K_s^{-1}}$ corresponds to Cert_HA, the last term corresponds to HA's signature.)

Message #4. $B \longrightarrow A$: $\{X_c, T_c, B, \{X_c, T_c, B, K_{ac}\}_{K_{ac}}\}$

(See also Figure 6.)

4.2 AUTHENTICATION GOALS

The goals that need to be achieved by the tripartite authentication protocol can be deduced from [1, 3, 4, 5]. Through authentication, the HA would know whether the Registration Request message it received is really produced by the MH if it can derive that the MH has the right key shared with itself and the Registration Request is freshly operated by the very key. Similarly, the MH would know whether the Registration Reply message it received is really produced by the HA if it can derive that the HA has the right key shared with itself and the Registration Reply is freshly operated by the very key. Moreover, the HA would know whether the FA is the authentic one by testing its certificate and its signature. It is the same with the FA. Finally, the MH would also know whether the FA is the authentic one and the FA would know whether the MH is a legal host. The goals are listed as follows with BAN logic notations:

(1) $A \mid\equiv C \mid\equiv A \overset{K_{ac}}{\leftrightarrow} C$ *(A believes that C believes the shared key K_{ac} between A and C)*

(2) $A \mid\equiv C \mid\equiv X_c$ *(A believes that C really sends the timely X_c)*

(3) $C \mid\equiv A \mid\equiv A \overset{K_{ac}}{\leftrightarrow} C$ *(C believes that A believes the shared key K_{ac} between A and C)*

(4) $C \mid\equiv A \mid\equiv X_a$ *(C believes that A really sends the timely X_a)*

(5) $B \mid\equiv \overset{K_c}{\mapsto} C$ *(B believes that K_c is the public key of C)*

(6) $B \mid\equiv C \mid\equiv \left\{X_c, T_c, B, A \overset{K_{ac}}{\leftrightarrow} C\right\}_{K_{ac}}$ *(B believes that C really sends the timely $\left\{X_c, T_c, B, A \overset{K_{ac}}{\leftrightarrow} C\right\}_{K_{ac}}$)*

(7) $C \mid\equiv \overset{K_b}{\mapsto} B$ *(C believes that K_b is the public key of B)*

(8) $C \mid\equiv B \mid\equiv \left\{X_a, T_a, A \overset{K_{ac}}{\leftrightarrow} C\right\}_{K_{ac}}$ *(C believes that B really sends the timely* $\left\{X_a, T_a, A \overset{K_{ac}}{\leftrightarrow} C\right\}_{K_{ac}}$ *)*

(9) $A \mid\equiv B$ *(A believes the identity of B)*

(10) $B \mid\equiv A$ *(B believes the identity of A)*

4.3 IDEALIZED PROTOCOL

The original protocol messages are transformed into the idealized form using BAN logic notations.

Message #1. *(Cleartext, omitted)*

Message #2. $B \longrightarrow C$:

$$\left\{X_a, T_a, A \overset{K_{ac}}{\leftrightarrow} C\right\}_{K_{ac}}, \{B, K_b\}_{K_s^{-1}}, \left\{ \begin{array}{c} \left\{X_a, T_a, \left\{X_a, T_a, A \overset{K_{ac}}{\leftrightarrow} C\right\}_{K_{ac}}\right\}, \\ \{B, K_b\}_{K_s^{-1}} \end{array} \right\}_{K_b^{-1}}$$

Message #3. $C \longrightarrow B$:

$$\left\{X_c, T_c, B, A \overset{K_{ac}}{\leftrightarrow} C\right\}_{K_{ac}}, \{C, K_c\}_{K_s^{-1}}, \left\{ \begin{array}{c} \left\{X_c, T_c, \left\{X_c, T_c, B, A \overset{K_{ac}}{\leftrightarrow} C\right\}_{K_{ac}}\right\}, \\ \{C, K_c\}_{K_s^{-1}} \end{array} \right\}_{K_c^{-1}}$$

Message #4. $B \longrightarrow A$: $\left\{X_c, T_c, B, A \overset{K_{ac}}{\leftrightarrow} C\right\}_{K_{ac}}$

4.4 ASSUMPTIONS

We formalize the assumptions for the improved authentication protocol with BAN logic notations as in Table 3:

Table 3: Improved Authentication Protocol Assumptions

A believes:	*B believes:*	*C believes:*
$A \overset{K_{ac}}{\leftrightarrow} C$ (1)	$\overset{K_b}{\mapsto} B$ (2)	$\overset{K_c}{\mapsto} C$ (3)
$\#(T_a)$ (4)	$\#(T_a)$ (5)	$\#(T_a)$ (6)
$\#(T_c)$ (7)	$\#(T_c)$ (8)	$\#(T_c)$ (9)
	$\overset{K_s}{\mapsto} S$ (10)	$\overset{K_s}{\mapsto} S$ (11)
	$S \Rightarrow \overset{K_c}{\mapsto} C$ (12)	$S \Rightarrow \overset{K_b}{\mapsto} B$ (13)
	$S \mid\equiv \overset{K_c}{\mapsto} C$ (14)	$S \mid\equiv \overset{K_b}{\mapsto} B$ (15)
		$A \overset{K_{ac}}{\leftrightarrow} C$ (16)

4.5 VERIFICATION

According to the rule of jurisdiction, with assumption (13) and (15) listed in Table 3, we get: $\frac{C|\equiv(S\Rightarrow\overset{K_b}{\mapsto}B),C|\equiv(S|\equiv\overset{K_b}{\mapsto}B)}{C|\equiv\overset{K_b}{\mapsto}B}$ .... *(Accords to goal 7)*

Thus, by the message meaning rule for public keys and message #2, we get:

$$\frac{C|\equiv\overset{K_b}{\mapsto}B,\ C\triangleleft\{\{X_a,T_a,A\overset{K_{ac}}{\leftrightarrow}C\}_{K_{ac}}\}_{K_b^{-1}}}{C|\equiv(B|\sim\{X_a,T_a,A\overset{K_{ac}}{\leftrightarrow}C\}_{K_{ac}})}$$

Thus, by the nonce verification rule and assumption (6), we get:

$$\frac{C|\equiv\#(T_a),C|\equiv(B|\sim\{X_a,T_a,A\overset{K_{ac}}{\leftrightarrow}C\}_{K_{ac}})}{C|\equiv(B|\equiv\{X_a,T_a,A\overset{K_{ac}}{\leftrightarrow}C\}_{K_{ac}})}$$ .... *(Accords to goal 8)*

Since message #2 contains T_a, we can derive: $C\ |\equiv\ \#(X_a)$

By the message meaning rule (1) for shared keys and assumption (16), we get: $\frac{C|\equiv A\overset{K_{ac}}{\leftrightarrow}C,\ C\triangleleft\{X_a,T_a,A\overset{K_{ac}}{\leftrightarrow}C\}_{K_{ac}}}{C|\equiv(A|\sim X_a)}$

Thus, according to the nonce verification rule, we get:

$$\frac{C|\equiv\#(X_a),C|\equiv(A|\sim X_a)}{C|\equiv(A|\equiv X_a)}$$ .... *(Accords to goal 4)*

By the message meaning rule (1) for shared keys, with assumption (16) and message #2, we get: $\frac{C|\equiv A\overset{K_{ac}}{\leftrightarrow}C,\ C\triangleleft\{X_a,T_a,A\overset{K_{ac}}{\leftrightarrow}C\}_{K_{ac}}}{C|\equiv(A|\sim A\overset{K_{ac}}{\leftrightarrow}C)}$

Since message #2 contains T_a, we can derive: $C\ |\equiv\ \#(A\ \overset{K_{ac}}{\leftrightarrow}\ C)$

Thus, by the nonce verification rule, we get:

$$\frac{C|\equiv\#(A\overset{K_{ac}}{\leftrightarrow}C),C|\equiv(A|\sim A\overset{K_{ac}}{\leftrightarrow}C)}{C|\equiv(A|\equiv A\overset{K_{ac}}{\leftrightarrow}C)}$$ .... *(Accords to goal 3)*

According to assumption (12) and message #3, by the rule of jurisdicton, we get: $\frac{B|\equiv(S\Rightarrow\overset{K_c}{\mapsto}C),B|\equiv(S|\equiv\overset{K_c}{\mapsto}C)}{B|\equiv\overset{K_c}{\mapsto}C}$ .... *(Accords to goal 5)*

Thus, by the message meaning rule for public keys and message #3, we get:

$$\frac{B|\equiv\overset{K_c}{\mapsto}C,\ B\triangleleft\{\{X_c,T_c,B,A\overset{K_{ac}}{\leftrightarrow}C\}_{K_{ac}},X_c\}_{K_c^{-1}}}{B|\equiv(C|\sim\{\{X_c,T_c,B,A\overset{K_{ac}}{\leftrightarrow}C\}_{K_{ac}},X_c\})}$$

Since message #3 contains T_c, we can derive: $B\ |\equiv\ \#(\{X_c,T_c,B,A\ \overset{K_{ac}}{\leftrightarrow}\ C\}_{K_{ac}},\ X_c)$

Thus, by the nonce verification rule, we get:

$$\frac{B|\equiv\#(\{X_c,T_c,B,A\overset{K_{ac}}{\leftrightarrow}C\}_{K_{ac}}),B|\equiv(C|\sim\{X_c,T_c,B,A\overset{K_{ac}}{\leftrightarrow}C\}_{K_{ac}})}{B|\equiv(C|\equiv\{X_c,T_c,B,A\overset{K_{ac}}{\leftrightarrow}C\}_{K_{ac}})}$$ ...... *(Accords to goal 6)*

By the message meaning rule for public keys, we get: $\frac{B|\equiv\overset{K_c}{\mapsto}C,\ B\triangleleft\{X_c\}_{K_c^{-1}}}{B|\equiv(C|\sim X_c)}$

Moreover, by the nonce verification rule, we get: $\frac{B|\equiv\#(X_c),B|\equiv(C|\sim X_c)}{B|\equiv(C|\equiv X_c)}$

From the content of X_c, B knows if C has refused A because of authentication failure. If not, B knows that A is trusted by C, thus we derive the new assertions: $B\ |\equiv\ (C\ |\equiv\ (A\ |\equiv\ A\ \overset{K_{ac}}{\leftrightarrow}\ C))$ *(B believes that C believes that A really have the key,* K_{ac}*,*

shared between A and C) .... *(Weaker than goal 10)*
and then go to message #4; otherwise, B stops here.

Since message #4 contains T_c, we can derive: $A \mid\equiv \#(A \overset{K_{ac}}{\leftrightarrow} C, X_c, B)$
By the message meaning rule (1) for shared keys and assumption (1), we get:
$$\frac{A|\equiv A \overset{K_{ac}}{\leftrightarrow} C, A \triangleleft \{X_c, B\}_{K_{ac}}}{A|\equiv (C|\sim \{X_c, B\})}$$
Thus, according to the nonce verification rule, we get:
$$\frac{A|\equiv \#(X_c), A|\equiv (C|\sim X_c)}{A|\equiv (C|\equiv X_c)}$$ .... *(Accords to goal 2)*
According to message #4 and assumption (1), by the message meaning rule (1) for shared keys, we get: $\frac{A|\equiv A \overset{K_{ac}}{\leftrightarrow} C, A \triangleleft \{X_c, T_c, B, A \overset{K_{ac}}{\leftrightarrow} C\}_{K_{ac}}}{A|\equiv (C|\sim A \overset{K_{ac}}{\leftrightarrow} C)}$
Thus, according to the nonce verification rule, we get:
$$\frac{A|\equiv \#(A \overset{K_{ac}}{\leftrightarrow} C), A|\equiv (C|\sim A \overset{K_{ac}}{\leftrightarrow} C)}{A|\equiv (C|\equiv A \overset{K_{ac}}{\leftrightarrow} C)}$$ .... *(Accords to goal 1)*

Finally, by the nonce verification rule, we get:
$$\frac{A|\equiv \#(B), A|\equiv (C|\sim B)}{A|\equiv (C|\equiv B)}$$ *(A believes that C believes the identity of B)* . *(Weaker than goal 9)*

5. DISCUSSION

From the verification results above, we achieved all of the authentication goals except goals (9) and (10). That is, the results showed that the MH and HA trust each other and the FA and HA trust each other. Since the direct authentication between the MH and the FA was omitted, we could not derive that the MH and FA trust each other, but that the MH trusts that the HA trusts the FA and the FA trusts that the HA trusts the MH. However, they can be viewed as having implicitly authenticated each other. The result was weaker than the other schemes [3, 4, 6] that perform direct authentication between the MH and FA, but we will show that it is enough for a secure registration process. In other words, the weaker beliefs do not compromise the security of the protocol, which we explain as follows.

While an illegal FA, say FA_{il}, masquerades as a legal one, say FA_l, it will fail to authenticate its identity to the HA. As we know, FA_{il} does not have FA_l's private key, so the HA will send back a Registration Reply with a code indicating that the FA failed authentication. Then the MH will try to find another FA. If FA_{il} concocts a Registration Reply and transfers it to the MH instead of the original Registration Reply, the MH will also find the truth since FA_{il} cannot concoct the right message digest which is supposed to compute with the key shared between the MH and HA. In such a case, the MH will try to find another FA. Note that, in this protocol, as long as an FA is not trusted at first, what it sends to the MH as Registration Reply will be ignored by the MH. That is, we deprived the right for the FA to refuse the MH, thus getting rid of the probability that an illegal FA performs denial-of-service attack by issuing Registration Reply stating the MH is denied. Meanwhile, when the MH received no reply within the timeout period and it has retransmitted the Registration Reply n times (n is a predesigned number of retransmission) in vain, the MH should intelligently try to find another FA.

Likewise, while an illegal MH, say MH_{il}, masquerades as a legal one, say MH_l, it will fail to authenticate its identity to the HA because it does not have the key shared between the HA and MH_l. Hence, the FA will know it from message #3 in which contained a code indicating that the MH failed authentication, so that the FA will not trust this MH_{il}.

In summary, our protocol meets the need for a secure Mobile IP authentication, i.e., it realizes the tripartite authentication.

6. CONCLUSIONS

This paper verified and analyzed the security of an improved authentication protocol for Mobile IP, which realizes an efficient and scalable tripartite authentication among the MH, the FA and the HA. The widely used BAN logic was used in the verification process. The result of the verification showed that desired beliefs were achieved for the MH-HA pair and the HA-FA pair. However, weaker beliefs were resulted from the MH and FA pair since we omitted the direct authentication between them. However, we showed that the weakness does not compromise the security of the protocol. Therefore, the proposed improved authentication protocol meets the security requirement of Mobile IP.

References

[1] Burrows M., Abadi, M., Needham R., (1990). "A Logic of Authentication". *http://www.research.digital.com/SRC/personal/Martin_Abadi/Papers/src39-revised.ps.*

[2] Gao, Q., Seah W., Lo A., (1999). "Secure and Improved Authentication in Mobile IP ". *Technical report 1999-NETWORKS-TR-001NR, Centre for Wireless Communications.*

[3] Jacobs, S., (1998). "Mobile IP Public Key Based Authentication". *draft-jacobs-mobileip-pki-auth-00.txt (work in progress).*

[4] Jacobs, S., (1999). "Mobile IP Public Key Based Authentication". *draft-jacobs-mobileip-pki-auth-01.txt (work in progress).*

[5] Perkins, C., (1996). "IP Mobility Support". *RFC2002.*

[6] Sanchez,L. A. and Troxel, G. D., (1997). "Rapid Authentication for Mobile IP". *draft-ietf-mobileip-ra-00.txt (work in progress).*

[7] Solomon, J. D., (1998). *Mobile IP, the Internet unplugged.* Upper Saddle River, N.J. : PTR Prentice Hall.

SESSION 8

TCP/IP Performance Evaluation

IP OVER ATM - PERFORMANCE EVALUATION, EXPERIMENTS AND RESULTS

Paulo C. A. Antunes, Walter Godoy Júnior, Eduardo Nisenbaum

Centro Federal de Educação Tecnológica do Paraná - Centro de Pós Graduação em Engenharia Elétrica e Informática Industrial - Curitiba – Paraná – Brazil

E-mail {pcaa, godoy, baum} @cpgei.cefetpr.br

Abstract This paper evaluates the performance of the IP protocol having ATM technology as the network support. Two models are approached: "The Classical IP and ARP over ATM" and "LAN Emulation". The theoretical performance was calculated based on the protocol overhead sources and compared to practical experiments. It is shown that: for applications working in IP over ATM environment, when we use short blocks of information transfer, the performance deteriorates considerably; and when we increase its size the performance, on the average, is limited about 86% of the line speed capacity. The influence of MTU size of the IP layer in the performance is also analyzed.

Keywords: ATM, IP, Performance

1. INTRODUCTION

The Asynchronous Transfer Mode (ATM) is a technology projected to integrate voice, video and computer data services, supplying individually to each service and application the Quality of Service (QoS) requested. The ATM is the connection-oriented service that uses small units of fixed size, denominated cells, in the transport of information, that allows fast hardware switching [1].

Due to wide use of the IP (Internet Protocol), dominant in the interconnection of networks and that presents continuous expansion in the world network (Internet), we will analyze the ATM as network support for the IP Protocol. To study objectifying the impacts of performance that this integration presents, in this paper we analyzed the proposed models by IETF "The Classical IP and ARP over ATM"[2] and by ATM Forum "LAN

Emulation"[3]. In both models the ATM is seen like a Data Link Layer, in the same way that it happens with Ethernet, Token Ring, FDDI, Frame-Relay, X.25, etc. The IP when uses ATM needs to adapt its packets, of variable data size, through the protocol in the ATM Adaptation Layer (AAL5) into cells fixed size small. However such procedure imposes overhead and processing cost. Such characteristics are focalized in this work.

This paper is organized in six sections. The second refers to the topology and equipments used in the experiment. In the third section we analyze performance limiting factors in the integration of the IP protocol environment ATM. In the fourth the proposed models by IETF and ATM Forum are analyzed and their overhead sources quantified. In the fifth section the realized experiments are shown and finally in the sixth the conclusions are written.

2. TOPOLOGY AND EQUIPMENT OF THE NETWORK

The tests were made at the laboratories of the high-speed network of Curitiba, Brazil, specially in the CEFET (Centro Federal de Educação Tecnológica do Paraná) and UFPR(Universidade Federal do Paraná) institutions (connected through 6.200 meters of single-mode optical fibers). The network topology is shown in *Figure 1.*

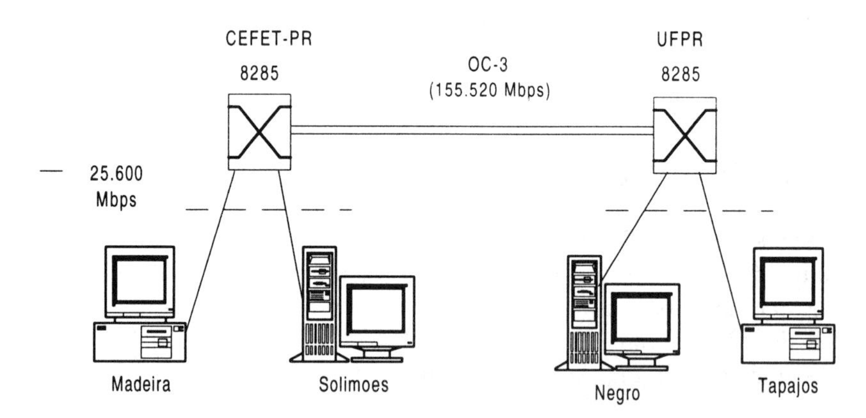

Figure 1: Network Topology

Switches ATM (IBM-8285) were used with backplane of 4.2 Gbps and it use the technique of EPD(Early Packet Discard)[4]. EPD is the technique, where in case of congestion, instead of discarding aleatory cells of each packet, all the cells belonging to the same packet but last are discarded. This avoids the unnecessary sending of cells that cannot be used in the receptor. The network interfaces used have PCI (Peripheral Component Interconnect) architecture with nominal capacity of 25.600 Mbps. The stations, CPU, operating systems and network interfaces used are described in *Table 1*.

Table 1: Stations, processors, operating systems and network adapters used

Station	Processor and Operating System	Network Adapter
Solimoes	Power PC 233 MHz, AIX 4.2.1	Turboways
Negro	Power PC 233 MHz, AIX 4.2.1	Turboways
Madeira	IBM-PC Pentium 166 MHz, NT 4.0	Turboways
Tapajos	IBM-PC Pentium 166 MHz, Linux	ForeRunner

In the physical interface local connection, between the stations and the switch, Unshielded Twisted Pairs UTP cabling (category 5) were used. In the remote connection of the Metropolitan Area Network (MAN), the switches were interconnected by optical link with single-mode fibers, with line rate of 155.520 Mbps. In the physic layer, the interface of 25.600 Mbps[5][6] doesn't have frame overhead, cells are transmitted in direct flow and the separation among them is done by the HEC field (Header Error Control) contained in each ATM cell[7]. The physical interfaces between switches were configured to use SONET (Synchronous Optical Network) frames. *Figure 2* [7] shows a SONET frame composed of three STS-1signals connected forming a STS-3c/OC-3c signal. Each SONET frame repeats itself to each 125 μs (8000 frames a second) = 8000 x [9 x (261+9)] x 8 bits], supplying the transmission rate of 155.520 Mbps.

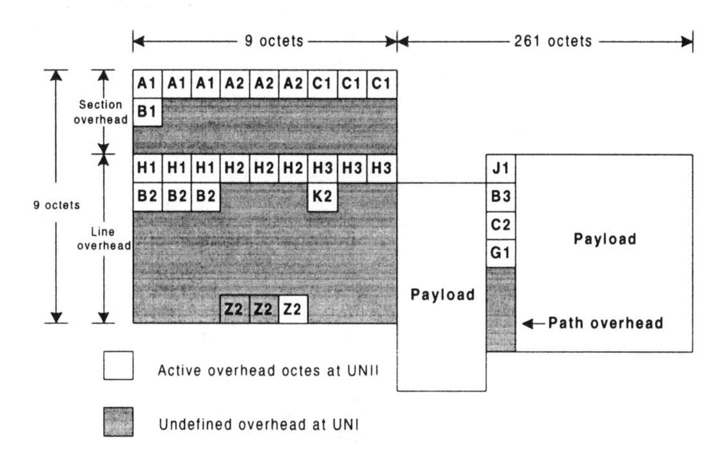

Figure 2: SONET frame STS-3C/OC-3C

The total size of the SONET STS-3c/OC-3c frame is 2430 octets [9x(9+261)]. Of which 90 octets are of overhead, divided in:
· The section overhead = 27 octets;
· The line overhead = 54 octets;
· The path overhead = 9 octets.

3. PERFORMANCE LIMITING FACTORS

The application, when transfers data to the network, involves the interaction of protocols in several layers. Each protocol can insert control informations to those data to identify it, in the header form, trailer or both. These information increase the reliability in the transport of the data, for instance, with the use of the fields: HEC (Header Error Control) of the ATM cell, CRC (Cyclic Redundancy Check) of AAL5, sum of integrity of the data (checksum) in the IP protocols, TCP (Transmission Control Protocol) and UDP (User Datagram Protocol). The information also identify the partners of the communication, as in the following case: in ATM the VPI (Virtual Path Identifier) and VCI (Virtual Channel Identifier); in IP the source and destination address; in TCP and UDP the indication of the communication port and PDU (Protocol Data Unit) size. These additional informations are denominated protocol overhead and it is one of the factors that limit the application to use at wire speed.

4. MODELS OF CLASSICAL IP AND LANE

The IP protocol in the Classical IP and LANE models use the protocol of the adaptation layer AAL5. The allowed maximum size of PDU(Protocol Data Unit) in CS(Convergence Sublayer) is of 65535 octets[8]. The format of the frames of the AAL5 and ATM layers is shown in *Figure 3*, where CPCS is the Common Part Convergence Sublayer and SAR is the Segmentation and Reassembly Sublayer.

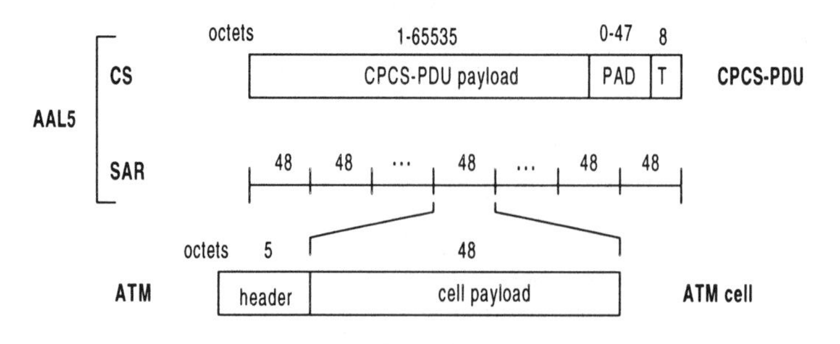

Figure 3: Frames of the AAL5 and ATM layers

In the transmission the SAR segments the PDU in blocks of 48 octets, no protocol identification is inserted in this sublayer. The SAR delivers the blocks, now denominated ATM-SDU(ATM-Service Data Unit) to the ATM layer and through the user-to-user indication parameter of primitive ATM-DATA.indication, informs the ATM layer which is the last block of the frame. The ATM layer inserts the header in each PDU, now denominated cell, marks the last unit of the frame, and delivers the physical layer for transmission. In the reception the SAR extracts the payload of the cells, convert them again in PDU and delivers to the superior protocols.

4.1 Classical IP

The default value of MTU (Maximum Transmission Unit) for the IP protocol in the layer AAL5 is of 9180 octets[2]. In the experiments beyond of that value, MTU of 1500 octets were used, too. The MTU indicates the maximum amount of data that can be inserted in one PDU without fragmentation. *Figure 4* shows the format of the PDUs in the layers of the transport protocols TCP and UDP for the model of Classical IP and ARP over ATM, as well as the amount of overhead and the number of corresponding octets and cells.

Each TCP or UDP segment has direct relationship with each IP packet, however the transport protocol can include several blocks of data in each segment[9].

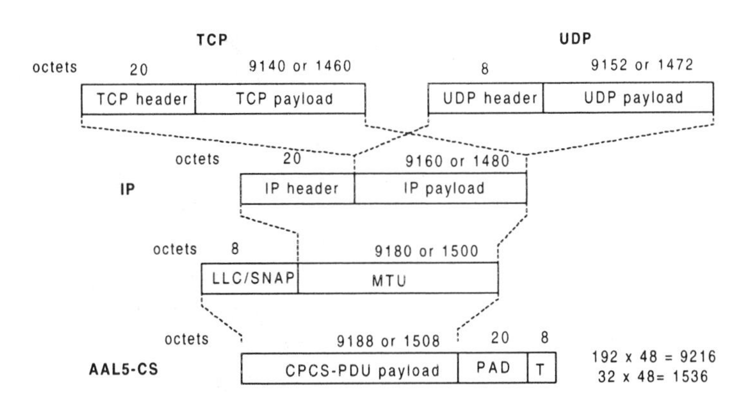

Figure 4: PDU format for the TCP,UDP, IP over AAL5 in Classical IP Model

4.2 LANE

The LAN Emulation (LANE) transports frames of the traditional Ethernet and Token-Ring networks; we concentrated experiments on Ethernet

LANES. The IEEE-802.3[10] standard for Ethernet networks defines that the minimum size of a frame should be of 64 octets, with 46 payload octets and 18 overhead octets. This standard also defines that the maximum size of the frame should be of 1518 octets, 1500 are payload octets and the rest, 18 octets, identify the frame.

When transporting Ethernet frames, the LANE excludes the field of CRC, because it is made in CPCS-PDU. The LANE model adds in each Ethernet frame one header of 2 octets. This header contains the LAN Emulation Client Identifier (LECID).

In the AAL5 adaptation protocol a trailer of 8 octets is added to each frame and, if necessary, it adds octets of padding (PAD) to turn the CPCS-PDU size a multiple of 48 octets (that are delivered to SAR). The trailer is only present in the last cell of each PDU. The recommendation of ITU-T[8] defines that the content of the trailer has to be constituted by fields CPCS-UU, used to transfer user to user informations of CPCS, of the Common Part Indicator(CPI) of CPCS, of the field length of CPCS-PDU, as well as of CRC that check the whole CPCS-PDU. The CPI field is still not defined, however actually it is used to turn the trailer to the size of 8 octets. *Figure 5* shows the trailer format of the CPCS-PDU.

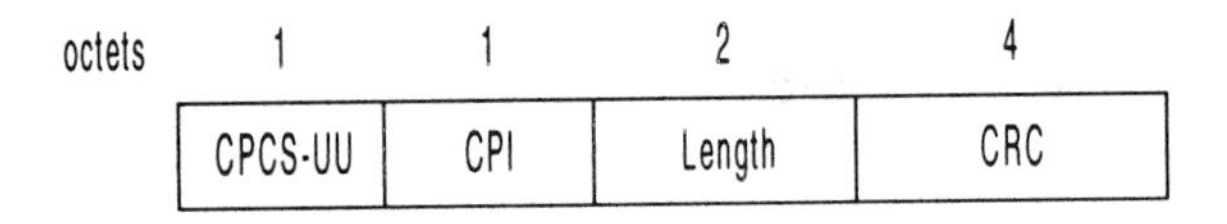

Figure 5: Trailer format to CPCS-PDU

The default size of a frame in AAL5-SDU in LANE is 1516 octets. This generates a PDU of 1536 octets in Ethernet, in which 1516 octets are data and the remaining 20 octets are overhead. One AAL5-SDU frame of 1536 octets corresponds to 32 cells in the ATM layer. The recommendation of ATM Forum[3] also allows to use larger sizes of AAL5-SDU(4544, 9234 and 18190) octets. In the experiment with standard size of Ethernet frames we used 1536 octets. *Figure 6* shows the format of the PDU on TCP, UDP layer for the LANE model using AAL5, as well as the overhead distribution in the layers and the number corresponding on octets and cells.

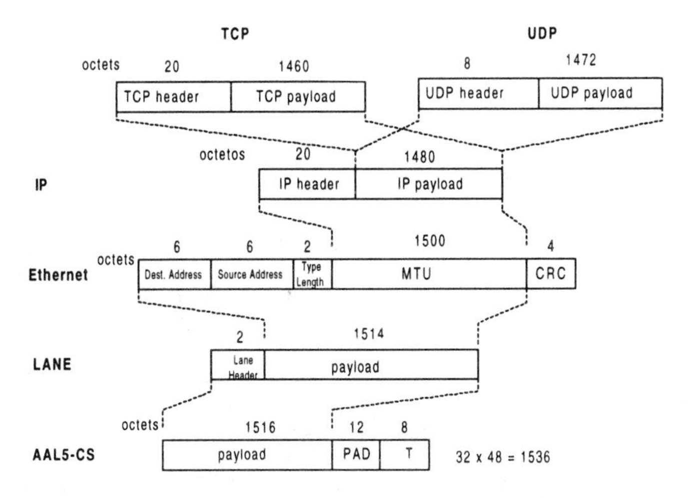

Figure 6: PDU format to TCP, UDP, IP, Ethernet over AAL5 in LANE

4.3 Theoretical Calculations of the Available Maximum Bandwidth for the Model of Classical IP

In order to evaluate the results obtained experimentally, initially the theoretical available maximum bandwidth was calculated to the application involving all the protocol layers. In the calculations were considered just the imposed overhead by the data unit of protocol in each layer.

Through the expressions below we calculated the available bandwidth(BW) and *Table 2* shows the calculated values for:

- ATM layer between switches

$$BW_{ATM} = \frac{frame\_payload}{frame\_length}.bit\_rate$$

- ATM adaptation layer

$$BW_{AAL} = \frac{cell\_payload}{cell\_legth}.BW_{ATM}$$

- IP

$$BW_{IP} = \frac{IP\_MTU}{CPCS\_PDU}.BW_{AAL}$$

- TCP and UDP

$$BW(TCP) = BW(UDP).\frac{IP\_payload}{IP\_MTU}.BW_{IP}$$

- Application when TCP is used

$$BW_{app} = \frac{TCP\_payload}{IP\_payload}.BW_{TCP}$$

- Application when UDP is used

$$BW_{appl} = \frac{UDP\_payload}{IP\_payload}.BW_{UDP}$$

Table 2: Bandwidth available after protocol overhead

(Mbps)	SONET OC-3c				UTP – 25.600	
Line rate	155.520				25.600	
To ATM	149.760				25.600	
To AAL	135.632				23.185	
	Classical IP MTU(octets)				LANE MTU(octets)	
	9180		1500		1500	
To IP	23.094		22.642		22.642	
To TCP	23.044		22.340		22.340	
To UDP	23.044		22.340		22.340	
	TCP	UDP	TCP	UDP	TCP	UDP
To Appl.	22.994	23.024	22.038	22.219	22.038	22.219

5. TRAFFIC MEASUREMENTS

In the measurements the program called Netperf developed by the Hewlett-Packard Company [11] was used. It uses the client/server architecture and utilizes socket in the communication among processes and between machines and processes. The tests consisted of sending continually prefixed size data blocks. To obtain index of reliability of 95%(±1,5%), periods of 60 seconds were necessary for each individual variation of parameter.

5.1 Local Test of Performance

The local test consisted of sending and receiving flows of tabulated data in the same communication port and in the same station. To estimate the station capacity of processing interacting with the protocol stack, when the TCP transport protocol is used, in agreement with the recommendation in [12] duplicates the value obtained in the transmission. *Table 3* shows the results obtained in each station for the TCP. The stations were configured with the following parameters: MTU of 1500 octets in the IP protocol;

maximum SDU of 1516 octets in AAL5 and socket buffers for transmission and reception using the same sizes = 4096 octets.

Table 3: Results of the obtained tests using the TCP protocol

Station	TX = RX (Mbps)	Estimated capacity (Mbps)
Solimoes	115.567	231.124
Negro	114.231	228.462
Madeira	41.480	82.960
Tapajos	111.782	223.564

5.2. Tests Among Stations

In the practical tests we analyzed the IP using the TCP transport protocol and the measures were collected in the transmitter. The environment was controlled, there wasn't competition for bandwidth for the analyzed models used UBR service (Unspecified Bit Rate) and in case of congestion they have preference in the discard.

5.3 Classical IP

Figure 7 shows the results obtained among the Solimoes and Negro stations; aimed at to evaluate the influence of socket buffer size in relation to variation of the size of the messages when MTU of 1500 octets is used. *Figure 8* is analogous to *Figure 7*, however MTU of 9180 octets was used. In both cases the socket buffers of transmission and reception were configured with the same sizes, beginning with 4096 octets and in the sequence increased by multiples of this value until 65536 octets. The messages began with 16 octets and were increased by the factor 4 in relation to the previous value and concluded in 65536 octets.

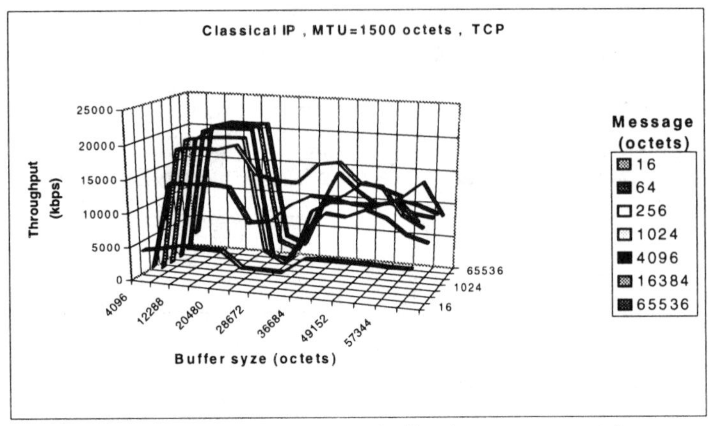

Figure 7: Throughput vs. socket buffer size vs. message size

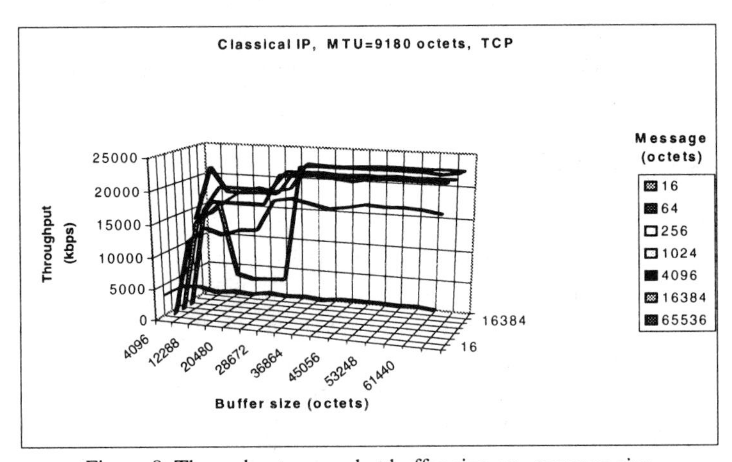

Figure 8: Throughput vs. socket buffer size vs. message size

It is specially observed that messages of small size (16 and 64 octets) in both cases presented a low performance, this is due to the overhead octets of the protocol layers being larger than the message size, still allied to the countless interruptions generated by the physical interface and to system calls, at the level of operating system. In the other messages the best found performance went to socket buffer size among 8192 octets and 16384 octets when it uses MTU(IP) of 1500 octets. To MTU 9180 octets the best performance went to socket buffers size larger of 28672 octets..

The result are shown in *Figure 9* and *10* involving the stations with smaller processing capacity, Madeira and Tapajos respectively. In them messages were tabulated with multiple size of the IP MTU default and socket buffers with multiple size of those values.

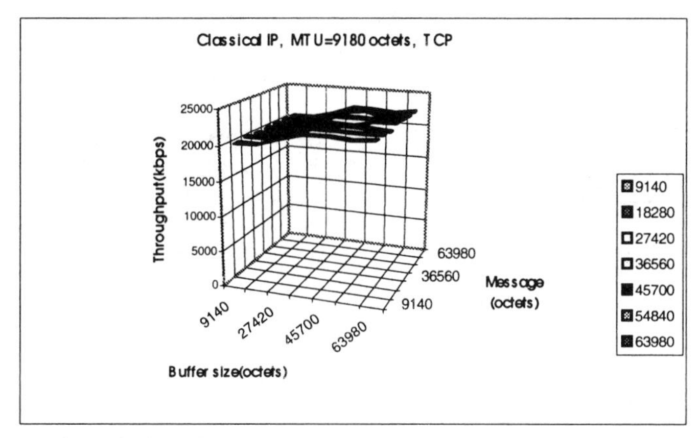

Figure 9: Experiment between the Madeira and Tapajos stations

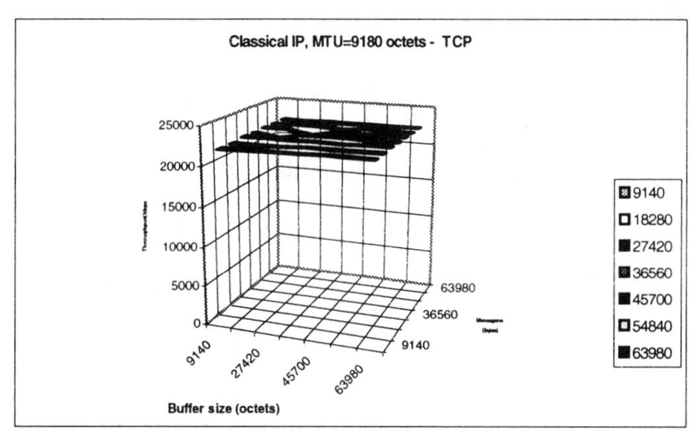

Figure 10: Experiment between the Tapajos and Solimoes stations

As the Tapajos as the Madeira stations demonstrated similar performances, obtaining maximum performance of 22.420 Mbps and 22.320 Mbps respectively that correspond to 97% of the calculated capacity.

5.4 LANE

In the experiments with LANE were pre-fixed socket buffer sizes in 16384 octets and messages beginning with 16 octets, leaving for 64 octets and proceeding for multiples of this value to reach 4096 octets. As it can be observed by *Figures 11* and *12*, the performance was considerably reduced for messages of small size, analogous to what was found in Classical IP model.

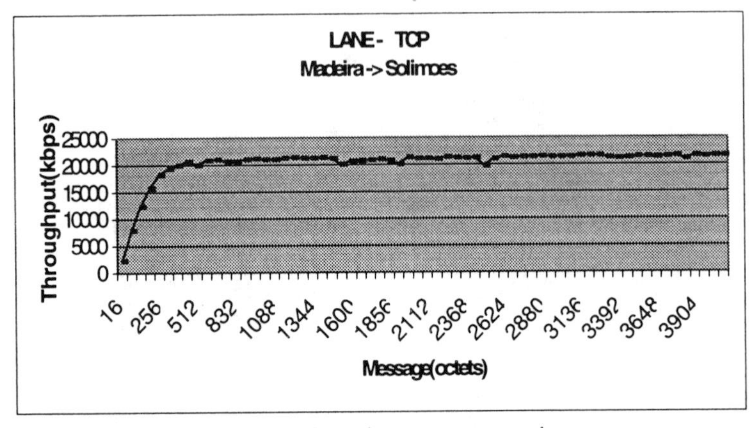

Figure 11: Throughput vs. message size

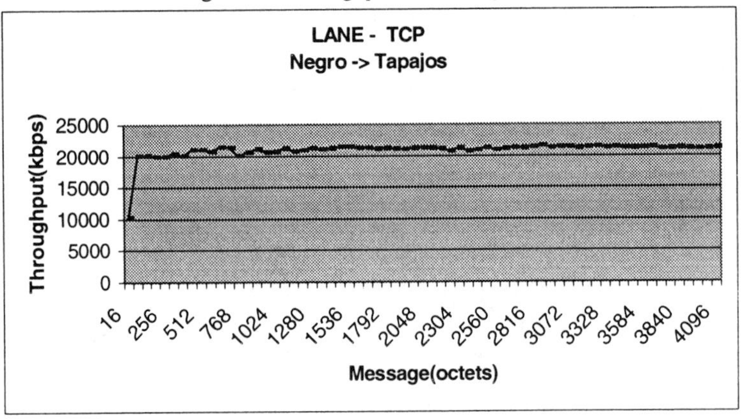

Figure 12: Throughput vs. message size

Table 4 shows the best values obtained in the Classical IP and LANE models.

Table 4: Maximum throughput obtained

Best Results	Theoretical (Mbps)	Measure (Mbps)	(%) of theoretical	(%) of line rate
Classical IP	22.038	21.618	98.09	86.08
LANE	22.038	21.579	97.91	86.08

6. CONCLUSIONS

The Classical IP model presents a lower level of encapsulation than the LANE model, however the performance approached itself with a difference less than 0,1% among them. This was due to the workstations having sufficient software and hardware resources for operating in the rate of 25.600 Mbps, without significant delay of processing. In both models it could be observed that for small messages the performance was degraded considerably, this penalty in using the lower messages is that the protocols overhead become a significant percentage of the overall transmission rate.

The identified parameters in the testbeds made possible to determine the factors that influence the performance, among them we highlighted: application message size, socket buffer size and MTU. The capacity of processing of the stations The capacity of processing, of the interfaces and devices of software(device drivers) in spite of being important in the performance, presented sufficiency of resources for maintaining the network in the available speed .

References

[1] PARTRIDGE, Craig, "Gigabit networking". Addison-Wesley, April 1995.

[2] LAUBACH, M., J. Halpern, "Classical IP and ARP over ATM", RFC 2225, Newbridge Networks, April 1998.

[3] ATM Forum, "LAN Emulation Over ATM version 1.0" ATM Forum af-lane-0021.000, Jan.1995.

[4] IBM 8285 Nways ATM Workgroup Switch, Installation and User's Guide,1996.

[5] ITU-T Recommendation I.432.5 : "B-ISDN user-network interface Physical layer specification: 25,600 kbit/s operation", June 1997.

[6] ATM Forum, "Physical Interface Specification for 25.6 Mb/s over Twisted Pair Cable", af-phy-0040.000. Nov. 1995.

[7] ATM Forum, "User-Network Interface Version 3.1 Specification", Sep. 1994.

[8] ITU-T Recommendation I.363.5 – "B-ISDN ATM Adaptation Layer specification: Type 5 AAL". Aug. 1996.

[9] STALLINGS, William, "High-speed networks - TCP/IP and ATM design principles", Prentice-Hall, Inc., 1998.

[10] IEEE 802-3: "Carrier sense multiple access method", New York: IEEE, 1985a.

[11] Hewlett-Packard Company, "Netperf: A network performance benchmark", revision 2.1, Information Network Division Hewlett-Packard Company, 1996.

[12] ANDRIKOPOULOS, T. Örs I et all, "TCP/IP throughput performance evaluation for ATM local area networks", Proceedings of 4th IFIP Workshop on Performance Modeling and Evaluation of ATM Networks, Ilkley, UK, July 1996.

BROADBAND SATELLITE NETWORK: TCP/IP PERFORMANCE ANALYSIS

Sastri Kota
Lockheed Martin Mission Systems
1260 Crossman Ave, MS:S40

e-mail: sastri.kota@lmco.com

Mukul Goyal, Rohit Goyal and Raj Jain
Computer and Information Science Department
The Ohio State University
2015 Neil Ave., Columbus, OH 43210
e-mail: {mukul,goyal,jain}@cis.ohio-state.edu

Abstract A number of satellite communication systems have been proposed using geosynchronous (GEO) satellites, as well as low earth orbit (LEO) constellations operating in the Ka-band and above. At these frequencies satellite networks are able to provide broadband services requiring wider bandwidth than the current services at C or Ku-band. As a consequence, some of the new services gaining momentum include mobile services, private intranets and high data rate internet access carried over integrated satellite-fiber ATM networks. Several performance issues need to be addressed before a transport layer protocol, like TCP can satisfactorily work over satellite ATM for large delay-bandwidth networks. In this paper, we discuss some of the architectural options and challenges for broadband satellite ATM networks. The performance results of TCP enhancements for Unspecified Bit Rate over ATM (ATM-UBR+) for large bandwidth-delay environments with various end system policies and drop policies for GEO satellite configurations for several buffer sizes are presented.

Keywords: Broadband Satellite, ATM, UBR, TCP, Performance Analysis

1. INTRODUCTION

The rapid globalization of the telecommunications industry and the exponential growth of the Internet is placing severe demands on global telecommunications. Satisfying this demand is one of the greatest challenges before telecommunications industry in 21st century. Satellite communication networks can be an integral part of the newly emerging national and global information infrastructures (NII and GII).

Satellite communication offers a number of advantages over traditional terrestrial point-to-point networks. Satellite networks can cover wide geographic areas and can interconnect remote terrestrial networks ("islands"). In case of damaged terrestrial networks, satellite links provide an alternative. Satellites have a natural broadcast capability and thus facilitate multicast communication. Finally, satellite links can provide bandwidth on demand by using Demand Assignment Multiple Access (DAMA) techniques.

The growing congestion of the C and Ku bands have increased the interest of satellite system developers in the Ka-band. Several factors influence the development of broadband satellite networks at Ka-band frequencies:

- *Adaptive Power Control and Adaptive Coding:* Adaptive power control and adaptive coding technologies have been developed for improved performance, mitigating propagation error impacts on system performance at Ka-band.
- *High Data Rata:* A large bandwidth allocation to geosynchronous fixed satellite services (GSO FSS) and non-geosynchronous fixed satellite services (NGSO FSS) makes high data rate services feasible over Ka-band systems.
- *Advanced Technology:* Development of low noise transistors operating in the 20 GHz band and high power transistors operating in the 30 GHz band have influenced the development of low cost earth terminals. Space qualified higher efficiency traveling-wave tubes (TWTAs) and ASICs development have improved the processing power. Improved satellite bus designs with efficient solar arrays and higher efficiency electric propulsion methods resulted in cost effective launch vehicles.

2. BROADBAND SATELLITE NETWORK

There are several options that drive the broadband satellite network architecture [8]:

- (GSO) Geosynchronous Orbits versus (NGSO) Non-Geosynchronous Orbits (e.g., LEOs, MEOs)
- No onboard processing or switching
- Onboard processing with ground ATM switching or “ATM like,” cell or fast packet switching
- Onboard processing and onboard ATM or “ATM like” fast cell/packet switching

However, most of the next generation broadband satellite systems have in common features like onboard processing, ATM or “ATM-like” fast packet switching, terminals, gateways, common protocol standards, and inter-satellite links [8]. Figure 1 illustrates a broadband satellite network architecture represented by a ground segment, a space segment, and a network control segment. *The ground segment* consists of terminals and gateways (GWs) which may be further connected to other legacy public and/or private networks. *The Network Control Station* (NCS) performs various management and resource allocation functions for the satellite media. Inter-satellite crosslinks in the *space segment* provide seamless global connectivity via the satellite constellation. The network allows the transmission of ATM cells over satellite, multiplexes and demultiplexes ATM cell streams for uplinks, downlinks, and interfaces to interconnect ATM networks as well as legacy LANs.

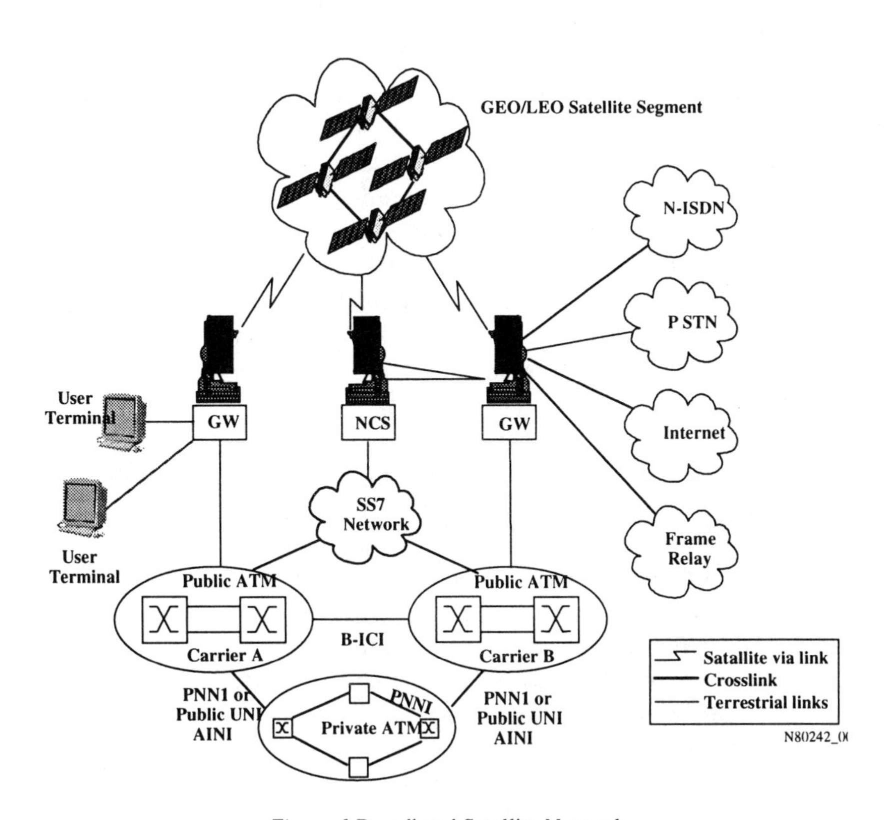

Figure 1 Broadband Satellite Network

The *gateways* support several protocol standards such as ATM User Network Interface (ATM-UNI), Frame Relay UNI (FR-UNI), Narrow-band Integrated Digital Network (N-ISDN), and Transmission Control Protocol/Internet Protocol (TCP/IP). The gateways interface unit provides external network connectivity. The number and placement of these gateways in both GEO and MEO systems depend on the traffic demand, performance requirements, and other international regulatory issues. The *user Terminals Interface Unit (TIU)* supports several protocol standards adapting to the satellite network interface. It includes the physical layer functionalities such as channel coding, modulation/demodulation, and other RF functions. The *space segment* consists of either a GEO or MEO constellation depending on the system design. Within payloads full onboard processing and ATM or "ATM-like" switching is assumed.

Interconnectivity to the external private or public networks is possible with the support of the standard protocol. For the satellite ATM case, the signaling protocols based on ITU-T Q.2931 can be used when necessary. For other networks, the common channel signaling protocol, e.g., Signaling System No. 7 (SS7), can be used. The other interconnection interfaces between public and private ATM networks are the ATM Inter-Network Interface (AINI), the Public User Network Interface (PUNI) or the Private Network-Network Interface (PNNI), and the default interface between two public ATM networks, namely, the B-ISDN Inter Carrier Interface (B-ICI). However, these interfaces require further modifications to suit the satellite interface unit development. There is a definite need for an integrated satellite-ATM network infrastructure and standards for interfaces and protocols are in development process.

Effective traffic management and media access protocols constitute main challenges for successful deployment of Satellite ATM networks. Limited bandwidth available on satellite links make it necessary to use DAMA techniques in order to support multimedia applications [7]. Congestion control is an essential part of traffic management. ATM-ABR service uses Explicit Rate Congestion Control where feedback from the network contains the explicit rate at which sources should send data. However, this scheme needs to be analysed in terms of the end-to-end delay requirements for satellite-ATM networks. In the long propagation delay satellite configurations, the feedback delay is the dominant factor in determining the maximum queue length. A feedback delay of 10 ms corresponds to about 3670 cells of queue for TCP over ERICA, while a feedback delay of 550 ms corresponds to 201,850 cells. Satellite switches can isolate downstream switches from such large queues by implementing Virtual Source/Virtual Destination (VS/VD) options [4].

3. TCP/IP TRAFFIC TRANSPORT OVER SATELLITE ATM

TCP/IP is the most popular network protocol suite and hence it is important to study how well these protocols perform on long delay satellite links. The main issue affecting the performance of TCP/IP over satellite links is very large feedback delay compared to terrestrial links. The inherent congestion control mechanism of TCP causes source data rate to reduce rapidly to very low levels with even a few packet loss in a window of data. The increase in data rate is controlled by ACKs received by the source. Large feedback delay implies a proportional delay in using the satellite link efficiently again. Consequently, a number of TCP enhancements (NewReno, SACK) have been proposed that avoid multiple reductions in source data rate when only a few packets are lost [2,9].The enhancements in end-to-end TCP protocol are called *End System Policies.*

Satellite ATM link performance can also be improved by using *intelligent switch policies*. The Early Packet Discard policy [10] maintains a threshold R, in the switch buffer. When the buffer occupancy exceeds R, then all new incoming packets are dropped. Partially received packets are accepted if possible. The Selective Drop policy [3] uses per-VC accounting, i.e., keeps track of current buffer utilisation of each active UBR VC. A UBR VC is called "active" if it has at least one cell currently buffered in the switch. The total buffer occupancy, X, is allowed to grow until it reaches a threshold R, maintained as a fraction of the buffer capacity K. A fair allocation is calculated for each active VC, and if the VC's buffer occupancy X_i exceeds its fair allocation, its subsequent incoming packet is dropped. Mathematically, in the Selective Drop scheme, an active VC's entire packet is dropped if

$$(X > R)\ AND\ (X_i > Z \times X/N_a)$$

where N_a is the number of active VCs and Z is another threshold parameter ($0 < Z <= 1$) used to scale the effective drop threshold.

4. END-SYSTEM POLICY VS SWITCH POLICY FOR SATELLITE-ATM

[5] discusses the relative impact of end system policies (TCP flavors: Vanilla, Fast Retransmit Recovery/Reno, NewReno, SACK), switch drop policies (Early Packet Drop and Selective Drop) and switch buffer sizes (0.5

RTT[1], 1 RTT, 2 RTT) on the performance of MEO and GEO links satellite UBR+ links for Internet traffic. The same issues have been studied earlier for *persistent/infinite* TCP traffic in [3]. Both studies establish that for long delay satellite links, end system policies are far more effective than switch policies in ensuring good performance.

4.1 Simulation Configuration and Experiments

Figure 2 shows the configuration used in all simulations. The configuration consists of 100 WWW clients being served by 100 WWW servers, one server for each client. Both WWW clients and servers use underlying TCP connections for data transfer. The WWW traffic model used in this study is an extension of that specified in SPECweb96 benchmark. [11] and is based on HTTP/1.1 standard [1]. The switches implement the UBR+ service with optional drop policies described before.

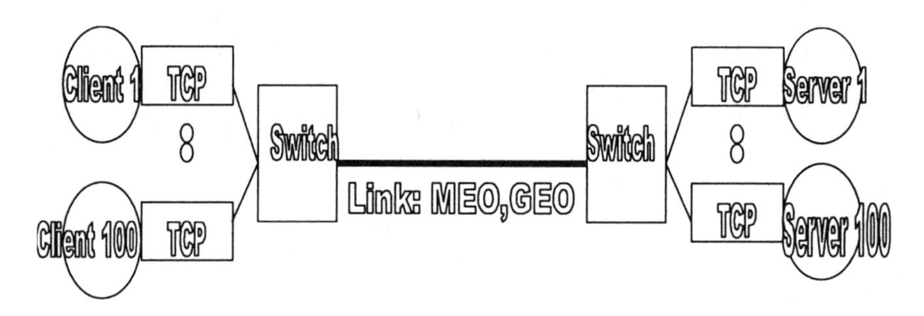

Figure 2 Simulation Configuration with 100 WWW Client-Server Connections

4.2 Configuration Parameters

Links connecting server/client TCPs to switches have a bandwidth of 155.52 Mbps (149.76 Mbps after SONET overhead), and a one way delay of 5 microseconds. The link connecting the two switches simulates MEO and GEO link respectively and has a bandwidth of 45Mbps (T3). The corresponding one-way link delays are 100 ms and 275 ms respectively. Since the propagation delay on the links connecting client/server TCPs to switches is negligible compared to the delay on the inter-switch link, the round trip times (RTTs) due to propagation delay are 200 ms and 550 ms for MEO and GEO respectively. All simulations run for 100 secs. TCP maximum segment size (MSS) is set to 9180. TCP timer granularity is set to

[1] A buffer size of 1 RTT means the round-trip time-bandwidth product of the link in terms of ATM cells.

100 ms. Using window scaling option, TCP maximum receiver window size is set to 2,097,120 and 4,194,240 bytes for MEO and GEO links respectively. The value of maximum receiver window is set so that it is greater than RTT-bandwidth product of the path. The TCP delay ACK timer is NOT set. Segments are ACKed as soon as they are received. The drop threshold R is 0.8 for both switch drop policies - EPD and SD. For SD simulations, threshold Z also has a value 0.8. We use three different values of buffer sizes corresponding to 0.5 RTT, 1 RTT and 2 RTT - bandwidth products of the end-to-end TCP connections for each of the propagation delays. The performance is measured in terms of the efficiency of link usage, i.e., the ratio of total throughput of all connections and the maximum possible throughput on the link.

4.3 Simulation Analysis Technique

We analyze the effects of 3 factors - TCP flavor, buffer size and drop policy - in determining the efficiency and for MEO and GEO links. The values a factor can take are called 'levels' of the factor. For example, EPD and SD are two levels of the factor 'Drop Policy'. The analysis consists of the calculating the following terms. A detailed description of analysis procedure is available in [6,5].

- **Overall mean**: This consists of the calculation of the overall mean 'Y' of the result (efficiency or fairness).
- **Total variation**: This represents the variation in the result values (efficiency or fairness) around the overall mean 'Y'. *The goal of the analysis to calculate, how much of this variation can be explained by each factor and the interactions between factors.*
- **Main effects**: These are the individual contributions of a level of a factor to the overall result. A particular main effect is associated with a level of a factor, and indicates how much variation around the overall mean is caused by the level. We calculate the main effects of 4 TCP flavors, 3 buffer sizes, and 2 drop policies.
- **First order interactions**: These are the interaction between levels of two factors. In our experiments, there are first order interactions between each TCP flavor and buffer size, between each drop policy and TCP flavor, and between each buffer size and drop policy.
- **Allocation of variation**: This is used to explain how much each factor contributes to the total variation.

4.4 Simulation Results

Following observations can be made about MEO and GEO links from Tables 1-2. TCP flavor explains 56.75% for MEO and 69.16% for GEO of the efficiency variation and hence is the major factor in deciding efficiency value. SACK results in substantially better efficiency than other TCP flavors. Thus, for long delay satellite links, SACK is the best choice in spite of complexity of its implementation. Buffer size explains 21.73% for MEO and 13.65% for GEO of the variation and interaction between buffer size and TCP flavors explains 13.42 for MEO and 7.54% for GEO of the variation. Efficiency values are largely unaffected as we increase buffer size from 0.5 RTT to 1 RTT. There is a marginal improvement in performance as buffer size is increased further to 2 RTT. Vanilla and Reno show substantial efficiency gains as buffer size is increased from 1 RTT to 2 RTT. Note that a buffer size of 0.5 RTT is sufficient for SACK. Further increase in buffer size brings very little performance improvement for SACK. Drop policy (EPD or Selective Drop) does not have an impact on efficiency as indicated by negligible allocation of variation to drop policy. From the observations above, it can be concluded that SACK with 0.5 RTT buffer is the optimal choice for MEO and GEO links with either of EPD and SD as switch drop policy.

Table 1 Simulation Results for MEO and GEO Links

Drop Policy	TCP Flavor	Buffer=0.5RTT Efficiency		Buffer=1RTT Efficiency		Buffer=2RTT Efficiency	
		MEO	GEO	MEO	GEO	MEO	GEO
EPD	Vanilla	0.848	0.791	0.879	0.792	0.899	0.848
	Reno	0.894	0.805	0.903	0.817	0.909	0.874
	NewReno	0.903	0.866	0.910	0.859	0.912	0.845
	SACK	0.908	0.902	0.912	0.909	0.916	0.921
SD	Vanilla	0.836	0.808	0.872	0.816	0.901	0.868
	Reno	0.876	0.810	0.898	0.781	0.902	0.863
	NewReno	0.892	0.790	0.892	0.832	0.898	0.851
	SACK	0.917	0.918	0.926	0.916	0.937	0.921

Table 2 Allocation of Variation for MEO and GEO Efficiency Values

Component	Sum of Squares		%age of Variation	
	Efficiency		Efficiency	
	MEO	GEO	MEO	GEO
Individual Values	19.3453	17.3948		
Overall Mean	19.3334	17.3451		
Total Variation	0.0119	0.0497	100	100
Main Effects:				
TCP Flavor	0.0067	0.0344	56.75	69.16
Buffer Size	0.0026	0.0068	21.73	13.65
Drop Policy	0.0001	0.0001	0.80	0.25
First-order Interactions:				
TCP Flavor-Buffer Size	0.0016	0.0037	13.42	7.54
TCP Flavor-Drop Policy	0.0007	0.0025	6.11	4.96
Buffer Size-Drop Policy	0.0001	0.0002	0.53	0.41

5. CONCLUSIONS

Broadband satellites networks are the new generation communication satellite systems that will use onboard processing and ATM and/or "ATM-like" switching to provide two-way communications. The proposed satellite or broadband satellite systems operate at Ka-band and above frequencies. Several technical challenges and issues, e.g., traffic management, Quality of Service (QoS) assurance, interoperability, efficient protocols, and standards. In this paper, we analysed design parameters based on end policies and switch parameters for efficient satellite ATM networks. In summary, as delay increases, the gains of end system policies are more important than the gains of drop policies and large buffers.

References

[1] R. Fielding, J. Gettys, J. Mogul, H. Frystyk, T. Berners-Lee, "Hypertext Transfer Protocol -HTTP/1.1", RFC 2068, January 1997.

[2] S. Floyd, T. Henderson, "The NewReno Modification to TCP's Fast Recovery Algorithm," Internet Draft, November 1998, Available from ftp://ftp.ietf.org/internet-drafts/drafts-ietf-tcpimpl-newreno-00.txt

[3] Rohit Goyal, Raj Jain, Shivkumar Kalyanaraman, Sonia Fahmy, Bobby Vandalore, "Improving the Performance of TCP over the ATM-UBR Service," *Computer Communications*, Vol 21/10, 1998.

[4] R. Goyal, R. Jain, M. Goyal, S. Fahmy, B. Vandalore, and S. Kota, "Traffic Management for TCP/IP over Satellite ATM Networks," *IEEE Communications Magazine,* March 1999, Vol. 37, No. 3, pp. 56-61

[5] Mukul Goyal, Rohit Goyal, Raj Jain, B. Vandalore, S. Fahmy, T. vonDeak, K. Bhasin, N. Butts, S. Kota, "Performance Analysis of TCP Enhancements for WWW Trafic using UBR+ with Limited Bufers over Satellite Links", ATM_Forum/98-0876R1, December 1998, http://www.cis.ohio-state.edu/~jain/atmf/a98-0876.htm

[6] R. Jain, The Art of Computer Systems Performance Analysis: Techniques for Experimental Design, Simulation, and Modeling, John Wiley & Sons Inc., 1991.

[7] S. Kota, J. Kallaus, H. Huey, and D. Lucantoni, "Demand Assignment Multiple Access (DAMA) for Multimedia Services – Performance Results," *Proc. MILCOM'97,* Monterey, CA, 1997.

[8] Sastri Kota, "Satellite ATM Networks: Architectural Issues and Challenges," *Proc. Conf. on Satellite Networks: Architectures, Applications and Technologies,* NASA Lewis Research Center, Cleveland, pp. 443-457, June 2-6, 1998.

[9] M. Mathis, J. Madhavi, S. Floyd, A. Romanow, "TCP Selective Acknowledgment Options," RFC 2018, October 1996.

[10] A. Romanow, S. Floyd, "Dymnamics of TCP Traffic over ATM Networks", *IEEE JSAC,* May 1995.

[11] SPEC, "An Explanation of the SPECweb96 Benchmark," Available from http://www.specbench.org/osg/web96/webpaper.html

A CUT-OFF PRIORITY FLOW CLASSIFICATION POLICY FOR DATA-DRIVEN IP/ATM SWITCHING SYSTEMS

Jun Zheng and Victor O. K. Li, Fellow, IEEE
Department of Electrical and Electronic Engineering
The University of Hong Kong
Pokfulam Road, Hong Kong
Email: {jzheng,vli}@eee.hku.hk

Abstract Flow classification is one of the key issues in the design of a data-driven IP/ATM switching system. To classify flows by applications is an effective approach in a real system due to its simplicity in implementation. In this paper, we propose a cut-off priority flow classification policy to improve the application-based flow classification policy. The basic idea is to assign priorities to the applications that are selected for ATM switching so that those comparatively longer-lived applications may have better performance under heavy traffic load. Performance analysis is also carried out to show the efficiency of this cut-off priority flow classification policy.

Keywords: Flow Classification, Cut-off Priority, IP, ATM Switching.

1. INTRODUCTION

Flow classification is one of the key issues in the design of a data-driven IP/ATM switching system [1]. To perform flow classification, a data-driven IP/ATM switching system must have a local flow classification policy/algorithm to decide whether or not an IP flow should be selected for ATM switching. Different policies may have different impacts on the system performance. For this reason, alternative flow classification policies have been proposed for better performance, including the protocol-based policy, the application-based policy, the X/Y algorithm [2] and the adaptive algorithm [3]. In this paper, we study flow classification policies and propose a cut-off priority policy to improve the application-based policy. The basic idea is to assign priorities to the applications that are selected for ATM switching. Higher priorities are assigned to those comparatively longer-lived applications so that they may have better performance under heavy traffic load. Performance analysis based on queuing theory is also carried out to show the efficiency of this cut-off priority flow classification policy.

The remainder of this paper is organised as follows. In Section 2, we give a brief review of flow classification policies. In Section 3, we present our proposed cut-off priority flow classification policy and analyse its performance. In Section 4, we give some numerical results to show its performance as compared with that of the application-based policy. In Section 5, we conclude our study.

2. FLOW CLASSIFICATION

The main function of flow classification in a data-driven IP/ATM switching system is to identify flows and decide whether they should be switched directly in the ATM hardware or forwarded by the routing software. This can be implemented by inspecting the values of the packet header fields and making a decision based upon a local policy. For this purpose, a data-driven IP/ATM switching system must have a flow classification policy/algorithm. Different policies may result in different system performance. For this reason, alternative flow classification policies have been proposed for better performance. Generally, long-lived flows with a large number of packets should be selected for ATM switching while short-lived flows with a small number of packets should be handled by normal hop-by-hop forwarding [1]. The designer may choose to implement a particular policy according to the traffic characteristics in the specific environment the system operates in.

One flow classification policy, the protocol-based policy [2], is to classify flows by protocols. With this policy, all TCP flows are selected for ATM switching while all UDP flows are forwarded by the routing software. The argument is that connection-oriented services generally last longer and have more packets to be sent over a short time than connectionless services. Similarly, flows can also be classified by applications (or port numbers), such as ftp (20), telnet (23) and http (80). This application-based policy [2] is dependent on the statistical measurement of the mean duration and the average number of packets of each flow. Only those applications that tend to generate long-lived flows and contain a large number of packets are selected for ATM switching. Both of the above two policies have the advantage of simplicity in implementation.

Another flow classification policy [4] is to count the number of packets received on each flow so that the first X packets of each flow are forwarded by the routing software while the subsequent packets are switched by the ATM hardware, independent of protocols and applications. A modified version of this policy is the X/Y policy [2]. This policy also counts the number of packets received on each flow. If the number of packets of a flow received within Y seconds exceeds X, further packets of the flow will be switched by the ATM hardware. The basis behind these two policies is that if X packets have already been received on a flow (within Y seconds), it is reasonable to expect that there will be more packets on this flow. The simulation studies in [2] and [4] showed that these two policies could separate short-lived flows from long-lived flows more effectively as compared with the protocol-based and the application-based policies.

All the above flow classification policies are static in that their criteria and control parameters (like X and Y) are statically set. However, due to the varying characteristics of IP traffic, it is difficult to predict the traffic characteristics exactly and set optimal criteria or control parameters. For this reason, a static flow classification policy that works well in one system and/or for one application today may perform poorly in other systems and/or for other applications in the future. To adapt to the varying traffic characteristics, an adaptive flow classification algorithm has been proposed in [3]. This adaptive algorithm dynamically adjusts the values of control parameters in response to the varying traffic load and tries to balance the utilization of the system resources. The simulation study showed that it could offer better performance than the static policies. However, the cost of such an adaptive algorithm in terms of the complexity in implementation and the additional load on the processor might be much higher than that of the static policies.

3. CUT-OFF PRIORITY POLICY

Even though a variety of flow classification policies have been proposed, to classify flows by applications is still an effective approach in a real data-driven IP/ATM switching system due to its simplicity in implementation, especially when real-time multimedia applications are introduced. In such a switching system, when a flow is selected for ATM switching, the system will request a virtual circuit identifier (VCI) for the flow from its VC space so that the flow can be switched directly in the ATM hardware. However, due to the limited VC space of a real system, a flow that is selected for ATM switching might be blocked at the VC space and thus could not be switched in the ATM hardware. From the viewpoint of resource utilization, the VC space should be used as fully as possible in order to have better performance. In this sense, it is desirable to select as many applications as possible for ATM switching. On the other hand, if too many applications are selected for ATM switching indiscriminately, those applications that last comparatively longer and contain a larger number of packets might be blocked at the VC space under heavy traffic load, which may affect the system performance. To take account of both the VC utilization and the system performance, we now propose a flow classification policy based on cut-off priority.

3.1 Policy Description

The proposed cut-off priority policy is based on the application-based policy, i.e. it classifies flows by applications. Normally, a set of applications is selected for ATM switching. These applications are assigned a certain number of priorities according to the statistical measurement of their mean duration and average number of packets. Higher priorities are assigned to those applications that last comparatively longer and contain a larger number of packets. Correspondingly, a certain number of thresholds are set for the usage of the VC space. In general, we assume k priorities and set $(k-1)$ thresholds for the usage of the VC space. When the VC usage exceeds threshold i $(i=1,2,\cdots,k-1)$, the flows with priority j $(1 \le j \le i)$ will no longer be selected for ATM switching. Only the flows with priority j $(i < j \le k)$ will be switched in the ATM hardware. For example, when the VC usage exceeds threshold 3, the flows with priority 1, priority 2 and priority 3 will not be selected for ATM switching. Only the flows with priority 4, priority 5, ..., priority k will be switched. It is conceivable that, with this cut-off priority policy, the number of higher priority flows switched can be efficiently increased under heavy traffic load as compared with that using the application-based policy. Meanwhile, the VC utilization will not decrease under light traffic load.

3.2 Performance Analysis

While there are many performance metrics for a data-driven IP/ATM switching system, we use the average number of packets switched in the system as the main metric in the following analysis. To analyse the performance, we use the performance model we proposed in [5], shown Figure 1. For comparison, we also discuss the performance of the application-based policy without cut-off priority.

3.2.1 Cut-Off Priority Policy

As mentioned earlier, we assume k priorities and $(k-1)$ thresholds. The probability that a flow is selected for ATM switching and it belongs to priority i is P_i and the average number of packets in a single priority-i flow is L_i $(i = 1,2,\cdots,k)$, which can be estimated from empirical data. The threshold values are C_i $(i = 1,2,\cdots,k-1)$, where $0 \le C_1 \le C_2 \le \cdots C_{k-1} \le C$ and C is the VC space size. According to the performance model, for the cut-off priority policy, the VC space of the system can be modelled as an $M/M/C/C$ loss system with k flow priorities. The arrival rate of priority-i flows to the VC space is $\lambda_i = P_i\lambda$, where λ is the total flow arrival rate to the system. The service rate for priority-i flows is μ_i, which can also be estimated from measurements.

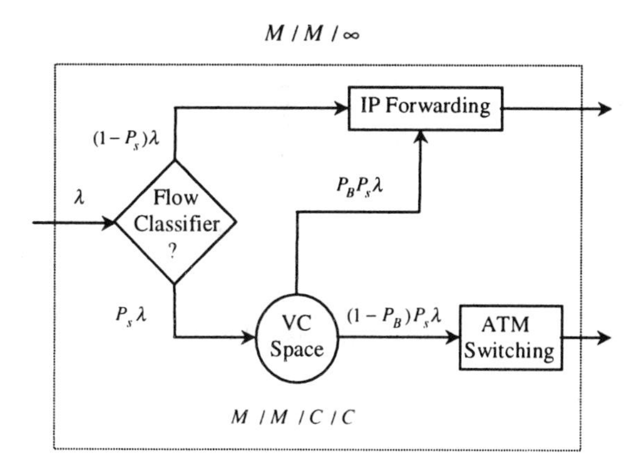

Figure 1 Performance Model

To perform the analysis, we use n to describe the state that is defined by

$$\bar{n} = (n_1, n_2, \cdots, n_k)$$

and Ω to describe a set of allowable states that is defined by

$$\Omega = \{\bar{n}: \ 0 \le n_i \le C_i, \ i = 1,2,\cdots k; \ 0 \le \sum_{i=1}^{k} n_i \le C\}$$

where n_i is the number of priority-i flows using the VC. Another notation needed is n_i^+ that is defined by

$$n_i^+ = (n_1, \cdots, n_{i-1}, n_i + 1, n_{i+1}, \cdots, n_k)$$

Then we can use the results obtained in [6]. The joint steady state distribution is given by

$$P(\bar{n}) = \prod_{i=1}^{k} \frac{a_i^{n_i}}{n_i!} \cdot G^{-1}(\Omega)$$

where $$G(\Omega) = \sum_{\bar{n} \in \Omega} (\prod_{i=1}^{k} \frac{a_i^{n_i}}{n_i!})$$

$a_i = \lambda_i / \mu_i$ is the offered load of priority-i flows and $1/\mu_i$ is the mean duration of priority-i flows. The blocking probability experienced by a priority-i flow is

$$P_{iB} = \frac{G(B_i^+)}{G(\Omega)}$$

where B_i^+ is defined by $B_i^+ = \{\bar{n} \in \Omega : n_i^+ \notin \Omega\}$.

The steady state distribution is given by

$$P(n_i) = \sum_{\substack{\bar{n} \in \Omega \\ n_i \equiv n_i}} P(\bar{n}) \qquad 0 \le n_i \le C_i$$

Therefore, the average number of priority-i flows switched is

$$N_i = \sum_{n_i=0}^{C_i} n_i P(n_i) \qquad 0 \le i \le k$$

and the total average number of packets switched can be calculated as

$$N_{pct} = \sum_{i=1}^{k} L_i N_i$$

3.2.2 Application-Based Policy

For the application-based policy, the VC space is completely shared by all priority flows. Compared with the analysis for the cut-off priority policy, the set of allowable states under the same conditions now becomes

$$\Omega = \{\bar{n}: \ 0 \le n_i \le C, \ i = 1, 2, \cdots k; \ 0 \le \sum_{i=1}^{k} n_i \le C\}$$

All other analytical steps are the same as that for the cut-off priority policy.

4. NUMERICAL RESULTS

In this section, we give some numerical results to illustrate the performance of the cut-off priority policy as compared with that of the application-based policy. For simplicity, we only discuss the case of $k=2$. For this purpose, we assume that $P_1=0.5$, $P_2=0.3$, $L_1=1000$ packets, $L_2=5000$ packets, $C=100$, $\mu_1=0.2$ flows/min, $\mu_2=1$ flow/min.

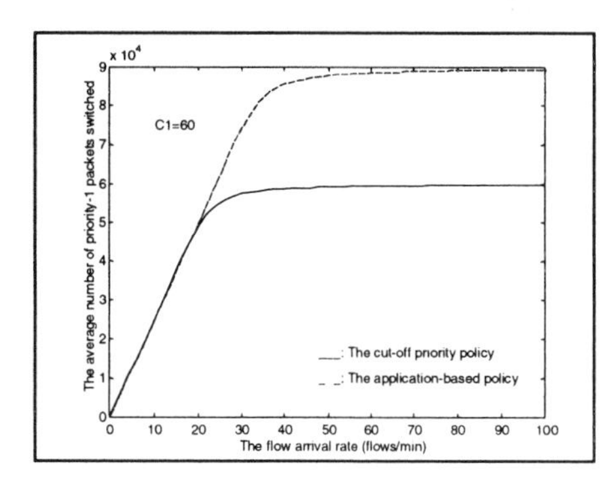

Figure 2 Avg. number of priority-1 packets switched versus the flow arrival rate

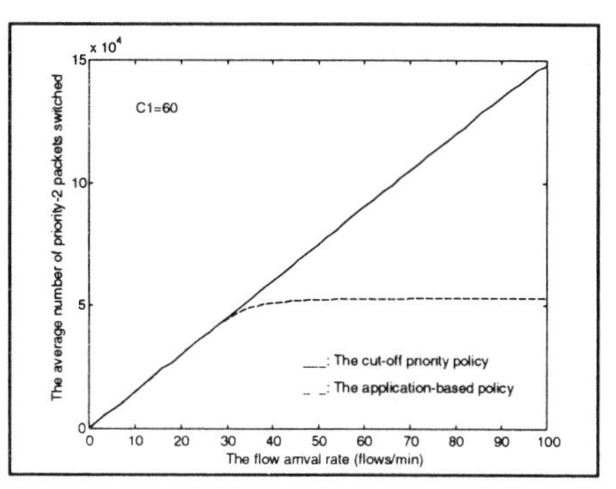

Figure 3 Avg. number of priority-2 packets switched versus the flow arrival rate

Figure 2 and Figure 3 respectively illustrate the comparison in the average number of priority-1 packets switched and the average number of priority-2 packets switched under a given threshold value $(C_1=60)$. When the flow arrival rate is very small, there is almost no difference in the performance for the two policies. This is because the VC usage will not reach the threshold under light traffic load. With the flow arrival rate increasing, the average number of priority-2 packets switched for the cut-off priority policy increases quickly while that for the application-based policy almost does not increase. On the contrary, the average number of priority-1 packets switched for the application-based policy is much larger than that for the cut-off priority policy. This is because, when the flow arrival rate is beyond a certain value, the VC usage will exceed the threshold. In this case, no more priority-1 flows will be selected for ATM switching, and only priority-2 flows will be switched. Therefore, with the cut-off priority policy, the performance of priority-2 applications in terms of the average number of packets switched is increased at the expense of decreased performance of priority-1 applications. However, this does not mean that the system performance in terms of the total average number of packets switched is also increased.

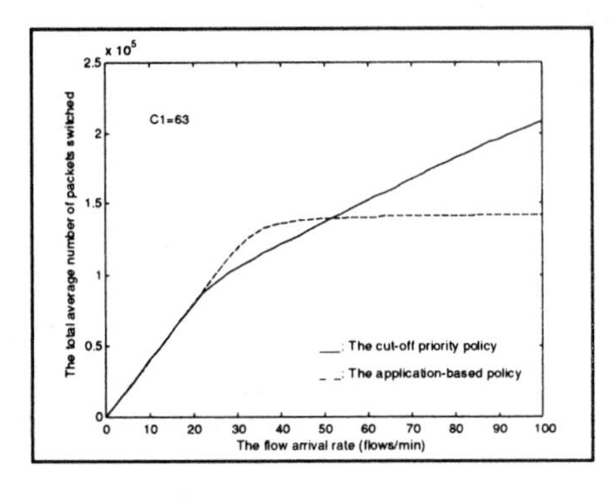

Figure 4 The total average number of packets switched versus the flow arrival rate

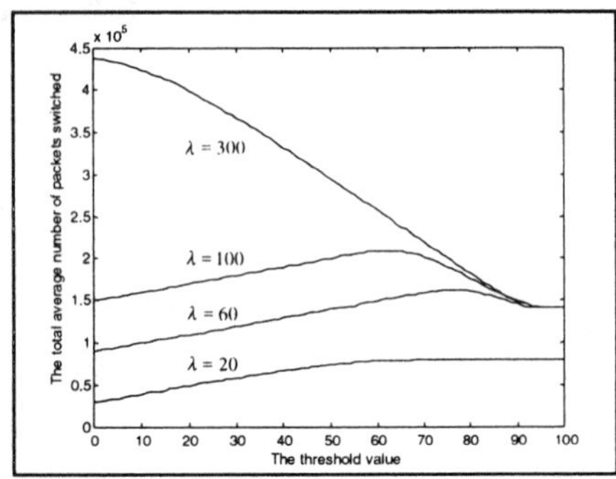

Figure 5 The total average number of packets switched versus the threshold value

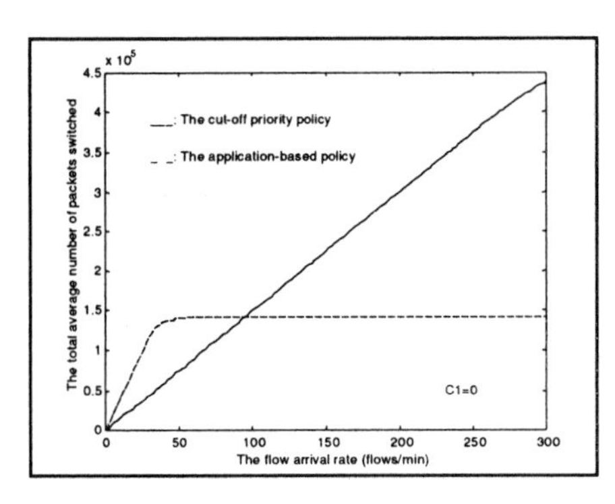

Figure 6 The total average number of packets switched versus the flow arrival rate

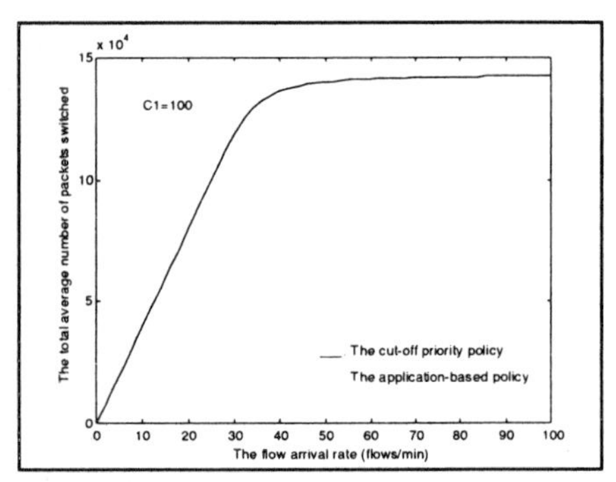

Figure 7 The total average number of packets switched versus the flow arrival rate

Figure 4 illustrates the comparison in the total average number of packets switched under a given threshold value $(C_1 = 63)$. It can be seen that the total average number of packets switched for the cut-off priority policy is not always larger than that for the application-based policy under different flow arrival rates. When the flow arrival rate is very small, there is almost no difference in the performance. With the flow arrival rate increasing, the performance for the application-based policy is better than that for the cut-off priority policy within a certain range of the flow arrival rate. Only when the flow arrival rate increases beyond a certain value will the performance for the cut-off priority policy become better than that for the application-based policy. The higher the flow arrival rate, the better the performance. Does this mean that, for some flow arrival rates, the total average number of packets switched for the cut-off priority policy is always smaller than that for the application-based policy?

To answer this question, let us see Figure 5, which illustrates the total average number of packets switched for the cut-off priority policy versus the threshold value under different flow arrival rates. For a given flow arrival rate, the threshold value may be adjusted to have better performance. No matter how large the flow arrival rate is, an optimal threshold value may be found that produces the best performance. Different flow arrival rates may have different optimal threshold values. For example, when the flow arrival rate is 60 flows/min and 100 flows/min, the optimal threshold value is 78 and 63 respectively. When the flow arrival rate is 20 flows/min, the optimal threshold value is 100, which is the VC space size. This means that, when the flow arrival rate is very small, the VC space is large enough to accommodate all the traffic. Accordingly, there is no need to set the threshold. When the flow arrival rate is 300 flows/min, the optimal threshold value is 0, which is another extreme. In this case, the flow arrival rate is very high. To provide better performance for priority-2 applications, the whole VC space should be dedicated to priority-2 applications.

Figure 4 and Figure 6 show the comparison in the total number of packets switched between the two policies under the optimal threshold value ($C_1 = 63$) for 100 flows/min and the optimal threshold value ($C_1 = 0$) for 300 flows/min respectively. Obviously, the performance for the cut-off priority policy is better than that for the application-based policy in both cases. The higher the flow arrival rate, the smaller the optimal threshold value and the better the performance. In Figure 7, the threshold value is 100, which corresponds to the optimal threshold value for 20 flows/min. Accordingly, no threshold is actually set and the performance for the cut-off priority policy is neither improved nor decreased.

Basically, given the VC space size and a flow arrival rate, the performance for the cut-off priority policy in terms of the total average number of packets switched under the corresponding optimal threshold value can be efficiently improved as compared with that for the application-based policy. The higher the flow arrival rate, the better the performance. The only exception occurs when the flow arrival rate is very small, in which case the performance is unchanged.

Based on the above observations, a further refinement of this cut-off priority policy is to adjust the thresholds to match the varying IP traffic. For this purpose, the system periodically measures the average flow arrival rate, calculates the corresponding optimal threshold values and then dynamically adjusts the threshold values.

5. CONCLUSION

In this paper, we proposed a cut-off priority flow classification policy for a data-driven IP/ATM switching system. The numerical results showed that, with the cut-off priority policy, the system performance in terms of the average number of higher-priority packets switched can be efficiently improved under heavy traffic load as compared with that using the application-based policy. Meanwhile, the optimal performance in terms of the total average number of packets switched in the system can be achieved by using the optimal threshold value of a given flow arrival rate. Even though the numerical analysis is discussed in the two-priority case, the results obtained apply to multiple priorities. Furthermore, by measuring the flow arrival rate and dynamically adjusting the threshold values, this cut-off priority flow classification policy will match the varying characteristic of IP traffic.

Acknowledgement

The authors would like to thank Dr. G. L. Li for his kind assistance in providing the valuable reference.

References

[1] Bruce Davie, Paul Doolan and Yakov Rekhter, "Switching in IP Networks," Morgan Kaufmann Publishers, 1998.
[2] Steven Lin and Nick Mckeown, "A Simulation Study of IP Switching," ACM SIGCOMM' 97, pp.15-24.
[3] Hao Che, San-qi Li and Arthur Lin, "Adaptive Resource Management for Flow-Based IP/ATM Hybrid Switching Systems," IEEE/ACM Transactions on Networking, Vol. 6, No. 5, Oct. 1998, pp. 544-557.
[4] Peter Newman, Greg Minshall and Thomas L. Lyon, "IP Switching—ATM under IP," IEEE/ACM Transactions on Networking, Vol. 6, No. 2, April 1998, pp. 117-129.
[5] Jun Zheng and Victor O. K. Li, "Performance Model for IP Switch," IEE Electronics Letters, Vol. 34, No. 21, Oct. 1998, pp. 2010-2011.
[6] Joseph S. Kaufman, "Blocking in a Shared Resource Environment," IEEE Transactions on Communications, Vol. Com-29, No. 10, Oct. 1981, pp. 1474-1481.

SESSION 9

Mobile Network Performance

ON BANDWIDTH RESERVATION POLICIES IN BROADBAND WIRELESS NETWORKS

Jelena Mišić and Tam Yik Bun
Department of Computer Science, Hong Kong University of Science and Technology
Clear Water Bay, Kowloon, Hong Kong
{jmisic,csben}@cs.ust.hk

Abstract Adaptive admission control reserves bandwidth for handoff calls by distributing bandwidth reservation values for each call to the neighboring base stations. The series of bandwidth reservation values distributed after k-th handoff determines the bandwidth reservation policy. This paper analyzes relationship between bandwidth reservation policies and call level QoS when user mobility is changing. We define classes of non-aggressive and aggressive bandwidth reservation policies. We show that bounded call level QoS under heterogeneous user mobilities can be achieved only by using aggressive bandwidth reservation policies. Examples of non-aggressive and aggressive bandwidth reservation policies are shown.

Keywords: multimedia wireless networks, adaptive call admission control, quality-of-service

1. INTRODUCTION

Third generation wireless networks should provide controlled call level QoS under the wide range of offered load and user mobilities. This should be provided through bandwidth reservation mechanism within wireless call admission control algorithm (WCAC). The WCAC algorithm must adapt to the traffic intensities in the surrounding cells, which can change when either new call arrival rates or user mobilities change. As the result, the handoff dropping probability and forced call termination probability must be bounded under different network conditions. Adaptive bandwidth reservation under uniform traffic conditions and under homogeneous user mobilities has been addressed through three main research directions:

Periodical, short-term handoff load estimation. New call admission decision is made according to the estimated short-term handoff load. Handoff load is estimated periodically based on channel occupancy in the neighboring cells [4].

Life-time handoff load estimation calculated over a number of equal time periods. Adaptive admission policy based on the *shadow cluster* concept has

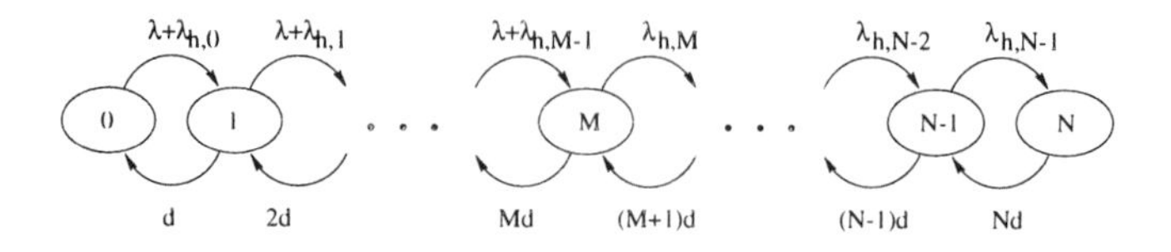

Figure 1 Markov chain for single type of traffic (where $d = h + \mu$)

been proposed in [1]. It is based on the calculation of probabilities that the mobile terminal will visit cells within its shadow cluster in consecutive, equal time periods during its life-time.

Event-based, short-term handoff load estimation. In [2], an adaptive WCAC algorithm is proposed in which the calculation of the estimated handoff load is triggered by certain events like handoff, origination and termination of calls in the surrounding cells.

Recently, adaptive WCAC with event-based handoff load estimation has been analyzed under non-uniform new call arrival rates and homogeneous mobilities in [3]. It determines the ranges of call level QoS when the call arrival rates are varying in the group of cells in the network.

This paper relates the bandwidth reservation policy within WCAC with the state probabilities of the underlying Markov chain, and derives its properties needed for the bounded call level QoS. The paper is organized as follows. Section 2 describes the Markov chain model for wireless cell with admission control. In Section 3 the bandwidth reservation process is described. Section 4 presents the performance analyses of the admission algorithm under uniform load with variable user mobility. In Section 5 we analyze the required properties of the bandwidth reservation policies and propose some of them. Section 6 concludes the paper.

2. SYSTEM MODEL

Consider a wireless cellular system with hexagonally tesselated cells. We use the assumptions that the probability to handoff to any of the six surrounding cells is equal to 1/6 and that the call duration and dwell time are exponentially distributed with parameters μ and h respectively. The cell capacity is N channels (in all further calculations we will use $N = 50$). Only narrow-band traffic is assumed consuming one channel per user under uniform traffic intensity in all the cells.

The Markov chain that represents system for the arbitrary admission algorithm for a single type of traffic is given in Fig. 1. The states in the chain represent the number of ongoing calls in the target cell. New call arrival rate is represented by λ. In this approach we address the observation that λ_h in the target cell should be decreasing function of the number of utilized channels in the target cell. Therefore we assume that the sum of current utilizations of the target cell and the cells from the surrounding ring is constant and equal to the sum of their average utilizations. In this case, for target cell with k ongoing calls, there are $7N\rho_{av} - k$ users in the surrounding ring with the "arriving parameter" h giving the handoff arrival rate: $\lambda_{h,k} = \frac{(7\rho_{av}N-k)h}{6}$.

The value $M = \rho_{max}N$ is the state where the call admission algorithm starts rejecting new calls. Beyond state M, only handoff calls are accepted. State probabilities of the Markov chain are then equal to:

$$P_k = \frac{(\frac{c}{6})^k\binom{A_1}{k}}{\sum_{i=0}^{M}(\frac{c}{6})^i\binom{A_1}{i} + \frac{\binom{A_1}{M}}{\binom{A_2}{M}}\sum_{i=M+1}^{N}(\frac{c}{6})^i\binom{A_2}{i}} \quad (0 \leq k \leq M) \tag{1}$$

$$P_k = \frac{(\frac{c}{6})^k\binom{A_2}{k}\frac{\binom{A_1}{M}}{\binom{A_2}{M}}}{\sum_{i=0}^{M}(\frac{c}{6})^i\binom{A_1}{i} + \frac{\binom{A_1}{M}}{\binom{A_2}{M}}\sum_{i=M+1}^{N}(\frac{c}{6})^i\binom{A_2}{i}} \quad (M < k \leq N) \tag{2}$$

where $c = h/(h+\mu)$, $A_1 = 7N\rho_{av} + \frac{6\lambda}{h}$ and $A_2 = 7N\rho_{av}$. The average utilization and probabilities of new call blocking and handoff call dropping for this system are therefore given by:

$$\rho_{av} = \frac{1}{N}\sum_{k=0}^{N} kP_k, \quad P_B = \sum_{k=M}^{N} P_k, \quad P_{hd} = P_N. \tag{3}$$

Since we consider systems with hard bounds on the handoff dropping probability $P_{hd} < 10^{-2}$, the number of new calls from both expressions can be approximated with $n_0 = \rho_{av}N\frac{\mu}{h+\mu}$. Furthermore, the number of calls in the target cell which have executed k handoffs is: $n_k = n_0(\frac{h}{h+\mu})^k$.

3. BANDWIDTH RESERVATION PROCESS

Every handoff call from the finite user population surrounding the target cell arrives independently from the others with the average rate h. In order to implement the bandwidth reservation to match the handoff arrival rate λ_h, we create the *bandwidth reservation rate* in the target cell using the whole user population in the first surrounding ring. Since the average dwell time $1/h$ may be different for various users and we don't know which mobile users will handoff to the target cell, mobile users contribute to the bandwidth reservation rate using some tentative numbers which we will denote as *bandwidth reservation values*. Every call that arrives to the cell either as a new call or handoff call will send bandwidth reservation value to all the surrounding cells. When the call leaves the cell that number is cleared from the surrounding cells. Every base station maintains the sum of those numbers associated with the bandwidth requirement of the call. By analogy to the handoff call arrival rate, the sum of bandwidth reservation values corresponds to the average bandwidth reservation rate of the *bandwidth reservation process*. Therefore, the bandwidth reservation rate in the target cell is:

$$\nu = \sum_{k=0}^{\infty} 6 n_k a_k \tag{4}$$

where n_k denotes the number of calls with k executed handoffs, and a_k denotes bandwidth reservation value sent after k-th handoff. After substituting value for n_k the average reservation rate becomes:

$$\nu = \rho_{av} N 6 \sum_{k=0}^{\infty} \frac{\mu}{\mu+h} (\frac{h}{\mu+h})^k a_k. \tag{5}$$

The average amount of bandwidth reservation values sent out by any connection from the ring surrounding the target cell is:

$$B = \frac{\nu}{\rho_{av} N} = 6 \sum_{k=0}^{\infty} \frac{\mu}{\mu+h} (\frac{h}{\mu+h})^k a_k. \tag{6}$$

Due to uniform cell capacity utilization and symmetry, connections from the target cell will send out the same average amount of bandwidth reservation values. In the text that follows we shall refer to B as the *bandwidth reservation parameter*. If we denote mobility parameter $\frac{h}{h+\mu} = c$, the bandwidth reservation parameter B becomes $B = 6(1-c)\sum_{k=0}^{\infty} c^k a_k$.

In case the value of mobility parameter c and bandwidth reservation policy a_k are known, the value of reservation rate can be calculated as: $\nu = \rho_{av} N B(c, a_k)$. The bandwidth reservation parameter B depends on the mobility parameter of the call, and on the policy for bandwidth reservation a_k.

3.1 TUNING THE BANDWIDTH RESERVATION

In order to do proper reservation of the bandwidth for various values of user's average dwell times, and for various bounds on handoff dropping probability, we choose bandwidth reservation values such that average bandwidth reservation rate is much larger than the handoff call arrival rate. In this case the regulating parameter is needed to determine which portion of the bandwidth reservation rate is needed to achieve the required QoS. We choose this regulating parameter to be the probability that "the number of the arrivals of the bandwidth reservation process" will exceed the currently unused capacity of the cell. This parameter is denoted as the *overload probability* and for the single traffic type it is defined as:

$$P_{ov} = e^{-\nu} \sum_{q=\lceil N(1-\rho_{curr.})\rceil}^{\infty} \frac{\nu^q}{q!} = \frac{1}{\Gamma(N(1-\rho_{curr.}))} \int_0^{\nu} e^{-x} x^{N(1-\rho_{curr.})} dx. \quad (7)$$

where $\rho_{curr.}$ denotes portion of the cell capacity currently occupied by the ongoing calls. By varying the parameter P_{ov}, it is possible to control the utilization in the cell, which means that handoff dropping probability can be controlled. Also, if ν changes due to the traffic fluctuations in the surrounding cells, for constant P_{ov}, $\rho_{curr.}$ will fluctuate in order to respond to different bandwidth reservation.

3.2 ADMISSION ALGORITHM BASED ON P_{OV}

We shall illustrate the algorithm using cell $(0, 0)$; but due to symmetry of the hexagonal network any other cell can be considered. The main steps of this algorithm are:

Step 1. Variables ν_j are initialized to zero and current utilization $\rho_{curr.}$ is set to zero. The values $a_{j,k}$ are chosen according to some QoS guaranteeing policy.

Step 2. When a new type j call arrives, the overload probability has to be calculated using expr. (7) in cell $(0, 0)$ and in the surrounding cells.

P_{ov} **checking in cell** $(0, 0)$: current utilization is incremented by α_j and the overload probability $P_{ov}(0, 0)$ is recalculated with the new value of $\rho_{curr.}$. If the updated value is larger than the predetermined limit, the new call is blocked.

P_{ov} **checking in surrounding cells :** In each cell $(1, i)$, $(0 \leq i \leq 5)$, the value of local ν_j is incremented by $a_{j,0}$ and the overload probability is recalculated. If the overload probability bound in any surrounding cell $(1, i)$ is violated, the new call must be blocked.

Step 3. The calls which are admitted in cell $(0, 0)$ can freely handoff to neighboring cells without invoking the call admission algorithm again. The handoff call is dropped only when there is insufficient bandwidth for serving the call. When a type j call executes k-th handoff from cell $(0, 0)$ to a neighbor cell $(1, i)$, and given that available bandwidth in the cell is sufficient for the call, the following actions must be performed:

Step 3.1. Update load PGF in cell $(1, i)$, which means that $\rho_{curr.}$ is incremented by α_j, ν_j is decremented by the bandwidth reservation value $a_{j,k}$.

Step 3.2. Update load PGF in the neighborhood of cells $(0, 0)$ and $(1, l)$. This action consists of the following:

Step 3.2.1. The new reservation values are distributed to cells surrounding cell $(1, i)$ and the corresponding values of ν_j are updated. Note that some cells not previously included in the neighborhood of cell $(0, 0)$ will receive the call's reservation factor for the first time. This reservation factor will be included in the execution of the admission algorithm for future new calls in these cells.

Step 3.2.2. The cells bordering the cell that was hosting the call before the handoff (0, 0), but which do not belong to the 1-ring neighborhood of the cell currently hosting the call, should delete the appropriate reservation factor.
Step 4. When a call terminates, all corresponding bandwidth reservation values in the neighboring cells are deleted and $\rho_{curr.}$ in the hosting cell is decremented by α_j.

4. QOS UNDER CHANGING MOBILITY

In this section we will relate the Markov chain presented in Fig. 1 with the admission algorithm from section 3.2. We use the threshold value of overload probability P_{ovT} to determine the threshold utilization ρ_{max} at which admission algorithm starts rejecting new calls.

Given the threshold value of overload probability P_{ovT} the admission equation derived from eqn. (7) is:

$$P_{ovT} = \frac{1}{\Gamma(N(1-\rho_{max}))} \int_0^{\rho_{av} N B} e^{-x} x^{N(1-\rho_{max})} dx. \tag{8}$$

The relationship between ρ_{av} and ρ_{max} through the clipping state in the Markov chain $M = \rho_{max} N$ is:

$$\rho_{av} = \frac{1}{N} \sum_{k=0}^{N} k P_k \tag{9}$$

With knowledge of the mobility parameters h, μ and B, and given the new call arrival rate λ, the pair (ρ_{max}, ρ_{av}) can be found by numerically solving the system of equations (9, 8).

Behavior of the QoS when the user mobility, or bandwidth reservation policy is changing (given constant new call arrival rate) can be obtained by considering either the handoff dropping probability or the forced call termination probability. The handoff dropping probability is determined by the expr. 2 for $k = N$. The forced call termination probability is determined by the following expression [5]:

$$P_{fct} = \frac{P_{hd}}{\mu/h + P_{hd}} \tag{10}$$

Note that P_{fct} as the call level QoS guarantee can be of more interest to the mobile user than P_{hd}, since the user is interested in the success of the entire call.

The problem of controlling call level QoS when the user mobility is changing can be therefore defined as the following:
Determine the bandwidth reservation policy a_k which will offer

$$P_{fct} = \chi(h, \mu) \tag{11}$$

(or $P_{hd} = \psi(h, \mu)$) when handoff rate is changing in the range $(h_{min}..h_{max})$ and call departure rate is changing in the range $(\mu_{min}..\mu_{max})$. The QoS shaping function $\chi(h, \mu)$ is determined by the network operator.

The task of finding the suitable bandwidth reservation policy requires solving of the system of equations (8, 9, 11) for the variables ρ_{av}, ρ_{max}, B. When the set of the values for the bandwidth reservation parameter are known, regression technique can be used to transform the set of points $[h, \mu, B]$ to the appropriate function $B(h, \mu)$. After that, $B(h, \mu)$ is transformed in series $B(c) = 6(1 - c)\sum_{k=0}^{\infty} c^k a_k$.

5. TYPES OF BANDWIDTH RESERVATION POLICIES

Better understanding of the role of the bandwidth reservation policy under changing handoff rates can be obtained if we look at the derivatives of the system of eqns. (9, 8, 11). Those derivatives can be obtained by considering state probabilities P_k given in expressions (1) and (2) as $P_k(\rho_{av}, \rho_{max}, h)$ and P_{ov} as $P_{ov}(\nu, \rho_{max})$:

$$\frac{\partial \rho_{av}}{\partial h} = \frac{1}{N}\sum_{k=0}^{N} k \frac{\partial P_k}{\partial \rho_{av}} \frac{\partial \rho_{av}}{\partial h} + \frac{1}{N} \frac{\partial(\sum_{k=0}^{N} k P_k)}{\partial \rho_{max}} \frac{\partial \rho_{max}}{\partial h} + \frac{1}{N}\sum_{k=0}^{N} k \frac{\partial P_k}{\partial h} \tag{12}$$

$$\frac{\partial P_{ov}}{\partial \nu}\frac{\partial \nu}{\partial h} + \frac{\partial P_{ov}}{\partial \rho_{max}}\frac{\partial \rho_{max}}{\partial h} = 0 \tag{13}$$

$$\psi'(h) = \frac{\partial P_N}{\partial \rho_{av}}\frac{\partial \rho_{av}}{\partial h} + \frac{\partial P_N}{\partial \rho_{max}}\frac{\partial \rho_{max}}{\partial h} + \frac{\partial P_N}{\partial h} \tag{14}$$

Here we again observe simplifying approximation applicable to the eqn. (13) similar to the one used in the analysis of QoS when new call arrival rate is changing [3]. The P_{ov} value corresponds to the area below the integrand function $f(x) = \frac{e^{-x} x^{N(1-\rho_{max})}}{\Gamma(N(1-\rho_{max}))}$ in the segment $x = [0, \nu]$, depicted in Fig. 2. The function $f(x)$ has the maximum at $x = N(1 - \rho_{max})$. When handoff rate h changes, the average bandwidth reservation rate will change due to the changes of average utilization ρ_{av} and bandwidth reservation parameter B, that is $\Delta\nu = N(\Delta\rho_{av}B + \Delta B\rho_{av})$. The character of the change $\Delta\nu$ depends on the particular bandwidth reservation policy, i.e. on the sign and absolute value of ΔB. Therefore, we have to consider two cases:

Aggressive bandwidth reservation policies. In such policies $\Delta\nu = N(\Delta\rho_{av}B + \Delta B\rho_{av}) > 0$ when handoff rate h grows by Δh. In this case, the end of the integrating segment ν tends to shift to the right towards the higher values of P_{ov}. However, P_{ov} regulation maintained by the admission algorithm has to keep the area below the function at the constant value P_{ovT}, so it will then shift the integrand function $f(x)$ to the right which is equivalent to the decrease of

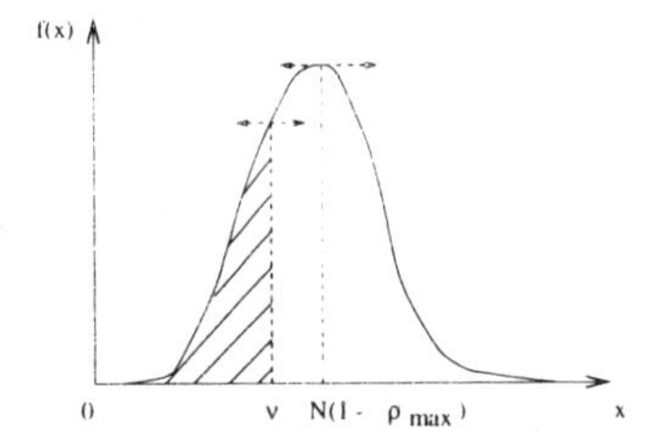

Figure 2 Representation of the utilization regulation property of the admission algorithm

the value of ρ_{max}.

Non-aggressive bandwidth reservation policies. In such policies $\Delta\nu = N(\Delta\rho_{av}B + \Delta B\rho_{av}) \leq 0$ when handoff rate h grows by Δh. In this case, the end of the integrating segment ν tends to shift to the left towards the lower values of P_{ov}. In order to keep the value of overload probability at level of P_{ovT}, the admission algorithm will increase the value of ρ_{max} and admit more calls.

Since $\Delta\nu \approx N\Delta(1 - \rho_{max})$ when handoff rate h changes the relationship between ν and ρ_{max} is:

$$\frac{\partial \nu}{\partial h} \approx -\frac{N\partial\rho_{max}}{\partial h}. \tag{15}$$

which shows that clipping utilization ρ_{max} is approximately linearly proportional to the bandwidth reservation rate ν. Since $\nu = \rho_{av}NB$, and by applying approximation (15) the equation (13) becomes:

$$\left(\rho_{av}\frac{\partial B}{\partial h} + \frac{\partial\rho_{av}}{\partial h}B\right) + \frac{\partial\rho_{max}}{\partial h} = 0 \tag{16}$$

By combining eqns. (12), (14) and (16) the following differential equations are obtained:

$$\frac{\partial\rho_{av}}{\partial h}\left(1 - \frac{1}{N}\sum_{k=0}^{N} k\frac{\partial P_k}{\partial\rho_{av}} + \frac{B}{N}\frac{\partial(\sum_{k=0}^{N} kP_k)}{\partial\rho_{max}}\right) =$$

$$-\rho_{av}\frac{1}{N}\frac{\partial(\sum_{k=0}^{N} kP_k)}{\partial\rho_{max}}\frac{\partial B}{\partial h} + \frac{1}{N}\sum_{k=0}^{N} k\frac{\partial P_k}{\partial h} \tag{17}$$

$$\frac{\partial\rho_{av}}{\partial h}\left(B\frac{\partial P_N}{\partial\rho_{max}} - \frac{\partial P_N}{\partial\rho_{av}}\right) = -\rho_{av}\frac{\partial P_N}{\partial\rho_{max}}\frac{\partial B}{\partial h} + \frac{\partial P_N}{\partial h} - \psi'(h) \tag{18}$$

From the equations (16) and (18), $B(h)$ function can be determined such to satisfy the required criterion. This is the system of two differential equations

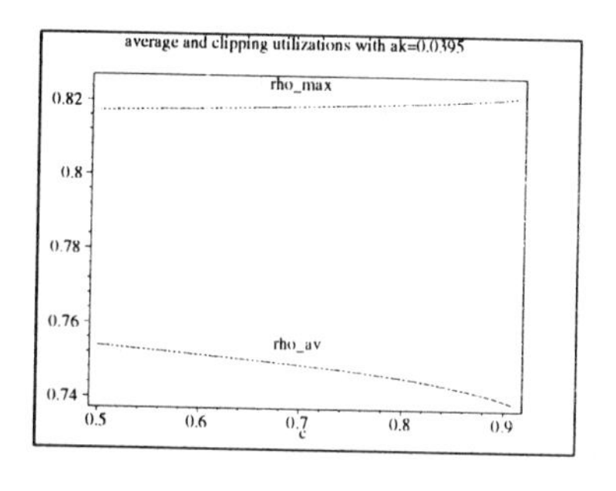

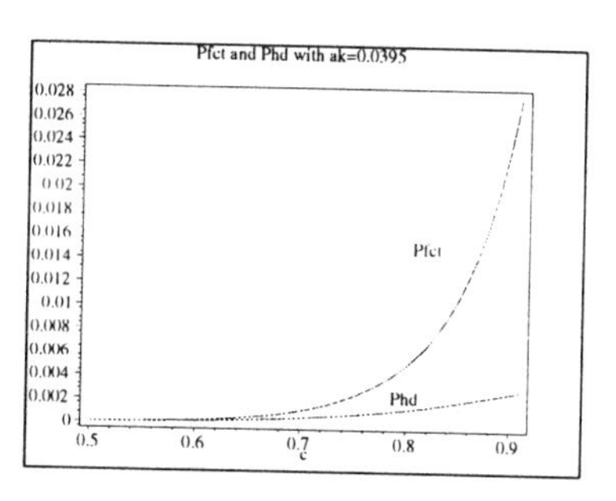

Figure 3 Call QoS behavior with constant bandwidth reservation parameter B, $\lambda = 0.1$, $c = \frac{h}{h+\mu}$, $\mu = 0.002$.

with changing parameters which again can be solved only numerically. Conversely, if we explicitly choose function $B(h, \mu)$, then we will obtain QoS characteristics $P_{fct} = \chi(h, \mu)$, as the result of bandwidth reservation policy.

In the following examples, we shall investigate explicit bandwidth reservation policies, i.e. the cases when the function $B(h, \mu)$ is known, and analyze the obtained P_{fct} and P_{hd}. In our analysis, the handoff rate h varies, while the call departure rate is fixed at the level $\mu = 0.002$. In all calculations we will assume that network operates at the nominal load $\frac{\lambda_{nom}}{N\mu} = 1$. Threshold for the admission algorithm is set to $P_{ovT} = 0.5$. The calibrating point for all policies is $B = 0.237$ under $h = 0.01$, $\mu = 0.002$ (i.e. $c = 0.833$).

Non-aggressive bandwidth reservation policy. If bandwidth reservation values are constant (independent of the number of previously executed handoffs) i.e. $a_k = a_0$, the bandwidth reservation factor becomes $B = 6a_0$. The value of $a_0 = 0.395$ is chosen to calibrate the admission system to give the desired tuple (B, h, μ). In this approach both handoff dropping probability and the forced call termination probability will increase with the growth of mobility parameter c as shown in Fig. 3 which prevents us from giving any guarantee on QoS as mobility changes. The consequence of the increasing forced call termination probability under fixed bandwidth reservation scheme is the decreasing average utilization as shown in Fig. 3.

Aggressive bandwidth reservation policies. Assume that a_k is proportional to the number of executed handoffs. Such approach will increase bandwidth reservation in two cases:

a) Calls with short dwell times and moderate duration times will increase bandwidth reservation according to their mobility dynamics.

b) Very long calls with moderate dwell times will implement the "call aging" concept to decrease the probability of forced termination when call has already executed a number of handoffs.

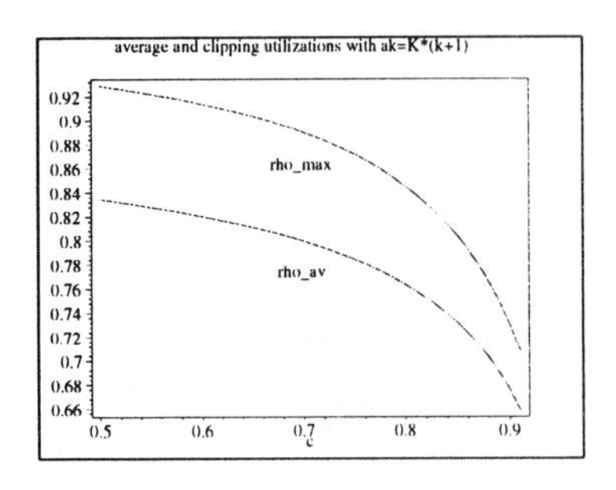

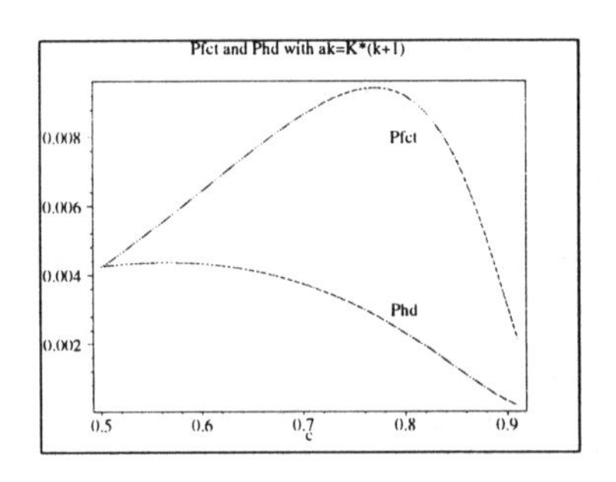

Figure 4 Call QoS behavior with bandwidth reservation parameter $B = \frac{6K}{1-c}$, $\lambda = 0.1, c = \frac{h}{h+\mu}$, $\mu = 0.002$.

As an example let us assume that $a_k = K(k+1)$. In this case, $B = 6K(1-c)\sum_{k=0}^{\infty}(k+1)c^k = \frac{6K}{1-c}$. This policy is easy to implement, since it does not involve computation, or storage at the mobile terminals.

6. CONCLUSION

In this paper we have analyzed the problem of controlling call level QoS under adaptive admission algorithm when the user mobility changes. We have shown that aggressive bandwidth reservation policies are needed to provide bounds on the forced call termination probability. In the future work, we will analyze QoS when both handoff rate and call departure rate are varying.

References

[1] D. A. Levine, I. F. Akyildiz and M. Naghshineh. "A Resource Estimation and Call Admission Algorithm for Wireless Multimedia Networks Using the Shadow Cluster Concept." *IEEE/ACM Trans. Networking*, vol. 5, no. 1, 1997, pp. 1–12., 1997.

[2] J. Mišić, S. T. Chanson and F. S. Lai. "Event Based resource Estimation in Admission Control for Wireless Networks with Heterogeneous Traffic." *Mobile Computing and Communications Review*, vol. 1, no. 4, 1997.

[3] J. Mišić and Y. B. Tam. "About the problem of hot-spots under adaptive admission control in the multimedia wireless networks." *to appear in proceedings of WCNC'99*, Sep. 1999.

[4] M. Naghshineh and M. Schwartz. "Distributed call admission control in mobile/wireless networks." *IEEE Journal on Selected Areas in Communications*, vol. 14, no. 4, pp. 711–717, May 1996.

[5] T.S. Rappaport. "Models for call handoff schemes in cellular communication networks." In *3rd WINLAB Workshop: Third Generation Wireless Information Networks*", Workshop Record, Apr. 28, 1992.

PERFORMANCE ANALYSIS OF CELLULAR SYSTEMS WITH DYNAMIC CHANNEL ASSIGNMENT

D.Tissainayagam and D.Everitt
Department of Electrical & Electronic Engineering
The University of Melbourne, Australia
{d.tissainayagam, d.everitt}@ee.mu.oz.au

Abstract We present a quasi-analytical method to compute the call blocking probabilities at a particular traffic density in a cellular system that employs dynamic channel assignment. The method is based on enumerating all the states in the entire state-space, a task that is beyond the capabilities of even the fastest computers when the cellular system has many cells and channels. We show how this task can be done by only focusing on the same system having just one channel. The results demonstrate that our combinatorial approach is very accurate.

Keywords: Dynamic Channel Assignment, State Space.

1. INTRODUCTION

By the end of the year 1998, the number of subscribers to a wireless communication system anywhere in the world was estimated to have been around 500 million. Presently, the average growth rate around the world is steady at just over 40%, which means that the wireless world would double in size in about 2 years' time. It has therefore become important to find means of increasing the capacity of cellular systems. This increase in capacity should not only absorb the expanding customer-base, but also maintain the present levels of quality of services. Many techniques have been employed to do just that: cell splitting [4], cell sectoring [7], frequency hopping [2], reuse partitioning [12], channel coding [9] are some of those that have already been implemented in cellular systems. A further option is to have in place a Dynamic Channel Assignment (DCA) algorithm.

Under DCA, channels are not distributed to cells in advance; all channels are deemed common to all the cells. A channel can be assigned to two callers in any two cells as long as they are separated by a 'buffer zone' of a cell's diameter. This is equivalent to saying that if a channel is used in one particular cell, then it cannot be used in any of the surrounding ring of six cells. This constraint is known as one-cell buffering. A call request in a cell is assigned a channel if and only if it is not in use anywhere in that cell's buffer zone.

Finding analytical estimations for the increase in capacity of cellular systems under DCA has proven to be a difficult task. Good approximations have been made in [1][3] [5][6][8][11]. However, an exact analysis has not been forthcoming. This paper attempts to do that in part.

We will name the cell in which a call request arrives to be the **call-cell**. In DCA, a channel that is in use within a cell is said to be **busy**. All other channels are considered **idle** in that cell. Not all idle channels in a cell, however, can be assigned to call attempts originating there. Some of them are **unavailable** in that cell for assignment because those channels are already busy in other cells that fall within its buffer zone. **Available** channels to a cell are those that are not busy anywhere within that cell and its buffer zone. Note that only available channels of a cell are assignable to call attempts there.

We denote the state of a channel in a cell by either a '1' (if it is busy) or a '0' (if it is idle). Throughout this paper, we let n be the number of channels at the disposal of the system. Then the state of a cell can be expressed in n bits. The state of the system is simply the collective states of all the cells. The set of all possible states of a system is its **state space**. In this paper, we first calculate the number of states in the state space $\mathcal{S}$ of a given cellular system. Then we calculate the number of those states in $\mathcal{S}$ that have at least one available channel in the call-cell - when the call-cell is specified. These are the states that permit a successful call attempt in the call-cell. The above computations are entirely mathematical. Furthermore, the enumeration of such states is independent of the traffic across the system. Finally, the probability of a state in $\mathcal{S}$ having at least one available channel in the call-cell is estimated via simulations. These traffic density dependent estimates are then made use of in computing the probability of a call attempt being blocked in a cell.

The **State Space Method**, as we call the above approach, is explained through an example and tested on two cellular systems. Later, it is generalised to a system containing any number of cells and channels.

2. EXAMPLE : THE 7-CELL SYSTEM

We take the cellular system shown in Figure 1 to explain our State Space method initially. The system has 7 cells, n channels and uses one-cell buffering. We shall treat cell 1 as the call-cell for this example. Note that a busy channel in the call-cell cannot be assigned to another user anywhere in the system.

First let us find the number of states in this cellular system's state space $\mathcal{S}$. Call that number S_n. Before doing that, a simpler problem is tackled: let S_1 be the number of states for the same system when it has just one channel - that

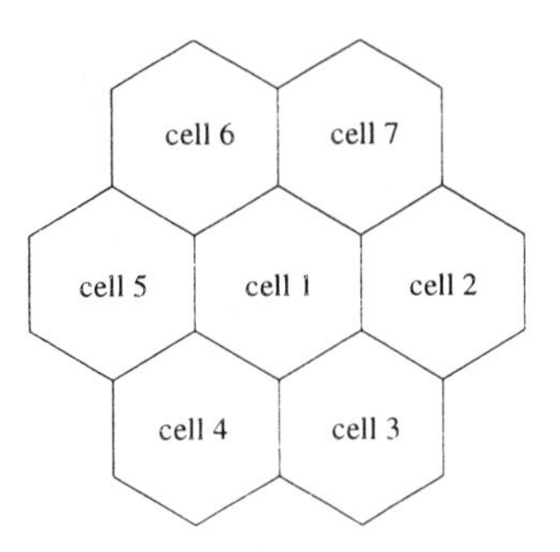

Figure 1 The 7-cell system

is, $n = 1$. S_1 can be found by direct enumeration. It turns out that there is 1 state with no calls, 7 states with one call, 9 states with two calls and 2 states with three calls. Hence,

$$S_1 = 19. \tag{1}$$

We also find in Figure 1 that for the first state, the number of calls in progress across the system is 0. For the next seven states, it is 1 and for the nine states thereafter it is 2. The number of calls in progress across the system is 3 for each of the last two states. There can be no state where 4 or more calls are in progress. We can express the above observations neatly by a **generator polynomial** which we define thus for the 7-cell, 1-channel system:

$$G(x) = 1 + 7x + 9x^2 + 2x^3. \tag{2}$$

The number of states with i calls in progress is simply the coefficient of x^i in the generator polynomial. Furthermore, $S_1 = G(1)$.

What if there are more than one channel in the system, though? The state of the 7-cell, n-channel system can always be broken down into a sum of the states of n identical 7-cell, 1-channel systems. An example is shown in Figure 2 for $n = 3$.

Figure 2(a) depicts the state of the 7-cell, 3-channel system. Channel #1 is busy in cell 1; channels #2 and #3 are busy in cell 2, and so on. This state can be decomposed into the states of three 7-cell, 1-channel systems as in Figures 2(b), 2(c) and 2(d). The channel that the 7-cell, 1-channel system possesses in Figure 2(b) is #1. In Figures 2(c) and 2(d), they are channels #2 and #3, respectively. We will call each of the the last three figures a *configuration*. The xth configuration only shows in which cells channel #x is busy (by a '1' in them) and idle (by a '0' in them). It will be realised that this decomposition into configurations is possible regardless of the value of n and regardless of the cellular topology of the system. Each configuration is totally independent of the rest. And each of them can exist in 19 different states - the number of states

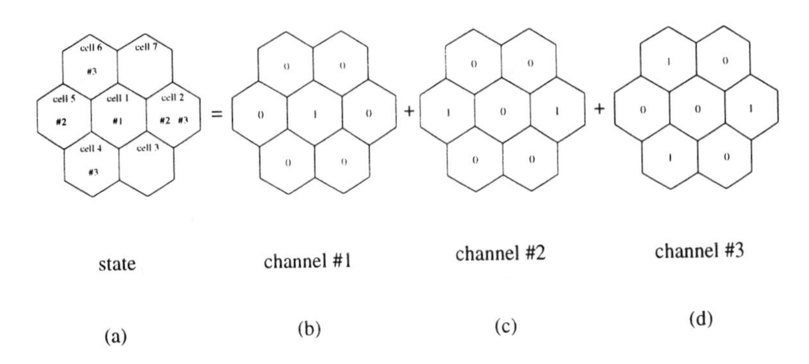

Figure 2 The break-down for a state in a 7-cell, 3-channel system

for the 7-cell, 1-channel system. Hence, the total number of possible states for the 7-cell, n-channel system is $19 \times 19 \times \ldots \times 19$ (n times). *i.e*

$$S_n = 19^n. \tag{3}$$

It will also be realised that the generator polynomial for the 7-cell, n-channel system is $G^n(x)$, where $G(x)$ has been defined in equation 2. It follows that the maximum number of calls that can be in progress, at a time, is $3n$, with each channel being busy in three cells. It also follows that $S_n = G^n(1)$.

Now, we calculate the number of states that have at least one available channel in cell 1 for the 7-cell, n-channel system. Let us call this number T_n. Focus back on the 7-cell, 1-channel system and its state space in Table 1. Out of the 19 possible states in it, only one state has the channel being available in cell 1 - namely column 1. If the system happens to be in this state when a call attempt is received in cell 1, then it will be successful. The rest of the 18 states have the channel being either busy in the call-cell (state 2) or unavailable to the call-cell (states 3 - 19). If the system were in any one of these 18 states, then a call attempt in cell 1 would be blocked. We note in passing that $T_1 = 1$, then.

Generalising to the n channel case, each of the n configurations in the decomposition of the state of the 7-cell, n-channel system can exist in 19 states, as was seen earlier. Out of which, only one state in nineteen in each configuration carries an available channel in the call-cell. Therefore, the number of states in the 7-cell, n-channel system which have *no* available channels in the call-cell is $18 \times 18 \times \ldots \times 18$ (n times). Hence,

$$T_n = S_n - 18^n = 19^n - 18^n. \tag{4}$$

Comparing equations 4 and 3, it is seen that the proportion of states in the state-space that have at least one available channel in the call-cell increases with n. For instance, just over half the states fall into this category when $n = 13$. Note that T_n can be derived from the generator polynomial $G(x)$: $T_n = G^n(1) - [G(1) - 1]^n$.

Now we move on to finding the probability that a cellular system is in a state where it has at least one available channel in the call-cell. This is an important quantity because it is also the probability that a call attempt in the call-cell will be successful. To begin with, turn again to the 7-cell, 1-channel system and its state space in Table 1. States 18 and 19 are equiprobable. This is so because under uniform traffic across the system, it does not matter in which three cells the channel is busy. Similarly, all of the states having two calls in progress are also equiprobable (states 9 to 17). Whether a state having two calls in progress is more probable than a state having three calls in progress depends on the traffic density. Under heavy traffic, the former state may be less probable than the latter state. Note also that only states 3 to 8 are equiprobable because cell 1 has six neighbours while the rest have only three; hence, the chances of a call attempt in cell 1 being successful (and being in state 2) are less than the chances of a call attempt elsewhere being successful.

Let the probability of this system having no calls in progress be π_0. Let the probability of cell 1 having a busy channel be π_a and the probability of any other cell alone having a busy channel be π_b. Also let the probabilities that the system has two and three calls in progress be π_2 and π_3, respectively. It must be noted that theses probabilities depend on the traffic density in the system. As the system has to be in one of the 19 states possible, we have that

$$\pi_0 + \pi_a + 6\pi_b + 9\pi_2 + 6\pi_3 = 1. \tag{5}$$

Now,

Prob (call attempt in cell 1 is successful) $=$ Prob (system in state 1) $= \pi_0$.

Therefore, the blocking probability for the call-cell in the 7-cell, 1-channel system is given by

$$B_1 = 1 - \pi_0. \tag{6}$$

With increased traffic density, π_0 decreases and B_1 increases.

Now, these probabilities are estimated when the system has n channels. Firstly, an assumption is made: the DCA algorithm used by the system is such that it does not consistently choose one particular available channel over another in the call-cell. This is necessary to maintain the independence assumption between configurations. Therefore, the responsibility of assigning an available channel to the offered traffic stream in the call-cell falls evenly on all n channels. Hence, it is seen that the probability that a channel is busy in 0, 1, 2 or 3 cells in the 7-cell, n-channel system under a uniform traffic density of ρ Erlangs per cell is the same as the probability that the channel is busy in 0, 1, 2 or 3 cells in the 7-cell, 1-channel system under a uniform traffic density of ρ/n Erlangs per cell, respectively. Having found $\pi_0, \pi_a, \pi_b, \pi_2$ and π_3 via this technique,

$$\begin{aligned} & \text{Prob (call attempt in cell 1 is successful)} \\ = & \text{ Prob (at least one available channel in cell 1)} \\ = & \text{ } 1 - \text{Prob (no available channels in cell 1)} = 1 - (1 - \pi_0)^n. \end{aligned}$$

Therefore, the blocking probability for the call-cell in the 7-cell, n-channel system is given by

$$B_n = (1 - \pi_0)^n. \tag{7}$$

3. APPLICATION TO THE 25-CELL SYSTEM

The State Space method is now applied to a larger cellular system shown in Figure 3. The number of channels at its disposal is n. One-cell buffering is used. We refrain from specifying a particular cell as the call-cell. The reason is that we want the estimated blocking probabilities to be applicable to any cell in the system - not just for cell 1 as in our example. To do that, first all cells must be made identical. Hence, we only analyse the system under uniform traffic density. Furthermore, all the cells are made to have exactly six neighbouring cells. This is achieved by juxtaposing replicas of the system to the top and bottom, and left and right of it. The resulting toroidal topology of the system is edgeless. Figure 3 depicts how cells 1 and 5, or cells 3 and 23 are really adjacent to each other.

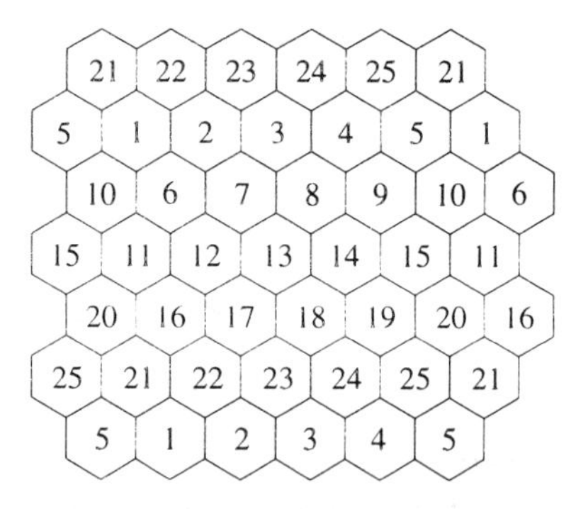

Figure 3 The 25 cell system

The example showed that all enumeration can be done from the generator polynomial of a cellular system with $n = 1$. Therefore, the generator polynomial for the 25-cell, 1-channel system is found first. It turns out to be

$$G(x) = 1 + 25x + 225x^2 + 900x^3 + 1600x^4 + 1100x^5 + 225x^6. \tag{8}$$

A short program was written to directly enumerate all possible states on a computer to determine the coefficients above. Therefore, the total number of states in the state space of the 25-cell, 1-channel system is given by

$$S_1 = G(1) = 4076. \tag{9}$$

Out of all those, how many states have the channel as available in one particular cell? It is not simply the first of the 4076 states; the one where no call is in progress in the system. This happened to be the case in our example because the call-cell and its buffer zone comprised the entire system. Now, there can be calls in progress on a channel outside a particular cell's buffer zone and still that channel would be available in that cell. Hence, we need to find the number of states in the state space where no calls are in progress within a particular cell and its buffer zone. The result is the following generator polynomial:

$$G_0(x) = 1 + 18x + 108x^2 + 256x^3 + 220x^4 + 54x^5. \tag{10}$$

This means that if, say, cell 13 happens to the call-cell, then there are 54 states where the call-cell and its buffer zone have no calls in progress and the rest of the cells collectively have five calls in progress - in cells 1, 4, 16, 19 and 23, for example. The point is that every state that contributes to $G_0(x)$, and there are $G_0(1) = 657$ of them, allows the call-cell to have the channel as available. And every state that does not contribute to $G_0(x)$ makes the channel unavailable to the call-cell. Generalising to the 25-cell, n-channel system, we have the following: the total number of state in the state space is

$$S_n = G^n(1) = 4076^n. \tag{11}$$

and number of states that have at least one available channel in a particular cell is

$$T_n = G^n(1) - [G(1) - G_0(1)]^n = 4076^n - 3419^n. \tag{12}$$

It takes just four channels to have a state space where half the elements in it would have at least one available channel to a given cell.

What is the probability that the 25-cell, 1-channel system is in a state where a call attempt in a cell will be successful? Note that there are a_i states, with i calls in progress in each, that have the channel as available in a cell. Here, a_i's are the coefficients of the generator polynomial $G_0(x)$. The total probability that the system is in one of these states is then given by $\sum_{i=0}^{5} a_i \pi_i$, where π_i is the probability that the channel is busy in i cells at some particular uniform traffic density ρ Erlangs per cell. The π_i's also have to satisfy the following equation (*c.f.* Equation 5):

$$\pi_0 + 25\pi_1 + 225\pi_2 + 900\pi_3 + 1600\pi_4 + 1100\pi_5 + 225\pi_6 = 1. \tag{13}$$

The probability of blocking for a call attempt in any cell is thus given by

$$B_1 = 1 - (\pi_0 + 18\pi_1 + 108\pi_2 + 256\pi_3 + 220\pi_4 + 54\pi_5). \tag{14}$$

Just as in the example, the probability of blocking in any cell in the 25-cell, n-channel system, at a traffic density of ρ Erlangs per cell, can be expressed as

$$B_n = [1 - (\pi_0 + 18\pi_1 + 108\pi_2 + 256\pi_3 + 220\pi_4 + 54\pi_5)]^n. \quad (15)$$

The π_i's have to be obtained from simulating the 25-cell, 1-channel system at a uniform traffic density of ρ/n Erlangs per cell.

4. RESULTS

The State Space method for calculating blocking probabilities is put to the test in the 25-cell system with $n = 45$. The traffic density ρ was varied from 8 to 20 Erlangs per cell. A simulation is run for the 25-cell, 1-channel system to obtain the probabilities $\pi_i, i = 0, 1, 2, 3, 4, 5$ and 6. Note that in the simulations the offered traffic would have to be in the range $\frac{8}{45} - \frac{20}{45}$ Erlangs per cell. Note also that no consideration need be given to the choice of DCA algorithm that has to be employed in the system - as there is only one channel. Once the π_i's are found, the blocking probabilities are estimated from equation 15. These estimates are compared against the blocking probabilities obtained by running the full simulation for the 25-cell, 45-channel system. Here, a DCA algorithm is necessary and the one chosen was called Random Channel Search algorithm [10]. This algorithm assigns any available channel, picked randomly, in the call-cell to a call attempt there and blocks it if none is available.

The comparison of blocking probabilities is shown in Figure 4. The State Space method approximates the actual call blocking probabilities very well across all traffic densities.

5. CONCLUSIONS

We proposed a novel method for estimating the call blocking probabilities for a cellular system with arbitrarily many cells and arbitrarily many channels. For a given system, by assuming that only one channel is available for users anywhere in the system, two generator polynomials $G(x)$ and $G_0(x)$ are found. This allows us to compute the total number of states in the state space when there are n channels in the system that can be assigned dynamically. The number of states that have at least one available channel in a particular cell can also be found. Both of the above computations are exact. The probabilities of call blocking at some traffic density can be estimated if the probabilities of i calls in progress in the system is known in advance. We approximate them from simulations. The simulations are confined to the simple case when the system has just one channel. The results demonstrate that the State Space method is accurate.

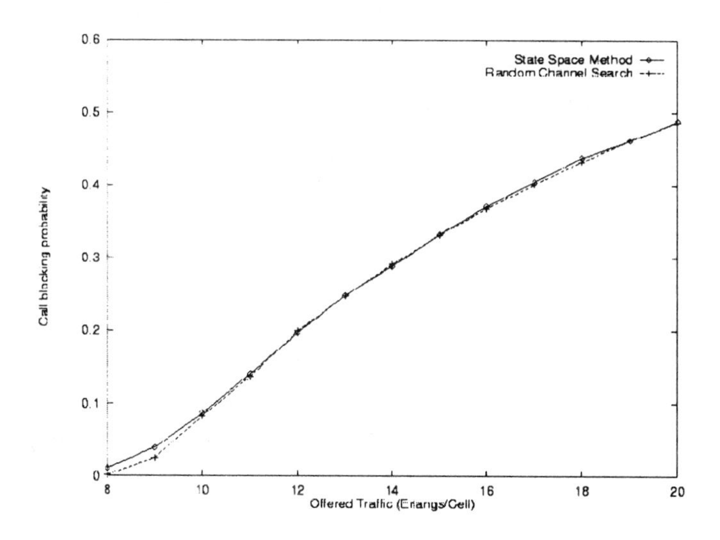

Figure 4 Call blocking probabilities for the 25-cell, 45-channel system

References

[1] D.Everitt and D.Mansfield, "Performance Analysis of Cellular Mobile Communication Systems with Dynamic Channel Assignment", *IEEE Journal on Selected Areas in Communications*, Vol. 7, pp. 1172-1180, 1989.

[2] M.Madfors, K.Wallstedt, S.Magnusson, H. Olofsson, P-O.Backman and S.Engstrom, "High Capacity with Limited Spectrum in Cellular Systems", *IEEE Communications Magazine*, Vol. 35, pp. 38-45, 1997.

[3] D.L.Pallant, "A Reduced Load Approximation for Cellular Mobile Networks Including Handovers", *Aust. Tel. Res.*, Vol. 26, pp. 21-29, 1992.

[4] T.Rappaport, "Wireless communications : principles and practice", IEEE Press, 1996.

[5] P.Raymond, "Performance Analysis of Cellular Networks", *IEEE Transactions on Communications*, Vol. 39, pp. 1787-1793, 1991.

[6] K.N.Sivarajan, R.J.McEliece and J.W.Ketchum, "Dynamic Channel Assignment in Cellular Radio", *Proc. of the 40th IEEE Vehicular Technology Conference*, pp. 631-637, 1990.

[7] G.Stuber, "Principles of Mobile Communication", Kluwer Academic, Boston, 1996.

[8] D.L.Pallant and P.G.Taylor, "Approximation of Performance Measures in Cellular Mobile Networks with Dynamic Channel Allocation", *Telecommunication Systems*, Vol. 3, pp. 129-163, 1994.

[9] J.Tisal, "GSM cellular radio telephony", Wiley, 1997.

[10] D.Tissainayagam, D.Everitt and M.Palaniswami, "Self Organising Feature Map Implementation of the Random Channel Search Algorithm", *Proc. of Asia-Pacific Conference on Communications*, Vol. 2, pp. 921-925, Dec. 1997.

[11] P.Whiting and S.Borst, "Performance Bounds for Dynamic Channel Assignment Schemes Operating under Varying Reuse Constraints", in *Proc. of INFOCOM '98*, Atlanta, USA, April 1998.

[12] J.Zander and M.Frodigh, "Capacity Allocation and Channel Assignment in Cellular Radio Systems Using Reuse partitioning", *Electronic Letters*, Vol. 28, No. 5, pp. 19-25, Feb. 1992.

AN EMERGENCY TOKEN BASED MULTIPLE ACCESS CONTROL SCHEME FOR WIRELESS ATM NETWORKS

Kun Pang[1],Zhisheng Niu[1],Junli Zheng[1] and Xuedao Gu[2]
[1]*Dept. of Electronic Engineering, Tsinghua University, Beijing,100084,China*
[2]*Academy of Telecommunication Technology, Beijing, 100083, China*
{pk,niu,zhengjl}@atm.mdc.tsinghua.edu.cn

Abstract This paper proposes an emergency token based MAC scheme for wireless ATM networks. Token buffer of MT(Mobile Terminal) are constructed in BS(Base Station) for future slot allocation to reduce overhead of access request. The scheduler in base station assigns slots to the sources by polling of token buffer. In order to deal with dynamic nature and stringent QoS of compressed video source, emergency message is defined to inform BS of the source's rate variation and buffer status. At the same time, voice sources are served without reservation. Also, handoff call of voice under micro-cellular environment can be accommodated timely. Emergency token buffer has higher priority over those of VBR and ABR. The emergency token scheduling scheme (ETSS) based on polling of token buffer is effective and easy to implement. Simulations are done to evaluate its performance. We found that the scheduling scheme proposed can significantly improve rtVBR's loss and delay performance while low signalling overhead is needed.

Keywords: Wireless ATM, Multiple Access Control, QoS(Quality of Service)

1. INTRODUCTION

Wireless communication today has evolved so dramatically that it is highly necessary for wireless system to have more capacity, better quality, and ability to support QoS of multimedia service. As a wireless extension to ATM networks, first proposed in [1], wireless ATM is still a very popular topic nowadays. How to provide wireless access to multimedia services with efficiency and QoS guarantees is one of the most important issues.

There are some typical MAC(Multiple Access Control) schemes such as DRAMA(Dynamic Resource Allocation Media Access)[2], DQRUMA (Distributed Queueing Request Update Multiple Access)[3], and PNP (Polling with Nonpreemptive Priority)[4]. DRAMA/DQRUMA are proposed to deal with integrated voice and data services, in which VBR services are regarded as data services. In DRAMA scheme, the access of data service is contention based. When contention occurs, the mobile station will retransmit after waiting for random time. DQRUMA is collision free through request access in control frame. These two schemes are well designed to accommodate wireless access, however, there are inherent limitations in them. First, they can not meet the QoS requirements of VBR service. VBR applications (especially real-time video) have relatively strict delay, jitter and loss requirements as well as burst nature over different time-scales. Second, the request access overhead is high. Polling with Nonpreemptive Priority(PNP) is a centralized MAC scheme with passive user. After admitted into the system, the sources will be served by polling. Polling scheme is seldom preferred in wireless system for it will be overburdened by large number of users. In PNP scheme of [4], fixed priorities are assigned to the sources. This implies that lower priority sources may be locked out completely under condition of heavy load [10]. Also, PNP does not consider the dynamic characteristics of video applications as well.

In this paper, we develop a scheduling scheme which can satisfy the dynamic nature involved in wireless multimedia. In stead of dividing the sources into fixed priorities, a dynamic priority scheme is implemented through emergency token. It means that the sources which have the urgent need to transmit will have higher priority. Although our token model based scheduling scheme is to be deployed in TDMA system, it is supposed to adapt to a CDMA one where the bandwidth used in CDMA system can be presented with token as well.

2. SYSTEM MODEL

Consider a cellular wireless system, where many mobile stations with multiple connections share a common wireless link. A base station is responsible for access of all mobile stations. It is assumed that user information is accommodated in slots. Since the downlink channel is a broadcasting channel where no contention takes place, the focus of our research is on uplink scheduling scheme in the base station.

It is assumed that each voice call may be modeled as 2-state Markov chain, alternating between ON and OFF states [11]. Let N, v_i and t_{i+1}denote total number of voice connection, number of voice packet arrivals in this frame and number of voice talk spurts in next frame. Then the probability for j talkspurts starting in the next frame, is given by

$$\Pr(t_{i+1} = j) = \binom{N - v_i}{j} (1 - p_{off})^j p_{off}^{N-v_i-j} \tag{1}$$

It is assumed that slot reordering and reassignment has been implemented to keep all the unused slots at the end of each frame[5]. When a CBR connection enters OFF period, the base station will be informed through a special bit pattern in the last cell of the connection in order to perform slot reordering. In our scheduling scheme, CBR sources are served by periodically feeding token. Therefore, our focus is on the dynamic scheduling of VBR sources. Like the analytical model proposed in [8], both the cell arrival process of nrt-VBR and that of MPEG are assumed to be 2-state MMBP. Though Poisson model isn't quite suitable for modeling MPEG video, MMBP model is exploited to simplify analysis and capture the bursty characteristics.

In order to develop a analytical model, we first assume that both the cell arrival process of nrt-VBR and that of MPEG are 2-state MMBP where the state transition probabilities $\omega_{11} = \omega_{12} = \alpha_v$ and $\omega_{11} = \omega_{12} = 1 - \alpha_v(v$ denote the class of VBR cell)[8]. A multi-server dedicated-buffer queueing model $MMBP1 + MMBP2/D/s(m1, m2)$,where is suitable for the analysis of the proposed MAC scheme.

(N_1, N_2, Y_1, Y_2) is defined as the system state during n-th slot, where N_1 is the sum of the number of occupied slots and the number of C-1(MPEG) cells waiting in Q_1, N_2 is the number of C-2(nrt-VBR) cells waiting in Q_2, and $Y_v$is the phase of the arrival process $MMBP_v$. By using Matrix-Geometric method,[9] the transition probability matrix P is solved and, therefore, the steady-state probability $\Pr\{N_1 = i, N_2 = j, Y_1 = l_1, Y_2 = l_2\}$ is acquired by solving following equation:

$$\mathbf{x}P = \mathbf{x} \tag{2}$$

$$\mathbf{xe} = 1 \tag{3}$$

Then the cell loss probability $P_{Lv}$of class v is given by

$$P_{Lv} = \frac{E[L_v]}{E[A_v]}, \qquad (v = 1, 2) \tag{4}$$

$$P_{L1} = \frac{\sum_{i=0}^{s+m_1} \sum_{l_1=1}^{r_1} E[L_1|N_1 = i, Y_1 = l_1] \Pr\{N_1 = i, Y_1 = l_1\}}{\sum_{k_1=0}^{\infty} k_1 \sum_{l_1=1}^{r_1} a_1(k_1, l_1)\pi_1(l_1)}$$

$$= \frac{\sum_{i=0}^{s+m_1} \sum_{k_1=s_1}^{\infty} (k_1 - s_1)\{\sum_{l_1=1}^{r_1} a_1(k_1, l_1)\, \bar{x}_{il_1}\}}{\lambda_1} \tag{5}$$

As for the loss probability of nrt-VBR traffic, however, it depends not only on the system state but on the parameters of both MPEG and nrt-VBR traffic. Following the similar probability arguments, we have

$$P_{L1} = \frac{\sum_{i=0}^{s+m_1} \sum_{j=0}^{m_2} \sum_{l_1=1}^{r_1} \sum_{l_1=1}^{r_2} E[L_2|N_1 = i, Y_1 = l_1, N_2 = j, Y_2 = l_2]x_{ij}(l_1, l_2)}{\sum_{k_2=0}^{\infty} k_2 \sum_{l_1=1}^{r_2} a_2(k_2, l_2)\pi_2(l_2)}$$

$$= \frac{\sum_{i=0}^{s+m_1} \sum_{j=0}^{m_2} \sum_{k_1=0}^{\infty} \sum_{k_2=s_2}^{\infty} (k_2 - s_2)\{\sum_{l_1=1}^{r_1} \sum_{l_2=1}^{r_2} a_1(k_1, l_1)a_2(k_2, l_2)x_{ij}(l_1, l_2)\}}{\lambda_2} \tag{6}$$

Since the regulated nrt-VBR's delay bound under PGPS service discipline, $D_i^* \leq \frac{\sigma}{\rho}$, is quite tolerable for bursty service[7]. We choose large buffer size for nrt-VBR regulated by leaky bucket(ρ_j, σ_j). To adapt to rt-VBR's dynamic characteristics and delay requirement, nrt-VBR is handled as class-2 traffic. However, the priority is a dynamic priority that benefits both fairness and efficiency.

3. FRAME STRUCTURE

We emphasize our research on slot allocation scheme of uplink channel. A TDMA based frame structure is shown in Figure 1 .The length of MAC frame is T. The control part of a frame is consisted of emergency minislots where emergency messages are transmitted through S_ALOHA and request minislots where access requests are made. The data part of a frame is consisted of data slots where user information packetized in wireless ATM cells are transmitted.. Our MAC structure is somewhat like that proposed in [8], however, reservation

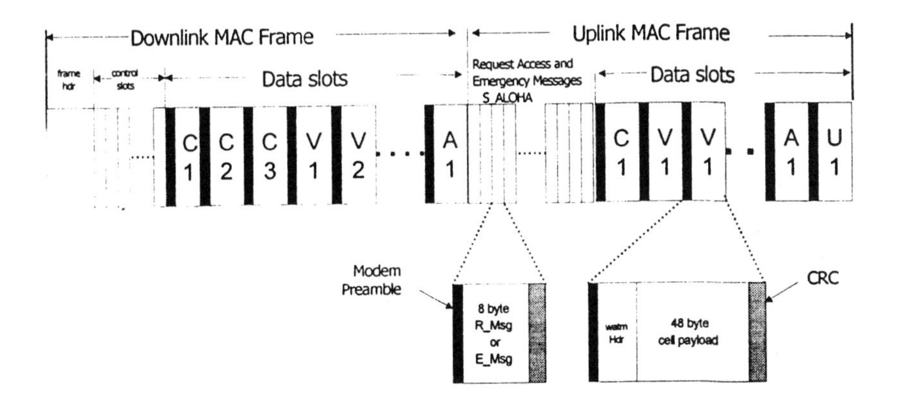

Figure 1 TDMA based MAC frame structure

slots is unneccessarry any more and MAC frame is not divided fixedly into CBR, VBR subframe to accommodate multi-rate CBR service. Also, we distinguish $Emergency\_Message$ from $Access\_Request$ in the control part of MAC frame, which is based on the idea that system will be overloaded when there are many emergency messages.

There are three types of 8-byte $Emergency\_Message$ defined: a)MPEG frame rate variation is over a certain threshold: $MPG\_Scene\_Message$ (We have known that VBR video is known to exhibit multiple time scale variations. Between scenes there is a significant change in the bit rate. For this reason, MPEG rate variation is emphasized in source modeling for accuracy.); b)rt-VBR source buffer is over threshold: $VBR\_Threshold\_Message$; c)CBR source enter ON period: $CBR\_Contend\_Message$. After receiving an emergency message, BS will put emergency tokens into emergency token buffer following the rule proposed in next section.

4. EMERGENCY TOKEN BASED SCHEDULING SCHEME(ETSS)

In wireless ATM scenario, the traffic sources send traffic descriptor and specified QoS to BS during connection setup period. After admitting access of sources, the scheduler in the base station will allocate slots to the connections according to the tokens got from token buffer where the tokens are generated with certain rule. The token polling shown in Figure 2 seems like mathematical model of reservation but with more flexibility. The token is generated with following rules:

i) For CBR source j with rate λ_j, its token is also generated with λ_j;

ii) For nrt-VBR source j regulated by leaky bucket(ρ_j, σ_j), its token is generated with sustained cell rate ρ_j;

iii) For MPEG-VBR source j, its token is generated variably according its frame rate $\rho_{I,j}$, $\rho_{P,j}$, $\rho_{B,j}$.

iv) For ABR service with specified MCR rate η_j, its token is generated with MCR rate η_j after t_a ms to meet its Poisson characteristics.

In uplink slot allocation, the scheduler polls the token buffers and assigns slots to the sources identified by tokens. If any $Emergency\_Message$ is received, the scheduler will first check out message type and change models for sources as following:

1) If it is a $CBR\_Contend\_Message$, put one token containing the connection-ID HOL into emergency buffer, then start generate token of the corresponding source into CBR token buffer. Therefore it doesn't impact other CBR's CDV.

2) If it is a $MPG\_Scene\_Message$, change token generating rate in model and put N_j^{i+1} emergency tokens into emergency buffer. Take statistical multiplexing gain into account, scheduler will stop sending N_j^{i+1} emergency tokens after a length of MPEG frame.

$$N_j^{i+1} = \Delta R_{frame\_change} * T \tag{7}$$

In equation above, if $\Delta R_{frame\_change} \leq 0$, then $N_j^{i+1} = 0$, only token generation rate is changed.

3) If it is a $VBR\_Threshold\_Message$, put N_{buffer_length} emergency tokens into emergency buffer. For sake of statistical multiplexing, stop sending token for $t = N_{buffer_length}/\rho_j$s after burst length.

When slot allocation begins, the scheduler scans the token buffer and assigns slots.

1) The scheduler scans CBR token buffer periodically. Any token found is removed and one slot is assigned to the corresponding CBR source.

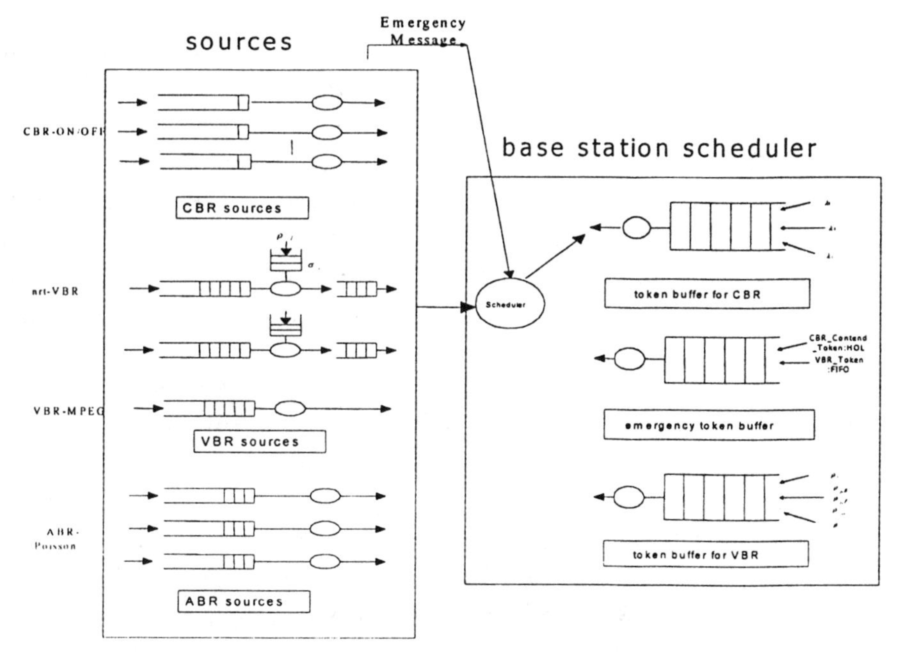

Figure 2 Dynamic token model based sheduling scheme

2) In order not to affect CBR's jitter requirement, emergency slot allocation is after CBR token buffer polling cycle. If there is no token found in CBR token buffer, the scheduler scans $EMERGENCY$ token buffer for token and assigns slots to corresponding sources.

3) When there is no token in the emergency tokens, the scheduler will enter VBR token polling cycle. The scheduler scans the VBR token buffer and assigns slots correspondingly until there is no token for VBR sources. The extraction of tokens is exhaustive so that slot allocation is continuous except that it will be interrupted by periodical CBR token polling.

4) After slot allocation for CBR and VBR is finished, the scheduler will assign remaining slots in a MAC frame to ABR and UBR sources with fair and efficient sharing.

5. SIMULATION RESULTS

It is assumed that link speed is 10Mbps in our system. The MAC subframe duration T is 2ms for delay and jitter requirements of CBR traffic. MPEG source's frame rate is 25Hz, therefore, video frame length(40 ms) is sufficiently lager than T_s so that information of video frame change is much meaningful. MPEG-I cells are generated because of limited wireless bandwidth, where average bit rate is from 0.8 to 1.6Mbps and peak-to-average ratio is 4. As for the ON/OFF parameter of voice will depend on coding technique. We use the parameters proposed in [11]. As for general VBR service, Poissonian model is enough for simulation. And $gamma\ beta\ autoregressive\ process$ (GBAR) proposed in [6] was used to generate MPEG video source traces in our simulation, which is considered as an appropriate model for VBR video source.

We write a MAC simulator in C/C++ to verify our scheduling scheme-ETSS and an error-free environment is assumed. The simulator is consisted of several modules of aloha manager, sources, slot allocation and token manager, etc. Since it is in microcellular environment, transmitting delay is ignored and only queueing delay is considered. From results in Figure 3, it is shown that rt-VBR's CLP(Cell Loss Probabilities) performance is improved under ETSS more than that of VBR round-robin. Emergency token may assign priority to corresponding source dynamically so that QoS of realtime and bursty source can be satisfied. The buffer size c can be expressed as the time required to empty a full buffer(buffer drain time). Buffer drain time $B_i = c/d_i$, where d_i is service rate of queue i. In Figure 4, we compare simulation results with analytical results in [8] of two kind of cells. As we expected, the performance curves of MPEG cells accord with each other under light MPEG load. But performance will deteriorate under heavy load. The reason is S_ALOHA transmission of emergency message will degrade greatly accompanied with emergency load's increase. The average queueing delay of MPEG cells in Figure 5 increasing drastically under heavy load is also due to the reason above. Number of emergency slots is the essential parameter of our model. From Figure 6, we can find that performance improvement is high when minislot number varies from 5 to 10(infinite buffer is used here). But improvement is rather low with the minislot number increased because system will be overloaded by emergency handling. And we can see that the scheduling scheme can work up to high utilization of 90% for the scheme exploits VBR's statistical multiplexing efficiently, however, the utilization is supposed to fall under error link, where retransmission is frequent.

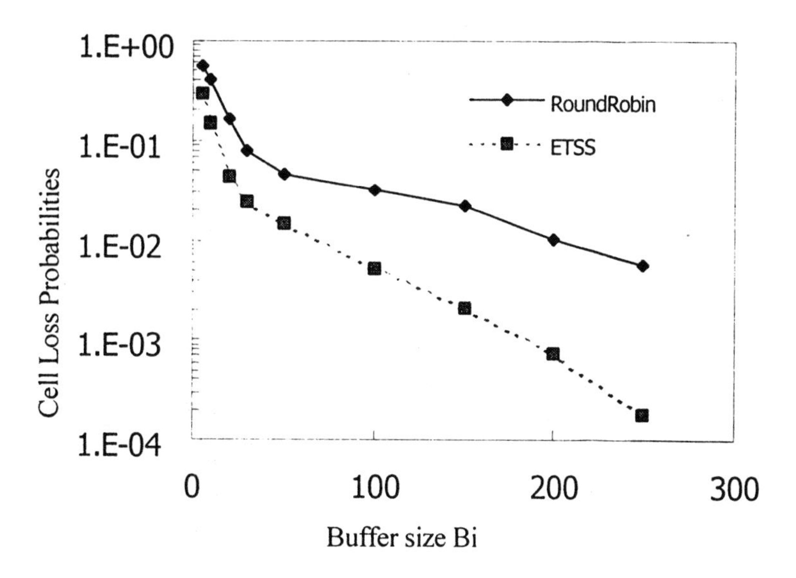

Figure 3 Buffer size vs cell loss probabilities (with utilization of 60%, Number of emergency slots=10)

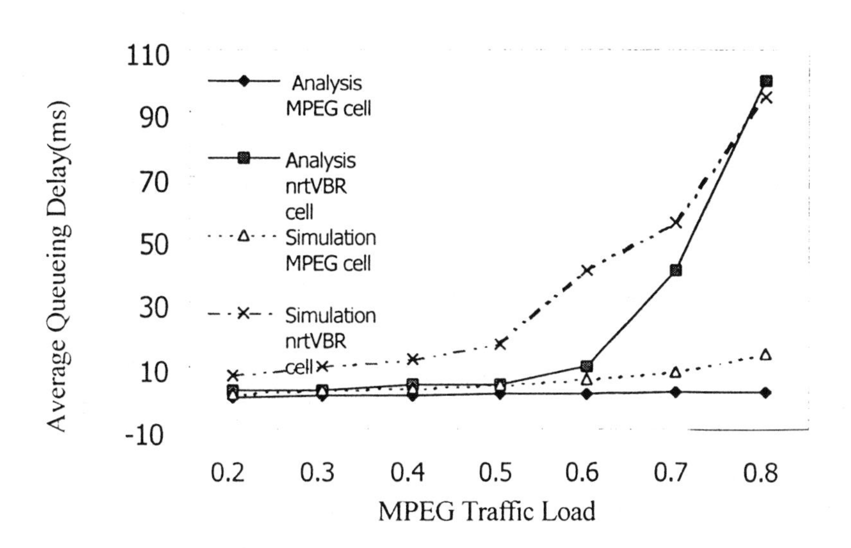

Figure 4 MPEG traffic load vs cell loss probabilities (with utilization of 60%, Number of emergency slots=10)

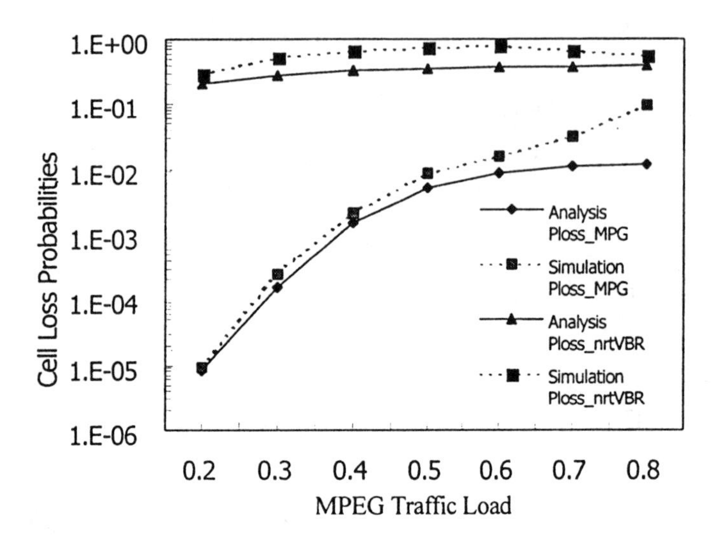

Figure 5 MPEG traffic load vs average queueing delay (with utilization of 60%, Number of emergency slots=10)

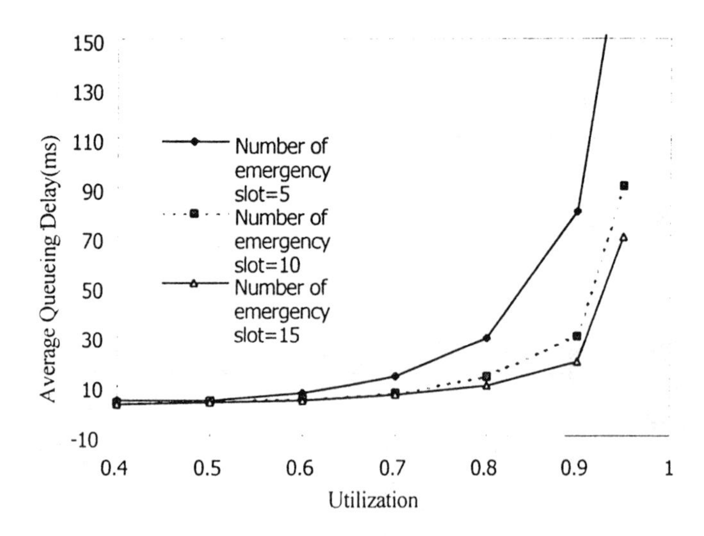

Figure 6 Average Queueing Delay of MPEG cell vs Utilization (60%MPEG load of VBR load)

6. CONCLUSION

A new multiple access protocol based on combination of source modeling and dynamic scheduling is developed. The performance of the new scheme has been evaluated by simulation. The results show that the delay and loss performance of dynamic MPEG sources can be improved greatly in designing wireless multiple access protocol. The future work is the combination of our scheduling scheme with a modified CAC strategy to meet QoS of multiple handoff connections.

Acknowledgments

This work is partially supported by Fujitsu Laboratory Ltd.

References

[1] D. Raychaudhuri and N.D. Wilson, ATM-based transport architecture for multiservices wireless personal communication networks, IEEE JSAC12 (1994), pp.1401-1414.

[2] B. A. Akyol D.C. Cox "A Dynamic Resource Allocation and Media Access Control Proposal for a Multi-tier Wireless ATM Network", Technical Report, Stanford University, June 1997.

[3] M. J. Karol, Z. Liu and K. Eng, "Distributed-queueing Request Update Multiple Access (DQRUMA) for Wireless Packet (ATM) Networks", IEEE International Conference on Communications (ICC'95), Seattle, USA. June 1995, pp.1224-1231.

[4] C-S. Chang "Guaranteed Quality-of-Service Wireless Access to ATM Networks", IEEE J. on Selected Areas in Communications, Vol.15, No 1, January 1997, pp.106 - 117.

[5] Xiaoxin Qiu, Victor O.K.Li, "A multiple access scheme for multimedia traffic in wireless ATM", IEEE/ACM Mobile Networks and Applications. Jan 1996, pp. 259-272

[6] Daniel P. Heyman, "The GBAR Source Model for VBR Videoconferences", IEEE/ACM Transactions on Networking, vol. 5. No. 4, August 1997, pp. 554-560.

[7] M. Schwartz, "Broadband Integrated Networks" Prentice Hall Inc.1996, pp.254-261.

[8] Z. Niu and KUBOTA "An Adaptive MAC Scheme for Wireless ATM and its Performance Evaluation", Chinese Journal of Electronics, Vol.7, No.4, 1998, pp341-348.

[9] M. F. Neuts, "Matrix-Geometric Solutions in stochastic Models" The John Hopkins University Press.1981, pp.254-261.

[10] A.S.Tanenbaum, "Computer Networks". EngleWood Cliffs, NJ: Prentice-Hall, 1981.

[11] S.Q.Li, "Dynamic Bandwidth Allocation on a Slotted ring with Integrated services". IEEE Transactions on Communication, Vol 36, No.7, pp.827-828.

SESSION 10

Bandwidth Allocation

EVALUATION OF ALLOCATION POLICIES ON HYBRID FIBER-COAX BROADBAND ACCESS NETWORKS FOR CONTENTION-TYPE TRAFFIC

César Fernàndez*

Computer Science Department, University of Lleida (UdL)

Ap. de correus 471, 25080, Lleida, Spain

cesar@eup.udl.es

Sebastià Sallent

Telematics Engineering Department, Polytechnical University of Catalonia (UPC)

Gran Capità, s/n. M. C3, 08034, Barcelona, Spain

matssr@mat.upc.es

Abstract Networks with centralized access mechanisms, *e.g.* HFC 802.14, are specified by a set of standards that in most cases does not cover internal operations of the central station, called Head-End Controller (HC) in HFC 802.14 networks. An example of this sort of operations could be the manner in which the HC allocates resources for contention-type traffic. In this paper we study two types of policies that the HC could use to allocate this kind of traffic. In both cases, we obtain computationally efficient analytical expressions of the mean *Contention Resolution Period* (CRP) using the methods described in [9] for a *Q-ary* resolution algorithm. Also we define two parameters of evaluation that in conjunction with the computations of the CRP could help to the HC to look for the best policy.

Keywords: Hybrid Fiber-Coax Networks, Medium Access Control Protocols, Contention Resolution Period Determination, Capacity Evaluation, Allocation Strategies.

1. INTRODUCTION

In recent years, it has been an obvious interest in offering higher capacity internet services to residential users. With this goal in mind, several alternatives have been heavily debated, from cooper loops as ADSL (Asymmetric

*Research partially supported by the project SMASH (TIC96-1038-C04-03) funded by the CICYT.

Digital Subscriber Line) to FTTH (Fiber To The Home). One of these viable alternatives that might be taken into account is HFC (Hybrid Fiber Coax) due to several reasons, such as the high bandwidth of fiber optics and the large number of existing networks of coaxial cable where an important amount of bandwidth is not being used.

In this sense, the IEEE 802.14 working group has done a considerable work on the definition of the Physical and Medium Access Control layer protocols of a bi-directional Cable TV network based on HFC technology. The last public release of the standard, see [1], presents the architectural view of a 802.14 network, detailing protocols and services involved.

HFC networks are systems where the access procedures are centralized on a single station, called Head-End Controller (HC). As a first approach, the HC tasks could be divided into two categories; those related with the HC-stations communication procedures, covered by the standard, and those internal to the HC devoted to allocate resources in a efficient fashion, out of the standard scope. One of these internal tasks is debated in this paper.

The motivation of this paper is a lack of analytical performance literature that studies the allocation procedures of the HC. Much work developed in this area focuses on simulation results or in particular access mechanisms not closely related with the standard [5; 3; 8; 10]. In this sense, we propose and analyze two alternatives for the allocation of mini-slots accessed by the stations, in order to acquire HC granting to transmit without collision. In both cases, we obtain computationally efficient analytical expressions of the mean *Contention Resolution Period* (CRP) using the methods described in [9] for a *Q-ary* resolution algorithm. Also we define two parameters of evaluation, namely the capacity for contention-type traffic and the rate of global bandwidth devoted to carry out such a traffic. Both parameters in conjunction with the computations of the CRP could help to the HC to look for the best policy; *i.e.* the policy that maximizes C under a given network load.

This paper is organized as follows: Section 2 introduces the basic aspects of the contention access mechanism for HFC 802.14 networks. Two possible strategies are defined. The first one, based on groups of mini-slots with fixed length and the second one with variable length groups. In Section 3 we obtain analytical results of the CRP for the fixed strategy. The same objectives are pursued in Section 4 for variable length groups. In Section 5 we define the evaluation parameters that will allow us to decide the better strategy to be applied by the HC under certain network scenarios. These parameters are numerically evaluated in Section 6 under Poisson traffic arrival. Finally, in Section 7 we conclude and discuss future lines of study.

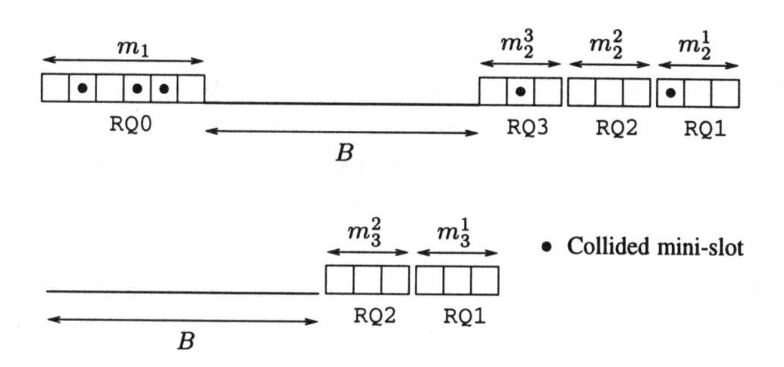

Figure 1 Schematic description of the HC contention mini-slot allocation policy.

2. OVERVIEW OF THE CONTENTION ACCESS MECHANISM

An HFC system can be represented as a group of stations linked to a Head-End Controller by means of two multiplexed frequency and unidirectional channels. The upstream channel, from stations to the HC, operating in the 5-50 MHz band, and the downstream channel, from the HC to stations, operating beyond 450 MHz until 1 GHz. Every station communicates with each other through the HC accessing the upstream channel. The HC echoes upstream information to the downstream channel, also using this channel to broadcast upstream channel status information.

The main goal for HFC Medium Access Control designers has been how to efficiently use the upstream channel over a transmission speed range from 3 to 100 Mbps. In this sense, the time in the upstream channel is split up into data-slots and contention mini-slots. The size of data-slots could be enough to allocate an ATM cell, meanwhile the contention slots are shorter. Contention mini-slots are used by the stations to send allocation requests to the HC using a determined contention algorithm (Stabilized Aloha, n-ary tree, XDQRAP, ...) For every successful request, the HC informs about the status of reserved data-slots employing the downstream channel.

Our analysis scopes in the operations described in the standard drafts, where the slot structure is defined by the HC in a dynamical manner. Nevertheless, some simplifications have been done in order to make feasible an analytical approach. In this sense, the guidelines of the protocol operation are the following:

As showed in Figure 1, the HC allocates groups of request mini-slot according to a certain policy. The first group of a contention resolution period is called `RQ0`. All the subsequent groups of mini-slots will be devoted to resolve contentions produced in `RQ0`. B mini-slots later, where B has to be larger than the round trip delay, the HC allocates as many groups of mini-slots as those collided in RQ0. This is called a cluster. In Figure 1, the first cluster is composed of three groups of mini-slots called RQ3, RQ2 and RQ1. Each station involved in a request is informed by the downstream channel. If no collision exists, the HC tells to the station where to transmit its information. Otherwise, the HC informs to the station where is located the following group of request mini-slots devoted to resolve its contention. The request mini-slot is randomly chosen by the station. This operation is repeated until no collission is produced in any group. At this moment, the HC is able to allocate a new RQ0.

Generally speaking, the size of RQ0 is m_1 mini-slots. This size can be modified by the HC at every round. The size of the groups belonging to cluster number i is m_i^j, where the superscript j denotes the group number j inside the cluster. Our assumption along the paper is that $m_i^j = m_i$ for all j.

At this point, two policies can be identified according to the size of the groups of mini-slots belonging to a cluster. The first one is an allocation policy where the group's length is the same for every cluster. That is, $m_i = m$ for all i. The second one allows a reduction of the group's length depending on the cluster considered. This group's length reduction pursues a best utilization of the upstream channel bandwidth.

The first goal is to find a computationally efficient expression, that gives us the mean length of the contention resolution period (CRP). This period is taken to be the elapsed time between two successive RQ0 clusters. Afterwards we define two kind of performance parameters in order to evaluate different policies.

3. CLUSTERS WITH GROUPS OF FIXED LENGTH

According to Figure 1, every group, independently of the cluster it belongs to, has the same length, say m mini-slots. Define the mean CRP given that N arrivals have occurred during last CRP as Y_N^m. In the same way that with others resolution mechanisms [9; 7; 4], Y_N^m can be recursively expressed as

$$Y_N^m = B + L_N^m = B + m + BP_N^m + \sum_{i_1 \cdots i_m}^{N} \binom{N}{i_1 \cdots i_m} \frac{1}{m^{i_1}} \cdots \frac{1}{m^{i_m}} \sum_{j=1}^{m} l_{i_j}^m, \tag{3.1}$$

where P_N^m is the probability that a collission occurs in RQ0 given that N arrivals exist, namely

$$\overline{P}_N^m = 1 - P_N^m = \prod_{i=1}^{N-1} \left(1 - \frac{i}{m}\right),$$

and $l_{i_j}^m$ can be expressed as

$$l_{i_j}^m = \begin{cases} 0 & i_j = 0,\ i_j = 1 \\ L_{i_j}^m & i_j \geq 2, \end{cases}$$

with $L_0^m = L_1^m = m$. In (3.1), $\sum_{i_1 \cdots i_m}^N$ is the sum over all possible combinations of $i_1 \cdots i_m$ such that equals N, and $\binom{N}{i_1 \cdots i_m}$ is the multinomial coefficient. Manipulating (3.1) conveniently, as detailed in [9; 6], yields

$$\begin{aligned} L_N^m &= m + \sum_{i=0}^{\infty} m^i \Bigg[m^2 + B - B(1-m^{-i})^N - m^2(1-m^{-i-1})^N \\ &\quad - B \sum_{j=2}^{\min(N,m)} \binom{N}{j} \overline{P}_j^m (1-m^{-i})^{N-j} m^{-ji} \Bigg] \\ &\quad - \sum_{i=0}^{\infty} \left[BN(1-m^{-i})^{N-1} + mN(1-m^{-i-1})^{N-1} \right]. \end{aligned} \quad (3.2)$$

4. CLUSTERS WITH GROUPS OF VARIABLE LENGTH

One could think, in contrast to the previous policy, an allocation policy that periodically reduces the length of the groups of mini-slots at every cluster delivered to the network. As we demonstrate later, a first approach could be to destinate the longest group (RQ0) to start the first collission resolution, where more arrivals remain to be solved, and to decrease progressively the length of the groups of every cluster until reaching a minimum.

Here we consider a policy based on the length reduction by a constant factor that divides the length, say r. In an extended version of this paper ([6]), we study a more general method, where the length of the groups does not necessarily obeys a determined rule. In both cases, we assume the same length in any group inside a cluster, that is, $m_i^j = m_i$ for all j, according to the notation introduced in Figure 1.

Let m_i be the length of any group of mini-slots in cluster number i. Take cluster number 1 as that which contains RQ0. The length for groups inside the next cluster is computed according a transformation $\sigma()$ as $m_{i+1} = \sigma_{r,m_\infty}(m_i)$, where

$$\sigma_{r,m_\infty}(x) = \begin{cases} \lfloor \frac{x}{r} \rfloor + 1 & \text{if } \lfloor \frac{x}{r} \rfloor + 1 \geq m_\infty \\ m_\infty & \text{otherwise} \end{cases}$$

and where $\lfloor y \rfloor$ denotes the integer part of y, $r \geq 2$ is an integer constant and $m_\infty \leq m_1$. The fact of adding 1 is not relevant, but it is maintained because formulation of $\sigma()$ for $r = 2$ was more compact on preliminary analysis.

In both strategies, constant factor and generic reduction, we will obtain expressions of $Y_N^{m_1}$ for $m_\infty = 2$. Other cases are left to further studies.

As done with the fixed length strategy, recurrence in N must be avoided if we intend to perform calculations for large N. So, we define a *two variant generating function*, indexing N and m. Resolution is detailed in [6]. Finally, we obtain

$$L_N^{m_1} = m + \sum_{k=2}^{N} \frac{N!}{(N-k)!} H_k^{m_1},$$

where

$$\begin{aligned} H_k^{m_1} &= \frac{1}{{m_1}^{k-1}} H_k^{\lfloor m_1/r \rfloor + 1} + B\frac{(-1)^k}{k!}(k-1) \\ &\quad + (\lfloor m_1/r \rfloor + 1) m_1 \frac{(-1/m_1)^k}{k!}(k-1) \\ &\quad - B \sum_{i=2}^{\min(k,m)} \overline{P}_i^{m_1} \frac{1}{i!} \frac{(-1)^{k-i}}{(k-i)!}. \end{aligned}$$

5. EVALUATION PARAMETERS

Once the mean CRP can be calculated numerically, we are able to obtain two important measures that describe the performance of the access mechanism when contention type traffic is considered.

The first parameter, subsequently called R, is the fraction of the total amount of available bandwidth dedicated to contention traffic. Namely, how many mini-slots are dedicated to solve contentions in contrast to those used to transfer information on them. R is a percentage.

The second parameter, say C, is the capacity of the access mechanism for contention type traffic. C is a throughput and is calculated in number of mini-slots by time of mini-slot.

In any case, both parameters depend on the mean CRP given than N arrivals are produced on previous CRP, Y_N, and on the probability that N arrivals were produced given a predefined tax λ. We call this probability $P_N(\lambda)$.

Let $Y_N(B)$ be the mean CRP for a particular strategy given that N arrivals must be allocated, considering that the HC has decided an inter-cluster gap of B mini-slots. It is clear that

$$R = \frac{\sum_{i=0}^{\infty} P_i(\lambda) Y_i(0)}{\sum_{i=0}^{\infty} P_i(\lambda) Y_i(B)}. \tag{5.3}$$

On the other hand, C can be calculated according to the definition of capacity of a *resolution contention algorithm* introduced by [7]

$$C = \sup_{\lambda} \frac{\lambda}{\sum_{i=0}^{\infty} P_i(\lambda) Y_i(B)}. \tag{5.4}$$

Clearly, the objective of our computations is to look for the best strategy used by the HC in order to allocate the contention clusters.

6. NUMERICAL RESULTS

This section shows several calculations performed under different network scenarios. The analyzed strategies have been basically two, clusters with groups of fixed length and clusters with groups of variable length with a constant factor reduction strategy. In the later case, many factors of reduction have been explored.

First of all, let us study the performance of the access mechanism when immediate feedback is considered ($B = 0$), in order to compare its capacity with others appeared in the literature. Afterwards, we show the throughput and the occupation rate for different values of B, helping this results the HC to look for the best policy. It is important to note that all the results are obtained considering arrival inputs as Poisson distributed, as well as those reported from alternate mechanisms.

The best capacity of HFC 802.14 algorithm for $B = 0$ when a fixed length policy is used is achieved for a group length $m = 2$. Its value is $C = 0.429512$. Variable length strategies do not introduce any improvement in C for $B = 0$.

Several proposals have been submitted to the committee during the elaboration of the 802.14 standard. These proposals include many contention access protocols, most of them studied some years ago when a lot of work was developed related to the evaluation performance of slotted contention networks. Some of these protocols and its corresponding capacities are *p-persistent* algorithms [2] (0.3679), *tree-based ternary* with *free* and *blocked access* [9] (0.4016,0.3662), and *first-come first-serve splitting (fcfs)* [2] (0.4878).

Let us now inspect the system performance for $B \neq 0$. At this moment, intensive calculations have been done for two values of B, $B = 20$ and $B = 100$. They corresponds to a network of 9,6 and 48 Km length respectively, considering 10 Mbps in the upstream channel and a mini-slot length of 6 bytes.

Figure 2 (left) shows the capacity versus contention rate bandwidth for a network of $B = 20$. As an example, fixed and variable constant rate ($r = 2$ and $r = 5$) strategies are plotted. Six points for fixed strategy are shown, corresponding to a range from $m = 3$ until $m = 8$. As can be seen, increasing m implies a better capacity but a larger bandwidth use, from 20% until more than 50% for $m = 8$. Also it is noted that a variable strategy with $r = 2$ outperforms in all cases than the fixed strategy. Moreover, the variable strategy with $r = 5$ outperforms $r = 2$, more clearly for large values of R. Note that any point of a plot corresponds to a certain value of m.

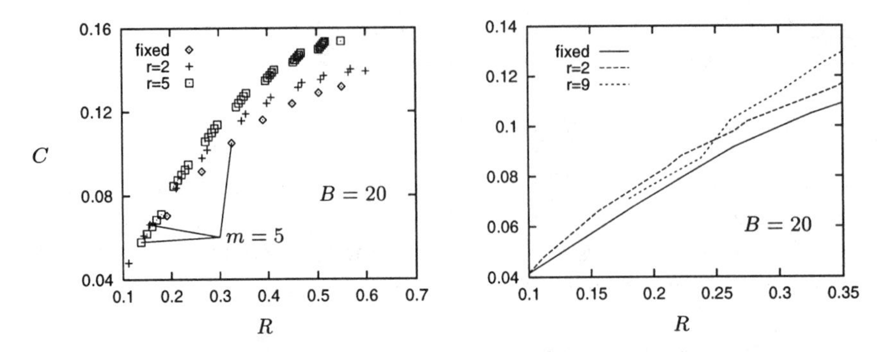

Figure 2 C versus R performance for $B = 20$

$B = 20$		$B = 100$	
R Range	Strategy	R Range	Strategy
$[0, 0.16]$	$r = 2$	$[0, 0.035]$	$r = 2$
$[0.16, 0.19]$	$r = 3$	$[0.035, 0.055]$	$r = 3$
$[0.19, 0.21]$	$r = 4$	$[0.055, 0.105]$	$r = 4$
$[0.21, 0.3]$	$r = 5$	$[0.105, 0.125]$	$r = 5$
$[0.3, 0.35]$	$r = 7$	$[0.125, 0.17]$	$r = 7$
$[0.35, 0.4]$	$r = 9$	$[0.17, 0.19]$	$r = 9$
$[0.4, 0.6]$	$r = 13$	$[0.19, 0.22]$	$r = 11, 13$

Table 1.1 Optimum strategy depending on R

But the question is: Which is the best choice of r for a variable constant rate strategy? The answer depends on the range of R considered. Figure 2 (right) plots the throughput of a system with $B = 20$ for contention traffic rates from 10% until 35%. At a first glance, this plot shows how the best strategy depends on R. More accurately, Table 1.1 lists the optimum strategy for a disjoint set of ranges of R, for $B = 20$ and $B = 100$ (only odd values of r beyond 5 have been used).

In this way, the HC is able to choose the optimum strategy calculating C versus R performance measurements for several values of B, namely, such strategy that maximizes throughput under a certain available bandwidth. Furthermore, if not enough bandwidth is available, the HC could increase B in order to find the first value of B that allows contention traffic allocation using $r = 2$.

7. CONCLUSIONS AND FUTURE WORK

In this paper we have studied two allocation policies used by the HC for the access of contention-type traffic. We have generalized the results of [9] and we have applied them to an allocation strategy where the splitting factor of the algorithm is variable and where delay is considered.

The obtained CRP expressions allow the computation of CRP for large values of N in short time. These expressions have been validated comparing the results against the recursive expressions for small values of N. Once calculated the CRP, this parameter is used to obtain the capacity (C) versus contention traffic bandwidth requirement (R), obtaining the best policy depending on R. As shown in Section 6, constant factor reduction strategies clearly outperforms the fixed policy when Poisson traffic is assumed, so look up tables such as Table 1.1 could help the HC to decide which is the best strategy under certain resource availability conditions for contention traffic.

Nevertheless, more work has to be done in this sense. First, expressions for $Y_N^{m_1}$ have to be generalized for $m_\infty > 2$. Second, some generic reduction strategies have to be tested in order to compare their performance against constant rate reduction. An example of generic reduction strategies could be a sequence m_i defined as a Fibonacci series. And third, other type of arrival processes could introduce variations in the look up tables of C versus R.

References

[1] *IEEE 802.14 Draft 2 Revision 2, 1997.* IEEE 802.14, 1997. Available at URL `http://texcat.com/users/chicago/draft2-rev2.pdf`.

[2] Dimitri P. Bertsekas and Robert Gallager. *Data Networks*. Prentice Hall, 1992.

[3] Chatschik Bisdikian, Bill McNeil, Rob Norman and Ray Zeisz. MLAP: A MAC Level Access Protocol for the HFC 802.14 network. *IEEE Communications Magazine*, 34(3):114–121, March 1996.

[4] John I. Capetanakis. Tree algorithms for packet broadcast channels. *IEEE Transactions on Information Theory*, 25(5):505–515, May 1979.

[5] James E. Dail, Miguel A. Dajer, Chia-Chang Li, Peter D. Nagill, Curtis A. Siller, Kotikalapudi Sriram and Norman A. Whitaker. Adaptative digital access protocol: A MAC protocol for multiservice broadband access networks. *IEEE Communications magazine*, 34(3):104–113, March 1996.

[6] Cèsar Fernàndez and Sebastià Sallent. Evaluation of allocation policies on hybrid fiber-coax broadband access networks for contention-type traffic (extended version). Technical Report DIEI-99-RT-2, Departament d'Informàtica i Enginyeria Industrial. Universitat de Lleida, April 1999.

[7] Huang Jian-Cheng and Toby Berger. Delay analysis of interval-searching contention resolution algorithms. *IEEE Transactions on Information Theory*, 31(2):264–273, March 1985.

[8] Ying-Dar Lin, Chia-Jen Wu and Wei-Ming Yin. PCUP: Pipelined Cyclic Upstream Protocol over Hybrid Fiber Coax. *IEEE Network*, 11(1):24–34, January/February 1997.

[9] Peter Mathys and Philippe Flajolet. Q-ary collision resolution algorithms in random-access systems with free or blocked channel access. *IEEE Transactions on Information Theory*, 31(2):217–243, February 1985.

[10] Dolors Sala and John O. Limb. Comparison of contention resolution algorithms for a cable modem MAC protocol. In *International Zurich Seminar on Broadband Communications*, February 1998.

JUST-IN-TIME OPTICAL BURST SWITCHING FOR MULTIWAVELENGTH NETWORKS

John Y. Wei, Jorge L. Pastor, Ramu S. Ramamurthy, and Yukun Tsai *
Telcordia Technologies
331 Newman Springs Road
Red Bank, NJ 07701-5699
U. S. A.
{wei,jorel,ramu,ykt}@research.telcordia.com

Abstract We describe the architecture, performance analysis and simulation result of a novel switching paradigm for optical WDM networks called Just-In-Time Optical Burst Switching (JIT-OBS) designed for ultra-low-latency transport of data-bursts across an optical WDM network. It combines the desirable features of circuit-switching and packet-switching, and features an out-of-band signaling scheme on a separate control channel with explicit feedback on delivery of data-bursts. We provide a performance analysis and simulation of the JIT-OBS approach, and compare its performance with those of circuit-switching and packet-switching approaches. We find that it has the best latency performance among the different switching mechanisms, and it has a better throughput performance than circuit-switching, and its performance is insensitive to network propagation delays.

Keywords: WDM, Optical Network, Optical Burst Switching Just-In-Time Signaling, Signaling Protocol, Packet Switching, Circuit Switching

1. INTRODUCTION

The emergence of broadband communications has increased the needs of bulk transport of high capacity signals and services. Multi-wavelength reconfigurable optical networks offer such a capability beyond current transport technologies such as SONET. The Multi-wavelength Optical Networking Program (MONET) [1] [2] [3] [4] sponsored by the U.S. Government's Defense Advanced Research Project Agency (DARPA) is a research consortium aimed at addressing the technology, architecture, and the management and control issues for this new emerging technology.

*This research was partially funded by the US Government DARPA Multi-wavelength Optical Networking (MONET) Project, contract number: MDA 972-95-3-0027.

In this paper we describe the Just-in-Time Optical Burst Switching (JIT-OBS) paradigm developed in MONET. The JIT-OBS paradigm is designed for ultra-low-latency unidirectional transport of data-bursts across an optical network. It combines the desirable features of circuit-switching and packet-switching, and features an out-of-band signaling scheme on a separate control channel and provides explicit feedback on delivery of data-bursts.

This paper is organized as follows. In Section 2, we outline a generic WDM switch architecture used for subsequent discussion. We then describe the different WDM switching paradigms, including the JIT switching paradigm we developed. Section 3 then presents an analysis of the performance of the JIT-OBS approach. Section 4 presents and discusses our initial simulation results. Section 5 concludes the paper, and comments on the MONET JIT signaling protocol implementation and the WDM switching experiments to be performed in the MONET Washington DC testbed.

2. WDM SWITCHING PARADIGMS

In optical WDM, the tremendous bandwidth of a fiber (potentially a few tens of terabits per second) are demultiplexed into many independent non-overlapping wavelength channels. Within certain restrictions, the wavelength channels are *transparent* in that they can transport data at different bit rates and modulation formats.

In this section, we outline a functional architecture for a generic WDM switch used for our subsequent discussion. We then describe the different switching paradigms for optical WDM in more detail. We note that each switching paradigm makes different assumptions on the WDM switch hardware, and requires different signaling schemes. The manner in which the header/control info is exchanged and the manner in which the path-setup and data-transfer are performed distinguishes the different schemes.

2.1 WDM SWITCH FUNCTIONAL ARCHITECTURE

A functional architecture of a generic WDM switch is illustrated in Fig. 1(a). Wavelength channels in an input fiber that enter the WDM switch are demultiplexed into individual wavelengths. They are then switched by the crossconnect to a specific output port. All wavelength channels destined to an output port are then multiplexed into the output fiber. The wavelength channels may be amplified and/or gain stabilized before they exit the switch. WDM switches [5] [6] differ in the extent to which they keep the optical signals transparent within the switch. In an all-optical WDM switch, the wavelength channels remain entirely in the optical domain. Such WDM switches may also perform wavelength conversion within the switch. Other WDM switches have optical-to-electronic (O/E), and electronic-to-optical (E/O) conversions performed on each wave-

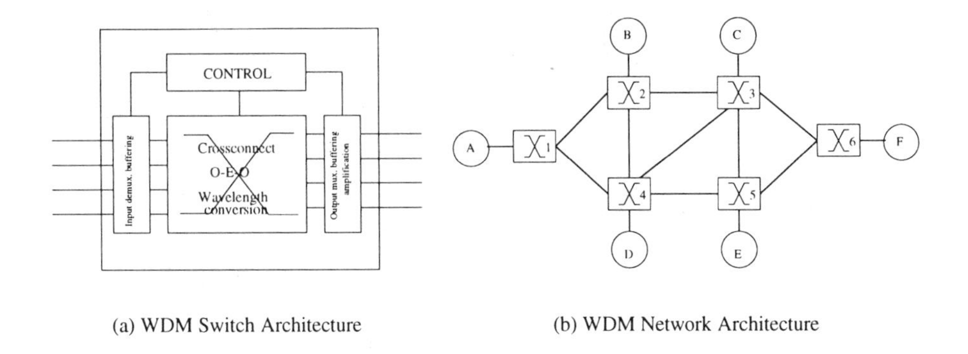

(a) WDM Switch Architecture

(b) WDM Network Architecture

Figure 1 Multiwavelength Optical Networking Architecture

length channel within the WDM switch. Input and/or output buffering may also be provided in an optical switch.

The control component in a WDM switch controls the state of the cross-connect, the wavelength converters, and output buffers. based on the control information present at the controller. An important parameter of a WDM switch is the *switching-time,* defined as the time it takes for the cross-connect to change state, and for the output channel to stabilize.

Fig. 1b depicts an optical network consists of WDM switches (labeled 1 through 6) interconnected by bidirectional fiber links. Technological constraints dictate that the number of WDM channels that can be supported in a fiber be limited to W (whose value is typical a few tens today.) Access stations are attached to a WDM switch via bidirectional fiber links. An access station is capable of sourcing data on any one of the available wavelengths. Data is transferred between access stations as unidirectional variable length optical bursts.

There are different ways to transmit the control information across the WDM network nodes. Control information can be transmitted along with the optical data as an in-band optical header (e.g. by utilizing a scheme such as subcarrier multiplexing.) In this mode, control information travels along with the data-burst, and is analogous to the packet switching. An implication is that the WDM switch must delay/buffer the data-burst until the header is decoded, and the data-path is established. As an alternative is to transmit control information independent of the data-burst on a separate signaling network. In this case, the control information (i.e. signaling) is utilized to setup the optical path prior to transmitting the data-burst, and is analogous to circuit-switching. Intuitively, a packet-switched scheme is suited for short data-bursts, and a circuit-switched scheme is efficient for large data-bursts.

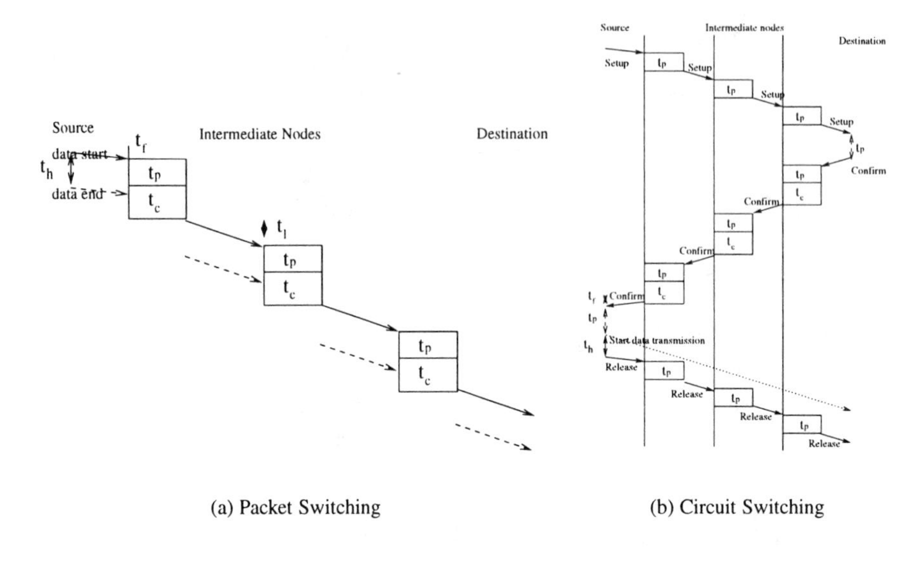

(a) Packet Switching (b) Circuit Switching

Figure 2 Conventional Switching Paradigms

Current optical technology provides very high bandwidth for transmission, but is limited in its ability to perform optical processing or buffering. Electronics on the other hand allows processing and buffering, but cannot match the bandwidth of optical transmission. In order to maximize the utilization of the optical network, we would like the control mechanism to minimize the time to setup the optical path for a data-burst, under the constraints imposed by the optical WDM switch technology. Just-In-Time switching is designed to combine the desirable features of packet-switching (small setup time) and circuit-switching (out-of-band signaling.)

2.2 PACKET-SWITCHING

In optical packet-switching as illustrated in Fig. 2(a), the control information associated with a data-burst travels with the data-burst as the packet header. At each intermediate node, the header is separated from the data-burst, and is processed to determine the output-port. A routing protocol may be used to determine the next-hop given the destination. The WDM switch controller sets up the crossconnect (along with wavelength conversion.) During the period of header processing, and cross-connect setup, the data-burst is buffered. If an output port is not available, the data-burst is dropped, and lost. No feedback is sent to the source access station.

2.3 CIRCUIT-SWITCHING

In optical circuit-switching, the data transfer path is setup prior to the transmission of the data burst. The access station that has a data-burst to transmit initiates an out-of-band distributed signaling procedure to determine the path, wavelengths, and setup cross-connects. When a data transfer path is available, the data-burst is transmitted by the access station.

An example of the circuit-switched signaling procedure is depicted in Fig. 2(b). A *SETUP* message is sent from source to destination. On its way, wavelengths are reserved at each link along the path. If at some intermediate node no wavelengths are available on the output port, then a *BLOCKED* message is sent back to the source. If the setup message reaches the destination access station successfully, then the destination responds with a *CONFIRM* message back to the source along the reverse path, and at each intermediate node, cross-connects are setup. When the confirm message reaches the source, the source transmits the data-burst. After the data-burst is transmitted, the source sends a *RELEASE* message which releases wavelengths along the path. We assume that the routing is performed by a routing control protocol, and is independent of the signaling protocol. In this work, we assume that the shortest-hop path is utilized for routing.

The setup time for a data-burst in circuit switching can be improved by pipelining the cross-connect setup times with the propagation time. Two variations of pipelining are possible:

- *Cut-at-Confirm:* where the crossconnect is installed (cut-through) after the CONFIRM message is sent (see Fig. 3(a)) on the reverse path.
- *Cut-at-Setup:* where the cut-through is performed right after the SETUP message is sent (see Fig. 3(b)) on the forward path.

2.4 JUST-IN-TIME OPTICAL BURST SWITCHING

Just-in-time optical burst switching combines desirable features of circuit-switching (explicit feedback, separate signaling network) while minimizing the setup time. The detailed design of the JIT signaling protocol is reported in [8]. Fig. 4 depicts a simplified abstract form of signaling for JIT-OBS that captures the salient features for our present purpose. An access station initiates JIT signaling by sending a *JIT_SETUP* message to its attached WDM switch. The WDM switch responds with a *JIT_CALL_PROCEEDING* to indicate that connection setup is on its way to the destination. In the reply message there is a *delay* parameter, which indicates how long the access station should wait before launching its data-burst. This delay parameter is estimated by the WDM switch (e.g. by a suitably developed routing algorithm) from the number of hops to the destination and associated setup time of the crossconnects along the

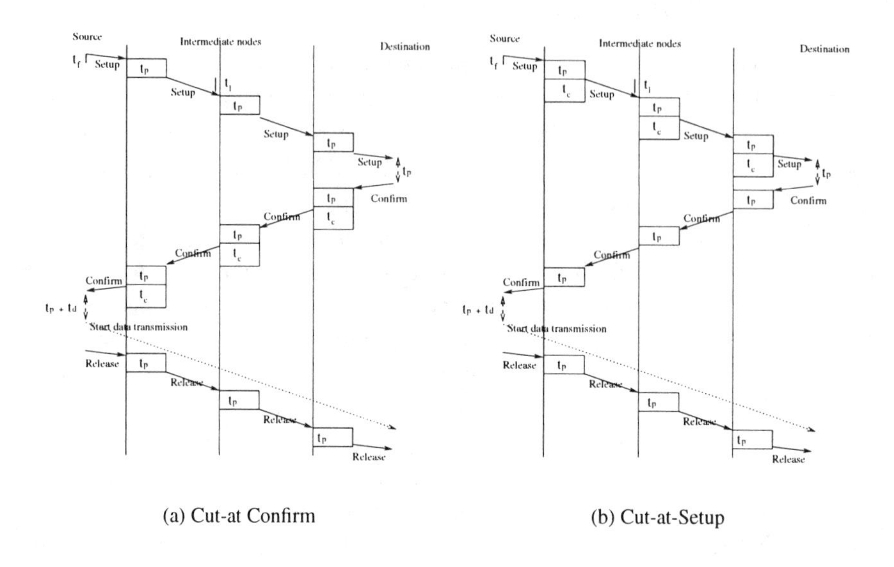

Figure 3 Pipelined Circuit Switching Paradigms

path. When the source access station receives the *JIT_CALL_PROCEEDING* message, it waits for the delay parameter, and then transmits its data-burst. When a WDM switch receives a *JIT_SETUP* message, it will attempt to reserve the wavelength on the output port and forward the *JIT_SETUP* message to its next hop. Cross-connect setup is performed in parallel with the next hop propagation. If the switch has no output wavelengths available on the next-hop output port, it sends back a *JIT_BLOCKED* message to the source access station. When a WDM switch gets the JIT_BLOCKED message it releases the reserved wavelength on the output port. When the destination access station receives the *JIT_SETUP* message, it responds with a JIT_CONNECT message which travels back to the source access station. After transmitting the data-burst, the source access station transmits a *JIT_RELEASE* to release all reserved wavelengths.

We note that the setup time for the circuit switched approach is significant, and therefore it is efficient only for data-bursts which are much longer than the setup time. The packet switched scheme on the other hand has shorter setup times since the control information travels with the data burst. However, in an all-optical WDM switch the header processing requirements are significant, and the data-burst has to be buffered during the time that the header is being processed, For example, a WDM packet switch must have the technology to separate the header from the optical data-burst at the input port and to reimpose it at the output port. Such technology is not required for the circuit-switched schemes. Furthermore, JIT-OBS and circuit switching schemes provide ac-

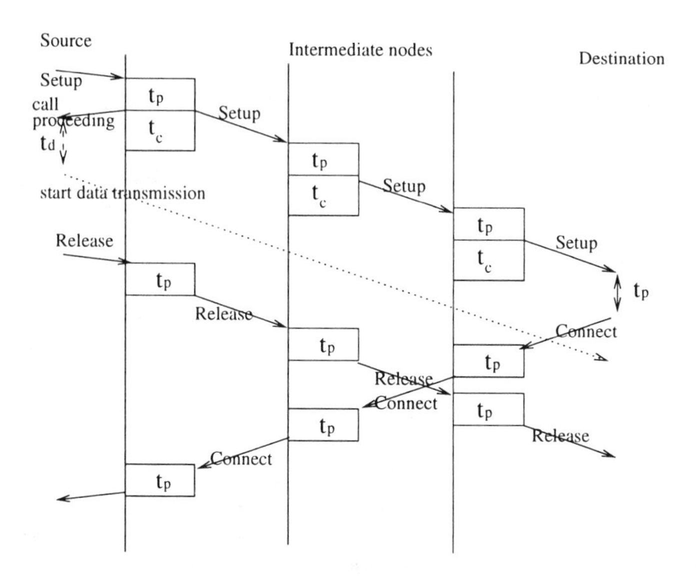

Figure 4 Just-In-Time Optical Burst Switching

knowledgement of delivery of the data-burst, while packet-switched schemes do not provide any such acknowledgements.

3. PERFORMANCE ANALYSIS

In this section we present a simple analytical model that examines the performance of each of the above schemes. In analyzing the performance of a WDM optical network, we are interested in the throughput, the latency, and their sensitivity to different network parameters.

3.1 ASSUMPTIONS AND NOTATION

We assume that the propagation delay on each link is identical (i.e., the fiber links are of the same lengths.) We assume that each WDM switch is capable of full wavelength conversion. We assume that the traffic distribution among node pairs is uniform. For each signaling scheme, we obtain the formulas for the channel holding time (i.e., the time duration from the instant a channel is reserved to the instant the channel is released) for a successful data-burst.

We now outline the notation utilized in the rest of this paper.

M: Number of access stations.

W: Number of wavelengths on the fiber that are used to transfer data-bursts.

t_f: Propagation delay from an access station to its attached WDM switch.

t_p: Protocol processing time at a WDM switch.

t_c: Crossconnect switching and stabilization time at a WDM switch.

t_l: Propagation delay on a fiber link between WDM switches.

t_h: Average burst duration.

γ: An upper bound on the throughput achievable on a single wavelength channel

C, S_1, S_2: Let there be C fiber links on a minimum cut of of the graph that represents the WDM network topology. Let the cut divide the network access stations into two sets, of sizes S_1 and S_2.

λ: Offered load to the network (in data-bursts/unit time)

λ_m: Maximum rate of traffic that can be sourced by any pair of access stations.

3.2 ANALYTICAL RESULTS

Channel Holding Time. We define the *channel reserve time* to be the instant at which an outgoing channel is reserved at a WDM switch for a call. *Channel release time* is the instant at which an outgoing channel that was reserved for a data-burst is released. *Channel hold-time* is the duration for which an outgoing channel is reserved by a WDM switch for a data-burst. Let n be the number of intermediate WDM switches along a path. From Figures 2, 3, and 4, we make the following observations.

For the circuit-switched scheme, a setup request arrives at the kth WDM switch on its path at time $t_f + kt_p + (k-1)t_l$. This is when an outgoing channel is reserved at the kth node. A release message arrives at the kth node at time $5t_f + (2n+2+k)t_p + nt_c + t_h + (2n-3+k)t_l$. Therefore the duration for which the channel is reserved for a call at the kth node is: $4t_f + (2n+2)t_p + nt_c + 2(n-1)t_l + t_h$

Similarly, for the Cut-at-Confirm circuit-switching scheme, we observe the following: (a) channel reserve time = $t_f + kt_p + (k-1)t_l$, (b) channel release time = $5t_f + (2n+2+k)t_p + t_c + t_h + (2n-3+k)t_l + t_d$, and therefore, (c) channel hold time = $4t_f + (2n+2)t_p + t_c + 2(n-1)t_l + t_h + t_d$, where the data delay, t_d, is determined as: $t_d \geq t_c - t_p - 2t_f, t_d \geq 0$.

Similarly, for the Cut-at-Setup circuit-switching scheme, we observe the following: (a) channel reserve time = $t_f + kt_p + (k-1)t_l$, (b) channel release time = $5t_f + (2n+2+k)t_p + t_h + (2n-3+k)t_l + t_d$, and therefore, (c) channel hold time = $4t_f + (2n+2)t_p + 2(n-1)t_l + t_h + t_d$, where the data delay, t_d, is determined as: $t_d \geq (t_c + nt_p) - (4t_f + 2(n-1)t_l + (2n+2)t_p), t_d \geq 0$.

For the packet switched scheme the channel holding time is $t_c + t_h$. For the JIT-OBS scheme, the data delay t_d is determined from the fact that, for any switch on path, its crossconnect must be setup before the data-burst arrives at the switch, i.e., for any $k \leq n$, where n is the number of WDM switches on the path,

$$t_d \geq (k-2)t_p + t_c - 2t_f \tag{1}$$

At k^{th} WDM switch on a n node path, (a) channel reserve time = $t_f + (k)t_p + (k-1)t_l$, (b) channel release time = $t_f + (n+k)t_p + (k-1)t_l + t_h + t_c$, and (c) channel hold duration = $nt_p + t_h + t_c$

Latency. The latency for a data-burst is defined as the duration from the instant the data-burst arrives at the source access-station, to the instant it arrives at the destination access station. For a route with n intermediate WDM switches (n+1 hops), the latency for the different schemes are the following:

- *Circuit Switching:* $6t_f + 3(n-1)t_l + (2n+2)t_p + nt_c$
- *Cut-at-Confirm Circuit Switching:* $6t_f + 3(n-1)t_l + (2n+2)t_p + t_c + t_d$
- *Cut-at-Setup Circuit Switching:* $6t_f + 3(n-1)t_l + (2n+2)t_p + t_d$
- *Packet Switching:* $2t_f + n(t_p + t_c) + (n-1)t_l$
- *JIT Optical Burst Switching:* $2t_f + nt_p + t_c + (n-1)t_l$

These equations are derived by summing up the different delay components in the end-to-end switching scenarios depicted in Figures 2, 3, and 4. We observe that JIT-OBS has the lowest latency.

Analysis of a single WDM switch. The switching time for a WDM switch imposes an upper bound on the achievable throughput on any channel.

$$\gamma = t_h/(t_h + t_c) \tag{2}$$

This arises from the simple fact that on any channel, data-bursts must be spaced apart by t_c.

Analysis of a WDM path. Let b_p be the probability that a data-burst arriving at a WDM switch is blocked due to output contention. Let there be k burst types arriving to the output link at a WDM switch with arrival rates, $\lambda_1, \lambda_2, \ldots, \lambda_k$ respectively, and with channel holding times $t_1, t_2, \ldots, t_k$, (which implies that the loads for each burst type is $\rho_1, \rho_2, \ldots, \rho_k$ where $\rho_i = \lambda_i/(1/t_i)$, respectively.) Then, the probability that all W wavelength channels are busy on the link is given by the M/M/m/K Markov chain with k customer arrival classes is determined from [7] as:

$$b_p = \sum_{n_1, n_2, n_k \leq W, \sum_{i=1}^{k} n_i = W} \prod_{i=1}^{k} (\rho_i)^{n_i}/G \tag{3}$$

where G is

$$G = \sum_{n_1,n_2,n_k \leq W, \sum_{i=1}^{k} n_i = W} \sum_{i=1}^{k} (\rho_i)^{n_i} \tag{4}$$

Let there be n WDM switches between two access stations. Then the probability that a burst from an access station destined to another access station is blocked, B_p, is given as: (assuming that each WDM NE blocks the burst independently.)

$$B_p = 1 - (1 - b_p)^n \tag{5}$$

Analysis of a WDM Network. The maximum load that the network can carry is determined by the bandwith of the minimum cut of the network as $C\gamma W$. Since the minimum cut separates the network into two sets of node with S_1, and S_2 nodes respectively, we obtain:

$$S_1 \times S_2 \times \lambda_m \leq C \times \gamma \times W \tag{6}$$

From the blocking probability of a path given in Eqn. 5, we can determine, B, the probability that a data-burst arriving to the network is blocked, B, by averaging the blocking probabilities of each node pair (assuming shortest path routing) weighted with the load between the node pair.

The normalized throughput ρ, when the offered load is λ is defined as follows:

$$\rho = \frac{\lambda}{(\lambda_m \times M \times (M-1))}(1 - B) \tag{7}$$

4. ILLUSTRATIVE SIMULATION RESULTS

Simulation Parameters. We simulate two representative networks, one with a 7 node bidirectional ring topology and another one with a 5×5 two-dimensional torus topology. We assume that the number of wavelengths was 8. For the simulations, we assumed that each node-pair was equally loaded, and the applied load was a fraction of the maximum load determined from Eqn. 6. We assumed Poisson arrivals and fixed data-burst sizes. We simulated two values of burst-sizes, one small value of 0.01ms (corresponding to 12500 bytes at 10Gbps), and a large value of 1000s. We simulated two values for propagation time on a link, one value of 0.27ms (corresponding to a link length of about 50 miles) and another value of 1ms. We assumed that the propagation delay to the first WDM switch is $0.0025ms$ corresponding to a fiber distance of about 1 mile. We assumed that the switching time at a WDM switch is $0.1ms$, and the protocol processing time at a WDM switch is $0.1ms$.

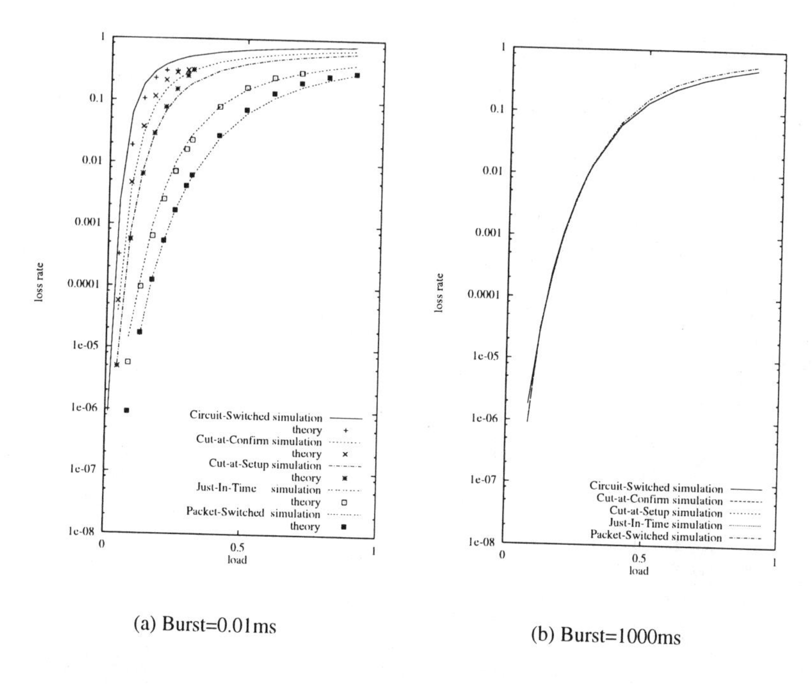

(a) Burst=0.01ms

(b) Burst=1000ms

Figure 5 Packet Loss versus Applied load

Simulation Results and Discussion. We observed similar performance behaviors for both topologies. Below we illustrate the results using those of the torus network. Fig. 5(a) illustrates the packet loss rate versus load when the data-burst is of size 0.01ms. First, we observe that the analytical values are in good agreement with the simulation values. Then, we observe that for a given load, packet-switching has the least blocking probability, followed by JIT-OBS, pipelined circuit-switching and circuit-switching. The intuitive reason is the channel-hold duration for the schemes increases in the same order. Fig. 5(b) illustrates the packet loss rate versus load when the data-burst is of size 1000ms. We observe that in this case, all the schemes have similar performance. This is because, since the data-burst transmission is large compared to the setup time, the channel hold duration for each scheme is approximately the same.

Figure 6 illustrates the normalized throughput against the applied load. We observe that for a given load, packet-switching achieves the highest throughput, followed by JIT-OBS and circuit-switching.

The average latency is the duration from the instant a data-burst arrives at an access station, to the instant that the data-burst arrives at the destination access station. Fig. 7 illustrates the network-wide average latency against the applied load. We find that JIT has the least latency followed by PKT switching, and pipelined circuit switching and circuit switching. JIT pipelines the crossconnect

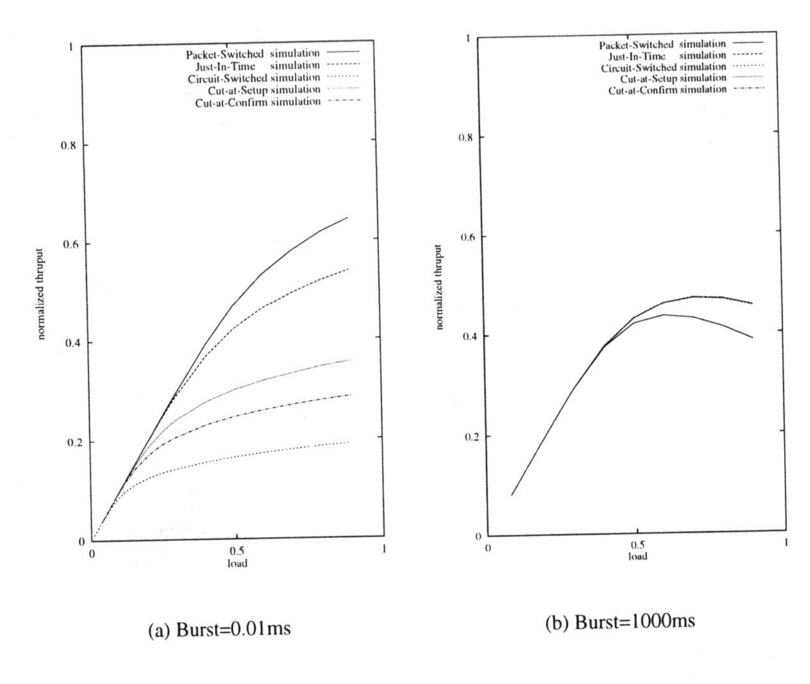

(a) Burst=0.01ms

(b) Burst=1000ms

Figure 6 Normalized Throughput versus Applied Load

setup time with data transmission and as a result is able to achieve the least latency.

Figure 8 illustrates the normalized throughput against load for two different values of the link propagation time: a value of 0.27 ms and a value of 1 ms. When the propagation delay increases, the throughput performance of circuit-switched schemes is adversely affected, while the performance of JIT and packet-switched schemes remain unaffected.

5. CONCLUSION

JIT-OBS combines the desirable features of circuit and packet switched schemes. We found that when the burst-size is small compared to the other parameters, throughput performance of JIT-OBS was better than circuit-switching and the latency performance of JIT-OBS was better than packet switching on representative network topologies. Furthermore, the performance of JIT-OBS is insensitive to the link propagation delay.

Under the MONET project, a signaling protocol for JIT-OBS has been designed and implemented [8]. This signaling implementation is currently being tested and validated against a laboratory network of Lucent optical WDM network elements. Additional testing and experimentation are scheduled in

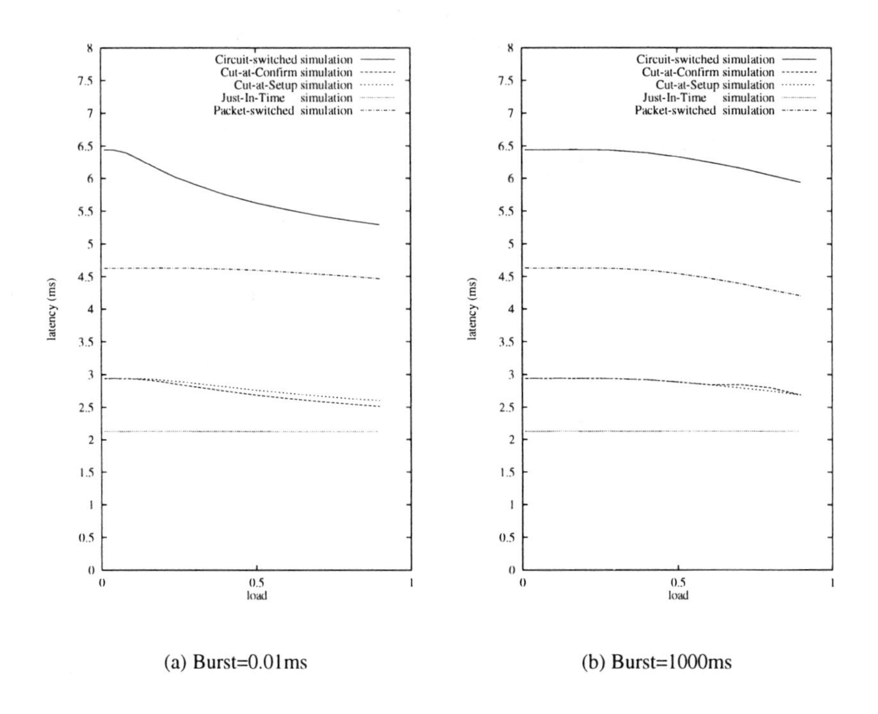

(a) Burst=0.01ms (b) Burst=1000ms

Figure 7 Latency versus Applied Load

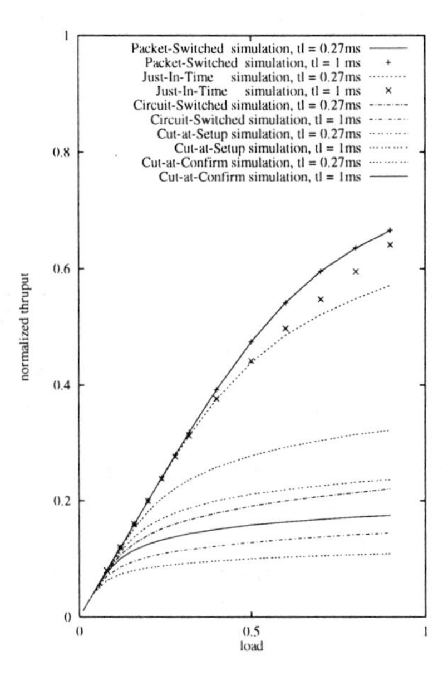

Figure 8 Normalized Throughput versus Applied Load

the upcoming MONET Washington DC testbed network demonstrations [9] with participation from several government agencies. These experiments will provide the opportunity to stress test our implementation and to validate our design.

As it was noted in [10], "...in the future, bandwidth will not be our problem. Latency will be the major challenge to overcome." Towards meeting this challenge, Just-In-Time Optical Burst Switching provides a good mechanism for ultra-low latency transport of variable sized bursts of data across an optical WDM network.

Acknowledgements

The authors wish to thank Dr. Ray McFarland from the National Security Agency's Laboratory for Telecommunications Science for introducing them to his notion of "Just-In-Time" signaling during the course of this research.

References

[1] R.E. Wagner, R.C. Alferness, A.A.M. Saleh, and M.S. Goodman, "MONET: Multiwavelength Optical Networking," *Journal of Lightwave Technology,* vol. 14, no. 6, pp. 1349–1355, June 1996.

[2] J.Y. Wei, C.-C. Shen, B.J. Wilson, and M.J. Post, "Network Control and Management of a Reconfigurable WDM Network," *MILCOM'96,* vol. 2, pp. 581–586, McLean, Virginia, October 1996.

[3] The MONET Research Team, "The MONET New Jersey Network Demonstration," *IEEE Journal on Selected Areas in Communications,* vol. 16, no. 7, pp. 1199-1219, September 1998.

[4] J.Y. wei, C.-C. Shen, B.J. Wilson, M.J. Post, and Y. Tsai, "Connection Management for Multiwavelength Optical Networking," *IEEE Journal on Selected Areas in Communications,* vol. 16, no. 7, pp. 1097–1108, September 1998.

[5] F. Masetti et. al., "High Speed, High Capacity ATM Optical Switches for Future Telecommunication Transport Networks," *IEEE Journal on Selected Areas in Communications,* vol. 14, no. 5, pp. 979-998, June 1996.

[6] D. Chiaroni et.al., "All-Optical Fast Packet-Switched Networks: Physical and Logical Limits of Operation," *SPIE Conference on All-Optical Networking: Architecture, Control and Management Issues,* vol. 3531, pp. 378–389, Boston, Massachusetts, November 1998.

[7] D. Bertsekas and R. Gallager, *Data Networks,* Prentice-Hall: Englewood Cliffs, New Jersey, 1992.

[8] J.Y. Wei, J.L. Pastor, R. McFarland, and J. Parker, "Design of The MONET Just-In-Time Signaling Protocol." In preparation.

[9] A.V. Lehmen, "Large-Scale Demonstration of Multiwavelength Optical Networking in MONET Washington DC Network," *OFC'99,* San Diego, California, February 1999.

[10] D. Farber, *Keynote Speech at the IEEE LEOS Summer Topical Workshop on Broadband Networking,* Monterey, California, July 1998.

CONSTRAINED MAX-MIN BANDWIDTH ALLOCATION FOR THE AVAILABLE BIT RATE ATM SERVICE

Seung Hyong Rhee and Takis Konstantopoulos *

Department of Electrical & Computer Engineering

University of Texas at Austin

Austin, Texas 78712, U.S.A.

{shrhee,takis}@alea.ece.utexas.edu

Abstract The available bit rate (ABR) is an ATM service category that provides an economical support of connections having vague requirements. An ABR session may specify its peak cell rate (PCR) and minimum cell rate (MCR), and available bandwidth is allocated to competing sessions based on the max-min policy. In this paper, we investigate the ABR traffic control from a different point of view: Based on the decentralized bandwidth allocation model studied in [9], we prove that the max-min rate vector is the equilibrium of a certain system of noncooperative optimizations. This interpretation suggests a new framework for ABR traffic control that allows the max-min optimality to be achieved and maintained by end-systems, and not by network switches. Moreover, in the discussion, we consider the constrained version of max-min fairness and develop an efficient algorithm with theoretical justification to determine the optimal rate vector.

Keywords: ATM Networks, Available Bit Rate, Decentralized Control, Max-Min Fairness, Equilibrium

1. INTRODUCTION

The available bit rate (ABR) is an ATM service category that provides an economical support of connections having vague requirements. An end-system may specify both a peak cell rate (PCR) and a minimum cell rate (MCR) on the establishment of the ABR session, and varies its input rate exploiting the available bandwidth along its path.

*The authors of this work were supported in part by Grant ARP 224 of the Texas Higher Education Coordinating Board. The first author is now with Digital Communications Lab, Samsung Advanced Institute of Technology, P.O.Box 111, Suwon, 440-600, Korea.

The ATM forum has proposed a rate-based closed-loop mechanism for the ABR traffic control: a network switch is responsible for computing a fair share of the available bandwidth among connections passing it and sending the information to end-systems. Available bandwidth is allocated to competing sessions based on the max-min policy, which has been widely used as a fairness criterion for bandwidth allocation [3]. Many algorithms have been suggested to implement this ABR control framework. For example, [11] proposed a scheme that uses Explicit Forward Congestion Indication (EFCI), which is a single bit indicator of congestion. To achieve a stable and speedy operation, explicit-rate feedback mechanisms were developed, for example, explicit rate indication [4] and uniform tracking [6].

In this paper, we investigate the ABR traffic control from a different point of view. Using the result on the decentralized bandwidth allocation model studied in [9], we prove that the max-min rate vector is the equilibrium of a certain system of noncooperative optimizations. This interpretation suggests a new framework for ABR traffic control that allows the max-min optimality to be achieved and maintained by end-systems, and not by network switches. In the framework, each session controller has a simple objective function that gives an optimal amount of bandwidth to be used for its connection, and dynamically adjusts its bandwidth responding to the link status along its path.

As ABR connections may specify their own minimum and maximum demands, the classical definition of max-min principle has been extended in the ABR traffic control literature to consider different MCR's or PCR's, e.g., max-min allocation above the MCR [6] or adding one extra link per session to impose PCR [7]. Recently, in [8], the conventional max-min policy was generalized to the case of connections with different constraints on their demands, and conditions for the optimality was driven in the same fashion with the classical one. We develop an efficient algorithm to determine the max-min rate vector for the constrained case: Given the sessions' PCR's and MCR's, the algorithm directly computes the max-min rate vector, and is proved to finish in $K \cdot L$ steps, where K and L are the numbers of sessions and links, respectively.

This paper is organized as follows. The constrained version of max-min policy is introduced in Section 2. In Section 3, a system of noncooperative optimization problems is formulated and it is proved that its equilibrium achieves the constrained max-min fairness. The centralized computing algorithm is developed in Section 4, while Section 5 proposes a new decentralized framework for achieving max-min fairness and shows some simulation results. Finally, Section 6 concludes the paper with some remarks.

2. CONSTRAINED MAX-MIN POLICY

Let $\mathcal{V}$ be a finite set of nodes and $\mathcal{L} \subseteq \mathcal{V} \times \mathcal{V}$ be a set of undirected links whose elements are unordered pairs of distinct nodes. Each link $\ell \in \mathcal{L}$ has an available bandwidth $C_\ell > 0$. A set $\mathcal{K} = \{1, \ldots, K\}$ of sessions is given and share the network $\mathcal{G} = (\mathcal{V}, \mathcal{L})$. Session $i \in \mathcal{K}$ has a set $\mathcal{L}^i \subseteq \mathcal{L}$ of links that it is associated with. The links in $\mathcal{L}^i$ constitute a fixed path between two end points of session i. Moreover, it is assumed that session i has Peak Cell Rate (PCR) M^i and Minimum Cell Rate (MCR) m^i such that $0 \leq m^i \leq M^i$. We denote the allocated rate (bandwidth) for session i by f^i, and the rate vector for all sessions in $\mathcal{K}$ by $\mathbf{f} = (f^1, \ldots, f^K)$. The set of sessions crossing link ℓ can be defined by

$$\mathcal{K}_\ell = \{\, i \in \mathcal{K} \mid \ell \in \mathcal{L}^i \,\},$$

and the total rate at link ℓ is denoted as $F_\ell = \sum_{i \in \mathcal{K}_\ell} f^i$. Then, a rate vector $\mathbf{f}$ should satisfy the following constraints:

$$\begin{aligned} m^i \leq f^i \leq M^i &\quad \text{for all} \quad i \in \mathcal{K}; \\ F_\ell \leq C_\ell &\quad \text{for all} \quad \ell \in \mathcal{L}. \end{aligned}$$

The constrained version of max-min policy, which is a natural extension of the conventional one in [3], can be stated as follows [8]. A vector $\mathbf{f}$ is said to be *feasible* if the above constraints are satisfied. Moreover, a rate vector $\mathbf{f}$ is said to be *max-min fair* if it is feasible and, for each $i \in \mathcal{K}$, f^i cannot be increased while maintaining feasibility without decreasing f^j for some session j for which $f^j \leq f^i$.

Definition 1 **[8]** Given a feasible rate vector $\mathbf{f}$, we say that link $\ell \in \mathcal{L}$ is a *bottleneck link* with respect to $\mathbf{f}$ for a session $i \in \mathcal{K}$ traversing ℓ if $F_\ell = C_\ell$ and $f^i \geq f^j$ for every session j traversing link ℓ for which $f^j > m^j$.

In [3], using the conventional model of max-min fairness, i.e., without constraints, a necessary and sufficient condition for a feasible rate vector $\mathbf{f}$ to be max-min fair is given. It can be extended to the constrained case as follows:

Theorem 1 **[8]** *A feasible rate vector* $\mathbf{f}$ *is max-min fair if and only if each session has either a bottleneck link with respect to* $\mathbf{f}$ *or a rate assignment equal to its maximum rate.*

3. ACHIEVING MAX-MIN FAIR RATE VECTOR

In this section, we formulate a decentralized bandwidth allocation model and prove that its unique equilibrium corresponds to the max-min fair rate vector of the network.

3.1 A SYSTEM OF NONCOOPERATIVE OPTIMIZATIONS

Assume that all sessions in $\mathcal{K}$ of network $\mathcal{G} = (\mathcal{V}, \mathcal{L})$ are active, and that each session independently decides an optimal rate for its ABR connection according to the following optimization problem:

$$\begin{array}{ll} \text{maximize} & \sum_{\ell \in \mathcal{L}^i} w_\ell^i \cdot U_\ell^i(\mathbf{f}), \quad i \in \mathcal{K} \\ \text{subject to} & m^i \leq f^i \leq M^i. \end{array} \tag{1}$$

The weighting factor w_ℓ^i is given by, for all i,

$$w_\ell^i = \begin{cases} 1 & , \quad \text{if } C_\ell^i \leq C_{\ell'}^i \text{ for all } \ell' \in \mathcal{L}^i \\ 0 & , \quad \text{otherwise,} \end{cases}$$

where

$$C_\ell^i = C_\ell - \sum_{j \neq i} f_\ell^j$$

is the remaining bandwidth seen by session i on link ℓ. U_ℓ^i is a real function that is strictly concave with respect to f^i given a remaining bandwidth C_ℓ^i and that has its optimal at

$$\arg\max_{f^i} U_\ell^i(\mathbf{f}) = \alpha C_\ell^i = \alpha(C - \sum_{j \neq i} f^j),$$

where α is a real number such that $0 < \alpha < 1$. The system of optimizations given by (1) is a special case of the decentralized bandwidth allocation model which was studied in [9]. According to Theorem 2 in [9], which establishes the uniqueness of the Nash equilibrium [5] of the decentralized model, (1) also has a unique equilibrium at which point no session can increase its objective without decreasing other sessions' objectives.

We now analyze the bandwidth allocation $\mathbf{f}^*$ achieved by the equilibrium. The set $\mathcal{K}_\ell$ can be divided into two disjoint subsets

$$\mathcal{K}_\ell^1 = \{\, i \in \mathcal{K}_\ell \mid w_\ell^{*i} = 1 \,\} \quad \text{and} \quad \mathcal{K}_\ell^0 = \{\, i \in \mathcal{K}_\ell \mid w_\ell^{*i} = 0 \,\}.$$

The equilibrium bandwidth of session $i \in \mathcal{K}_\ell^1$ can be figured out by the lemma below.

Lemma 1 *For a session $i \in \mathcal{K}$ that passes through link $\ell \in \mathcal{L}$ and $w_\ell^i = 1$ at the equilibrium, its bandwidth is determined such as*

$$f^{*i} = \begin{cases} M^i & , \quad \text{if } M^i < FS_\ell \\ m^i & , \quad \text{if } m^i > FS_\ell \\ FS_\ell & , \quad \text{otherwise,} \end{cases}$$

where the unique point FS_ℓ is given by

$$FS_\ell = \frac{\alpha}{1-\alpha}(C_\ell - F_\ell^*). \tag{2}$$

Proof. Due to space limitations, we only sketch the idea of proof; Refer to the proof of Lemma 2 in [10] for details. Let $FS_\ell = \frac{\alpha}{1-\alpha}(C_\ell - F_\ell^*)$. Then, by the Kuhn-Tucker conditions [2] at the equilibrium, we obtain the following implications:

$$\begin{array}{l} M^i < FS_\ell \;\Rightarrow\; f^{*i} < FS_\ell \;\Rightarrow\; \lambda^{*i} > \mu^{*i} \;\Rightarrow\; f^{*i} = M^i \\ m^i > FS_\ell \;\Rightarrow\; f^{*i} > FS_\ell \;\Rightarrow\; \lambda^{*i} < \mu^{*i} \;\Rightarrow\; f^{*i} = m^i, \end{array}$$

where λ^{*i} and μ^{*i} are Lagrange multipliers. The remaining case $f^{*i} = FS_\ell$ can be proved by removing the users of the first two cases and computing the unconstrained equilibrium. □

3.2 MAX-MIN FAIRNESS OF THE EQUILIBRIUM

All sessions in $\mathcal{K}_\ell$ can be divided further into five disjoint sets as Figure 1. A session with $w_\ell^i = 1$ belongs to one of three sets $\mathcal{A}_\ell$, $\mathcal{B}_\ell$, and $\mathcal{K}'_\ell$, which are defined by

$$\begin{aligned} \mathcal{A}_\ell &= \{\, i \in \mathcal{K}_\ell^1 \mid m^i > FS_\ell \,\} \\ \mathcal{B}_\ell &= \{\, i \in \mathcal{K}_\ell^1 \mid M^i < FS_\ell \,\} \\ \mathcal{K}'_\ell &= \mathcal{K}_\ell^1 - \mathcal{A}_\ell - \mathcal{B}_\ell. \end{aligned}$$

Session i in $\mathcal{K}_\ell^0$ has a link $\ell' \neq \ell$ along its path $\mathcal{L}^i$, and its equilibrium bandwidth is determined in link ℓ' by Lemma 1. However, in link ℓ, sessions in $\mathcal{K}_\ell^0$ can be divided into two sets, $\{\, i \in \mathcal{K}_\ell^0 \mid f^{*i} \geq FS_\ell \,\}$ and $\{\, i \in \mathcal{K}_\ell^0 \mid f^{*i} < FS_\ell \,\}$, depending on the value of FS_ℓ.

Then, the total flow in link ℓ at the equilibrium is

$$F_\ell^* = \sum_{i\in\mathcal{A}_\ell} m^i + \sum_{i\in\mathcal{B}_\ell} M^i + |\mathcal{K}'_\ell| \cdot FS_\ell + \sum_{i\in\mathcal{K}_\ell^0} f^{*i}. \tag{3}$$

If we substitute (3) into (2) and let $\alpha \to 1$, the limiting value of FS_ℓ is

$$\lim_{\alpha\to 1} FS_\ell = \frac{1}{|\mathcal{K}'_\ell|}(C_\ell - \sum_{i\in\mathcal{A}_\ell} m^i - \sum_{i\in\mathcal{B}_\ell} M^i - \sum_{i\in\mathcal{K}_\ell^0} f^{*i}), \quad \ell \in \mathcal{L}. \tag{4}$$

Thus putting (4) in (3) yields

$$\lim_{\alpha\to 1} F_\ell^* = C_\ell, \quad \ell \in \mathcal{L}.$$

In the theorem below, it turns out that the allocation $\mathbf{f}^*$ by (1) becomes the max-min fair rate vector as $\alpha \to 1$.

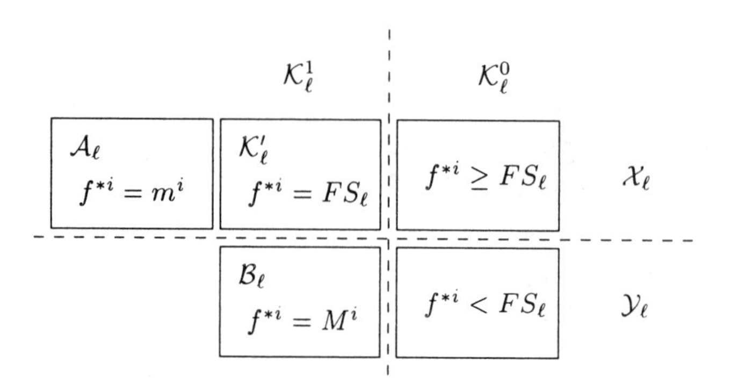

Figure 1 Five disjoint sets of sessions passing link ℓ at the equilibrium.

Theorem 2 *As $\alpha \to 1$, the limit of bandwidth allocation $\mathbf{f}^*$ in the system of optimizations given by (1) is the max-min fair rate vector of network $\mathcal{G} = (\mathcal{V}, \mathcal{L})$ for the sessions $\mathcal{K}$.*

Proof. Clearly $\mathbf{f}^*$ is feasible, since $\mathbf{f}^* \in \mathcal{S}$ and $F_\ell^* \leq C_\ell$ for all ℓ [9]. We consider a link ℓ and examine all five disjoint subsets in Figure 1. First, for a session $i \in \mathcal{K}'_\ell$ or $i \in \mathcal{A}_\ell$, $f^{*i} = FS_\ell \vee m^i \geq FS_\ell$. For any session j for which $f^{*j} > m^j$, $j \in \mathcal{K}_\ell$,

$$f^{*j} \begin{cases} = FS_\ell \wedge M^i \leq FS_\ell & , \quad \text{if} \quad j \in \mathcal{K}_\ell^1 \\ < FS_\ell & , \quad \text{if} \quad j \in \mathcal{K}_\ell^0. \end{cases}$$

The second inequality comes from the fact that for a session in $\{ i \in \mathcal{K}_\ell^0 \mid f^{*i} \geq FS_\ell \}$, $f^{*i} = m^i$, by Lemma 2 below. Thus for session i in $\mathcal{K}'_\ell \cup \mathcal{A}_\ell \cup \{ i \in \mathcal{K}_\ell^0 \mid f^{*i} \geq FS_\ell \}$,

$$f^{*i} \geq FS_\ell \geq f^{*j} \text{ for any } j \text{ for which } f^{*j} > m^j, \quad j \in \mathcal{K}_\ell.$$

At the limit of $\mathbf{f}^*$ as $\alpha \to 1$, $F_\ell^* \to C_\ell$. Then by Definition 1, ℓ is a bottleneck link with respect to $\mathbf{f}^*$ for a session in $\mathcal{K}'_\ell \cup \mathcal{A}_\ell \cup \{ i \in \mathcal{K}_\ell^0 \mid f^{*i} \geq FS_\ell \}$. Now consider a session $i \in \{ i \in \mathcal{K}_\ell^0 \mid f^{*i} < FS_\ell \}$. i has at least one link $\ell' \neq \ell, \ell' \in \mathcal{L}^i$, for which $w_{\ell'}^i = 1$. By the same reasoning, the session belongs to $\mathcal{K}'_{\ell'}$, $\mathcal{A}_{\ell'}$ or $\mathcal{B}_{\ell'}$. In case it belongs to $\mathcal{K}'_{\ell'}$ or $\mathcal{A}_{\ell'}$, ℓ' is its bottleneck link, otherwise $f^{*i} = M^i$. Thus each session passing link ℓ has a bottleneck link or a rate equal to its PCR. Thus by Theorem 1, $\mathbf{f}^*$ is a max-min fair rate vector as $\alpha \to 1$. □

Lemma 2 *Session i in $\{ i \in \mathcal{K}_\ell^0 \mid f^{*i} \geq FS_\ell \}$ has a rate equal to its MCR at* $\mathbf{f}^*$, *i.e.,* $f^{*i} = m^i$.

Proof. As the session belongs to $\mathcal{K}_\ell^0$, there is a link $\ell' \neq \ell$, $\ell' \in \mathcal{L}^i$, such that $w_{\ell'}^i = 1$, i.e. $C_\ell^{*i} > C_{\ell'}^{*i}$. Then for the session $i \in \mathcal{K}_\ell^0$ for which $f^{*i} \geq FS_\ell$,

$$\begin{aligned} f^{*i} &\geq FS_\ell = \frac{\alpha}{1-\alpha}(C_\ell - F_\ell^*) = \frac{\alpha}{1-\alpha}(C_\ell^{*i} - f^{*i}) \\ &> \frac{\alpha}{1-\alpha}(C_{\ell'}^{*i} - f^{*i}) = \frac{\alpha}{1-\alpha}(C_{\ell'} - F_{\ell'}^*) = FS_{\ell'} \end{aligned}$$

Thus $f^{*i} = m^i$ at link ℓ' by Lemma 1. □

4. COMPUTING ALGORITHM

We develop a simple algorithm for computing the max-min fair rate vector $\mathbf{f}$. The algorithm is formulated in the following way. First, using Algorithm 1 below which determines the fair share solution of a single link, we compute the solution for each link in the network. Then, it is iterated until the max-min rate vector of the network is found, by removing bottleneck link(s) at each iteration.

We are given the link capacity C_ℓ and information on sessions in $\mathcal{K}_\ell$, i.e., M^i's and m^i's. Algorithm 1, which is a modified version of Algorithm 1 in [10], computes FS_ℓ of link ℓ, so that each sessions rate can be determined by Lemma 1. It is easy to see that FS_ℓ is determined in at most $|K_\ell|$ iterations.

Algorithm 1 (Computation of FS_ℓ***)***
REPEAT
$FS_\ell := \frac{C_\ell}{|\mathcal{K}_\ell|}$;
$\delta_\mathcal{A} := \sum_{i \in \mathcal{A}_\ell}(m^i - FS_\ell)$ ***and*** $\delta_\mathcal{B} := \sum_{i \in \mathcal{B}_\ell}(FS_\ell - M^i)$;
IF $\delta_\mathcal{A} > \delta_\mathcal{B}$
THEN $\mathcal{K}_\ell := \mathcal{K}_\ell - \mathcal{A}_\ell$, $C_\ell := C_\ell - \sum_{i \in \mathcal{A}_\ell} m^i$;
ELSE IF $\delta_\mathcal{A} < \delta_\mathcal{B}$
THEN $\mathcal{K}_\ell := \mathcal{K}_\ell - \mathcal{B}_\ell$, $C_\ell := C_\ell - \sum_{i \in \mathcal{B}_\ell} M^i$;
OTHERWISE ($\delta_\mathcal{A} = \delta_\mathcal{B}$)***, STOP.***
UNTIL $\mathcal{K}_\ell = \emptyset$.

We now develop an algorithm for computing the max-min rate vector for network $\mathcal{G} = (\mathcal{V}, \mathcal{L})$. At each iteration of the algorithm, FS_ℓ is independently computed for all links using Algorithm 1. The idea is to find link ℓ such that $FS_\ell \leq FS_{\ell'}$ for all $\ell' \in \mathcal{L}$. If we let p be such a link, then by virtue of Proposition 1 below, we can assure that $\mathcal{K}_p^0 = \emptyset$ at the link. That is, FS_p can be determined by only the local information at link p, and the sessions in $\mathcal{K}_p$ have rate assignments regardless of other links.

Proposition 1 *For a network $\mathcal{G} = (\mathcal{V}, \mathcal{L})$ and sessions in $\mathcal{K}$, let p be the link for which $FS_p = \min_\ell FS_\ell$ for all $\ell \in \mathcal{L}$, where each FS_ℓ, $\ell \in \mathcal{L}$, is independently computed at each link by Algorithm 1. Then*

$$w_p^i = 1, \quad \textit{for all } i \in \mathcal{K}_p.$$

Proof. Consider a session i in $\mathcal{K}_p$, and let q be a link such that $q \neq p$ and $q \in \mathcal{L}^i$. Then by (2), for any $\alpha \in (0, 1)$,

$$\frac{\alpha}{1-\alpha}(C_p - F_p) = FS_p \leq FS_q = \frac{\alpha}{1-\alpha}(C_q - F_q).$$

Thus for any i in $\mathcal{K}_p$,

$$C_p^i = C_p - \sum_{\substack{j \neq i \\ j \in \mathcal{K}_p}} f^j = C_p - F_p + f^i < C_q - F_q + f^i = C_q - \sum_{\substack{j \neq i \\ j \in \mathcal{K}_q}} f^j = C_q^i. \qquad \square$$

At next iteration, the link(s) which has the smallest FS_ℓ is removed from $\mathcal{L}$, and the sessions crossing the link(s) is regarded as constant rate sessions; i.e., for session i in $\mathcal{K}_p$, where p is removed from the set of links, let $M^i = m^i = f^i$. The algorithm is iterated until all links are removed. Then clearly every session has a bottleneck link or rate assignment equal to its PCR. The max-min fairness follows by Theorem 1.

Algorithm 2 (max-min fair bandwidth allocation of network $\mathcal{G}$)
REPEAT
For each $\ell \in \mathcal{L}$, compute FS_ℓ using Algorithm 1;
$\mathcal{P} := \{ \ell \in \mathcal{L} \mid FS_\ell \leq FS_{\ell'}$ for all $\ell' \in \mathcal{L} \}$;
/* For sessions passing link(s) in $\mathcal{P}$, f^i is determined by (1). */
Let $m^i := M^i := f^i$ for $i \in \mathcal{K}_\ell, \ell \in \mathcal{P}$, and $\mathcal{L} := \mathcal{L} - \mathcal{P}$;
UNTIL $\mathcal{L} = \emptyset$.

Since at least one link is removed form the set of links, Algorithm 2 terminates in at most $|\mathcal{L}|$ iterations. For a faster computation, $\mathcal{L}$ can be reduced further in step 3 by removing the link(s) that is not in $\mathcal{P}$ but has all constant rate sessions. This feature is not implemented in Algorithm 2, and its possible effect can be checked in the following example.

Example 1 The network configuration in Figure 2, which is borrowed from [1], consists of four switches connected via three links with capacity 100 each. Four constrained sessions are established and each link is shared by at least two sessions.

Table 1 shows the procedure for computing $\mathbf{f} = (f^1, f^2, f^3, f^4)$ using Algorithm 2. First, FS_ℓ, $\ell = 1, \ldots, 4$, is computed and FS_3 is the smallest.

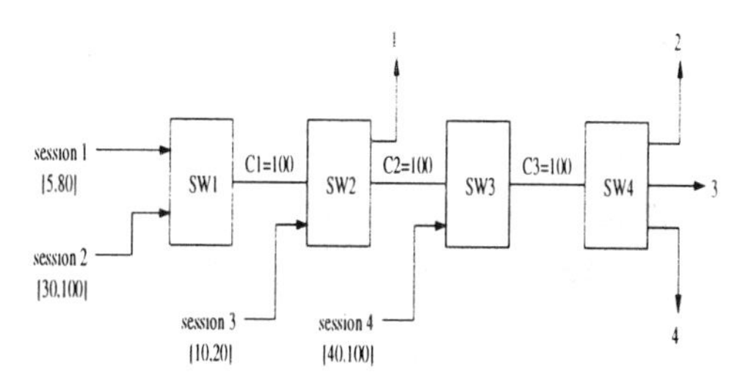

Figure 2 Network configuration of four switches connected via three links. Each session rate is constrained by its MCR and PCR which are denoted by two numbers in the bracket.

Thus we can assign rates to session $2, 3$, and 4 according to Lemma 1. Those assigned rates will not be changed during the computation. At the next iteration, we consider only link 1 and link 2, and $FS_1 > FS_2$. However since all sessions in $\mathcal{K}_2$ were assigned rates at the first iteration, nothing is changed in the second iteration except the set $\mathcal{L}$. If we modify the algorithm so that link 2 is removed at the first iteration, then the algorithm terminates in two iterations. Finally, only link 1 is remained and $\mathbf{f}$ is determined as $(60, 40, 20, 40)$. Its max-min fairness is asserted by Theorem 1.

Table 1 Computation of max-min fair vector of the four-switch network using Algorithm 2.

iter	*step 1*			*step 2*	*step 3*				
	FS_1	FS_2	FS_3	$\mathcal{P}$	f^1	f^2	f^3	f^4	$\mathcal{L}'$
1	50	80	40	$\{3\}$		40	20	40	$\{1,2\}$
2	60	50		$\{2\}$		40	20	40	$\{1\}$
3	60			$\{1\}$	60	40	20	40	$\emptyset$

5. DECENTRALIZED FRAMEWORK FOR ABR TRAFFIC CONTROL

In this section we suggest a new decentralized framework to achieve the max-min fair bandwidth allocation based on the results of Section 3. We consider a dynamic scheme in which each ABR session adjusts its rate dynamically in response to network status. Let $\mathbf{f}(t)$ be a K-dimensional vector whose components are session rates at a discrete time t. Then the dynamics of the system can be formulated by the following iterative equation:

$$\mathbf{f}(t+1) = T(\mathbf{f}(t)), \quad t = 0, 1, \ldots, \tag{5}$$

where T is some continuous mapping. Sessions change their rates iteratively until $\mathbf{f}$ converges to the fixed point solution of (5), according to the following rule.

Session behavior If session $i \in \mathcal{K}$ has a chance to update its rate at time t, it changes its rate such as, given remaining capacities of links in $\mathcal{L}^i$,

$$f^i(t) = M^i \wedge [\alpha C_p^i(t) \vee m^i], \tag{6}$$

where $p \in \mathcal{L}^i$ is a link such that $C_p^i(t) \leq C_q^i(t)$ for all $q \in \mathcal{L}^i$.

Note that response of a session given by (6) is the optimal of the optimization problem given in (1). Thus, if α is close to 1, the fixed point solution of the above system corresponds to the equilibrium of (1) which has the max-min property.

We adopt the *Gauss-Seidel* type iteration by which only one component of $\mathbf{f}$ can be updated at a time and the most recent information is available. Thus, in our implementation, $\mathbf{f}(t-1)$ and $\mathbf{f}(t)$ can be different only in their i^{th} element. A couple of assumptions are used in this framework. First, we do not specify how sessions obtain the information on the link status: We assume that sessions are able to acquire the information from the switches or by measurement. Second, it is assumed that there is no information delay.

The distinct feature of this scheme is that it achieves the max-min fairness by the sessions' independent and simple computations as in (6). Network switches do not perform any traffic control, nor maintain any local information. Moreover, the rate adjust mechanism given by (6) guarantees that there is no overflow at any time as long as MCR's are all zeros.

The Gauss-Seidel type algorithm can be implemented in two different ways: In a *synchronous* implementation, sessions update their rate in a specified order. On the other hand, there is no specified order in an *asynchronous* one. In the following subsection, we present simulation results using the asynchronous iteration.

5.1 ASYNCHRONOUS ITERATIONS

There are many ways to select a session at each time, for example, first-come-fist-serve based on requests of sessions or random selection among sessions.

Figure 3 shows the changes in individual and total rates in a single link network with $C = 100$. With $\alpha = 0.99$, the individual rates converge to the fair share rate of the link. Each session has an opportunity to change its rate

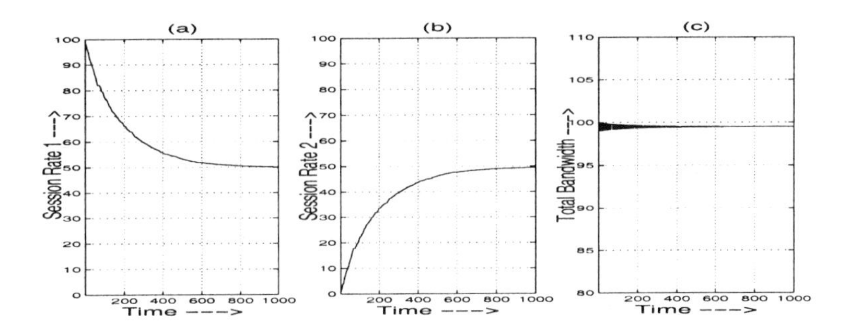

Figure 3 Asynchronous iterations in the single link with two unconstrained sessions: (a) session rate 1 (b) session rate 2 (c) total rate

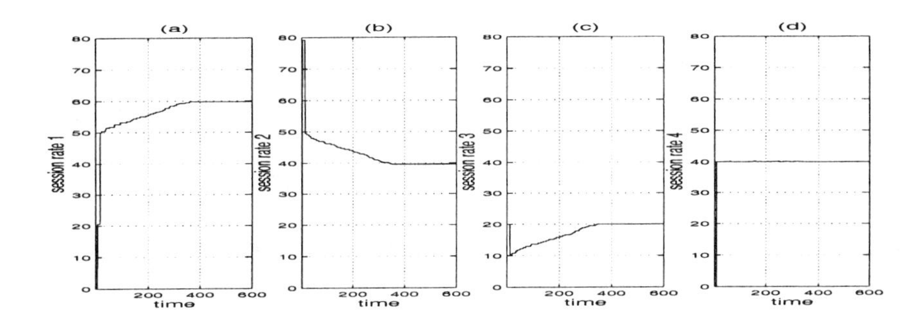

Figure 4 Changes of four individual session rates in the four-switch network by the asynchronous iterations

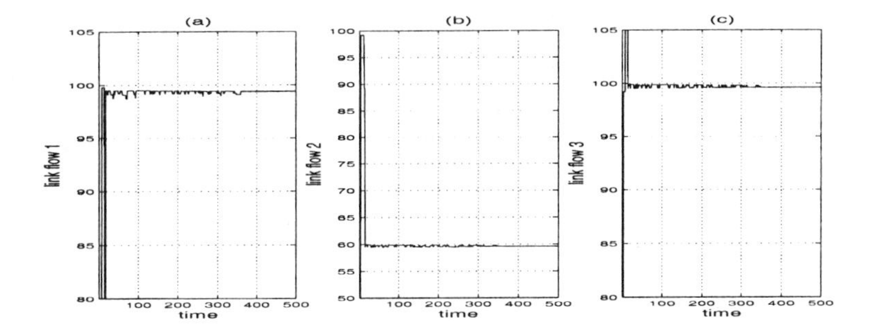

Figure 5 Total link rates in the asynchronous iterations

with equal probability. As session rates are changed in non-periodic manners, graphs of Figure 3 show irregular fashions.

Similarly, the asynchronous iteration can be applied to general topology networks with constrained sessions. Figure 4 and Figure 5 depict the rate changes of individual sessions and links, respectively, in the four switch network of Figure 2. With $\alpha = 0.99$, four individual rates converge to $(60, 40, 20, 40)$ as computed in Example 1.

5.2 SPEEDING UP THE CONVERGENCE RATE

If K unconstrained sessions share a single link, at any value of α they equally share the capacity at the steady-state of the iterations. From (2) the rate of a session will be

$$f^i = \frac{\alpha}{1-\alpha}(C - F) = \frac{\alpha}{1-\alpha}(C - Kf^i) = \frac{\alpha}{1+(K-1)\alpha}C$$

and the total flow should be

$$F = Kf^i = \frac{K\alpha}{1+(K-1)\alpha}C. \tag{7}$$

In Figure 3, as we set $\alpha = 0.99$, the actual rates at the steady-state are $f^1 = f^2 = 49.75$ and $F = 99.5$. In order to let the residual capacity shrink to zero, α should be very close to 1. The problem here is that, as $\alpha \to 1$, it takes very long time for the rates to converge to the steady-state. Figure (6) shows the effect of α: As α goes to 1, total rate in the link increases, however it takes much more time to be converged.

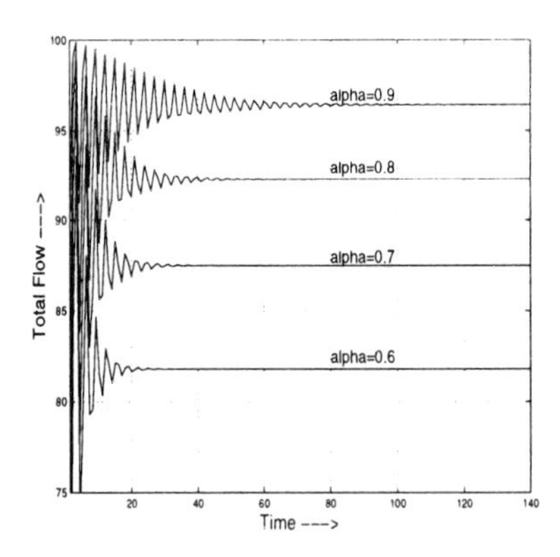

Figure 6 The effect of α on the convergence of the algorithm. As α goes to 1, the total flow increases with lower converging speed.

One approach to avoid the difficulty is to give a value of α so that a target utilization of the link is achieved. Given target utilization ρ and capacity $\overline{C}$, to achieve the total rate $\rho\overline{C}$, the value of α is determined from (7) such as

$$\alpha = \frac{F}{K\overline{C} - (K-1)F} = \frac{\rho}{K - (K-1)\rho}. \tag{8}$$

Note that the value of α to achieve ρ, depends only on K, and is independent of $\overline{C}$. Figure 7 depicts this approach for $\rho = 0.9$ in a single link with $\overline{C} = 100$. As $K = 2$, α is set to $\frac{\rho}{2-\rho} = 0.8182$ by (8). Individual session rates achieve the equal share rate 45 very fast, and the exact link utilization is attained as desired.

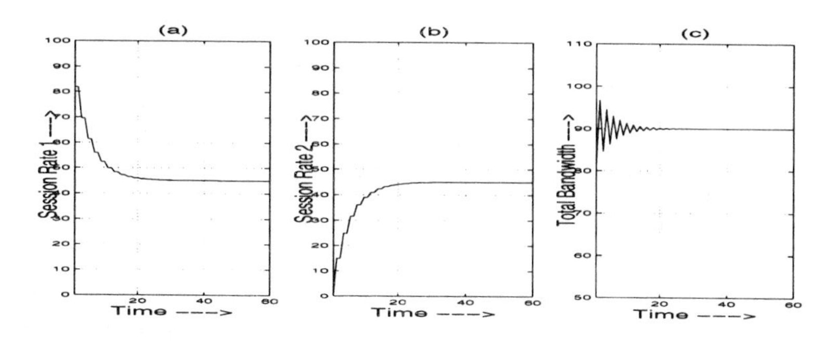

Figure 7 Achieving max-min bandwidth allocation by link utilization ρ. Target link utilization $C = 90$ is attained at the steady state: (a) session rate 1 (b) session rate 2 (c) total flow

6. CONCLUSIONS

We have considered the constrained version of max-min bandwidth allocation for the ABR traffic control in ATM networks. Based on the decentralized bandwidth allocation model of [9], we proved that the equilibrium of a certain system of noncooperative optimizations is the max-min rate vector, and develop an efficient algorithm to determine the vector. The above work suggested a new framework for ABR traffic control such that the max-min fairness can be achieved by simple interactions of ABR sessions not by network switches. Simulation results were presented using asynchronous Gauss-Seidel type iteration.

It should be noted that the decentralized framework proposed in this paper awaits investigation on the effect of information delay to the convergence properties of the algorithm.

References

[1] Arulambalam, A., Chen, X., and Ansari, N. (1996). Allocating fair rates for available bit rate service in ATM networks. *IEEE Communications Magazine*, pages 92–100.

[2] Bertsekas, D. (1995). *Nonlinear Programming*. Athena Scientific, Belmont, MA.

[3] Bertsekas, D. and Gallager, R. (1992). *Data Networks.* Prentice Hall, London, 2nd edition.

[4] Charny, A., Clark, D., and Jain, R. (1995). Congestion control with explicit rate indication. In *Proceedings of IEEE ICC*, pages 1954–1963.

[5] Fudenberg, D. and Tirole, J. (1992). *Game Theory.* MIT Press, Cambridge, MA.

[6] Fulton, C., Li, S.-Q., and Lim, C. S. (1997). An ABR feedback control scheme with tracking. In *Proceedings of IEEE INFOCOM*, pages 805–814.

[7] Gafni, E. and Bertsekas, D. (1984). Dynamic control of session input rates in communication networks. *IEEE Trans. on Automatic Control*, 29(11):1009–1016.

[8] Hou, Y., Tzeng, H., and Panwar, S. (1997). A generalized max-min network capacity assignment policy with a simple ABR implementation for an ATM LAN. In *Proceedings of IEEE GLOBECOM*, pages 503–508.

[9] Rhee, S. H. and Konstantopoulos, T. (1999a). A decentralized model for virtual path capacity allocation. In *Proceedings of IEEE INFOCOM*, pages 497–504, NY.

[10] Rhee, S. H. and Konstantopoulos, T. (1999b). Decentralized optimal flow control with constrained source rates. *IEEE Communications Letters*, 3(6):188–190.

[11] Yin, N. and Hluchyj, M. (1994). On closed-loop rate control for ATM cell relay networks. In *Proceedings of IEEE INFOCOM*, pages 99–108.

SESSION 11

Switching Systems

INTEGRATION OF IP-PACKET/ATM/CIRCUIT-SWITCHING TRAFFIC AND OPTIMUM CONDITIONS OF THE RELATIVE COST FUNCTION FOR TYPICAL TRAFFIC MODES

Noriharu Miyaho
NTT Service Integration Laboratories
miyaho@rdh.ecl.ntt.co.jp

Abstract An integrated switching-system architecture is proposed in which IP-packet, X.25-packet, ATM, and circuit-switching traffic can be handled simultaneously. It is based on a hierarchical memory system. The optimum boundary condition for frames sent over a TDM line is discussed from the viewpoint of minimum-cost requirements. Effectively utilizing transmission lines in an international-connection environment is particularly important when integrating various kinds of multimedia traffic to be sent over one line. Typical examples with key traffic-characteristic parameters are discussed and evaluated for determining optimum bandwidth allocation.

Keywords: Integrated switching, IP, ATM, Circuit switching, TDM boundary optimization Cost function

1. INTRODUCTION

A previously proposed mechanism [1] enables packet-switching (PS) and circuit-switching (CS) functions to be integrated in a switching-system architecture by using a hierarchical memory system. Various types of data traffic with different bearer rates can then be time division multiplexed (TDM) onto a single transmission line by setting an appropriate boundary in each frame to be sent over a TDM line to separate the two switching functions.

Communication services are generally specified as either guaranteed or best effort in terms of communication quality and routing mechanisms [2][3]. The QoS service classifications for circuit, packet, ATM, and frame-relay switching are shown in Table 1.

Table 1 QoS Service classification

Switching type		CO/CL	QoS guarantee	Delay characteristics
Circuit switching		CO	Guaranteed	Excellent (guaranteed to be constant and small)
Packet switching	X.25	CO	Guaranteed (virtual circuit)	Poor (not guaranteed)
	IP	CL	Best effort	Very poor (not guaranteed)
ATM switching	CBR, GFR	CO	Guaranteed	Good (not guaranteed)
	UBR,VBR, ABR	CO	Best effort	Poor (not guaranteed)
Frame relay		CO	Best effort	Good (not guaranteed)

CBR:Constant Bit Rate GFR:Guaranteed Frame Rate UBR:Unspecified Bit Rate VBR:Variable Bit Rate ABR:Available Bit Rate
CO: Connection-Oriented CL:Connectionless

Connection-oriented packet switching has traditionally been used to provide reliable communication services, but with the introduction of optical fibers, conventional packet switching is being replaced by frame-relay switching, particularly for transmitting large files and interconnecting LANs. Using frame-relay switching eliminates the need for retransmission control and hence achieves more efficient packet-data transmission. In Japan, the number of subscribers to packet-switching services, excluding DDX-TP (which is available via conventional telephone networks), was more than 300,000 at the end of 1998, while that to frame-relay services was only 50,000. However, the latter increased more than twice from the previous year, while demand for the former is subsiding.

Connection-less (CL) IP packet-switching service is deemed a best-effort service because it does not handle retransmission in the network layer when an error occurs; instead, error recovery is handled in an upper transport layer, such as the transmission-control-protocol (TCP) layer. Therefore, the CL function is considered to be useful for accommodating the multimedia databases that will be used in access networks.

The demand for Internet protocol (IP) services is increasing. These services currently include non-real-time and comparatively low-speed communication services, such as electronic mail (using SMTP), file transfer (using SMTP), and WWW access (using HTTP). Higher-speed and QoS-guaranteed communication services using IP should become available in the near future. In Japan, the number of subscribers to IP services over the Internet now exceeds ten millions; however, service quality has been degraded by the increasing traffic volume and limited router-processing capacities. To handle multimedia traffic with a wide range of communication speeds with QoS control, ATM switching is generally recognized as an efficient solution;

however, when ATM switching is used, it is a little bit difficult to ascertain the real-time traffic characteristics, such as for high-quality-voice and hi-fi stereophonic-sound traffic. This means that even if most multimedia traffic is handled using packet, ATM, or frame-relay switching to support efficient utilization of the transmission line and to ensure sufficient communication quality, genuine real-time traffic should sometimes be handled as such by using comparatively narrow-band circuit switching. This narrow-band switching should be appropriately applied to international connections because of the high cost of using international transmission lines. High-quality hi-fi stereophonic sound and highly bandwidth compressed voice are considered to be examples of genuine real-time traffic.

Taking the above discussion into account, we should handle non-real-time traffic, such as packet-switching (PS) traffic, and strict real-time traffic, such as high-quality voice and sound traffic, simultaneously. While ATM-mode traffic, such as real-time variable-bit-rate traffic, can currently be handled as real-time traffic, ATM mode is still inferior to CS mode.

In this paper I propose an integrated switching-system architecture in which PS, ATM, and CS traffic are integrated based on a hierarchical memory system. I also present a method for determining the optimum boundary conditions for a TDM transmission line. I focus on the conditions for a 2-Mb/s TDM line from the viewpoint of cost.

Integration of the IP-packet processing with layer 4 control and of ATM traffic processing with QoS control by using a hierarchical memory system concept is presented first. This paper also discusses optimum traffic multiplexing conditions from the viewpoint of minimum cost requirements.

2. PROPOSED INTEGRATED SWITCHING SYSTEM ARCHITECTURE

The basic concept of the proposed integrated switching-system architecture is presented in Fig. 1. The ATM traffic-handling function is included in addition to the IP-packet (including X.25)/CS traffic-handling functions by defining the following seven features.

(1) The boundary in a TDM packet is initially set assuming circuit-switching and other types of traffic. It is adjusted to match the variations in traffic types by using a movable boundary scheme [4]-[5].

(2) IP/X.25-packet/ATM-cell and circuit-switching functions are performed by sharing the same hierarchical memory structure that is conventionally used for program storage and execution in a general-purpose computer. The ATM traffic and PS traffic are handled by making use of a specific hardware mechanism inside the high-speed integrated switching storage (ISS). Switching program execution is performed so that instructions stored in the high-speed

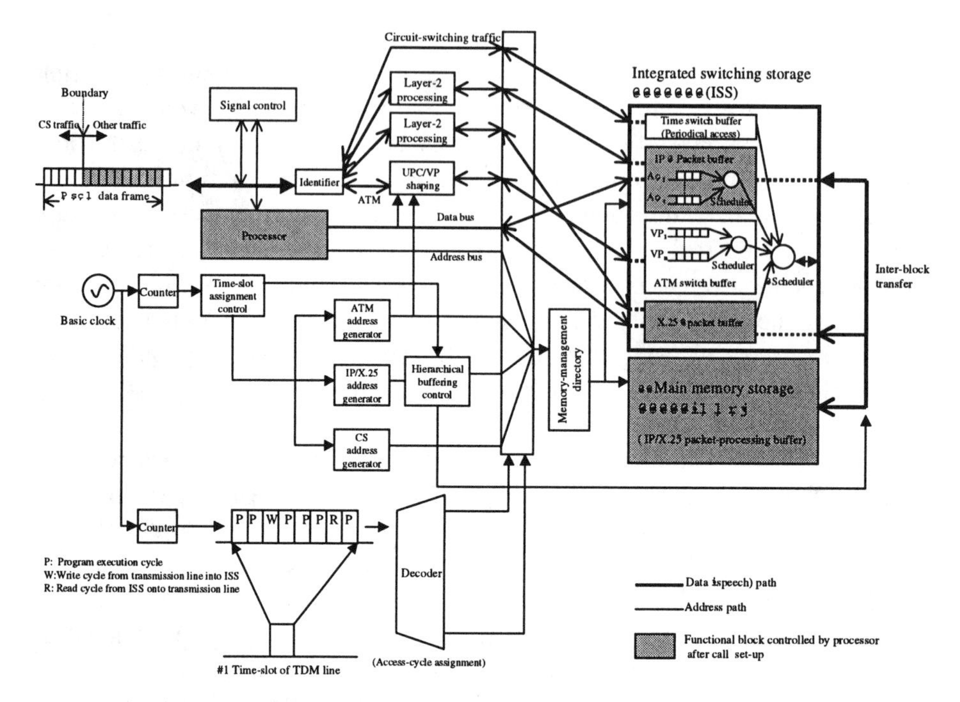

Fig. 1 Proposed integrated switching-system architecture.

ISS are read out and processed. When the required information (instructions or data) is not stored in the ISS, the appropriate data block is fetched from the large-capacity main memory storage, processed, and stored in the ISS so as to be available for subsequent processing. This process is the same as cache processing in a general-purpose computer.

(3) The communication channels (with multiple time-slots) assigned for PS, ATM, and CS traffic are multiplexed, and each time slot is identified during call set-up as containing a PS, CS, or ATM call. Doing this requires that the calling terminal send the call-identification signaling information in advance to the switching node. This information is sent through a control-signal channel, as in the ISDN user-network interface.

(4) The IP/X.25 traffic-processing function processes the appropriate data stored in the ISS. When the IP-packet /ATM traffic density is low and the data buffer areas for these traffic types in the ISS have vacancies, the processor handles these types of traffic on a first-in first-out (FIFO) basis for each service-type queue, sending the data into the outgoing TDM transmission line.

(5) When IP/X.25 packets arrive with a long service time, such as for a mail-box service, or when there is a shortage of buffer space for these packets in the ISS due to a large number of packets arriving simultaneously, the data packets are transferred to the MMS on a FIFO basis. In contrast, when there is a shortage of ATM-traffic buffer space, ATM cells are lost in the ISS,

which leads to degradation in the ATM QoS.
(6) A CS call periodically accesses the same area of the CS buffer space (which is equivalent to a time switch) in the ISS until a disconnection signal is received by the signal control block and processed by the processor. The CS function is thus achieved by this time-slot interchanging action. Therefore, it can easily guarantee a short and constant delay, with real-time characteristics. In contrast, PS traffic is processed continuously by the processor according to the corresponding communication protocol, such as error recovery/re-transmission and virtual-to-real address conversion for the corresponding program execution.
(7) ATM traffic can be handled in a manner similar to that for CS traffic, except it also needs usage parameter control (UPC), VP shaping control, and required-QoS control for each virtual path (VP). Also, the time-slot assignment for ATM traffic in a TDM frame for each specific call is not fixed. Only ATM traffic needs these mechanisms ; other types of traffic do not need them. ATM traffic also follows the Q.2931 protocol.

In the proposed switching architecture, IP packets can be handled in both the IP layer and the TCP/UDP layer by making use of the common processor. A conventional router can normally handle only IP addresses (by referring to a routing table), but in this proposed architecture, it can also handle the port numbers included in the TCP/UDP header. For each IP packet call, after identifying the port number (which normally determines a specific application server; WWW, FTP, etc.), the processor stores the IP packet in the corresponding service queue buffer (A1-Am). Because this priority-control mechanism can be further enhanced by combining the IP address, the type of service (which is included in the IP header), and the port number, it is effective for handling application- or protocol-specific priority control. This differs substantially from the end-to-end user connection-oriented QoS control normally used in ATM switch networks, in which several priority-specific VP buffers (VP_1-VP_n) are used.

This proposed design concept is also applicable to a single-stage memory system if the ISS buffer is large enough to handle the packet traffic and the required packet-switching speed is comparatively low. Figure 2 shows the ISS access cycle flow based on the above discussion.

Integrated PS/CS/ATM communications can be made available by selectively offering packet- or circuit-switching services that correspond to the user's rapidly changing requests. To adequately respond to these requirements, the PS/CS/ATM integrated switching concept needs to be carefully examined from the perspectives of economy and flexibility.

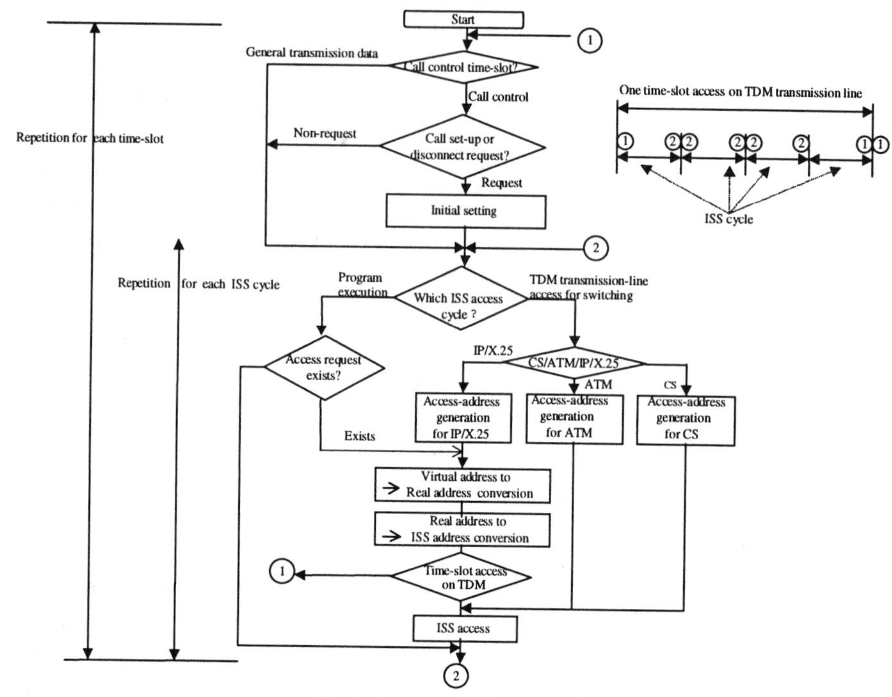

Fig.2 Proposed ISS access flow.

3. DETERMINING THE OPTIMUM BOUNDARY CONDITION

To determine the optimum boundary condition, we define the costs of transferring B bits of information per day by circuit switching and by packet switching as Cc and Cp, respectively. Assuming the parameters defined in Table 2, we can derive the following equations.

$$C_c = A_1 \bullet B / D + A_2 \bullet B/ (R_1 \rho) \tag{1}$$

$$C_p = A_3 \bullet B/P + A_4 \bullet B/D + A_5 \bullet B/\lambda \tag{2}$$

By using these formulae, we can minimize newly defined relative cost Zrel. Let Ncs and Nps be the number of channels assigned to circuit and packet switching, respectively, and let q be the basic channel bearer rate (e.g., 64 kb/s). If a primary-rate E1 transmission line is assumed, we get Ncs + Nps = m (constant=30).

The parameters R_1, R_2, and λ in Table 1 are defined as follows.

$$R_1 = q \bullet N_{cs} \tag{3}$$

$$R_2 = q \bullet N_{ps} \tag{4}$$

$$\lambda = q \bullet N_{ps}/P \tag{5}$$

Thus, the relative cost Z_{rel} is expressed as follows.

$$\begin{aligned} Z_{rel} &=(C_c +C_p)/ B = A_1/D+A_2/(R_1\rho)+A_3/P+A_4/D + A_5/\lambda \\ &=[(A_1+A_4)/D+A_3/P]+A_2/(q\bullet N_{cs}\bullet\rho)+(A_5\bullet P)/q\bullet(m-N_{cs}) \end{aligned} \tag{6}$$

Although Ncs is actually an integer (≥0), we assume it is a real number and that Zrel is a continuous function of Ncs. The minimum value of Zrel is found by setting the derivative of Zrel equal to zero (i.e., Zrel'=0) and calculating (Ncs)opt, the corresponding value of Ncs for $0 \le Ncs \le m$.

The (Ncs)opt that gives the minimum Zrel under these conditions is as follows.

$$(N_{cs})_{opt} = (m/q) \bullet [(A_2 \bullet A_5 \bullet P/\rho)^{1/2} - A_2/\rho]/[A_5 \bullet P/q - A_2/(q \bullet \rho)] \qquad (7)$$

Table 2 Parameters for determining for optimum boundary condition.

A_1	Circuit-switching processing cost per call (unit/call)
A_2	Circuit-switching speech-path channel cost per sec (unit/s)
A_3	Packet-switching processing cost per packet (unit/packet)
A_4	Packet-switching processing cost per virtual call (unit/call)
A_5	Packet data buffer occupation cost (unit/ bit ·s)
B	Total information amount per day (bit/day)
C_c	Relative cost of circuit switching for transmission of B bits per day
C_p	Relative cost of packet switching for transmission of B bits per day
D	Total information amount per call (bit/call)
P	Packet size (bit/packet)
R_1	Line speed for circuit switching (bit/s)
R_2	Line speed for packet switching (bit/s)
λ	Reciprocal of holding time for buffer occupation for packet switching
ρ	Line utilization rate for circuit switching (time required for information transmission divided by line holding time)

4. EVALUATION

To evaluate the effectiveness of the method proposed for determining the optimum boundary condition, let us consider practical examples in which we assume appropriate traffic parameters for Eq. (6). We also assume that either PS or CS traffic is dominant. Table 3 shows seven typical traffic-model cases in which PS and CS traffic are appropriately combined, taking current communication-service trends into account.

Based on the results for these cases, in which either the CS or PS traffic is dominant depending on certain conditions, the optimum boundary conditions can be determined.

Table 3 Models for evaluating method for determining optimum boundary condition

Case	Model	PS characteristics	CS characteristics
1	X.25 + CS	Short packet (128 B)	Low traffic density (0.1) Medium speed (64 Kb/s)
2	X.25 + CS	Long packet (4096 B)	Low traffic density (0.1) Medium speed (64 Kb/s)
3	IP + CS	IP packet (576 B)	Low traffic density (0.1) Medium speed (64 Kb/s)
4	IP + CS	IP packet (576 B)	High traffic density (0.8) Low speed (8 Kb/s)
5	IP + CS	IP packet (576 B)	High traffic density (0.8) High speed (128 Kb/s)
6	ATM + CS	ATM cell (53 B)	Low traffic density (0.1) Medium speed (64 Kb/s)
7	ATM + CS	ATM cell (53 B)	High traffic density (0.9) Medium speed (64 Kb/s)

Accurate values for parameters A_1 to A_5 are not needed to evaluate the relative costs or to obtain the conditions for minimizing transmission costs. We can set the same approximate value for parameters A_1 and A_4 based on the author's experience with many types of packet, circuit, and ATM switching-system implementations. Generally, A_2 is of the same order as A1. The important thing is setting the values for A_3 and A_5 compared to those for A_1 and A_2. The absolute value of A3 is normally one order of magnitude lower than that for A_4, which depends on the application service and traffic characteristics. The absolute value of A_5 is probably one order of magnitude lower given recent trends in the development of economical high-speed random access memory. Therefore, the values of parameters A_1 to A_5 were set based on the above reasoning and were the same for the seven examined cases, enabling extraction of the essential requirements.

To evaluate cost function Zrel, a utilization of 2 Mb/s was assumed for an international dedicated line, whose running cost is far higher than that of other communication facilities. Assuming a frame structure of 2.048 Mb/s, 30-B (64 Kb/s) channels were assumed because one frame (= 125μs) is composed of 32 64-Kb/s time-slots (time-slot #0 - #31); #0 is used for synchronization and #16 is used for signaling (D-channel).

Relative cost Zrel is approximately expressed as follows for the seven traffic models, assuming the parameter settings shown in Table 3. In cases 1 to 3 and 6 for normal CS-application services, real-time voice and interactive large-file transmission applications are assumed. The corresponding required average transmission speed is 64 Kb/s and the traffic density is 0.1.

<u>Case 1</u>: X.25 short packet + medium-speed CS

A_1=100 (unit), A_2=100 (unit), A_3=10 (unit), A_4=100 (unit), A_5=0.1 (unit),

D=1 (Mb), P=128 (B), q=64 (Kb/s),ρ = 0.1, Ncs+Nps=30

Zrel = 2 + 200/(64Ncs) + 20/{64•(30-Ncs)} (8)

Case 2: X.25 long packet + medium-speed CS

A_1=100 (unit), A_2=100 (unit), A_3=10 (unit), A_4=100 (unit), A_5=0.5 (unit),

D=1 (Mb), P=4096 (B), q=64 (Kb/s), ρ = 0.1, Ncs+Nps=30

Zrel = 0.1 + 200/(64Ncs) + 51/(30-Ncs) (9)

Here, 576 Byte size of IP packet is assumed according to the specification on RFC879 on TCP/IP case.

Case 3: IP packet + medium-speed CS

A_1=100 (unit), A_2=100 (unit), A_3=10 (unit), A_4=100 (unit), A_5=0.5 (unit),

D=1 (Mb), P=576 (B), q=64 (Kb/s), ρ = 0.1, Ncs+Nps=30

Zrel = 0.44 + 200/(64Ncs) + 92/64(30-Ncs) (10)

Case 4: IP packet + low-speed CS

A_1=100 (unit), A_2=100 (unit), A_3=10 (unit), A_4=100 (unit), A_5=0.5 (unit),

D=1 (Mb), P= 576 (B), q=8 (Kb/s), ρ = 0.8, Ncs+Nps=240

Zrel = 0.44 + 200/(64Ncs) + 740/64(240-Ncs) (11)

For Case 4, because the basic time-slot speed in CS is 8 Kb/s, the total number of equally available time-slots is assumed to be 240 (= 30 x 8). Because an 8-Kb/s-bandwidth efficiently compressed-voice technology[6] (conforming to ITU-T G.729, Conjugate Structure Algebraic CELP) is assumed, the corresponding traffic-density parameter should be set to a large value (such as 0.8). To make this case comparable with the other cases, the optimum boundary range of 0 to 240 is converted to 0 to 30.

Case 5: IP packet + high-speed CS

A_1=100 (unit), A_2=100 (unit), A_3=10 (unit), A_4=100 (unit), A_5=0.5 (unit),

D=1 (Mb), P= 576 (B), q=64 (Kb/s), ρ = 0.8, Ncs+Nps=15

Zrel = 0.44 + 200/(1024Ncs) + 46/84(15-Ncs) (12)

For Case 5, since the basic time-slot speed in CS is 128 Kb/s, the total equivalently available number of time-slots is assumed to be 15(=30 / 2). Because a 128-Kb/s time-slot is assumed to be used for high-quality stereophonic sound, the corresponding traffic-density parameter should be set to a large value (such as 0.8). To make this case comparable with the other cases, the optimum boundary range of 0 to 15 is converted to 0 to 30.

Case 6: ATM cell + medium-speed CS

A_1=100 (unit), A_2=100 (unit), A_3=10 (unit), A_4=100 (unit), A_5=0.5 (unit),

D=1 (Mb), P= 53 (B), q=64 (Kb/s), ρ =0.1, Ncs+Nps=30

Zrel = 4.7 + 200/(64Ncs) + 0.66/(30-Ncs) (13)

For Case 6, the ATM traffic was assumed to have a fixed packet size for evaluation simplicity.

Case 7: ATM cell + medium-speed CS

A_1=100 (unit), A_2=100 (unit), A_3=10 (unit), A_4=100 (unit), A_5=0.5 (unit),

D=1 (Mb), P= 53 (B), q=64 (Kb/s), $\rho = 0.9$, Ncs+Nps=30

$$Zrel = 4.7 + 200/(576Ncs) + 0.66/(30-Ncs) \tag{14}$$

The equations for the seven cases can be illustrated in Figs. 3 (a) to (g) for evaluating the optimum boundary. Several points should be noted.

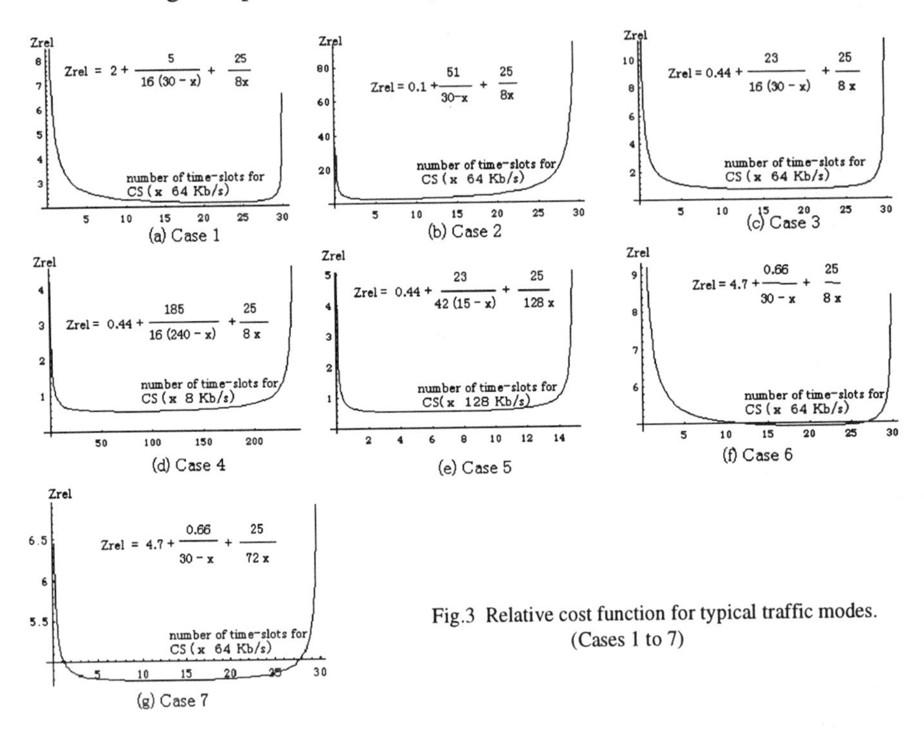

Fig.3 Relative cost function for typical traffic modes. (Cases 1 to 7)

(1) When the CS processing cost exceeds that of PS, i.e., in the X.25 short-packet case, the initially assigned boundary range for CS is comparatively wide, and the required number of time-slots for CS is large (17 to 27).
(2) When the PS processing cost exceeds that of CS, i.e., in the X.25 long-packet case, the initially assigned boundary range for CS is comparatively narrow (3 to 10), and the required number of time-slots for CS is small.
(3) When the required speed for CS is high, the initially assigned boundary range is comparatively wide; however, the optimum value assigned for CS traffic is approximately the same as in the low-speed (8-Kb/s) case if the corresponding traffic density is the same.
(4) When the traffic density of CS increases, the assigned number of time-slots for CS can be increased to achieve minimum cost on the condition that ATM traffic is considered to be a fixed size short packet.

After the initially assigned channel allocation between the switching functions has been determined based on the proposed method, it can be adjusted to match the variations in traffic types by using a movable boundary scheme [4].

The proposed concept can also be applied to a mixture of frame-relay traffic. In this paper, ATM traffic of UBR class (best-effort type) was assumed for analytical simplicity because ATM traffic is considered to be equivalent to fixed-length-packet traffic.

5. CONCLUSION

A new concept was proposed for a switching system architecture in which packet-switching, ATM, and circuit-switching traffic with QoS control are integrated based on a hierarchical memory system. The optimum boundary condition for frames sent over an expensive TDM transmission line on which PS, ATM, and CS traffic are integrated was evaluated by setting several appropriate cost-related parameters. The PS processing cost parameter was found to be more sensitive than the CS one. This proposed method can also be applied to a mixture of CS, PS, ATM (UBR class), and frame-relay traffic.

Future studies should consider the allowable call-blocking ratio in the CS case and the allowable maximum buffering delay in the PS case in addition to the ATM traffic characteristics during burst mode in order to precisely define the optimum boundary condition. In addition, the effective ATM cell handling technology for versatile cell transfer mode such as ABR, VBR, and GFR should also be investigated.

References

[1] Ishikawa, H., Kosuge, Y., and Miyaho, N., "A Study on Hybrid Switch Architecture Using Hierarchical Memory System", IEICE Trans.Commun., Vol. 68-B, No. 1, pp. 30-37, Jan., 1985.

[2] Lee, W. C., Hluchy, M.G., and Humblet, P.A., "Routing Subject to Quality of Service Constraints in Integrated Communication Networks", IEEE Network, July/Aug., pp. 46-55, 1995.

[3] Mase, K. and Kimura, "Quality of Service Issues in Communication Networks," IEICE B-1, Vol. J80-B-I, pp. 283-295, June 1997.

[4] Niitsu, Y., "Evaluation of the Movable Boundary Scheme in Circuit and Packet Switched Networks", Transactions of IEICE, Vol. J68-B, No. 10, 1985.

[5] Miyaho, N. and Miura, A., "Integrated Switching Architecture and its Traffic Handling Capacity in Data Communication Networks", IEICE Trans. Commun., Vol. E-79-B, No. 12, pp.1887-1899, Dec., 1996.

[6] R. Salami, et.al., “Design and description of CS-ACELP : Toll quality 8Kb/s speech coder,” IEEE Trans. Speech on Audio Processing, Vol.6, No.2, pp116-130,1998.

PERFORMANCE ANALYSIS OF SPEEDED-UP HIGH-SPEED PACKET SWITCHES

Aniruddha S. Diwan[1], Roch A. Guérin[2], and Kumar N. Sivarajan[1] *

[1]*Indian Institute of Science*
Electrical Comm. Engg. Dept.
Information & Acoustics Bldg.
Bangalore 560 012, India
{diwan,kumar}@ece.iisc.ernet.in

[2]*University of Pennsylvania*
Dept. Elec. Eng., Rm. 367 GRW
200 South 33rd Street
Philadelphia, PA 19104
guerin@ee.upenn.edu

Abstract In this paper, we study the performance of high-speed packet switches, where the switch fabric operates at a slightly higher speed than the links, i.e., a speeded-up switch. As link speeds keep increasing, the speedup of N (number of links) needed for pure output queueing (which is the best) becomes a significant technical challenge. This is one of the main reasons for the renewed interest in *moderately speeded-up switch fabrics*. The aim of this paper is to highlight the result that only a moderate speed-up factor (less than two) is sufficient to achieve full input link utilization. In particular, we emphasize that this holds, even without relying on a central switch controller making intelligent decisions on which packets to schedule through the switch. As shown in recent works, i.e., [10, 12, 8, 2, 11], there are clearly benefits to using intelligent controllers, but they do come at a cost. Instead, in this paper we focus on what can be achieved by relying *simply* on switch speedup. We do so by means of analysis and simulations for average queue length in switches with speedup. We also present simulation results on delay performance.

Keywords: Packet Switch, Speedup Factor, HOL Blocking, Matrix Analytic

1. BASIC SWITCH ARCHITECTURE

Consider the packet switch of figure 1. This switch has N input ports and N output ports and employs a combination of input and output queueing. We assume a *nonblocking* switch fabric so that all contentions inside the switch for packets destined to *different* output ports are avoided. Note that contention remains inevitable among packets addressed to the *same* output port. Fixed-length packets, or cells, arrive at the input ports of the packet switch. This cell structure is only internal to the switch, so that the links could carry variable size packets. However, in the rest of the paper, we focus on the case of fixed-size packets. Each packet

*A part of this work was done when R. A. Guérin and K. N. Sivarajan were at IBM T. J. Watson Res. Ctr., NY.

contains an identifier that indicates which output (we limit ourselves to the case of unicast flows) j, $1 \leq j \leq N$, it is destined for. In this paper, we are concerned with the performance analysis of such a switch fabric when it is *speeded-up*. That is, the switch fabric runs at a speed greater than that of the input and output links. For simplicity, we assume in our analysis that all input and output links run at the same speed.

Another important characteristic of the switch we consider is that each input makes independent decisions on which packet to send. As a result, the queueing structure at the inputs is a simple FIFO queue. This is in contrast to the switch fabrics of [10, 12, 8, 2, 11], which assume that packets are sorted according to their destination, and possibly priority, at the inputs, so that a central controller can make a selection based on the best possible combination of packets to send through the switch.

The operation of the simple switch fabric which we consider assumes, therefore, independent transmission attempts from all inputs. The basic packet transmission time through the switch fabric is called a *switch-slot*, and switch-slot boundaries at all the links are synchronized. For our analysis, we assume that packet arrivals on all N links are statistically identical, and packet destinations are independent from packet to packet and uniformly distributed across all N outputs. We denote as a *link-slot* the time it takes for a packet to arrive on the link. We also assume that the statistics of packet arrivals on all input links are identical. Because of potential contentions, it may take several switch slots to transfer a packet through the switch, as only packets with distinct destinations can be transferred within a given switch-slot. However, the impact on the inputs depends on the relation between switch-slots and link-slots, or in other words on the speedup of the switch. In switches with a speedup less than N, some queueing will occur on the inputs, and in our analysis we assume infinite buffers on inputs (and outputs), so that no packets are lost. In cases when packets from multiple inputs contend for the same destination in a switch-slot, one packet is chosen according to a *contention resolution policy*.

2. SWITCH THROUGHPUT

Let us denote the transmission time of a packet on any of the links by $1/R_l$ and the transmission time of a packet inside the switch by $1/R_s$. Thus R_s and R_l are the transmission rates (in packets per second) of the switch and input links, respectively. When the switch fabric is speeded up, $R_l < R_s$, and the ratio R_s/R_l is termed the *speed-up factor*. Let us denote the expected number of packet arrivals on each input link per link-slot by p. We term p the *input link utilization*. The expected number of packet arrivals on each input link per switch-slot is then $q = \frac{R_l}{R_s}p$. We say that a switch is *saturated* if it has a packet available for transmission through the switch fabric, in every switch-slot. We define the *saturation throughput* of a switch fabric as the expected number of packets per input link that is transmitted by the switch in each switch-slot, when all switch inputs are saturated. Clearly the saturation throughput will depend on the distribution of the packet destinations and the speed-up factor. At one extreme, for a speed-up factor of 1, if all packets on all input links are destined to a single output port i, the saturation throughput is $1/N$. When packets on input link $i, 1 \leq i \leq N$, are always destined for output port i, the saturation throughput is 1. The saturation throughput has been widely studied in the case when, for each packet, each output port is equally likely to be the destination. As stated in section 1, this is the case we assume in this paper. In this case, the saturation throughput can also be interpreted as the *output link utilization when the input links are saturated.* (Also recall our assumption of independent and statistically identical input links.)

When the switch is not saturated we have to specify the process by which packets arrive on the links, in addition to the distribution of their destinations. The packets that cannot be transmitted through the switch immediately upon arrival, due to HOL blocking, are queued in an infinite buffer at each input link. We define the *stability throughput* as the maximum link utilization for which these input queues are stable. It has been implicitly assumed in much of the literature on input queueing that, as long as the arrival rate of packets on each input link is less than the saturation throughput of the switch, these input queues are stable [5]. However, that this is an assumption requiring proof has been recognized by Jacob and Kumar and their proof for the case of Bernoulli arrivals (successive packets have independent destinations) with identical rates on all input links may be found in [6].

Saturation-Stability Property: For a specified distribution of the packet destinations, we will say that an arrival process satisfies the *Saturation-Stability Property* if the input link queues are stable whenever the expected number of arrivals on each input link per switch-slot is less than the saturation throughput of the switch. We will assume that the packet arrival processes arising in our discussion satisfy the Saturation-Stability property.

We now ask: *For a given input link speed R_l what is the switch speed R_s required in order to achieve an input link utilization of unity, i.e.,* 100% *(stability) throughput?* Note that, due to our assumptions, specifically the symmetry among the output links and the equality of the transmission speeds on the input and output links, the stability of the output queues is assured. Our answer is embodied in the following proposition.

Proposition 1 *The input link utilization of an input queueing switch fabric running at a rate $R_s > R_l/\gamma$, where R_l is the rate on the input links and γ is the saturation throughput of the switch with the same distribution for the packet destinations as that of the packets arriving on any input link, can be made arbitrarily close to one, provided the arrival process satisfies the Saturation-Stability Property.*

Proof: The expected number of arrivals on each input link *per switch-slot* is $q = \frac{R_l}{R_s}p$. If $R_s > R_l/\gamma, q < \gamma p < \gamma$ for all p, and the input queues are stable since the arrival process satisfies the Saturation-Stability Property by assumption. Thus the input link utilization p can be made arbitrarily close to 1 and the switch achieves a (stability) throughput of 100%.

3. QUEUE LENGTH ANALYSIS

In this section we analyze the queue lengths at various queueing points in a speeded-up switch fabric. The time it takes to transmit a packet fully on any link is called a *link-slot*. A link-slot consists of r switch-slots. The time it takes to transfer a packet from an input port to an output port, i.e., inside the switch fabric, is s switch-slots. In other words, the link speed is less by a factor of r/s than the switch fabric speed and we say that the speed-up factor of the switch fabric is r/s. Note that this type of model is applicable to any speed-up factor of the form r/s and is thus a general model. It can be seen that pure input-queued switches $(r = s = 1)$ and pure output-queued switches $(r = N, s = 1)$ are also special cases of the above model.

3.1 PACKET ARRIVAL PROCESS

We now describe the packet arrival process which is considered for analyzing the speeded-up switch fabric. This arrival process is a suitable adaptation of the uniform i.i.d. Bernoulli arrival process. Further, for analytical simplicity, the arrival process considered here is over synchronized switch-slots. Such a switch-slot arrival process leads to the Markov chain of figure 2. An ON switch-slot corresponds to an arrival. An OFF switch-slot corresponds to no arrival. The state transition probabilities are given in figure 2. The destination output port of a packet is chosen uniformly randomly out of N output ports. The packet arrival rate, λ, in packets per switch-slot can be obtained by solving for the steady-state vector of this Markov chain and we get,

$$\lambda = \frac{p}{1 + (r-1)p} \tag{1}$$

and the rate of packet arrivals in packets per link-slot is $r\lambda$.

3.2 ASYMPTOTIC ANALYSIS OF SWITCH PERFORMANCE

The packet switch with the input and output queues forms a complex queueing network. It seems that the exact queueing analysis of this complex queueing network is intractable for finite N. In order to get a handle on the problem, we follow the approach outlined in [7], and try to analyze the different parts of the network separately. For this, consider the travelogue of a tagged packet from the moment it arrived at an input queue to the moment it is transmitted fully on its destination output link. The tagged packet encounters queueing at three points in the network. It is first queued in the input queue where it has arrived. The packets in each input queue are transfered on a FIFO (first in first out) basis. When all the packets in front of this tagged packet in its input queue are transmitted across the switch, the tagged packet enters the HOL position of this input queue. At this point, there may be packets in the HOL position of other input queues whose destination is the same as that of the tagged packet. There may also be a packet which is already in the transfer process to the destination of the tagged packet, in which case the destination is busy. The switch can start transferring only one of the former type of packets in a switch slot to the destination and then *only if* the destination is free. This is called *HOL contention*. This is the second queueing point in the network. Note that the HOL packets destined for a given output form a contention (or virtual) queue corresponding to that output. Therefore the tagged packet is queued in the HOL position, i.e., in its contention queue. Which packet from a given contention queue is to be transferred to the corresponding output (if it is free) in the next s switch-slots is decided by a *contention resolution policy*. The third queueing occurs in the output queue because the tagged packet is transmitted on the output link only when all the packets that arrived before it are transmitted fully on the output link.

We analyze the queue-length of these three types of queues for infinite input and output queues, and for the asymptotic case ($N \to \infty$) in the following subsections. Contention at the HOL positions is resolved thus: If k HOL packets contend for a particular output in a switch-slot, one of the k packets is chosen uniformly randomly from those k packets. The other packets have to wait until that switch-slot in which the output becomes free and a new selection is made among the packets that are then waiting. The arrival process of section 3.1 is assumed to satisfy the saturation-stability property of section 2. For this, the arrival rate, λ, should be less than the saturation throughput of the switch. Observe that a switch with only input queues is exactly the same as a pure input-queued switch on a switch-slot basis. The saturation throughput is 0.586 for the uniform i.i.d. Bernoulli arrival model [7] and tends to 0.5 for bursty arrivals [9] over the switch-slots. Therefore as long as $\lambda < 0.5$, the switch input queues are stable and, due to the symmetry among the output links and the equality of the transmission speeds on the input and output links, the switch output queues are also stable.

3.2.1 Contention Queue Analysis. The destination output of a packet is selected uniformly randomly among the N outputs as mentioned before. The situation at the HOL positions of the input queues is similar to that at the HOL positions of a pure input-queued switch. That is, in both cases packets encounter the HOL blocking phenomenon. We assume that each output contention process tends to be independent asymptotically (as $N \to \infty$). This is indeed the case for a pure input-queued switch with uniform i.i.d. Bernoulli traffic [7]. With this assumption the contention queues can be analyzed separately.

We say that an input queue is unbacklogged in a given switch-slot, say mth, if and only if, either a packet transfer ended in the $(m-1)$th switch-slot or it was empty during the $(m-1)$th switch-slot. A packet occupying HOL position in mth switch-slot is said to be unbacklogged if the corresponding input queue is unbacklogged. Otherwise the packet is said to be backlogged. Now consider any one contention queue, say corresponding to output i. In a given switch-slot, say mth, the contention queue contains backlogged as well as unbacklogged packets. A packet whose transfer is in progress in the mth switch-slot is also considered as a backlogged packet. We denote the number of backlogged packets by B^i_m and the number of unbacklogged packets by A^i_m. We assume that, as $N \to \infty$, the steady-state number of packets moving to the head of unbacklogged input queues in each switch-slot, and destined for output i, (A^i), becomes Poisson at rate, $\overline{F}/N = \rho_0$, where $\overline{F}$ is the mean steady-state number of new packets at the HOL positions. In other words, $\lim_{N\to\infty} \Pr\{A^i = k\} = e^{\rho_0} {\rho_0}^k / k\,!$, and $\rho_0 = \lambda$ below saturation. That this "Poisson process" assumption is correct, is substantiated by the simulated performance [4].

(B^i_m) can be thought of as the system queue length in mth switch-slot. The above Poisson process assumption and the form of B^i_m suggest that the contention queue can be modeled as a discrete-time $BP/D_s/1$ queue with random order of service. BP represents discrete-time batch-Poisson process. D_s represents deterministic service time which, in this case, is s switch-slots (the transmission time of a packet inside the switch). This queueing model can be analyzed to get the steady-state queue-length distribution in a random switch-slot and the delay distribution of the packets in the queue. Each output contention process is assumed to be independent as $N \to \infty$ and all the output contention processes are identical. Hence it is sufficient to analyze any one of them. Note that when the speed-up factor is of the form $r/1$, i.e., integer speed-up factor r, the contention queue is modeled as $BP/D_1/1$. This is the same queueing model as that of the contention queue in [7]. There this model is referred to as discrete $M/D/1$ queue. Thus the analysis of this discrete $M/D/1$ developed in [7] is directly applicable to the case of *integer* speed-up factors. The analysis of the contention queue in [7] is a special case of the analysis developed below.

Distribution of Queue-length in a Random Switch-Slot. We cannot model the queue-length of the contention queue *alone* by a DTMC. If we know how many switch-slots of service time have been completed for the current packet in transfer out of the deterministic s switch-slots of service time, we can model the queue-length by a DTMC. The switch status is T_0 at the beginning of a switch-slot if the switch is not busy at a switch-slot boundary. In this status the switch is ready to serve any packet in this switch-slot. If a packet arrives when the status is T_0 and the contention queue is empty, the packet starts service immediately from this switch-slot. If say, n packets are in the contention queue waiting for service, one of them starts service from this switch-slot according to the contention resolution policy if no packet arrives in this switch-slot. If a packet arrives in this switch-slot, one of the total $(n+1)$ packets starts service from this switch-slot. The status T_i, where $1 < i \leq s$, at the beginning of a switch-slot indicates that there is a packet transfer in progress and that first i switch-slots of service have been completed. A packet is eligible to get service in the same switch-slot in which it arrives.

We club the queue-length in a switch-slot and the switch-status of that switch-slot to form a state. This state can be modeled by a 2-D DTMC over the state space $\{(i,j) : i \geq 0, j \in \{T_0, T_1, \cdots, T_{s-1}\}\}$. Then queue-lengths in successive switch-slots follow this 2-D DTMC. It is assumed that the number of arrivals in the contention queue during each switch-slot has Poisson probabilities. Therefore the state transition probability matrix takes the form,

$$\mathbf{P} = \begin{bmatrix} B_0 & B_1 & B_2 & B_3 & \cdots \\ C & A_1 & A_2 & A_3 & \cdots \\ \tilde{0} & A_0 & A_1 & A_2 & \cdots \\ \tilde{0} & \tilde{0} & A_0 & A_1 & \cdots \\ \vdots & \vdots & \vdots & \vdots & \ddots \end{bmatrix} \quad (2)$$

where B_0 is an 1×1 matrix, $B_i,\ i > 0$ are $1 \times s$ matrices, C is an $s \times 1$ matrix and $A_i,\ i \geq 0$ are $s \times s$ matrices. Note that the $\mathbf{P}$ matrix is a structured $M/G/1$-type of matrix. Please refer to [4] for calculation of the entries of the matrices. Algorithms to solve for the steady-state probabilities vector of such matrices by matrix analytic methods have been implemented in a software package called TELPACK [1]. We have used TELPACK to compute the steady-state probability vector of the $\mathbf{P}$ matrix. The steady-state probability vector of the $\mathbf{P}$ matrix is of the form, $\pi = [\pi^0 \pi^1 \pi^2 \cdots \pi^i \cdots]$, where $\pi^0 = [q_{0,T_1}]$ and, $\pi^i = [q_{i,T_0} q_{i,T_2} \cdots q_{i,T_s}], : i > 0$. q_{i,T_j} is the steady-state probability of the queue-length being i and the switch status being T_j in a random switch-slot. We now can compute the steady state probability vector of queue-length in a random switch-slot which is, $\tilde{l} = [l_0 l_1 l_2 \cdots l_i \cdots]$, where l_i is the probability that the queue-length is i in a random switch-slot. The l_is are given by, $l_0 = q_{0,T_0}$ and $l_i = \sum_{j=0}^{s} q_{i,T_j}$.

Distribution of Delay. Karol *et al* [7] have developed a simple numerical method for computing the delay distribution of a discrete-time $BP/D_1/1$ queue, with packets served in random order (Appendix III of [7]). In this section we extend this scheme to the case of a discrete-time $BP/D_s/1$ queue, with packets served in random order. The number of packet arrivals at the beginning of each switch-slot is Poisson distributed with rate λ and each packet requires s switch-slots of service time. We focus our attention on a particular "tagged" packet in the system, during a given switch-slot. Let p^m_{k,T_j} denote the probability, conditioned on there being a total of k packets in the system and the switch status being T_j during the given switch-slot, that the remaining delay is m switch-slots until the tagged packet completes service.
The p^m_{k,T_j} can be obtained by recursion on m as follows,

$$p^m_{1,T_0} = \begin{cases} 1: & m = s \\ 0: & m \neq s \end{cases} \quad (3)$$

$$p^s_{k,T_0} = \frac{1}{k}: \quad k \geq 1 \quad (4)$$

$$p^{m}_{k,T_0} = (k-1)p^{s}_{k,T_0} \cdot \sum_{j=0}^{\infty} p^{m-s}_{k-1+j,T_0} \cdot \frac{e^{-s\lambda}(s\lambda)^j}{j\,!} \quad : \begin{cases} m = x \cdot s \\ x > 1, \\ k > 1 \end{cases} \tag{5}$$

Averaging over k, the delay D has probabilities as follows:
For $m = s \cdot i$ such that $i \geq 1$ and i integer,

$$\Pr\{D = m\} = \sum_{k=1}^{\infty} p^{m}_{k,T_0} \cdot \sum_{n=0}^{\infty} q_{n,T_0} \cdot \frac{e^{\lambda}\lambda^{k-n-1}}{(k-n-1)\,!} \tag{6}$$

For $m = s \cdot i + j$ such that $i \geq 1$ and $1 \leq j \leq (s-1)$ and i, j integers,

$$\Pr\{D = m\} = \sum_{k=1}^{\infty} p^{m-j}_{k,T_j} \cdot \sum_{n=1}^{\infty} q_{n,T_j} \cdot \frac{e^{(s-j+1)\lambda}((s-j+1)\lambda)^{k-n}}{(k-n)\,!} \tag{7}$$

For all the remaining m, the delay probability is zero. The q_{i,T_j} are obtained in section 3.2.1. With (6)–(7) we get the delay distribution of a packet in the system. The moments of the delay distribution are determined numerically from the delay probabilities in (6)–(7).

3.2.2 Input Queue Analysis. As $N \to \infty$, successive packets in an input queue i experience the same service time distribution because their destination addresses are independent and are equiprobable. The number of switch-slots elapsed between the entry of a tagged packet in the HOL position of its input queue and the exit of that packet from the input queue, is equal to the delay of the packet in the contention queue. It is as if the tagged packet is served for that many switch-slots in its input queue. In other words, the service time distribution of a packet in an input queue is the delay distribution of a packet in the contention queue. With the arrival process of section 3.1 and the service time distribution of section 3.2.1, we can model the input queue as a $G/G/1$ queue. The transition matrix $\mathcal{P}$ of the 2-D DTMC has the form,

$$\mathcal{P} = \begin{bmatrix} B_0 & B_1 & B_2 & B_3 & \cdots \\ A_0 & A_1 & A_2 & A_3 & \cdots \\ \tilde{0} & A_0 & A_1 & A_2 & \cdots \\ \tilde{0} & \tilde{0} & A_0 & A_1 & \cdots \\ \vdots & \vdots & \vdots & \vdots & \ddots \end{bmatrix} \tag{8}$$

Please refer to [4] for calculation of the entries of the matrices. To get the steady-state probability vector of the $\mathcal{P}$ matrix, we need to solve the set of equations, $\tilde{\pi} = \tilde{\pi}\mathcal{P}$ and $\|\tilde{\pi}\| = 1$. where $\tilde{\pi}$ is a vector of the form, $\tilde{\pi} = [\pi^0\pi^1\pi^2 \cdots \pi^i \cdots]$ and π^i is a vector of the form, $\pi^i = [\pi_{i,0}\pi_{i,1}\pi_{i,3} \cdots \pi_{i,r-1}]$. $\pi_{i,j}$ is the steady-state probability of the state (i, j) of the 2-D DTMC. Note that the $\mathcal{P}$ matrix is a structured $M/G/1$-type of matrix. We use TELPACK [1] to compute the steady-state probability vector $\tilde{\pi}$ of our $\mathcal{P}$ matrix. The steady-state probability vector $\tilde{\pi}$ enables us to compute the steady-state distribution of queue-length as seen by a departing packet. The steady-state queue-length vector is of the form, $\tilde{q} = [q_0 q_1 q_2 \cdots q_i \cdots]$ where, $q_i = \sum_{j=0}^{r-1} \pi_{i,j}$. The moments of the distribution of the queue-length as seen by a departing packet are then numerically computed using the above equations. The transition rates into and out of each state in a discrete-state stochastic process must be identical. By using the time-average interpretations for the steady-state queue-length probabilities, we see that the distribution of queue-length as seen by a departing packet is the same as the distribution of queue-length as seen by an arriving packet [13, pp. 387–388]. Note that in [13, pp. 387–388], the Markov chain is one-dimensional. In our case it is two-dimensional. But as we are interested in the steady-state probability vector of levels (i.e., queue-lengths), the arguments are applicable in this case also. We compare the results of the above analysis with the results of simulations in section 5. As it is not possible to compute the distribution of queue-length in a random slot, we cannot analyze the delay of a packet in the input queue. In section 5, we show simulation results for the delay of a packet in the input queue.

3.2.3 Output Queue Analysis. We have assumed that, as $N \to \infty$, all contention queue processes are identical and independent. It then follows that all output queue systems are also identical and independent. It is sufficient to analyze only one output queue. It is difficult to characterize the departure process of a contention queue due to its general queueing model structure. The exact arrival process at the output queue is then unknown. We use an approximate ON-OFF type of arrival model to analyze the output queue. An exact analysis like that of input queues may be difficult in this case. A packet arrives in the output queue after s switch-slots in a busy period. This is because the contention queue service time is s switch-slots. Therefore we model the arrival process as follows: A packet arrives at output queue i in a switch-slot only if the switch status is T_{s-1} at the beginning of the switch-slot (section 3.2.1). There are always $(s-1)$ idle switch-slots (corresponding to the switch status T_0 to T_{s-2}) before an arrival in the output queue. The group of all these switch-slots is called a *busy cluster.* In the busy period of the contention queue, packets depart regularly at the end of the busy clusters. When the contention queue is empty in a switch-slot, no packet arrives in output queue i in that switch-slot. We construct the ON-OFF arrival model such that, $E[B]$, the mean busy period of the contention queue is equal to s times the mean number of busy clusters. The mean OFF period is equal to, $E[I]$, the mean idle period of contention queue i. The ON-OFF Markov process is depicted in figure 3. The ON state and the series of associated $OFF_j, 1 \leq j \leq s-1$ form a busy cluster. It is easy to see that $p = 1 - s/E[B]$ and $p = 1 - 1/E[I]$ since the mean busy period of the contention queue is equal to s times the mean number of busy clusters. The mean OFF period is equal to the mean idle period of contention queue i. Note that for *integer* speed-up factors $s = 1$. There are no idle slots in a busy cluster and a busy cluster consists of only ON slot. The ON-OFF Markov chain of figure 3 reduces to the simple ON-OFF Markov chain of figure 4.

The service time of the output queue is r switch-slots and is deterministic (the transmission time of a packet on the output link). We analyze this output queue model. The queue-length in a random switch-slot cannot be modeled by a DTMC. We look at the departure-switch-slots. The output queue-length as seen by a departing packet in a switch-slot together with the ON-OFF chain status in that switch-slot form a 2-D DTMC over the state space $\{(i,j) : i \geq 0, j \in \{ON, OFF_1, OFF_2, \cdots, OFF_{s-1}, OFF\}\}$ where i is the queue-length in a departure-switch-slot and j is the state of the ON-OFF process. Let $ON \equiv s$ and $OFF \equiv 0$ and $OFF_i \equiv i$. We say that i is the level of the 2-D DTMC and j is the phase of the 2-D DTMC. Let $m = \lceil r/s \rceil$. The state transition matrix is of the form,

$$\mathcal{P} = \begin{bmatrix} B_0 & B_1 & B_2 & \cdots & B_m & \tilde{0} & \tilde{0} & \tilde{0} & \cdots \\ C & A_1 & A_2 & \cdots & A_m & \tilde{0} & \tilde{0} & \tilde{0} & \cdots \\ \tilde{0} & A_0 & A_1 & \cdots & A_{m-1} & A_m & \tilde{0} & \tilde{0} & \cdots \\ \tilde{0} & \tilde{0} & A_0 & \cdots & A_{m-2} & A_{m-1} & A_m & \tilde{0} & \cdots \\ \vdots & \vdots & \vdots & \vdots & \vdots & \vdots & \vdots & \vdots & \ddots \end{bmatrix} \tag{9}$$

Note that the $\mathcal{P}$ matrix is a structured $M/G/1$-type matrix. Please refer to [4] for calculation of the entries of the matrices. The steady-state probability vector of the $\mathcal{P}$ matrix is computed by using TELPACK. The steady-state probability vector is of the form, $\pi = [\pi^0 \pi^1 \pi^2 \cdots \pi^i \cdots]$ where, $\pi^i = [\pi_{i,0} \pi_{i,1} \cdots \pi_{i,s}]$ The steady-state probability vector of queue-lengths as seen by a departing packet is of the form, $\tilde{q} = [q_0 q_1 q_2 \cdots q_i \cdots]$ where $q_i = \pi_{i,0} + \pi_{i,1} + \cdots + \pi_{i,s}$. The results of the above analysis are compared with the results of simulations in the next section of this chapter. We are unable to analyze the delay performance of this output queue. Nevertheless, in section 5 of this chapter, we show the results of simulations for the delay of a packet in the output queue.

4. MEAN DELAY SIMULATIONS

In this section we study the simulation results of the mean delay of the speeded-up packet switches for the arrival process of section 3.1. We have considered a 64×64 nonblocking speeded-up switch fabric. We plot the total average delay of a packet in the switch obtained through simulations in figure 5. The average delays for the cases of pure input-queued and pure-output queued switches are also plotted in figure 5. It can be seen from the plot that the delay curves for speed-up factors 2 and 3 are close to the delay curve of the pure output-queued switch. So even a moderate speed-up of 2 quickly overcomes the HOL blocking phenomenon of pure input-queueing. To illustarte this, we have plotted the input and output queue delays separately for a speed-up factor of 2 in figure 6. We also consider bursty traffic for delay simulations with the following model. A burst of successive packets is destined to the same output port but the destinations of each burst are chosen uniformly from among the N output ports. The burst sizes have the following *modified geometric distribution*: If the mean burst size is b, the size of a burst is $1 + B$ where B is a geometric r. v. with mean $b - 1$. The mean queueing delays achieved by output queueing and input queueing with a speedup factor of 2 are shown in Fig 7. Fig 8 shows the mean queueing delays for an average burst size $b = 50$. This emphasizes the fact that speed-up of 2 is sufficient even for bursty traffic.

5. QUEUE LENGTH ANALYSIS VS SIMULATION

Results for three speed-up factors are studied: $1.5(= 3/2)$, $2(= 2/1)$ and $3(= 3/1)$. The analysis of the input and output queues gives the mean queue-length as seen by a departing packet. We obtain the same data from simuation for comparison. We plot the mean input queue-length obtained by analysis and simulation in figure 9. It can be seen that the results are in good agreement. Note that even though the input queue analysis is exact, it is asymptotic, and thus we compare it with simulations to judge the effect of finite switch size (64×64). The mean output queue-length as seen by a departure is plotted in figure 10. The reason for a slight discrepancy here is the approximation of the arrival process to the output queue by the ON-OFF process. Observe that the discrepancy is more pronounced for the speed-up factor 1.5. For speed-ups of 2 and 3 the results of both cases are in good agreement. Thus the approximation works better for higher speed-up factors. Overall, the asymptotic analysis results conform well with the simulation results.

6. CONCLUSIONS

In this paper, we have used matrix-geometric techniques to investigate the impact of speedup on the performance of combined input and output queueing packet switches. The important observation is that only a moderate speedup factor (about two) is necessary to approach not only the throughput but also the delay performance of pure output-queued switches. Moreover, these results hold for random scheduling of the head-of line packets; no complex scheduling and/or matching algorithms are required within the switch. This is practically significant for the design and operation of high speed switch fabrics, especially with the continued increase in link speeds and switch sizes.

References

[1] A. Agrawal and N. Oguz.
Telpack Software, http://www.cstp.umkc.edu/org/tn/telpack/home.html, ftp://ftp.cstp.umkc.edu/telpack/software/.

[2] A. Charny, P. Krishna, N. Patel and R. Simcoe. Algorithms for providing bandwidth and delay guarantees in input-buffered crossbars with speedup. Proceed-

ings of the Sixth International Workshop on Quality of Service (IWQoS'98), 1998.

[3] W. E. Denzel, A. P. J. Engbersen, and I. Illiadis. A flexible shared buffer switch for ATM at Gb/s rates. *Computer Networks and ISDN Systems*, 27(4):611-624, Jan 1995.

[4] A. S. Diwan. *Performance analysis of speeded-up high-speed packet switches.* MS thesis, Indian Institute of Science, Bangalore (INDIA), Feb 1998.

[5] I. Illiadis, W. E. Denzel. Analysis of packet switches with input and output queueing. *IEEE Transactions on Communications*, 41(5):731-740, May 1993.

[6] L. Jacob. *Performance analysis of scheduling strategies for switching and multiplexing of multiclass variable bit rate traffic in an ATM network.* PhD thesis, Indian Institute of Science, Bangalore (INDIA), Dec 1992.

[7] M. Karol, M. Hluchyj, and S. Morgan. Input versus output queueing on a space division switch. *IEEE Transactions on Communications*, 35(12):1347–1356, Dec 1987.

[8] P. Krishna, N. S. Patel, A. Charny and R. Simcoe. On the speedup required for work-conserving crossbar switches. Proceedings of the Sixth International Workshop on Quality of Service (IWQoS'98), 1998.

[9] S. Q. Li. Performance of a non-blocking space-division packet switch with correlated input traffic. In *IEEE GLOBECOM*, pages 1754–1763, 1989.

[10] N. McKeown, V. Anantharam, and J. Walrand. Achieving 100% throughput in an input-queued switch. In *IEEE INFOCOM, San Francisco, CA*, pages 296–302, March 1996.

[11] B. Prabhakar and N. McKeown. On the speedup required for combined input and output queued switching. Computer Systems Lab, Technical Report CSL-TR-97-738, Stanford University.

[12] I. Stoica and H. Zhang. Exact emulation of an output queueing switch by a combined input output queueing switch. Proceedings of the Sixth International Workshop on Quality of Service (IWQoS'98), 1998.

[13] R. W. Wolff. *Stochastic modeling and the theory of queues.* Prentice Hall, Englewood Cliffs, New Jersey, 1989.

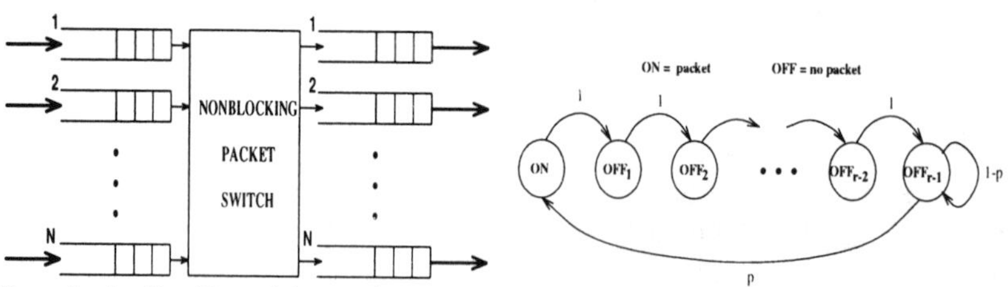

Figure 1 An $N \times N$ speeded-up packet switch with input and output queues.

Figure 2 Markov chain modeling the arrival process in the input queues of the packet switch.

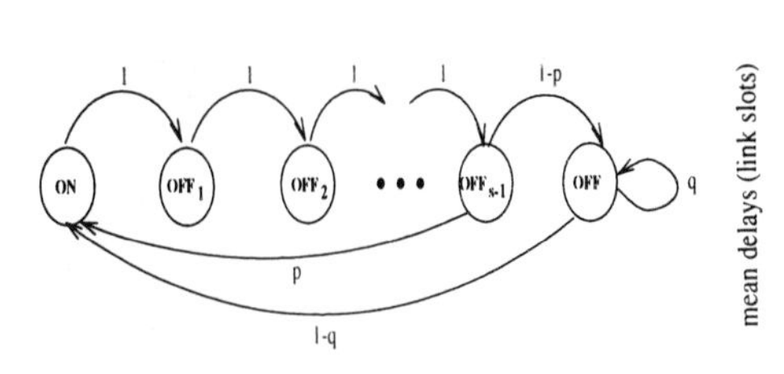

Figure 3 Approximate model of the arrival process in the output queues for a rational speed-up factor r/s.

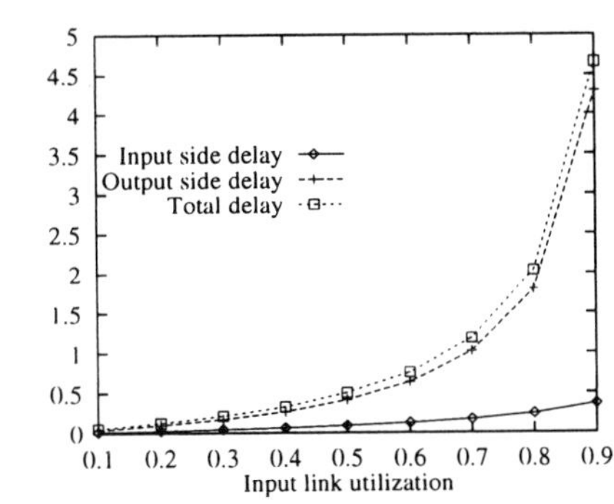

Figure 6 Mean delay in input queue and the mean delay in output queue and total mean delay of a packet for a speed-up factor of 2.

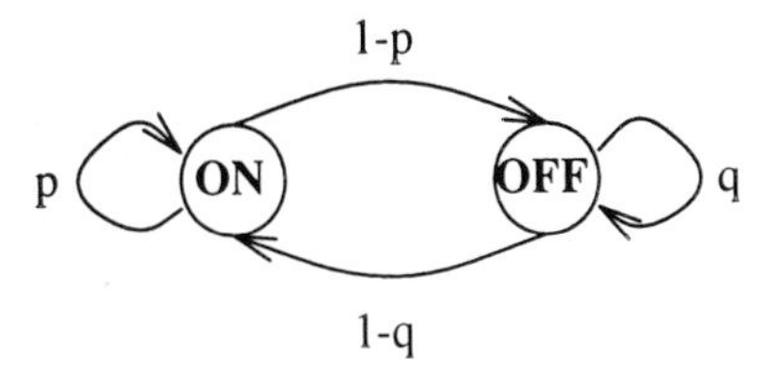

Figure 4 Approximate model of the arrival process in the output queues for an integer speed-up factor.

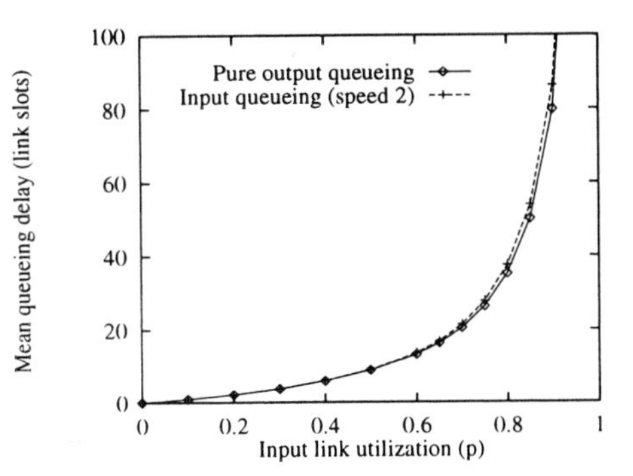

Figure 7 Delay performance of a switch with speed-up of 2 and delay performance of pure output-queued switch. mean burst size $b = 10$.

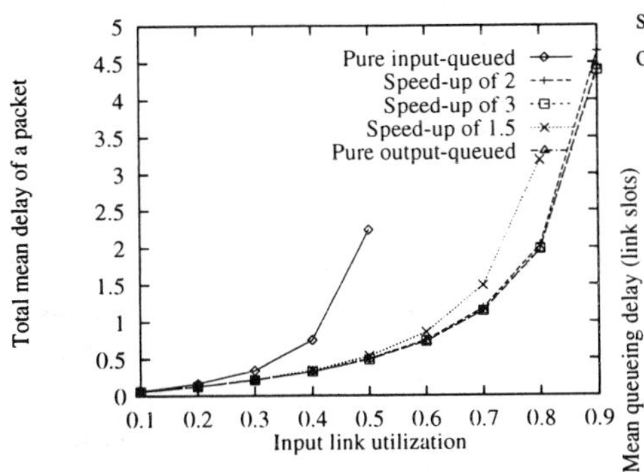

Figure 5 Mean total packet delay results for switches with speed-up factors of 1.5, 2 and 3 and delay results of pure input-queued and pure output-queued switches.

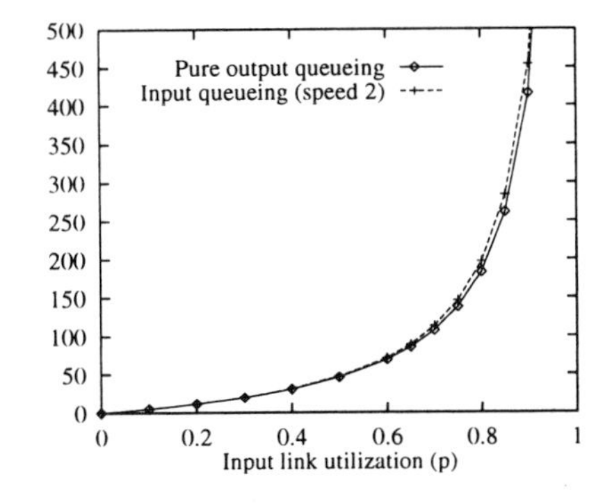

Figure 8 Delay performance of a switch with speed-up of 2 and delay performance of pure output-queued switch. mean burst size $b = 50$.

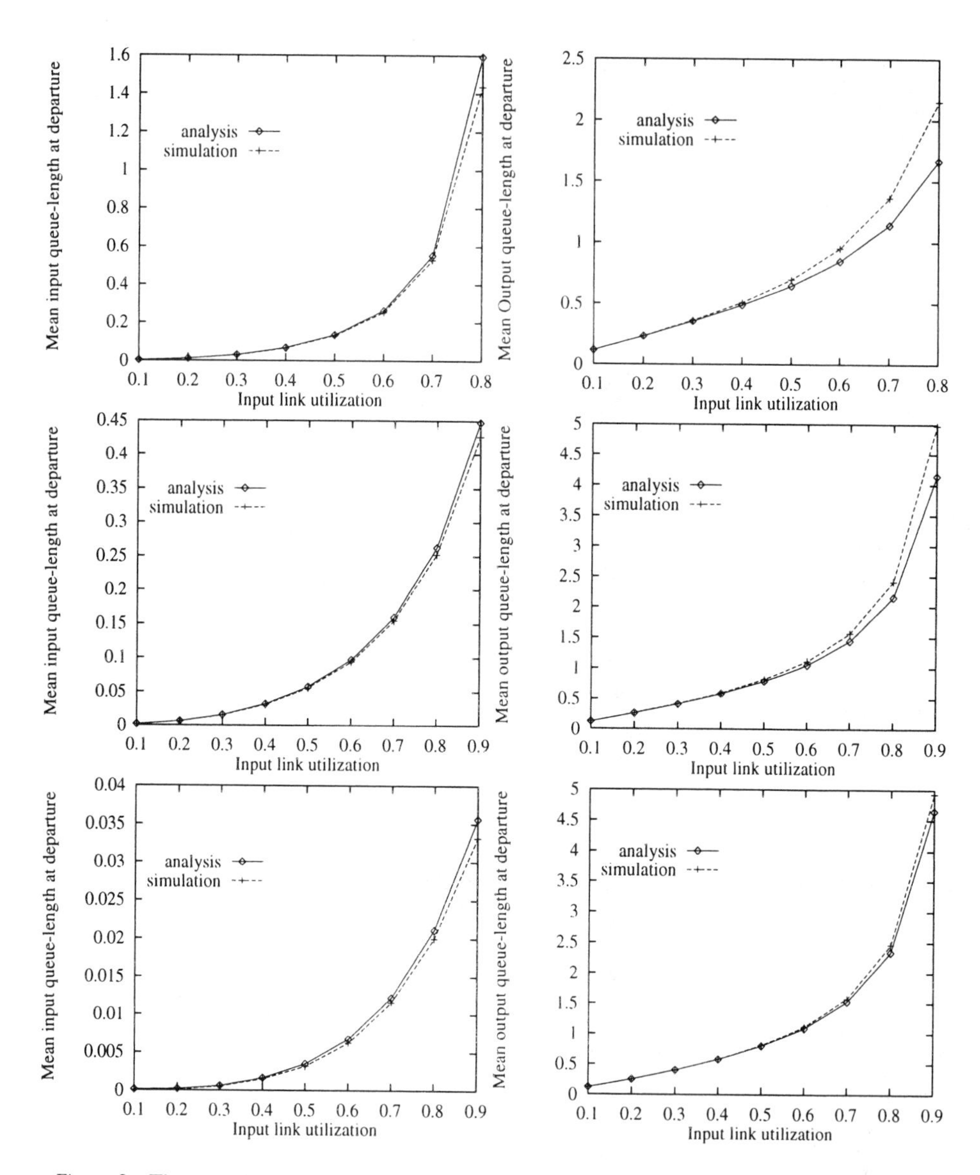

Figure 9 The mean input queue-length as seen by a departure obtained by analysis is compared to that obtained by simulation, (top) speed-up of 1.5, (middle) speed-up of 2, (bottom) speed-up of 3. The input link utilization is $r\lambda$. see (1). The arrival process is of section 3.1.

Figure 10 The mean output queue-length as seen by a departure obtained by analysis is compared to that obtained by simulation, (top) speed-up of 1.5, (middle) speed-up of 2, (bottom) speed-up of 3. The input link utilization is $r\lambda$. see (1). The arrival process is of section 3.1

INTERWORKING SUPPORT IN A MULTI-DISCIPLINE SWITCH

K. P. T. Raatikainen
VTT Information Technology, Telecommunications
P.O.Box 1202, FIN-02044 VTT, Finland
pertti.raatikainen@vtt.fi

Abstract The evolving multimedia services are foreseen to be delivered to customers over heterogeneous transport networks. The ATM technology has been considered as the most obvious choice, but other transporting concepts have also been studied and entirely new ones suggested, especially, for IP based communications. Diversity in transporting concepts implies that network nodes, such as switches and routers, should connect to different kinds of networks. This introduces the need to support multiple switching disciplines in a single switching fabric and, additionally, a clear need for interworking. As a solution to the emerging problem, this paper introduces the concept of multidiscipline switching to integrate different switching systems into a compact fabric and to offer interworking between the connected networks.

Keywords: Switching, Interworking, Multimedia, Call Control

1. INTRODUCTION

Transport networks for delivery of future multimedia services are foreseen to be based on heterogeneous transport and switching technologies. Relevant alternatives include, e.g., ATM (Asynchronous Transfer Mode), conventional PSTN (Public Switched Telephone Network), various IP (Internet Protocol) based concepts and new label switching concepts. ATM is a mature concept, capable of offering support for a large variety of services, and has often been envisaged as a unifying transmission and switching technology. Today it is mainly a trunk network solution, but is expanding also to the access network side. This is based on the new access technologies, such as the various Digital Subscriber Loop (DSL) concepts aimed at extending usage of the traditional copper based local loops that can offer high bit rate ATM connections to residential users.

CATV network operators are introducing different cable modem solutions for interactive service delivery. Although standardisation is not settled, estimates for usage of CATV networks for multimedia delivery are expected to show quite high growth rates in the very near future [1]. The conventional telecommunications network has a stabile position and it can be envisioned to hold this position for quite a many years and will be used for delivering most varying set of services. This is even more obvious when taking ISDN into consideration. The narrow-band ISDN, especially, is becoming more and more popular for private access.

An interesting ongoing trend is the rapid growth of Internet which has boosted popularity of the TCP/IP (Transport Control Protocol/Internet Protocol) protocol suite. The IP based data networks are foreseen to continue growing and the need for IP routing and switching increases. Consequently, a number of concepts to carry IP traffic have been suggested. Transmission of IP over ATM has been a standardisation target [2, 7] and has inspired vendors to develop competing solutions, e.g., IP switching, IP/ATM, ARIS and Tag switching [7, 8]. IETF has started to develop a harmonising concept to be called Multi-Protocol Label Switching (MPLS) and, additionally, has founded a working group to study service differentiation mechanisms to accelerate routing and switching of IP packets [6, 14]. Progress of these techniques is in the early stage, but great expectations are placed on them.

IP over ATM introduces unnecessary overhead and duplication of functionality [14] lowering the available payload bandwidth [1]. As a result, other transport concepts have been proposed. Examples are concepts that convey IP traffic directly in physical level transport frames, e.g., in SDH/ SONET (Synchronous Digital Hierarchy/Synchronous Optical Network) [6] frames. Concepts called IP over photons are also under intense research [5].

Referring to the above, it can be concluded that in the near future the numerous multimedia services will be offered to users over a diverse set of networks, deploying different transport and switching technologies. Due to this, there will be an increasing need for equipment capable of supporting multiple switching schemes and offering interworking between the different networks. This paper introduces a multidiscipline switching concept to support switching and routing of multimedia traffic carried over heterogeneous networks while offering interworking between the dissimilar networks. Chapter 2 presents the concept of multidiscipline switching highlighting the major implementation issues related to interworking. Chapter 3 introduces an experimental switching solution to offer interworking between circuit switched PSTN, packet switched IP and cell switched ATM networks. Chapter 4 has the concluding remarks.

2. MULTIDISCIPLINE SWITCHING

A conventional telecommunications switch connects to a homogeneous transport network while a multidiscipline switch operates in a heterogeneous network environment. Therefore, a multidiscipline switch connects to different kinds of networks and has to support parallel switching concepts. In order to cope with different switching disciplines, the switch must have a common way to route the dissimilar data units through the switching fabric and implement sophisticated software solutions to manage connections and the physical fabric.

2.1 Switching Bus

The switching bus often implements special containers (internal to the switch) to convey the different size data units (coming from the dissimilar networks) through the physical bus (see Fig. 1). In order to share its capacity between the supported systems, the switch can have a separate routing and switching mechanisms for each transport system or, alternatively, a common mechanism for all systems. Depending on the selected solution, the switching resources are shared either permanently or dynamically between the connected systems. Interworking is performed on a higher level, so the bus itself does not need to know about possible interworking needs. The problem of allocating switching capacity for the parallel switching schemes and dimensioning of a ring shaped switching bus is discussed in [3, 4, 13].

2.2 Network Interfaces

Since a multidiscipline switch interfaces to different sorts of networks, it has to implement a diverse set of line interfaces. The main function of a line interface is to adapt the switch to an external network and convert incoming data streams to the form accepted by the switch.

Since the physical switching bus does not normally include much intelligence, the interfaces should be equipped with capabilities to support interworking. The switch can also include units that perform specialised interworking functions, such as audio and video coding conversions and synchronisation of service data streams. Provided that the data to be switched do not require any data conversions, necessary interworking functions can be implemented in the line interfaces. In this elementary case, interworking functions include processing of incoming data units to the form accepted by the receiving network. An example is the case when a conventional 64 kbps PDH channel is converted into an ATM channel. The PDH line-card collects time-slots of individual channels to assemble AAL packets and then ATM cells. These are further encapsulated into ATM containers, internal to the switch, for passing them trough the switching bus. In the opposite direction, a similar process can take place in the ATM interface or ATM cells are directly routed to the PDH interface which performs the relevant interworking functions. This means that either both or just the other one of the interworking interfaces must be able to process the switch specific ATM and PDH containers.

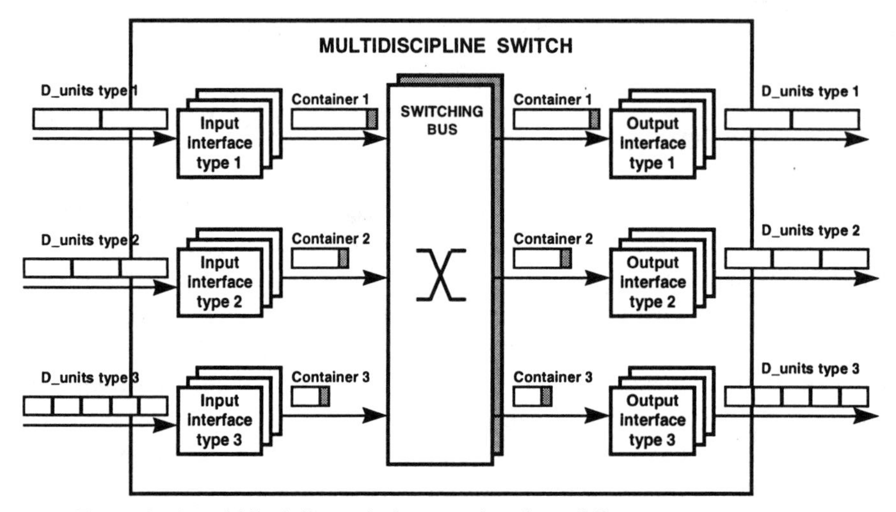

Figure 1 A multidiscipline switch supporting three different transport systems.

If manipulation of the incoming data is required, special units are implemented to perform the necessary processing. The input line interface directs the received data first to the special processing unit which later sends the processed data to the destination interface. An example of such a case is a voice over IP application where one of the communicating parties is connected to an IP network and the other to a PSTN network. Voice in the 64 kbps PDH channel is PCM coded using ITU-T standardised A- or μ-low. In the IP network, the voice coding can be based on ITU-T's H.323 recommendation. Now, interworking means despite processing of data units to fit into the form of the other network also the relevant voice coding conversion.

2.3 Switch and Connection Control

Due to the different nature of the associated transport technologies, their signalling and connection control procedures, connection types and quality of service (QoS) measures may deploy entirely different control disciplines. Therefore, a multidiscipline switch must combine several control and service architectures into a compact control system. Each architecture requires its own set of protocols and thus a straightforward solution is to implement the protocols (of each virtual switch) separately. In this way, each protocol stack and associated call control functions work as if they were run on top of a physical switch of that particular network architecture.

Interworking between the connected systems can be accomplished by implementing a common call control application which is placed on top of the control stacks. A general control block is needed to manage resources of the physical switch (e.g. configuration of the switch), convert application level connection commands to physical connections and perform physical level routing of the data units.

3. AN EXPERIMENTAL SWITCH

An experimental multidiscipline switch SCOMS (Software Configurable Multidiscipline Switch) is being developed at VTT Information Technology and Helsinki University of Technology. It integrates (64 kbps PDH) time-slot, (ATM) cell and (IP) packet switching into a single fabric. The switching platform comprises the switching bus, interfaces to connect the switch to trunk and access networks and the switch control unit (see Fig. 2). Additionally, special plug-in-units can be implemented to perform coding and alike interworking functions.

3.1 Switching Bus

The physical switching platform is based on the Frame Synchronized Ring (FSR) concept [10], developed and patented by VTT Information Technology. This platform has characteristics necessary to implement the various fundamental features of a multidiscipline switch, e.g., inherent support for real-time multicasting, flexible addressing capabilities and support for simultaneous switching of different size data units.

Basically, FSR is a slotted ring consisting of nodes connected together with unidirectional point-to-point links. Due its sophisticated MAC (Medium Access Control) and the use of destination release policy, FSR supports spatial reuse of the switching bus enabling multiple simultaneous connections. Data is conveyed through the bus in fixed size containers that can be dimensioned individually. This allows implementation of multiple virtual switches into a single fabric. FRS's performance has been analysed, e.g., in [9] and [10], and compared to other interconnection networks in [11]. A bridging solution, utilising FSR as the backbone technology for delivering multimedia services, is introduced in [12]. Dimensioning of the FSR bus in multidiscipline applications has been analysed in [3] and [13].

On the physical level, each "virtual switch" occupies selected FSR nodes which are equipped with associated line-cards. Individual FSR nodes can be configured freely to support either of the three switching schemes. Switch configuration can be changed dynamically by replacing a line-card by another type of line-card and modifying configuration registers in the switch control unit.

3.2 Containers for Data Unit Switching

Since the data units, i.e., time-slots, cells and packets, are of different size, special containers are assigned for each traffic type. In the PDH networks, the basic unit of data is an 8-bit byte and time-slot switching refers here to switching of these 8-bit bytes. A PDH container is fixed to convey 16 time-slots (belonging to an individual 64 kbps connection) at a time. The payload size of 16 octets keeps the container's assembly delay low enough to guarantee required delay performance of conventional digital PSTN traffic.

Switching of ATM traffic is supported by another container size that can hold exactly one cell at a time. Bit rates and quality measures of individual ATM connections may differ from each other and even the bit rate of a single connection may vary with time. Thus, the nature of cell switching differs greatly from time-slot switching and it was justifiable to assign separate containers for the PDH and ATM traffic.

In IP networks, the packet lengths vary from a few dozens of octets to 64k octets and to manage with the long and varying packet sizes, the IP packets must be segmented into mini-packets. These are the data units to be encapsulated into IP containers. Since IP traffic is bursty in nature, the process of switching IP packets resembles that of ATM cells. However, manipulation of IP packets differs greatly from manipulation of cells.

Routing of the PDH, ATM and IP containers through the switch is based on FSR's internal addressing mechanism. In a line interface, the PDH channel numbers, ATM channel identifiers or IP destination addresses are mapped to FSR addresses. These are further used inside the switch to route the containers from the source interface to the destination one. In the destination interface, the containers are disassembled and the received time-slots, cells or packets are reassembled and directed onto the outbound line.

3.3 Interworking and Network Interfaces

Each of the three virtual switches are equipped with associated line-cards to have connections to the trunk and access networks as shown in Fig. 2. ATM traffic is carried over STM-1 links in the trunk network and over E1 or ADSL (Asynchronous Digital Subscriber Line) links on the access network side. The PDH virtual switch is equipped with (2 Mbps) E1-cards to connect both to the access and trunk networks. The narrow-band ISDN and nx64 kbps cards are used to offer additional links to customers. IP traffic is carried over STM-1 links in the trunk network and over Ethernet links on the access network side.

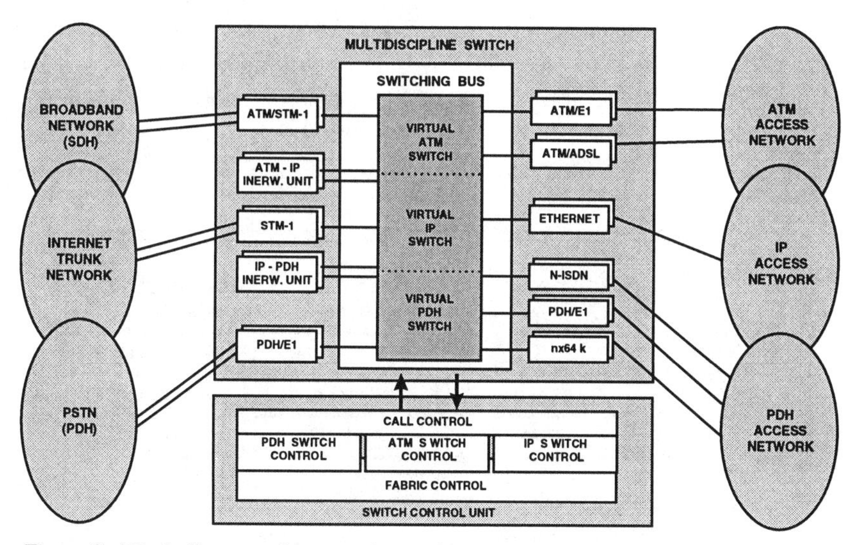

Figure 2 Block diagram of an experimental multidiscipline switch with interworking capabilities.

Since each of the three virtual switches has its own container type, interworking between the connected systems on the switching bus level is not possible. Conversion from any incoming system to any other outgoing system has to be performed on the line interface level. If heavier processing is needed, interworking units designed for specific service applications are attached to the switch. In Fig. 2 the switch is equipped with interworking units which allow voice connections between PDH and IP as well as between ATM and IP networks. All interworking voice connections are directed first to the associated interworking units and then to the destination line-cards.

3.4 Switch Control

Operation of the switch is controlled by a workstation which for the moment implements call control and physical fabric control functions. The control workstation is connected to the switch with an ATM/STM-1 link, and all protocols receive and transmit their signalling messages in reserved ATM virtual channels. Therefore, a non-ATM line interface must map the incoming signalling messages to a specific ATM virtual channel, assigned for transferring the signalling to the control workstation. ATM/STM-1 was chosen for the workstation connection, because it is a cost-effective way to have a high-speed signalling link which offers a straightforward way to manage various control and signalling connections. An Ethernet interface would be another alternative to connect the workstation to the switch.

Since the switch implements three totally separate and incompatible switching schemes, the call control block includes separate signalling software packages for each system (see Fig. 3). If interworking capabilities were not implemented, calls originating from the PDH network must terminate at the PDH interfaces. Likewise ATM connections could be set up only between the ATM interfaces and IP connections between the IP interfaces.

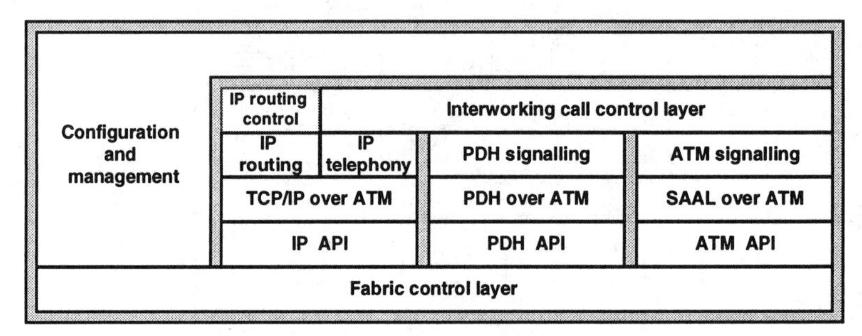

Figure 3 Interworking call control architecture.

Interworking between the different networks is performed by a shared interworking call control layer. Since the originating and terminating ends of the connections may reside in different kinds of networks, the call control layer decodes, processes and forwards connection setup requests from each protocol stack. The call control interworking function maps the connection setup parameters between the different networks and maintains status information of all the established connections.

At the bottom layer, all protocol stacks use services of the common fabric control layer. The fabric control layer offers network specific Application Programming Interfaces (APIs) that implement functions necessary to reserve switching resources and manage connections through the physical switch.

4. CONCLUSIONS

Diversity in networking technologies and lack of a unifying transport solution is leading to a situation that evolving multimedia services will be delivered to users over a diverge set of networks. The ATM technology has been envisaged to become the prevailing technology due to its flexibility. However, ATM faces unavoidable transmission overhead which leads to inefficient use of transport capacity. Thus, other transport and switching concepts have also been considered. Obviously, there will be several alternatives to carry multimedia services which implies that switching nodes will connect to networks that utilise different transport and switching technologies. Due to this, there will be an increasing need for equipment capable of supporting multiple switching schemes and offering interworking between the different networks.

In this article, a multidiscipline switching concept SCOMS is introduced as a compact solution to tackle with the incompatible transport and switching systems. The benefit from multidiscipline switching is to share the physical switching resource between different networks. Additionally, more complex services can be implemented if interworking between the different networks is supported. A generic solution, housing two or more virtual switches in a compact unit, would effectively cut down installation and maintenance costs of the future broadband networks.

Implementation of several switching schemes into a single fabric poses a number of demanding research and development topics. Problems to be solved include, e.g., switching of different size data units, non-compliant call control and signalling methods, varying buffering needs and different QoS requirements. Innovations, related to distributed switching and separation of physical fabric control from call connection and service control, enable development of such a switching equipment.

An experimental switching solution integrating 64 kbps based circuit, ATM cell and IP packet switching with interworking capabilities is presented. The physical switching platform is based on the Frame Synchronized Ring (FSR) concept. FSR's ring-shaped switching bus is dimensioned to perform effectively by implementing separate transport containers (internal to the switch) for circuit, cell and packet switching. Control of the switch applies multi-layered control architecture, dividing the control functions into fabric, network and service control layers. Signalling protocols of the different networks are implemented separately, although, they use services of the common fabric control layer. Interworking between the networks is supported by a shared interworking call control module that is able to manage calls that originate and terminate at different sorts of networks.

References

[1] H. Armbtüster: *"Information Infrastructures and Multimedia Communications - Different Approaches of the Telephone, Data, and Radio/TV World"*. IEEE Comm. Mag., Vol. 35, No. 9, Sep. 1997, pp.92 - 101.

[2] ATM-Forum: *"LAN Emulation over ATM Specification"*, Version 1, 1995.

[3] K. Kaario, P. Raatikainen: *"Dimensioning of a Multimedia Switching Bus"*. Proc. of 24th Euromicro Conference, Västerås (Sweden), IEEE Computer Society Press, Aug. 1998, Vol. II, pp. 567 - 573.

[4] K. Kaario, P. Raatikainen: *"Sharing of payload capacity in a multimedia switching bus"*. Proceedings of ITC-16, Teletraffic Engineering in a Competitive World, IEE, Elsevier Science B.V., Edinburg (UK), June 1999, Vol. 3a, pp. 551 - 560.

[5] E. Livermore, R.P. Skillen, M. Beshai, M. Wernik: *"Architecture and Control of an Adaptive High-Capacity Flat Network"*. IEEE Comm. Mag., Vol. 36, No. 5, May 1998, pp. 106 - 112.

[6] J. Manchester, J. Anderson, B. Dosi, S. Dravida: *"IP over SONET"*. IEEE Comm. Mag., Vol. 36, No. 5, May 1998, pp. 136 - 142.

[7] P. Newman, G. Minshall, T. Lyon, L. Huston: *"IP Switching and Gigabit Routers"*. IEEE Comm. Mag., Vol. 35, no. 1, Jan. 1997, pp. 64 - 69.

[8] P. Newman, G. Minshall, T.L. Lyon: *"IP Switching - ATM Under IP"*. IEEE/ACM Transactions on Networking, Vol. 2, No. 2, April 1998, pp. 117 - 129.

[9] T. Pyssysalo, P. Raatikainen, J. Zidbeck: *"FSR - a Fair Switching Architecture"*. Proc. of the 7th Euromicro Workshop on Real-time Systems, Odense (Denmark), IEEE Computer Society Press, June 1995, pp. 116 - 123.

[10] P. Raatikainen: *"Analysis and Implementation of a High-Speed Packet Switching Architecture - the Frame Synchronized Ring"*. Doctor's Theses, Helsinki University of Technology, 1996, 184 pages.

[11] P. Raatikainen, J. Zidbeck: *"Performance Comparison of Experimented Switching Architectures for ATM"*. Proc. of the 22nd Euromicro Conference, Prague (Czech Republic), IEEE Computer Society Press, Sep. 1996, pp. 405 - 411.

[12] P. Raatikainen, J. Zidbeck: *"A Bridging Solution for Delivering Multimedia Services in CATV Networks"*. Proc. of the 23rd Euromicro Conference, Budapest (Hungary), IEEE Computer Society Press, Sep. 1997, pp. 223 - 230.

[13] P. Raatikainen: *"Switching Bus for Multidiscipline Switching"*. Proc. of ICT'98, Porto Carras (Greece), June 1998, Vol. II, pp. 278 - 282.

[14] A. Viswanathan, N. Feldman, Z. Wang, R. Callon: *"Evolution of Multiprotocol Label Switching"*. IEEE Comm. Mag., Vol. 36, No. 5, May 1998, pp. 165 - 173.

AN ATM SWITCH CONTROL INTERFACE FOR QUALITY OF SERVICE AND RELIABILITY

Rahul Garg
Indian Institute of Technology, Delhi, India
rahul@cse.iitd.ernet.in

Raphael Rom
Technion, Haifa, Israel and Sun Microsystems, Palo Alto CA, USA
rom@ee.technion.ac.il

Abstract Traditional communication switches include an embedded processor that implements both the switch control and network signalling. Such an architecture is failure-prone due to the complexity of the control software of modern network. Recently, a new approach of controlling switches through an external controller is gaining momentum, due the flexibility and reliability it affords. In addition, open control protocols and interfaces for controlling and managing networks, are now emerging as an alternative to specifications and standards. For instance, in the OpeNet project Sun Labs has designed and implemented an open, high performance ATM network control platform.

In this paper we describe the OpeNet Switch Control Interface (ONSCI)–an open local switch control protocol, for controlling an ATM switch. An ATM switch controller uses this protocol to setup or tear down virtual circuits and perform other control and management functions in an ATM switch. Important and distinguished features of this protocol are primitives for Quality of Service (QoS) management in the switch and support for fault-tolerant operation in case of failure of a switch controller. The design of these primitives is based on conceptual modeling of the switch architecture. The model is generic enough to cover a wide range of ATM switches.

The protocol was implemented and integrated with the OpeNet platform. Without going into specifics of the protocol, we describe its design principles and show how it has affected our protocol.

Keywords: ATM, GSMP, QoS, Fault Tolerance, Availability, Reliability, IP-Switching, Open Interfaces, Admission Control, GCAC, PNNI.

1. INTRODUCTION

Communication switches are built from two major components: a hardware component where the data switching takes place, and a software component which provides the control mechanisms and the integration of the individual switch into a complete network. In traditional switches the control software is implemented on a dedicated processor embedded into the switch's hardware. This integration is typically tightly coupled and is tailor-made for the specific hardware and network application. Such an architecture suffers from many disadvantages such as inability to cope with the progress of processor and software technology and the inability to re-use the software.

A new trend has started recently and is gaining momentum: to devise open network architectures based on distributed systems principles. Several research groups as well as network equipment manufacturers are engaged in active pursuit of the problem. Moreover, standardization efforts have recently started by the IEEE (e.g., the P1520 working group on application programming interfaces for networks [10] and its ATM sub-working group) with a goal of defining APIs for future multimedia networks. With such an interface in place, the controller is typically detached from the rest of the hardware, need not be tightly coupled with it, the entire design allows for easy upgrading of software and lends itself to better fault tolerance.

As part of its activities, Sun Microsystems Laboratories designed and implemented OpeNet [3], an open, non-proprietary, high performance, switch independent ATM network control platform. To achieve the goal of switch independence, none of the mechanisms deployed by OpeNet rely on any particular switch. However, to be deployed for operation, the OpeNet must be interfaced with an actual ATM switch. Thus the need for a generic interface to control ATM switches was felt. This lead to the development of the OpeNet Switch Control Interface (ONSCI) [7] which is the subject of this document.

Our work has two basic distinguishing features from other works on open ATM switch control interfaces. Firstly, we provide support for managing Quality of Service (QoS) in the switch in a manner compatible with the ATM Forum specifications. Secondly, primitives for fault tolerant operation are provided to assist recovery of a switch controller after its failure. In the event of failure of a switch controller, a backup controller (or the recovering failed controller) may assume the control of the switch and after some recovery operations, starts functioning as its primary controller. Other operations supported by ONSCI are setting up and tearing down VCs, configure and manage its ports, manage its VPs and get other statistics and information related to performance monitoring. The interface uses a message exchange protocol which is client/server in nature; the switch controller sends requests to the switch which responds after processing the request.

The Generalized Switch Management Protocol (GSMP) [14] was the first protocol developed to control ATM switches. GSMP supports only basic operation like setting up and tearing down virtual circuits, monitoring performance, and lacks the more advanced QoS functions needed in an ATM network. The second version of GSMP [15] addressed the problem of QoS in the framework of Class Based Queuing (CBQ) [6]. This model is appropriate for the Internet suite of protocols like RSVP [19], but is still inadequate (and was not intended) for ATM networks. A GSMP extension was proposed [1], in the context and framework of Xbind [11] that uses the model of schedulable region [9] to carry out admission control, but is not compatible with the specifications of the ATM Forum. The work by Evans et._al. [5] define a reasonable model of managing an ATM switch that provides QoS guarantees, but proposes only an approach and not a complete specification. None of these addresses fault tolerance issues.

The rest of the paper is organized as follows. Section 2 contains a brief overview of ONSCI design. To provide QoS support, the model of switch's resources and how the controller manages these resources is very important. This is discussed in Section 3, which is the major contribution of this paper. Section 4 details our assumptions and approach to support fault tolerance. The paper concludes in Section 5.

2. ONSCI DESIGN

A traditional *switch* of an ATM network is shown in Figure 1. It has two parts: A switching component, sometimes called the fabric or the cross-connect, a set of tables containing the forwarding information and configuration data, and a processing unit that embodies the control functions.

The ATM switch forwards its incoming cells to different output ports. This forwarding is done at very high speed using the cross-connect. The entries in the forwarding table dictate which cell should be forwarded to which port. Most of the flow control, monitoring, statistics collection is jointly done done by the hardware and software components. Other network wide operations like routing etc., are done solely by the processor. In a typical network, decisions are made by the software running in the processor and then stored in the tables to be used by the cross-connect. Such manipulation of the data structures is typically done by a tight integration between the processor, the cross-connect, and the specific manner in which the data is stored in the tables.

To overcome the deficiencies of a traditional ATM switch, a new architecture is proposed as depicted in Figure 2. An *ATM node* is divided into two major parts: The node controller (or switch controller) and the nodal switching subsystem (also referred to as "the switch"). The switching subsystem is essentially a very simple traditional switch where the processor need have extremely limited computational power and a small interconnect-protocol (software) module

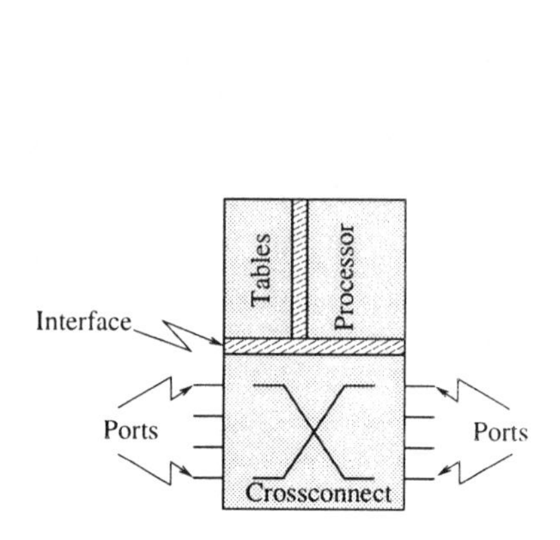

Figure 1 A traditional ATM switch

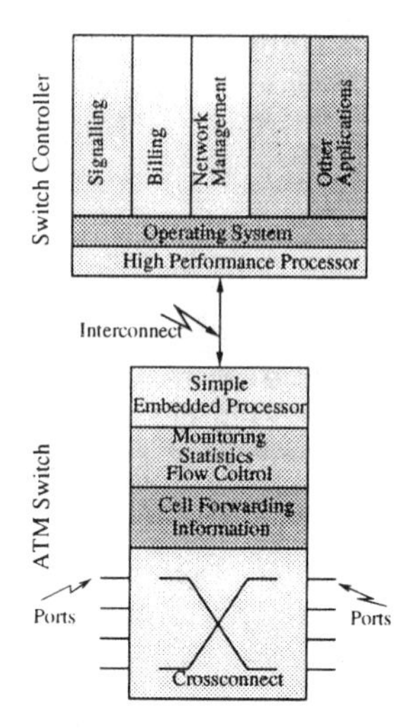

Figure 2 An ATM Node

with which it communicates with the controller. The switch is expected to have a limited capability processor, and a minimal operating system. The design of ONSCI is influenced by this assumption.

The switch controller is a generic high performance computing platform. No restrictions are imposed on its architecture. For instance, it may be a high end workstation with a generic operating system.

The switch controller sends messages to the ATM switch through a duplex interconnect. The ATM switch processes the message and sends back the response to the controller. ONSCI uses a regular ATM port for the interconnection. This makes the ATM switch uniform and enables a simpler and richer fault recovery mechanism (Section 4).

ONSCI primitives allow programming of the switch forwarding table. Thus, the controller may send and receive (signalling) messages on special VCs (VCI between 0 and 31 on VPI 0) of each port by diverting them to specific VCs on the ATM interconnect. The switch tables can be programmed using ONSCI to reserve resources for connections (in a manner compatible with the ATM Forum). ONSCI is not tied to one particular signalling protocol. For instance, in one configuration of the network, all the nodes may use PNNI signalling, whereas in another configuration OpeNet signalling may be used. ONSCI also supports functions that are used for operations such as monitoring, management, configuration control, link failure indications, and processing of OAM cells.

To maintain the reliability, the processor that is implemented in the switch is a very simple one, not running any complex operation but just responding to requests from the controller. Therefore, it is clear that the controller is the susceptible point with respect to reliability. Our main design philosophy is that the switch should be able to continue operating, as much as possible, even with the failed controller and providing a mechanism with which a standby controller can gain the control of the switch. When a standby controller takes over the control of switch, it needs to recover the state of primary controller just before its crash. Primitives of these operations are included in ONSCI. Our design also allows to take the controller down in a graceful manner so as to allow upgrading or fixing the control software.

3. MODELS AND STRUCTURE

The switch interface design is based on conceptual modeling of the switch behavior as well as the way it supports different QoS requirements. These models cover most of the switch implementations regardless of the way the basic switching function is implemented, as well as the ways buffers are managed across different connections, and the way QoS is provided. It also captures a broad range of the switch core control architecture.

3.1 THE SETTING

ONSCI is defined in terms of AAL5 message exchanges between the switch and a controller. This interconnection may use pre-configured ATM VCs on a regular ATM port, or a virtual circuit between the switch and the controller as shown in Figure 3. This allows a single controller to control multiple switches as depicted in Figure 3. The switch controller controls switch-1 directly (i.e., the control VC indicated by the dot-dashed line) and controls switch-2 indirectly (i.e., the virtual circuit indicated by the dashed line).

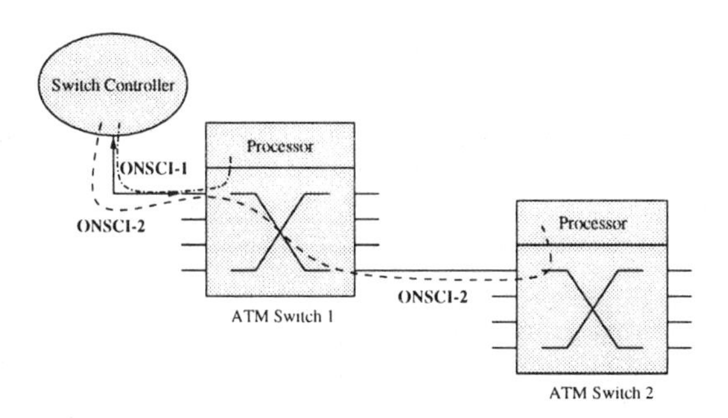

Figure 3 Control of multiple switches

3.2 PROTOCOL NATURE

The interaction between the switch controller and the switch is master-slave. The controller issues a command to the switch by sending a request packet which is encapsulated in an AAL5 frame. The switch performs the request and send a response message with a similar structure. Every packet contains a 32-bit transaction identifier which is used by the controller to match a response to a request. Most of the request messages are idempotent, and it is the responsibility of the controller to re-issue requests in case of packet losses. The resource management messages are not idempotent and carry another identifier. The state of one resource management packet is remembered by the switch to detect retransmissions.

Other than these messages, the switch asynchronously sends event messages to the controller to report certain conditions like a link going down, in a manner similar to that described in [14].

3.3 CONNECTIONS

The switch and its controller refer to individual connections by means of an abstract data structure which is referred to as generalized forwarding table. An entry in the generalized forwarding table consists of a set of input designators, a set of output designators, a traffic descriptor and some additional parameters. An I/O designator is a triplet of port number, VP and VC identifiers and is the means by which the switch identifies individual data flows.

Every cell arriving at any input designator are switched to *all* the corresponding output designators. The traffic descriptor follows the ATM Forum's specification and includes traffic type and other parameters (see [7]). An entry in the generalized forwarding table defines a flow in one direction only. A bidirectional connection includes two entries, one in each direction. The port to which the controller is attached treated as other ports and cells can be forwarded to and from it just like other ports.

3.4 SWITCH PARAMETERS AND QOS MODEL

The switch can support a set of predefined QoS classes. The set can vary from the minimum set defined by the ATM Forum (CBR, VBR-RT, VBR-NRT, ABR, UBR) [18] to a larger set which is proprietary to an individual switch. The switch has its own proprietary way to translate the class identity and call rate into its internal switch tables. This translation is only known to the switch and may be hidden from the controller. Sometimes this translation is easy as mapping the given QoS class into a internal priority class and including rate parameters for switch policing functions. Other more sophisticated switches need to translate the rate parameters and insert these translation into switch internal buffer management table. The translation to the internal format is done by the embedded processor on the switch.

3.5 SWITCH ACCOMMODATION CONTROL MODEL

In addition to the ability of the switch to carry calls of certain QoS and rates, we assume that an accommodation test mechanism also exists by which the ability of the switch to accommodate an additional connection can be checked.

At the source of a connection a routing decision must be made, and the source node must be able to estimate the ability of every switch along the computed route to accommodate the requested connection (we refer to this as the *remote* CAC test). During connection set-up, every node along the path must verify that the resources required by the connection are available (we refer to this as the *local* CAC test). Furthermore, there are two phases during the set-up time: first the node checks for resource availability and sets the resource aside tentatively, and then, when all nodes along the path have confirmed the availability of the resources a commitment is made.

The local CAC test described above, is done by the local controller in cooperation with the local switch. The accommodation test must be conservative in the sense that whenever it indicates that a call can be accommodated, it should be able to obey the QoS contract in the physical switch. On the other hand, the accommodation control test should be fairly accurate and not overly conservative to allow for efficient utilization of all the physical resources available.

In order to approximate remote CAC, a generic call admission control (GCAC) procedure has been defined which takes the new call rate parameters (SCR, PCR, MBS) along with the three parameters (ACR, CRM, VF) which are advertised by remote node for each of its link. The advertised GCAC parameters are switch dependent (VF parameter in PNNI GCAC), and the controller should be able to interrogate the local switch regarding the availability of resources (VF parameter).

The above observations have led to a CAC model whose major components are shown in Figure 4. The translator block converts commands which are expressed in terms of a generalized VC table entry to the switch internal formats and parameters. In the reverse direction, the translator will make some of its internal settings available to the controller. The accommodation test and availability checks are performed with the accommodation oracle whose components are depicted in Figure 5.

At the core of the accommodation oracle is a function or a performance model which can estimate quite accurately whether a new call of given QoS and rate parameters can be accommodated. While such a procedure is an essential block in any ATM system, it is the most switch dependent one. Even if we use known

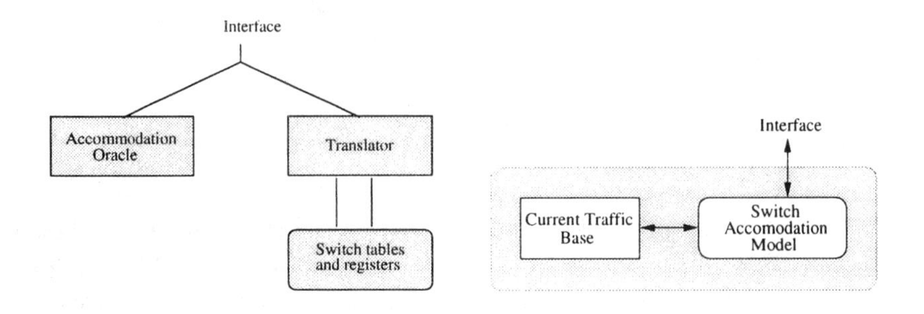

Figure 4 Components of accommodation mechanism

Figure 5 Structure of accommodation oracle

models (such as one of the equivalent capacity models of [18], [8] or [4]) there are necessarily some parameters in these models which are switch specific and capture, for example, the amount and structure of buffering in the switch. It is quite possible that the implementation of the oracle might contain a proprietary switch model. In addition the oracle maintains a current traffic base, whose structure is undefined but which embodies the oracle's view of existing traffic base. To make an accommodation decision the oracle combines its notion of the current traffic with the requested new connection.

With no further assumptions, the traffic base will consist of the list of connections with their traffic descriptors and QoS parameters. This will render the test and translate operations very complex and time consuming, adversely impacting the connection set-up procedure as we described earlier. To facilitate these operations we further assume that the accommodation control function is additive (or cumulative). This means that if A is the set of existing calls over a link of the switch and c_1 and c_2 are two given calls, and if c_1 is feasible under link state $A \cup c_2$, then c_2 is feasible under the link state $A \cup c_1$. This implies that the list of individual connections need not be maintained but that some form of aggregation would suffice and that allocating and de-allocating resources can be easily incorporated into the base. We also assume that the accommodation control function is monotonic which means that if c_1 is feasible under state A and c_2 has a rate description (and QoS) that is smaller than that of c_1, then c_2 is also feasible under state A. With this assumption one availability check would suffice to quickly determine the availability of resources for several (small) connections.

If implemented in the switch, the accommodation oracle and the translator have direct access to all the switch registers and can make the most accurate calculations. If implemented in the controller, the accommodation test function can be more efficient as the oracle can make full use of the computational resources of the controller and does not need to communicate with the switch for each decision. In this case the oracle may have to retrieve switch-specific information from the switch, for which the *private* primitive is provided in the interface.

4. SWITCH RELIABILITY SUPPORT

The switch controller which has a large software component is more prone to failure than the ATM switch. ONSCI primitives allow a secondary controller located anywhere in the network, to replace the failed controller of a switch, enhancing the controller's availability. The controller is not involved in normal data transfer operation, and therefore the connections successfully established before its failure are not affected by its failure.

As soon as the failure is detected, a secondary controller sets-up VCs to the switch and modifies its forwarding table (using a pre-established virtual circuit) to divert the special VCs at each of its ports. It then queries the ATM switch to get a list of connections in progress and gathers information from its neighbours to reconstruct the state of primary controller and starts working as the primary controller of the switch. The following sections describe these operations in more detail.

4.1 MULTIPLE CONTROLLERS OF A SWITCH

A *control point* is a special I/O designator on which incoming messages are treated as ONSCI requests by the switch and the outgoing messages contain the corresponding switch response. Every switch has at least one default control point, which is used by the switch to locate its primary controller when the switch is powered up.

Secondary controllers may be attached to the switch by dynamically creating new control points (using ONSCI primitives) and creating a VC from the new control point to the secondary controller. Thus, a controller can, in principle control, more than one ATM switch at a time as indicated earlier in Figure 3. In a realistic setting, a primary controller of a switch may also act as a secondary controller for another switch.

4.2 REBUILDING STATE INFORMATION

When a secondary controller takes over the control of an ATM switch, it first rebuilds the state of the primary controller, then it creates an additional secondary controller (if possible) and finally advertises its presence in the network and starts working as the primary controller of the switch. Of these, the first one is the hardest. The relevant state of the primary controller includes list of connections (in the generalized VC table), traffic descriptor and QoS of individual connections, and other information like link status of individual link, the load on ATM switches etc. The secondary controller gathers the relevant information by three methods: from periodic broadcasts, by querying neighbors, and by querying switch.

The list of active connections on a port along with the VCI, traffic descriptor and QoS of the connections can be obtained from neighbors. This along with the forwarding information from the switch's cross-connect is sufficient for building the generalized forwarding table. State of the traffic base of accommodation control oracle is constructed by accumulating the traffic descriptor and QoS of each connection to the traffic base. Recall that in Section 3.5 we assumed that the accommodation control procedure is cumulative. Therefore the connections can be re-accumulated in any order to reconstruct the traffic base.

The state information thus constructed reflects the state of the controller at the time it crashed. Some events (like connection tear-down) may have occurred between the crash and subsequent recovery. Thus, the recovering controller needs to carry out corresponding update in its state and possibly some additional operations, described next.

4.3 PARTIAL AND ZOMBIE CONNECTIONS

A partial connection is a connection such that there is one neighboring controller which believes that the connection exists and there is another which believes it has been closed. A zombie connection is one where none of the adjacent controllers believes that the connection exists. A connection may become partial or zombie if its tear-down is initiated after crash of a controller in its path.

The recovering controller identifies the deleted input and output branches, traffic descriptor and QoS of partial connections, using the neighbour and switch information. It then removes these branches from the switch forwarding table, updates the current traffic base and initiates a connection tear-down if needed.

Zombie connections are more difficult to deal with. Their existence can be identified but their traffic descriptors and QoS parameters may not be recoverable. In order to reclaim the resources allocated to zombie connections, the entire traffic base may be rebuilt from scratch.

The overall availability of the network to incoming calls mainly depends upon the number of secondary controllers, the mean time taken by a secondary controller in rebuilding its state and the mean time taken to repair the faulty controller. From these parameters, using the standard reliability theory techniques [2], [17] it should be possible to figure out the (approximate) number of secondary controllers needed to ensure a given availability.

The primitives in ONSCI protocol provide support needed by the controller from the switch in order to make the network tolerant to faults in the controller. For the details of the protocol, the reader is referred to [7]. With proper design of protocols at higher layer, and support from ONSCI it is possible to build a fault tolerant network.

5. DISCUSSION AND CONCLUSION

We have proposed the OpeNet Switch Control Interface (ONSCI), a new open ATM switch control interface based upon principles of distributed systems. ONSCI is switch independent and supports a variety of primitives to allow many network control platforms to be implemented on top of it. We chose the ATM Forum's traffic management model and the OpeNet signalling as an example.

Two innovative contributions are included in ONSCI: (1) support for a resource management scheme for the provision of general QoS that is both switch and signaling platform independent (2) support for a fault tolerant operation in the vein of increasing availability. Our design assumes limited switch capability in terms of the amount of memory and particularly CPU power. The idea behind this design is to allow simple (read: stable and reliable) processor as part of the switch.

One of the major problem in designing this interface was how to identify resources of the switch allocated to a particular connection. Imagine a situation when a branch is to be added to an existing multicast connection. In order to compute the amount of additional resources needed for this purpose, the resources already allocated (buffers, bandwidth, etc.) must be known. The CPU of the switch performs only very simple operations and is incapable of storing this resource mapping. The controller cannot store this information either as the amount and structure of switch resources is switch specific which cannot be generalized.

Resource identification has other benefits also. If the switch resources allocated to a particular connection were identifiable, all the commands in the resource management group could have been made idempotent, simplifying the protocol. The problem of reclaiming resources from zombie connections would have become much easier.

Explicit resource identification requires enormous book-keeping in the switch (and a more complicated switch model). It also needs significant software on the switch. This conflicts with our original goal of decoupling software and hardware components of a traditional ATM switch, distributed implementation of these two components, and open switch control interface. Despite the advantages of explicit resource identification, we decided not to opt for it. The tradeoff in favor of simple switch software is desirable.

We solved the resource identification issues in a variety of ways. In terms of the resource management commands, we chose to implement those in a more reliable manner which includes specific ACKs for each request. In terms of resource searching, we require that all messages that refer to resources should carry enough information to allow the switch to infer, rather than compute or search, the resources involved (e.g., including all I/O designators in identifying resources of a multicast connection). This is, in addition to the assumption that resource accommodation is monotonic and cumulative.

The inability to identify resources impacts the recovery process after failures. To that end we provided primitives that will allow to reconstruct the state of a failed controller, considering the fact that a large amount of information can be acquired from neighboring controllers.

ONSCI was implemented and integrated with the OpeNet control platform. Pure ONSCI implementation on an ATM switch was not available, but we had access to GSMP-capable ATM switch. We wrote a thin layer of software which converts ONSCI message to GSMP messages and GSMP response to ONSCI response. Using this basic functionality of ONSCI, like connection management, statistics etc., were successfully tested.

Acknowledgments

The authors thank Israel Cidon and Moshe Sidi for their valuable comments and suggestions while designing the protocol. Thanks to Amit Gupta who provided much needed help in design as well as implementation phase.

References

[1] C. M. Adam, A. A. Lazar, and M. Nandikesan. QOS extensions to GSMP. Technical Report 471-97-05, Center for Telecommunications Research, Columbia University, New York, April 1997.

[2] E. Balaguruswamy. *Reliability Engineering.* Tata McGraw Hill, New Delhi, India, 1984.

[3] I. Cidon, T. Hsiao, P. Jujjavarapu, A. Khamisy, A. Parekh, R. Rom, and M Sidi. The OpeNet architecture. Technical Report SMLI-TR-95-37, Sun Microsystems Laboratories, Mountain View, CA, USA, December 1995. http://www.sunlabs.com/technical-reports/1995/smli_tr-95-37.ps.

[4] A. Elwalid and D. Mitra. Effective bandwidth of general markovian traffic and admission control of high speed networks. *IEEE/ACM Transactions on Networking*, 1:329–343, 1993.

[5] Joe Evans. ATMF extensions to GSMP. In *Proceedings of OPENSIG Fall '97 Workshop: Open Signalling for ATM, Internet and Mobile Networks*, Columbia University, New York, NY, October 1997.

[6] S. Floyd and V. Jacobson. Link-sharing and resource management models for packet networks. *IEEE/ACM Transactions on Networking*, 3(4):365–386, August 1995.

[7] R. Garg and R. Rom. The OpeNet switch control interface specification. Technical Report SMLI TR-97-0543, Sun Microsystems Laboratories, Mountain View, CA, USA, November 1997.

[8] L. Georgiadis, R. Guerin, V. Peris, and K. Sivarajan. Efficient QoS provisioning based on per node traffic saping. *IEEE/ACM Transactions on Networking*, pages 482–501, 1996.

[9] J. Hyman, A. Lazar, and C. Pacifi. Real-time scheduling with quality of service constraints. *IEEE Journal on Selected Areas of Communications*, pages 1052–1063, September 1991.

[10] IEEE. Proposed IEEE standard for application programming interfaces for networks. http://www.iss.nus.sg/IEEEPIN/.

[11] A. A. Lazar, S. Bhonsle, and K.S. Lim. A binding architecture for multimedia networks. *Journal of Parallel and Distributed Systems*, 30(2):204–216, November 1995.

[12] A. A. Lazar and Franco Marconici. Towards an API for ATM switch control. Technical Report CU/CTR/TR 441-96-07, Center for Telecommunications Research, Columbia University, April 1996. http://www.ctr.columbia.edu/comet/xbind/xbind.html.

[13] K. Van Der Merwe and Ian Leslie. Switches and dynamic virtual ATM networks, July 1996. Computer Laboratory, University of Cambridge.

[14] P. Newman, W. Edwards, R. Hinden, E. Hoffman, F. Ching Liaw, T. Lyon, and G. Minshall. Ipsilon's genaral switch management protocol, August 1997. RFC 1987.

[15] P. Newman, W. Edwards, R. Hinden, E. Hoffman, F. Ching Liaw, T. Lyon, and G. Minshall. Ipsilon's general switch management protocol specification version 2.0, March 1998. RFC 2297.

[16] P. Newman, G. Minshall, and T.L. Lyon. IP switching-ATM under IP. *IEEE/ACM Transactions on Networking*, 6(2):117–29, April 1998.

[17] Charles O Smith. *Introduction to Reliability in Design*. McGraw Hill, NY, 1976.

[18] The ATM Forum. Private network-network specification interface v1.0 (PNNI 1.0). Technical Report af-pnni-0055.000, The ATM Forum, March 1996.

[19] L. Zhang, S. Deering, D. Estrin, S. Shenker, and et. al. RSVP: a new resource reservation protocol. *IEEE Network*, 7(5):8–18, September 1993.

SESSION 12

Traffic Flow Control

THE VIRTUAL BANDWIDTH BASED ER MARKING ALGORITHMS FOR FLOW CONTROL IN ATM NETWORKS

Tao Yang, Ping Wang, and Wengang Zhai
Ascend Communications, Inc., One Robbins Road, Westford, MA 01886
{tyang,pwang,zhai}@ascend.com

Abstract ABR service is designed for a wide range of applications that do not require bounded delay and loss ratio, but rather prefer low loss ratio and high throughput with only a minimum cell rate guaranteed. In this paper, we introduce the concept of *virtual bandwidth* and discuss its application to ABR flow control. We develop a new explicit rate marking algorithm that adopts a *traffic-driven* measurement-based approach to track the available bandwidth and the virtual bandwidth concept to achieve both fairness and high utilisation. This algorithm exhibits O(1)-computational complexity, O(1)-storage complexity, fast responsiveness, and quick convergence. It does not require special information from end systems nor *any information* from other switches, such as bottleneck link and bottleneck connection indication. Simulation results show that the proposed scheme adapts well to the distributed dynamic network environment and converges to max-min fairness allocation quick.

Keywords: ATM Networks, Available Bit Rate Service, Flow Control, ER Marking, Virtual Bandwidth, Traffic-Driven.

1. INTRODUCTION

ATM networks offer the ABR service as one of its five service classes. The ABR service does not require bounded delay and loss ratio for a given connection, and guarantees only a minimum cell rate (MCR) specified by the user. To support ABR service, each switch in the network adopts a closed-loop control mechanism to control the source rate of each connection according to availability of bandwidth. The objective is to achieve high bandwidth utilization, fairness, and low cell loss ratio.

In a typical rate-based ABR flow control model, sources adapt their rates to network conditions. Information about the network condition is conveyed to the sources through *resource management* (RM) cells, periodically generated by sources and turned around by the corresponding destinations.

RM cells travelling from a source to destination are called forward RM (FRM) cells and those travelling the opposite direction are called backward RM (BRM) cells. Each RM cell contains several fields including the source minimum cell rate (MCR), current cell rate (CCR), and the explicit rate (ER). When a RM cell traverses along its path, its fields can be updated by a switch to reflect its congestion state. When the RM cell returns to the source, the values of these fields are used by the source to adjust its rate.

The rule by which a switch updates the content of a RM cell and/or a control bit in the header of a data cell is called the *switch behavior*. The ATM Forum requires that each switch implement at least one of three algorithms: explicit forward congestion indicator (EFCI) marking, relative rate (RR) marking, and explicit rate (ER) marking. For details of these marking algorithms, the reader is referred to [20]. Amongst the three, ER marking is the most effective and is also the most challenging and complicated one. In this paper, we will study only ER marking.

There are several ER marking algorithms proposed in the literature (see [1, 4, 18] for related surveys). These algorithms can be classified into two categories: *direct marking* and *progressive marking*. Examples of direct marking algorithms include [2, 5, 6, 11, 12, 15, 24, 25, 26]. This approach allows the link to achieve both fairness and high utilization. A typical implementation requires a certain form of per-connection accounting and needs extra processing time to perform necessary bookkeeping. At high speed, these requirements can be very expensive. Examples of progressive marking algorithms include [3, 7, 8, 9, 10, 13, 14, 16, 17, 19, 21, 22, 23, 27]. The advantage of progressive ER marking algorithms is simplicity and lower implementation complexity. But it is generally considered as being slow to converge to max-min fairness allocation.

Therefore, there is a compelling demand to develop a new ER scheme with the following objectives:

- *O(1)-Processing Complexity*. This implies that the processing time required to do rate allocation should be independent of the number of ABR connections traversing the link.
- *O(1)-Storage Complexity*. The O(1)-storage complexity rules out the costly per-connection accounting, found in most direct marking schemes [1,12,15,26] and some progressive marking algorithms [16].
- *Fast Responsiveness*. The responsiveness of the new algorithm should be comparable to that of direct marking algorithms.
- *Max-min Fairness*. In steady state, the new algorithm should converge to the max-min fair allocation of the chosen fairness criterion.
- *High Bandwidth Utilization*. Capable of achieving high utilization when not all connections are bottlenecked elsewhere.
- *Robustness*. Congestion indicators based on only *'Local Bound'* information, which is insensitive to either end system behavior, other switch behavior, or the network topology. This is essential for the

switch being scalable and successfully performing in the future heterogeneous networking environment.

In this paper, we introduce the concept of virtual bandwidth and discuss how it can be applied to develop a simple progressive ER marking algorithm, called the Traffic-driven, Virtual-bandwidth based ER Marking Algorithm (T-VERMA), that aims to achieve the above objectives. In the next section, we introduce the concept of virtual bandwidth and an approach to achieve max-min fairness. In section 3, we describe how to apply the concept to do distributed rate allocation and present details of T-VERMA. In section 4, we present simulation results and compare T-VERMA with two algorithms. Finally, we conclude our work in section 5.

2. VIRTUAL BANDWIDTH

2.1 Network Model and Notations

We consider a network consisting of a set of links, $\mathbf{J} = \{1, 2, \ldots, M\}$, with C_j being the capacity (bandwidth) of link $j \in \mathbf{J}$. Let $\mathbf{I} = \{1, 2, \ldots, N\}$ be a set of connections competing for network resources (bandwidth) and MCR_i be the minimum cell rate to be guaranteed by the network for connection $i \in \mathbf{I}$. Define $\mathbf{J}(i)$ to be the set of *links* traversed by connection $i \in \mathbf{I}$ and $\mathbf{I}(j)$ to be the set of *connections* traversing link $j \in \mathbf{J}$. To guarantee MCR for each connection, we assume that

$$\sum_{i \in \mathbf{I}(j)} MCR_i \le C_j \tag{1}$$

holds for all $j \in \mathbf{J}$.

For $i \in \mathbf{I}(j)$, we denote by $R_{i,j}$ the *rate allocated* to connection i by link j, also called the *local rate allocation* for connection i at link j. The matrix $\lfloor R_{i,j} \rfloor$ is called the *allocation matrix.* $\lfloor R_{i,j} \rfloor$ is *feasible* if it satisfies:

$$R_{i,j} \ge MCR_i, \quad i \in \mathbf{I} \text{ and } j \in \mathbf{J}(i) \text{; and } \sum_{i \in \mathbf{I}(j)} R_{i,j} \le C_j, \quad j \in \mathbf{J}. \tag{2}$$

It is *work conserving* if it satisfies (2) *and* $\sum_{i \in \mathbf{I}(j)} R_{i,j} = C_j$ for any $j \in \mathbf{J}$. The allowed rate for connection $i \in \mathbf{I}$ is given by $R_i = \min\{R_{i,j} \mid j \in \mathbf{J}(i)\}$, also called the *global rate allocation*. The vector $(R_1, R_2, \cdots, R_N)$ is called the *allocation vector* derived from the allocation matrix $\lfloor R_{i,j} \rfloor$.

For a given network, it is possible that there are multiple feasible allocation matrices. The questions are: which one is optimal and how to find it? This is the focal point of all ER marking algorithms and their primary function is to compute the *right* local allocation. Here, the objectives are two folds: *fairness* and *high utilization.*

2.2 Fairness Criteria

From (1), we observe that the quantity $C_j - \sum_{i\in \mathbf{I}(j)} MCR_i$, or the *excessive bandwidth* at link $j \in \mathbf{J}$, is greater than or equal to zero. The fairness issue is on how to distribute the excessive bandwidth amongst all traversing connections. There are two well-known fairness criteria: *equal share* and *proportional share.* In equal share, the excessive bandwidth is equally shared by all traversing connections. In proportional share, it is shared in proportion to their respective minimum cell rates. Let $S_{i,j}$ be the fair share of connection $i \in \mathbf{I}(j)$ at link $j \in \mathbf{J}$. Then,

$$S_{i,j} = \begin{cases} MCR_i + (C_j - SUM_j)/|\mathbf{I}(j)|, & \text{for equal share criterion,} \\ MCR_i + (C_j - SUM_j) \times MCR_i / SUM_j, & \text{for proportional share criterion,} \end{cases} \quad (3)$$

where $SUM_j = \sum_{i\in \mathbf{I}(j)} MCR_i$ and $|\mathbf{A}|$ is the total number of elements in set $\mathbf{A}$. It can be shown that if the condition in (1) is satisfied then $S_{i,j} \geq MCR_i$ for any $j \in \mathbf{J}$ and $i \in \mathbf{I}(j)$. In this paper, we assume that *all* links in the network adopt either the equal share criterion or the proportional share criterion.

2.3 Bottleneck Links and Bottleneck Connections

One approach to rate allocation is to set $R_{i,j} = S_{i,j}$. This achieves fairness but may cause low link utilization. To improve, the concept of bottleneck link was introduced. For a given allocation matrix $\lfloor R_{i,j} \rfloor$, a link $j \in \mathbf{J}$ is said to be a *bottleneck link* if $\sum_{i\in \mathbf{I}(j)} R_i = C_j$ holds. A connection $i \in \mathbf{I}$ is said to be a *bottleneck connection* if it traverses at least one bottleneck link. It can be shown that if an allocation matrix $\lfloor R_{i,j} \rfloor$ is work conserving, then there exists at least one bottleneck link in the network and, for every link $j \in \mathbf{J}$, there exists at least one connection $i \in \mathbf{I}(j)$ that is bottlenecked.

2.4 Max-min Fair Allocation

An optimal allocation should be the one that maximizes the network resource utilization and maintains fairness amongst all connections. This is called the *max-min fair allocation.* A procedure has been proposed to compute the max-min fair allocation [5]. In Table 1, we rephrase this procedure for a more general scenario described in the previous subsections.

Table 1. Max-Min Fair Allocation – The Traditional Approach

1. At each link j, allocate each connection its fair share $S_{i,j}$;
2. At each bottleneck link j, mark each connection i and assign it the rate R_i;
3. Decrease capacities of all links by the total capacity consumed by the marked connections traversing these links.
4. Consider a reduced network with all link capacities adjusted as

above and with marked connections removed. Repeat the procedure until all connections are assigned their rates.

2.5 The Virtual Bandwidth

We see that, in the traditional procedure, each link simply removes traversing bottleneck connections, decreases its capacity by the total bandwidth consumed by these bottleneck connections, and then redistribute the remaining bandwidth among other (non-bottleneck) connections using the selected fairness criterion. Here, we describe a procedure that takes the *opposite direction* for computing the max-min fair allocation. Instead of decreasing the link capacity, we *increase* the link capacity to a level that is *above* its actual or real value, hence the term *virtual bandwidth*.

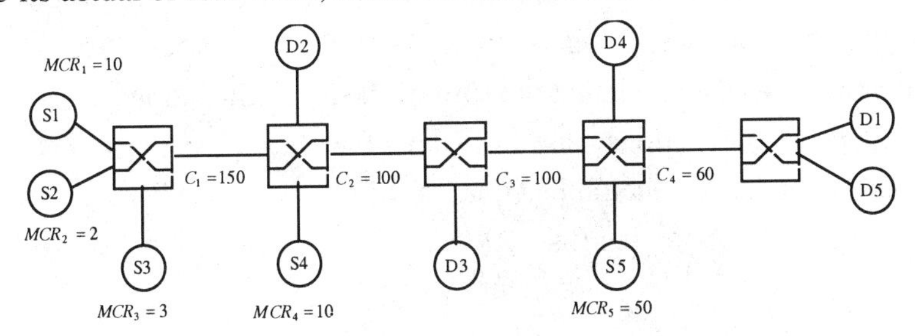

Figure 1. An Example.

To explain, let us consider Link 1 in Figure 1. According to proportional share criterion, it allocates 100, 20, and 30 to connections 1, 2, and 3, respectively. Suppose that connection 1 is bottlenecked elsewhere (link 4) in the network and is allowed to transmit only at the rate of 10. In the traditional procedure, the link would decrease the total link capacity of 150 by 10 and then redistribute the remaining 140 between connections 2 and 3. This gives 10 to connection 1, 56 to connection 2, and 84 to connection 3.

In the virtual bandwidth approach, we *increase* the link capacity to its virtual bandwidth, which is 420 in this case, and then redistribute the *total* virtual bandwidth of 420 among *all* connections, *including* the one that is bottlenecked elsewhere. This will give 280 to connection 1, 56 to connection 2, and 84 to connection 3. We notice that since connection 1 is bottlenecked elsewhere at the rate of 10, assigning a rate *greater* than its previous allocation *will not* cause over allocation.

In principle, the virtual bandwidth should be at such a capacity level that the sum of the local rates of all non-bottleneck connections plus the sum of the global rates of all bottleneck connections is equal to the real link capacity, hence achieving 100% link utilization. For the above simple example, let $\tilde{C}_1$ be the virtual bandwidth of Link 1. Then, $2\tilde{C}_1/15$ and $3\tilde{C}_1/15$ are the new allocations to connections 2 and 3 (both unbottlenecked), respectively. Hence, $\tilde{C}_1$ should satisfy

$$2\tilde{C}_1/15 + 3\tilde{C}_1/15 + 10 = C_1 = 150. \quad (4)$$

Solving the above for $\tilde{C}_1$ yields $\tilde{C}_1 = 420$.

As in the traditional method, computing the max-min fair allocation using virtual bandwidth is also done in an iterative manner. For $j \in \mathbf{J}$, let $\tilde{C}_j$ be the virtual bandwidth of link j and $\tilde{C}_j(k)$ be the virtual bandwidth computed at the kth iteration. Let $R_{i,j}(k)$ and $R_i(k)$ be, respectively, the local and the global allocations for the kth iteration. Also, let $\mathbf{I}_B(j,k)$ ($\mathbf{I}_U(j,k)$) be the set of connections traversing link j that are bottlenecked elsewhere (unbottlenecked anywhere) at the kth iteration.

Given the results of the kth iteration, the virtual bandwidth $\tilde{C}_j(k+1)$ is computed as follows. If all connections traversing link j are bottlenecked either elsewhere or at this link, then $\tilde{C}_j(k+1) = \tilde{C}_j(k)$. Otherwise, $\tilde{C}_j(k+1)$ should assume such a value that the sum of the local allocations $R_{i,j}(k+1)$ for non-bottleneck connections plus the sum of global allocations $R_i(k+1)$ of bottleneck connections should be equal to the real link capacity C_j, i.e.,

$$\sum_{i \in \mathbf{I}_U(j,k)} R_{i,j}(k+1) + \sum_{i \in \mathbf{I}_B(j,k)} R_i(k+1) = C_j. \quad (5)$$

Since

$$R_{i,j}(k+1) = \begin{cases} MCR_i + \left(\tilde{C}_j(k+1) - SUM_j\right) / |\mathbf{I}(j)|, & \text{for equal share,} \\ MCR_i + \left(\tilde{C}_j(k+1) - SUM_j\right) \times MCR_i / SUM_j, & \text{for proportion al share,} \end{cases} \quad (6)$$

for any $i \in \mathbf{I}(j)$, and $R_i(k+1) = R_i(k)$ for any $i \in \mathbf{I}_B(j,k)$, we can solve (5) and (6) simultaneously for $\tilde{C}_j(k+1)$ and obtain

$$\tilde{C}_j(k+1) = \begin{cases} \left(C_j - \sum_{i \in \mathbf{I}_U(j,k)} MCR_i - \sum_{i \in \mathbf{I}_B(j,k)} R_i(k) \right) \times N_j / |\mathbf{I}_U(j,k)| + SUM_j, & \text{equal share,} \\ \left(C_j - \sum_{i \in \mathbf{I}_B(j,k)} R_i(k) \right) \times SUM_j / \sum_{i \in \mathbf{I}_U(j,k)} MCR_i, & \text{prop. share.} \end{cases} \quad (7)$$

The iterative procedure for computing the max-min fair allocation using virtual bandwidth is outlined in Table 2.

Table 2. Max-Min Allocation: A Virtual Bandwidth Approach

1. Set $k = 0$ and $\tilde{C}_j(0) = C_j$ for all $j \in \mathbf{J}$.
2. Compute $R_{i,j}(k)$ using formula (6) for all $j \in \mathbf{J}$ and all $i \in \mathbf{I}(j)$.
3. For each link j, determine whether it is a bottleneck. If so, mark any unmarked traversing connections.
4. For each link j, determine whether it has both marked and unmarked connections traversing it. If so, increase its current virtual bandwidth $\tilde{C}_j(k)$ to $\tilde{C}_j(k+1)$ using (7). Otherwise, set $\tilde{C}_j(k+1) = \tilde{C}_j(k)$.
5. Terminate if all connections have been marked. Otherwise, increase k by one and go to step 2.

$\lfloor R_{i,j}(k) \rfloor$ may not be feasible with respect to the real capacity C_j. However, it is always feasible and work conserving with respective to the $\tilde{C}_j(k)$. Furthermore, if a connection is bottlenecked at iteration k, then it remains to be bottlenecked in iteration (k+1) and its global allocation will not be affected by the re-distributions of bandwidth occurred at various links in the network. These properties ensure that the iterative process will converge to the max-min fair allocation. Once $\tilde{C}_j$ has been obtained, then the local allocation $R_{i,j}$ for connection $i \in \mathbf{I}(j)$ can be computed as follows:

$$R_{i,j} = \begin{cases} MCR_i + \left(\tilde{C}_j - SUM_j\right) / |\mathbf{I}(j)|, & \text{for equal share criterion,} \\ MCR_i + \left(\tilde{C}_j - SUM_j\right) \times MCR_i / SUM_j, & \text{for proportional share criterion,} \end{cases} \quad (8)$$

regardless of whether connection $i \in \mathbf{I}(j)$ is bottlenecked or not.

3. VIRTUAL BANDWIDTH BASED ER MARKING ALGORITHM

We now present the proposed new algorithm T-VERMA, aiming at achieving the objectives outlined in the Introduction. T-VERMA consists of two major operations: *rate allocation* and *virtual bandwidth estimation*.

3.1 Rate Allocation

This operation is executed each time the link processes a backward RM cell. Here, it first computes the local allocation for the connection using formula (8). It then compares the local allocation with the value in the ER field of the RM cell and overwrites the ER field with the local allocation if the former is larger than the latter. This operation is summarized in table 3 for equal share criterion and table 4 for proportional share criterion.

Table 3. The Rate Allocation for the case of Equal Share Criterion.

```
If (A BRM Cell Departs) {
      R ← BRM_MCR + (VB * Adjust(ABR_Q) - SUM) / N;
      If (R < BRM_ER) BRM_ER = R;
}
```

Table 4. Rate Allocation - Proportional Share Criterion.

```
If (A BRM Cell Departs) {
      R ← VB * Adjust(ABR_Q) * BRM_MCR / SUM;
      If (R < BRM_ER) BRM_ER = R;
}
```

In the above algorithm, we use the function `Adjust()` to adjust the virtual bandwidth, a technique used by a number of ER marking algorithms (see [12] and [16]). The value of `Adjust()` depends on the current queue length. The longer the queue, the smaller the value of `Adjust()`. The idea here is to reserve some bandwidth to drain the queue when its length exceeds a certain threshold [1]. A simple example is given as follows:

$$Adjust(x) = \begin{cases} c - \dfrac{c-1}{T_0}x, & 0 \le x \le T_0, \\ 1, & T_0 < x \le T_1, \\ 1 - \dfrac{1-d}{T_2 - T_1}(x - T_1), & T_1 < x \le T_2, \\ d, & x > T_2, \end{cases} \tag{9}$$

where $c > 1 > d > 0$ and T_0, T_1, and T_2 are queue thresholds. The interval $[0, T_0]$ defines the underload region, $[T_0, T_1]$ is the steady-state region, $[T_1, T_2]$ is the overload region, and $[T_2, \infty]$ is the heavy-load region. More details on this topic and some excellent examples of this type of queue control functions can be found in [28].

3.2 Virtual Bandwidth Estimation

In this operation, T-VERMA estimates the virtual bandwidth of the link. Here, the virtual bandwidth is progressively updated. Each time, it is unchanged, increased, or decreased from its present value, depending on the link congestion state, the available bandwidth, and the total ABR traffic volume. In steady-state, the estimated virtual bandwidth is expected to converge to the true virtual bandwidth of the link after some iterations.

T-VERMA periodically monitors the available bandwidth for ABR connections and the total ABR traffic load. Let $\{t_0, t_1, t_2, \cdots, t_k, \cdots\}$ be the sequence of time epochs at which the measurements are made. The time interval $(t_k, t_{k+1}]$ is called the kth *measurement interval.*

3.2.1 Basic Updating Rules

First, for the link concerned, we define the following variables:

$ABR\_Capacity(t)$: the measured available capacity for ABR connections;

$ABR\_Load(t)$: the measured total ABR traffic load at time t;

$ABR\_U(t)$: the ABR utilization at time t and is equal to $ABR\_Load(t)$ / $ABR\_Capacity(t)$;

$ABR\_Q(t)$: the length of the ABR queue at time t;

$\tilde{C}(t)$: the true virtual bandwidth of the link, calculated in theory based on values of $ABR\_Capacity(t)$ of all links; and

$VB(t)$: the estimated virtual bandwidth at time t.

The rules for updating the estimated virtual bandwidth $VB(t)$ are as follows. If $ABR\_U(t) = 1$, $VB(t)$ is unchanged. If $ABR\_U(t) > 1$, $VB(t)$ is decreased to avoid sustained congestion. If $ABR\_U(t) < 1$, $VB(t)$ is increased to avoid under-utilization.

3.2.2 The Updating Function

The magnitude of the change (decrease or increase) of $VB(t)$ depends on the values of $ABR\_U(t)$ and $ABR\_Q(t)$. Suppose that an update is to be carried out at time t. Then, we update the estimated virtual bandwidth using the following formula:

$$VB(t^{+}) = f(ABR\_Q(t), ABR\_U(t)) \cdot VB(t), \qquad (10)$$

where $VB(t^{+})$ is the new virtual bandwidth and $f(x, y)$ is a non-negative function, called the *virtual bandwidth updating function*. The form of the function $f(x, y)$ is not strictly specified. However, it should be non-increasing in x and in y; below 1.0 as either x, or y, or both approach infinity; and above 1.0 as either x, or y, or both approach zero. As an example, we present the following virtual bandwidth updating function:

$$f(x, y) = \begin{cases} \max(1, g(x)), & 0 \le x < \infty;\ 0 \le y < U_0, \\ g(x), & 0 \le x < \infty;\ U_0 \le y \le U_1, \\ \min(1, g(x)), & 0 \le x < \infty;\ U_1 < y < \infty, \end{cases} \qquad (11)$$

where $g(x) = 1 - \alpha + \alpha \cdot q(x)$, α is a smoothing factor, and $q(x)$ is a standard queue control function. Typical values of α should be around 0.1. An example of $q(x)$ is the piece-wise hyperbolic function given in [16] as:

$$q(x) = \begin{cases} \dfrac{b \cdot Q_0}{(b-1) \cdot x + Q_0}, & 0 \le x \le Q_0, \\ 1.0, & Q_0 < x \le Q_1, \\ \dfrac{a \cdot (Q_2 - Q_1)}{(1-a) \cdot x + a \cdot Q_2 - Q_1}, & Q_1 < x \le Q_2, \\ a, & x > Q_2, \end{cases} \qquad (12)$$

where a (<1.0) and b (>1.0) are respectively the minimum and maximum values of $q(x)$. The parameters U_0 and U_1 specify a targeted utilization band and the parameters Q_0, Q_1, and Q_2 specify a number of congestion zones.

3.2.3 RM Cell Driven vs. Traffic Driven

We now discuss the issue of *when* to update $VB(t)$. First, we point out that the timing of updating control variables is crucial to an ER marking algorithm. Most existing ER marking algorithms update at least some of their control variables at the time of processing RM cells and hence are said to be of *RM cell driven*. For these algorithms, updating control variables requires some information carried in RM cells and therefore *has to* be done when processing RM cells. The major disadvantage of being RM cell driven is that it can cause the control variables and the link congestion state out of synchronization, especially for progressive ER marking algorithms.

Table 5. Virtual Bandwidth Estimation.

```
If (End of Measurement Interval) {
    Update ABR_Capacity according to traffic measurements;
    Update ABR_Load according to traffic measurements;
    ABR_U ← ABR_Load / ABR_Capacity;
```

```
        VB ← VB * f(ABR_Q, ABR_U);
        If (VB < ABR_Capacity) VB ← ABR_Capacity;
    }
```

Here we adopt a different approach: to have the update of control variables driven by traffic. This approach ensures that the values of control variables can more accurately represent the current congestion-state of the link. A necessary condition for being traffic-driven is that the update does not need information carried by RM cells. From the previous subsection, we can see that updating the estimated virtual bandwidth $VB(t)$ requires only $ABR\_U(t)$ and $ABR\_Q(t)$, and can be done without any information from RM cells. Indeed, *this is the most important advantage of combining virtual bandwidth with progressive marking*! We choose the updating time to be at the end of each measurement interval $(t_k, t_{k+1}]$. The value of $ABR\_Q(t)$ is always accurate. To summarize, we present the pseudo code for the second (and last) part of the virtual bandwidth algorithm (T-VERMA) in Table 5.

4. SIMULATION RESULTS

We will now show simulation results for three algorithms: ERICA with Weighted Fairness (ERICA-WF) [27], EDERA [12], and our new algorithm T-VERMA. We have focused on three scenarios. In the first, we test the responsiveness of the three algorithms when there is a sudden increase in bandwidth and a sudden decrease in bandwidth. We also test in this scenario whether or not these algorithms converge to the max-min fair allocation after each change. In the second scenario, we test how the algorithms perform when some of the sources are non-persistent, meaning that they may transmit at a rate below their allowed cell rates and sometimes even below their MCRs. In Scenario 3, we compare the performance of the three algorithms in the presence of VBR sources.

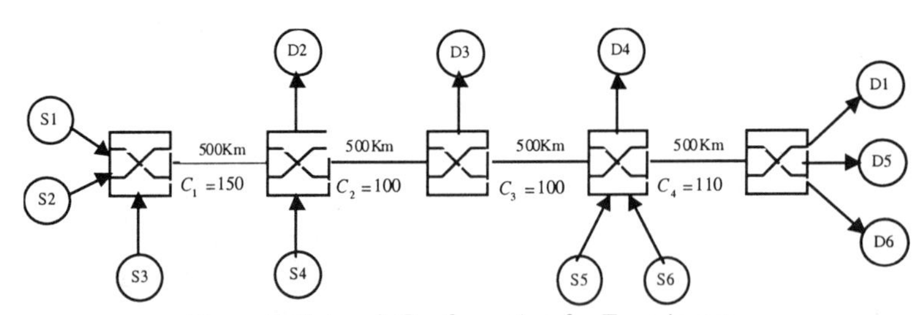

Figure 2. Network Configuration for Experiments.

All experiments are designed based on the network configuration shown in *Figure 2*. Link capacity is in Mbps. The distance between a switch and a source or a destination is 10 kilometers.

Due to space limitation, only brief simulation results (ACR plots) of scenario 1 are shown here (see *Figure 3*). It is observed that both EDERA and T-VERMA response to the sudden significant change in bandwidth very well, with EDERA being slightly better. ERICA-WF reacts well (so do the other two) to the sudden decrease in bandwidth but its response to the sudden increase in bandwidth is relatively slow.

Simulation results for scenario two indicates that in T-VERMA, non-persistent sources are treated fairly and are allocated a rate proportional to their respective MCR at the bottleneck link. However, it is observed that both EDERA and ERICA-WF seem to penalize non-persistent sources by allocating them a rate that is smaller, in proportion, than the rates received by their persistent peers at a bottleneck link.

In scenario 3, it is observed that all three algorithms respond to traffic disruptions caused by the VBR source very well!

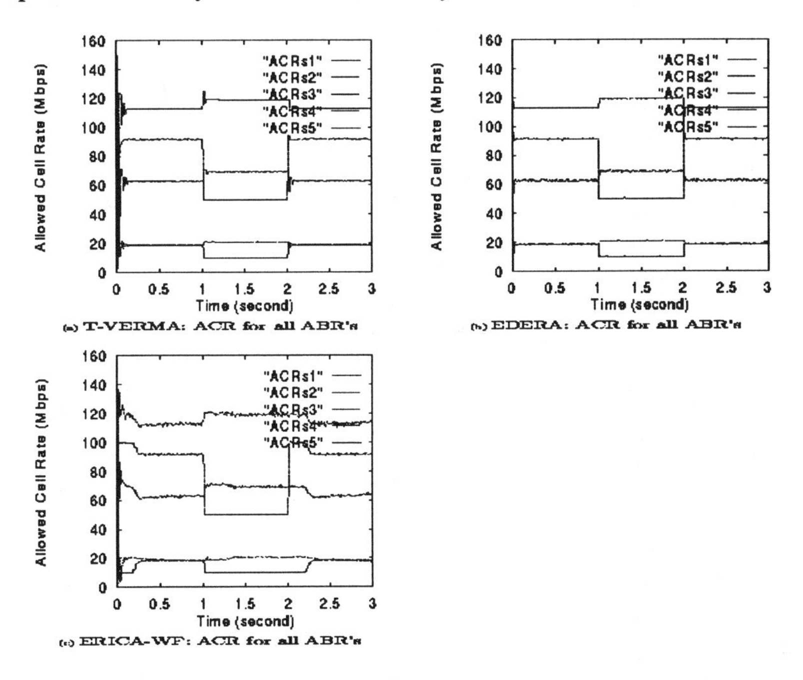

Figure 3. Comparison of T-VERMA, EDERA and ERICA-WF under scenario 1.

5. CONCLUSIONS

In this paper, we introduced the concept of virtual bandwidth and showed how it can be used to do global calculation of the max-min fair allocation for ABR connections in an ATM network. Based on the concept of virtual bandwidth, we developed a simple ER marking algorithm, called T-VERMA, that has O(1) storage complexity and O(1) computational complexity.

We have conducted extensive simulation studies on T-VERMA and compared it with two fairly recent algorithms, namely EDERA [12] and ERICA-WF [27]. From these studies, we can conclude that the responsiveness of T-VERMA is comparable to that of the more sophisticated

EDERA and ERICA-WF. Furthermore, T-VERMA is capable of treating all ABR connections, including those that are non-persistent, in a fair fashion and can converge to max-min fair allocation in steady state. Finally, T-VERMA has shown remarkable robustness in presence of VBR connections.

References

[1] Arulambalam, A., X. Chen, and N. Ansari, "Allocating Fair Rates for Available Bit Rate Service in ATM Networks," IEEE Communication Magazine, vol. 34, 92-100, 1996.

[2] Arulambalam, A., X. Chen, and N. Ansari, "An Intelligent Explicit Rate Control Algorithm for ABR Service in ATM Networks," ICC'97, 200-204, 1997.

[3] G. Bianchi, L. Fratta, and L. Musumeci, "Congestion Control Algorithms for the ABR Service in ATM Networks," IEEE GLOBECOM, 1996.

[4] F. Bonomi and K.W. Fendick, "The Rate-Based flow Control Framework for the Available Bit Rate ATM Service," IEEE Network Magazine, March/April 1995.

[5] Charny, A., D. Clark, and R. Jain, "Congestion Control with Explicit Rate Indication," Proc. ICC'95, June 1995.

[6] Chiussi, F., A. Arulambalam, Y. Xia, and X. Chen, "Explicit Rate ABR Schemes Using Traffic Load as Congestion Indicator," IEEE GLOBECOM'97, 76-84, 1997.

[7] Chiussi, F., T. Wang, "An ABR Rate-Based Congestion Control Algorithm for ATM Switches with Per-VC Queueing," IEEE GLOBECOM'97, 2108-2117, 1997.

[8] Chiussi, F., Y. Xia, and V. Kumar, "Dynamic Max Rate Control Algorithm for Available Bit Rate Service in ATM networks", IEEE GLOBECOM'96, 2108-2117, 1996.

[9] F.M. Chiussi, Y. Xia, and V.P. Kumar, "Virtual Queueing Techniques for ABR Service: Improving ABR/VBR Interaction," IEEE INFOCOM'97, 1997.

[10] S. Fahmy, R. Jain, S. Kalyanaraman, R. Goyal, and B. Vandalore, "On Determining the Fair Bandwidth Share for ATM Connections in ATM Networks," Proceedings of the IEEE International Conference on Communications (ICC) 1998, June 1998.

[11] N. Ghani and J.W. Mark, "Dynamic Rate-Based Control Algorithm for ABR Service in ATM Networks", IEEE GLOBECOM'96, Vol. 2, Nov. 1996, London, UK, pp.1074-1079.

[12] N. Ghani and J.W. Mark, "An Enhanced Distributed Explicit Rate Allocation for ABR services", 15th International Teletraffic Congress (ITC-15), Washington D.C., June 1997.

[13] Jain, R., S. Kalyanaraman, R. Goyal, S. Fahmy, and R. Viswanathan, "ERICA Switch Algorithm: A Complete Description," ATM Forum 96-1172, August, 1996.

[14] R. Jain, S. Kalyanaraman, and R. Viswanathan, "The OSU Scheme for Congestion Avoidance in ATM Networks: Lessons Learnt and Extensions," Performance Evaluation (North-Holland), Special Issue on

Traffic Control in ATM Networks, Vol. 31, No.1-2, November 1997, pp. 67-88.

[15] L. Kalampoukas, A. Varma, K.K. Ramakrishnan, An Efficient Rate Allocation Algorithm for ATM Networks Providing Max-Min Fairness," High Performance Networking VI. IFIP sixth Int. Conf. on High Performance Networking, 1995.

[16] S. Kalyanaraman, R. Jain, S. Fahmy, R. Goyal, and B. Vandalore, "The ERICA Switch Algorithm for ABR Traffic Management in ATM Networks," Submitted to IEEE/ACM Trans. on Networking, Nov. 1997, http://www.cis.ohiostate.edu/~jain/papers/erica.htm.

[17] S. Muddu, F.M. Chiussi, C. Tryfonas, and V.P. Kumar, "Max-Min Rate Control Algorithm for Available Bit Rate Service in ATM Networks", Proc. ICC'96, June 1996.

[18] H. Ohsaki, M. Murata, H. Suzuki, C. Ikeda, and H. Miyahara, "Rate-Based Congestion Control for ATM Networks," Computer Comm. Review, ACM SIGCOMM, 1995.

[19] Roberts, L., "Enhanced PRCA (Proportional Rate-Control Algorithm)," ATM Forum 94-0735R1, 1994.

[20] Shirish S. Sathaye, "ATM Forum Traffic Management Specification 4.0," ATM Forum af-tm-0056.000, April 1996.

[21] K.-Y. Siu and H.-Y. Tzeng, "Adaptive Proportional Rate Control (APRC) with Intelligent Congestion Indication," ATM Forum 94-0888, September 1994.

[22] K.-Y. Siu and H.-Y. Tzeng, "Limits of Performance in Rate-Based Control Schemes," ATM Forum Contribution 94-1077, November 1994.

[23] Siu, K. and T. Tzeng, "Intelligent Congestion Control for ABR Service in ATM Networks," Computer Communication Review, 24(5), 81-106, October 1995.

[24] Wei K. Tsai, Yuseok Kim and Lee Hu, "ASAP: A Non-per-VC Accounting Max-Min Protocol for ABR Flow Control with Optimal Convergence Speed," IEEE SICON'98, Singapore, June 30 - July 3, 1998.

[25] Tsang, D., W. Wong, S. M. Jiang and E. Liu, "A fast Switch Algorithm for ABR Traffic to Achieve Max-Min Fairness," IEEE 1996 International Zurich Seminar on Digital Communications, February 19-23, 1996.

[26] Tsang, D. and W. Wong, "A New Rate-Based Switch Algorithm for ABR Traffic to Achieve Max-Min Fairness with Analytical Approximation and Delay Adjustment, INFOCOM'96, 1174-1181, 1996.

[27] B. Vandalore, S. Fahmy, R. Jain, R. Goyal, and M. Goyal, "A Definition of General Weighted Fairness and its Support in Explicit Rate Switch Algorithms," submitted to ICNP'98, May 1998, http://www.cis.ohiostate.edu/~jain/papers/icnp98_bv.htm.

[28] B. Vandalore, R. Jain, R. Goyal, and S. Fahmy, "Design and analysis of queue control functions for explicit rate switch," Submitted to the IC3N'98, May 1998. Also available at http://www.cis.ohio-state.edu/~jain/papers.html.

CONVERGENCE OF ASYNCHRONOUS OPTIMIZATION FLOW CONTROL

Steven H. Low and David Lapsley *
Department of Electrical & Electronic Engineering
University of Melbourne
Australia
{slow,lapsley}@ee.mu.oz.au

Abstract We proposed earlier an optimization approach to reactive flow control where the objective of the control is to maximize the total source utility over their transmission rates. The source utility functions model their valuation of bandwidth and can be different for different sources. The control mechanism is derived as a gradient projection algorithm to solve the dual problem. In this paper we generalize the algorithm and the convergence result to an asynchronous setting where the computations at and the communications among the links and sources are uncoordinated and based on possibly outdated information.

Keywords: Optimization flow control, congestion control, congestion pricing, asynchronous algorithm.

1. INTRODUCTION

We have proposed previously an optimization approach to flow control where the control mechanism is derived as a gradient projection algorithm to solve (the dual of) a global optimization problem [10; 11]. The purpose of this paper is to show that the basic algorithm converges in both synchronous and asynchronous settings.

Specifically consider a network that consists of a set L of unidirectional links of capacity c_l, $l \in L$. The network is shared by a set S of sources, where source s is characterized by a utility function $U_s(x_s)$ that is concave increasing in its transmission rate x_s. The goal is to calculate source rates that maximize

*The first author acknowledges the support of the Australian Research Council under grant S499705, the second author acknowledges the Australian Commonwealth Government for their Australian Postgraduate Awards, and both acknowledge the financial support of Melbourne IT, Australia.

the sum of the utilities $\sum_{s \in S} U_s(x_s)$ over x_s subject to capacity constraints. Solving this problem centrally would require not only the knowledge of all utility functions, but worse still, complex coordination among potentially all sources due to the coupling of sources through shared links. The key to a distributed solution is to consider the dual problem that decomposes the task into simple local computations to be executed at individual links and sources.

The algorithm takes the familiar form of reactive flow control. Based on the local *aggregate* source rate each link $l \in L$ calculates a 'price' p_l for a unit of bandwidth. A source s is fed back the scalar price $p^s = \sum p_l$, where the sum is taken over all links that s uses, and it chooses a transmission rate x_s that maximizes its own benefit $U_s(x_s) - p^s x_s$, utility minus the bandwidth cost. These individually optimal rates $(x_s(p^s), s \in S)$ may not be socially optimal for a general price vector $(p_l, l \in L)$, i.e., they may not maximize the total utility. The algorithm iteratively approaches a price vector $(\hat{p}_l, l \in L)$ that aligns individual and social optimality such that $(x_s(\hat{p}_l), s \in S)$ indeed maximizes the total utility. In order words, the price $\hat{p}^s$ represents the complete congestion information source s needs for its control decision.

The basic algorithm is presented in [10] and a preliminary prototype is briefly discussed in [11]. The basic algorithm requires communication of link prices to sources and source rates to links. This requirement is greatly simplified in [13,12], as follows. In [13] we prove that a link can simply set its price to a fraction of its buffer occupancy, thus eliminating the need for explicit communication from sources to links. This can be seen as have the links estiamte the gradient using local information in carrying out the gradient projection algorithm. In [12], we describe a marking scheme, inspired by [6], that achieves the communication from links to sources using only binary feedback. The result is a variant of Random Early Detection (RED) scheme [4], that not only stabilizes network queues, as RED does, but does so in a way that optimizes a global measure of performance.

Optimization based flow control have also been proposed in [5; 7; 3; 8; 9; 6]. All these works, as ours, motivate flow control by an optimization problem and derive their control mechanisms as a solution to the optimization problem. They differ in their choice of objective functions or their solution approaches, and result in rather different flow control mechanisms to be implemented at the sources and the network links. In particular both [8; 9] and our work solve the same optimization problem of maximizing aggregate utility over source transmission rates. The two works however differ in their solution approach, which lead to different algorithms and their implementation through marking [6; 12]. See [14] for a detailed comparison.

The present paper is structured as follows. In Section 2 we present the optimization problem and its dual that motivate our approach. In Section 3 we briefly derive a synchronous solution and present its convergence property.

In Section 4 we extend the synchronous algorithm and its convergence to an asynchronous setting. All proofs are omitted due to space limitation and can be found in [14].

2. MODEL

Consider a network that consists of a set $L = \{1, \ldots, L\}$ of *unidirectional* links of capacity c_l, $l \in L$. The network is shared by a set $S = \{1, \ldots, S\}$ of sources. Source s is characterized by four parameters $(L(s), U_s, m_s, M_s)$. The path $L(s) \subseteq L$ is a subset of links that source s uses, $U_s : \Re_+ \to \Re$ is a utility function, $m_s \geq 0$ and $M_s \leq \infty$ are the minimum and maximum transmission rates, respectively, required by source s. Source s attains a utility $U_s(x_s)$ when it transmits at rate x_s that satisfies $m_s \leq x_s \leq M_s$. We assume U_s is increasing and strictly concave in its argument. Let $I_s = [m_s, M_s]$ denote the range in which source rate x_s must lie and $I = (I_s, s \in S)$ be the vector. For each link l let $S(l) = \{s \in S \mid l \in L(s)\}$ be the set of sources that use link l. Note that $l \in L(s)$ if and only if $s \in S(l)$.

Our objective is to choose source rates $x = (x_s, s \in S)$ so as to:

$$\textbf{P:} \quad \max_{x_s \in I_s} \sum_s U_s(x_s) \tag{1}$$

$$\text{subject to} \quad \sum_{s \in S(l)} x_s \leq c_l, \quad l = 1, \ldots, L. \tag{2}$$

The constraint (2) says that the total source rate at any link l is less than the capacity. A unique maximizer, called the primal optimal solution, exists since the objective function is strictly concave, and hence continuous, and the feasible solution set is compact.

Though the objective function is separable in x_s, the source rates x_s are coupled by the constraint (2). Solving the primal problem (1–2) directly requires coordination among possibly all sources and is impractical in real networks. The key to a distributed and decentralized solution is to look at its dual, e.g., [2, Section 3.4.2, 15]:

$$\textbf{D:} \quad \min_{p \geq 0} \ D(p) = \sum_s B_s(p^s) + \sum_l p_l c_l \tag{3}$$

where

$$B_s(p^s) = \max_{x_s \in I_s} U_s(x_s) - x_s p^s \tag{4}$$

$$p^s = \sum_{l \in L(s)} p_l. \tag{5}$$

The first term of the dual objective function $D(p)$ is decomposed into S separable subproblems (4–5). If we interpret p_l as the price per unit bandwidth at link l then p^s is the total price per unit bandwidth for all links in the path of s. Hence $x_s p^s$ represents the bandwidth cost to source s when it transmits at rate x_s, and $B_s(p^s)$ represents the maximum benefit s can achieve at the given price p^s. We shall see that this scalar p^s summarizes all the congestion information source s needs to know. A source s can be induced to solve maximization (4) by bandwidth charging. For each p, a unique maximizer, denoted by $x_s(p)$, exists since U_s is strictly concave.

In general $(x_s(p), s \in S)$ may not be primal optimal, but by the duality theory, there exists a $p^* \geq 0$ such that $(x_s(p^*), s \in S)$ is indeed primal optimal. Hence we will focus on solving the dual problem (3). Once we have obtained the minimizing prices p^* the primal optimal source rates $x^* = x(p^*)$ can be obtained by individual sources s by solving (4), a simple maximization (see below). The important point to note is that, given p^*, individual sources s can solve (4) *separately without the need to coordinate with other sources.* In a sense p^* serves as a coordination signal that aligns individual optimality of (4) with social optimality of (1). Note that despite the notation, a source s does not require the vector price p, but only a scalar $p^s = \sum_{l \in L(s)} p_l$ that represents the sum of link prices on its path.

Indeed the unique maximizer $x(p)$ for (4) can be given explicitly, from the Kuhn–Tucker theorem, in terms of the marginal utility:

$$x_s(p) \quad = \quad [U_s'^{-1}(p)]_{m_s}^{M_s} \tag{6}$$

where $[z]_a^b = \max\{a, \min\{b, x\}\}$. Here $U_s'^{-1}$ is the inverse of U_s', which exists over the range $[U_s'(M_s), U_s'(m_s)]$ since U_s' is continuous and U_s *strictly* concave. It is indeed the demand function in economics. It is illustrated in Figure 1. Let $x(p) = (x_s(p), s \in S)$.

In this paper, we abuse notation and use $x_s(\cdot)$ both as a function of scalar price $p \in \Re_+$ and of vector price $p \in \Re_+^{|L|}$. When p is a scalar, $x_s(p)$ is given by (6). When p is a vector, $x_s(p) = x_s(p^s) = x_s(\sum_{l \in L(s)} p_l)$. The meaning should be clear from the context.

3. SYNCHRONOUS DISTRIBUTED ALGORITHM

In [10; 14] we propose to solve the dual problem using the gradient projection algorithm where link prices are adjusted in opposite direction to the gradient $\nabla D(p)$ whose l-th component is given by:

$$\frac{\partial D}{\partial p_l}(p) \quad = \quad c_l - x^l(p) \tag{7}$$

where $x^l(p) := \sum_{s \in S(l)} x_s(p)$ is the aggregate source rate at link l. The synchronous algorithm there takes the following form. In each iteration, each

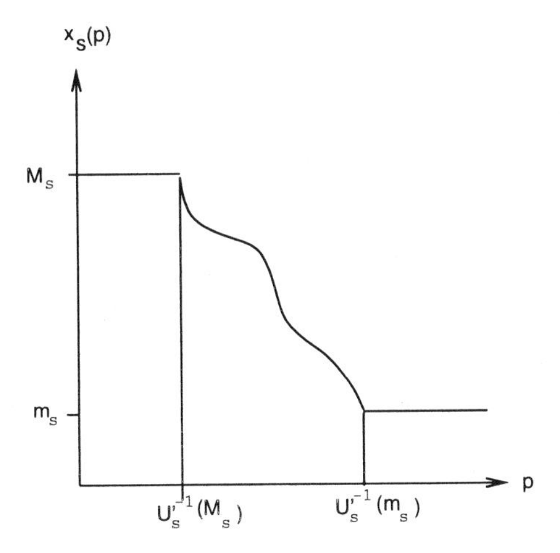

Figure 1 Source rate $x_s(p)$ as a function of (scalar) price p.

link l adjusts its price according to:

$$p_l(t+1) \quad = \quad [p_l(t) + \gamma(x^l(p(t)) - c_l)]^+ \tag{8}$$

where $[z]^+ = \max\{z, 0\}$. This is consistent with the law of supply and demand: if the demand $x^l(p(t))$ for bandwidth at link l exceeds the supply c_l, raise price $p_l(t)$; otherwise reduce price $p_l(t)$. Each source adjusts its rate to:

$$x_s(t+1) \quad = \quad [U_s'^{-1}(p^s(t))]_{m_s}^{M_s} \tag{9}$$

They then exchange their results: each link receives from all sources using that link their sources rates and each source receives a scalar price equal to the sum of the link prices along its path. The iteration is repeated until it converges.

We prove in [13; 14] that the algorithm generates a sequence that approaches the optimal rate allocation, provided the following conditions are satisfied:

C1: On the interval $I_s = [m_s, M_s]$, the utility functions U_s are increasing, strictly concave, and twice continuously differentiable.

C2: The curvatures of U_s are bounded away from zero on I_s: $-U_s''(x_s) \geq 1/\overline{\alpha}_s > 0$ for all $x_s \in I_s$.

These conditions imply that the dual objective function is Lipschitz which guarantees the convergence of gradient projection algorithms. Define $\overline{L} := \max_{s \in S} |L(s)|$, $\overline{S} := \max_{l \in L} |S(l)|$, and $\overline{\alpha} := \max\{\overline{\alpha}_s,\ s \in S\}$. In words $\overline{L}$ is the length of a longest path used by the sources, $\overline{S}$ is the number of sources sharing a most congested link, and $\overline{\alpha}$ is the upper bound on all $-U_s''(x_s)$.

Theorem 1 *Suppose assumptions C1–C2 hold. Provided that the stepsize* γ *satisfies* $0 < \gamma < 2/\overline{\alpha}\overline{LS}$, *starting from any initial rates* $m \leq x(0) \leq M$ *and prices* $p(0) \geq 0$, *every accumulation point* (x^*, p^*) *of the sequence* $(x(t), p(t))$ *generated by algorithm (8–9) are primal–dual optimal. That is,* x^* *gives the source rates that maximize total utility and* p^* *the shadow bandwidth prices.*

4. ASYNCHRONOUS DISTRIBUTED ALGORITHM

The synchronous model of the last section assumes that updates at the sources and the links are synchronized to occur at times $t = 1, 2, \ldots$. In this section we will extend the model to an asynchronous setting which better resembles the reality of large networks. In such networks sources may be located at different distances from the network links. Network state (prices in our case) may be probed by different sources at different rates, e.g., the Resource Management (RM) cells in an ATM networks are sent at different rates by different sources. Feedbacks may reach different sources after different, and variable, delays. These complications make our distributed computation system consisting of links and sources asynchronous. In such a system some processors may compute faster and execute more iterations than others, some processors may communicate more frequently than others, and the communication delays may be substantial and unpredictable.

Let $T_l^1 \subseteq \{1, 2, \ldots, \}$ be a set of times at which link l adjusts its price based on its current knowledge of source rates and $T_s^2 \subseteq \{1, 2, \ldots, \}$ a set of times at which source s updates its rate. The asynchronous algorithm is similar to the synchronous algorithm, except that computations and communications by sources and links are not coordinated and the computations are carried out using possibly outdated information.

Algorithm: Asynchronous Gradient Projection

Link l's algorithm:

1. *From time to time link l receives source rates from sources that go through link l. Link l replaces the oldest rates in its local memory with the newly received rates.*

2. *At each update time* $t \in T_l^1$, *link l computes an estimate* $\lambda_l(t)$ *of* $\partial D/\partial p_l(p(t))$ *(see (11–14) below) and adjusts its price according to*

$$p_l(t+1) \quad = \quad [p_l(t) - \gamma\lambda_l(t)]_0^{\bar{p}_l}$$

where $\bar{p}_l$ *is a large constant satisfying* $\bar{p}_l > \max_{s \in S(l)} U_s'(m_s)$ *and* $[a]_0^{\bar{a}} = \min\{\max\{0, a\}, \bar{a}\}$. *At times* $t \notin T_l^1$, $p_l(t+1) = p_l(t)$.

3. *From time to time link l communicates the current price to sources that go through link l.*

Source s's algorithm:

1. *From time to time source s receives bandwidth prices fedback from links in its path. Source s replaces the oldest prices in its local memory with the newly received ones.*

2. *At each update time $t \in T_s$ source s chooses a new rate based on its current estimate $\hat{p}^s(t)$ of prices (see (16–18) below):*

$$x_s(t+1) = x_s(\hat{p}^s(t))$$

It then transmits at this rate until the next update, i.e., $x_s(t+1) = x_s(t)$ for $t \notin T_s$.

3. *From time to time source s communicates the current source rate to links in its path.*

We now describe more precisely the update steps in the above algorithm. Link l updates its price at times $t \in T_l^1$ according to

$$p_l(t+1) = [p_l(t) - \gamma \lambda_l(t)]_0^{\bar{p}_l} \tag{10}$$

where (cf. (7))

$$\lambda_l(t) = c_l - \hat{x}^l(t) \tag{11}$$

$$\hat{x}^l(t) = \sum_{s \in S(l)} \hat{x}_{ls}(t) \tag{12}$$

$$\hat{x}_{ls}(t) = \sum_{t'=t-t_0}^{t} a_{ls}(t', t)\, x_s(t'), \quad s \in S(l) \tag{13}$$

with

$$\sum_{t'=t-t_0}^{t} a_{ls}(t', t) = 1, \quad \forall t, \forall l, s \text{ with } s \in S(l). \tag{14}$$

In (10) the projection onto the range $[0, \bar{p}_l]$, instead of $[0, \infty)$, can be motivated by the fact that in practice a price must be represented by a finite number of bits. Moreover, since $\bar{p}_l > \max_{s \in S(l)} U_s'(m_s)$, the projection imposes no restriction on the source rates. In (11–12), $\hat{x}^l(t) = \sum_{s \in S(l)} \hat{x}_{ls}(t)$ is the aggregate estimated source rates. Note that the estimate $\hat{x}_{ls}(t)$ depends on (l, s, t) and can be different for different link–source pairs and at different times. It is obtained by 'averaging' (convex sum) over the past source rates (see (13–14)). This model is very general and includes in particular the following two popular types of policies:

- **Latest data only:** only the last received rate $x_s(\tau)$, for some (possibly *unknown*) $\tau \in \{t - t_0, ..., t\}$, is used to estimate $\hat{x}_{ls}(t)$, i.e., $a_{ls}(t', t) = 1$ if $t' = \tau$ and 0 otherwise.

- **Latest average:** only the average over the latest k received rates is used in the estimate $\hat{x}_{ls}(t)$, i.e., $a_{ls}(t', t) > 0$ for $t' = \tau - k + 1, \ldots, \tau$ and 0 otherwise, for some (possibly *unknown*) $\tau \in \{t - t_0, ..., t\}$.

The interpretation in both cases is that rates $x_s(t')$ for $t' > \tau$ have not been received at link l by time t, and rates $x_s(t')$ for $t' < \tau$ or for $t' \leq \tau - k$ have been discarded.

In the following, we abuse notation and use $x_s(\cdot)$ both as a function of time, to denote source rate at time t under the algorithm, and as a function of price given by (6). The meaning should be clear from the context.

Source s updates its rate at times $t \in T_s^2$ according to

$$x_s(t) \quad = \quad x_s(\hat{p}^s(t)) \tag{15}$$

where $x_s(\cdot)$ is given by (6), and

$$\hat{p}^s(t) \quad = \quad \sum_{l \in L(s)} \hat{p}_{ls}(t) \tag{16}$$

$$\hat{p}_{ls}(t) \quad = \quad \sum_{t'=t-t_0}^{t} b_{ls}(t', t)\, p_l(t'), \qquad l \in L(s) \tag{17}$$

with

$$\sum_{t'=t-t_0}^{t} b_{ls}(t', t) \quad = \quad 1, \qquad \forall t, \forall l, s \text{ with } l \in L(s). \tag{18}$$

In (15–16) the source computation is the same as in the synchronous case except that it is based on its current estimate $\hat{p}^s(t)$ of link prices. As in the link algrithm the estimated link price $\hat{p}_{ls}(t)$ is obtained by 'averaging' over the past available prices (see (17–18)), and can depend on (l, s, t). Again the 'averaging' model is very general and include the policy of using only the last received price or the average over the last k prices; see above.

Note that (13) and (17) above tacitly assume that the one-way delay between any (l, s) pair is no more than t_0.

Our main result states that the difference between the various estimates and their true values converges to zero and that the algorithm yields the optimal rate allocation, provided the following additional assumption is satisfied:

C3: For all links l and sources s, the time between consecutive updates (i.e., the difference between consecutive elements of T_l^1 or T_s^2) is bounded.

Theorem 2 *The conclusion of Theorem 1 holds provided assumptions C1–C3 are satisfied and the stepsize γ is sufficiently small.*

5. CONCLUSION

We have presented an asychronnous model for optimization flow control proposed in [10], where the objective of the control is to maximize total user utilities. We have shown that both the synchronouns and the asynchronous algorithms converge under very mild conditions on the utility functions.

We close with extensions on the work presented in this paper. The basic algorithm presented here requires communication between links and sources. In [13] we prove that it is possible to eliminate the need for explicit communication from sources to links. In [12], we describe a marking scheme that simplifies the communication in the reverse direction to binary feedback. These simplifications combine to yield a variant of RED scheme, called REM (Early Random Marking), that is applicable to Interent using the proposed explicit congestion notification bit in IP header. The optimization framework in which REM is derived has two advantages. First, though it may not be possible, nor critical, that optimality is exactly attained in a real network, the optimization framework offers a means to explicitly steer the *entire* network towards a desirable operating point. Second it makes possible a systematic method to design and refine practical flow control schemes, which can be treated simply as implementations of a certain optimization algorithm, where modifications to the flow control mechanism is guided by modifications to the optimization algorithm. For instance, it is well known that Newton algorithm has much faster convergence than gradient projection algorithm. By replacing the gradient projection algorithm presented in [10; 14] by the Newton algorithm we derive in [1] a practical Newton–like flow control scheme that can be proved to maintain optimality, has the same communication requirement as the original scheme but enjoys a much better convergence property.

References

[1] Sanjeewa Athuraliya and Steven Low, "Newton–like algorithm for optimization flow control," Submitted for publication, 1999.

[2] Dimitri P. Bertsekas and John N. Tsitsiklis. *Parallel and distributed computation*. Prentice-Hall.

[3] Costas Courcoubetis, Vasilios A. Siris, and George D. Stamoulis. Integration of pricing and flow control for ABR services in ATM networks. *Proceedings of Globecom'96*, November 1996.

[4] S. Floyd and V. Jacobson. Random early detection gateways for congestion avoidance. *IEEE/ACM Trans. on Networking*, 1(4):397–413, August 1993.

[5] R. G. Gallager and S. J. Golestani. Flow control and routing algorithms for data networks. In *Proceedings of the 5th International Conf. Comp. Comm.*, pages 779–784, 1980.

[6] R. J. Gibbens and F. P. Kelly. Resource pricing and the evolution of congestion control. *Automatica*, 35, 1999.

[7] Jamal Golestani and Supratik Bhattacharyya. End-to-end congestion control for the Internet: A global optimization framework. In *Proceedings of International Conf. on Network Protocols (ICNP)*, October 1998.

[8] F. P. Kelly. Charging and rate control for elastic traffic. *European Transactions on Telecommunications*, 8:33–37, 1997.

[9] Frank P. Kelly, Aman Maulloo, and David Tan. Rate control for communication networks: Shadow prices, proportional fairness and stability. *Journal of Operations Research Society*, 49(3):237–252, March 1998.

[10] David E. Lapsley and Steven H. Low. An optimization approach to ABR control. In *Proceedings of the ICC*, June 1998.

[11] David E. Lapsley and Steven H. Low. An IP Implementation of Optimization Flow Control. In *Proceedings of the Globecom'98*, November 1998.

[12] David Lapsley and Steven Low, "Random early marking: An optimisation approach to internet congestion control," in *Proceedings of IEEE ICON '99*, September 1999.

[13] Steven H. Low. Optimization flow control with on-line measurement. In *Proceedings of the ITC*, volume 16, June 1999.

[14] Steven H. Low and David E. Lapsley. Optimization flow control, I: basic algorithm and convergence. IEEE/ACM Transactions on Networking, to appear 1999.

[15] David G. Luenberger. *Linear and Nonlinear Programming, 2nd Ed.* Addison-Wesley Publishing Company, 1984.

[16] John N. Tsitsiklis and Dimitri P. Bertsekas. Distributed asynchronous optimal routing in data networks. *IEEE Transactions on Automatic Control*, 31(4):325–332, April 1986.

PERFORMANCE ANALYSIS OF RATE-BASED FLOW CONTROL UNDER A VARIABLE NUMBER OF SOURCES

Y.-C. Lai[1], Y.-D. Lin[2]

[1]*National Cheng Kung University*
Dept. Computer Science and Information Engineering
No. 1, Ta Hsueh Road, Tainan, Taiwan
laiyc@locust.csie.ncku.edu.tw

[2]*National Chiao Tung University*
Dept. Computer and Information Science
No. 1001 Ta Hsueh Road, Hsinchu, Taiwan
ydlin@cis.nctu.edu.tw

Abstract Rate-based flow control plays an important role for efficient traffic management of ABR service in ATM networks. In this paper, a performance analysis of a rate-based flow control mechanism is presented. In our analytical model, the number of active sources is variable. A new source arrives when a connection is established, and an existing source departs when it has transmitted its data. Hence our model not only reflect the real scenes, but also correctly estimate the effect of the rate-based flow control.

Due to this variation, the analysis of the steady state is not enough. Therefore the analysis of transient cycles is also developed. Using the results of both analyses, we derive the equations of cell loss probability and utilization.

Keywords: ATM Networks, ABR(Available Bit Rate), Rate-based Flow Control

1. INTRODUCTION

ATM (Asynchronous transfer mode) is the most promising transfer technology for implementing B-ISDN. It supports applications with distinct QoS requirements such as delay, jitter, and cell loss and with distinct demands such as bandwidth and throughput. To provide these services for a wide variety of applications, in addition to CBR (Constant bit rate), rt-VBR (Real-time variable bit rate), nrt-VBR (Non-Real-Time VBR), and UBR (Unspecified bit Rate) services, the ATM forum defined a new service class known as ABR (Available bit rate) service to support data applications economically. Also an end-to-end adaptive control mechanism called closed-loop rate-based flow control is applied to this service. In this control scheme, the allowed cell transmission rate of each ABR connection is dynamically regulated by feedback information from the network [1-3]. If the network is congested, the source end decreases its cell transmission rate when it receives the congestion indication. Also, the source end increases its cell transmission rate when congestion is relieved. The rate-based control mechanism could efficiently control the connection flows and utilize the network bandwidth.

Recently several analyses and simulations have been conducted for rate-based control schemes. First Bolot and Shankar used differential equations to model the rate increase and decrease [4]. Yin and Hluchyj proposed analytical models for early versions of ABR control with a timer-based approach [5,6]. Ramamurthy and Ren developed a detailed analytical model to capture the behavior of a rate-based control scheme and obtain approximate solutions in closed forms [7]. Ohsaki et al. made an analysis and comparison between different switches in the steady state and initial transient state [8,9,10]. Ritter derived the closed form expression to quickly estimate the buffer requirements of different switches [11]. We derived the equations of cell loss probability and utilization for the rate-based control, and provided some rules to reduce cell loss probability and raise utilization [12]. These papers provide much insight into the effect of using the rate-based flow control. However, none of these papers consider the variable number of sources.

These researches under the assumption of a fixed number of sources are inaccurate. First, in real conditions, the number of sources is variable; a new connection may be established and an existing connection may be released. Second, using the rate-based control, the most obvious oscillations in buffer size happen at the time of a new source arrival or an existing source departure. Ignoring the fact of the variation about the number of sources does not correctly show the impact of rate-based control.

In this paper, we assume the number of sources to be variable to reflect the real conditions. An existing source may depart and a new source may arrive. We analyze the rate-based flow control under a variable number of sources, and deduce the equations of cell loss probability and utilization. In order to get the precise values of these parameters, besides the analysis for the steady state, the analysis for a transient state is developed. This is because the dynamic behavior is very different from the stable one when an existing source departs or a new source arrives.

2. ANALYTICAL MODEL

First we briefly introduce the basic operation of a close-loop rate-based control mechanism [3]. When a connection is established, the source end system (SES) sends the cells at the allowed cell rate (ACR) which is set as initial cell rate (ICR). In order to probe the congestion status of the network,

The SES sends a forward Resource Management (RM) cell every N_{RM} data cells. The destination end system (DES) returns the forward RM cell as a backward RM cell to the SES. Depending on the received backward RM cell, the SES adjusts its allowed cell rate, which is bounded between Peak cell rate (PCR) and Minimum cell rate (MCR).

The RM cell contains a 1-bit congestion indication (CI) which is set to zero, and an explicit rate (ER) field which is set to PCR initially by the SES. Depending on the different ways to indicate the congestion status, two types of switches are implemented. One is the Explicit Forward Congestion Indication (EFCI) switch, the other is the Explicit Rate (ER) switch. In the EFCI type, the switch in the congestion status sets the EFCI bit to one (EFCI=1) in the header of each passing data cells. The DES, if a cell with EFCI=1 has been received, marks the CI bit (CI=1) to indicate congestion in each backward RM cells. In the ER type, the switch sets the EFCI bit of the RM cells to indicate whether there is congestion or not, and sets the ER field to indicate the bandwidth the connection should use. The performance results and comparisons between the two types of switches are shown in [10] in detail.

When the SES receives a backward RM cell, it modifies its ACR using additive increase and multiplicative decrease. Depending on CI, the new ACR is computed as follows:

$$ACR = \max(\min(ACR + N_{RM} \cdot AIR, ER), MCR), \quad \text{if CI=0},$$

$$ACR = \max(\min(ACR \cdot (1 - \frac{N_{RM}}{RDF}), ER), MCR), \quad \text{if CI=1},$$

where AIR is the additive increase rate and RDF is the rate decrease factor. AIR and RDF are defined in the traffic management specification version 4.0 [3]. Although the new version of the specification has made some changes about AIR and RDF [13], the analysis for the new notations is easily translated from this paper.

In this paper, we focus on the EFCI switch and use a simple model as shown in Fig. 1. There are some ABR sources sharing a bottleneck link where the bandwidth is BW. We assume that these sources are homogeneous; that is, they all have the same parameters ICR, PCR, MCR, AIR and RDF. The number of sources is variable. Source arrival is according to a Poisson process and the duration of a source is according to an exponential process. Let λ be the source arrival rate and $1/\mu$ be the mean source duration. Thus, the distribution of the number of sources is the same as the number in the system for an M/M/∞ queue. Also we assume that each source always has cells to send, i.e. it has infinite backlog. This assumption allows us to investigate the performance of an EFCI switch in the most stressful situations.

The buffer size at the switch is denoted by Q_B. The switch determines the congestion condition according to its queue length. There are two values, high

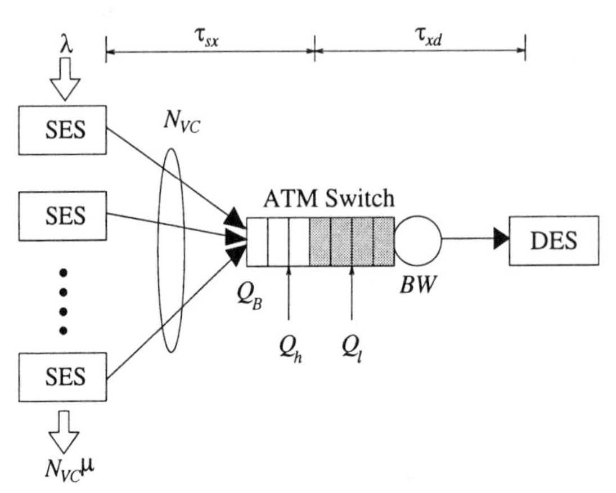

Figure 1 Analytic model for the rate-based flow control.

threshold Q_h and low threshold Q_l, which decide whether congestion occurs or not. When the queue length exceeds Q_h, the EFCI bit of passing data cells is set to one to indicate congestion. The congestion is relieved when the queue length drops below Q_l.

We define τ_{sx} as the propagation delay between the SES and the switch, and τ_{xd} as the propagation delay between the switch and the DES. Also, the feedback propagation delay from the switch to the SES is denoted by τ_{xds} and the round trip propagation delay is denoted by τ. Thus we get the relation $\tau_{xds} = \tau_{sx} + 2\tau_{xd}$ and $\tau = 2(\tau_{sx} + \tau_{xd})$. The propagation delay is a critical parameter of system performance.

3. TRANSIENT CYCLES

For lack of space, in this paper we skip the analysis of the case that the number of sources is fixed. However, those results are the base for the analysis of the case that the number of sources is variable. Hence the readers ought to reference the paper

[12] to know the meaning of the notations and the derivation of the analysis of steady state.

As we know, the cell loss probability and utilization during a transient cycle are determined by the time instant when the number of sources is changed. The worst case to cell loss probability happens as a new source arrives at the time instant when the high threshold is reached, $t_{Q_h}^-$. The reason is as follows. If a new source arrives after $t_{Q_h}^-$, this new source does not send its cells during the time interval between congestion detection and this new source arrival. Hence, cell loss probability is smaller. On the other hand, the congestion is detected earlier if a new source arrives before $t^-_{Q_h}$. Therefore, the maximum ACR is smaller, which causes cell loss probability to be smaller [11]. Thus, the closer the new source arrival to $t^-_{Q_h}$, the higher cell loss probability we got. Similarly the worst case of utilization is that an existing source departs at the time instant when the low threshold is attained, $t^-_{Q_l}$. In this section, we only consider these worst cases. Therefore, we shall obtain the upper bound of cell loss probability and lower bound of utilization during transient cycles.

3.1 SOURCE DEPARTURE

Without loss of generality, we assume that the number of active sources is i. When an existing source departs at the time instant $t^-_{Q_l}$, the speed of adjusting ACR is not changed immediately. After the switch has sent all the cells in the buffer, then the SES speeds up its adjustment of the ACR. That is, the rate of SES receiving the backward RM cells is changed from BW/iN_{RM} to $BW/(i-1)N_{RM}$ after the time $Q_l/BW + \tau$. Now we derive utilization at worst case.

The minimum rate, ACR_{min}, in a transient cycle is the same as in the steady state. Thus we get the time interval, $\Delta\tilde{t}^-_{Q_{min}}[i\text{-}1]$, during a transient cycle where the number of active sources changes from i to i-1, by solving the equation,

$$(ACR_{min}[i] + \frac{BW \cdot AIR}{i} \cdot \frac{Q_l}{BW})(i-1) + BW \cdot AIR \cdot (\Delta\tilde{t}^-_{Q_{min}}[i-1] - \frac{Q_l}{BW}) = BW.$$

The minimum queue length, $\tilde{Q}_{min}[i\text{-}1]$, during a transient cycle where the number of active sources changes from i to i-1, is given by

$$\tilde{Q}_{min}[i-1] = Q_l - \int_{\Delta t_{Q_l}}^{\Delta t_{Q_l}+\tau} (BW - (i-1) \cdot ACR_{t_0}(t))dt$$

$$- \int_0^{\frac{Q_l}{BW}} (BW - (i-1)(ACR_{min}[i] + \frac{BW \cdot AIR}{i} t))dt$$

$$- \int_0^{(\Delta\tilde{t}^-_{Q_{min}}[i-1] - \frac{Q_l}{BW})} (BW - (i-1)(ACR_{min}[i] + \frac{Q_l AIR}{i}) - BW \cdot AIR \cdot t)dt.$$

Then we get

$$\tilde{Q}_{min}[i-1] = Q_l - \tau \cdot BW + (i-1) \cdot RDF \cdot e^{-\frac{BW}{i \cdot RDF} \cdot \Delta t_{Q_l}} \cdot (1 - e^{-\frac{BW}{i \cdot RDF} \cdot \tau})$$

$$- \frac{(2BW - (i-1)(2ACR_{min}[i] + \frac{Q_l AIR}{i}))\frac{Q_l}{BW}}{2} - \frac{(BW - (i-1)(ACR_{min}[i] + \frac{Q_l AIR}{i}))^2}{2 \cdot BW \cdot AIR}. \tag{1}$$

After the queue length reaches the minimum, the system enters the steady state where N_{VC} is equal to i-1. Hence cell loss probability and utilization during this transient cycle are calculated as

$$\tilde{\rho}_1[i-1] = 1 - \frac{\tilde{N}_{waste1}[i-1]}{\tilde{T}_{cycle1}[i-1] \cdot BW}, \quad \text{and}$$

$$\tilde{P}_{loss1}[i-1] = \frac{N_{loss}[i-1]}{\tilde{T}_{cycle1}[i-1] \cdot \tilde{\rho}_1[i-1] \cdot BW + N_{loss}[i-1]}. \tag{2}$$

where

$$\tilde{N}_{waste1}[i-1] = \max(0, 0 - \tilde{Q}_{min}[i-1]),$$

$$\tilde{T}_{cycle1}[i-1] = 2\tau + \Delta t_{Q_l}[i-1] + \Delta \tilde{t}_{Q_{min}}^{-}[i-1] + \Delta t_{Q_h}[i-1] + \Delta t_{Q_{max}}^{-}[i-1].$$

3.2 SOURCE ARRIVAL

In the steady state or the transient cycle of a source departure, because the ACR of each source is the same, the speed of adjusting the ACR is the same among all sources. Therefore, we do not need to take care the scenarios that different sources bring. In contrast, in the transient phase of a new source arrival, the ACR of the new source is different from the existing sources. Then, for this new source, the speed of adjusting the ACR is also different from the existing sources. It depends on a '*rate ratio*' a queueing delay before, i.e., right before the RM cell joins the switch queue. The rate ratio, R, is the ratio between the ACR of the new source and the ACR of the existing sources.

Since the rate ratio and queueing delay are changed continuously, analyzing the speed of adjusting the ACR during this transient cycle is very difficult. We introduce the concept of '*average interval*' to alleviate this difficulty. The forepart of this transient cycle is divided into some average intervals, whose lengths are not fixed, as shown in Fig 2. In the nth average interval ΔI_n, the switch sends all the cells which have been sent by the sources in the previous average interval ΔI_{n-1}. Also the variable behavior of rate ratio is approximated by a constant, which is the ratio between the average ACR of the new source and the average ACR of the existing sources during this average interval. Hence the speed of adjusting ACR in the ΔI_n completely depends on the rate ratio, R_{n-1}, in the ΔI_{n-1}. The approximation works well when the length of average intervals is not large.

The evolution of the number of active sources.

When a new source arrives at the time instant $\Delta t_{Q_h}^{-}$, the SES sends its cells at the rate ICR. The new source keeps this rate until the first RM cell returns. Let the first average interval begin at the time instant $t_{Q_h}^{-}$, and end at the time instant when the first RM cell of this new source returns. Hence the length of

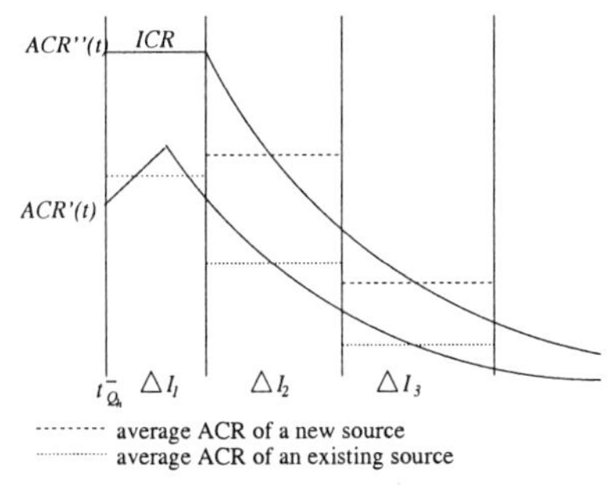

Figure 2 Transient cycle of a new source arrival.

the first average interval is

$$\Delta I_1 = \frac{Q_h}{BW} + \tau.$$

For simplicity of presentation, let $ACR''_j(t)$ and $ACR'_j(t)$ be the dynamic behavior of the new source and of those existing sources in the jth average interval, respectively. Note that t is the escaped time from the beginning time of the average interval. Now we get the dynamic behavior of the new source and the existing sources in the first average interval as

$$ACR''_1(t) = ICR, \quad 0 \le t < \Delta I_1,$$

$$ACR'_1(t) = \frac{BW}{i}(1 + AIR(\Delta t_{Q_h} + t)), \quad 0 \le t < \tau,$$

$$ACR'_1(t) = \frac{BW}{i}(1 + AIR(\Delta t_{Q_h} + \tau))e^{-\frac{BW}{iRDF}(t-\tau)}, \quad \tau \le t < \Delta I_1.$$

The rate ratio R_1 is given by

$$R_1 = \frac{\int_0^{\Delta I_1} ACR''_1(t)}{\int_0^{\Delta I_1} ACR'_1(t).}$$

The system is in phase II or IV after the first average interval, so the dynamic behavior is as

$$ACR''_n(t) = ACR''_n(0)e^{-\frac{BW \cdot R_{n-1}}{(i+R_{n-1})RDF}t}, \quad 0 \le t < \Delta I_n,$$

$$ACR'_n(t) = ACR'_n(0)e^{-\frac{BW}{(i+R_{n-1})RDF}t}, \quad 0 \le t < \Delta I_n.$$

where $ACR''_n(0)$ and $ACR'_n(0)$ equal to $ACR''_{n-1}(\Delta I_{n-1})$ and $ACR'_{n-1}(\Delta I_{n-1})$, respectively. The length of the second and nth average intervals is

$$\Delta I_2 = \frac{\int_0^{\Delta I_1}(iACR'_1(t) + ACR''_1(t))dt}{BW} - \tau,$$

$$\Delta I_n = \frac{\int_0^{\Delta I_{n-1}} (iACR'_{n-1}(t) + ACR''_{n-1}(t))dt}{BW}, \quad n > 2.$$

The rate ratio of the nth average interval is

$$R_n = \frac{\int_0^{\Delta I_n} ACR''_n(t)dt}{\int_0^{\Delta I_n} ACR'_n(t)dt}.$$

Without loss of generality, we assume that the queue length reaches maximum at the mth average interval.

$$iACR'_m(\tilde{t}[i+1]) + ACR''_m(\tilde{t}[i+1]) = BW.$$

Thus we get

$$\tilde{Q}_{max}[i+1] = (\Delta I_m - \tilde{t}[i+1])BW + \int_0^{\tilde{t}[i+1]} (iACR'_m(t) + ACR''_m(t))dt.$$

We assume that the system enters the steady state after the time instant when the queue length reaches the maximum. Hence cell loss probability and utilization during this transient cycle is calculated as

$$\tilde{\rho}_2[i+1] = 1 - \frac{N_{waste}[i+1]}{\tilde{T}_{cycle2}[i+1] \cdot BW},$$

$$\tilde{P}_{loss2}[i+1] = \frac{\tilde{N}_{loss2}[i+1]}{\tilde{T}_{cycle2}[i+1] \cdot \tilde{\rho}_2[i+1] \cdot BW + \tilde{N}_{loss2}[i+1]}, \tag{3}$$

where

$$\tilde{N}_{loss2}[i+1] = \max(0, \tilde{Q}_{max}[i+1] - Q_B),$$

$$\tilde{T}_{cycle2}[i+1] = 2\tau + \Delta t_{Q_h}[i+1] + \sum_{j=1}^{m-1} \Delta I_j[i+1] + \tilde{t}[i+1] + \Delta t_{Q_l}[i+1] + \Delta t_{\bar{Q}_{min}}[i+1].$$

4. COMBINATION OF STEADY STATE AND TRANSIENT CYCLES

Now we derive the equations of cell loss probability and utilization. Remind that source arrival is according to a Poisson process with parameter λ, and the source existence duration is according to an exponential process with parameter $1/\mu$. The distribution of the number of sources is the same as the number in the system for an M/M/∞ queue. Hence we get

$$P_i = \frac{(\frac{\lambda}{\mu})^i}{i!} e^{-\frac{\lambda}{\mu}},$$

$$T_i = \frac{1}{\lambda + i\mu},$$

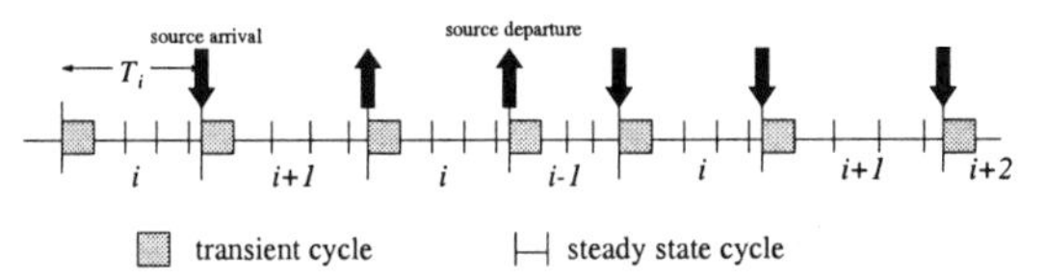

Figure 3 The evolution of the number of active sources.

where P_i is the probability that i sources are active in the system, and T_i is the mean duration when i sources are active.

Because the event that a new source arrives or an existing source departs happen not very often, λ and μ are small. Therefore many cycles may pass by between two events. There are few opportunities that an event happens during a transient cycle. So the behavior of system looks like Fig. 3. When there are $i+1$ sources in the system and an existing source departs, the system passes the time interval of a transient cycle, and then enters the steady state. Hence the time of the system staying in the transient cycle is $\tilde{T}_{cycle1}[i]$, and the time of the system staying in the steady state is T_i-$\tilde{T}_{cycle1}[i]$. We use the probability method to obtain cell loss probability and utilization as

$$P_{loss1}[i] = \tilde{P}_{loss1}[i]\frac{\tilde{T}_{cycle1}[i]}{T_i} + \bar{P}_{loss}[i]\frac{T_i - \tilde{T}_{cycle1}[i]}{T_i},$$

$$\rho_1[i] = \tilde{\rho}_1[i]\frac{\tilde{T}_{cycle1}[i]}{T_i} + \bar{\rho}[i]\frac{T_i - \tilde{T}_{cycle1}[i]}{T_i}. \tag{4}$$

When there are $i-1$ sources in the system and a new source arrives, similarly we get $P_{loss2}[i]$ and $\rho_2[i]$.

As described above, there are two ways that the system enters the condition that i sources are active. The departure rate of an existing source when there are $i+1$ sources is $P_{i+1}(i+1)\mu$. On the other hand, the arrival rate of a new source when there are $i-1$ sources is $P_{i-1}\lambda$. Hence the probability of first way is $\frac{P_{i+1}(i+1)\mu}{P_{i+1}(i+1)\mu+P_{i-1}\lambda}$, and the probability of second way is $\frac{P_{i-1}\lambda}{P_{i+1}(i+1)\mu+P_{i-1}\lambda}$. Also according to the local-balance equation of the queueing theory, the rate $P_{i+1}(i+1)\mu$ is equal to $P_i\lambda$, and the rate $P_{i-1}\lambda$ is equal to $P_i i\mu$. Therefore the probabilities of first and second ways are $\frac{\lambda}{\lambda+i\mu}$ and $\frac{i\mu}{\lambda+i\mu}$, respectively.

Finally we obtain cell loss probability and utilization as

$$P_{loss} = \sum_{i=0}^{\infty} P_i(\frac{\lambda}{\lambda+i\mu}P_{loss1}[i] + \frac{i\mu}{\lambda+i\mu}P_{loss2}[i]),$$

$$\rho = \sum_{i=0}^{\infty} P_i(\frac{\lambda}{\lambda+i\mu}\rho_1[i] + \frac{i\mu}{\lambda+i\mu}\rho_2[i]). \tag{5}$$

Although the summation is infinite, we have to limit the number of sources. We can set a reasonable bound for N_{VC} so that boundary states have probability very close to 0.

5. CONCLUSION

When the variation of the number of sources is neglected, the derived cell loss probability and utilization do not correctly reflect the effect of the rate-based flow control. Meanwhile, although the variation in the number of sources is considered, the results of ignoring the transient cycles are still unsatisfactory. In this paper, an accurate analysis for rate-based flow control under a variable number of sources is provided.

References

[1] B. Flavio and W. F. Kerry, *The Rate-Based Flow Control Framework for the Available Bit Rate ATM service*, IEEE Network Magazine, pp. 25-39, March/April 1995.

[2] P. Newman, *Backward Explicit Congestion Notification for ATM Local Area Networks*, Proc. IEEE GLOBECOM, vol. 2, pp. 719-723, Houston, TX, Dec. 1993.

[3] ATM Forum Technical Committee TMWG, *ATM Forum Traffic Management Specification Version 4.0*, ATM Forum/95-0013R8, Oct. 1995.

[4] J. C. Bolot and A. U. Shankar, *Dynamical Behavior of Rate-Based Flow Control Mechanisms*, Computer Communication Review, pp. 25-39, April 1990.

[5] N. Yin and M. G. Hluchyj, *On Close-Loop Rate Control for ATM Cell Relay Networks*, IEEE INFOCOM'94, pp. 99-108, Toronto, 1994.

[6] N. Yin, *Analysis of a Rate-Based Traffic Management Mechanism f or ABR Service*, IEEE GLOBECOM'95, pp. 1076-1082, Singapore, 1995.

[7] G. Ramamurthy and Q. Ren, *Analysis of the Adaptive Rate Control for ABR Service in ATM Networks*, IEEE GLOBECOM'95, pp. 1083-1088, Singapore, 1995.

[8] H. Ohsaki, M. Murata, H. Suzuki, C. Ikeda and H. Miyahara, *Analysis of Rate-Based Congestion Control Algorithms for ATM Networks, part 1: Steady State Analysis*, IEEE GLOBECOM'95, pp. 296-303, Singapore, 1995.

[9] H. Ohsaki, M. Murata, H. Suzuki, C. Ikeda and H. Miyahara, *Analysis of Rate-Based Congestion Control Algorithms for ATM Networks, part 2: Initial Transient State Analysis*, IEEE GLOBECOM'95, pp. 1095-1101, Singapore, 1995.

[10] H. Ohsaki, M. Murata, H. Suzuki, C. Ikeda and H. Miyahara, *Rate-Based Congestion Control for ATM Networks*, ACM SIGCOMM Computer Communication Review, pp. 60-72, April 1995.

[11] M. Ritter, *Network Buffer Requirement of the Rate-Based Control Mechanism for ABR Services*, IEEE INFOCOM'96, pp. 1190-1197, San Francisco, California, 1996.

[12] Y. C. Lai and Y. D. Lin, *Performance Analysis of Rate-Based Congestion Control and Choice of High and Low Thresholds*, IEEE ICCCN'97, pp. 70-75, Las Vegas, Nevada, 1997.

[13] ATM Forum Technical Committee TMWG, *ATM Forum Traffic Management Specification Version 4.0*, ATM Forum/96-tm-0056.000, Apr. 1996.

SESSION 13

Routing

MULTICAST ROUTING IN ATM NETWORKS

Chi-Chung Cheung[1], Hon-Wai Chu[2], Danny H.K. Tsang[3] and Sanjay Gupta[4]

[1] *Department of Computer Science*
City University of Hong Kong
csccc@cityu.edu.hk

[2] *School of Science and Technology*
Open University of Hong Kong
wchu@ouhk.edu.hk

[3] *Department of Electrical and Electronic Engineering*
Hong Kong University of Science and Technology
eetsang@ee.ust.hk

[4] *GSM Products Division*
Motorola, USA
guptasn@cig.mot.com

Abstract In this paper, we study the problem of multicast routing in ATM based networks. We propose a dynamic multicast algorithm in ATM networks called Least Load Multicast Routing (LLMR) algorithm and the algorithm with statistical multiplexing is considered. A numerical example is presented to show the characteristics of LLMR. Moreover, we find that the connection blocking probabilities decrease with increasing destination size.

Keywords: ATM Networks, Multicast Routing.

1. INTRODUCTION

Multicasting refers to the ability of a set of more than two nodes or end-users in a communication network to communicate simultaneously with each other. Applications that require multicast capability (either point-to-multipoint (PTM) as in distributional video or multipoint-to-multipoint (MTM) as in video conferencing, online collaboration and others) will be an integral part of future broadband services. *Asynchronous Transfer Mode* (ATM) has been almost universally accepted as the multiplexing and switching technique for *Broadband Integrated Services Digital Networks* (B-ISDN). ATM — a *high-speed, Virtual Circuit (VC) oriented packet-switching technique* that uses fixed length packets called *cells*— provides an acceptable and cost effective means for meeting the requirements of future broadband networks [8].

Given the popularity of ATM based broadband networks *we study the problem of multicast routing in ATM based networks.* Recent research efforts in the area of multicast routing (few of them have been conducted keeping in perspective the requirements of ATM networks) have followed the following broad approach: the underlying network is modeled as a graph (directed or undirected) with links as edges and switches as nodes; and each edge is assigned a "cost" (for instance the cost could be a measure of expected delay). The multicast routing problem is then reduced to finding a "tree" that spans the nodes that wish to participate in a given multicast connection that has minimal cost [1][4][7][10][11][12]. The above problem is known in the literature as the *Steiner tree* problem (an NP-complete [9] problem) for which several heuristics have been proposed. Recently, [3] proposed dynamic algorithms to handle the multicast routing problem in circuit switching networks and the bandwidth of each connection is peak rate allocated. In this paper, we proposed a dynamic multicast algorithms in ATM networks called Least Load Multicast Routing (LLMR) algorithm.

In this paper we make a distinction between ATM networks operating in the *statistical* and *non-statistical multiplexing modes.* An ATM network operates in the statistical multiplexing mode if, for any VP, the VCs routed over the VP are permitted to have an aggregate offered bit rate that exceeds the bandwidth allocation of the VP. During these excess periods, the buffer before the first link of the VP will accumulate with cells and perhaps overflow. The ATM network is operated in the non-statistical multiplexing mode if, for each VP, the aggregate instantaneous bit rate offered by the VCs does not exceed the VP's bandwidth allocation[1].

In ATM based networks there is also a distinction between *point-to-multipoint* (PTM) and multipoint-to-multipoint (MTM) VCs. In a PTM VC, there exists a single source (henceforth referred to as *root*) and a set of destination nodes, and information is sent from the source to the destinations. We would like to emphasize that no information is sent from the destinations to the source or other destinations. On the other hand in an MTM VC all participating nodes communicate with each other.

The paper is organized as follows. We describe the relevant features of ATM networks and their architecture briefly in Section 2. In Section 3 we propose multicast routing policies for various classes of ATM networks. In Section 4 we present the simulation results and compare it with other dynamic multicast routing algorithms. Finally, we conclude in Section 5 by summarizing the paper and outlining avenues for future research.

2. MULTICAST ROUTING POLICIES

Consider the VP network to be a graph $G = (V, E)$ where V and E are the set of nodes and VPs respectively. Denote by $C(e), e \in E$, the bandwidth allocated to VP e. The routing policies that we propose in the paper is employed for routing PTM VC connection requests and thus the underlying graph is assumed to be directed. For routing MTM VC connection requests, since ATM supports PTM VC connections with asymmetric bandwidth, we can employ multi PTM VC connections to support MTM VC connection requests. Each node in a group that wishes to communicate can establish a PTM VC connection to all of the other nodes in the group. For a group of N nodes, this requires N PTM VC connections.

We consider the case where statistical multiplexing is performed within VP subnetworks. For reasons of notational simplicity, we consider a case of homogeneous VCs, (i.e., all VCs have identical traffic characteristics and QOS requirements). In order to simplify the discussion, we assume that the QOS requirement only involves cell loss. Specifically, for each VC in progress, suppose that the fraction of cells lost are not permitted to exceed ϵ (typical values of ϵ are between 10^{-6} and 10^{-9}). Let $p_e(l)$ be the fraction of cells lost at the input buffer of VP e when l VCs are being routed through VP e, these probabilities can be determined by a cell-level analysis; for example, see [2][5][13].

Consider an VC request with destination set $S(c)$ and a *connected, acyclic* graph (tree) $T = (V(c), E(c))$, $S(c) \subseteq V(c)$. Nodes $s \in V(c) - S(c)$ are referred to as *alternate* nodes. A VP, $e \in E(c)$, is referred to as an alternate VP if it connects to an alternate node. Let $t(e)$ denote the reservation threshold for VP e, i.e., VP e can be used as an alternate VP only if its residual capacity exceeds $t(e)$. Let $\pi^T(s_1, s_2), s_1, s_2 \in V(c)$ be the set of VPs in the path between the nodes s_1, s_2. The tree T can be used for routing VC request c, if and only if

(a) the QOS requirement of the VC being routed is satisfied, i.e.,

$$\prod_{e \in \pi^T(s_1, s_2)} [1 - p_e(l_e + 1)] > 1 - \epsilon \text{ for all } s_1, s_2 \in S(c)$$

(we have implicitly assumed that the cell loss probabilities on distinct VPs are independent) and

(b) the QOS requirement of any pre-existing VC is not violated after addition of VC c on tree T. (At the very least, all existing VCs that are routed through VP $e, e \in E(c)$ will have to be (re)examined.)

Checking for condition (b) above is computationally prohibitive, even for small networks, as it involves examining VPs that are NOT in $V(c)$. It can be demonstrated that for a large fraction of VC requests, before a multicast VC can be assigned a routing tree, the status of all the VPs in the VP subnetwork will be required. (For a detailed discussion refer to [6].) We would like to remind that QOS requirements often involve bounds on delay as well. In this case step (a) and step (b) will involve convolution of random variables – a computationally intensive exercise. Further, the overhead, due to the increased information exchange required, on the signaling network, increases substantially. In light of the above discussion, it is easy to see that determining the acceptability of an VC request on any given tree places significant stress on the network management and control protocols and is therefore unacceptable.

In order to reduce the computational requirements placed on the switches and the burden on network management and control functions, it is critical that a VP be able to decide if it can be used as a part of the routing tree without requesting any information from other VPs in the VP subnetwork or the tree. Therefore we propose the following (more restrictive) admission rules for determining if a VC request c can be routed on a tree $T = (V(c), E(c)), S(c) \subseteq V(c)$:

Revised Admission Rules

Rule 1 $p_e(l_e + 1) < \epsilon_e^*$, $\epsilon_e^* \leq \epsilon$ for all $e \in E(c)$, have been chosen a priori; and

Rule 2 $\prod_{e \in \pi^T(s_1, s_2)} [1 - \epsilon_e^*] > 1 - \epsilon$ for all $s_1, s_2 \in S(c)$.

When operating under the scope of the revised admission rules, the admissibility of a VC on any routing tree can be determined after considering the state of VPs that are a part of the routing tree only, thus reducing the complexity of the admissibility criterion. However, the above simplicity in admissibility criterion is achieved at the expense of denying VC requests when there exist feasible trees (possibly minimal) to route them.

The revised admission rules transform (for purposes of routing VCs only) a VP subnetwork with statistical multiplexing to an equivalent VP subnetwork with no statistical multiplexing. Note that in this case the capacity of VP e is defined as

$$C(e) := \max\{l : p_e(l_e) \leq \epsilon^*\}.$$

The revised admission rules require that in networks with statistical multiplexing the diameter of the routing trees not exceed ϵ/ϵ^*. Complexity of algorithms that construct maximum residual capacity trees with constraints on their maximum diameter are substantial. Therefore, we propose that routing policies that limit the maximum number of hops from a given node to others in the destination (henceforth referred to as the *distance* between destination nodes) set be employed. For instance, if the distance from the source node to any other destination node can not exceed h^*, the maximum diameter of the tree will certainly be less than $2h^*$. Hence, for MTM VCs we propose using h^* — the maximum distance between the source and any other destination node — to be $\frac{\epsilon}{2\epsilon^*}$; however, for PTM VCs where data flows from one node to the remaining nodes in the destination set choosing $h^* = \frac{\epsilon}{\epsilon^*}$ is clearly sufficient. We now propose the following routing policy for ATM networks with statistical multiplexing for a VC request with destination set $V(d)$, source node s_r and a maximum distance of h^* from the source node to any other destination node:

Least Loaded Multicast Routing (LLMR)

1. *Initialization:* Let $T_1 = (V_1 + V_t, E_1)$ where $V_1 = s_r, V_t = \phi$, and $E_1 = \phi$. Set $i = 1$ and proceed to Step 2 (VP selection).

2. *VP Selection:* If $S(c) \subseteq V_i + V_t$ the route selection process is complete and the tree $T = (V_i + V_t, E_i)$ is used to route the VC request. If $i < d(c)$ choose VP $e^* = (s_1^*, s_2^*), s_1^* \in V_i + V_t, s_2^* \in S(c) - V_i - V_t$, such that $R(e^*) \geq R(e)$ where $e = (s_1, s_2)$ is a VP such that $s_1 \in V_i + V_t, s_2 \in S(c) - V_i - V_t$, (any ties are broken randomly). If $R(e^*) \geq b(c)$ and the distance between s_r and s_2^* is $\leq h^*$ then (i) $E_{i+1} = E_i + e^*$ and $V_{i+1} = V_i + s_2^*$ and (ii) set $i = i + 1$ and Step 2 is repeated (VP selection); Otherwise search for an alternate node is initiated as in Step 3.

3. *Choose an alternate node*: If $V_t \neq \phi$ the VC request is denied; *otherwise* (if $V_t = \phi$) define an alternate path, ψ, as the set of VPs $e_1 = (s_1, s_3), e_2 = (s_3, s_2)$ such that (i) $s_1 \in V_i, s_2 \in S(c) - V_i$,, (ii) $s_3 \in V - S(c)$, and (iii) distance between s_r and s_2 is $\leq h^*$. Let $R(\psi) = \min\{R(e_1) - t(e_1), R(e_2) - t(e_2)\}$ be the residual capacity of the alternate path ψ. Denote by ψ^* the alternate path with maximum residual capacity. If $R(\psi^*) \geq b(c)$ then (i) $V_t = \{s_3^*\}$; (ii) $V_{i+1} = V_i + s_2^*$; and $E_{i+1} = E_i + e_1^* + e_2^*$; (ii) set $i = i + 1$ and (iii) proceed to Step 2 (VP selection). If $R(\psi^*) < b(c)$, the *VC request is denied.*

Observe that in LLMR a minimal tree with the largest residual capacity is chosen for routing the VC. In case no minimal tree are available, from the set of trees with no more than one alternate nodes the tree having the largest residual capacity is chosen. In case no such tree exists with the required bandwidth available on each component VP, the VC request is denied. Note that we only allow at most one alternate node added to establish any given VC request. It is because there is only marginal improvement in performance when we use more than one alternate node (further if the load on the network is sufficiently high the performance of the system degrades by using increasing number of alternate nodes). We will show this fact through the simulation result in Section 3.

To illustrate the routing algorithm outlined above consider a PTM VC request with the destination set $\{2, 3, 4\}$ and source node 1. The diameter of the routing tree is not allowed to exceed 3. The steps in the connection establishment procedure are as shown in Figure 1 — the numbers next to a VP denote its residual capacity and VPs that are included in the set $E(c)$ are drawn with "thick" lines. alternate paths and alternate nodes are not shown for clarity. Note that node 3 is the first node to be added as the VP $(1, 3)$ has the largest residual capacity (3) among VPs, $(1, 4)$ and $(1, 2)$, that are connected to node 1 – the root node. Subsequently, nodes 2 and 3 (in this order) are added to the routing tree and the VPs included are $(3, 2)$ and $(2, 4)$, respectively.

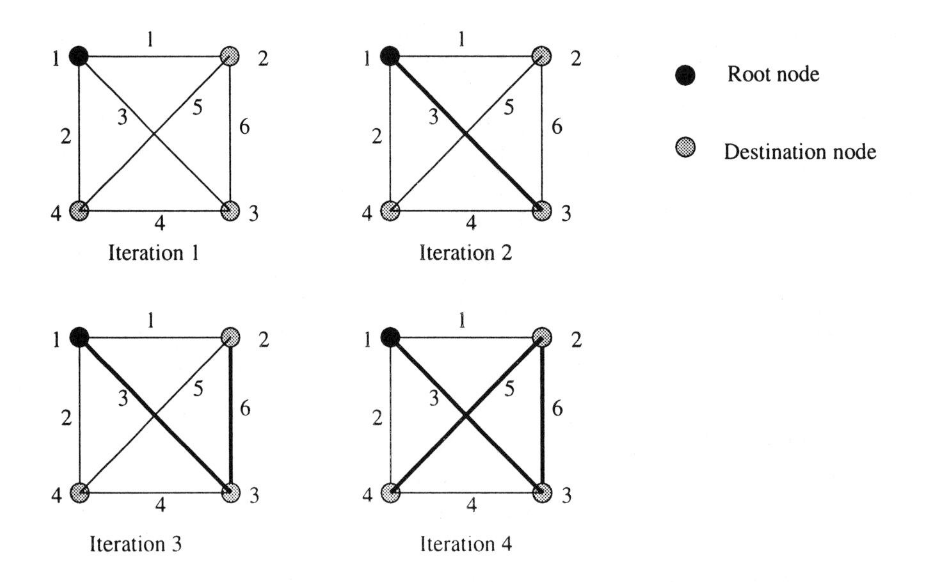

Figure 1 Example to illustrate LLMR without alternate node

Figures 2 and 3 consider a 6-node network and a VC request with destination set $\{1, 3, 5\}$ with node 6 as the source node. In figure 2, ϵ^* is chosen such that the diameter of the routing tree is not allowed to exceed 4. Node 3 is the first one to be added as VP $(6, 3)$ has a residual capacity of 6 that is larger than the residual capacities of the VPs $(6, 1)$ and $(6, 5)$ which is followed by node 5 (VP $(6, 5)$ is chosen due to its larger residual capacity over VP $(3, 5)$ which has a residual capacity of 0). Search for a VP to node 1 from the nodes 3,5, and 6 reveals that there are no available VPs with non-zero residual capacities and therefore the search for an alternate node is initiated. For the case at hand, node 2 is selected as the alternate node (the alternate path chosen is shown by dashed lines). In figure 3, the diameter of the routing tree is limited to be not greater than 3. The connection establishment procedures of destination nodes 3 and 5 are the same as the example in Figure 2. However, the alternate path selected for node 1 is different to the previous example since the diameter of the routing tree of the previous example is 4, which is not allowed in this case.

Thus the second best is selected and the diameter of the routing tree is limited not to exceed 3.

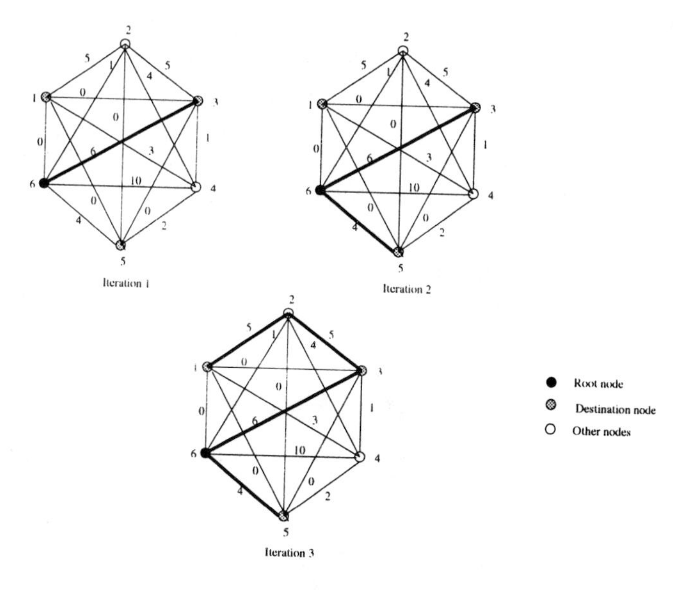

Figure 2 Example to illustrate LLMR with an alternate node

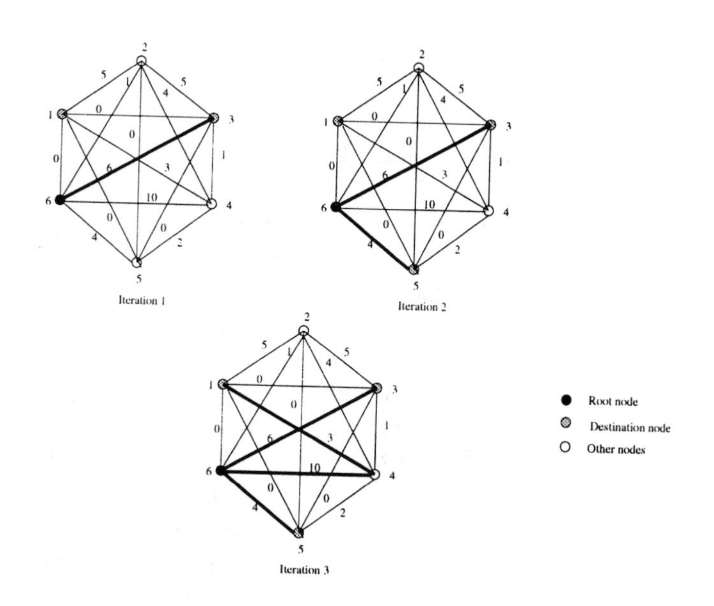

Figure 3 Example to illustrate LLMR with an alternate node

We now determine the complexity of the LLMR routing algorithm. When no alternate nodes are employed the number of VPs queried are $d(d-1)/2$ and the total number of comparisons needed is equal to that required in sorting $d(d-1)/2$ numbers which is $O(d(d-1)\log_2 d)$. The *time complexity* of the above is $d-1$, i.e., the time required to set-up a VC request with a destination set of size d is $d-1$ times the time required to set-up a point-to-point VC request. The use of an alternate node increases the time complexity to d. The number of VPs whose status is requested depends on which intermediate stage the alternate node is added. If the alternate node is added at stage i, i.e., when there are i destination nodes in the routing tree constructed thus far, the number of VPs queried for their status are $i(d-i)$ direct VPs and $2i(d-i)(N-d)$ alternate VPs. In the worst case the number of VPs whose status is requested is less than $d(d-1)/2 + \frac{d^2}{2}(N-d)$.

3. SIMULATION RESULTS

We consider a fully connected network with 12 nodes and each link have a capacity of 30 bandwidth units. The maximum destination size of a connection request is 6. In order to maintain the same link load of each type of VC requests, the arrival rate, λ_d, of a VC request with destination size d is assumed to be inversely proportional to the minimum number of links, $d-1$, used to establish the VC, i.e. $(d-1)\,\lambda_d = \lambda$, where λ is a constant. For convenience we define the normalized network load, L, as the ratio of total offered load to total network capacity, i.e.,

$$L = \frac{\sum_{d=2}^{6} L_d}{\sum_{e \in E} C(e)}.$$

For each set of simulation point, 10 independent runs are carried out and the length of each run is 10^6 units of mean VC request interarrival time. For each run, the initial 10% of each run was discarded to avoid the effect of transient state. The symbols around each simulation point are to show the upper and lower bounds of 95% confidence interval.

The performance of LLMR algorithm is shown in Figure 4. When the loading of a network is high (larger than 100%), the normalized revenue losses is also high since the network resources are insufficient and no algorithm can help under this situation. However, when the loading is moderate, the normalized revenue loss of LLMR algorithm decreases rapidly because LLMR select least loading path to avoid congestion. Figure 5 shows that the normalized revenue loss with at most alternate node is much smaller than that without any alternate node. Furthermore, the revenue loss changes only marginally when the maximum number of alternate nodes allowed per VC request increases. It is because more resources are used to make connection over alternate paths and less resources are for future VC requests. It is therefore prudent to restrict the maximum number of alternate nodes to at most one for multicast routing.

Figure 6 shows an interesting phenomenon: the larger the destination set size of the VC request the smaller the blocking probabilities. It is because there are two factors affecting the VC blocking probability: the number of links required and the number of possible choices to establish a multicast VC. A VC request with a large destination set requires more links so that its blocking probability should be higher than that with smaller destination set. However, the number of possible links to build a connected tree is also larger and thus there is higher chance to avoid congestion and blocking. For example, in a fully connected network, there are 3 choices to establish a connected tree if the destination size is 3, but for a destination size of 4, 16 choices are available to establish a VC. That means a VC request for a destination size of 4 has 13 more choices to successfully establish than that for a destination size of 3. It turns out the latter factor dominates in a VC request with larger destination set.

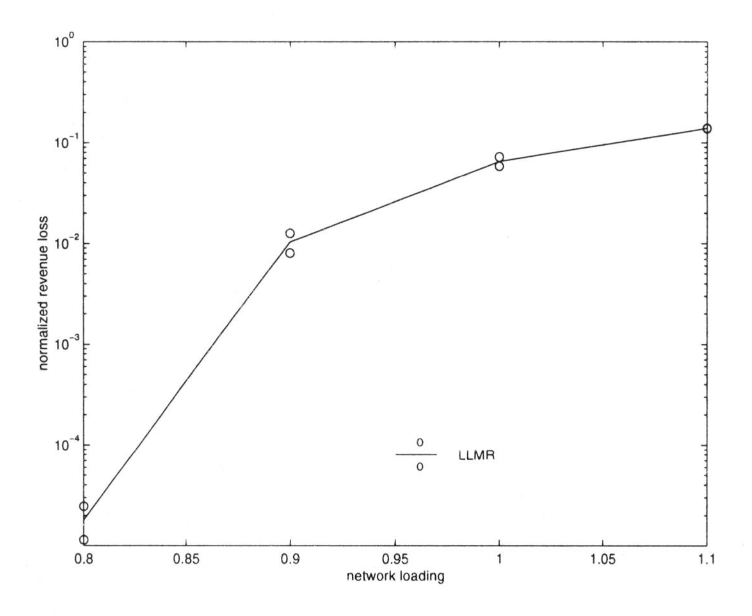

Figure 4 Normalized revenue loss vs. network loading

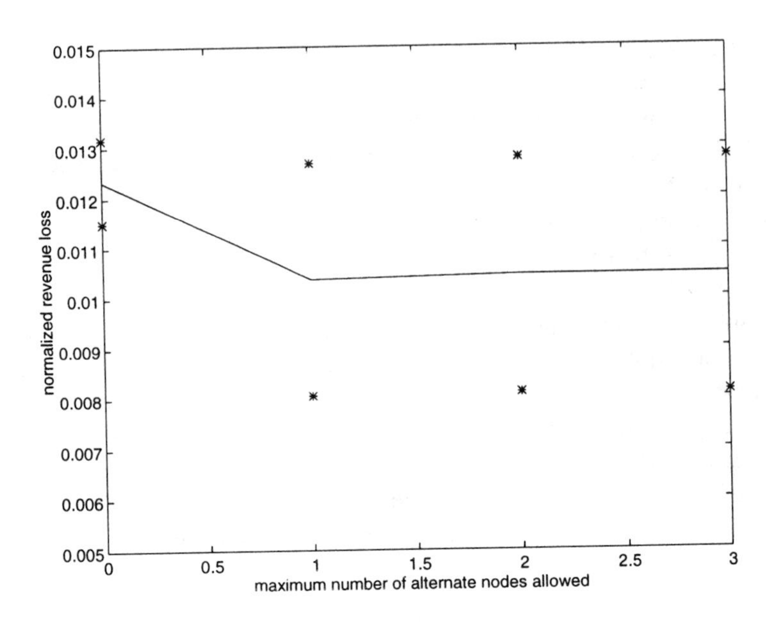

Figure 5 Normalized revenue loss vs. maximum number of alternate nodes allowed (offered load = 0.9)

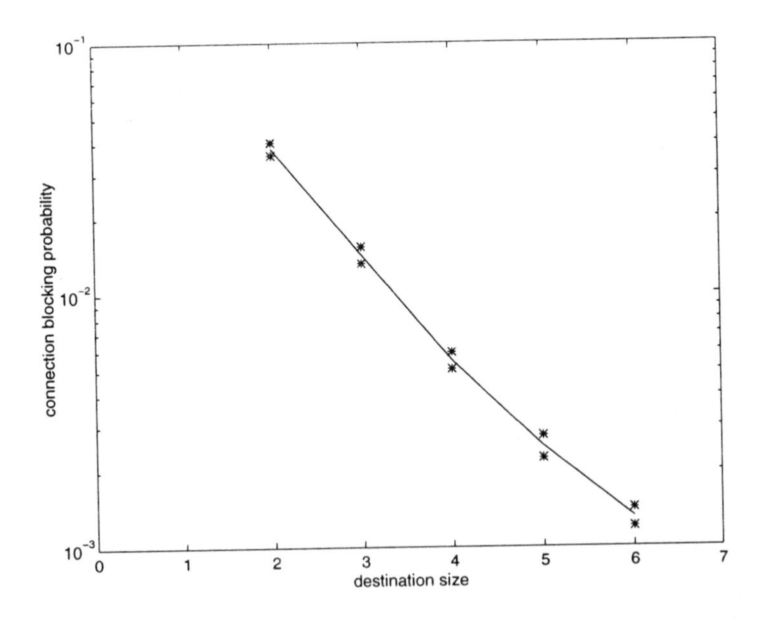

Figure 6 VC blocking probability vs. destination size (offered load=0.9)

4. CONCLUSIONS

In this paper, we propose new dynamic multicast routing algorithms for ATM networks and the algorithms with statistical multiplexing are considered. The simulations results shows the characteristics of LLMR. Moreover, we find that the connection blocking probabilities decrease with increasing destination size.

Notes

1. The wording *non-statistical multiplexing* has been chosen for consistency with the terminology in the ATM literature. We stress, however, that even for non-statistical multiplexing the cells propagating on a VP are "randomly" allocated to the various VCs established over the VP.

References

[1] M. H. Ammar, S. Y. Cheung, and C. M. Scoglio. Routing multipoint connections using virtual paths in an ATM network. In *Proceedings of INFOCOM*, pages 98–105, 1993.

[2] D. Anick, D. Mitra, and M. M. Sondhi. Stochastic theory of a data-handling system with multiple sources. *The Bell Systems Technical Journal*, vol. 61:1871–1894, 1982.

[3] Chi-Chung Cheung, D.H.K. Tsang, S. Gupta and H-W Chu. Least loaded first multicast routing in single rate loss networks. To appear in ICC '96, 1996.

[4] M. Doar and I. Leslie. How bad is naive multicast routing? In *Proceedings of INFOCOM*, pages 82–89, 1993.

[5] R. Guerin, H. Ahmadi, and M. Naghshineh. Equivalent capacity and its application to bandwidth allocation in high-speed networks. *IEEE Journal on Selected Areas in Communications*, vol. 9:968–981, 1991.

[6] S. Gupta, K. W. Ross, and M. El Zarki. *On Routing in ATM Networks.* Prentice-Hall and Manning, 1995. Routing in Communications Networks.

[7] V. P. Kompella, J. C. Pasquale, and G. C. Polyzos. Multicasting for multimedia communication. In *Infocom*, pages 2078–2085, 1992.

[8] J. B. Lyles and D. C. Swinehart. The emerging gigabit environment and the role of local ATM. *IEEE Communications Magazine*, vol. 30:52–58, April 1992.

[9] M. Minoux. A class of combinatorial problems with polynomially solvable large scale set covering/partitioning relaxations. *R.A.I.R.O Operations Research*, vol. 21(2):105–134, 1988.

[10] H. Tode, Y. Sakai, M. Yamamoto, H. Okada, and Y. Tezuka. Multicast routing algorithm for nodal load balancing. In *Infocom*, pages 2086–2095, 1992.

[11] B. M. Waxman. Routing of multipoint connections. *IEEE Journal on Selected Areas in Communications*, vol. 6:1617–1622, December 1988.

[12] B. M. Waxman. Performance evaluation of multipoint routing algorithms. In *Infocom*, pages 980–986, 1993.

[13] T. Yang and D.H.K. Tsang. A novel approach to estimating the cell loss probability in an ATM multiplexer loaded with homogeneous on-off sources. *IEEE Transactions on Communications*, vol. 43:117–126, Jan. 1995.

THE FLOODING MECHANISM OF THE PNNI ROUTING PROTOCOL

Performance Aspects

P. Jocher[1], L. Burgstahler[2], and N. Mersch[3]

1. *LKN, Technische Universität München, Arcisstr. 21, D-80290 München, Germany*
   E-mail: jocher@ei.tum.de
2. *IND, Universität Stuttgart, Pfaffenwaldring 47, D-70569 Stuttgart, Germany*
   E-mail: burgstahler@ind.uni-stuttgart.de
3. *Siemens AG, Hofmannstr. 51, D-81359 München, Germany*
   E-mail: norbert.mersch@icn.siemens.de

Abstract The Private Network Node Interface (PNNI) provides a flexible and scaleable routing architecture for ATM networks comprising a routing protocol and a signaling protocol. To obtain more experiences about PNNI, we developed a PNNI Emulator. We investigated the simple flooding mechanism of the PNNI routing protocol used to distribute topology information through the network. Beside theoretical results, the paper also presents some measurements of example networks.

Keywords: ATM, emulation, flooding, performance, PNNI, routing

1. INTRODUCTION

The *Private Network Node Interface* (PNNI), standardized by the ATM Forum (see [1]), provides a flexible and scaleable routing architecture for ATM networks comprising a routing protocol and a signaling protocol. PNNI routing includes mechanisms for the autonomous exchange of aggregated topology information to form a hierarchical representation of the network. Moreover, Quality of Service parameters are supported as required by ATM. PNNI signaling is based on a subset of UNI 4.0 signaling.

Our earlier investigations (see [2] and [3]) on principle aspects of the PNNI performance showed, that one of the crucial points is the load from topology information packet processing.

This paper is focusing on those performance aspects of the PNNI routing protocol concerning the flooding mechanism used to distribute topology information through the network. First, we look at the *PNNI Topology State Elements* (PTSEs) before we investigate the simple flooding mechanism. Besides theoretical results, this paper also presents measurements in example networks using our emulation tool developed at the Institute of Communication Networks/TUM in cooperation with Siemens AG.

The remaining part of the paper is organized as follows: Section 2 describes the basic characteristics of the distribution of topology information within PNNI. Section 3 concerns with the performance aspects of the PNNI routing protocol focusing on the flooding mechanism. Finally, section 4 concludes the paper and gives an outlook on future work.

2. PNNI TOPOLOGY INFORMATION DISTRIBUTION

2.1 Overview

PNNI uses source routing to determine a path through a network. Hence, every node needs a complete description of the topology to perform the necessary computations. However, when first turned on each node has only information about its own state. To complete a node's view of the network, the distribution of information must be provided by the routing protocol.

Section 2.2 describes the structure of the PNNI topology information groups; section 2.3 describes the distribution methods. These explanations are limited to a single peer group network.

2.2 Topology Information

The PNNI protocol provides a three leveled data structure for topology information. On the first level are the information groups (IG). Each IG only covers one specific part of a node, e.g. one port and its resources. On a second level, the IGs are bundled in PNNI Topology State Elements (PTSEs). Each PTSE contains IGs of only one type, so there are many different PTSEs describing each node.

PTSEs are the units of flooding and retransmission. As they do not contain information about the originating node, they need some kind of envelope when being sent to a neighbor. The PNNI Topology State Packet (PTSP) is such an envelope that transports PTSEs including information about the node's identity.

2.2.1 Information Groups

IGs can be divided into three classes:

- *Nodal information*: Nodal information includes the identity of a node, its capabilities, and information about the hierarchy. As long as there is no change in the hierarchy or no need to re-elect the peer group leader, the information in this group are static.
- *Topology state information*: Both, link and nodal state parameters, describing the characteristics of a link and a node respectively, belong to this group. Some topology state information are highly dynamic (e.g. the available bandwidth), while others are more static (e.g. the administrative weight). To keep a node's topology information up-to-date, the dynamically changing IGs have to be distributed frequently.
- *Reachability information*: End-system addresses are contained in these IGs. Their number depends on the node's role in the network. Nodes in access area may have many end systems attached (thus having many addresses in their databases), while nodes in the core network only might have a small number of attached end systems. As long as there is no mobility involved, their content is rather static and does not need to be distributed very often.

2.2.2 PTSE

PTSEs are used to bundle different IGs covering a certain aspect of a topology. While IGs only carry values that describe this topology aspect, the PTSEs also contain administrative information like a remaining lifetime or an IG identifier. PTSEs do not carry any information about the originating node.

Each PTSE can contain any number of IGs, provided they are of the same type and the PTSE does not exceed the maximum packet size. It is not necessary that all IGs of one type are bundled in one PTSEs. Rather, IGs can be bundled in a way that expresses a certain logical relation.

2.2.3 PTSP

To transmit PTSEs to a neighbor PTSPs are used. PTSPs contain at least one PTSEs of any type from a single originating node. Note, that only the PTSP reveals the source of the PTSEs in its header. For the receiving node, all information in a PTSP belongs to the same node. While it is recommended to transmit as many PTSEs in one PTSP as possible, the size of the PTSP must not exceed the maximum packet size.

2.3 Information Distribution

Two methods of distributing topology information are specified in PNNI: Database synchronization and flooding.

2.3.1 Database Synchronization

Database synchronization should happen rarely. Whenever two nodes learn for the first time that they belong to the same peer group, they exchange their complete database. They do this by announcing their database contents to the newly found neighbor. Then, the missing PTSEs are requested form the neighbor and finally exchanged.

2.3.2 Flooding

Flooding is a reliable method to distribute information within a network. Its main advantage, but also its main drawback is redundancy. On receipt of a PTSE that is not yet in its database, a node forwards this PTSE to *all neighbors*, except the one the PTSE was received from. If there is more than one path between any two nodes, a PTSE will be forwarded over each of them. Therefore, appropriate measures have to be taken, to prevent redundant PTSEs from consuming to much processing power at the receiving node.

Each received PTSE is checked, whether it is already installed in the nodal database. Following, there are two actions possible:

- Discarding the PTSE, if it is already installed in the database.
- Forwarding the PTSE via flooding, installing it in the database and then acknowledging of the PTSE.

There are two major reasons, why a node originally floods a PTSE:

- *Triggered Update*: Triggered flooding happens if a completely new PTSE is originated by a node or if there is a significant change in an IG within an existing PTSE (e.g. new end system addresses are added, the available bandwidth changed beyond a threshold etc.).
- *Aging*: Aging causes flooding if either the remaining lifetime of a PTSE reaches zero or if the remaining lifetime of the PTSE reached a certain threshold in its originating node. To prevent the PTSE from being deleted the originating node floods an update, even if the contents did not change.

Summarizing, while database synchronization is limited to the moment where two neighbors learn about their existence, flooding lasts as long as the network is up and running.

3. PERFORMANCE ASPECTS

In earlier investigations (see [2]) on PNNI performance we measured the processing load of a typical node[1]. Results showed, that with the given protocol stack about 80% of the load were due to PTSP processing while only 10% were caused by route computation. Moreover, in a hierarchical multi peer group network additional processing capacity is necessary for nodes representing their peer group at the higher network levels.

In the first section of this chapter we will give theoretical estimates of PTSE rates and their influence on the processing performance of a typical node. Following we are focusing on the simple flooding mechanism used to distribute topology information within the peer groups. Based on theoretical considerations and additional measurements we will show that - depending on the topology - a not to be neglected percentage of PTSEs is redundant.

3.1 Routing Protocol Processing

When a PNNI switch receives a new PTSE, it sends this PTSE to all neighbor nodes except the one the PTSE was received from. This is independent from the fact that some neighbors might already have flooded the same PTSE. It is also not altered by jittering the refresh interval, since jittering only influences the time a new PTSE is originated.

Therefore, we can follow, that if adj_i is the number of adjacent nodes of node i, every PTSE originated by node i must be

- sent: adj_i times
- received: 0 times

and every PTSE not originated by node i must be

- sent: $adj_i - 1$ times
- received: $adj_i - t$ times

t refers to the fact that, due to the nature of flooding, a particular PTSE is not flooded upstream on those links, which form the shortest path tree (SPT) with the originating node as the root. Thus the value of t depends on the considered originating node and receiving node. What follows is:

- The necessary routing protocol performance capacity of a PNNI switch increases linearly with the number of adjacent nodes, i.e. with the meshing of the network.
- Nodes, which own a number of PTSEs, which is above the average of a PNNI network, need less processing power for the flooding than nodes, with a smaller number of PTSEs.

[1] Square mesh topology with 25 nodes (single peer group) and 60% mean offered load.

This has to be taken into account when adding a node with low performance to a PNNI network. Based on this an estimation of the routing protocol processing capacity is possible and will be performed subsequently. However, we should take the following effects into account:

- Receiving PTSEs is more expensive than sending PTSEs. This is due to the big effort needed to decode the PTSEs and if necessary incorporate them into the database.
- The insertion of new PTSEs into the database is done only for the first new PTSE and not for the multiple duplicates received additionally.
- The processing time for a database insertion or check depends on the filling grade of the database.

It is clear that an estimation of the routing protocol processing capacity as performed below yields only a lower bound for the necessary performance capacity of a switch. Additional capacity must be provided due to the following reasons:

- PNNI timers are jittered but flooding is still bursty as our measurements in section 3.2.2 confirm. More capacity can prevent long queues.
- PTSE retransmissions due to bit errors must be taken into account.
- In case of the failure of a node or the insertion of a new node into the peer group database synchronization is required by the neighbor nodes of the failed or new node respectively.
- These considerations apply per peer group. Logical group nodes demand for additional capacity.

3.1.1 Flooding of Address PTSEs

This paragraph shows how the address PTSE processing performance of a PNNI switch depends on various parameters. The main factors influencing the address PTSE rate in a network are:

- Number of address PTSEs in the network: *NoAddrPTSE*
- Time between the refresh flooding of PTSEs: *PTSERefreshInt*
- Network Meshing (average number of neighbors): *NoNeighbors*
- The average value of t (in the network): T

Hence the average rate for received address PTSEs per node (*AddressPTSERate*) can be calculated as follows[2]:

$$AddressPTSERate \leq \frac{NoAddrPTSE \cdot (NoNeighbors - T)}{PTSERefreshInt}$$

[2] The '≤' relation refers to the fact that not every node necessarily originates address PTSEs.

Reserve capacity is needed, since the following parameters may change:

- The meshing of the network may be increased.
- The number of addresses within the network may increase with the consequence of an increased number of address PTSEs.

Within PNNI networks address summarization is applied to reduce the number of PTSEs to be flooded. Hence the address structure, i.e. the association of addresses to PNNI switches has a very big influence on the number of address PTSEs. Moving addresses within a network might deteriorate the summarization of addresses within peer groups, resulting in additional PTSE traffic.

3.1.2 Flooding of Link State PTSEs

Concerning the planning of the routing protocol processing capacity it is difficult to estimate the maximum rate of significant changes and thus the flooding rate, since this heavily depends on the dynamic behavior of the network and on the PNNI parameters. As an upper bound only the minimum interval between the flooding of PTSEs (*MinPTSEInt*) can be taken.

This results in the following formula for the average maximum rate of received link state PTSEs (*MaxLsPTSERate*) in a node within a PNNI network consisting of l links[3]:

$$MaxLsPTSERate < (2 \cdot l) \cdot \frac{NoNeighbors - T}{MinPTSEInt}$$

Extensive simulation is needed to evaluate the different influence factors of the update rate and to find a sensible upper bound (or increase the *MinPTSEInt*).

3.2 Flooding

3.2.1 Theoretical Results

We demonstrate in this section, that - depending on the topology - the simple flooding mechanism generates a not to be neglected percentage of redundant information in the network.

Figure 1 shows the flooding of an information element in an example network. Based on a significant event, *node a* originates a new information

[3] The '<' relation refers to the fact that some of the link state PTSEs are originated by the nodes themselves.

element and forwards it to all neighboring peers (see figure 1-1). On receipt of this element, *node b* as well as *node d* checks if an instance is already in the respective database. If not, both nodes update their databases and forward the information to all neighbors, except the one it was received from (see figure 1-2). Thus, *node c* gets the same information twice. Because of sequential processing, one information element - in our example form *node b* - will be checked first, installed in the database and forwarded (see figure 1-3). Then, the second one will be processed and discarded, as an instance is already in the database. Thus, five information elements had to be processed in the example network. Only three of them would have been sufficient to update the databases of the respective nodes.

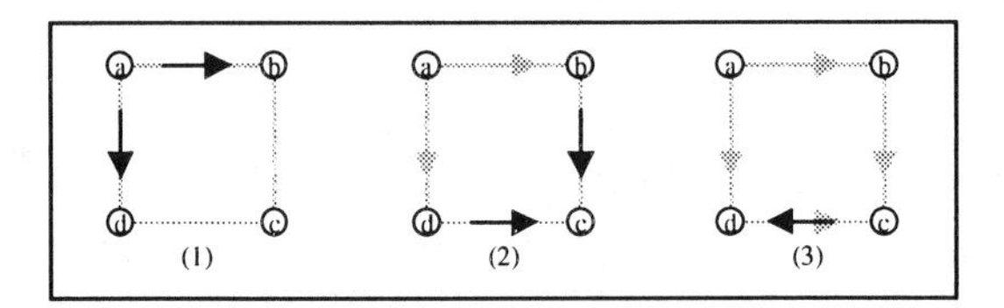

Figure 1. Example of the simple flooding mechanism

Due to the nature of flooding 'overlapping' may not occur on those links, which form the SPT with the originating node as the root. Hence, when a new information element is originated in a connected network consisting of n nodes and l bi-directional links, the following can be derived:

- Connectivity C:

$$C = \frac{2 \cdot l}{n}$$

- Number of information elements distributed on the SPT:

$$P_{spt} = n - 1$$

- Number of redundant information elements:

$$P_{redundant} = 2 \cdot (l - (n - 1))$$

- Number of total distributed information elements:

$$P_{total} = P_{spt} + P_{redundant} = (n - 1) + 2 \cdot (l - (n - 1)) = 2 \cdot l - n + 1$$

– Ratio between P_{total} and P_{spt}:

$$\frac{P_{total}}{P_{spt}} = \frac{(n-1)+2\cdot(l-(n-1))}{n-1} = 2\cdot\frac{l}{n-1} - 1$$

The connectivity C is inappropriate to describe the behavior of a network with reference to the flooding mechanism, as table 1 shows.

Table 1. Properties of linear chains

Topology	No. of nodes	No. of links Total	No. of links SPT	C	$P_{redundant}$
Linear chain	16	15	15	**1.88**	**0**
Linear chain	25	24	24	**1.92**	**0**

Therefore, we define a redundancy coefficient *R*, which implies information about the flooding behavior of the network:

$$R = \frac{l}{n-1}$$

The smallest possible value $R = 1$ only appears in connected networks consisting of a topology with a simple tree structure. Moreover, it follows:

$$\frac{P_{total}}{P_{spt}} = 2\cdot R - 1$$

Table 2 shows the properties of some regular graphs. Already in the simple square mesh topology with 25 nodes 57% of the processed information elements are redundant. In the full mesh topology, this share increases to over 95%.

Table 2. Properties of regular graphs

Topology	No. of nodes	No. of links Total	No. of links SPT	R	P_{total}	P_{spt}	$P_{redundant}$	P_{total} / P_{spt}
Linear chain	25	24	24	1.00	24	24	0	1.00
Ring	25	25	24	1.04	26	24	2	1.08
Square mesh	25	40	24	1.67	56	24	32	2.33
Full mesh	25	300	24	12.50	576	24	552	24.00

Summarizing, because of the simple flooding mechanism, the number of redundant information elements caused by a single event strongly depends on the network topology. Consequently, paying attention to this fact in the network planning process would be one possibility to minimize the distribution of redundant information. Another approach is to modify the flooding mechanism. [8] proposes an interesting flooding method, delivering network updates faster than conventional mechanisms, while at the same time using significantly less bandwidth. However, an adaptation to our specific problem is necessary.

3.2.2 Measurements

To verify the theoretical results, we performed measurements on two emulated PNNI networks: A full mesh topology with 8 nodes and a square mesh topology with 16 nodes.

Both networks formed a single peer group with one end-system per node. Between any two neighbor nodes, there was only one bi-directional link. Each of them had a capacity of 155 Mbit/s (STM-1). The PNNI specific parameters had been set to values recommended in the annex of [1].

To simplify the measurements, we admitted only CBR (Constant Bit Rate) connections, though VBR (Variable Bit Rate) traffic could be easily supported by the use of equivalent bit rates (see [4], [5], [6], and [7]).

Bi-directional calls had been generated according to a Poisson process using three different call classes: 60% requested an 848 kbit/s connection, 30% requested a 4 Mbit/s connection and 10% a 12 Mbit/s connection. The link costs used for path computation were equal for all links. Hence, we varied the call arrival rate to adjust the mean offered load to 75%. The calls were equally distributed over the network according to a uniform random distribution of sources and destinations. The mean call holding time was 480 s. Table 3 contains additional properties of both networks.

Table 3. Network properties

Topology	No. of nodes	No. of links		R	P_{total}	P_{spt}	$P_{redundant}$	P_{total} / P_{spt}
		Total	SPT					
Full mesh	8	28	7	4.00	49	7	42	7.00
Square mesh	16	24	15	1.60	33	15	18	2.20

Figure 2 shows the total number of measured PTSEs with a time granularity of 1 s in the full mesh network. Due to the small number of end-system addresses, most of the PTSEs are generated because of link state changes. This is also the reason why the refresh of address information is nearly invisible in figure 2.

One of the interesting characteristics is the plateau on the level of about 98 PTSEs/s and (in a weaker form) also on the level of about 196 PTSEs/s. Since all connections are bi-directional and a PTSE is generated for each direction of a link, every significant change causes $2 \cdot 49 = 98$ PTSEs (see table 3) to be processed in the example network. Additional measurements with reference to single PTSEs confirmed this.

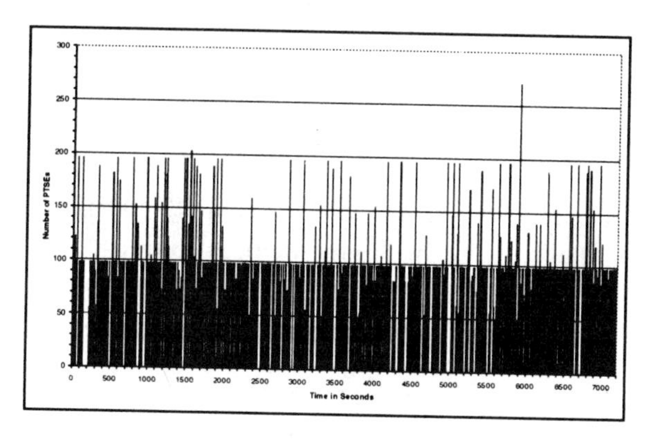

Figure 2. Total number of flooded PTSEs in the full mesh network

Consequently, the plateaus in figure 2 refer to significant change events. They occurred either on one link (98 PTSEs to be processed) or on two links simultaneously (196 PTSEs to be processed). In the latter case, it can be assumed that either two connections each spanning one link or one connection spanning at least two links has caused the significant change.

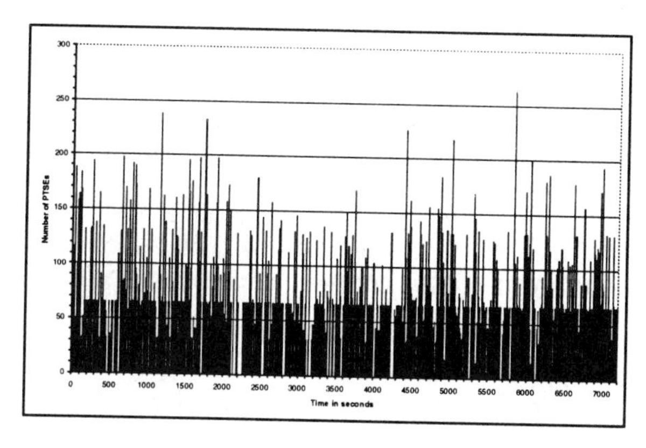

Figure 3. Total number of flooded PTSEs in the square mesh network

Figure 3 shows the number of measured PTSEs in the square mesh network. Again, a significant change generates a PTSE for each direction of a link. Thus, plateaus may occur at a multiple of $2 \cdot 33 = 66$ PTSEs/s (see table 3), as figure 3 confirms. The partly uneven formed plateaus are only caused by the time interval of 1 s used to measure the PTSEs.

Summarizing, our measurements confirmed the theoretical results. Additional investigations of several load scenarios have confirmed that the PTSE traffic is characterized by short peaks of high processing activity followed by periods of silence.

4. CONCLUSION

In this paper, performance aspects of the PNNI routing protocol have been presented focusing on the PNNI topology information elements and their flooding mechanism. The main factors determining the minimum routing protocol processing capacity of a PNNI switch are:

- the addresses structure (association of addresses to switches),
- the rate of significant changes on the links,
- the meshing of the network,
- the values of the various timers and thresholds defined in [1], Annex E.

The number of redundant PTSEs in a peer group caused by a single event strongly depends on the network topology. Consequently, paying attention to this fact in the network planning process is one possibility to minimize the distribution of redundant information. Another approach is the modification of the flooding mechanism. Therefore, further investigations are necessary and part of our future work.

References

[1] ATM Forum Technical Committee, Private Network-Network Interface Specification Version 1.0 (PNNI 1.0). ATM Forum af-pnni-0055.000, March 1996.

[2] U. Gremmelmaier, P. Jocher, J. Püschner and M. Winter, Performance Evaluation of the PNNI Routing Protocol using an Emulation Tool. 16th Int. Switching Symposium, Toronto, September 1997.

[3] P. Jocher, J. Frings, U. Gremmelmaier and M. Winter, Planning Aspects of ATM Networks using the PNNI Routing Protocol. NOC'98, Manchester, June 1998

[4] J. Roberts, U. Mocci, and J. Virtamo, Broadband Network Teletraffic. Springer, Berlin, 1996.

[5] International Telecommunication Union, Framework for traffic control and dimensioning in B-ISDN. ITU Recommendation E.735, May 1997.

[6] International Telecommunication Union, Methods for cell level traffic control in B-ISDN. ITU Recommendation E.736, May 1997.

[7] International Telecommunication Union, Dimensioning methods for B-ISDN. ITU Recommendation E.737, May 1997.

[8] Y. Huang, P. K. McKinley, Switch-Aided Flooding Operations in ATM Networks. IEEE INFOCOM `97, Kobe, April 1997.

A MODEL FOR EVALUATING THE IMPACT OF AGGREGATED ROUTING INFORMATION ON NETWORK PERFORMANCE

J.L.Rougier[1]*, A.R.P.Ragozini[2], A. Gravey[3], D. Kofman[1]

[1] *Ecole Nationale Supérieure des Télécommunications. France.*

[2] *Università Federico II di Napoli, Dipartimento di Ingegneria Elettronica e delle Telecomunicazioni. Italy.*

[3] *France Telecom — CNET. France.*

Corresponding Authors: rougier@email.enst.fr, ragozini@unina.it

Abstract This paper presents a mathematical model for evaluating the impact of information aggregation on the performance of PNNI-driven ATM networks. The routing aggregation scheme affects both the performance and scalability of these networks. However, to date, little is still known on how the choice of PNNI configuration parameters influences the network behavior. A generic model for the aggregation process is proposed in order to study its impact on the network utilization and routing overhead. Random geometric considerations are used to obtain closed-form approximations of various performance measures.

Keywords: QoS Routing, Hierarchy, aggregation, PNNI, Performance Analysis, Random Geometry.

1. INTRODUCTION

1.1 MOTIVATIONS

In this paper, we analyze the trade-off between performance and scalability in PNNI-driven ATM networks. The PNNI protocol, defined by the ATM-Forum, provides dynamic, hierarchical and *QoS-sensitive* routing capabilities. Unfortunately, the design and dimensioning of networks running PNNI is still not well known. For instance, the performance and scalability of PNNI are both affected by the routing aggregation scheme. However, very little is still known

*This author is supported by a France Telecom-CNET grant (project CNET/ENST PE96-7672)

on how this scheme impacts the network performance (e.g., call blocking rate) or scalability (i.e., routing overhead).

1.2 PNNI BASIS

QoS routing. The PNNI standard [1] defines a routing protocol adapted for the specific characteristics of ATM networks. PNNI is a *QoS-sensitive* routing protocol, which means that the choice of the paths is based on the knowledge of network state and connection QoS requirements. QoS routing potentially increases the likelihood of accommodating calls with diverse QoS requirements while optimizing the network usage.

However, dynamic QoS routing requires that the switches exchange link attributes or metrics, such as link load information, which can consume considerable network resources. The routing computation based on this information can also become very expensive in terms of processing and memory consumption [6; 9; 17]. This means that the routing information (such as link loads) should only be exchanged on a regular basis (with a small frequency) or after significant events. However, the routing information updating policy is critical to network performances; a careful trade-off between accuracy and complexity must be found. In this paper, we shall neglect the discrepancies caused by the routing update scheme — thus assuming that all the nodes have up-to-date routing information. The impact of stale routing information on network performance has been studied in a companion paper [15].

Hierarchy. The link information exchange, and the processing that it induces, can become prohibitive as the network size grows. PNNI ensures scalability by dividing the network into routing domains, called *Peer-Groups* — as depicted in Figure 1(a). A node in a Peer-Group (PG) has a complete view of the topology and routing information of the PG but only an aggregated view of the topology and routing information of the rest of the network. As information exchanged between different Peer-Groups is *aggregated* (i.e., summarized), the routing overhead is reduced. The impact of hierarchy on the routing overheads (e.g. routing table size, routing computation complexity) has been studied in [10; 12; 14]. In this paper, we analyze the impact of this hierarchical organization on network utilization.

Aggretation. The aggregation scheme aims at representing a whole peer-group, which potentially contains many links and nodes, with a small number of parameters. For instance, if the *simple node representation* [1] is used, the whole peer-group is seen as a single node[1]. In this case, the available bandwidth announced represents how much residual bandwidth could be expected by calls which have to cross the Peer-Group. A more precise alternative, is to use the *complex node representation* [1], where a peer-group is represented by an

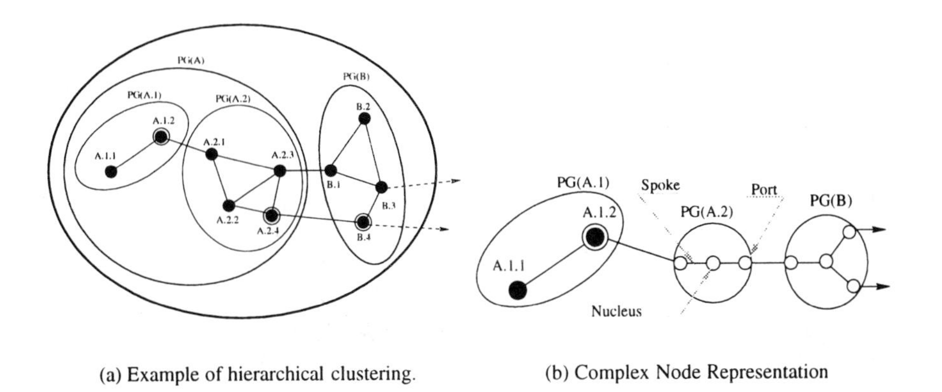

(a) Example of hierarchical clustering.

(b) Complex Node Representation

Figure 1 Hierarchy in PNNI

asymmetric star — thus announcing one set of link attributes and metrics for each entry point to the PG (see Figure 1(b)). The design of efficient and scalable aggregation schemes is difficult [7; 8; 11], as it requires a delicate trade-off between scalability and routing accuracy (network optimization).

Source-Routing. PNNI is a *source-routing* protocol, which means that the path is chosen once at the source switch[2]. As a specific routing computation could not possibly be run at each call arrival, a set of precomputed paths for any possible destination is usually stored in each switch [1; 8; 17]. The source selects one of these available paths which fits the call QoS requirements using GCAC (Generic Call Admission Control). If none of the precomputed paths is feasible, then an on-demand routing computation may be used or the call can be rejected.

In a hierarchical network, some topological information is hidden by the aggregation scheme, making it impossible for the source to determine a complete path. However, a logical path through the Peer Groups is computed by the first switch based on aggregated routing information. The route is then refined at each entrance in a new Peer-Group in order to determine how to cross the selected domain.

Crankback. When selecting a path, a node uses the currently available information about resources and connectivity. That information may be inaccurate because of hierarchical aggregation for instance. Therefore, a call may be blocked along its specified route. A *crankback* procedure is then used for partially releasing the reserved resources. An alternate path over which to route the call may then be chosen.

1.3 FOCUS OF THIS WORK

PNNI offers network designers a considerable amount of latitude in limiting routing overhead — at the expense of performances. In particular, both routing information aggregation and distribution can be controlled. Some recent studies, based on simulations, have analyzed the tradeoff between scalability and performance in PNNI [7; 16; 17]. Unfortunately, most studies analyzes on the routing update policy and do not consider the aggregation scheme. To the authors knowledge, no analytical results are available today.

In this paper, an analytical model is provided which allows to study the impact of aggregation on network performance (e.g. call blocking probability). Our results are based on the random-geometric model proposed in the next section. In section 3 various performance measures are expressed for a network with two hierarchical levels. Some results are analyzed in section 4. Conclusions and future work are presented in section 5. Technical considerations have been placed in appendices for the sake of clarity.

2. THE MODEL

In this paper, a random-geometric representation is used to model the network. Random geometry [5] has been recently applied with success to the performance evaluation of various networking problems [2; 14].

Physical Topology. The topology of the network is generated by stationary *Poisson processes* in the plane. At the physical level, the switches are represented by points of a Poisson process π_0 of intensity λ_0. In order to get a finite network, only the points which lie in a finite area W will be considered. Without loss of generality, we shall assume that W is a disc of unitary surface ($|W| = 1$), so that the average number of switches in the network is λ_0.

The connectivity between these nodes will be constructed by the so-called *Delaunay Graph* (see appendix A.1). A weighted graph is obtained by marking each link i with its available bandwidth X_i — other metrics or attributes such as the delays or jitter will not be considered. In order to limit the routing overhead, the values $\{X_i\}$ are normally broadcasted periodically or when a significant event occurs, so that the other nodes should only be able to observe estimations of the real bandwidths. In this paper, we shall concentrate on hierarchy and aggregation, and we shall neglect this phenomenon. The impact of stale routing information on network performances has been studied in a companion paper [15].

Hierarchy. In this paper, we shall consider a network wth two levels of hierarchy. A new stationary Poisson process π_1, of intensity λ_1, is introduced to divide the network into peer-groups. More precisely, the network is clustered using Voronoi tessellations [2; 5] (see appendix A.1) built with this new process. For each $u \in \pi_1$ we define a Peer-Group as a set containing all the nodes of π_0 which fall in the Voronoi cell $C_u(\pi_1)$.

The nodes at the higher level (the "logical nodes") correspond to the points of π_1. The logical connectivity between nodes at the higher level is also modeled by a Delaunay Graph. Such a model is depicted in Figure 2(a).

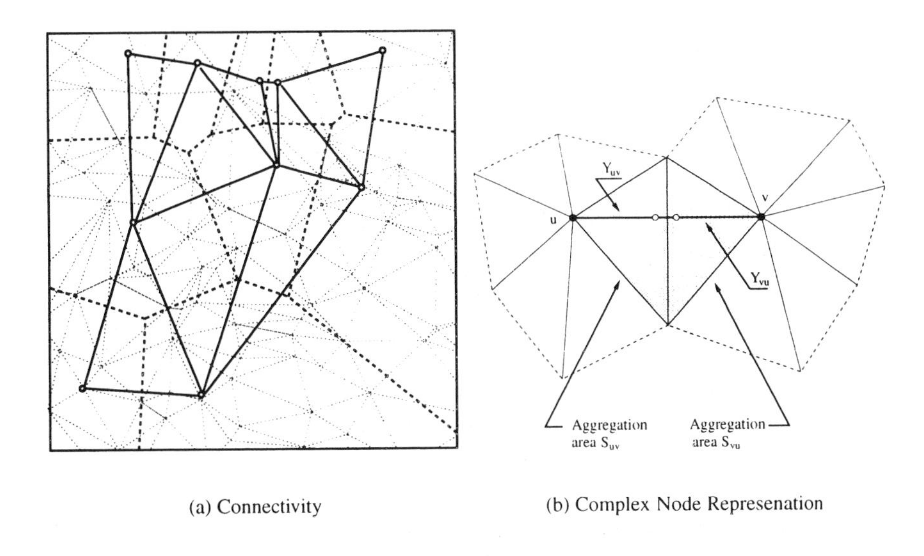

(a) Connectivity

(b) Complex Node Represenation

Figure 2 The hierarchical model.

Aggregation model.. In this paper, a complex node representation as defined in [1] is used[3]. The nuclei will be chosen as the centroids of the Voronoi cells (see Figure 2(b)). The links joining each port to the nucleus are marked by the aggregated bandwidth that one could expect when trying to reach nodes at the "center" of the peer-group.

Aggregation strategies are not specified in the standards. In the following we propose a generic aggregation model[4], as a function of its "aggressiveness". The cell is divided into non-overlapping areas (one for each port) as shown in Figure 2(b). The different aggregated values $Y_{u,.}$ are then computed independentely in each area: Given $\beta \in [0, 1[$, the aggregated value $Y_{u,v}$ will be chosen as the β-sample quantile[5] of the set $\{X_i\}$ in the corresponding area. β can be seen as the "aggressiveness" of the aggregation strategy: For $\beta = 0$, $Y_{u,v}$ is the minumum of all the available bandwidths, for $\beta \to 1$, the aggregated value is very optimistic as it corresponds to the maximum bandwidth available in the set.

A logical link of level 1, say uv, will then be marked by a single quantity $X^1_{u,v} = \min\{Y_{u,v}, Y_{v,u}\}$. We shall assume that the $\{Y_{\cdot}\}$ are independent and identically distributed[6]. This assumption implies that $\{X^1_{\cdot,\cdot}\}$ is a set of independent and identically distributed random variables. Finally, we can introduce $\mathbf{p}_1 = \mathbb{P}\left(X^1 > b\right)$ and $\mathbf{q}_1 = \mathbb{P}\left(Y > b\right)$ (omitting the link indications from the i.i.d. assumption) where b is the amount of bandwidth[7] requested by the used. It should be noted that, by definition, $\mathbf{p}_1 = \mathbf{q}_1^2$.

Routing algorithm. We shall assume that the routing algorithm at the source switch is recursive (As discussed in [13]). Furthermore, we shall use precomputed paths. At each call arrival, a list of precomputed paths for the destination is examined. A list of "feasible" paths (i.e. paths that can *a priori* accommodate the requested QoS) is then deduced using the GCAC function. If no feasible path exists, the call is rejected by the source switch. Otherwise, the selected route is the shortest one (in terms of number of hops).

Delaunay graphs are highly connected, which allows us to build many alternate routes. As highlighted in Figure 3, it is always possible to avoid a blocking link in two ways using a two hop bypass. In our model the set of precomputed paths is built considering the default shortest-path [8] and all the possible bypasses in case a link on this direct path does not have enough bandwidth, i.e., whenever $X < b$ (if b is the amount of bandwidth requested in the call set-up). It should be noted that each blocking link introduces a penalty of one hop in the total path length[9]. In order to control the 'cost' of routes, the source will limit the number of penalties, rejecting the call whenever the path uses too many ressources.

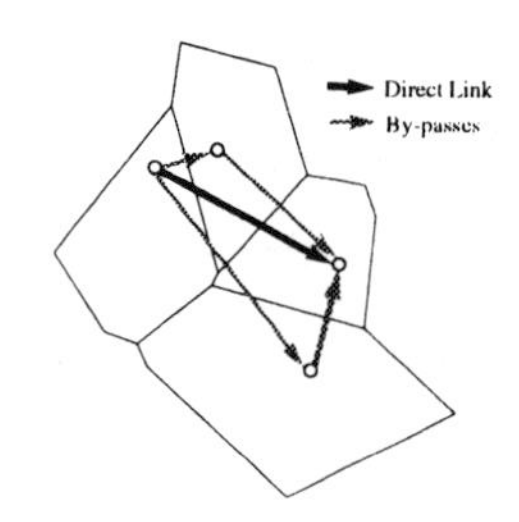

Figure 3 Alternate by-passes

3. PERFORMANCE MEASURES

In the following, the performance measures are first computed as conditional probabilities or averages, given a particular topology. An approximation for averaging these quantities is then discussed at the end of this section.

Passing through a spoke. Let $\mathbf{q}_{0,1}$ denote the probability of passing a spoke (in the complex node representation). The corresponding event implies that :

(i) at level-1, GCAC finds enough bandwidth on the spoke (i.e. $X^1 > b$);

(ii) at the physical level, it is actually possible to find a feasible route.

Let N_0 denote the number of links in the aggregation region considered and let $m_0 = \lfloor \beta_0 N_0 \rfloor + 1$ be the sample quantile index. Let Ω_s denote the event "s out of N_0 links fail" (i.e. s out of N_0 links have $X < b$). Let now M_0 denote the number of links needed to cross the peer-group (on the shortest path), and let Υ_k denote the event "k links out of M_0 are eliminated by GCAC" (as $X < b$). In order to control the overhead of the paths, we impose $k \le K_0$ with $K_0 = \lfloor \alpha_0 M_0 \rfloor$. It is easy to see that $\mathbb{P}(\Upsilon_k|\Omega_s)$ is hyper-geometrically distributed,

$$\mathbb{P}(\Upsilon_k|\Omega_s) = \frac{\binom{N_0-s}{M_0-k}\binom{s}{k}}{\binom{N_0}{M_0}} \tag{1}$$

Thus, it is easy to see that :

$$\mathbf{q}_{0,1} = \sum_{s=0}^{m_0-1} \sum_{k=0}^{K_0 \wedge s} \mathbb{P}(\Omega_s)\, \mathbb{P}(\Upsilon_k|\Omega_s)\, \mathbf{q}_{0,1|\Omega_s,\Upsilon_k} \tag{2}$$

where $a \wedge b = \min\{a, b\}$ and $\mathbf{q}_{0,1|\Omega_s,\Upsilon_k}$ denotes the probability of passing through the spoke, given Ω_s and Υ_k. This probability describes whether a path with k alternate by-passes can be found or not, i.e. wether the failing links can be by-passed. Denoting $T \in \{0, 1, 2, 3, 4\}$ as the number of links for which $X < b$ around a failing link, it is easy to see thet T has a hypergeometric distribution:

$$\mathbb{P}(T = t) = \frac{\binom{N_0-s-(M_0-k)}{4-t}\binom{s-k}{t}}{\binom{N-M}{4}}$$

For $t = 0$ and $t = 1$, it is always possible to by-pass the failing link. For $t = 2$, the probability of by-passing the failing link is $\frac{1}{2}$. This reasoning can be extended for the k blocking links by introducing the set $t_0, \ldots, t_{k-1}$:

$$\mathbf{q}_{0,1|\Omega_s,\Upsilon_k} = \sum_{t_0=0}^{2} \cdots \sum_{t_{k-1}=0}^{2} \frac{\binom{N_0-M_0-(s-k)}{4-t_0\ \ldots\ 4-t_{k-1}}\binom{s-k}{t_0\ \ldots\ t_{k-1}}}{\binom{N_0-M_0}{4\ \ldots\ 4}} 2^{-\sum_{i=0}^{k-1} \mathbf{1}_{\{t_i=2\}}} \tag{3}$$

Blocking. Let M_1 denote the number of hops needed to cross the network at level-1 (using the shortest path algorithm). The probability of passing through a link of level-1 is, by construction, $\mathbf{p}_{0,1} = \mathbf{q}_{0,1}^2$. Thus, it is possible to pass directly through a link of level-1 with probability $\mathbf{p_d} = \mathbf{p}_{0,1}$. If not, two alternate by-passes could be tried instead. The probability of passing through one of the two alternate by-passes is then $\mathbf{p_a} = \mathbf{p}_{0,1}^2 + \left(1 - \mathbf{p}_{0,1}^2\right)\mathbf{p}_{0,1}^2$ which can be restated as $\mathbf{p_a} = \left(2 - \mathbf{p}_{0,1}^2\right)\mathbf{p}_{0,1}^2$. The probability of blocking is then given by this simple expression :

$$1 - \mathcal{P}_b = \mathbf{p}_{0,1} \sum_{k=0}^{K_1} \binom{M_1}{k} \mathbf{p_d}^{M_1-k}(1 - \mathbf{p_d})^k \mathbf{p_a}^k \tag{4}$$

where $K_1 = \lfloor \alpha_1 N_1 \rfloor$ controls the length of the path in the level-1 graph. The first term $\mathbf{p}_{0,1}$ comes from the first and the last spokes which cannot be bypassed.

Rejection. Rejection occurs whenever the source switch is not able to find any feasible route which leads to the destination. Hence the expression,

$$\mathcal{P}_r = 1 - \mathbf{p}_{0,1} \sum_{k=0}^{K_1} \binom{M_1}{k} \mathbf{p}_1^{M_1-k} (1-\mathbf{p}_1)^k \left(2-\mathbf{p}_1^2\right)^k \mathbf{p}_1^{2k} \tag{5}$$

where $\mathbf{p}_1$ is expressed in appendix A.2 . The term $\mathbf{p}_{0,1}$ comes from the first and the last spokes GCAC which cannot be by-passed.

Crankback. The probability of crankback is defined as the probability that the call is crankbacked at least once. On a single link, the probability of crankback is then $1 - \mathcal{P}_c^{(1)} = \mathbf{p}_{0|1} = \frac{\mathbf{p}_{0,1}}{\mathbf{p}_1}$. Taking by-passes into account, the expression becomes

$$1 - \mathcal{P}_c^{(1)} = \frac{\mathbf{p_d p}_{0|1} + (1-\mathbf{p_d})\mathbf{p_a p}_{0|1}^2}{\mathbf{p_d} + (1-\mathbf{p_d})\mathbf{p_a}}$$

where $\mathbf{p_d}$ and $\mathbf{p_a}$ are defined in the previous paragraphs.

Blocking with no re-routing.. When re-routing is not allowed after a crankback, the expression of the blocking probability becomes:

$$\mathcal{P}_b' = 1 - (1-\mathcal{P}_r)(1-\mathcal{P}_c) \tag{6}$$

Blocking without GCAC at level-1.. For comparative purposes, we have analyzed the network when the source node does not run any GCAC on the logical links. The probability of crossing a "spoke" can then be expressed as :

$$\mathbf{q}_{0,1}^* = \sum_{k=0}^{K_0} \binom{M_0}{k} \mathbf{p}_0^{M_0-k} (1-\mathbf{p}_0)^k \left(2-\mathbf{p}_0^2\right)^k \mathbf{p}_0^{2k} \tag{7}$$

Posing $\mathbf{p}_{0,1}^* = {\mathbf{q}_{0,1}^*}^2$, the blocking probability can be expressed as :

$$\mathcal{P}_b^* = 1 - \mathbf{p}_{0,1}^* \sum_{k=0}^{K_1} \binom{M_1}{k} {\mathbf{p}_{0,1}^*}^{M_1-k} (1-\mathbf{p}_{0,1}^*)^k \left(2-{\mathbf{p}_{0,1}^*}^2\right)^k {\mathbf{p}_{0,1}^*}^{2k} \tag{8}$$

Crankback without GCAC at level-1.. In this case, the crankback probability becomes :

$$\mathcal{P}_c^* = 1 - \mathbf{p}_{0,1}^* {\mathbf{p}_{0,1}^*}^{M_1} \tag{9}$$

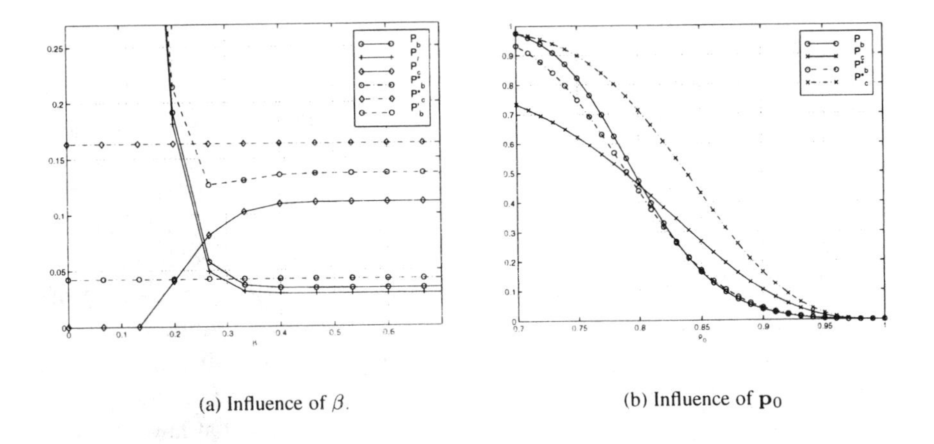

(a) Influence of β.

(b) Influence of $\mathbf{p}_0$

Averaging.. The previous quantities have to be averaged with respect to the random graphs π_0 and π_1. However the distributions of the aggregation areas and of the path lengths (in terms of hops) are not known to date. We shall assume that $M_0 = \langle \frac{1}{2}\sqrt{\frac{\lambda_0}{\lambda_1}} \rangle$ and $N_0 = \langle \frac{1}{4}\frac{\lambda_0}{\lambda_1} \rangle$, where $\langle . \rangle$ is the round operation and $\frac{\lambda_0}{\lambda_1}$ is the average number of nodes in a cell[10]. Furthermore, we shall assume $M_1 = \langle \frac{2}{\pi}\sqrt{\lambda_1} \rangle$. where $\frac{2}{\pi}$ is the diameter of the network. Those assumptions are discussed in appendix A.1 .

4. NUMERICAL RESULTS AND ANALYSIS

In Figure 4(a), the network behavior is represented as a function of the "aggressiveness" of the aggregation strategy for $\lambda_0 = 2000$ (2000 nodes in average) and $\lambda_1 = 33$ (33 peer-groups in average), $\mathbf{p}_0 = 0.9$, $K_1 = 1$ and $K_0 = 2$. We observe that, for conservative strategies (for small values of β) the model presents an high blocking rate caused by a high source rejection probability. Increasing β, the rejection and the blocking rates ($\mathcal{P}_r$ and $\mathcal{P}_b$) decrease but the crankback rate increases (leading ot a larger routing overhead).

As expected, $\mathcal{P}'_b$ is largely outperformed by $\mathcal{P}_b$ for any value of β unlighting the usefullness of rerouting the calls after crankback. we note that $\mathcal{P}'_b$ has a minimum for $\beta \approx 0.25$, where an optimal trade-off between rejection at the source and set-up failures(due to aggregation inaccuracy) is found.

The blocking and crankback rates when GCAC is disabled at the higher level can also be investigated. In Figure 4(a), it can be seen that this routing strategy is slightly less efficient for aggressive values of β. Futhermore, a higher crankback rate is induced by this scheme. We can conclude that performing

GCAC (even on rough aggregated information) can help to preserve the network resources while decreasing the routing overhead.

Given the aggressiveness of the aggregation algorithm, it is also interesting to observe the network behavior as a function of the required QoS. In Figure 4(b) the probabilities of blocking and crankback for the two routing strategies are presented as functions of $\mathbf{p}_0$. We have chosen the same parameters of the previous plot and set β to 0.35. As in the previous case, performing GCAC at level-1 allows a lower crankback rate for every $\mathbf{p}_0$ and, then, a better network utilization. We observe that for high required bandwidths, i.e., for low values of $\mathbf{p}_0$, using GCAC gives worse blocking rates than the second strategy. This can be explained by the fact that performing GCAC on aggregated information can exclude logical links, with insufficient declared bandwidth, that could have accepted the call. Instead, the second strategy, not using aggregated information, avoids only the logical links that have actually caused a crankback.

5. CONCLUSIONS AND WORK IN PROGRESS

We obtained analytical expressions for the crankback and blocking probability as a function of the routing aggregation policy. It turned out that :

- aggressive aggregation policies perform better with respect to the blocking probability[11] because of call rejection at the source;

- the GCAC function helps in selecting appropriate paths, even with very rough routing information ($\beta \to 1$), the crankack probability is significantly reduced (w.r.t. the case where GCAC is disabled);

- however, *GCAC* performs poorly for calls requesting a large amount of bandwidth, it could be better not to use GCAC for these calls.

Work is still going on in order to incorporate other phenomena (such as stale routing information caused by the routing update policy) in the model and for averaging more precisely the performance measures.

Appendix

A.1 RANDOM GEOMETRY

This appendix is devoted to the presentation of the Random Geometry concepts on which our analysis is based. For a more formal and exhaustive presentation of this concepts, see [5; 2; 3].

In this section, π denotes a stationary Poisson process in $I\!R^2$ of intensity λ. For all $u \in I\!R^2$, $||u||$ denotes the Euclidean norm of u.

Voronoi Cells.. The *Voronoi cell* with centroid $u \in \pi$, denoted by $C_u(\pi)$, is the region of the plane whose points are closer to u than to any other point of

π. A Poisson process can then be used to randomly partition the plane $\mathbb{R}^2$. The resulting mosaic is called a *Poisson-Voronoi Tessellation*.

Delaunay Graph.. Two points $u, v \in \pi$ are directly connected in the Delaunay Graph, if and only if their Voronoi cells $C_u(\pi)$ and $C_v(\pi)$ have a common edge (i.e. the cells are adjacent). Such a graph and its underlying Poisson-Voronoi tessellation are represented in Figure 2(a).

Path "lengths" in the Delaunay graph.. In this paper, we are interested in the number of hops used in a path. Let us denote this measure by $\nu(\text{path}(s,t,\pi))$ for two arbitrary points s and t. For a specific paths (the "markov paths" in [3]), this measure is known explicitely as : $\nu(\text{path}(s,t,\pi)) = frac4\pi\sqrt{\lambda}||s-t||$. More generally, scaling argument show that, for for sub-addtive routing algorithms (see [3]),$\mathbb{E}\left[\nu(\text{path}(s,t,\pi))\right] \sim \kappa\sqrt{\lambda}\,||s-t||$ when $||s-t||$ is large, where κ is a constant characterizing the routing algorithm. In this paper we shall pose $\kappa = 1$ for simplicity.

A.2 ORDER STATISTICS

Let $\{X_i\}_{i \in \mathcal{I}}$ be a set of random variables. Let then define $X_{(1)} < X_{(2)} < \ldots < X_{(i)} < \ldots X_{(N)}$, where $N = |\mathcal{I}|$. Given $\beta \in [0,1[$ and $m = \lfloor \beta N \rfloor + 1$, the random variable $X_{(m)}$ is called the *β sample quantile*.

Let us denote $Q_b = \mathbb{P}\left(X_{(m)} > b\right)$, $P_b = \mathbb{P}\left(X > b\right)$ and $\Omega_s = \{X_{(s)} \leq b < X_{(s+1)}\}$ for any $s < m$. It is easy to see that $Q_b = \sum_{s=0}^{m-1} \mathbb{P}(\Omega_s)$. Furthermore, assuming that $\{X_i\}_{i \in \mathcal{I}}$ are i.i.d., we get :

$$\mathbb{P}(\Omega_s) = \binom{N}{s} P_b^{N-s} \left(1 - P_b\right)^s \qquad \text{(A.1)}$$

Notes

1. More precisely, the Peer Group is seen as a symmetric star. All the links of the star have the same attributes and metrics.

2. Source switch: the switch which is directly connected to the host originating the call.

3. Except that no "exception-bypasses" are supported.

4. The limit of this model is that it does not consider any topological information (as explained in the following). Any real aggregation strategy should rely on the Peer-Group internal topology to compute aggregated values.

5. This choice implies that β is the fraction of aggregated links whose bandwidth is less than $Y_{u,v}$ in the area of interest.

6. The independence relation is false, as the numbers of links aggregated in adjacent areas are correlated.

7. The parameter b can represent either a peak, average or effective bandwidth, depending on the admission policy.

8. We consider the number of hops as length of a path; we shall assume for this algorithm $\kappa = 1$ (see also [3]).

9. It is a worst case assumption, as an alternate path with the same number of hops could be found as the graph is highly connected.

10. In computing the average value of N_0 two terms appear: the former is $\frac{3}{2}$, it represents the ratio between the intensities of the link and node processes at physical level (see [5]), the latter is the term $\frac{1}{6}$ and represents the fraction of the cell aggregated in a "spoke".

11. through conservative policies are preferable in order to limit crankbacks

References

[1] The ATM Forum[1]. *Private Network-Network Interface Specification version 1.* document: af-pnni-0055.000. March 1996.

[2] F.Baccelli, S.Zuyev. *Poisson-Voronoi Spanning Trees with applications to the Optimization of Communication Networks.* INRIA Research Report No. 3040. Nov.1996.

[3] F.Baccelli, K.Tchoumatchenko, S.Zuyev. *Markov Paths on the Poisson-Delaunay Graph.* INRIA Research Report No. 3420. Mai.1998.

[4] H.A.David.*Order Statistics.* Wiley Interscience, Second Edition, 1981.

[5] A.Frey, V.Schmidt, *Marked Point Processes in the Plane — Part I. A survey with applications to Spatial Modeling of Communication Networks.* Adv. Perf. Anal. 1, pp. 65-110, 1998.

[6] R.Guérin, A.Orda. *QoS-based Routing in Networks with Inaccurate Information: theory and Algorithms.* Proc. IEEE Infocom 97, 1997.

[7] F.Hao, E.W.Zegura, S.Bhatt. *Performance of the PNNI Protocol in Large Networks.* Proc. IEEE ATM Workshop 98, 1998.

[8] A.Iwata, R.Izmailov et al. *QoS Aggregation Algorithms in Hierarchical ATM Networks.* Proc. of IEEE ICC'98, Atlanta, June 1998.

[9] A.Iwata, R.Izmailov et al. *Routing Algorithms for ATM Networks with multiple QoS requirements.* Proc. IFIP WATM'95, Paris, Dec. 1995.

[10] L.Kleinrock, F.Kamoun. *Hierarchical Routing for Large Networks.* Computer Networks, vol.1. 1977.

[11] W.C.Lee. *Topology Aggregation for Hierarchical Routing in ATM Networks.* ACM Computer Communication Review, vol.25, April 1995.

[12] P.Van Mieghem. *Estimation of an Optimal PNNI Topology.* Proc. IEEE ATM'97, Lisboa. 1997.

[13] P.Van Mieghem. *Routing in a Hierarchical Structure.* Proc. IEEE ICATM'98, Colmar. 1998.

[14] J.L.Rougier, D.Kofman, A.Gravey. *Optimization of Hierarchical Routing Protocols.* Proc. IFIP ATM'98, Ilkley (UK). July 1998.

[15] J.L.Rougier, A.Ragozini, A.Gravey, D.Kofman. *Estimation of Crankback Probability in Hierarchical PNNI network.* Proc. IEEE ICATM'99, Colmar (France). June 1999.

[16] A.Shaikh, J.Rexford, K.G.Shin. *Efficient Precomputation of Quality-of-Service Routes.* Proc. IEEE NOSSDAV 98, July 1998.

[17] A.Shaikh, J.Rexford, K.G.Shin. *Evaluation of Overheads of Source Directed Quality of Service Routing.* Proc. IEEE ICNP98, October 1998.

SESSION 14

Congestion and Admission Control

CALL ADMISSION CONTROL FOR PREEMPTIVE AND PARTIALLY BLOCKING SERVICE INTEGRATION SCHEMES IN ATM NETWORKS

Ernst Nordström

Department of Computer Systems, Information Technology, Uppsala University, Box 325, S–751 05 Uppsala, Sweden
Email: ernstn@docs.uu.se

Abstract This paper evaluates a Markov decision approach to single–link Call Admission Control for CBR/VBR and ABR/UBR services. Two different schemes that support integration of narrow-band ABR/UBR and wide-band CBR/VBR services are evaluated: the standard preemptive scheme and the modified partial blocking scheme. The structure of the Markov decision policy shows an "intelligent blocking" feature, which implements bandwidth reservation for wide-band calls. The numerical results show that the Markov decision method yields higher long-term reward than the complete sharing method when the ability to create sufficient capacity for wide-band calls through partial blocking/preemption is limited. The results also show that the modified partial blocking scheme, which allows total preemption, gives the highest average reward rate.

Keywords: ATM networks, Call admission control, Service integration, Preemption, Partial blocking, Markov decision theory

1. INTRODUCTION

Call Admission Control (CAC) in Asynchronous Transfer Mode (ATM) networks should support an efficient integration of the Variable Bit Rate (VBR), Constant Bit Rate (CBR), Available Bit Rate (ABR) and Unspecified Bit Rate (UBR) service classes. One of the main design issues is how to share the capacity between guaranteed services (CBR and VBR) and best effort services (ABR and UBR). The design must utilize that fact that best effort calls have the ability to reduce their bandwidth in case of congestion. Two methods that meet this constraint are the standard preemptive scheme and the standard partial blocking scheme.

In the standard preemptive scheme, best effort calls are preempted when guaranteed service calls arrive to a busy link. In this paper, the best effort calls that are chosen for preemption are selected at random. When calls depart from the link such that sufficient free capacity becomes available, a preempted best effort call enters service again. The preemptive scheme was analyzed in [3] in the case when all calls enter a queue before service. It was found that the scheme is capable of improving the link utilization at the expense of fairness. The common FIFO policy was shown to maintain fairness at some expense of link utilization.

In the standard partial blocking scheme [1, 2], the best effort services adapt their bandwidth requirement to the available capacity such that the bandwidth - holding time product remains constant. Each best effort call can specify a minimal accepted service ratio, $r_{min} \in (0,1]$ (along with the bandwidth requirement, b) which is used in the call negotiation process. A best effort call is accepted only if the available bandwidth b_a fulfills the criteria: $r_{min} b \leq b_a \leq b$. Throughout the lifetime of a call, the instantaneous service rate $r(t)$, defined as $b_a(t)/b$, may fluctuate according to the current load and available capacity on the link. The standard partial blocking scheme was analyzed in [1, 2] were it was found that the scheme gives low blocking probability and efficient link utilization for best effort calls.

The standard preemptive and partial blocking scheme was evaluated in [9] using optimal call admission control policies derived from Markov decision theory [10]. The two methods were shown to yield high average reward rates for different mixes of narrow-band and wide-band traffic. Several alternative methods to the Markov decision approach have been proposed in the literature, e.g. class limitation, trunk reservation and dynamic trunk reservation. The comparison presented in [6] indicates that for many cases, the trunk reservation and dynamic trunk reservation policies can provide fair, bandwidth efficient solutions, having performance close to the optimal Markov decision policy.

This paper evaluates the efficiency of Markov decision based call admission control policies for the standard preemptive scheme and a modified version of the partial blocking scheme. The modified partial blocking scheme is controlled by two different minimal service ratios. The first ratio, $r_{min,dim} \in (0,1]$, controls the access of best effort calls and limits the number of accepted best effort calls. The second ratio, $r_{min,user} \in [0,1]$, controls the access of guaranteed service calls. Using two minimal service ratios it is possible to both limit the time spent in the system for best effort calls and to allow a zero instantaneous service ratio. Note that the preemption occurring with $r_{min,user}=0$ is fair since all best effort calls will have their bandwidth reduced to zero upon preemption, which is not the case with the standard preemptive scheme.

Markov decision theory provides a computationally efficient technique to find the optimal CAC policy in terms of long-term reward. The Markov decision policy maps states to admission decisions (actions), i.e. to accept or reject a new call. The Markov decision approach evaluates the long-term reward of each action in each state, and chooses the action which maximizes the reward. The evaluation is based on a Markov model of the decision task, which comprises the state transition probabilities and the expected reward delivered at each state transition. The decision task model is parameterized by the call arrival and departure rates, which are assumed to be measured on line.

The Markov decision technique has been applied to the link access control problem [7] and the network routing problem [4] assuming that blocked calls are lost. The technique has also been applied to link allocation [8] and routing problems [5] in the context of blockable narrow-band and queueable wide-band call traffic.

This paper is organized as follows. In the next section, the CAC problem is introduced. Section 3 presents a Markov decision model for the CAC task for the standard preemptive scheme and for the modified partial blocking scheme. Section 4 describes the policy iteration technique of Markov decision theory in which the value determination problem is handled by solving a sparse linear equation system. Section 5 presents the numerical results. Finally, section 6 concludes the paper.

2. THE CAC PROBLEM

In the CAC problem, a link with capacity C [units/s] is offered calls from K traffic classes of CBR[1] and ABR calls. Calls belonging to class $j \in J=\{1, 2, ... K\}$ have the same bandwidth requirements and similar arrival and holding time dynamics. For ease of presentation, we consider K=2 traffic classes throughout the rest of this paper. The two classes consists of a narrow-band ABR class and a wide-band CBR class, indexed by 1 and 2, respectively.

We assume that class-j calls with peak bandwidth requirement b_j arrive according to a Poisson process with average rate λ_j [s^{-1}], and that the CBR call holding time is exponentially distributed with average $1/\mu_2$ [s]. The ABR call holding time for the preemptive scheme and the partial blocking schemes is exponentially distributed with average $1/\mu_1$ in the case when the call experiences no preemption and no partial blocking, respectively. If the ABR calls are partially blocked, the call holding time can be calculated by techniques from Markov driven workload processes, see [2].

The task is to find a CAC policy π that maps *request states* $(j,x) \in J \times X$ to *admission actions* $a \in A$, $\pi: J \times X \rightarrow A$, such that the long-term reward is maximized. The set A contains the possible admission actions, $\{ACCEPT, REJECT\}$. The set X contains all feasible *system states*. For the preemptive scheme it is given by:

1. VBR calls can be modelled the same way adopting the notion of effective bandwidth.

$$X_1 = \left\{(n_1, n_2, p) : p = 0,\ n_j \geq 0,\ \sum_{j \in J} n_j b_j \leq C\right\} \cup \tag{1}$$

$$\left\{(n_1, n_2, p) : p \in \{1, 2, ..., p_{max}\},\ n_j \geq 0,\ \sum_{j \in J} n_j b_j = C\right\},$$

where n_j is is the number of class-j calls accepted on the link, and p is the number of preempted ABR calls, which can take on the values $p \in P = \{0, 1, ..., p_{max}\}$. For later use, we also introduce the set of feasible *link states* for the preemptive scheme:

$$N = \left\{(n_1, n_2) : n_j \geq 0,\ \sum_{j \in J} n_j b_j \leq C\right\}. \tag{2}$$

For the partial blocking scheme, the set of feasible system states to enter when admitting best effort calls is given by:

$$X_{2,dim} = \left((n_1, n_2) : n_j \geq 0,\ n_1 b_1 r_{min,dim} + n_2 b_2 \leq C\right) \tag{3}$$

The set of feasible system states to enter when admitting guaranteed service calls is given by:

$$X_{2,user} = \{(n_1, n_2) : 0 \leq n_1 \leq \lfloor C/(b_1 r_{min,dim}) \rfloor,$$

$$0 \leq n_2 \leq \lfloor C/b_2 \rfloor, n_1 b_1 r_{min,user} + n_2 b_2 \leq C\} \tag{4}$$

where $r_{min,dim} \in (0,1]$ is a minimal accepted service ratio used for dimensioning purposes, i.e. to control the number of ABR calls in the system, and $r_{min,user} \in [0,1]$ is the minimal service ratio acceptable for the user when admitting guaranteed service calls. Note that $r_{min,dim} \geq r_{min,user}$.

3. A MARKOV DECISION MODEL FOR CAC

This section presents a Markov decision model for CAC for the standard preemptive scheme and the modified partial blocking scheme. The Markov decision model specifies a Markov chain which is controlled by actions in each state. The actions result in state transitions and reward delivery to the system. The control objective is to find the actions that maximize the average reward accumulated over time. In the current application, the Markov chain evolves in continuous time, and we therefore face a semi-Markov decision problem (SMDP).

The SMDP state x corresponds to the system state in the previous section, i.e. $x=(n_1,n_2,,p)$ for the preemptive scheme, and $x=(n_1,n_2)$ for the partial blocking scheme. The SMDP action a is represented by a vector $a=(a_1,a_2)$, corresponding to admission decisions for presumptive call requests. The action space for both the preemptive and the partial blocking scheme becomes:

$$A = \{(a_1,a_2) : a_j \in \{0,1\}, j \in J\}. \tag{5}$$

were $a_j=0$ denotes call rejection and $a_j=1$ denotes call acceptance. The permissible action space in state x is a state-dependent subset of A. For the preemptive scheme, the permissible action space becomes:

$$\begin{aligned} A_1(x) = \{(a_1, a_2) \in A : a_1 &= 0 \text{ if } n + \delta_1 \notin N, \\ a_2 &= 0 \text{ if n} + \delta_2 - \Delta(n_1, n_2)\delta_1 \notin N \text{ or p} + \Delta(n_1, n_2) \notin P\} \end{aligned} \tag{6}$$

where $n=(n_1,n_2)$, δ_j denotes a vector with zeros except for a one at position j, and

$$\Delta(n_1, n_2) = \theta\left(\frac{1}{b_1}\left[\sum_{j \in J} n_j b_j - C + b_2\right]\right) \tag{7}$$

where $\theta(s)=0$ if $s \leq 0$ and $\theta(s)=\lfloor s \rfloor$ if $s>0$. The quantity $\Delta(n_1,n_2)$ denotes the number of ABR calls that should be preempted in link state (n_1,n_2) in order to reserve capacity for a new CBR call. For the partial blocking scheme, the permissible action space becomes:

$$A_2(x) = \{(a_1,a_2) \in A: a_1 = 0 \text{ if } n+\delta_1 \notin X_{2,dim},\ a_2 = 0 \text{ if } n+\delta_2 \notin X_{2,user}\} \tag{8}$$

The Markov chain is characterized by state transition probabilities $p_{xy}(a)$ which expresses the probability that the next state is y, given that action a is taken in state x. For the preemptive scheme, the state transition probabilities for $j \in J$ become:

$$p_{xy}(a) = \begin{cases} \lambda_j a_j \tau(x,a), & n_y = n_x + \delta_j \in N, \\ & p_y = p_x = 0, \\ \lambda_2 a_2 \tau(x,a), & n_y = n_x + \delta_2 - \Delta(n_1, n_2)\delta_1 \in N, \\ & p_y = p_x + \Delta(n_1, n_2) \in P, n_x + \delta_2 \notin N \\ n_{xj} \mu_j \tau(x,a), & n_y = n_x - \delta_j + min(b_j/b_1, p_x)\delta_1 \in N, \\ & p_y = max(p_x - b_j/b_1, 0) \in P, \\ n_{xj} \mu_j \tau(x,a), & n_y = n_x - \delta_j \in N, \\ & p_y = p_x = 0, \\ 0 & \text{otherwise} \end{cases} \tag{9}$$

where the quantity $\tau(x,a)$ denotes the average sojourn time in state x:

$$\tau(x,a) = \left[\sum_{j \in J} n_{xj} \mu_j + a_j \lambda_j \right]^{-1} \tag{10}$$

The first term in the state transition probability expression above gives the state transition probability for a CBR or ABR call arrival to a link with some free capacity without any preemption of ABR calls. The second term gives the state transition probability for a CBR call arrival to a link with sufficient free capacity after preemption of ABR calls. The third term gives the state transition probability for CBR or ABR call departures when the preemption queue is non-empty. The fourth term gives the state transition probability for CBR or ABR call departures when the preemption queue is empty.

For the partial blocking scheme, the state transition probabilities become:

$$p_{xy}(a) = \begin{cases} \lambda_1 a_1 \tau(x,a), & n_y = n_x + \delta_1 \in X_{2,dim}, \\ \lambda_2 a_2 \tau(x,a), & n_y = n_x + \delta_2 \in X_{2,user}, \\ n_{x1} \mu_1 r(x) \tau(x,a), & n_y = n_x - \delta_1 \in X_{2,user}, \\ n_{x2} \mu_2 \tau(x,a), & n_y = n_x - \delta_2 \in X_{2,user}, \\ 0, & \text{otherwise} \end{cases} \tag{11}$$

where $r(x)$ denotes the instantaneous service ratio in state x:

$$r(x) = \begin{cases} 1, & \sum_{j \in J} n_j b_j \le C, \\ [C - n_{x2} b_2]/(n_{x1} b_1), & \sum_{j \in J} n_j b_j > C, \end{cases} \tag{12}$$

The average sojourn time in state x is given by:

$$\tau(x, a) = \left[n_{x1}\mu_1 r(x) + n_{x2}\mu_2 + \sum_{j \in J} a_j \lambda_j \right]^{-1} \tag{13}$$

The expected accumulated reward in state x is given by $R(x,a)=q(x)\tau(x,a)$. For the preemptive scheme the reward accumulation rate is given by $q(x)=\Sigma_{j \in J} r_j n_{xj} \mu_j$. For the partial blocking scheme the reward accumulation rate is given by $q(x)=r_1 n_{x1} \mu_1 r(x)+r_2 n_{x2} \mu_2$. The quantity r_j, which specifies the reward for carrying a type-j call, can be written $r_j=r_j' b_j/\mu_j$, where r_j' denotes the normalized reward parameter. In this paper, we let the normalized reward parameter depend on the pricing model used for call charging.

4. MARKOV DECISION COMPUTATIONS

This section describes a method for solving the CAC task, formulated as a SMDP. The method of choice is *policy iteration*, which is one of the computational techniques within Markov decision theory to determine an optimal policy.

The admission to the link is controlled by the so-called gain function, $g_j(x,\pi)$. This function simply measures the increase in long-term reward due to acceptance of a class j call in state x under policy π. Calls are accepted if the gain function is positive and rejected otherwise. The gain function can be expressed in terms of the relative value function, $v(x,\pi)$, as $g_j(x,\pi)=v(x+\delta_j,\pi)-v(x,\pi)$. The difference $v(x,\pi)-v(y,\pi)$ can be interpreted as the expected difference in accumulated reward over an infinite interval starting in state x instead of in state y under policy π. The relative value function is computed by the policy iteration algorithm.

The policy iteration algorithm computes a series of improved policies in an iterative manner. The computation of an improved policy π_{k+1} from the current policy π_k involves three steps:

- task identification
- value determination
- policy improvement

The first step involves determining the Markov decision model, i.e. the state transition probabilities and the expected rewards. These quantities are parameterized by link call arrival rates λ_j and call departure rates μ_j, see section 3. The arrival/departure rates are obtained from measurements to make the Markov decision model adaptive to actual traffic characteristics. The measurement period corresponds to the policy improvement period. The measurement period should be of sufficient duration for the system to attain statistical equilibrium.

The second step involves computing the relative value function for the current policy. The value determination step consists of solving the set of linear equations:

$$\begin{aligned} v(x,\pi) &= R(x,a) - g(\pi)\tau(x,a) + \sum_{y\in X} p_{xy}(a)v(y,\pi) \qquad ;x \in X \\ v(x_r,\pi) &= 0 \end{aligned} \tag{14}$$

where x_r is an arbitrary chosen reference state (e.g. the empty state) and $g(\pi)$ denotes the average reward rate. The solution involving all the $v(x,\pi)$ and $g(\pi)$ can be obtained by any standard routine for sparse linear systems.

The third step is the actual policy improvement. This step consists of finding the action that maximizes the relative value in each state:

$$\max_{a\in A(x)} \left\{ R(x,a) - g(\pi)\tau(x,a) + \sum_{y\in X} p_{xy}(a)v(y,\pi) \right\} \qquad ;x \in X \tag{15}$$

Policy iteration can be proved to converge to an optimal policy in a finite number of iterations in the case of finite state and action space [10].

The proposed method can be summarized as follows. Choose an initial admission policy π and a relative value function $v(x,\pi)$. During a finite period, allocate calls according to the gain function associated with the chosen relative value function. At the same time, measure traffic statistics (call arrival rates and call departure rates) in order to determine the Markov decision task for the current policy. Evaluate the applied policy in the context of the current Markov decision task, by solving a sparse linear equation system, and improve the policy. Apply the new policy during the next period, measure the traffic statistics and repeat the policy evaluation and the policy improvement step and so forth.

5. NUMERICAL RESULTS

This section evaluates the performance of two CAC methods for the preemptive scheme and the partial blocking scheme: the Markov decision (MD) method and the complete sharing (CS) method. Performance measures of interest are the average reward rate and the average time an ABR call spends in the system (the call holding time). For the preemptive scheme, the preemption probability is also evaluated.

The results are based on simulations for a single link with capacity C=48 [units/s], which is offered different mixes of ABR (class 1) and CBR (class 2) traffic. The bandwidth requirements are b_1=1, b_2=6 [units/s], and the mean call holding times $1/\mu_1=1/\mu_2=1$ [s], assuming that the ABR calls experiences no preemption and no partial blocking.

The arrival rates λ_1 and λ_2 were varied so that the average offered traffic equalled the link capacity:

$$\frac{b_1\lambda_1}{C\mu_1}+\frac{b_2\lambda_2}{C\mu_2} = 1.0 \tag{16}$$

A step size of 0.2 in the arrival rate ratio λ_1/λ_2 has been used when plotting all the figures. Moreover, the curves presented in the figures are obtained after averaging over 30 simulation runs and 95% confidence intervals, computed assuming normally distributed values, are also shown for each curve.

Figure 1 and 2 shows the average reward rate for the preemptive scheme for different arrival rate ratios, different maximal sizes of the preemption queue, and different normalized reward parameters for the ABR class.

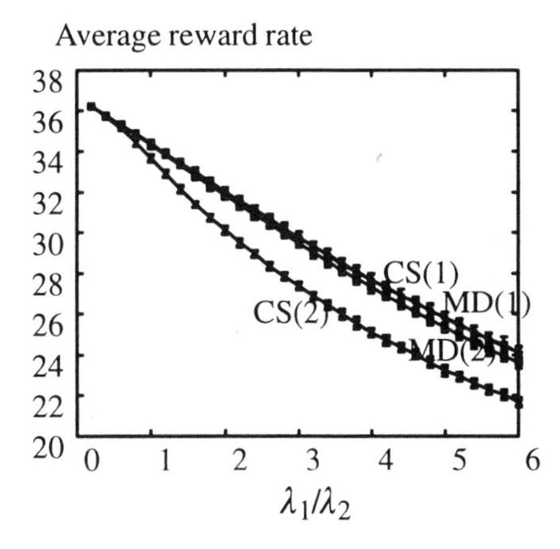

Figure 1: Average reward rate for different arrival rate ratios for the preemptive scheme with r_1'=0.05. Case 1 has p_{max}=24. Case 2 has p_{max}=6.

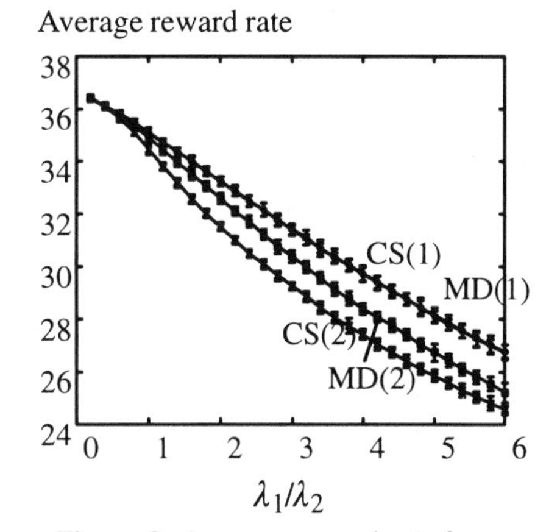

Figure 2: Average reward rate for differentarrival rate ratios for the preemptive scheme with r_1'=0.20. Case 1 has p_{max}=24. Case 2 has p_{max}=6.

When the maximal size of the preemption queue is large (p_{max}=24), the average reward rate of the MD and CS method are similar. When the maximal queue size is small (p_{max}=6), the MD method gives a larger average reward rate compared to the CS method. The reason is that for small maximal queue sizes the MD method implements so called "intelligent blocking" in individual states. By rejecting narrow-band call requests, typically when the free capacity equals the size of a wide-band call, bandwidth is reserved for the wide-band class, which increases the long-term reward.

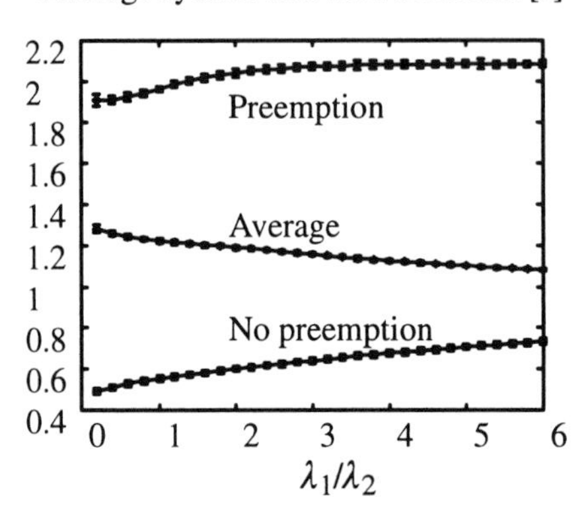

Figure 3: Average system time for ABR calls for different arrival rate ratios for the preemptive/MD scheme with p_{max}=24 and r_1'=0.05.

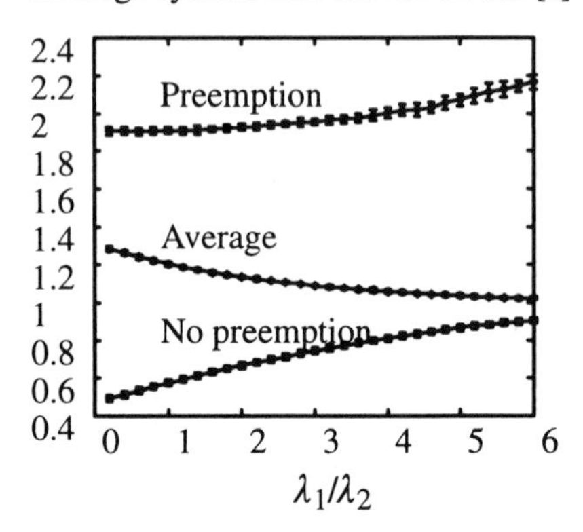

Figure 4: Average system time for ABR calls for different arrival rate ratios for the preemptive/MD scheme with p_{max}=6 and r_1'=0.05.

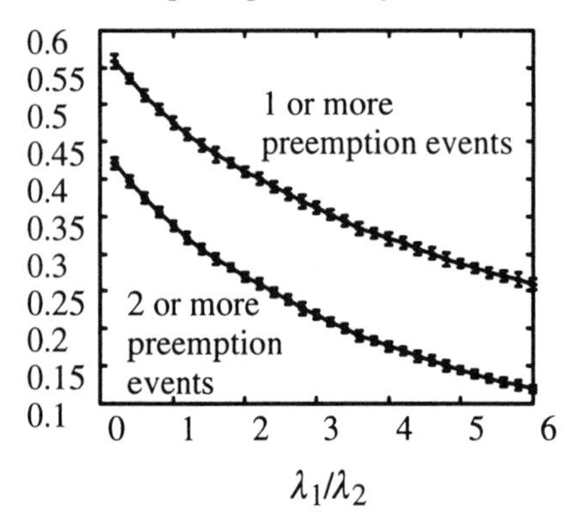

Figure 5: Preemption probability for ABR calls for different arrival rate ratios for the preemptive/MD scheme with p_{max}=24 and r_1'=0.05.

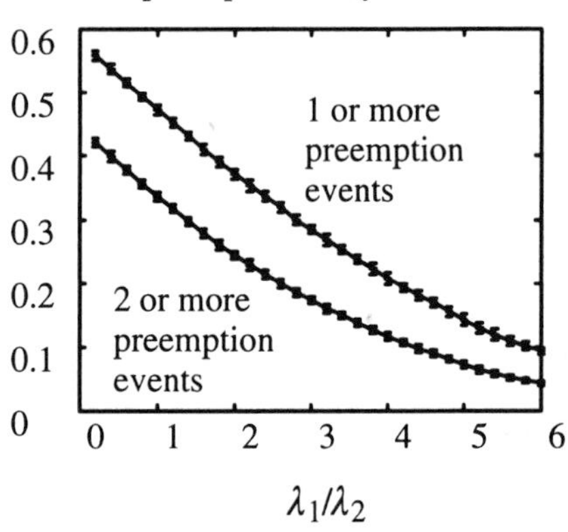

Figure 6: Preemption probability for ABR calls for different arrival rate ratios for the preemptive/MD scheme with p_{max}=6 and r_1'=0.05.

Figure 3 and 4 shows the average time an ABR call spends in the system in the preemptive scheme with the MD method for different maximal sizes of the preemption queue. Three different curves are shown in each figure. The lower curve shows the average system time for calls that are not preempted. The middle curve shows the average system time taking all calls (preempted and not preempted) into account. The upper curve shows the average system time for calls that are preempted. The lower curve is below 1 since short calls are more likely not to be preempted.

Figure 5 and 6 show the preemption probability for the preemptive scheme with the MD method for different arrival rate ratios and different maximal queue sizes. Two different curves are shown in each figure. The upper curve shows the probability of preemption occurring 1 or more times during the lifetime of an ABR call. The lower curve shows the probability of preemption occurring 2 or more times.

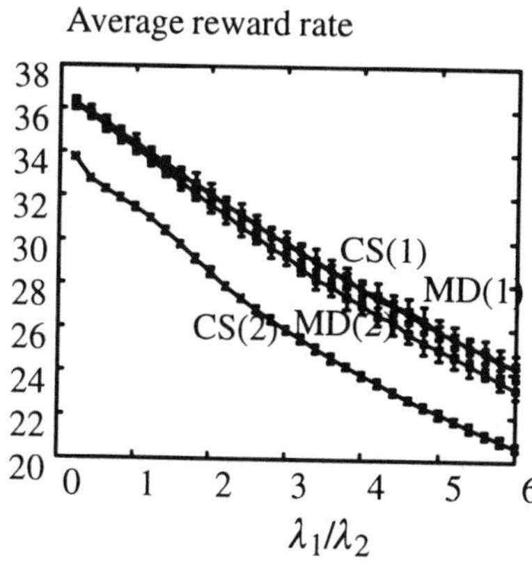

Figure 7: Average reward rate for different arrival rate ratios for the partial blocking scheme with price factor r_1'=0.05.
Case 1 has $r_{min,dim}$=0.5, $r_{min,user}$=0.
Case 2 has $r_{min,dim}$=0.5, $r_{min,user}$=0.5.

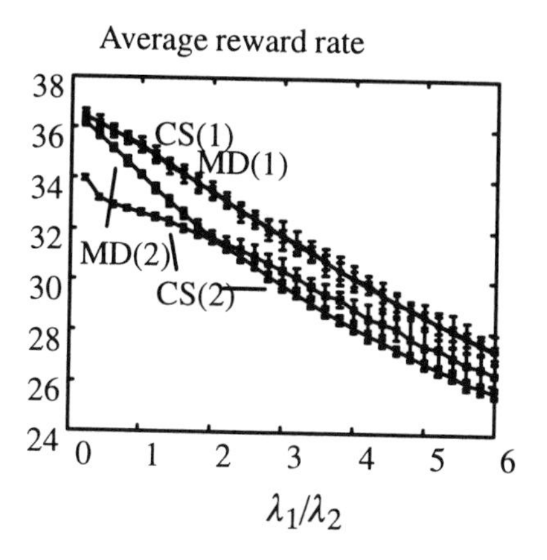

Figure 8: Average reward rate for different rrival rate ratios for the partial blocking scheme with price factor r_1'=0.20.
Case 1 has $r_{min,dim}$=0.5, $r_{min,user}$=0.
Case 2 has $r_{min,dim}$=0.5, $r_{min,user}$=0.5.

Figure 7 and 8 shows the average reward rate for the partial blocking scheme for different arrival rate ratios, different values of $r_{min,dim}$ and $r_{min,user}$, and different values of the normalized reward parameter for the ABR class. When $r_{min,user}$ =0, i.e. when total preemption is allowed, there is no performance difference between the MD and CS method. When $r_{min,user}$ =0.5, the intelligent blocking feature of the MD method results in a higher average reward rate. The narrow-band ABR class is blocked in all link states when the normalized reward parameter for the ABR class is low (r_1'=0.05). When the value is higher (r_1'=0.20) the MD policy blocks ABR calls in all link states when $\lambda_1/\lambda_2<2$, and in individual link states when $\lambda_1/\lambda_2>2$.

When the narrow-band ABR class is completely blocked, we face a severe fairness problem. However, the complete blocking can be avoided by increasing the normalized reward parameter r_1' for the ABR class. Of course, we can not expect the average reward rate to be as high as when the narrow-band ABR class is completely blocked since the blocking probability for the wide-band CBR class will increase. Nevertheless, changing the normalized reward parameters is a simple way to control the distribution of blocking probabilities among different call classes [4].

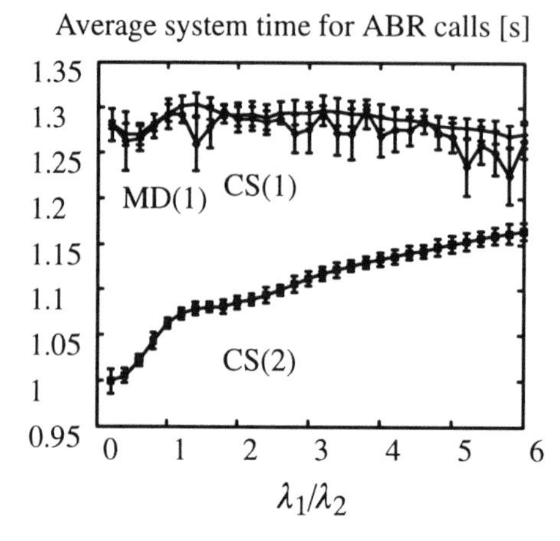

Figure 9: Average system time for ABR calls for different arrival rate ratios for the partial blocking scheme with r_1'=0.05.
Case 1 has $r_{min,dim}$=0.5, $r_{min,user}$=0.
Case 2 has $r_{min,dim}$=0.5, $r_{min,user}$=0.5.

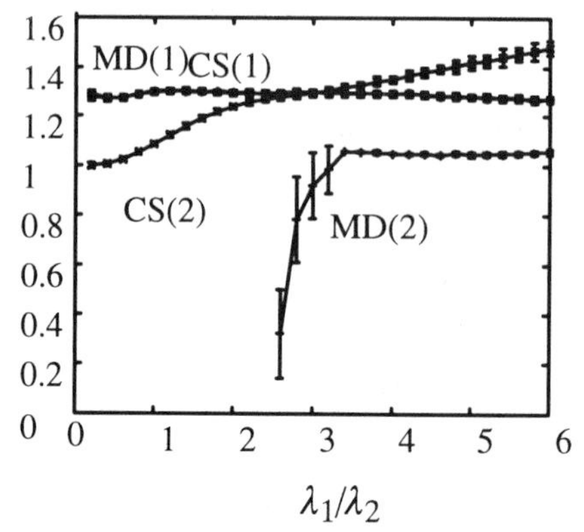

Figure 10: Average system time for ABR calls for different arrival rate ratios for the partial blocking scheme with r_1'=0.20.
Case 1 has $r_{min,dim}$=0.5, $r_{min,user}$=0.
Case 2 has $r_{min,dim}$=0.5, $r_{min,user}$=0.5.

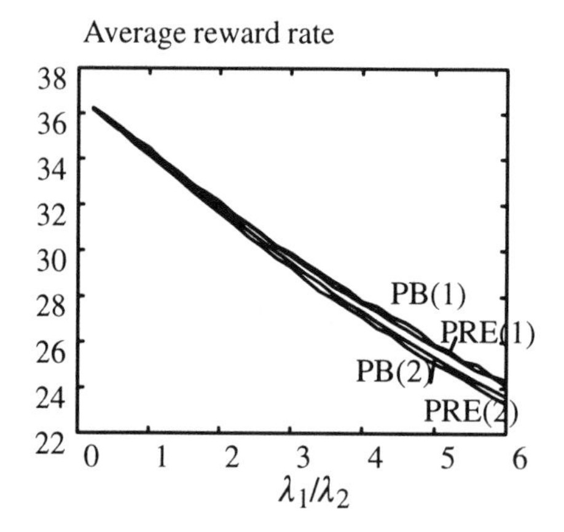

Figure 11: Average reward rate comparison between the preemptive/MD and partial blocking/MD scheme for different arrival rate ratios with r_1'=0.05.
PB case 1: $r_{min,dim}$=0.5, $r_{min,user}$=0.
PB case 2: $r_{min,dim}$=0.5, $r_{min,user}$=0.5.
PRE case 1: p_{max}=24.
PRE case 2: p_{max}=6.

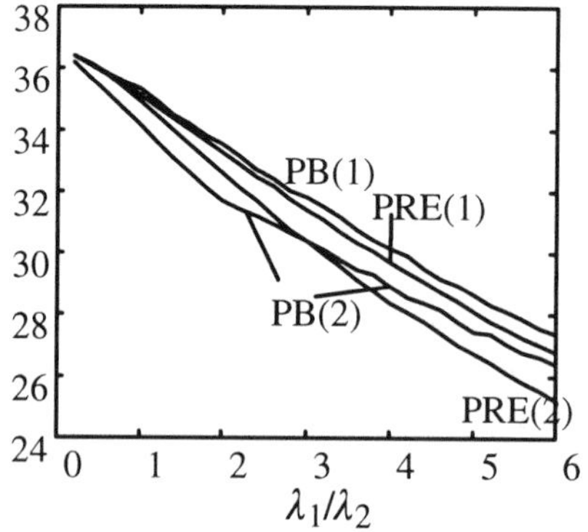

Figure 12: Average reward rate comparison between the preemptive/MD and partial blocking/MD scheme for different arrival rate ratios with r_1'=0.20.
PB case 1: $r_{min,dim}$=0.5, $r_{min,user}$=0.
PB case 2: $r_{min,dim}$=0.5, $r_{min,user}$=0.5.
PRE case 1: p_{max}=24.
PRE case 2: p_{max}=6.

Figure 9 and 10 shows the average system time for ABR calls in the partial blocking scheme with the MD method for different arrival rate ratios and different values of $r_{min,dim}$ and $r_{min,user}$. No curve is shown for the case when ABR calls are blocked in each link state. In figure 9, the case 1 curve for the Markov decision method has larger confidence intervals than the case 1 curve for the complete sharing method.

For comparison, figure 11 and 12 shows the average reward rate for different realizations of the preemptive and the partial blocking scheme with CAC based on the MD method. The method with highest average reward rate is obviously partial blocking with $r_{min,user}$=0, i.e. when total preemption is allowed. Confidence intervals are not shown in order to improve the readability of the figures.

The results presented in the figures were obtained after 10 adaptation epoches with the policy iteration method. Each adaptation period contained 100 000 simulated call events. The performance values in the figures are based on measurements of 400 000 call events after policy convergence.

6. CONCLUSION

This paper has evaluated the efficiency of Call Admission Control (CAC) based on Markov decision theory for two schemes that supports integration of guaranteed services and best effort services: the standard preemptive scheme and the modified partial blocking scheme. The Markov decision technique can be used to compute CAC policies that are optimal in terms of long-term reward. The optimality is achieved by intelligent blocking of narrow-band ABR/UBR calls, either completely, or at link states where typically the free capacity equals the size of a wide-band call.

The presented numerical results show that the Markov decision method yields higher long-term reward than the complete sharing method when the ability to create sufficient capacity for wide-band CBR calls through partial blocking/preemption is limited. The results also show that the modified partial blocking scheme, which allows total preemption ($r_{min,user}$=0), gives the highest average reward rate.

ACKNOWLEDGEMENTS

The author would link to thank Jakob Carlström, Søren Blaabjerg and Gábor Fodor for stimulating discussions. This work was financially supported by NUTEK, the Swedish National Board for Industrial and Technical Development.

References

[1] Blaabjerg S and Fodor G, A Generalization of the Multirate Circuit Switched Loss Model to Model ABR Services in ATM Networks, in Proc. of the IEEE International Conference on Communications, Singapore (1996).

[2] Blaabjerg S, Fodor G and Andersen A, A Partially Blocking-Queueing System with CBR/VBR and ABR/UBR Arrival Streams", in Proc. of 5th International Conference on Telecommunications Systems, Nashville, USA (1997).

[3] Kraimeche B and Schwartz M, Bandwidth Allocation Strategies in Wide-Band Integrated Networks, IEEE Journal on Selected areas in Commun., vol. SAC-4, no. 6, pp. 869-878 (1986).

[4] Dziong Z and Mason L, Call Admission and Routing in Multi-Service Loss Networks, IEEE Trans. on Commun., vol. 42, no. 2, pp. 2011-2022 (1994).

[5] Dziong Z, Liao K and Mason L, Flow Control Models for Multi-Service Networks with Delayed Call Set Up, in proc. of INFOCOM'*90,* pp. 39-46, San Francisco, USA, (1990).

[6] Dziong Z and Mason L, Fair-Efficient Call Admission Control Policies For Broadband Networks - A Game Theoretic Framework, *IEEE/ ACM Trans. on Networking.*, vol. 4, no. 1, pp. 123-136, (1996).

[7] Ross K and Tsang D, Optimal Circuit Access Policies in an ISDN Environment: A Markov Decision Approach, IEEE Trans. on Commun., vol. 37, no. 9, pp. 934-939, (1989).

[8] Nordström E, Near-Optimal Link Allocation of Blockable Narrow-Band and Queueable Wide-Band Call Traffic in ATM Networks, in Proc. of 15th International Congress on Telecommunications, ITC'15, Washington D.C., USA (1997).

[9] Nordström E, Blaabjerg S and Fodor G, Admission Control of CBR/ VBR and ABR/UBR Call Arrival Streams: A Markov Decision Approach, in proc. of IEEE ATM'97 Workshop, Lisboa, Portugal (1997).

[10] Tijms H, Stochastic Modeling and Analysis - a Computational Approach, Wiley (1986).

PREDICTIVE RESOURCE ALLOCATION FOR REAL TIME VIDEO TRAFFIC IN BROADBAND SATELLITE NETWORKS

H.O.Awadalla, L.G.Cuthbert and J.A.Schormans
Department of Electronic Engineering, Queen Mary & Westfield College, Mile End Road, London E1 4NS, UK. {h.o.awadalla,l.g.cuthbert,,j.a.schormans}@elec.qmw.ac.uk

Abstract Satellites are attractive for broadband networks because they provide fairly high bit-rate connections over a wide area. However, satellite networks are characterised by having a fixed amount of shared bandwidth between all users, and by relatively large propagation delays. These problems pose a challenge when implementing a high performance dynamic resource allocation scheme to effectively share the bandwidth among a large population of Earth terminals, with heterogeneous mix of traffic and different QOS requirements. This paper proposes the implementation of a predictive resource allocation scheme for the real time video component of the traffic in a broadband satellite network environment. Two predictive implementations are investigated: one based on time series analysis, and the other based on neural networks.

Keywords: ATM Networks, Broadband Satellite, Resource Allocation, Time Series Analysis, Neural Networks.

1. INTRODUCTION

Variable bit rate (VBR) video traffic is expected to be one of the major traffic types that need to be supported by broadband satellite networks. Broadband satellite networks will use ATM and most of the video encoding will be done using the MPEG standard (ISO Moving Picture Expert Group) [1]. Hung et al [2], have shown that for the up-link access, delay sensitive VBR connections should use both variable-rate and fixed-rate demand assignments. This is because variable-rate demand assignment on its own can cause excessive cell transfer delay (CTD) resulting in unacceptable cell losses at the receiving end. In Hung's scheme (which we refer to as the classical scheme), the Earth terminals send the value of the occupancy of their buffers to the satellite. The occupancy of the terminal buffer can be defined by the following recursion [2]:

$$occupancy(n)=[occupancy(n-1)+a(n)-o^{fixed}(n)-o^{variable}(n)-o^{free}(n)]+ \quad (1)$$

Where a(n) is the number of cell arrivals to the terminal queue of a particular connection at the end of the nth TDMA frame, $o^{fixed}(n)$, $o^{variable}(n)$ and $o^{free}(n)$ are the number of slots assigned by fixed-rate DA, variable-rate DA, and free assignment, respectively, to the connection on the uplink during the nth frame. Here, x+ = max(x,0). At time (n+z), the satellite on-board scheduler makes the assignments for the up-link frame that will be received by the terminals at time (n+2z). z is the propagation delay expressed in frames.

In this paper, we propose an implementation for a predictive resource allocation scheme where the assignment of the TDMA frame slots to the real-time VBR connections is performed according to the predicted MPEG frame values. One predictive implementation is based on time series analysis that is easy to incorporate, does not require any modification to the scheduler and does not require great computational complexity at the terminals. The other predictive implementation studied is based on neural networks and this requires greater computational complexity at the terminals, but like the time series analysis does not require modification to the scheduler.

Our results show that the time series analysis implementation is found to be comparable to the neural networks implementation in terms of the utilisation of both the ground and space segments resources.

1.1 The Broadband Satellite Network

This paper considers a broadband TDMA satellite communication system that uses an on-board packet switch in conjunction with an on-board scheduler to provide broadband services to a large population of Earth terminals. The Earth terminals can be both fixed and portable, with peak bit rates (up to 8Mbit/s) as shown in figure 1. TDMA is used for the up-link while TDM is used for the down-link. Residential and business applications that may use video encoding based on the MPEG standard include among other applications; direct-to-home video, telecommuting, distance learning and corporate training, collaborative group work and telemedicine.

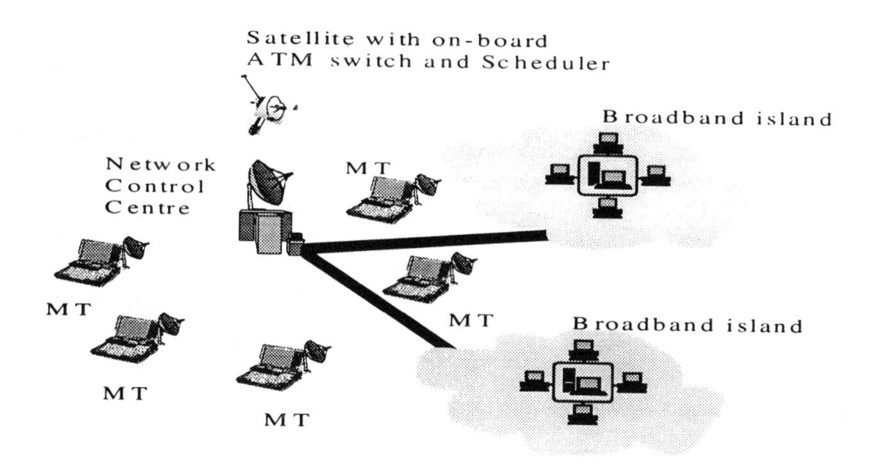

Figure 1 The broadband satellite network envisaged.

2. PROPERTIES OF MPEG VIDEO DATA STREAMS

The MPEG coding algorithm is now widely used in multimedia-video communications. An MPEG stream contains deterministic periodic sequences of frames referred to as a Group of Pictures (GOP) [3].

Frames can be encoded in three modes: intra-frames (I-frames), forward predicted frames (P-frames) and bi-directional predicted frames (B-frames) as shown in figure 2.

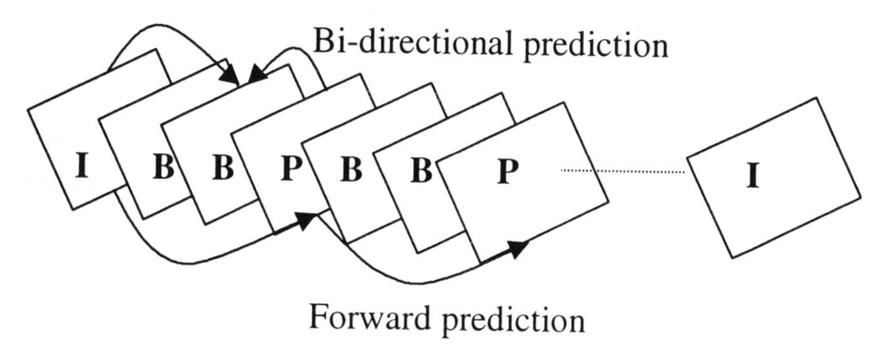

Figure 2 MPEG Group of Pictures (GOP) pattern

An I-frame is encoded as a single image, with no reference to any past or future frames. The encoding scheme used is similar to JPEG compression for still images. A P-frame is encoded relative to the past reference frame, the future frame (which is the closest following I- or P-frame), or both. The encoding algorithm for B-frames is similar to P-frames, except that the motion vectors may refer to areas in the future reference frames.

In our work, we used two empirical MPEG-1 video traces from the movie Jurassic Park and a German "talk show". The traces vary in the degree of GOP-by-GOP correlation and frame-by-frame correlation as well as the scene activity. The traces are each about half an hour long and are available from the internet [4]. The sequence of MPEG I, P and B frames that was used is IBBPBBPBBPBB.......(12 frames/GOP) and there are 40,000 frames in total per trace (each frame time is 40 ms).

The statistical properties of the traces are shown below in Table 1:

Sequence	**Mean [Mbit/s]**	**Peak [Mbit/s]**
Jurassic Park	0.33	1.01
Talk show	0.36	1.00

Table 1 Simple statistics of the encoded sequences

3. IMPACT OF THE PREDICTIVE IMPLEMENTATION ON THE GROUND SEGMENT

Implementing a predictive resource allocation scheme for the real time video connections at the ground segment (Earth terminals) entails modifications to the terminal design. Instead of generating the bandwidth demands of all connections at the terminal using Hung's recursion (equation 1), the demands of the real time video connections are generated using the predicted MPEG frames values. Hence the bandwidth generation process will have two flows as inputs: a continuous flow indicating the queue occupancy and another flow that is short lived (at connection set-up) from the real time video sources as shown in figure 3. This latter flow contains information about the MPEG properties of the video streams, e.g. the frames durations and GOP pattern. These properties are required by the predictive function which resides at the "Bandwidth requests generation" as shown in figure 3.

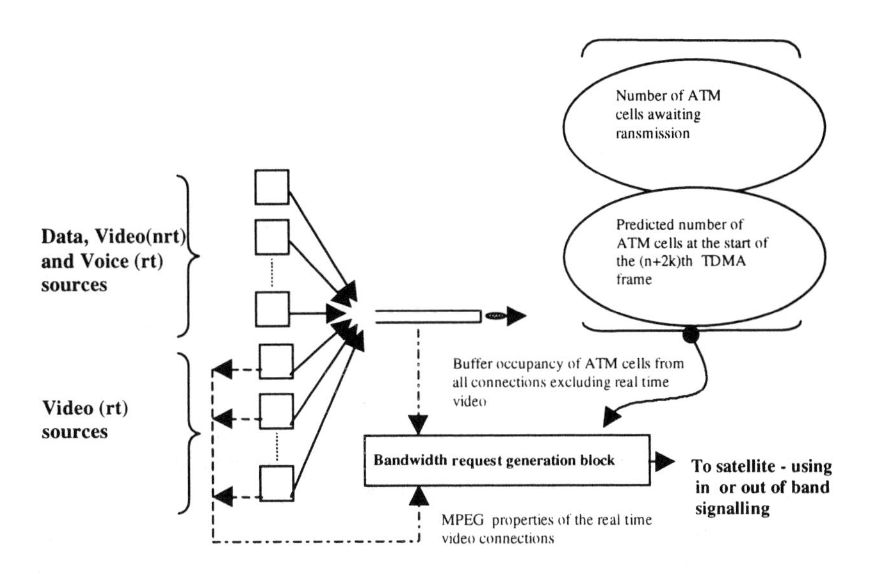

Figure 3 A conceptual design of the Earth terminal

4. PREDICTION OF MPEG TRAFFIC USING TIME SERIES ANALYSIS

In our proposed scheme, we assume that the MPEG stream is a time series $X(t)$ (*that has seasonal effects as well as trends*) generated by an additive model of the form:

$$X(t)=T(t)+S(t)+R(t) \quad t=1,2,.... \tag{2}$$

Here $T(t)$ is the trend, $S(t)$ the seasonal term, and $R(t)$ is the irregular or random term. We regard each frame of the GOP pattern as a unique seasonal term so that we have the following seasonal effects S_1, S_2, ...S_{12}. Each seasonal effect, has period s; that is, it repeats after s time periods:

$$S(t+s) = S(t) \qquad \forall \ t \tag{3}$$

No trend was observed in the three traces, hence we set $T(t)$ to zero. We also set $R(t)$ to zero. The jth seasonal is denoted as $S(j)$ and the period is taken as s. Then for the additive model we use an exponentially weighted moving average (EWMA) and, following [5], we have two smoothed series:

$$M(t) = a.(X(t -S(t-s)) + (1-a).M(t-1) \ \ |a| < 1 \tag{4a}$$

$$S(t) = b.(X(t) - M(t)) + (1-b).S(t-s) \quad |b| < 1 \tag{4b}$$

Where a and b are the discounting parameters. The forecast of $X(t+k)$ made at time t is:

$$F(t+k) = M(t) + S(t+k-s) \tag{5}$$

5. PREDICTION OF MPEG TRAFFIC USING NEURAL NETWORKS

The earth terminal must be able to predict at least 250 ms ahead to account for the round trip delay. A feed-forward three layered (30-15-10) Neural Network was trained on each trace. Fifteen successive MPEG frames were used as inputs to the Neural Network to predict the twenty third successive frame which starts after (40*7) = 280 ms from the end of the fifteenth frame, as shown in figure 4. The extra 30 ms ensures that processing delays are catered for.

As there are no rules for choosing the number of inputs as well as the number of hidden layers and neurons, a number of experiments were required to achieve a near optimum configuration for the neural network. It was found that the more inputs the neural network has, the better the

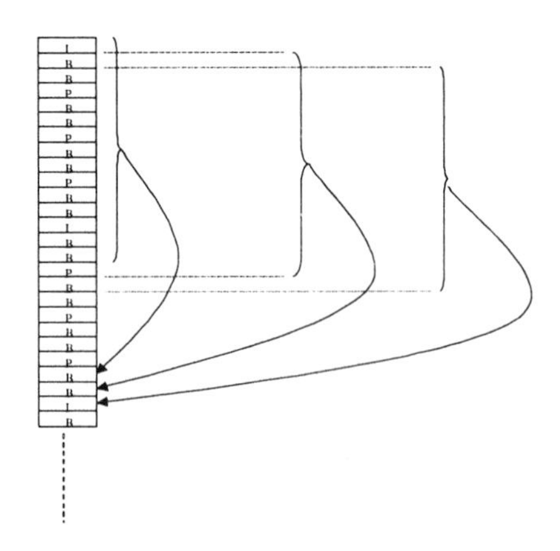

Figure 4 Prediction using Neural Networks

prediction. However, it was also noticed that the improvement is not substantial for more than ten inputs. Also the more inputs and hidden layers, the greater the complexity that makes the implementation of the neural network at the earth terminal unattractive.

The learning algorithm used by the neural network was the Back Propagation [6] algorithm and the activation function was the sigmoid function, which is a commonly used activation function for back propagation neural networks. A training file of 5000 samples (1/9 of the whole MPEG trace) was used to train the neural network (*this was carried out for each trace*). All inputs were scaled so that they lay in the range (0-1). This was achieved by dividing all samples of each training file by the size of the largest frame in the respective trace.

6. SIMULATION STUDIES AND RESULTS

In the simulations, we assume that each MPEG trace starts at the same time as the start of a TDMA frame and that the MPEG video traffic is deterministically smoothed over the MPEG frame period, i.e. cells emitted during a frame period are evenly spaced. We can then write the forecast value of the integer number of cells that will arrive during the frame starting at (*t+ k+tdma frame duration*) as a result of the predicted MPEG frame as:

$$E = \left\lceil F(t+k) \cdot \textit{tdma frame duration/mpeg frame duration} \right\rceil \tag{6}$$

k in the case of the time series analysis is equal to 440 ms *(11*40)* and for the neural network case, it is equal to 280 ms *(7*40)*.

To allow for under-prediction, we allocate ϕ number of fixed slots in every TDMA frame (or multiple of frames) to the real time video connection. The earth terminal signals a demand of:

$$D=E+\phi \tag{7}$$

for bandwidth to the satellite through inband or out-of-band signalling as opposed to signalling the buffer occupancy as is the case in the classical scheme. For simplicity we set the TDMA frame duration equal to the MPEG frame duration and size the TDMA frame slot to hold one ATM cell.

The values of the different parameters in the simulation are all shown below in table 2:

Uplink capacity	8 Mbit/s
TDMA frame duration	40 ms
Round trip propagation delay (RTD)	250 ms
Discount parameter “a”	0.075
Discount parameter “b”	0.75

Table 2 Simulation parameters values

When using time series analysis, large values of “b” and small values of “a” make the smooth series react rather quickly to changes, hence the choice of “a” and “b”. Also, in the time series analysis, when estimating the seasonal effects (S_1, S_2, ...S_{12}), we consider the first two GOP pattern data and following [5] we take the seasonal minus the overall mean.

6.1 Performance Objective

The ground segment and space segment resources are the Earth terminal’s buffer and the TDMA frame slots respectively.

To compare the proposed predictive and the classical schemes, a number of simulations were performed with different values of o^{fixed} for the classical scheme and different values of ϕ for the predictive schemes. The performance objective in all three schemes was the same: a cell transfer delay of less than (RTD+TDMA frame duration), i.e. a cell delayed by more than one frame duration, at the sending terminal’s buffer is discarded at the receiving terminal.

We define the relative mean allocation rate as the ratio of the mean allocation rate to the mean cell rate. Figures 5,6 and figures 7,8 show the cell discard probability and the mean buffer size when each of the three traces is considered (plotted against the relative mean allocation rate). It is apparent that the predictive schemes proposed here, requires less TDMA frame slots to achieve a specific performance objective as well as less buffer space than the classical scheme.

The time series analysis predictive implementation is less complicated than the neural network one and compares well with it, bearing in mind that the neural network predicts frames 280 ms ahead while the time series analysis scheme predicts frames 440 ms ahead.

A Neural Network has to be trained on a portion of the traffic to be able to predict accurately. If however, the traffic characteristic changes, the neural network may not be able to achieve reasonable performance.

For all the above reasons, we suggest that it will be more economical and feasible to implement the predictive scheme using time series analysis.

7. CONCLUSIONS

Two predictive implementations for resource allocation for real time video traffic in a satellite network, were presented. The implementations are based on time series analysis and Neural Networks.

Both implementations outperform the classical assignment scheme in the usage of both the ground and space segment resources of the satellite network. Moreover, they do not require extra computational effort from the satellite, which is vital if the scheduler is to be housed on board the satellite. However, the time series analysis implementation is more attractive than the Neural one as it needs minimum computational support at the Earth terminals. A conceptual design for the Earth terminal to cope with a predictive implementation for resource allocation has also been outlined.

References

[1] D.LE GALL: 'MPEG: A video compression standard for multimedia applications'. Communications of ACM, 34(4), April 1991, pp. 46-58

[2] A..HUNG, M.J.MONTPETITT, G.KESIDIS: ' A Framework for ATM via Satellite', IEEE GLOBECOM, London, Nov.1996, vol.2., pp.1020-1025

[3] O.ROSE: 'Statistical properties of MPEG video traffic and their impact on traffic modelling in ATM systems'. Research Report Series, Report No.101, February 1995,Institute of Computer Science,University of Wuerzburg, Germany.

[4] ftp-info3.informatik.uni-wuerzburg.de Directory: /pub/MPEG/traces

[5] G.JANACEK, L.SWIFT: 'Time Series: Forecasting,simulation,applications", Ellis Horwood series in mathematics and its application, Chichester, pp.14 –32

[6] D.E.RUMELHART, G.E.HINTON, R.J.WILLIAMS: 'Learning representations By Back-Propagating Error.' Nature, 1986, 323:533-536. Reprinted in Anderson & Rosenfeld, 1988, pp. 696-699

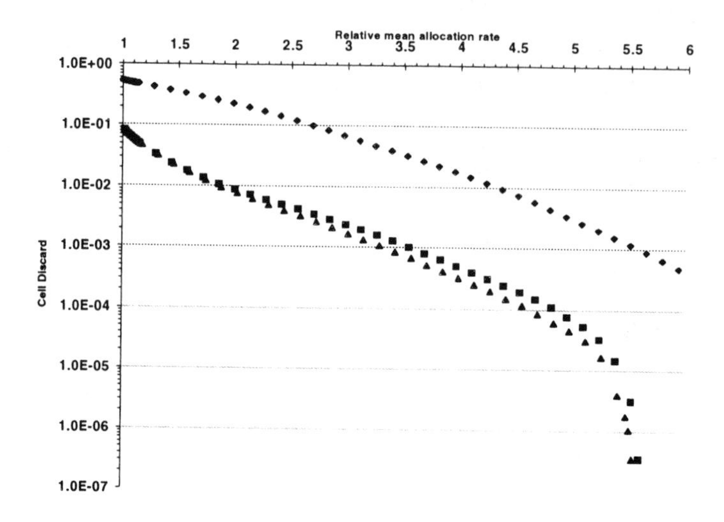

Figure 5 Cell discard (Jurassic Park trace)

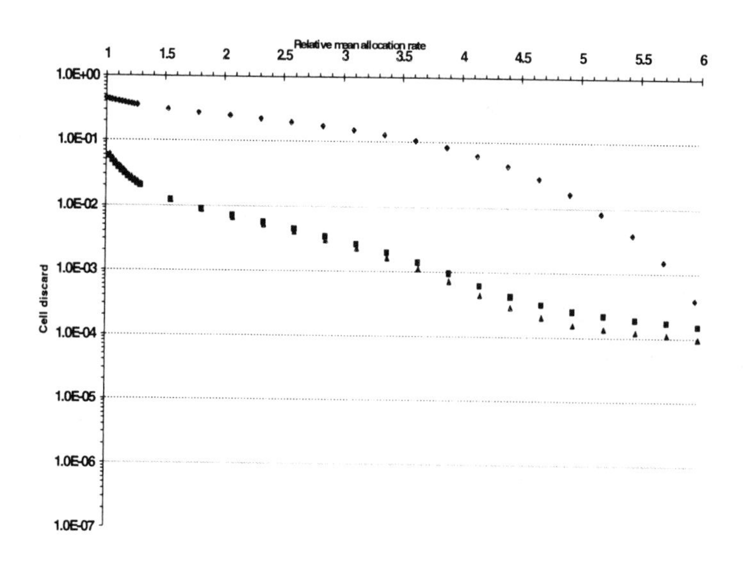

Figure 6 Cell discard (Talk show trace)

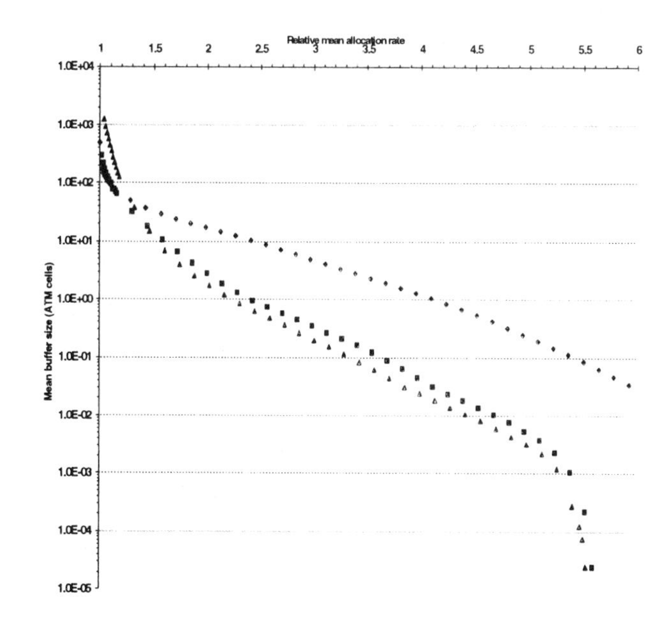

◆ Classical scheme ■ Predictive "Time Series Analysis – EWMA" scheme
▲ Predictive "Neural" scheme

Figure 7 Mean buffer size (Jurassic Park trace)

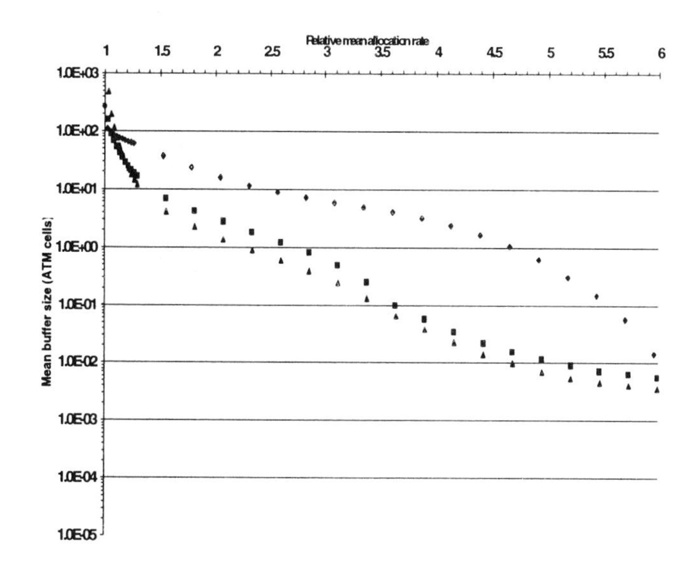

◆ Classical scheme ■ Predictive "Time Series Analysis – EWMA" scheme
▲ Predictive "Neural" scheme

Figure 8 Mean buffer size (Talk show trace)

ADAPTIVE NEURAL CONGESTION CONTROLLER FOR ATM NETWORK WITH HEAVY TRAFFIC

Ng Hock Soon, N. Sundararajan, and P. Saratchandran
School of Electrical and Electronic Engineering,
Nanyang Technological University, Singapore.
ensundara@ntu.edu.sg

Abstract This paper presents an adaptive control scheme using a newly developed Minimal Resource Allocation Network (MRAN) to solve the traffic congestion problem in ATM networks. MRAN generates a minimal radial basis function neural network by adding and pruning hidden neurons based on the input data and is ideal for on-line adaptive control for fast time varying nonlinear systems. The ATM traffic modeling is carried out using the well-known network simulation software OPNET for multiplexed traffic (combining both speech and video signals). Performance of MRAN controller is compared with conventional method and Back-Propagation (BP) neural network controller with the aim of minimizing the congestion episodes and maintaining the quality. Simulation results indicate that MRAN controller performs better than both conventional and BP controller in reducing the congestion and maintaining a better quality of the traffic.

Keywords: Congestion control, Neural networks, ATM

1. INTRODUCTION

Congestion control is a traffic management mechanism to protect the network and the end-system from congestion in order to achieve network performance objectives, while at the same time promoting the efficient use of network resources. Congestion control refers to the set of actions taken by the network to minimize the intensity, spread and duration of congestion. Feedback flow control is one of the solutions that have been extensively studied in the literature [1]. In feedback control schemes, when possible traffic congestion is detected at any network element, feedback signals are sent to all the sources and the traffic submitted to ATM connections is then regulated by modifying the source coding rates. Recently, use of artificial neural networks (ANN) in traffic management of ATM networks is gaining momentum [2]-[3]. ANNs have several valuable properties that are quite useful when implementing ATM traffic control. First, ANNs can implement direct adaptive control tailored to the actual characteristics of the cell and /or call traffic. No explicit model of the traffic is needed as in traditional methods. ANNs can learn the relationships between many inputs and outputs and can explicitly consider propagation delay. Second, the parallel structure of ANNs can be exploited in hardware implementations, which provide short and predictable response times.

Recently, an adaptive controller using neural networks for congestion control in ATM multiplexers has been developed [4]. The motivation to use neural networks is to utilize their learning capabilities to adaptively control a non-linear dynamic system without having to define an accurate analytical model of the system. The neural network learns the dynamics of the system from input/output examples. Another motivation is to use the adaptive capabilities of neural networks to handle unpredictable time varying and statistical fluctuations of ATM traffic, which can not be described by theoretical models.

In this scheme [4], the control signal is generated based on the real time measurement of arrival rate process and queuing processes which are indicative of the congestion episodes. This control signal is then fed back to the traffic sources to dynamically modulate the arrival rates by changing the source coding rates. The number of cells waiting in the multiplexer buffer is used as an indicator of congestion. During periods of buffer overloads, the source coding rates will be decreased at the expense of quality, since decreasing the coding rate will decrease the signal to noise (SNR) ratio of the traffic. The sources coding rate for the Adaptive Differential Pulse Code Modulation (ADPCM) scheme considered lie between 4 bits/sample, 3 bits/sample or 2 bits/sample. This involves a trade-off. The control law should try to strike a balance between minimizing the cell loss rate on one hand and maximizing the coding rate on the other hand. To achieve this, a performance index function which consists of two error terms are defined, one the difference between the desired and actual number of cells waiting in the buffer and the second error term which is the difference between the original uncontrolled coding rates of the coders and the controlled rate after applying the control signal. Maximizing the performance index involves in minimizing these two error terms and this is used to adjust the weights of the neural network. The neural network used is the well known back propagation feed forward network and the results indicate that the proposed neural adaptive control scheme can reduce the congestion in the network in a significant manner.

Recently, a new minimal radial basis function (RBF) neural network called Minimal Resource Allocation Network (MRAN) has been developed by the authors [5], which uses a sequential learning scheme for adding and pruning RBF hidden layer neurons, so as to achieve a minimal network with better approximation accuracy. When no neurons are added or removed, the algorithm uses an Extended Kalman Filter (EKF) to update the centers, widths and weights of each of the hidden neurons. This paper presents the application of MRAN for adaptive congestion control scheme for ATM networks. In comparison to the adaptive controller in [4] where the neural network had a fixed structure i.e. fixed number of neurons and only its weights were adjusted, in the proposed scheme the network builds up the hidden neurons from the input data and it does this in an efficient manner to realize a compact RBF network with better approximation accuracy. Also, instead of adjustments of only the weights as in [4] the proposed MRAN adaptive control scheme provides for adjustments of the centers, widths and also the weights which result in better approximation for the input -output nonlinear functions.

The paper is organized as follows. Section 2 describes the proposed adaptive control scheme using MRAN for congestion control of ATM traffic, which is similar to that of [4]. Section 3 describes briefly MRAN algorithm. Section 4 describes the adaptive neural control for ATM networks under heavy traffic using OPNET while section 5 show MRAN controller performs to clear the heavy congestion in the network. Conclusions from this study are summarized in Section 6.

2. ADAPTIVE CONGESTION CONTROLLER FOR ATM NETWORKS

Figure 1 shows the adaptive congestion control scheme using MRAN and is similar to the scheme in [4] except that the neural controller is based on MRAN instead of BP network. In Fig.1, the controlled source coding rate is defined by the equation:

$$C(k) = C_o u(k) \tag{1}$$

where
C(k) = controlled coding rate at sample k
$C_o(k)$ = maximum uncontrolled coding rate of the source
u(k) = feedback control signal produced by the controller at sample k
n(k+1) = number of cells in the buffer at sample (k+1)
$n_d(k+1)$ = desired number of cells in the buffer at sample (k+1),

$n_d(k+1) \leq n_{max}$ (maximum length of the buffer)
u(k+1) = feedback control signal at the sample (k+1)
$u_d(k+1)$ = desired value of the feedback control signal which is also the maximum value of the feedback control signal: $u(k+1) \leq u_d(k+1)$
z^{-1} represents a unit delay.

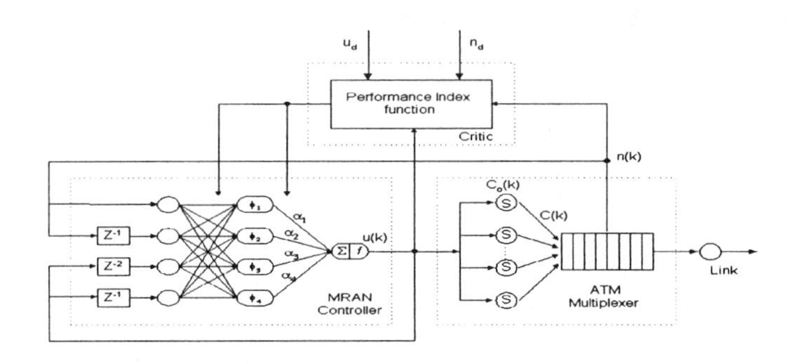

Figure 1 Adaptive Congestion Controller Using MRAN

The congestion control system consists of a critic part and a neural networks controller part. The inputs to the control algorithm are taped delay values of the number of cells in the multiplexer buffer (which is a measure of potential congestion problem) and the taped delay of the feedback control signal. The controller's output is a predicted optimal control signal that is fed back to the input sources to alter their coding rates. This will directly control the traffic arrival rate. During overflow condition, the control signal will reduce the packet arrival rate by decreasing the coding rate of the ADPCM for both bursty and VBR sources. On the other hand, the coding rate is switched back to higher level to maintain the traffic quality.

The critic part involves the performance index of the system (cost function) to be minimized. According to this cost function, the critic part evaluates the system performance and generates an evaluation signal that is a function of the deviation of the system performance from the desired optimal level and is used to change the weights of the neural network controller. Hence, if the control signal is driving the system toward the desired objectives, it is reinforced. Otherwise, the weights are changed to generate a correct control signal. The control signal value will keep updating to minimize two error signals over the measurement period: the difference between the original uncontrolled coding rate of the coders and the controlled rate after applying the control signal; and the difference between the desired and actual number of cell in the buffer. There is a tradeoff between these two objectives, means minimizing the former will at the same time increase the second error signal term. The performance index function (J) is given as below:

$$J(P) = \sum_{k=1}^{L} R_n S_n(k+1)\varepsilon_n^2(k+1) + R_u \varepsilon_u^2(k+1)) \tag{2}$$

P = trial number
L = length of the measurement period
$S_n(k+1)$ = Reward Signal to reset the control signal as long as the number of cells in the buffer is less than the desired level.
$= 0$ if $n(k+1) < n_d(k+1)$
$= 1$ if $n(k+1) \geq n_d(k+1)$
R_n = weight value on the buffer overflow performance index
R_u = weight value the on the the coding rate performance measure.
$\varepsilon_n^2(k+1) = (n_d(k+1) - n(k+1))^2$
$\varepsilon_u^2(k+1) = (u_d(k+1) - u(k+1))^2$
ϵ_u = deviation from voice/video quality from its maximum coding rate.
ϵ_n = cell loss term.

The term ϵ_n represents the cell loss and the term ϵ_u represents the deviation of the traffic quality from its maximum value. Thus, the feedback control signal is determined such that it minimizes both the cell loss rate and the deviation of the traffic quality from its original uncontrolled value.

3. MINIMAL RESOURCE ALLOCATION NETWORK(MRAN)

The MRAN is a minimal Radial Basis Function Neural Network (RBFNN) which is a sequential learning algorithm recently developed by Yingwei et al [5] which combines the growth criterion of RAN with a pruning strategy to realize a minimum RAN. The hidden layer consists of an array of neurons (ϕ_1 to ϕ_n) connected to the output by n connection weights (α_1 to α_n). The output of the hidden layer is the vector $\phi_k(\mathbf{x})$ with m inputs $x(x_1$ to $x_m)$. The second layer of the RBF network is essentially a linear combiner. The overall network response is:

$$f(x) = \alpha_0 + \sum_{k=1}^{K} \alpha_k \phi_k(\mathbf{x}) \tag{3}$$

where $\phi_k(x)$ is the response of the kth hidden neuron to the input x, and α_k is the weight connecting the kth hidden unit to the output unit. α_0 is the bias term. Here, K represents the number of hidden neurons in the network. $\phi_k(x)$ is a Gaussian function given by,

$$\phi_k(x) = exp(-\|x - \mu_k\|^2/\sigma_k^2) \tag{4}$$

where μ_k is the center and σ_k is the width of the Gaussian function. $\| \;\|$ denotes the Euclidean norm.

In the MRAN algorithm, the network begins with no hidden units. As each input-output training data (x_n, y_n) is received, the network is built up based on certain growth criteria. The algorithm adds hidden units, as well as adjusts the existing network, according to the data received. The criteria that must be met before a new hidden unit is added are :

$$\|x_n - \mu_{nr}\| > \epsilon_n \tag{5}$$

$$e_n = y_n - f(x_n) > e_{min} \tag{6}$$

$$e_{rmsn} = \sqrt{\sum_{i=n-(M-1)}^{n} \frac{e_i^2}{M}} > e_{min1} \tag{7}$$

where μ_{nr} is the center (of the hidden unit) which is closest to x_n, the data that was just received. ϵ_n , e_{min} and e_{min1} are thresholds to be selected appropriately. Equation (5) ensures that the new node to be added is sufficiently far from all the existing nodes. Equation (6) decides if the existing nodes are insufficient to obtain a network output that meets the error specification. Equation (7) checks that the network has not met the required sum squared error specification for the past M outputs of the network. Only when all these criteria are met, is a new hidden node added to the network. Each new hidden unit added to the network will have the following parameters associated with it :$\alpha_{K+1} = e_n$, $\mu_{K+1} = x_n$, $\sigma_{K+1} = \kappa\|x_n - \mu_{nr}\|$.

The overlap of the responses of the hidden units in the input space is determined by κ, the overlap factor. When an input to the network, does not meet the criteria for a new hidden unit to be added, the network parameters $w = [\alpha_0, \alpha_1, \mu_1^T, \sigma_1, ..., \alpha_K, \mu_K^T, \sigma_K]^T$ are adapted using the EKF as follows :

$$w_n = w_{n-1} + e_n k_n \tag{8}$$

where k_n is the Kalman gain vector given by,

$$k_n = [R_n + a_n^T P_{n-1} a_n]^{-1} P_{n-1} a_n \tag{9}$$

where a_n is the gradient vector (for details, see[5]),R_n is the variance of the measurement noise and P_n is the error covariance matrix which is updated by,

$$P_n = [I - k_n a_n^T] P_{n-1} + QI \tag{10}$$

where Q is a scalar that determines the allowed random step in the direction of the gradient vector. If the number of parameters to be adjusted is N, P_n is a $N \times N$ positive definite symmetric matrix.

The algorithm also incorporates a pruning strategy, which is used to prune hidden nodes that do not contribute significantly to the output of the network, or are too close to each other. The former is done by observing the output of each of the hidden nodes for a period of time, and then removing the node that has not been contributing a significant output for that period. Consider the output, o_k of the kth hidden unit :

$$o_k = \alpha_k exp(-\|x - \mu_k\|^2 / \sigma_k^2) \tag{11}$$

If α_k or σ_k in the above equation is small, o_k might become small. Also, if $\|x - \mu_k\|$ is large, the output will be small. This would mean that the input is far away from the center of this hidden unit. To reduce inconsistency caused by using the absolute values of the outputs, their values are normalized to that of the highest output. This normalized output of each node is then observed for M consecutive inputs. A node is pruned, if the output of that node falls below a threshold value for M consecutive inputs. The dimensions of the EKF are then reduced to suit the reduced network.

4. OPNET SIMULATION OF ATM WITH HEAVY TRAFFIC

The ATM traffic system is simulated using OPNET Modeler[6]. OPtimized Network Engineering Tools (OPNET) is a comprehensive engineering system capable of simulating communications networks with detailed protocol modeling and performance analysis. OPNET features include graphical specification of models; a dynamic, event-scheduled Simulation Kernel; integrated data analysis tools; and hierarchical, object-based modeling. OPNET's hierarchical modeling structure accommodates special problems such as distributed algorithm development.

The traffic model is shown in Figure 2. There are 3 kind of input sources: bursty, VBR and custom traffic. Two bursty sources with average arrival rate of 29 packets/sec and VBR sources are multiplexed into a FIFO queue. The custom traffic is set to randomly add in some heavy traffic load to overload the network, from there, observations of how the

MRAN and BP controller perform to overcome the congestion under heavy traffic condition can be made. Here, the custom traffic with constant arrival rate of 50 packet/sec was fed into the queue at the period of 50-100 sec, 200-230 sec, 350-400 sec and 500-550 sec. Each bursty source

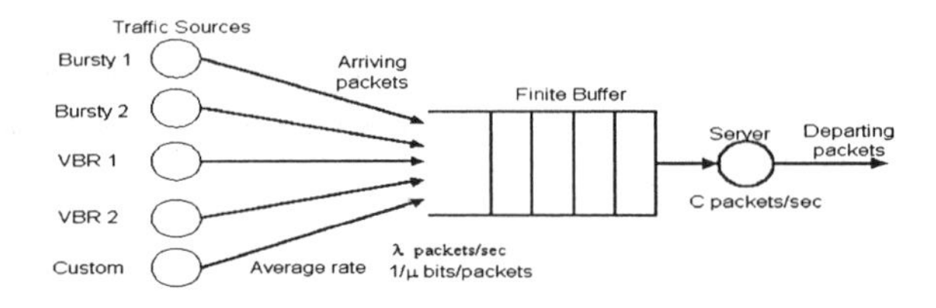

Figure 2 G/D/1/50 Traffic Model

is simulated using ON/OFF binary-state model. In this case, 29 cells are generated during ON period while no cell is generated during the OFF period. Both periods are exponentially distributed random variables with means $1/\beta = 0.35$ sec and $1/\alpha = 0.65$ sec. At the same time, the service capacity is set to 100 packets/sec which will lead to utilization over 100%. Consequently, severe traffic congestion will be occurred. The traffic control scheme will handle this problem by decreasing the source-coding rate. As a result, packet arrival rate will be reduced to avoid the congestion episode. On the other hand, compression made (by reducing the coding rate) will affect the quality of the traffic sources.

This G/D/1/50 queue consists of a first-in-first-out (FIFO) buffer with packets arriving randomly and a server, which retrieves packets from the buffer at a constant service rate. Its performance depends on three parameters: packet arrival rate, packet size, and service capacity. If the combined effect of the average packet arrival rate and the average packet size exceeds the service capacity, the queue size will be fill up immediately. In order to assess the performance of the controller, first the simulation is carried out without any controller and this results in a severe congestion. Figure 3 present the typical simulation results based on OPNET simulation for this case. Figure 3(a) shows the traffic situation without any controller and it can be clearly seen that the traffic condition is heavily congested especially for the period where custom traffic with a constant arrival rate of 50 packets/sec was pumped into the queue. The number of cell in the buffer is concentrated at the top of the capacity of the buffer as shown in Figure 3(b). This leads to a serious congestion problem where as we can observe from Figure 3(c), that overflow occurs more than 15,000 times. Thus, the buffer is said to be severely overloaded. As a result, a very high Cell Loss Rate at about 0.25 occurred all the time according to Figure 3(d). The cost function

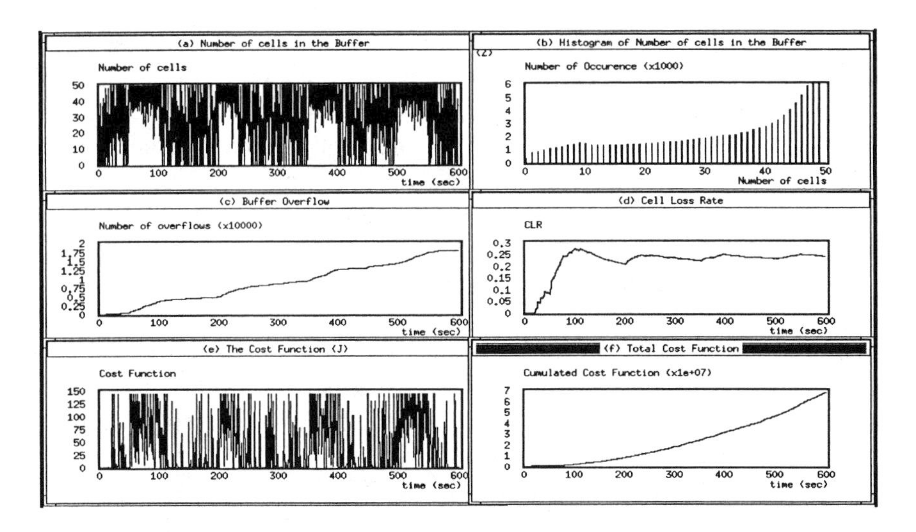

Figure 3 Multiplexed Traffic without Control

and the its accumulated sum are also shown in Figure 3(e) and Figure 3(f) respectively as a deviation measurement from its optimal condition.

5. PERFORMANCE OF MRAN CONGESTION CONTROLLER

After knowing that in the above scenario the congestion problem is severe and noting that it is quite hard to overcome congestion for a dynamically changing multiplexed traffic, the MRAN controller is integrated into the loop and Figure 4 shows the performance of the MRAN controller. In the adaptive traffic control system, MRAN will intelligently adjust the coding rate so that it can optimize the traffic quality and congestion. The time interval rate about 0.01 sec is small enough to obtain the significant changes in the queuing system. The length of the measurement period will affect the sensitivity for the neural network control system to overcome the congestion. However, too frequent updates may result in possible instabilities in the controller. Besides, the weight parameter R_u and R_n give the priority either for achieving good traffic quality or minimized cell loss rate.

The MRAN controller is allowed to operate and its efficiency in removing the congestion is assessed. Figure 4(a) shows the buffer size after applying the MRAN control. As we can observe, there is only a few short period of congestion. Initially, in the case without control, the buffer size is always full which lead to serious traffic congestion. Figure 4(b) shows the buffer occurrence histogram in which the buffer size is always below the maximum buffer capacity and it is concentrated in the range of 1–30 cells. This can definitely avoid overflow. Figure 4(c) shows the overflow vs. time. Figure 4(d) shows the Cell Loss Rate (CLR) which is the ratio of lost cells to the total number of transmitted cells to the buffer. Again, it is clear that MRAN works to keep the CLR as low as $2.5x10^{-3}$. At the same time, Figure 4(e) shows the cost function which is minimized by the traffic control system. The MRAN control tackles the congestion very well and tries to minimize the cost function immediately by adaptively changing the source-coding rate through feedback control signal. Figure 4(f) is the total cost function cumulated over the time. In the case without control, the cost function will keep increasing exponentially. However, MRAN keeps a control of it although there are a lot of dynamically traffic fluctuation condition.

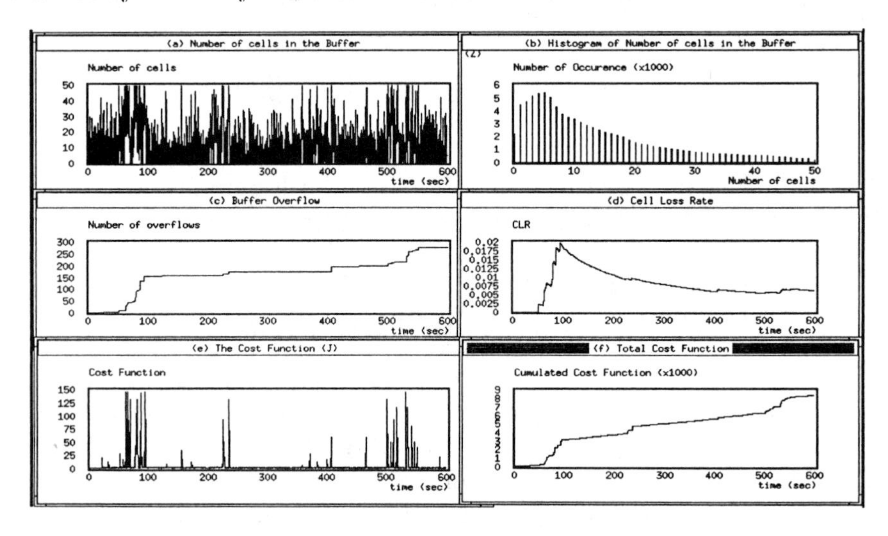

Figure 4 Multiplexed Traffic with MRAN Control

Some performance comparisons between the conventional, BP and MRAN congestion controller have been done. Conventional control which is a modified simple ERICA congestion control scheme is being used to reduce the packet generation rate to the queue when congestion is detected. This conventional control scheme will decrease the source coding rate by a factor of 0.10 during congested period and increased the source coding rate by a factor of 0.01 when the congestion is over. This decreasing rate chosen is 10 times larger than increasing rate, this is to ensure that congestion can be avoided effectively. At the same time, by increasing the traffic with 0.01 any immediate congestion situation that occured can be recovered faster. There are three most important criteria to be highlighted: Traffic Quality, Cell Loss Rate and the total cost function as a measure of the overall performance of the system. Figure 5 below shows the Cell Loss Rate for the conventional, BP and MRAN controller which have been plotted in the same graph.

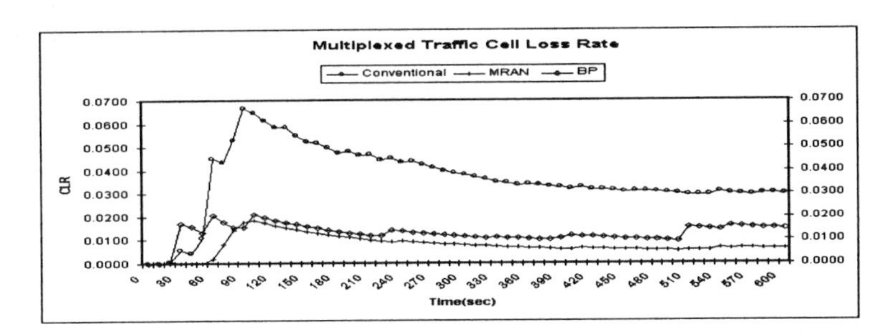

Figure 5 Cell Loss Rate

It is obvious that MRAN does a much better job to reduce and maintain the CLR as low as possible. At the same time, it manages to maintain the traffic quality as shown in Figure 6 although under severe congestion period when the Custom Source sent a large amount of packet for certain durations. On the other hand, the cumulated cost function is used to compare the overall performance of the controllers to do the best optimization job for maintaining the traffic quality and keep the CLR at very low level. This is shown in Figure 6 where MRAN controller is much better than BP along the simulation time. Again, the graph indicates that MRAN tackles the congestion problem faster and more efficiently than conventional or BP.

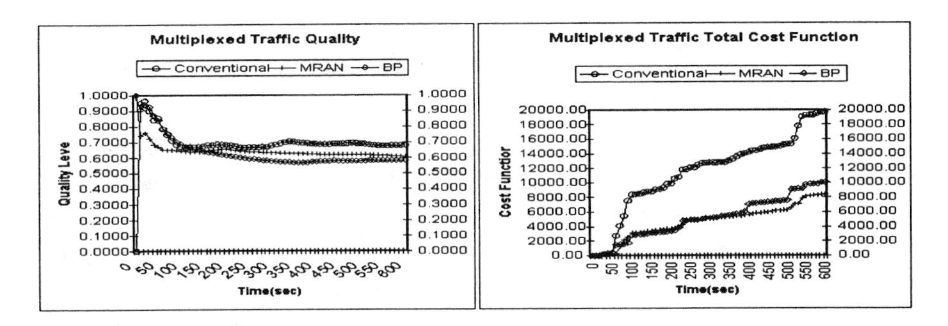

Figure 6 Traffic Quality

From the above simulation results, MRAN performed better where it can reduce CLR four times lower and at the same time maintain the traffic quality 5% higher than BP controller. Furthermore, MRAN can react faster with its most optimized network structure.

6. CONCLUSION

In this paper, three adaptive congestion control schemes using conventional method, BP neural networks and the recently developed MRAN are compared using OPNET simulation of ATM networks with heavy traffic. The neural network controllers generate feedback control signal in accordance to the traffic congestion situation and try to reduce the congestion episodes while maintaining the quality of the traffic. The performance index used as a measure of the traffic performance consists of two parameters with different weights, one concerning the cell loss rate while the other is related to the quality of multiplexed traffic. It is shown that MRAN can adapt and control the system more effectively as compared to conventional or BP controller even under heavy congestion. Based on a detailed comparison based on several simulation studies, it is shown that MRAN controller responds faster and is more efficient than conventional and BP schemes. This is due to the minimal network structure of MRAN which is suitable for fast sequential learning and application to the time-varying nonlinear dynamic system.

References

[1] P.Newman "Backward explicit congestion notification for ATM local area network". *Proc. IEEE GLOBECOM'93*, Houston, TX,1993, pp. 719-723.

[2] "Computational and Artificial Intelligence in High Speed Networks". *IEEE Journal on Selected Areas in Communication*, Vol. 15, No. 2, February, 1997.

[3] C. Douligerisand G. Develokos "Neuro-fuzzy Control in ATM Networks". *IEEE Communication Magazine*,Vol. 35, No. 5, May, 1997, pp. 154-162.

[4] I. Habib, A. Tarraf, and T. Saadawi "A neural network controller for congestion control in ATM multiplexers". *Computer Networks and ISDN Systems*, Vol . 29, No. 3, 1997, pp.325-334.

[5] Y. Lu, N. Sundararajan and P. Saratchandran "A Sequential Learning Scheme for Function Approximation Using Minimal Radial Basis Function Neural Networks". *Neural Computation*,Vol. 9, 1997, pp. 461 - 478.

[6] R. Goyal, R. Jain, S. Fahmy and S. Narayanaswamy "Modeling Traffic Management in ATM Networks with OPNET". *Proc. of OPNETWORK'98*,Washington DC, May 1998.

SESSION 15

Multicast Protocols

PERFORMANCE COMPARISON OF BRANCH POINT ALGORITHMS FOR MULTICAST ABR FLOW CONTROL

Dong-Ho Kim*, Jang-Kyung Kim *, Byung-Chul Kim**, and You-Ze Cho**
**Network Equipment Test Center, ETRI, Taejon, KOREA*
Phone: +82-42-860-6668, Fax: +82-42-961-5404
{dhkim, jkkim}@netc.etri.re.kr
***School of Electronic & Electrical Engineering, Kyungpook National University, KOREA*
bckim@palgong.kyungpook.ac.kr, yzcho@ee.kyungpook.ac.kr

Abstract This paper compares the performance of feedback consolidation algorithms with/without a fast overload indication function at a branch point switch for multicast (or point-to-multipoint) available bit rate (ABR) flow control in ATM networks. A new backward resource management (BRM) cell-discarding policy is proposed which controls additional BRM cell overhead due to fast overload indication function. The performance of various consolidation algorithms with the proposed fast overload indication function is also compared using simulations. The simulation results show that a fast overload indication function is very effective in a severe overload situation, particularly in an initial period with a higher initial cell rate. The fast overload indication function can be also combined with any feedback consolidation algorithm. However, its performance is highly dependent on the underlying basic consolidation algorithm employed.

Keywords: Feedback Consolidation Algorithm, BRM Cell Discarding Policy, Multicast ABR Flow Control, ATM.

1. INTRODUCTION

Multicast (or point-to-multipoint) available bit rate (ABR) service capabilities are essential for asynchronous transfer mode (ATM) networks to efficiently support many data applications including IP multicasting. The current version of the ATM Forum Traffic Management Specification defines ABR multicast capabilities, but it only specifies the fundamental behaviour of branch points of a multicast tree [1]. The source and destination behaviours of a multicast ABR connection are the same as those of a unicast (or point-to-point) ABR connection, except that data cells must not be transmitted in a backward direction. A branch point switch replicates each data cell and forward resource management (FRM) cell received from the source onto each branch that leads to a destination. Each destination returns the received FRM cells to the source, a branch point therefore receives backward RM (BRM) cells as many as the number of its downstream branches for each FRM cell. The key issue at a branch-point switch for multicast ABR services is how to consolidate BRM cells returning from downstream branches to avoid feedback implosion problem [1].

A number of feedback consolidation algorithms have been proposed in [3,7,9,10,11]. Consolidating BRM cells at a branch point, however, can cause undesirable effects such as *consolidation noise*, *consolidation delay,* and *consolidation loss* [3,7]. These problems can incur a slower response, inferior fairness, longer queue length, or lower link utilization for a multicast ABR connection compared with a unicast ABR connection. Recently, Kim et al. have proposed various solutions for resolving consolidation problems and a scalable feedback consolidation algorithm [7]. Their algorithm can eliminate consolidation noise and loss. Although, this algorithm can also limit the total consolidation delay within the longest round-trip time (RTT) of the multicast tree regardless of the number of branch points, it still exhibits a higher queue length due to consolidation delay particularly when an initial period with high initial cell rate (ICR) or a severe overload condition. Recently, fast overload indication functions have been proposed by many researchers in order to reduce consolidation delay [1,2,4,6,8]. The main idea behind this is that a severe overload situation should be reported to the source as soon as it is detected.

Most previous researches on multicast ABR flow control have focused on the development of basic consolidation algorithms and their performance evaluation [3,7,9,10,11]. Some papers have evaluated performance according to various fast overload indication functions for a particular algorithm [2,4,6,8]. However, relatively little has been reported on the performance comparison of basic consolidation algorithms with the same fast overload indication function.

The remainder of this paper is as follows: Section 2 introduces the existing feedback consolidation algorithms with/without fast overload indication function briefly. Section 3 presents feedback consolidation algorithm considered in this paper and proposes a new BRM cell discarding policy to control additional BRM cell overhead due to fast overload indication function. Section 4 compares the performance of feedback consolidation algorithms with/without a fast overload indication function by simulation. Finally, Section 5 makes conclusions.

2. RELATED WORKS

2.1 Basic Feedback Consolidation Algorithms

This paper classifies basic feedback consolidation algorithms according to two components as followings; (1) *Feedback information storing method for consolidation:* how to store feedback information extracted from BRM cells received from downstream branches in local variables and then consolidate them. (2) *BRM cell returning condition:* when to return the consolidated feedback information to the source. Table 1 shows this classification of basic feedback consolidation algorithms.

Table 1 Classification of basic feedback consolidation algorithms

Feedback storing method for consolidation	Per-VC	Most existing algorithms
	Per-branch for each multicast VC	Cho [3], Kim [7]
BRM cell returning condition	Wait for FRM	Roberts [10], Tzeng [11]
	Wait for BRM after FRM received	Ren's the first algorithm [9],
	Wait for BRM from all branches	Ren's the second algorithm [9]
	Wait for BRM from the farthest destination	Cho [3], Kim [7]

To date, a number of basic feedback consolidation algorithms have been proposed in [3,7,9,10,11]. A simple basic feedback consolidation algorithm for a multicast ABR service was proposed by Roberts [10]. In his algorithm, for every BRM cell received, the branch point first consolidates the congestion information to local variables maintained on a per-VC basis, and then discards the received BRM cell. Also, whenever a branch point receives an FRM cell, it returns a BRM cell along with consolidated information to the upstream node. However, this BRM cell may only contain congestion information about the current switch and not include any from the branches. In order to reduce such consolidation noise, Tzeng et al. in [11] proposed a slightly more conservative scheme. Their main idea was that a branch point would return a BRM cell to its upstream node only after receiving at least one BRM cell from its branches subsequent to the previous feedback.

Both algorithms discussed above require switches to generate BRM cells at the branch points, incurring higher complexity in switch implementation. Ren et al. proposed two consolidation algorithms that do not require switches to generate BRM cells [9]. In their first algorithm, a BRM cell that is received from a branch immediately after an FRM cell has been received by the branch point is passed back to the source. However, this algorithm may exhibit consolidation noise, similar to that experienced in scheme [11], if the branch point has not received BRM cells from one or more branches. Accordingly, to avoid such consolidation noise, in their second algorithm a BRM cell is only passed when BRM cells have been received from all branches. Ren's the first and second algorithms have been also included in the traffic management specification of the ATM Forum as sample branch-point switch algorithms [1].

However, in most existing algorithms, consolidating the BRM cells at a branch point can cause undesirable effects such as *consolidation noise*, *consolidation delay,* and *consolidation loss*. *Consolidation noise* means that the BRM cells returned to the source contain incorrect congestion information about some downstream branches. That is, this noise occurs due to the loss of congestion information about some branches and the loss of the latest feedback information which lead to a higher allowed cell rate (ACR) oscillation and a longer ramp-up delay respectively. Thus, it will introduce an inaccuracy into the computation of the ACR of the source. *Consolidation delay* indicates an additional delay at each branch point, since a branch point usually has to wait for an FRM cell from upstream node or BRM cells from other branches in order to relay the congestion information carried by a BRM cell to its upstream node. This additional delay can cause a slower response to congestion, resulting in a longer queue length or lower link utilization. *Consolidation loss* occurs when a branch point does not return as many BRM cells as the number of FRM cells received. Such loss of BRM cells can cause a slower response for a multicast VC [3,7].

Recently, Kim et al. have proposed various solutions for resolving those consolidation problems [7]. They have also proposed a scalable basic feedback consolidation algorithm by combining the proposed solutions. In their algorithm, a branch point stores feedback information on a per-branch basis for each VC and returns the BRM cells received from the farthest destination only. As a result, this algorithm can eliminate consolidation noise and loss, and also bounds the total consolidation delay within the longest RTT of the multicast tree regardless of the number of branch points. However, it still exhibits a larger queue length in some situations such as an initial period with a higher ICR or severe overload condition.

2.2 Feedback Consolidation Algorithms with Fast Overload Indication Function

Recently, a number of consolidation algorithms with fast overload indication functions have been proposed in [2,4,6,8] to facilitate an immediate warning about congestion when a severe overload condition occurs in a network. Table 2 summarises the characteristics of the existing consolidation algorithms with fast overload indication.

The fast overload indication was originally proposed by Jang et al. [6]. This algorithm uses Roberts' algorithm as its basic consolidation algorithm. All the other algorithms in [2,4,8] use Ren's the second algorithm as their basic consolidation algorithm.

Table 2 Feedback consolidation algorithms with fast overload indication function

Algorithms	Jang [6]	Moh [8]	Fahmy [4]	Chen [2]
Basic consolidation algorithm	Roberts' algorithm	Ren's the second algorithm	Ren's the second algorithm	Ren's the second algorithm
Overload condition detection	Last_ER, Dynamic threshold	Last_ER, Static threshold, Timer	Last_ER, Static threshold	Last_ER, Static threshold, Probability

In all the algorithms, a branch point detects an overload situation using a rate comparison with the last returned ER, Last_ER, and the current received ER, ER. If the ratio of the Last_ER and ER is larger than a specified threshold, a branch point will immediately send an extra BRM cell to the upstream node. Jang's algorithm uses a dynamic threshold using a load factor and BN (backward notification) bit in the BRM cells to detect the condition for sending an extra BRM cell. All the other algorithms use a static threshold, a timer, or a probability. However, their differences in performance are marginal, particularly when the threshold value is high [2].

Table 3 Characteristics of consolidation algorithms considered

Algorithms	1	2	3	4
Feedback storing method	Per-VC	Per-branch for each multicast VC	Per-VC	Per-branch for each multicast VC
BRM cell returning condition	Wait for all feedback	Wait for BRM from the farthest destination	Algorithm 1 + fast overload indication	Algorithm 2 + fast overload indication
Comments	Ren's the second algorithm [9]	Kim [7]	Modified from Algorithm 1	Modified from Algorithm 2

3. CONSOLIDATION ALGORITHMS CONSIDERED

Table 3 summarises the characteristics of the consolidation algorithms considered for comparison in this paper. Four consolidation algorithms are considered. Algorithms 1 and 2 are basic consolidation algorithms without fast overload indication. Algorithm 1 is Ren's the second algorithm, which has been widely used, as a basic consolidation algorithm in many papers [2,4,8,9]. Algorithm 2 is a counter-based basic feedback consolidation algorithm proposed by the authors in [7]. Algorithms 3 and 4 are the improved versions of Algorithms 1 and 2, respectively, with an added fast overload indication for reducing consolidation delay. In Algorithms 1 and 3, a branch point stores feedback information received from downstream branches on a per-VC basis. In contrast, in Algorithms 2 and 4, a branch point stores feedback information on a per-branch basis for each multicast VC.

3.1 Algorithm 1

The main idea of this algorithm is that a BRM cell is passed to the source only when BRM cells have been received from all branches [9]. The pseudo code is as follows.

A branch point switch maintains resisters MER, XER, flags MCI, MNI, XNI, and M flags for each multicast VC.

```
On the receipt of an FRM (ER, CI, NI) cell:
        Let XER = ER, XNI = NI;
        Multicast this FRM cell to all participating branches;

On the receipt of a BRM (ER, CI, NI) cell from branch i:
        Set M(i) = 1 for branch i;
        Let MER = min(ER, MER), MCI = OR(CI, MCI), and MNI = OR(NI, MNI);
        If (M(j) = 1 for all the other branches j,  j ≠ i ) then
                Send this BRM (MER, MCI, MNI) cell back to the source;
                Reset M(j) = 0 for all participating branches j;
                Reset MER = XER, MCI = 0, and MNI = XNI;
        Else
                Discard this BRM cell;
```

3.2 Algorithm 2

The main ideas of this algorithm are that each branch point stores the feedback information on a per-branch basis for each VC and only passes the BRM cells returning from the farthest destination among its all branches [7]. The pseudo code of Algorithm 2 is presented below.

A branch point switch maintains a resister MER, flags MCI, MNI, and a counter CTR for each branch.

```
On the receipt of an FRM (ER, CI, NI) cell:
        Multicast this FRM cell to all participating branches;

On the receipt of a BRM (ER, CI, NI) cell from branch i:
        Let CTR(i) = CTR(i)+1 for the corresponding branch i;
        Let MNI(i) = NI, MCI(i) = CI, and MER(i) = ER for the corresponding branch i;
        If (CTR(j) > 0 for all other branches j,  j ≠ i ) then
                Let ER = min(MER(j)), CI = OR(MCI(j)), NI = OR(MNI(j)) for all par-
                   ticipating branches j;
                Let CTR(j) = CTR(j)-1 for all participating branches j;
                Send this BRM (ER, CI, NI) cell back to the source;
        Else
                Discard this BRM cell;
```

3.3 Algorithm 3

Some variations of this algorithm were presented in [2,4,8]. The normal operation of Algorithm 3 is the same as in Algorithm 1. That is, in an under-load situation a branch point can pass a BRM cell to the upstream node only when it has received feedback from all branches. A difference occurs only when an overload situation has been detected. An overload situation is detected when the current feedback, ER, received from a branch is *much less* than the last feedback, Last_ER, returned by the branch point to the source [2,4,6,8]. Normally, a threshold is used for detecting the "much less" condition as follows:

$$\frac{ER}{Last\_ER} \leq \alpha \text{,}$$

where α is a threshold and $0 < \alpha < 1$. This threshold can be determined as a static value [2,4,8] or a dynamic parameter considering the load factor [6].

In order to avoid an additional increase of BRM overhead due to fast overload indication function, a Skip_Num counter is used for each multicast VC and initialised to zero [4]. The Skip_Num is increased whenever a BRM cell is sent before the normal BRM cell returning condition is satisfied. When a normal BRM cell returning condition is satisfied and the value of the Skip_Num is greater than zero, this particular feedback can be ignored and the Skip_Num is decreased by one if the following BRM cell discarding condition is satisfied:

$$Last\_ER \leq MER \leq \beta \cdot Last\_ER \text{,}$$

where β is a scaling factor and $\beta > 1$, and MER represents a consolidated ER value when the normal BRM cell returning condition is satisfied. The main idea of the proposed discarding policy is that a sudden change to an underload in the network status should be notified to the source even though the Skip_Num value is greater than zero and the normal BRM cell returning condition is satisfied. That is, the additional BRM overhead due to fast overload indication function are only controlled when the network is in a steady state so as to avoid any degradation in utilization by the indiscriminate discarding of a BRM cell indicating an underload. The pseudo code of Algorithm 3 is presented below.

A branch point maintains resisters MER, XER, Last_ER, flags MCI, XNI, MNI, Send_Flag, Reset_Flag, a counter Skip_Num, and M flags for each multicast VC.

```
On the receipt of an FRM (ER, CI, NI) cell:
        Let XER = ER, XNI = NI;
        Multicast this FRM cell to all participating branches;

On the receipt of a BRM (ER, CI, NI) cell from branch i:
        Let Send_Flag = 0;
        Let Reset_Flag = 0;
        Set M(i) = 1 for branch i;
        Let MER = min(ER, MER), MCI = OR(CI, MCI), and MNI = OR(NI, MNI);
        If (M(j) = 1 for all the other branches j, j ≠ i ) then
                Set Send_Flag = 1;
```

```
        Set Reset_Flag = 1;
If ( β *Last_ER ≥ MER ≥ Last_ER AND Skip_Num>0 AND Send_ Flag=1) then
        Let Skip_Num = Skip_Num-1;      /* β =1.1*/
        Reset Send_ Flag = 0;
If (MER < α *Last_ER) then              /* Threshold α = 0.9 */
        If (Send_Flag = 0) then
                Let Skip_Num = Skip_Num+1;
                Set Send_Flag = 1;
If (Send_Flag = 1) then
        Let Last_ER = MER;              /* Initially set to ICR */
        Send this BRM (MER, MCI, MNI) cell back to the source;
Else
        Discard this BRM cell;
If (Reset_Flag = 1) then
        Reset M(j) = 0 for all participating branches j;
        Reset MER = XER, MCI = 0, and MNI = XNI;
```

3.4 Algorithm 4

The normal operation for Algorithm 4 is the same as in Algorithm 2. That is, in an underload situation a branch point passes a BRM cell to the upstream node only when it receives feedback from the farthest destination among all its branches. The abnormal operation for Algorithm 4 in an overload situation is the same as in Algorithm 3. The pseudo code of Algorithm 4 is presented below.

A branch point maintains a resister MER, flags MCI, MNI, and a counter CTR for each branch. And it also maintains a resister Last_ER, a flag Send_Flag, and a counter Skip_Num for multicast VC.

```
On the receipt of an FRM (ER, CI, NI) cell:
        Multicast this FRM cell to all participating branches;

On the receipt of a BRM (ER, CI, NI) cell from branch i:
        Let Send_Flag = 0;
        Let ER(i) = ER, CI(i) = CI, and NI(i) = NI for branch i;
        Let CTR(i) = CTR(i)+1 for branch i;
        If (CTR(j) > 0 for all the other branches j, j ≠ i ) then
                Set Send_Flag = 1;
                Let CTR(j) = CTR(j)-1 for all participating branches j;
                Let MER = min(ER(j)), MCI = OR(CI(j)), and MNI = OR(NI(j)) for all
                    participating branches j;
        If ( β *Last_ER ≥ MER ≥ Last_ER AND Skip_Num>0 AND Send_Flag=1) then
                Let Skip_Num = Skip_Num-1;      /* β =1.1*/
                Let Send_Flag = 0;
        If (ER(i) < α *Last_ER) then            /* Threshold α = 0.9 */
                If (Send_Flag = 0) then
                        Let Skip_Num = Skip_Num+1;
                        Set Send_Flag = 1;
        If (Send_Flag = 1) then
```

```
        Let MER = min(ER(j)), MCI = OR(CI(j)), and MNI = OR(NI(j)) for all
            participating branches j;
        Send this BRM (MER, MCI, MNI) cell back to the source;
        Let Last_ER = MER;              /* Initially set to ICR */
Else
        Discard this BRM cell;
```

4. PERFORMANCE EVALUATION

4.1 Parameter Values and Network Model

Table 4 presents the ABR parameter values used for performance evaluation in this paper. All ATM switches are assumed to be ERICA switches with a target link utilization of 90% [5]. It is also assumed that all links had a bandwidth of 155.52 Mbps.

Table 4 ABR parameter values considered

Parameter	Description	Value
PCR	Peak cell rate	155.52 Mbps
ICR	Initial cell rate	4 Mbps
MCR	Minimum cell rate	0.1 Mbps
Nrm	Number of cells between FRM cells	32
RIF	Rate increase factor	1
RDF	Rate decrease factor	1/32,768
CRM	Missing RM cell count	32
CDF	Cutoff decrease factor	1/16
ADTF	ACR decrease time factor	0.5 sec

A simple network configuration is used to investigate the scalability of consolidation algorithms with and without fast overload indication function. Figure 1 illustrates the network model. In this model, the link distance between the end stations and their access switches is assumed to be 1 km, except for the distance between the switch S_N and destination A_N that determined as 3,000 km. The link distance between the switches S_1 and S_N is maintained at 1,000 km regardless of the number of branch points N. The source A is a multicast ABR source passing through N branch points, and source B is a unicast variable bit rate (VBR) source as background traffic. It is assumed that the ABR source A is a persistent source with 20 independent VCs and the VBR source B is a periodic high(120 Mbps)/low(30 Mbps) source with a 100 ms duration for each. In this configuration, the link between S_N and S_{N+1} is always a bottleneck point.

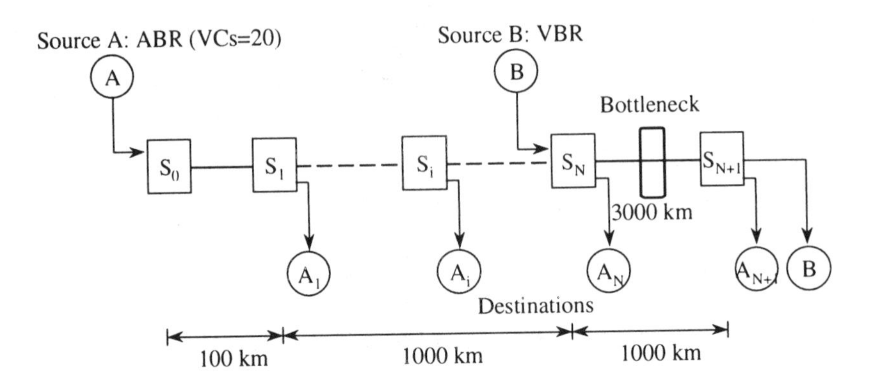

Figure 1 Network model.

4.2 Simulation Results

Figure 2 shows the ACR changes of source A according to the rate changes of the VBR source B for a different number of branch points N. All algorithms considered can eliminate ACR oscillation problem. However, Algorithms 1 and 3 suffer from ramp-up delay problem due to consolidation noise since they sometimes lose the latest feedback information due to storing it on a per-VC basis. This consolidation noise can be eliminated if each branch point stores feedback information on a per-branch basis for each multicast VC [3,7].

From this figure, the impact of the number of branch points N on ACR changes can also be observed. In the case of an overload situation, Algorithms 3 and 4 exhibit a faster ramp-down than Algorithms 1 and 2 by virtue of the fast overload indication function, regardless of N. In the case of an underload situation, Algorithms 1 and 3 exhibit a much slower ramp-up of the ACR when N=10, compared with their behaviours when N=2. In contrast, Algorithms 2 and 4 demonstrate their insensitiveness to an increase of N in terms of the ACR changes of source A. This is due to the fact that, in Algorithms 1 and 3, the total accumulated consolidation delay increases proportionally with N, whereas, in Algorithms 2 and 4 the total accumulated consolidation delay is bounded within the longest RTT in the multicast tree, regardless of N.

Figure 3 illustrates the impact of the number of branch points N on the queue length of the bottleneck switch S_N. Algorithms 3 and 4 exhibit a smaller queue length than Algorithms 1 and 2.

Table 5 compares the utilization of the bottleneck link between switches S_N and S_{N+1}. Algorithm 2 exhibits a higher link utilization than the other algorithms. Among algorithms with a fast overload indication function, Algorithm 4 produces a better performance than Algorithm 3, and is superior to Algorithm 1 which does not use a fast overload indication function. Algorithms 1 and 3 are both sensitive in link utilization to the number of branch points N. On the other hand, Algorithms 2 and 4 consistently provide a higher link utilization regardless of the number of branch points.

Table 5 Link utilization at the bottleneck link

Algorithms	1	2	3	4
$N = 2$	0.85	0.88	0.82	0.85
$N = 10$	0.78	0.88	0.76	0.85

Table 6 compares the fairness index among the multicast VCs at the bottleneck link. All of the algorithms seem to be fair when N=2, however, this is not true for Algorithms 1 and 3 when N=10 due to consolidation problems. Algorithms 2 and 4 show a good fairness regardless of N.

Table 6 Fairness index at the bottleneck link

Algorithms	1	2	3	4
$N = 2$	0.99	0.99	0.98	0.99
$N = 10$	0.90	0.98	0.89	0.98

Table 7 shows the ratio of BRM cells to FRM cells at the source. Algorithms 1 and 3 suffer from a severe consolidation loss, thereby the number of BRM cells returned to the source is much lower than that of FRM cells generated by the source. However, this ratio in Algorithms 2 is consistently maintained at one without any consolidation loss. In addition, the overhead of BRM cells of Algorithms 3 and 4 are almost the same as those of Algorithms 1 and 2, respectively, since they control the additional increase of BRM cells due to a fast overload indication function.

Table 7 Ratio of BRM cells to FRM cells at the source

Algorithms	1	2	3	4
$N = 2$	0.69	1	0.69	1
$N = 10$	0.67	1	0.68	1

From these results, it can be concluded that Algorithms 3 and 4 with fast overload indication function are very effective in terms of queue length at the expense of a lower link utilization than Algorithms 1 and 2, respectively. However, in an underload situation, the behaviour of consolidation algorithms with fast overload indication function is the same as that of their basic consolidation algorithms without fast overload indication function, since fast overload indication function only works when a severe overload situation has been detected. Therefore, the overall performances are highly dependent on the underlying basic consolidation algorithms.

5. CONCLUSIONS

This paper has proposed a new discarding policy for controlling excessive BRM cell overhead due to fast overload indication function. In addition, the performance of consolidation algorithms with/without fast overload indication function were compared using simulations.

Simulation results showed that fast overload indication function was very effective in a severe overload situation, particularly in an initial period with a higher ICR. The queue length at a bottleneck switch can be significantly decreased at the expense of the link utilization. In an underload situation, however, the behaviour of consolidation algorithms with fast overload indication function is the same as that of their basic consolidation algorithms without fast overload indication function, since fast overload indication function only works when a severe overload situation has been detected. Accordingly, the overall performance of consolidation algorithms with fast overload indication function is highly dependent on their underlying basic consolidation algorithms employed.

References

[1] ATM Forum Technical Committee, "The ATM traffic management specification draft version 4.1," *STR-TM-41.01 Straw Ballot*, Mar. 1999.

[2] H-S. Chen and K. Nahrstedt, "Feedback consolidation and timeout algorithms for point-to-multipoint ABR service," *In Proc. of IEEE ICC'99*, June 1999.

[3] Y-Z. Cho, S-M. Lee, and M-Y. Lee, "An efficient rate-based algorithm for point-to-multipoint ABR service," *In Proc. of IEEE GLOBECOM'97*, Dec. 1997.

[4] S. Fahmy, R. Jain, R. Goyal, B. Vandalore, S. Kalyanaraman, S. Kota, and P. Samudra, "Feedback consolidation algorithms for ABR point-to-multipoint connections," *In Proc. of the IEEE INFOCOM'98*, Apr. 1998.

[5] R. Jain, S. Kalyanaraman, R. Goyal, S. Fahmy, and F. Lu, "ERICA+: Extensions to the ERICA switch algorithm," *ATM Forum/95-1346*, Oct. 1998.

[6] W-S. Jang, Y-Z. Cho, and M-Y. Lee, "Performance analysis of branch-point switch behaviors for point-to-multipoint ABR flow control in ATM networks," *In Proc. of the 23rd Korea Information Science Society Fall Conference (KISS-FC)*, Sept. 1996.

[7] D-H. Kim, Y-Z. Cho, Y-Y. An, and Y. Kwon, "A scalable consolidation algorithm for point-to-multipoint ABR flow control in ATM networks," *In Proc. of IEEE ICC'99*, June 1999.

[8] W. M. Moh, "On multicasting ABR protocols for wireless ATM networks," *In Proc. of the International Conference on Network Protocols (ICNP) '97*, Oct. 1997.

[9] W. Ren, K-Y Siu, and H. Suzuki, "On the performance of congestion control algorithms for multicast ABR service in ATM," *In Proc. of IEEE ATM'96*, Aug. 1996.

[10] L. Roberts, "Rate based algorithm for point to multipoint ABR service," *ATM Forum/94-0772R1*, Nov. 1994.

[11] H-Y. Tzeng and K-Y. Siu, "On max-min fair congestion control for multicast ABR service in ATM," *IEEE JSAC*, Vol. 15, No. 3, pp. 545-555, Apr. 1997.

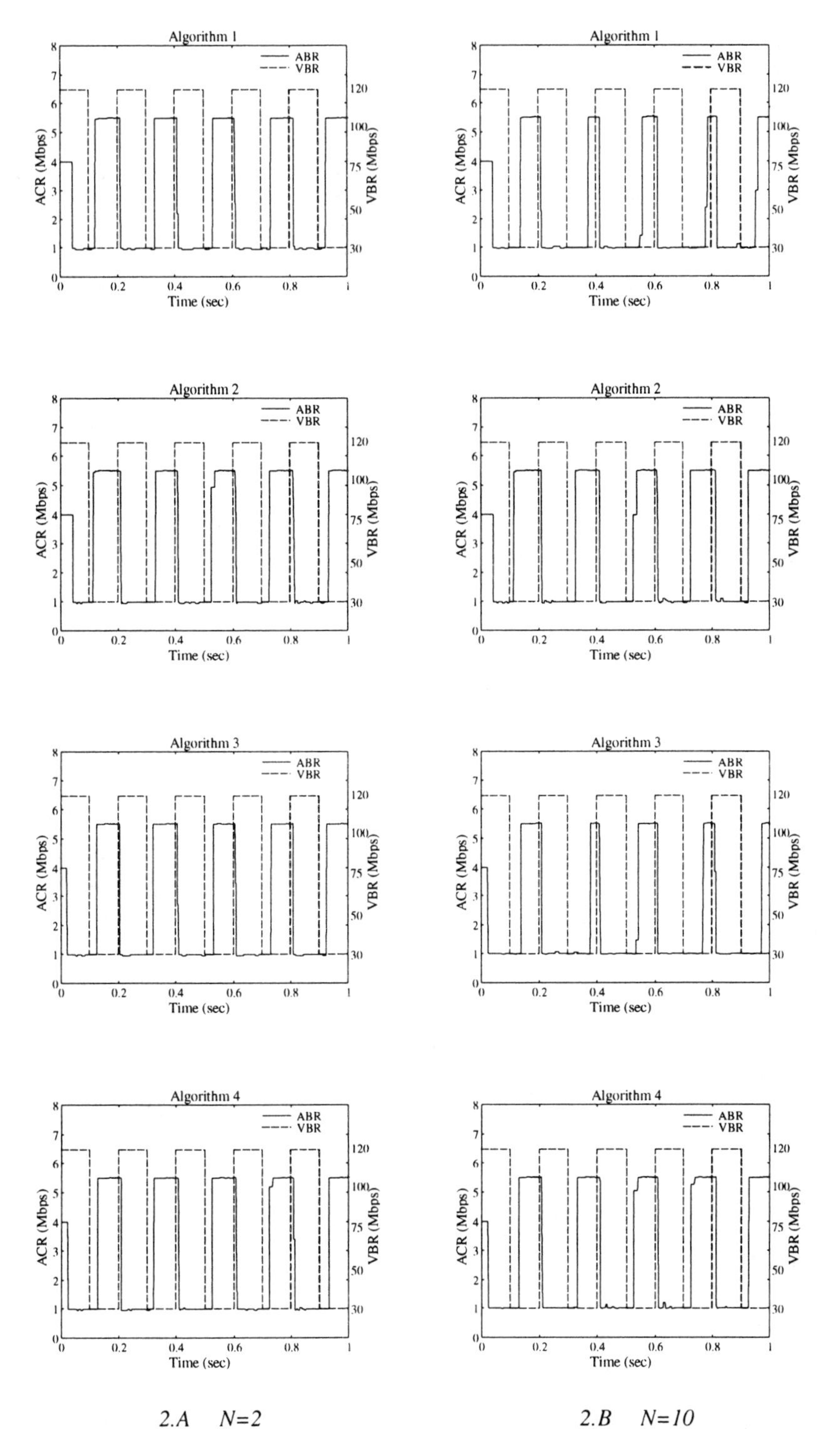

2.A N=2 *2.B N=10*

Figure 2 ACR changes according to the number of branch point *N*.

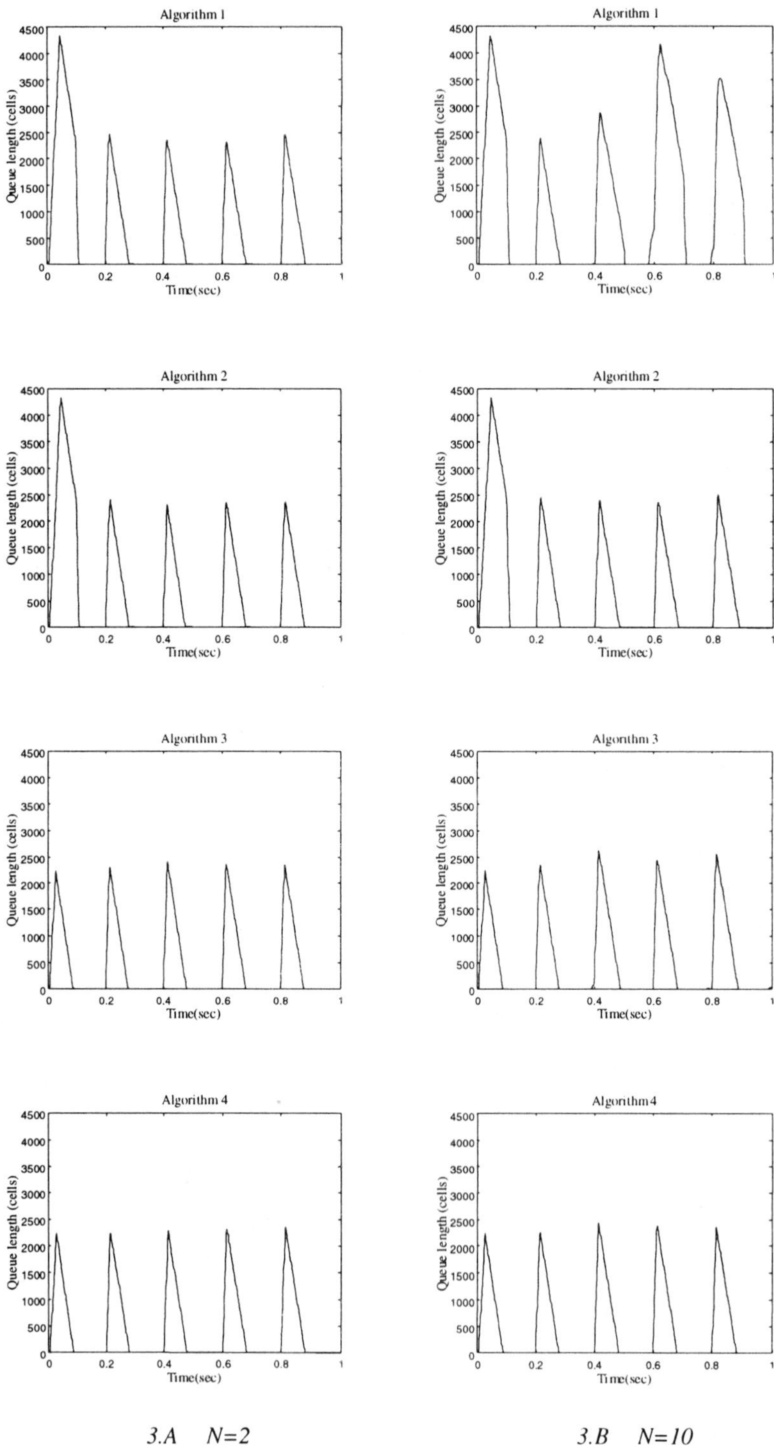

3.A *N=2* *3.B* *N=10*

Figure 3 Queue lengths according to the number of branch point *N*.

A QOS-BASED RESOURCE RESERVATION PROTOCOL FOR PRIORITY-DIFFERENTIATED DELAY-BOUNDED MULTICAST

Longsong Lin[1], Ming-Shou Liu[2], Lih-Chyau Wuu [1] and Chun-Yeh Tsai[1]

[1] *Dept. of Elec. and Info. Engr.*
Natl. Yunlin Univ. Sci. & Tech.
Touliu, Yunlin, Taiwan
{ lin, wuulc } @el.yuntech.edu.tw

[2] *Dept. of Elec. Engr.*
Natl. Chinyi Inst. of Tech.
Taiping, Taichung, Taiwan.
{ liu } @chinyi.ncit.edu.tw

Abstract Discerning the diverging developments of Internet routing and resource reservation signaling protocols, we argue that when it comes to the overall efficiency and utilization of the network resources, the reservation signaling must collaborate with underly ing routing protocol. This paper presents a multicast QoS-based routing and reservation protocol (MQRP) that integrates resource reservation algorithm with multicast routing to guarantee end-to-end delay bounds for priority-differentiated multicast. MQRP allows each receiver to prioritize its selection on multiple hierarchical layer-encoded media streams and bounds the end-to-end delays for the selected flows. Given the priority and delay constraints, this receiver-initiated protocol employs RSVP signaling messages and OSPF-based routing mechanism to establish a multicast tree and reserve link bandwidth, aiming at maximizing the gain and utilization of the netw ork. To compare protocols for solving the *priority-differentiated, delay-bounded multicast problems*, simulations are conducted on a class of *random networks*. It is shown that MQRP has best scalability to the number of network nodes as well as offered source streams. Furthermore, in case that the network resource is insufficient to convey all the requests, MQRP can admit the request with higher priorities so as to leverage the overall network utilization.

Keywords: Multicast algorithms, Resource reservation protocols, Multimedia multicast, QoS-based routing

1. INTRODUCTION

Motivated by the increasing demand on broadband multimedia applications as well as by the enriched functionality at users desktops nowadays, there is a need to provide multiple-point [1], guaranteed quality of services (QoS) [2][3]

in a heterogeneous networking environment. Multicast promises the efficient use of the network bandwidth for multiple-point communication; nevertheless, delivering guaranteed multicast service is more than a simple capacity issue. It involves not only a routing protocol to establish QoS-assured paths but also a signaling procedure to secure the bandwidth for each individual, thus engineering towards optimal usage of network resources. To provide efficient resource usage while guar anteeing QoS routes, multicast routing and resource reservation protocols thus come to orchestrate naturally [1][4].

Resource reservation protocols and QoS-based routing, however, have evolved into separated efforts and neither an integrated discussion nor multicast context has been much discussed. ST-II and RSVP are among two resource reservation protocols proposed for supporting guaranteed services [5][6]. For both, the guaranteed quality of service is characterized by approximating a "fluid model" of service [5][7][8] in which a stream effectively sees a dedicated wire of the assigned bandwidth between the source and the de stination. The major difference between ST-II and RSVP lies in how multicast is supported within reservation signaling. In addition, RSVP is not equipped with any mechanism to instruct routing on QoS attributes; it is still an open issue how QoS is attained, and yet none of the current researches have yet sufficiently dealt with the interaction of routing and QoS guarantees [4]. Apart from ST-II and RSVP, QoS-based routing [9] approach, which does not include a resource reservation mechanism, allows the determination of a path that will possibly accommodating the requested QoS. Thereupon, paths for flows would be determined based on some knowledge of resource availability in the network, as well as the QoS requirement of flows.

The issues surrounding resource reservation and QoS routing for multicast have been centered at how to discover the status of resource usage in the current multicast. This status information could improve the feasibility of finding a path for a new receiver joining to a multicast flow during path computation. QOSPF [10] handles this problem by having routers broadcast reserved resource information to other routers. With the global information on residual bandwidth, each node in QOSPF can compute the path that satisfies a QoS request. However, in m ulticast case, this approach may not generate an optimal routing tree as it is not constructed by considering the receivers' requirements. To avoid broadcasting the information of resource availability as with QOSPF, alternate path routing [11] deals with this issue by using probe messages to find a path that has sufficient resources to fulfill the QoS requirements. Path QoS Computation (PQC) method, proposed in [12], uses RSVP PATH messages to propagate bandwidth allocation information. However, a router receiving the PATH message gets the knowledge of the resource allocation only on those links along the path from the source to itself. Allocation for the same flow on other remote branches of the multicast tree is not available. Thus, the PQC method may not be sufficient to find feasible paths to all receivers.

In this paper we present a multicast QoS-based routing and resource reservation protocol (MQRP) to support the *priority-differentiated, delay-guaranteed multicast*. The protocol allows a customer to specify different priorities on a set of selected streams from a source and set an end-to-end delay bounds for the requested streams. It integrates a link-state routing protocol with the RSVP bandwidth reservation protocol to establish the multicast tree and reserve tree bandwidth simultaneously. The goal in designing MQRP is to leverage the overall efficiency and utilization of the network.

Our approach distinguishes from previous works in the following respects. 1.) ***Dynamic determination of feasible paths initiated from receivers***: The receiver-initiated MQRP avoids the broadcast storm as with QOSPF and adopts alternative path approach by dynamically establish the multicast tree from all receivers to the source. requirements is a more difficult 2.) ***Optimization of resource usage***: MQRP scheme can improve the utilization and efficiency of network resources. It attempts to optimize the network resource utilization by minimizing the reserved bandwidth of the tree as well as by maximizing the total number of streams admitted to the ne twork, and to optimize the efficiency by admitting streams with higher gains. 3.) ***Path selection based on per-flow priority***: MQRP allows a user to prioritize the selected streams and to specify a delay bound for these streams. The prioritized scheme not only allows a user to specify priority on multiple selected streams but also enables traffic engineering routing paths for optimal usage of network bandwidth. 4.) ***Application supports for service differentiation*** MQRP allows a receiver to accept a partial allocation of the full stream bandwidth. The *partial bandwidth allocation* strategy not only reflects the heterogeneity of host capability but also provides the flexibility for network provider to maximize the overall utilization.

2. PRIORITY-DIFFERENTIATED DELAY-BOUNDED MULTICAST

Formulation of the Problem. Given a network with a multicast group G, the source node s announces a set of streams with specification T_{spec} to all network nodes. A potential receiver d in G requests multiple streams by specifying a priority vector $P(d)$ and an associated delay bound $\Delta(d)$. The problem is to find a multicast tree T connecting all the requesting receivers and to allocate bandwidth $R^j(u,v)$ for each stream j on each link $(u,v) \in T$ so as to satisfy $\Delta(d)$ for the admitted streams. On establishing such tree, it is the objectives to maximize the gain over these requests and minimize the tree cost, in order to leverage the utilization and efficiency of the network resources. Specifically, the utilization of the network resource (bandwidth)

Υ is defined as $\Upsilon = \frac{TS}{Cost}$, where $TS = \sum_{(u,v)\in T,\forall j} 1_{[R^j(u,v)\neq 0]}$, is the total number of streams that has been admitted to the multicast. (The indication function $1_{[x]}$ returns one if x =TRUE). The tree cost is defined as the total amount of bandwidth that has been allocated to the multicast tree T, i.e., $Cost = \sum_{(u,v)\in T,\forall j} R^j(u,v)$, for all $R^j(u,v) \neq 0$. Likewise, the efficiency is defined as $\Theta = \frac{Gain}{MG}$, where $Gain$, the sum of the priorities of all the admitted streams, i.e., $\sum_{d\in G,\forall j} P^j(d), R^j(u,d) \neq 0$, and $MG = \sum_{d\in G,\forall j} P^j(d)$, is the maximal gain that can be possibly obtained from all the requests.

An Overview of the Protocol. MQRP assumes that the link latency information is propagated over the network via a link state routing protocol (e.g., OSPF); thereby, each node has the information of global network topology and hence can derive the minimum path latency (MPL) to all othe r nodes. The signaling procedure of MQRP is depicted in Figure 1. The protocol proceeds in three phases, with respect to the signaling phases of RSVP. First, during the stream announcement phase, the source announces the stream specification T_{spec} by sending a RSVP PATH message via the minimum latency paths downstream to each destination.

Next, during the tree establishment and resource reservation phase, each node willing to participate the multicast session specifies the R_{spec} in a RSVP request message RESV, indicating the priorities and delay bounds requested. These RESV messages are sent upstream towards the source node. On receiving the messages, intermediate nodes merges the requests and connect to an upstream node using the link latency information offered by link state routing table and the information contained in R_{spec} and T_{spec}. The phase ends when the source receives all the RESV messages, when the tree is established with bandwidth R reserved on each link for each stream.

Finally, during the admission control and resource allocation phase, the source exerts admission control to the requesting streams and starts to allocate bandwidth for each stream by sending RSVP PATH Refresh messages toward all receivers via the tree tha t has been established in previous phase. On receiving the PATH Refresh message, each node extracts the information in the Refresh message and calculate the amount of bandwidth R to reserve for each stream to be admitted to the multicast.

3. MULTICAST QOS-BASED ROUTING AND RESERVATION

The algorithm for MQRP consists of two parts: one for multicast tree establishment and resource reservation and the other for admission control and bandwidth allocation. The former reserves bandwidth when establishing the

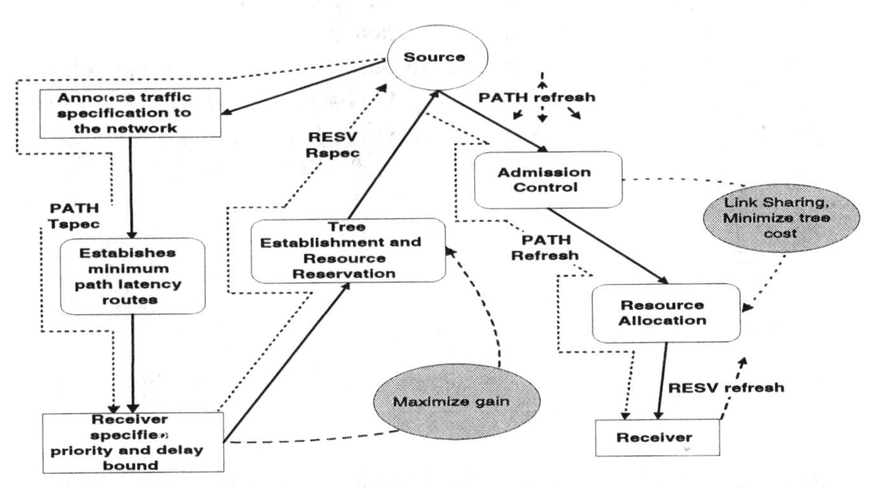

Figure 1 Signals in the integrated reservation and multicast protocol

tree, whereas the latter performs admission control and allocates the bandwidth for each stream.

3.1 MULTICAST FLOW AND DELAY MODEL

The operation of MQRP is based on the admission control policy of multicast flows and on a delay model. The algorithm for multicast tree establishment starts with merging requests from destinations with distinct flow requirements.

a peak rate of 35 via a by successfully. In the example,

eive the stream in peak rate so that it acking of bandwidth, but they still can participate the multicasting by

Delay model from node s to nodes d_1 and

The delay model used in MQRP is approximated by the parameters in the traffic specification and the delay error terms incurred in each router, as specified in the perfect fluid model [7][8]. According to the fluid model [13], the end-to-end queuing delay can be computed via the traffic parameters in T_{spec}, which include the peak rate of the flow (p), the token bucket depth (b) and rate (r), the minimum policed unit (m), and the maximum datagram size (M) [4]. Assuming that a fixed amount of latency is associated with each transmission link, then the end-to-end delay from source s to a destination d, $D(s,d)$, is equal to the end-to-end queuing delay $Q(s,d)$ plus the fixed amount of path latency $PL(s,d)$ from s to d, i.e.,

$$D(s,d) = Q(s,d) + PL(s,d)$$

While the path latency is fixed, the queuing delay will vary as the bandwidth assigned to the path changes. This fluid model explicitly shows that a stream conforming to a token bucket will increase its queuing delay as the allocated path bandwidth decreases [4]. If a bandwidth $R(u,d)$ from node u to a destination d is reserved and is greater than or equal to the peak rate p, i.e., $R(u,d) \geq p$, then the queuing delay $Q(u,d)$ can be estimated by

$$Q(u,d) = \frac{M + C_{tot}}{R(u,d)} + D_{tot}. \tag{1}$$

Otherwise, for $r \leq R(u,d) < p$, it is computed as

$$Q(u,d) = \frac{(b-M)(p-R)}{R(u,d)(p-r)} + \frac{M + C_{tot}}{R} + D_{tot}, \tag{2}$$

where D_{tot} and C_{tot} account for the sum of error terms in each routers along the path.

be admitted to the multicast

3.2 TREE ESTABLISHMENT AND BANDWIDTH RESERVATION

The MQRP protocol is represented as a distributed algorithm operating at each node. As MPL is derived from PATH messages, each node then initialize two parameters: aggregated priority $\bar{P}$ and admissible path delay $\bar{\Delta}$. The aggregate priority $\bar{P}(i)$ at node i is the sum of the priority values of all streams in the current tree. The admissible path delay $\bar{\Delta}(i)$ represents the latest time a packet should arrive at this node i. If a packet arrives at this node with a delay larger than $\bar{\Delta}(i)$, then it is impossible for the packet to be delivered to the destination within specified delay bound.

PROCEDURE Multicast QoS Routing and Reservation Protocol

1 **If** receive the PATH message
  Compute $MPL(d), \forall d \in V$

2 **For** each destination $d \in G$
  $\bar{P}(d) = P(d),\ \ \bar{\Delta}(d) = \Delta(d)\ ,\ R(d) = p(d)$

3 **If** receives the RESV message
  $\bar{\Delta}(v) = Min_d\{\bar{\Delta}_d(v)\},\ \bar{P}(v) \leftarrow \bar{P}(v) + \bar{P}(w)$
  $R(v) = Max_w\{R(v,w)\}, \forall w \in Child(v)$

4 **For** each u, $MPL(u) < MPL(v)\ , MPL(u) + L(u,v) < \bar{\Delta}(v)$
  Calculate $\rho(u,v) = Min(\frac{C(u,v)}{S_{sum}(v)}, 1)$
    $S_{sum}(v) = \sum_j R^j(v,w),\ \ \{j|\bar{P}^j(v) \neq 0\}$
  Estimate $R(u,v) = R(v)\rho(u,v)$
  For each destination d

$R(u,d) = Min(R(u,v), R(v,d))$
$Q(u,d) \leftarrow Compute\_Q\_Delay(R(u,d))$
$D(u,d) = Q(u,d) + PL(u,d)$
If ($D(u,d) < \Delta(d)$)
$\bar{\Delta}_d(u) \leftarrow Min_j\{(\Delta(d) - D^j(u,d))\}\ \forall j$, s. t. $\{j|P^j(d) \neq 0\}$
Else If ($u \neq s$) $\bar{\Delta}_d(u) = 0$
5 $u^* = Arg_u Max_u\{\bar{\Delta}_d(u)\}$, $Parent(v) \leftarrow u^*$
6 send RESV($\bar{\Delta}_d(u^*), R(u^*, v), \bar{P}(v)$) message to u^*.
END PROCEDURE

The tree establishment phase is initiated from the receivers by sending RESV messages to the source. A node v extracts $\bar{\Delta}_d(v)$, $R(v,w)$ and $\bar{P}(w)$ carried in the RESV message from a downstream node w, where $\bar{\Delta}_d(v)$ is the admissible delay at node v for destination d and $R(v,w)$ is the bandwidth reserved on link (v,w). For a number of destinations requesting the stream via node v, the smallest admissible delay must be selected for v. That is, $\bar{\Delta}(v) = Min_d\{\bar{\Delta}_d(v)\}$. Also, node v must select the largest bandwidth among these requests merging to v as the bandwidth $R(v)$ to serve on the upstream link. That it, $R(v) = Max_w\{R(v,w)\}$, for all the children w's of node v.

To reserve the bandwidth, node v needs to find an optimal nodes u^* to connect, among the neighboring nodes. The neighboring node u must satisfy two conditions: $MPL(u) < MPL(v)$ and $MPL(u) + L(u,v) < \bar{\Delta}(v)$. The minimum path latency from source to v via node u must satisfy, $MPL(u) + L(u,v) \leq \bar{\Delta}(v)$; otherwise, it is impossible to deliver the streams to node v in time. Besides, if v request the bandwidth $R(v)$ on link (u,v) that is greater than the capacity $C(u,v)$, then it is impossible to deliver all the streams at $R(v)$. In such case, in order to accommodate as many streams as possible on link (u,v), the bandwidth requirement must be reduced by a fraction $\rho(u,v)$ such that the bandwidth requirement for link (u,v) becomes $R(u,v) = R(v)\rho(u,v)$.

According to the fluid delay model, the queueing delay (and thus the end-to-end delay) is determined by the path bandwidth (the bandwidth of smallest link among a path), $R(u,d)$. Hence, queuing delay $Q(u,d)$ can be estimated according to Equation 1 or 2. As a result, the end-to-end delay from node u to destination d can be obtained through $D(u,d) = Q(u,d) + PL(u,d)$.

Now, the admissible delay $\bar{\Delta}_d(u)$ at node d, where $D(u,d) < \Delta(d)$, is updated to $Min_j\ (\Delta(d) - D^j(u,d))$, for all j's in d such that $P^j(d) \neq 0$. Having all the admissible delays computed at all neighboring node, node v selects the optimal neighboring u^* by

$$u^* = Arg_u\ Max\ \{\bar{\Delta}_d(u)\},$$

where "$Arg_u\ Max$" returns the corresponding index u that produces the largest $\bar{\Delta}_d(u)$ value. Finally, node v will sends upstream node u^* a RESV message that contains $\bar{\Delta}_d(u)$, $R(u^*, v)$, and $\bar{P}(v)$ to node u^*. When receiving the RESV message from node v, node u^* will reserve the amount of bandwidth $R(u^*, v)$ on link (u^*, v) and use $\bar{\Delta}_d(u)$ and $P_d(v)$ to select a neighboring node to connect, repeating the above procedure.

3.3 ADMISSION CONTROL AND RESOURCE ALLOCATION

The admission control and resource allocation algorithm allocates bandwidth to each tree link. This algorithm starts from the source node when it received all the RESV messages from the destinations and then iterates downstream to the destination nodes. When a node v receives a PATH Refresh from its upstream node, it extracts the $PL(s, d)$ and estimate the queueing delay according to $Q(s, d) = \Delta(d) - PL(s, d)$. Then, it estimates the amount of bandwidth required for this queueing delay using Equation 1 or 2. This bandwidth, $\tilde{R}$, will be further redistributed among the streams in order to maximize the total gain. The function $Band\_Share()$ returns the optimal distribution of the bandwidth $R(v, d)$ for maximal gain. Suppose that w is a child of v which lead the streams from v to all destinations. For each stream, node v reserves the amount of bandwidth that is maximum of the bandwidths required by all destination along the link (v, w) so that all the destinations can receive the stream. Specifically, v reserves $R(v, w) \leftarrow Max_d\{R(v, d)\}$. Finally, node v allocates a bandwidth on the link (v, w) for each stream by first selecting the stream with largest priority and repeating until the link capacity is exhausted.

PROCEDURE Admission Control and Bandwidth Allocation
If receives a PATH Refresh,
  For each child w
    $R(v, d) \leftarrow Band\_Share(v, d)$ for each d
    $R(v, w) \leftarrow Max_d\{R(v, d)\}$
  Repeat $j = Arg\ Max_{j \notin A} \{\bar{P}^j(v)\}$
      Allocate bandwidth $R^j(v, w)$
      $A = A + \{j\}$
  Until ($\sum_{j \in A} R^j(v, w) > C(v, w)$)
  $R^j(v, w) = 0$, for $j \notin A$
  send PATH Refresh($R(v, w)$) to w
END PROCEDURE

4. SIMULATION

The performance of the protocol will be evaluated in terms of two measures: utilization and efficiency, as defined earlier. During simulations, focused are the performance impacts due to change of the metrics including the network size, percentage of the multicast membership, total number of streams, delay offset, and bandwidth sufficiency. The *delay offset* is defined as the specified delay bound minus the calculated path latency, $\Delta - MPL$, which accounts for the time interval that can be used to trade for more streams to join. The *bandwidth sufficiency*, a quantity that indicates how sufficient the network capacity is, compared to the stream bandwidth specification, is defined as $\frac{\frac{1}{|E|}\sum_{(u,v)\in E} C(u,v)}{\frac{1}{q}\sum_{j=1,\dots,q} S_j}$, where $|E|$ is the number of links in the network.

Several protocols are compared in the simulations. In addition to the MQRP protocol, we also derive two its variants that allocate bandwidth along the path without trimming the tree on phase three: the MQRP-Total and the MQRP-Partial. The partial reservation scheme allows reserving a portion of the peak rate with reduced media quality. The P-RSVP is a modified version of RSVP for accommodating the priority on receiver request. The NHP3 is the third procedure of the four proposed procedures in [2], but the authors do not specify the tree structure in the work. We modify the third procedure in their work into two variants: maximum capacity tree (NHP3-1) and minimum cap acity tree (NHP3-2).

sign the link bandwidth in

4.1 RESOURCE UTILIZATION AND SCALABILITY

To study the robustness and scalability, protocols are evaluated on a class of random network [14] that have dynamically changed topology and are relatively large scale. In the experiment, we gradually scale the network to 200 nodes while keeping the average node degree to 3, and measure the utilization of the network with a given fixed bandwidth sufficiency. It means that the ratio of network capacity to stream bandwidth remains constant regardless that the network grows. Of the network nodes, 40 percent are designed as the receivers. Figure 2 shows the result that MQRP protocol is more scalable than the other algorithms. The scalability of MQRP protocol is about four times that of the P-RSVP. In Figure 3, we find that the utilization increases as the number of streams increases. The utilization produced by the MQRP scheme increases in a rate about four times that of the NHP3-1 scheme and about two times that of the P-RSVP scheme. This demonstrates the capability of MQRP protocol on accommodating more number of streams. The bandwidth reservation scheme in MQRP can admit more steams to the multicast when all the protocols running in a network with scarce resource.

est path delay observed at o that the overall utilization and efficiency are leveraged.

4.2 GAIN AND BANDWIDTH EFFICIENCY

The efficiency is determined by the total gain that can be anticipated. As the network scale from 20 nodes to 200 nodes, the MQRP and P-RSVP protocols are very competitive, as shown in Figure 4. This shows that the P-RSVP is also an efficient protocol because the receiver always adopts the shortest path delay to the source without considering the priority in establishing the multicast tree. Although it accommodates more streams to enter; however, it will not guarantee the best streams (streams with high priority) are admitted. Hence, as shown in Figure 2 the utilization degrades. The protocols with the worst efficiency are NHP3-2 and MQRP-Partial. The NHP3-2, although establishing the tree by considering the priority, finds a minimum capacity tree, which limits the number of streams to be admitted into the multicast tree, resulting in poor utilization.

. As a result, it may not be able to tial has the worst efficiency. But, as shown in Figure 3, it still In Figure 5, we compare the efficiency by varying the bandwidth sufficiency from 0.2 to 1.2. MQRP protocol has the best efficiency among them, whereas the NHP3-2 has the worst when the bandwidth sufficiency is below 2. But the MQRP-Partial become s the worst when the bandwidth sufficiency is above 2. This shows that MQRP is suitable for the network with relatively heavy-load or the network with scarce capacity relatively to the total requested bandwidth of the receivers.

5. CONCLUSION

In this article, we have shown that the proposed MQRP is able to optimize the bandwidth utilization and efficiency for network resources whilst satisfying user preferences and constraints. From the simulation results, it is also shown that MQRP is able to scale best with the number of network nodes as well as offered streams. To sum up, MQRP as a QoS-based routing and reservation multicast protocol is able to accommodate heterogeneous application requirements to support prioritized delay-bounded multicast service, enabling the network provider to optimize the resource utilizat ion and to gain the revenue efficiently from the provisioning.

References

[1] G.N. Rouskas, and I.Baldine, "Multicast Routing with End-To-End Delay and Delay Variation Constrains", *IEEE Journal on selected areas in communications,* Vol. 15 , No. 3, April 1997, pp.346-356.

[2] N. Shacham and James S. Meditch, "An Algorithm For Optimal Multicast of Multimedia Stream,", in Proceedings of *IEEE INFOCOM'94,* Vol. 2, pp.586-64.

[3] V.P. Kompella, J.C. Pasqale, and G.C. Polyzos, "Multicast Routing for Multimedia Communication", *IEEE/ACM Transactions on Networking,* 1993.
[4] L. Zhang, S. Deering, D. Estrin, S. Shenker and D. Zappala, "RSVP:A New Resource ReSer Vation Protocol", *IEEE Network,* Sept. 1993.
[5] S. Schenker, C. Partridge, and R. Guerin, "Specification of Guaranteed Quality of Service", *IEEE Communications Magazine,* May 1997.
[6]D.J. Mitzel, D. Estrin, S Shenker and L. Zhang. , "An Architectural Comparison of ST-II and RSVP", IEEE 1994.
[7]A. Parekh and R. Gallagher, "A Generalized processor Sharing Approach to Flow Control - The Single Node Case," *IEEE/ACM Trans. Networking,* vol. 1, no. 3, 1993, pp. 366-57.
[8]A. Parekh and R. Gallagher, "A Generalized processor Sharing Approach to Flow Control - The Multiple Node Case," *IEEE/ACM Trans. Networking,* vol. 2, no. 2, 1996, pp. 137-50.
[9]E. Crawley, R. Nair, B. Rajagopalan, and H. Sandick, " A Framework for QoS-based Routing in the Internet," RFC 2386, August 1998.
[10]Zhang, Z., Sanchez, C., Salkewicz, B., and E. Crawley, "QoS Extensions to OSPF", Work in Progress.
[11]Zappala, D., Estrin, D., and S. Shenker, "Alternate Path Routing and Pinning for Interdomain Multicast Routing", *USC Computer Science Technical Report # 97-655*, USC, 1997.
[12]Y. Goto, M. Ohta and K. Araki, "Path QoS Collection for Stable Hop-by-Hop QoS Routing", *Proc. INET '97*, June, 1997.
[13] S. Shenker, C. Partridge, and R. Guerin, “Specification of Guaranteed Quality of Service,” RFC 2212, September 1997.
[14] B. M. Waxman, "Routing of Multiple Connections," *IEEE Journal on selected areas in communications,* Vol. 6, No. 9, December 1988, pp.1617-1622.

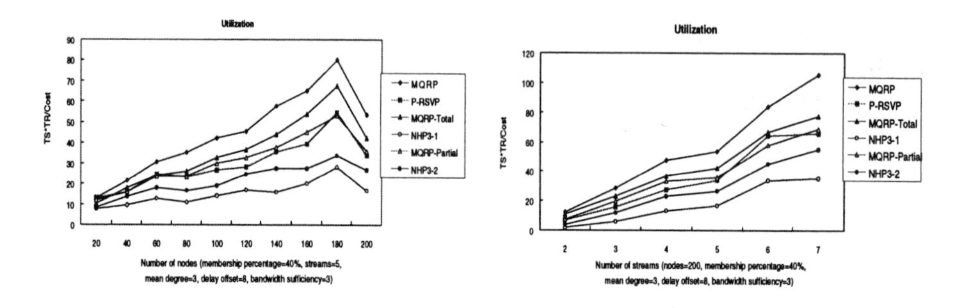

Figure 2 Utilization versus the network size

Figure 3 Utilization versus the number of source streams

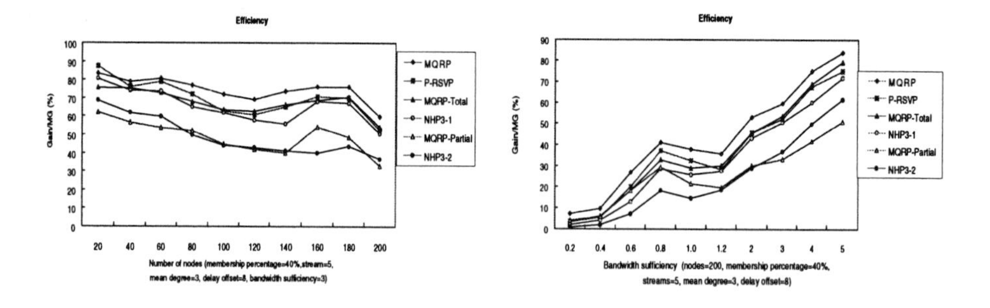

Figure 4 Efficiency versus the network size

Figure 5 Efficiency versus bandwidth sufficiency

MULTIRATE RESOURCE SHARING FOR UNICAST AND MULTICAST CONNECTIONS

Khaled Boussetta and Andre-Luc Belyot
Laboratoire PRiSM
University of Versailles Saint-Quentin
45, Avenue des Etats-Unis
78035 Versailles, France
{boukha,beylot}@prism.uvsq.fr

Abstract Recent advances in network equipment and the important progress in video, audio and data compression techniques has given rise to new multimedia group applications, such as Video-conferencing or Webcasting. These applications need guarantees on the QoS (Quality of Service) and require group communication services. ATM seems to be well adapted to support these applications, since it provides several service classes and QoS guarantees. However, these guarantees are possible only if the network is able to allow efficient resource sharing. Therefore, resource sharing for multirate traffic is a fundamental issue, specially for multicast connections with a huge number of recipients. In this paper, we generalize the mathematical model of Kaufman [6] to compute the exact values of call blocking probability for unicast and multicast multirate multiclass traffics sharing a single link. We show that the model has a product form solution. Moreover, we give a recursive algorithm to obtain the normalization constant and the call blocking probability.

Keywords: Multicast, resource sharing, product solution, recursive solution, ATM.

1. INTRODUCTION

Over the last few years, several multimedia group applications have emerged. In particular, the extension of the Internet to support the multicast within the MBONE [2] makes the use of multimedia group applications over wide area networks a commonplace reality. Some of these applications, such as CSCW (Computer Supported Cooperative Work) or video-conferencing, are very popular within the Internet community. These applications facilitate group communication and allow members of a group to work together or to share resources. We believe that there will be an increasing need for developing this kind of applications in the future and so, there will be an absolute necessity to provide the multicast capability in the networks.

The support of multicast leads to many problems [4] such as call admission control, multicast routing [3], group management, multicast reliability [1], flow control [8], etc. However all these questions are a consequence of one fundamental problem : the resource sharing problem. Until now, most of the answers given to these problems aim at supporting unicast connections with heterogeneous requirements. Thus, multirate multiclass loss models for unicast traffic, have been widely studied. It was shown that they have a product form solution, and exact blocking probability can be computed. Moreover, recursive solutions have been developed to bypass numerical problems that may arise in large systems.

Efficient Sharing of resources for multicast connections is required to allow the support of a huge number of recipients. However, the performance of a link supporting multicast connection has not been extensively investigated. Therefore, most of these research works concern the switch call blocking probability [7]. [5] gives an interesting solution to compute multicast call blocking in a single link, that is, the authors use generalized Engset system to deduce the call blocking probability. Nevertheless, only multicast traffics have been considered, while the effect of unicast connections sharing the same link was not discussed. Moreover, difficulties in computing will arise for large systems since no recursive solution is given.

In this paper, we use a mathematical model to compute the exact solution of the call blocking probability for both unicast and multicast multirate traffics sharing a single link. Our work differs from [5] in that we generalize the model of Kaufman [6] to compute the exact steady-state blocking probability for multicast and unicast multirate multiclass connections. To obtain the normalization constant, we use a new recursion algorithm which gives the resource distribution for multicast connections. Finally, the call blocking probabilities for multicast and unicast traffics are obtained with our recursive algorithm.

The rest of the paper is organized as follows. In the next section, we present the problem of resource sharing for unicast and multicast traffic, then we present our analytical model and its product solution form. We also describe our recursive solution to compute the call blocking probability . In section 3, an example is given to illustrate the use of our recursive algorithm. A summary of this paper and possible extensions are discussed in section 4.

2. CALL BLOCKING FOR MULTIRATE TRAFFIC

2.1 MODEL DESCRIPTION

We study the call blocking probability in a link with C resource capacity that allows the support of unicast and multicast multirate multiclass streams. All arrival call connections and holding times are independent. We suppose that the other links in the network have infinite capacity, so that no blocking that may affect our system occurs.

Let U be the number of unicast customer types. Class i unicast customers arrive according to a Poisson process of rate $\lambda_{u,i}$ and require simultaneously $b_{u,i}$ resource units. We assume that the holding time of class i call is exponentially distributed with mean $\frac{1}{\mu_{u,i}}$. Thus, the offered load is $\rho_{u,i} = \frac{\lambda_{u,i}}{\mu_{u,i}}$. A Unicast customer whose requirement cannot be satisfied is blocked.

Suppose that there is M multicast customer types, each class corresponds to a multicast session. Thus, an infinite user population can join each multicast session. That is, it is reasonable to consider that multicast customers of class i arrive according to a Poisson process with rate $\lambda_{m,i}$ and has exponential residency time with parameter $\mu_{m,i}$. Thus, the offered load is $\rho_{m,i} = \frac{\lambda_{m,i}}{\mu_{umi}}$. All the members of a given multicast session will require the same bandwidth capacity. That is, multicast customers of class i, will require $b_{m,i}$ resource units.

Definition 1 *i is an active class if there is at least one multicast customer of class i being served in the link.*

As calls of multicast class i will not be blocked if i is already active, then $b_{m,i}$ units of the system are allocated to a new arriving multicast customer of class i, only if i is not already an active class and there is at least $b_{m,i}$ free resources in the system.

2.2 PRODUCT FORM SOLUTION

Under the preceeding assumptions, it is easy to see that the state of our system is described by the state vector $n = (u, m) = (u_1, \cdots, u_U, m_1, \cdots, m_M)$. Here u is the vector number of calls in progress for unicast class and m is the state vector for multicast ones.

Let Ω be the set of allowed states. Ω depends on the sharing policy P. In this paper, we address Complete Sharing policy (CS). However, join distribution will also have a product form solution for Upper Limit (UL) and Guaranteed Minimum (GM) policies.

Let $1_{\{m_i>0\}} = 1$ if i is an active class. Then, the set of allowed states Ω for CS policy can be written as :

$$\Omega = \left\{ (u,\, m) \,/\, \sum_{i=1}^{U} b_{u,i} \cdot u_i + \sum_{i=1}^{M} b_{m,i} \cdot 1_{\{m_i>0\}} \le C \right\} \tag{1}$$

Theorem 1 *The state probability corresponding to a resource sharing policy CS is given by :*

$$P\,(u, m) = \frac{1}{G\,(\Omega)} \cdot \prod_{i=1}^{U} \frac{\rho_{u,i}^{u_i}}{u_i!} \cdot \prod_{i=1}^{M} \frac{\rho_{m,i}^{m_i}}{m_i!} \tag{2}$$

Where $G\,(\Omega)$ is the normalization constant for Ω.

$$G\,(\Omega) = G\,(C,\, U,\, M) = \sum_{n \in \Omega} \left(\prod_{i=1}^{U} \frac{\rho_{u,i}^{u_i}}{u_i!} \cdot \prod_{i=1}^{M} \frac{\rho_{m,i}^{m_i}}{m_i!} \right) \tag{3}$$

The proof is trivial since the state probability verifies the global equilibrium balance equation.

2.3 CALL BLOCKING PROBABILITY

Let B_i^+ denotes the sets of blocking states for unicast or multicast class i. Then, the call blocking probability for class i can be expressed as :

$$P_{bi} = \sum_{n \in B_i^+} P\,(n) = 1 - \sum_{n \notin B_i^+} P\,(n) = 1 - \sum_{n \in \Omega - B_i^+} P\,(n) \tag{4}$$

Thus,

$$P_{bi} = 1 - \frac{1}{G\,(\Omega)} \sum_{n \in \Omega - B_i^+} \left(\prod_{i=1}^{U} \frac{\rho_{u,i}^{u_i}}{u_i!} \cdot \prod_{i=1}^{M} \frac{\rho_{m,i}^{m_i}}{m_i!} \right) = 1 - \frac{G\,(\Omega - B_i^+)}{G\,(\Omega)} \tag{5}$$

Furthermore, when i is a unicast class, then Kaufman [6] shows that the previous quantity can be expressed as :

$$P_{bi,u} = 1 - \frac{G\,(C - b_{u,i},\, U,\, M)}{G\,(C,\, U,\, M)} \tag{6}$$

Let $M-i$, denotes M multicast classes without the class i, which also means that the class i will never be active. Then, a similar expression for the multicast call blocking probability $P_{bi,m}$ can be found. See the Appendix for the proof.

$$P_{bi,m} = 1 - \frac{e^{\rho_{m,i}}\, G\left(C - b_{m,i},\, U,\, M - i\right)}{G\left(C,\, U,\, M\right)} \tag{7}$$

2.4 UNICAST CONNECTIONS

In the previous section, we gave the exact solution to compute the call blocking probability for both multicast and unicast classes. However, the state system has as many dimensions as the number of traffic classes. Therefore, in a large system capacity C and a huge number of classes $M+U$, numerical difficulties may arise when computing both the normalization constant and the call blocking probability. Fortunately, Kaufman [6] found a solution to map the multi-dimensional state space into a one dimensional state space. It has been shown that the distribution of the number of resource units allocated for unicast connections has a recursive expression. Moreover, this formula is only linearly dependent on U.

These results can also be applied in our model to compute the resource distribution for unicast connections. In fact, since the state probability has a product form, it can be easily verified that the multicast traffic has no consequences on the validity of the recursive solution. However, since the normalization constant in our model depends on both unicast and multicast connections, we have to consider first an unormalized state probability for resource distribution for the unicast traffic. Thus, let $q_U\left(.\right)$ be this quantity. Then :

$$q_U\left(j\right) = \sum_{\{u:u.b_u=j\}} \prod_{i=1}^{U} \frac{\rho_{u,i}^{u_i}}{u_i!} \tag{8}$$

Here j denotes the number of resource units allocated to the unicast connections and $u.b_u = \sum_{i=1}^{U} b_{u,i} u_i$. Kaufman [6] shows that $q_U\left(j\right)$ can be computed recursively as follow :

$$q_U\left(j\right) = \begin{cases} 0 & , j < 0 \\ 1 & , j = 0 \\ q_U\left(j\right) = \frac{1}{j} \sum_{i=1}^{U} \rho_{u,i}\, b_{u,i}\, q_U\left(j - b_{u,i}\right) & , j > 0 \end{cases} \tag{9}$$

2.5 MULTICAST CONNECTIONS

The recursive solution given in the previous section cannot be applied for the multicast connections. However, in the last case, we find another solution to reduce the state space. Thus, let us define $q_M(.)$ as the unormalized state probability for resource distribution in the multicast traffic. Remembering that a class i request will never be blocked if i is an active class, an infinite population customer of the same class i can join an active class i. Thus, defining $1_m.b_m = \sum_{i=1}^{M} b_{m,i}.1_{\{m_i>0\}}$ we can write :

$$q_M(j) = \sum_{\{m:1_m.b_m=j\}} \prod_{i=1}^{M} 1_{\{m_i>0\}} \left(\sum_{m_i=1}^{+\infty} \frac{\rho_{m,i}^{m_i}}{m_i!} \right) \tag{10}$$

Thus, we find

$$q_M(j) = \sum_{\{m:1_m.b_m=j\}} \prod_{i=1}^{M} 1_{\{m_i>0\}} \left(e^{\rho_{m,i}} - 1\right) \tag{11}$$

Now let $q_k(.)$ be the unormalized resource distribution probability for multicast classes whose index vary from of 1 to k. Let b_m^k denotes a k dimension capacity vector for multicast calls. Thus, $1_m^k.b_m^k = \sum_{i=1}^{k} b_{m,i}.1_{\{m_i>0\}}$

$$q_k(j) = \sum_{\{m:1_m^k.b_m^k=j\}} \prod_{i=1}^{k} 1_{\{m_i>0\}} \left(e^{\rho_{m,i}} - 1\right) \tag{12}$$

Since this quantity can be divided into two parts, the first one represents the case where no customers of class k are being served, and the second one for the case where class k is active, then :

$$\begin{aligned} q_k(j) = & \sum_{\{m:1_m^k.b_m^k=j/m_k=0\}} \prod_{i=1}^{k} 1_{\{m_i>0\}} \left(e^{\rho_{m,i}} - 1\right) \\ & + \sum_{\{m:1_m^k.b_m^k=j/m_k>0\}} \prod_{i=1}^{k} 1_{\{m_i>0\}} \left(e^{\rho_{m,i}} - 1\right) \end{aligned} \tag{13}$$

Consider the left-side of the sum of equation 13, then this quantity can be written as follows :

$$\sum_{\{m:1_m^{k-1}.b_m^{k-1}=j\}} \prod_{i=1}^{k-1} 1_{\{m_i>0\}} \left(e^{\rho_{m,i}} - 1\right) = q_{k-1}(j) \tag{14}$$

The right-hand side of the sum of equation 13 can be written as follows:

$$\left(e^{\rho_{m,k}}-1\right)\sum_{\left\{m:1_m^{k-1}.b_m^{k-1}=j-b_{m,k}\right\}}\prod_{i=1}^{k-1}1_{\{m_i>0\}}\left(e^{\rho_{m,i}}-1\right) \tag{15}$$

We find the recursive formula for the multicast traffic :

$$q_k\left(j\right)=\begin{cases}0 & ,j<0,\forall k\\ 1 & ,j=0,\forall k\\ 0 & ,j>0,k=0\\ q_{k-1}\left(j\right)+\left(e^{\rho_{m,k}}-1\right)q_{k-1}\left(j-b_{m,k}\right) & ,j>0,k>0\end{cases} \tag{16}$$

Finally, using 9 and 16 we can compute the normalization constant $G\left(\Omega\right)$:

$$G\left(\Omega\right)=\sum_{i=0}^{C}q_M\left(i\right)\left(\sum_{j=0}^{C-i}q_U\left(j\right)\right) \tag{17}$$

2.6 MULTICAST CALL BLOCKING PROBABILITY

A multicast call of class r is blocked only if r is not an active class and there are less than $b_{m,r}$ free resource units in the link. Thus, let $q_{M-r}\left(.\right)$ be the unormalized resource distribution probability for multicast connections such that the class r is not active :

$$q_{M-r}\left(j\right)=\sum_{\left\{m:1_m.b_m=j/m_r=0\right\}}\prod_{i=1}^{M}1_{\{m_i>0\}}\left(e^{\rho_{m,i}}-1\right) \tag{18}$$

Given the recursive formula given in 16 we obtain the following :

$$q_M\left(j\right)=q_{M-r}\left(j\right)+\left(e^{\rho_{m,r}}-1\right)q_{M-r}\left(j-b_{m,r}\right) \tag{19}$$

Thus we have a recursive formula that will give the exact value of $q_{M-r}\left(j\right)$

$$q_{M-r}\left(j\right)=\begin{cases}0 & ,j<0\\ 1 & ,j=0\\ q_M\left(j\right)-\left(e^{\rho_{m,r}}-1\right)q_{M-r}\left(j-b_{m,r}\right) & ,j>0\end{cases} \tag{20}$$

Finally, the call blocking probability for unicast and multicast connections can be computed using the following equations :

$$P_{bi,u} = \frac{1}{G(\Omega)} \sum_{j=0}^{C} \left(q_M(j) \sum_{k=0}^{b_{u,i}} q_U(C-j-k) \right) \tag{21}$$

$$P_{bi,m} = \frac{1}{G(\Omega)} \sum_{j=0}^{C} \left(q_U(j) \sum_{k=0}^{b_{m,i}} q_{M-i}(C-j-k) \right) \tag{22}$$

3. EXAMPLE

We consider a link with a bandwidth capacity equal to 10 Mbps. Let the basic bandwidth unit be 64 Kbps. Thus, $C = 160$. We choose two unicast classes : $U = 2$ and four multicast classes : $M = 4$. For each class we take the following capacity requirements :

- $b_{u,1} = 32, b_{u,2} = 16\lambda$
- $b_{m,k} = 32, \forall k/1 \leq k \leq M$

We suppose that for both unicast and multicast mean call duration time are 120 seconds. Thus, to analyze the link occupancy at different loads, we introduce λ as the inter-arrival time of a customer (unicast and multicast) in the link such as :

- $\lambda_{u,1} = \lambda$
- $\lambda_{u,2} = 2\lambda$
- $\lambda_{m,k} = 0.8^k\lambda, \forall k/1 \leq k \leq M$

Varying λ from 0 to 900, we compute using equations 21 and 22 the call blocking probability for unicast and multicast traffics. The results are plotted in figure 1. The figure shows that the call blocking probabilities for unicast traffics increase when the traffic load grows. However, multicast call blocking probabilities for all classes are always under a certain threshold. Thus, these probabilities increase until a certain value of λ, after which they decrease. This is due to the fact that after a certain value of λ, these classes will always be active. At this point, the traffic load is so high that there is at any time at least one multicast customer being served; thus, for these multicast sessions a new customer will never be blocked.

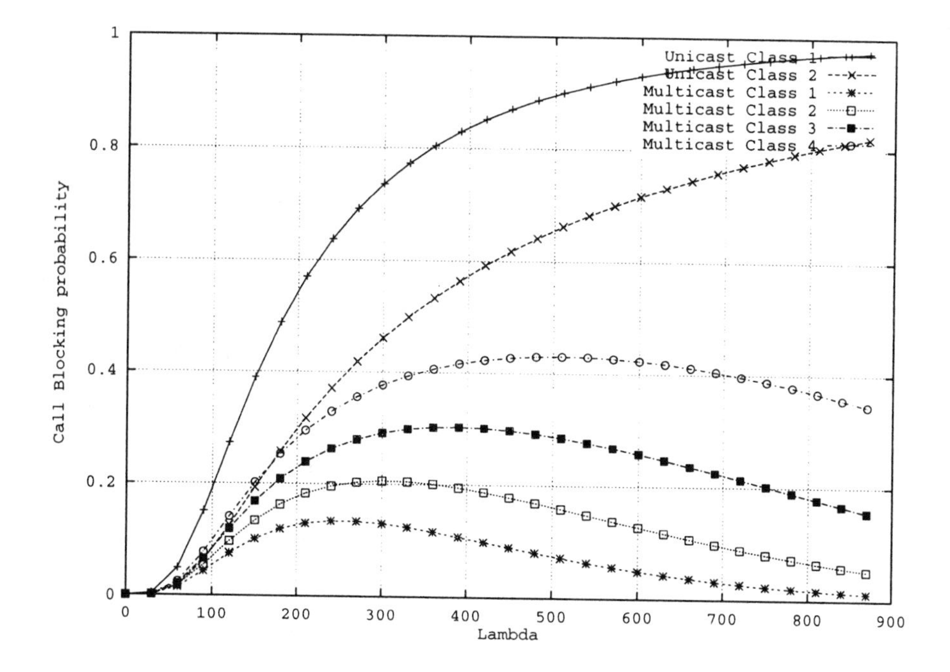

Figure 1 Call Blocking Probability

4. CONCLUSION

In this paper, we addressed the multirate resource sharing problem for unicast and multicast traffics. We show that in Complete Sharing policy the model have a product form solution. Thus, we deduce the exact call blocking probability values for both unicast and multicast connections. Moreover, to avoid the numerical problems that may arise when computing these quantities, we extend the recursive solution of Kaufman [6] to the multicast case.

Our future prospects will concern the adaptation of these results to other sharing policies, such as Upper Limit (UL) and Guaranteed Minimum (GM) policies. We also investigate the problem of multicast resource sharing in a wireless environment.

Appendix

Let us write $G\left(\Omega - B_i^+\right)$. This quantity can be divided into two parts, the first one represents the case where there is no customer of class i being served, thus, no blocking occurs for this class if the sum of the allocated resource units in the system is under $C - b_{m,i}$. The second part of the equation represents the case where the class i is active, that is no blocking for this class will occur. Therefore, We obtain that :

$$G\left(\Omega - B_i^+\right) = \sum_{j=0}^{C-b_{m,i}} q_{M-i}(j) \left(\sum_{k=0}^{C-b_{m,i}-j} q_U(k)\right) + \left(e^{\rho_{m,i}} - 1\right) \sum_{j=b_{m,i}}^{C} q_{M-i}(j - b_{m,i}) \left(\sum_{k=0}^{C-j} q_U(k)\right) \tag{A.1}$$

Thus, taking $J = j - b_{m,i}$ in the second sum of A.1 we find that :

$$G\left(\Omega - B_i^+\right) = e^{\rho_{m,i}} G\left(C - b_{m,i}, U, M - i\right) \tag{A.2}$$

We can now deduce that the multicast call blocking probability for i is :

$$P_{bi,m} = 1 - \frac{e^{\rho_{m,i}} G\left(C - b_{m,i}, U, M - i\right)}{G\left(C, U, M\right)} \tag{A.3}$$

References

[1] C. Diot. Reliability in multicast services and protocols ; a survey. In *International Conference on Local and Metropolitan Communication Systems*, December 1994.

[2] Hans Eriksson. MBone: The Multicast Backbone. *Communications of the ACM*, 37:54–60, 8 1994.

[3] D. Reeves H. Salama and Y. Viniotis. Evaluation of multicast routing algorithms for real-time communication on high-speed networks. *IEEE Journal on Selected Areas in Communication*, February 1997.

[4] G. C. Polyzos J. C. Pasquale and G. Xylomenos. The multimedia multicasting problem. *ACM Multimedia Systems Journal*, 6(1):43–59, 1998.

[5] S. Aalto J. Karvo, J. Virtamo and O. Martikainen. Blocking of dynamic multicast connections in a single link. In *International Broadband Communications Conference, Future of Telecommunications*, pages 473–483. Paul J. Kühn and Roya Ulrich Edition, April 1998.

[6] D. Kaufman. Blocking in a shared resource environment. *IEEE Transactions on communications*, 29(10):1474–1481, October 1981.

[7] J. S. Turner. An optimal nonblocking multicast virtual circuit switch. Technical Report wucs-93-30, March 1994.

[8] H. A. Wang and M. Schwartz. Performance analysis of multicast flow control algorithms over combined wireless/wired networks. In *IEEE INFOCOM'97*, April 1997.

SESSION 16

Traffic Control

ON THE PCR POLICING OF AN ARBITRARY STREAM PASSING THROUGH A PRIVATE ATM NETWORK BEFORE REACHING THE PUBLIC UNI

Lorenzo Battaglia
Digital Communication Systems Dept.
Technical University of Hamburg-Harburg
Germany
Battaglia@TU-Harburg.de

Ulrich Killat
Digital Communication Systems Dept.
Technical University of Hamburg-Harburg
Germany
Killat@TU-Harburg.de

Abstract Any Call Admission Control algorithm and any Network Dimensioning depend on the reliability of the chosen Usage Parameter Control algorithm. The standardized ATM UPC algorithm is the GCRA. An ideal ATM PCR policer should guarantee a very low cell loss probability (10^{-9}) for the cells of a well-behaved user and police contract violation by the source as rigidly as possible. All this independently of the characteristics of the monitored stream and of the background traffic. When we started our investigations no acceptable GCRA parameters dimensioning had been presented and no alternative policing solution could claim to be more general and simpler than the GCRA. We already provided an analytic GCRA parameters dimensioning which transforms the GCRA into an ideal PCR policer, also in the case of online PCR renegotiation, and into an ideal ACR policer of ABR services. In this paper we generalize our results to the PCR policing of a *GCRA* $(T_{PCR}\,;\ \tau_0)$ complying stream passing through an ATM network before reaching the public UNI. The GCRA dimensioning we present allows the policer to discard no cell of a well-behaved user and to rigidly police contract violation. Our results hold for any kind of background traffic.

Keywords: ATM, Traffic Monitoring, Network Dimensioning

1. INTRODUCTION

ATM is the switching and multiplexing technology chosen by the ITU for the operation of B-ISDN. Basically, ATM technology is designed to combine the reliability of circuit switching with the efficiency and flexibility of packet switching technologies [1]. ATM provides 5 service categories with a rich set of related QoS parameters and guarantees an end-to-end QoS to individual connections [3].

Call Admission Control (CAC), Network Dimensioning and Usage Parameter Control (UPC) are very important issues in ATM technology. They all depend on how input traffic is described, on how cell streams' characteristics change when passing an ATM queueing system. CAC and Network Dimensioning strongly depend on how reliable the chosen UPC algorithm is. The standardized ATM UPC algorithm is the Generic Cell Rate Algorithm (GCRA). The GCRA depends on two parameters: the Increment I and the Limit L. When just one GCRA is used, Peak Cell Rate (PCR) policing, I and L are respectively denoted by T and τ. If two GCRA's are used, PCR and Sustainable Cell Rate (SCR) policing, the four parameters are respectively denoted by T, τ, T_s and τ_s. The ATM Forum has established that streams at the input of ATM networks should be described by means of the GCRA parameters used to police them. A recent internet draft on Differentiated Services [14] has proposed to assign drop precedence levels of packets within the same Assured Forwarding class by using a leaky bucket policer. Hence the GCRA concept has received increasing attention for both, ATM-based and IP-based networks.

One of the main criteria to dimension an ATM policer is to minimize cell loss. The loss of a single ATM cell would cause the loss of a higher layer's packet, resulting in wasted bandwidth and loss of QoS. At the same time it is very important for the network provider that any malicious user is policed as rigidly as possible. Any charging system as well as any network dimensioning to guarantee a desired QoS is valid only if all cell streams entering the network do comply with the traffic contract. An ideal ATM policer should then guarantee a very low cell loss probability (10^{-9}) for the cells of a well-behaved user and police contract violation as rigidly as possible. All this independently of the characteristics of the monitored stream and of the cell delay variations due to the other ATM applications, whose cells pass through the same ATM queueing systems (see ATM UPC definitions in [3] and Figure 4-2 therein). To be able to protect all network resources, the policing function must be located as close as possible to the actual traffic source [20] and obviously be under direct control of the network provider. Since this function must be available for every connection during the entire active phase and must operate in real time, the policing method used must be fast, simple and cost effective.

Since the dimensioning of GCRA parameters was left to the user without clear guidelines, several alternative policing methods have been introduced in the literature. Most of them have been evaluated against their capability to enforce the PCR of ON-OFF sources. All have been compared to the GCRA. Window-based control methods have been proposed [12], [Rathgeb91]. Fuzzy Logic [10]-[13], Neural Networks [22] and Finite State Automata [18] control methods have been presented. Even "improved" versions of the GCRA [2], [4], [16], [17] have been tested. No solution could however claim to be more general and simpler than the GCRA.
Many other authors investigated the rejection probability p_r of the GCRA when monitoring a CBR source passing an ATM multiplexer. Some, e.g. [15], tried to relate the statistical parameters of the background traffic to the chosen GCRA parameters. Others, e.g. [9], saw the whole as a $(M + D)/D/1$, FIFO. All came to the conclusion that in order to ensure a sufficiently low value for p_r it is necessary to police a higher rate than negotiated, in some cases [8] even significantly higher. Such a choice is surely not the ideal to rigidly police contract violation by the source. We started our investigations with the task to dimension the GCRA in such way to minimize the rejection probability for the cells of a well-behaved user and to rigidly police contract violation by the source, and this independently of the characteristics of the monitored stream and of the background traffic. We started with the rejection probability of a GCRA when monitoring a well-behaved CBR stream passing an ATM multiplexer. We extended our results also to the more general case of an ATM multiplexer equipped with a global traffic shaper which guarantees a minimum distance of T_{shape} time slots between two consecutive cells [5]. We treated then the PCR policing of any stream passing an ATM multiplexer, also in the case of on-line PCR renegotiation, and the Available Cell Rate (ACR) policing of ABR services [6]. In [7] we dimensioned two sets of GCRA parameters in such way to fairly police PCR and SCR of an arbitrary stream passing through an ATM multiplexer. In this paper we treat the PCR policing of an arbitrary stream which complies with a $GCRA(T_{PCR}, \tau_0)$ passing through a private ATM network before reaching the public User Network Interface (UNI), where the policing function is performed. According to the PCR Reference Model [3] we refer to a "private" and to a "public" network to distinguish two networks with different providers. Also in this case we provide a GCRA parameters dimensioning which transforms the GCRA into an ideal PCR policer: No cell of a well-behaved user is discarded whereas contract violation by the source is policed as rigidly as possible. Our results hold for any kind of background traffic (even self similar) generated by the other ATM applications running at the user side, whose cells pass through the same ATM queueing systems in the private ATM network. This paper is organized as follows. In Section 1 we

model the problem. In Section 2 some important concepts of the GCRA are recalled. Section 3 provides the framework for GCRA parameters dimensioning. In Section 4 and 5 the policing issue respectively of streams complying with a $GCRA(T_{PCR}, 0)$ and with a $GCRA(T_{PCR}, \tau_0)$ is treated.

2. MODELLING THE PROBLEM

We treat the PCR policing of an arbitrary stream passing through a private network before reaching the public UNI. We refer to Fig.2. Let S be the source we want to monitor. The monitored stream passes through M queues (multiplexers and switches) before entering the public ATM UNI. We assume each queue, e.g. the kth, to have N_k input lines and one output link and to be equipped with a buffer of finite length B_k (see also example in Fig. 3). Time slots on each of the input lines and on the output link are of the same length and synchronized to each other. Every new cell is assumed to arrive at the beginning of a slot and can be transmitted at the end of the same time slot if no predecessor exists (see Fig.1).

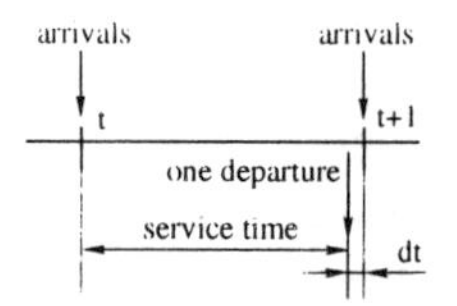

Figure 1 Queue Discipline: Departure occurs immediately before the beginning of the next time slot.

About the background traffic and the topology of the rest of the network we don't make any hypotheses. This choice may at first glance be surprising

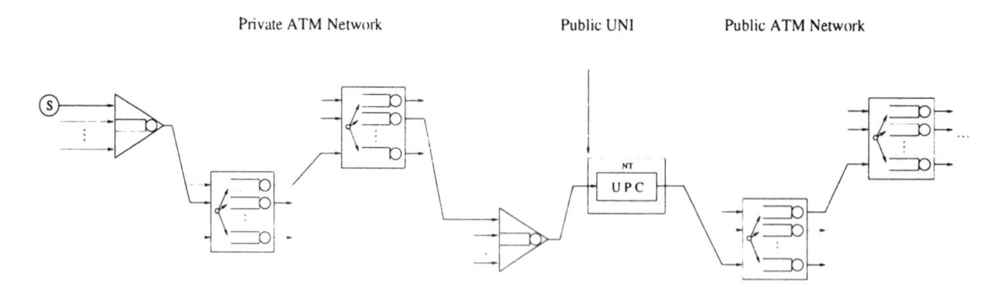

Figure 2

because the delay behaviour of the monitored cells will, of course, depend on the characteristics of the background traffic and on the topology of the rest of the network. However, since B_k time slots is the upper-bound to the delay any monitored cell can experience passing through an ATM queueing system equipped with a buffer of length B_k and a server, it is possible to provide an ideal GCRA dimensioning once the negotiated PCR and the lengths of the crossed buffers $\{B_k : k = 1, ..., M\}$ are known. In the ensuing discussion we denote by $GCRA(T, \tau)$-stream ($GCRA(T, \tau)$-source) an arbitrary stream (source) which complies with a $GCRA(T, \tau)$.

3. THE GENERIC CELL RATE ALGORITHM

The GCRA depends on two parameters the Increment, I, and the Limit, L. In order to get familiar with it we refer to a $GCRA(I, L)$. Let's consider a possible realization of the departure process of a monitored stream from the last ATM queueing system. Let $t_a(i)$ be the discrete time instant in which the ith cell of the monitored stream arrives at the policer and $TAT(i)$ its Theoretical Arrival Time. Whenever the ith cell arrives after its TAT, i.e. $t_a(i) > TAT(i)$, it is accepted and the following TAT is given by

$$TAT(i+1) = t_a(i) + I. \tag{1}$$

Else it is accepted only if

$$TAT(i) - L \leq t_a(i)$$

holds and the new TAT is given by

$$TAT(i+1) = TAT(i) + I. \tag{2}$$

4. HOW TO TUNE I AND L

In order to rigidly police contract violation by the source we set I to the reciprocal of the negotiated rate. In order to reject no cell of a well-behaved user for any background traffic, i.e. to properly dimension L, we introduce the new concept of τ coverage area (Subsection 3.1) and apply it to the worst case sequence into which the monitored stream can be transformed by the background traffic (Subsection 3.2). Since for the PCR policing L is denoted by τ, in the ensuing discussion we coherently refer to τ.

4.1 THE τ COVERAGE AREA

Any stream of m cells arriving at the GCRA is uniquely identified by the sequence of the discrete time instants $t_a(i)$, $i = 1, 2, ..., m$, in which the ith monitored cell enters the policer. Let TAT(i), $i = 2, 3, ..., m$, be the corresponding sequence of the Theoretical Arrival Time of the ith cell at the GCRA. Whenever the ith cell arrives at the policer before its TAT any

$$\tau \geq TAT(i) - t_a(i)$$

allows to accept that cell. We define τ coverage area, τ_{CA}, for the ith cell as

$$\tau_{CA}(i) = [TAT(i) - t_a(i)]^+$$

i.e. the smallest value of τ that allows the policer to accept it. Obviously a τ_{CA} of 0 time slots is already enough to accept any cell arriving after its TAT. Considering the whole sequence and defining

$$\tau_{CAmax} = max\{\tau_{CA}(2), \tau_{CA}(3), ..., \tau_{CA}(m)\}$$

any $\tau \geq \tau_{CAmax}$ allows then to accept all the cells without losses. Such an a posteriori dimensioning of τ will be valid for any sequence $t_a(i)$, $i = 1, 2, ...$, if derived from the worst case sequence into which the monitored stream can be transformed by the background traffic.

4.2 THE WORST CASE SEQUENCE

We define as the worst case sequence into which the monitored stream passing through a private ATM network can be transformed by the background traffic the sequence $t_a(i)$, $i = 1, 2, ...$, for which each $\tau_{CA}(i)$, $i = 2, 3, ...$, is the biggest possible. Whenever the ith monitored cell enters the policer before its TAT

$$\tau_{CA}(i) = TAT(i) - t_a(i) \tag{3}$$

holds. The biggest $\tau_{CA}(i)$ will then occur when $TAT(i)$ and $t_a(i)$ are respectively the biggest and the smallest allowed for i. We denote the latter quantity by $t_{a\,min}(i)$. To simplify the results, in the ensuing discussion we denote as time slot 0 that in which the first of the monitored cells enters the first queue of the private ATM network. Moreover, since we are interested in the difference $TAT(i) - t_a(i)$ we don't consider in the delay the service time (1 time slot) which all cells experience passing through each ATM server. Since with the previous assumption the delay that any monitored cell will experience passing through all the M queues is at most given by

$$\sum_{k=1}^{M} B_k \tag{4}$$

time slots, the biggest TAT allowed for i is

$$TAT(i) = \sum_{k=1}^{M} B_k + (i-1)I. \tag{5}$$

We assume lines to be short and routing operations to be performed fast enough not to introduce any further delay. As far as the $t_{a\,min}(i)$ is concerned, it is the first time slot in which the ith monitored cell can at the earliest leave the Mth queue. Formally

$$t_{a\,min}(i) = max\{t_{a\,Q}(i, M), t_{a\,min}(i-1) + 1\} \tag{6}$$

with

$$t_{a\,Q}(i, j) = max\{t_{a\,Q}(i, j-1), t_{a\,Q}(i-1, j) + 1\}$$

and $j = 1, 2, ..., M$. We denote by $t_{a\,Q}(i, j)$ the earliest arrival time of the ith monitored cell at the jth queue and by $t_{a\,Q}(i, 0)$ the earliest departure instant of the ith cell from the monitored source. Clearly $t_{a\,min}(i)$, as in general $t_{a\,Q}(i, j)$ is the latest of the two events: the earliest arrival time of the ith cell at the previous queue (if the $(i-1)^{th}$ monitored cell has already left the previous queue, the ith monitored cell can leave it at the earliest immediately) or right after the departure of the $(i-1)^{th}$ monitored cell from the previous queue (if another previous monitored cell is in the previous queue, the ith monitored cell can leave it at the earliest right after the $(i-1)^{th}$). The worst case sequence occurs then when the background traffic delays the first of the monitored cells by $\sum_{k=1}^{M} B_k$ time slots without delaying any of the following ones. The τ_{CA} sequence is then in the worst case given by

$$\tau_{CA}(i) \quad = \quad \sum_{k=1}^{M} B_k + (i-1)I - t_{a\,min}(i) \tag{7}$$

with $i = 2, 3, ...$. In order to rigidly police contract violation by the source we set I to the reciprocal of the negotiated rate. We will show that for the worst case sequence the biggest $\tau_{CA}(i)$, $i = 2, 3, ...$, is a finite quantity. We consistently denote the last quantity by τ_{CAmax}.

This τ_{CAmax} will then be the ideal τ to police the PCR of any stream, passing through a private ATM network guaranteeing that for any background traffic no cell of a well-behaved user will be rejected. Any smaller τ would lead to a positive probability to reject a cell of a well-behaved user. Any bigger τ would tolerate cell delay variations which wouldn't ever occur in the stream of a well-behaved user.

5. PCR POLICING ISSUE OF A $GCRA(T_{PCR}, 0)$- -STREAM PASSING THROUGH A PRIVATE ATM NETWORK

We consider the PCR policing of a $GCRA(T_{PCR}, 0)$-stream passing through M queues at the user's premises before reaching the public UNI (see example in Fig.3). Passing through the M queues the monitored cells will in general experience cell delay variations that can make appear the monitored source to send at a higher rate than negotiated. We use a $GCRA(T, \tau)$ to police this stream. Since in the worst case any $GCRA(T_{PCR}, 0)$-source is a well-behaved CBR source with period T_{PCR}, GCRA parameters can a fortiori be dimensioned so to fairly police such a source. To rigidly police contract violation by the source we set $T = T_{PCR}$. To evaluate the τ_{CAmax} two cases have to be considered: $\sum_{k=1}^{M} B_k < T_{PCR}$ and $\sum_{k=1}^{M} B_k \geq T_{PCR}$.

5.1 FIRST CASE: $\sum_{K=1}^{M} B_K < T_{PCR}$

In this case, since no cell can be delayed of more time slots than its period

$$t_{a\,min}(i) = t_{a\,Q}(i, M) = t_{a\,Q}(i, M-1) = ... = t_{a\,Q}(i, 0) \tag{8}$$

holds. Since

$$t_{a\,Q}(i, 0) = (i-1)T_{PCR} \tag{9}$$

then

$$\tau_{CA}(i) = \sum_{k=1}^{M} B_k + (i-1)T_{PCR} - (i-1)T_{PCR} = \sum_{k=1}^{M} B_k \tag{10}$$

with $i = 2, 3, ...$, for the worst case sequence holds. Clearly $\tau_{CAmax} = \sum_{k=1}^{M} B_k$.

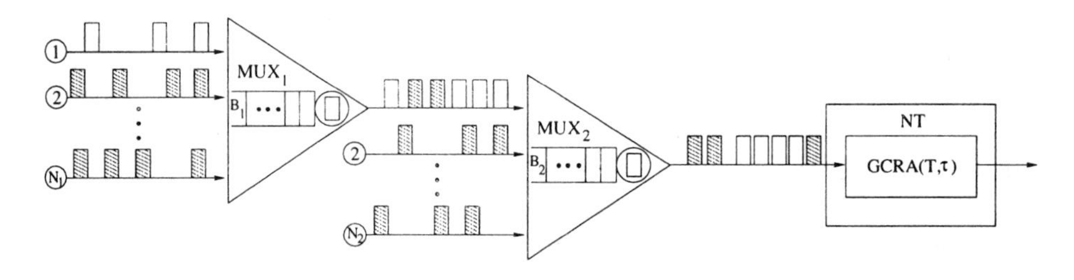

Figure 3 The PCR policing issue of a $GCRA(T_{PCR}, \tau_0)$-stream feeding input 1 of the first multiplexer: The background traffic can make appear the monitored source to send at a higher rate than negotiated.

5.2 SECOND CASE: $\sum_{K=1}^{M} B_K \geq T_{PCR}$

The worst case sequence has at first a burst of

$$b = \left\lfloor \frac{\sum_{k=1}^{M} B_k}{T_{PCR} - 1} \right\rfloor + 1$$

cells spaced at one time slot, followed by all the other cells which don't experience any delay. The τ_{CA} sequence is then

$$\tau_{CA}(i) = \begin{cases} (i-1)(T_{PCR}-1) & \text{for } 2 \leq i \leq b \\ \sum_{k=1}^{M} B_k & \text{for } i > b \end{cases} \tag{11}$$

since in this case

$$t_{a\ min}(i) = \begin{cases} i + \sum_{k=1}^{M} B_k - 1 & \text{for } 2 \leq i \leq b \\ t_{a\,Q}(i,0) & \text{for } i > b. \end{cases} \tag{12}$$

Thus

$$\tau_{CAmax} = max\{\tau_{CA}(2), \tau_{CA}(3), ..., \tau_{CA}(b), \sum_{k=1}^{M} B_k\}.$$

Since the sequence $\tau_{CA}(2), ..., \tau_{CA}(b)$ is monotonically increasing and

$$\tau_{CA}(b) \leq \sum_{k=1}^{M} B_k, \tag{13}$$

$\tau_{CAmax} = \sum_{k=1}^{M} B_k$. Hence setting T to T_{PCR} and choosing $\tau = \sum_{k=1}^{M} B_k$ allows to ideally police the PCR of any stream passing through M buffers having lengths $\{B_k : k = 1, ..., M\}$, so that for any background traffic contract violation by the source is rigidly policed and so that no cell of a well-behaved user is discarded.

6. PCR POLICING ISSUE OF A $GCRA(T_{PCR}, \tau_0)$--STREAM PASSING THROUGH A PRIVATE ATM NETWORK

We consider now the PCR policing of a $GCRA(T_{PCR}, \tau_0)$-stream passing through M queues at the user's premises before reaching the public UNI. We use a $GCRA(T, \tau)$ to police this stream. We analyze at first the case of $M = 1$ and then generalise our results. In order to rigidly police contract violation by the source we set $T = T_{PCR}$. In order to dimension τ we evaluate the τ_{CAmax}. Clearly each $\tau_{CA}(i)$, $i = 2, 3, ...$, is the biggest possible if the first monitored cell of the $GCRA(T_{PCR}, \tau_0)$-stream experiences the longest possible delay in the queue before being served and if the following cells arrive at the earliest (see definition of the $t_{a\,Q}(i,0)$) and are not delayed by any further background traffic. We turn now our attention to each $t_{a\,Q}(i,0)$, $i = 1, 2, ...$, which we already defined as the earliest departure instant of the ith cell from the monitored source. Two cases have to be considered: $\tau_0 < T_{PCR}$ and $\tau_0 \geq T_{PCR}$.

6.1 CASE OF $\tau_0 < T_{PCR}$

In this case

$$t_{aQ}(i,0) = \begin{cases} 0 & \text{for } i = 1 \\ (i-1)T_{PCR} - \tau_0 & \text{for } i \geq 2. \end{cases} \tag{14}$$

6.2 CASE OF $\tau_0 \geq T_{PCR}$

In this case there's at first a burst of N back-to-back cells with

$$N = \left\lfloor \frac{\tau_0}{T_{PCR} - 1} \right\rfloor + 1 \tag{15}$$

and

$$t_{aQ}(i,0) = \begin{cases} i-1 & \text{for } i = 1, 2, ..., N \\ (i-1)T_{PCR} - \tau_0 & \text{for } i \geq N+1. \end{cases} \tag{16}$$

The issue we address can be simply solved by means of an equivalence. Let's consider the PCR monitoring of a well-behaved source passing through an ATM queueing system equipped with a buffer of length B. Let T_{PCR}=1/PCR. For the worst case sequence into which such a stream can be transformed by the background traffic [5] two cases have to be considered: $B < T_{PCR}$ and $B \geq T_{PCR}$.

6.3 CASE OF $B < T_{PCR}$

In this case

$$t_{aQ}(i,0) = \begin{cases} 0 & \text{for } i = 1 \\ (i-1)T_{PCR} - B & \text{for } i \geq 2. \end{cases} \tag{17}$$

6.4 CASE OF $B \geq T_{PCR}$

In this case there's at first a burst of b back-to-back cells with

$$b = \left\lfloor \frac{B}{T_{PCR} - 1} \right\rfloor + 1 \tag{18}$$

and

$$t_{aQ}(i,0) = \begin{cases} i-1 & \text{for } i = 1, 2, ..., b \\ (i-1)T_{PCR} - B & \text{for } i \geq b+1. \end{cases} \tag{19}$$

The PCR monitoring of a $GCRA(T_{PCR}, \tau_0)$-stream is then equivalent to the PCR monitoring of a $GCRA(T_{PCR}, 0)$-source, whose cells pass through an ATM queueing system equipped with a buffer of length τ_0. The results of the previous Section can be then generalized to this case. Setting T to T_{PCR} and choosing

$$\tau = \sum_{k=1}^{M} B_k + \tau_0 \tag{20}$$

allows to ideally police the PCR of any $GCRA(T_{PCR}, \tau_0)$-stream passing through M buffers having lengths $\{B_k : k = 1, ..., M\}$ so that for any background traffic contract violation by the source is rigidly policed and so that no cell of a well-behaved user is discarded.

Discussion

We have presented a worst-case GCRA parameters dimensioning. Setting T to T_{PCR} and τ to τ_{CAmax}, the maximum cell delay variation a cell can experience when passing through an ATM queueing system, allows the GCRA to rigidly police contract violation by the source and to discard no cell of a well-behaved user. All this for any kind of background traffic. Any smaller τ would lead to a positive probability to reject cells of a well-behaved user. At the same time a smaller τ would decrease the potential cell clumping allowed at the UNI. The biggest cell-burst a source policed by a $GCRA(T, \tau)$ is allowed to send at a (higher) rate $1/\delta$ (cells/time slot) is of

$$N = \left\lfloor 1 + \frac{\tau}{T - \delta} \right\rfloor \tag{21}$$

cells. In order to achieve a smaller N, it could be thought to choose by means of simulations a $\tau < \tau_{CAmax}$, which still guarantees a sufficiently low discard probability to the cells of a well-behaved user. The main question is then if this is always necessary. If it is possible in some cases to take advantage of this general solution. Simulations have big disadvantages. They are time consuming (even with speed-up algorithms) and, above all, very complex models are needed to simulate "real" traffic. According to Eq. 21 the potential cell clumping becomes less significative the smaller τ is with respect to T. According to real buffer lengths, to real network speeds and to the real rate requirements of the most used applications, τ_{CAmax} comes out to be far smaller than T so that in most of the cases no further optimization of τ is necessary.

Conclusions

In this paper we have addressed the PCR policing issue of a stream which complies with a $GCRA(T_{PCR}, \tau_0)$ passing through a private ATM network before reaching the public UNI.

We have provided an ideal dimensioning of GCRA parameters which transforms the GCRA into an ideal PCR policer. Independently of the characteristics of the monitored stream and of the background traffic, no cells of a well-behaved user will be discarded, whereas any malicious user will be policed as rigidly as possible.

With our policing solution no well-behaved user will experience loss of QoS due to the unjustified discard of an ATM cell. Any CAC and any charging system will work and any network dimensioning will be valid since cell streams allowed to enter the ATM network will comply with the traffic contract.

References

[1] R. Ahmad, *Next Generation of Broadband Network Architectures-Traffic Flow Control and End-to-End Performance*, IFIP ATM '98, 20th-22nd July 1998, Craiglands Hotel, Ilkley, UK.

[2] A. Atlasis, G. Stassinopulos, A. Vasilakos, *Leaky bucket mechanism with learning algorithm for ATM traffic policing*, IEEE Symp. on Comp. and Commun. - Proc. 1997. IEEE Los Alamitos, CA, pp 68-72

[3] ATM Forum, Traffic Management Specification, Version 4.0, af-tm-0056.000, April 1996.

[4] W. Bao, S. Cheng, *Parameter adaptation mechanism for ATM traffic policing*, Chinese Journal of Electronics, vol. 6, no. 1, Jan. 1997.

[5] L. Battaglia, Z. Bažanowski, U. Killat *Dimensioning GCRA Parameters to Fairly Police a CBR Stream Passing an ATM Multiplexer Equipped with a Traffic Shaper*, IFIP ATM '98, 20th-22nd July 1998, Craiglands Hotel, Ilkley, UK.

[6] L. Battaglia, U. Killat *QoS of IP over ATM: The PCR and ACR Policing Issue*, IWS '99, 18th-20th Feb. 1999, Osaka University Convention Center, Suita, Japan.

[7] L. Battaglia, U. Killat *How to Dimension Two Sets of GCRA Parameters to Fairly Police PCR and SCR of Connections Passing an ATM Multiplexer*, AEÜ, International Journal of Electronics and Communications, 53 (1999) No. 3, 129-134, Urban & Fischer Verlag.

[8] P. Boyer, F.M. Guillemin, M.J. Servel, J.P. Coudreuse *Spacing Cells Protects and Enhances Utilization of ATM Network Links*, IEEE Network Magazine, Vol. 6, No. 5, September 1992, pp 38-49.

[9] P. Castelli, A. Forcina, A. Tonietti *Dimensioning Criteria for Policing Functions in ATM Networks*, INFOCOM '92.

[10] V. Catania, G. Ficili, S. Palazzo, D. Panno, *A fuzzy decision maker for source traffic control in high speed networks*, Proc. 1995 Int. Conf. on Network Protocols (ICNP 95), Tokyo, Nov. 7-10, 1995.

[11] V. Catania, G. Ficili, S. Palazzo, D. Panno, *A fuzzy expert system for Usage Parameter Control in ATM networks*, Proc. GLOBECOM '95, Singapore, Nov. 13-17, 1995.

[12] V. Catania, G. Ficili, S. Palazzo, D. Panno, *A Comparative Analysis of Fuzzy Versus Conventional Policing Mechanisms for ATM Networks*, IEEE/ACM Transactions on Networking, Vol. 4, no. 3, June 1996.

[13] C. Douligeris and G. Develekos, *A Fuzzy Logic Approach to Congestion Control in ATM Networks*, Proc. IEEE ICC '95, Seattle, WA, June 1995.

[14] J. Heinanen, F. Baker, W. Weiss and J. Wroklawski, *Assured Forwarding PHB Group*, Internet Draft, January 1999.

[15] F. Huebner, *Dimensioning of a Peak Cell Rate Monitor Algorithm Using Discrete-Time Analysis*, ITC 14 / J. Labetoulle and J.W. Roberts (Editors).

[16] Z. Jiang, Z. Liu *Improved algorithm of usage parameter control in ATM networks*, Int. Conf. on Commun. Tech. Proc., ICCT vol. 1, 1996. IEEE, Piscataway, NJ, pp 24-27.

[17] Z. Jiang, Z. Liu *Fuzzy leaky bucket for policing mechanism in ATM networks*, Neural Networks for Signal Processing. Proc. IEEE Workshop 1996, Piscataway, NJ, pp 510-517.

[18] L. Mason, A. Pelletier, J. Lapointe *Toward optimal policing in ATM networks*, Computer Communications, vol. 19, no. 3, Mar 1996, pp 194-204.

[19] R. O. Onvural, *Asyncronous Transfer Mode Networks: Performance Issues*, Norwood, MA: Artech House, 1994.

[20] E. P. Rathgeb, *Modeling and Performance Comparison of Policing Mechanisms for ATM Networks*, IEEE J. Select. Areas Commun. vol. 9, no. 3, pp. 325-334, Apr. 1991.

[21] J. Sairamesh and N. Shroff, *Limitations and pitfalls of leaky bucket, a study with video traffic*, Proc. 3rd Int. Conf. Comput. Commun. Networks, Sept. 1994.

[22] A. Tarraf, I. Ibrahim, and T. Saadawi, *A novel neural network traffic enforcement mechanism for ATM networks*, IEEE J. Select. Areas Commun., vol. 12, no. 6, pp. 1088-1096, Aug. 1994.

MANAGEMENT AND CONTROL OF DISTRIBUTED MULTIMEDIA DEVICES AND STREAMS THROUGH OBJECT-ORIENTED MIDDLEWARE

Reinhold Eberhardt, Christian Rueß, Jochen Metzler
DaimlerChrysler AG, Research and Technology 3
PO Box 2360, 89013 Ulm, Germany
Phone: ++49 731 505 2103
Fax: ++49 731 505 4110
{Reinhold.Eberhardt, Christian.Ruess, Jochen.Metzler}@DaimlerChrysler.com

Abstract In this paper an architectural concept for the management of a distributed multimedia environment based on object-oriented middleware is presented. The system enables remote control of distributed multimedia devices such as MPEG encoders, decoders, video servers and file systems with multimedia content. Through this approach, the multimedia streams between the devices can be bound to an appropriate communication infrastructure, such as native ATM, RSVP or another network which may or may not be QoS aware. The system is easily extendable and provides a generic framework for the support of a variety of multimedia devices and communication infrastructures. Experiences using a prototype of the concept are given.

Keywords: Object-oriented distributed networking, multimedia communications, application and network QoS, ATM and IP QoS

1. INTRODUCTION

The provisioning of multimedia content is gaining increased importance in many areas of our daily life. Product presentation and visualization as well as learning environments are only a few examples. Furthermore, multimedia products such as hardware encoders and decoders are now available for everyone at reasonable prices. The speed of current computing systems is sufficient so as to permit real-time decoding and encoding of high-quality videos. On the other hand, computer networks provide the speed required to support a distribution of the multimedia devices and their sharing.

In order that a distributed multimedia environment may be supported, an architecture for the control and the management of such systems is required. Not only is there a need to manage the multimedia devices but there is also a requisite to control the QoS-aware communication system between the devices.

This paper sets forth a system architecture geared for the management and control of distributed multimedia devices and streams through object-oriented middleware. The system proposed is largely derived from the CORBAtelecoms specification [2], yet has the added benefit of reducing or extending it to fit our requirements as needed. The system is based on a CORBA platform with Java language mapping [1], [3], [4]. However, native code is integrated in the Java platform to support native ATM and device-specific libraries. For performance reasons, multimedia flows are directly transferred within the native code without propagation to the higher Java or CORBA level, thereby creating an interface to control the multimedia traffic from Java instead of a Java ATM API [7].

Section 2 shows how the user could make use of such a platform. In Section 3 a short overview of the architecture is presented. Section 4 goes into detail on several of the possible concepts for implementation of the system. The last two sections summarize related and future work.

2. THE SYSTEM AS SEEN BY THE USER

The users of a distributed multimedia system would like to be able to deploy the available resources, e.g. a remote encoder and the intermediate network. If the resources can be accessed remotely via a well-defined CORBA interface, several possible scenarios are generated:

Client-driven control: Should a user at a client PC with an MPEG-2 decoder desire to see MPEG-2 encoded material from a camera or VCR located at a remote location, the client could be utilized to set up the network connection and then configure and control the remote and local devices. In this scenario, the management of the system is driven solely by the client.

Server-driven control: The management and control of the devices and streams could also be driven by a server. For example, when using a camera to observe a security area and a motion sensor to detect any movement in this area, the motion sensor could trigger the transmission of pictures taken by the camera to a specific client, e.g. the security inspector responsible.

Manager-initiated control: In the third scenario, management and control is performed by a management application outside the client and server. Consider a telelearning scenario with an MPEG encoder in a central studio and several decoders in classrooms interconnected by a heterogeneous network infrastructure: when the session starts, a manager configures and controls all of the devices and sets up the intermediate network connections.

The objective of the system is to provide the flexibility needed to interconnect arbitrary source and sink devices across an arbitrary network infrastructure. The users can express their requirements and choose from a list of devices and communication systems those they want to use. The mapping of user requirements with available services is carried out via a trading service.

3. THE ARCHITECTURE

The architecture of our system reduces the CORBAtelecoms specification [2], on one hand, and, on the other hand, also extends it to meet our specific requirements. In our approach, the devices only support one single, unidirectional data flow at a time. This may change when the system is extended. A device used in a connection is classified as either a source or a sink object. A device can support both directions, e.g. a fileserver; however, during the setup of a connection, the direction of the flow is selected. In the same manner, streams are simplified to only support a single, unidirectional data flow.

In the initial state of the system, the servers running on the hosts provide factory objects for the different devices and StreamEndPoints (SEP). The factory objects are registered within the CORBA naming service and provide a CORBA interface.

For portability reasons, we use Java as a programming language for the implementation of the different services, as many of the APIs that come with encoder, decoder and ATM network interface card merely provide support for C/C++. For this reason, as well as for performance reasons, some parts of the system have to be programmed in C/C++ and integrated in Java via the Java Native Interface (JNI) [9]. This leads to a layered architecture, consisting of a high-level CORBA interface, accessing a lower level Java implementation that uses native code to achieve access to the lower level resources.

4. THE PROTOTYPE

4.1 CORBA Interfaces

In order that different devices may be included in the system, interface inheritance was used at the CORBA level. A base interface is utilized for the joint operations of all of the devices. The interfaces of those devices which should be included have to inherit from this basic interface "device":

```
enum DeviceOperation {start, stop, pause};
enum VideoFormat {MPEG1, HalfD1, FullD1};
enum StreamDirection {Source, Sink};
interface Device {
  void operation(in DeviceOperation op);
  void bind(in StreamEndPoint sep);
};
interface Encoder : Device {
  void set_format(in VideoFormat format);
  void set_bitrate(in double kbps);   ...
};
interface Decoder : Device { ... };   ...
```

The same approach was deployed for the StreamEndPoints: a single base interface provides the common operations for all of the SEPs:

```
enum StreamDirection {Source, Sink};
interface StreamEndPoint {
    void set_other(in string other);
    void set_direction(in StreamDirection d);
    void connect();
  };
  interface SEP_ATMCBR : StreamEndPoint {
    void set_bitrate(in double kbps);
  };
```

4.2 Java Implementations

As a rule, the CORBA interfaces are implemented in Java classes. Should it become necessary to access specific native libraries in order to access the hardware in turn, the Java native interface is used (see next section). Nevertheless, the CORBA interface maps to a Java implementation class in the first step. Certain methods of the Java class may be implemented as native methods.

The CORBA interfaces of the different classes represent the control interface of the system. The actual data transmission is performed beneath this surface. When using native SEPs and devices, data can be transmitted direct at native level. If the SEPs and devices are pure Java objects, the transmission can be carried out in Java. A mixture of different kinds of objects necessitates data propagation from native code to Java and vice versa. Any combination has to be supported.

Two major aspects have to be considered for direct data transmission:

- The implementation of the SEP must have a direct reference to the implementation of the local bound device.
- Devices and SEPs must provide methods for a direct hand-over of data to each other.

The first aspect is realized through a table which maps a CORBA object reference (an IOR) to a Java object which implements this interface. The same approach is used for the StreamEndPoint objects. When a new SEP is instantiated from a factory, a corresponding entry is made in the table. When a StreamEndPoint IOR is passed to a device object via the bind method, the device consults the table and acquires the implementation object and then registers its own implementation at the SEP, thereby ensuring that the implementation of the SEP knows the implementation object of the device and vice versa.

Two modes are taken into consideration for the direct data transmission between the implementation objects:

- Push: data can be pushed from an originating device into an SEP for transmission and from a destination SEP to a device.
- Pull: an SEP can retrieve data from a device at the sender, or the receiving device could poll the data from the sink SEP.

Exactly which mode is used depends on whether or not a specific bandwidth is prescribed by the SEPs and devices. For example, an encoder is configured for a specific bandwidth, and the connection to the sink SEP supports this bandwidth. In this case it is much more effective for the encoder to push the data to its associated SEP. If a FileSender is used as source and data is to be transmitted at the maximum data rate supported by the connection to the other SEP, then the SEP should pull the data from the FileSender.

To support push/pull models, both the SEPs and devices support a push and a pull method for the forwarding of data at the Java level. The methods may be implemented using native methods. In addition, these methods may be called direct at native level via a similar mechanism.

4.3 JNI Interfaces

Generally, the Java programming language is used to implement the CORBA objects with the given interfaces. Hardware-specific native code is integrated via the Java Native Interface. Special JNI wrapper classes are utilized to provide an interface to the lower level services in Java.

In the first step, the transmission of MPEG-2 material from an Optibase encoder to an Optibase decoder over ATM CBR is implemented in a prototype. Decoder, encoder and the ATM networks interface card are accessed through native code. Consequently, we decided not to propagate any multimedia data to the Java level. Instead, we hand the data from the encoder over direct to the ATM library, and, at the decoder side, forward the data direct from the ATM library to the decoder board. The JNI therefore provides a Java interface for the control of the data flow on the native layer.

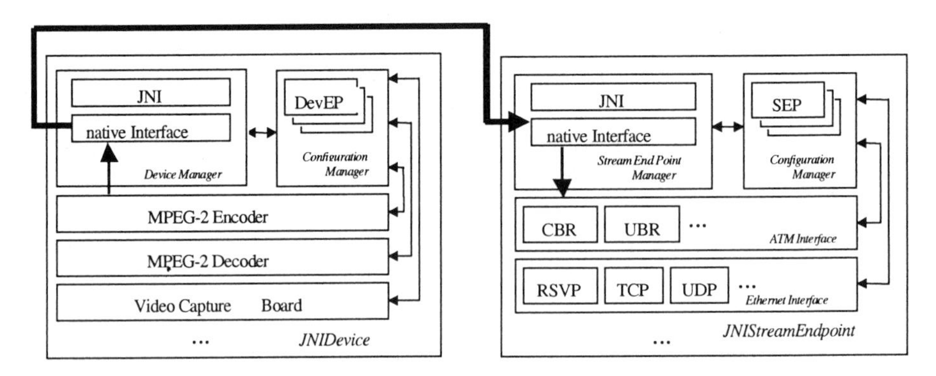

Figure 1 Integration of Devices with JNI

One major task of the JNI other than the provision of access to a hardware device is to manage how the multiple virtual devices at the CORBA level access the joint hardware and how the link between a multimedia device and a StreamEndPoint is achieved. Figure 1 roughly illustrates the components within the JNI layer in one of the host systems.

Each time a new StreamEndPoint is to be created, a new SEP record is inserted in the configuration manager. These records are globally available within the server process for all kinds of SEPs. The records include all the relevant information for a StreamEndPoint, such as the type of the SEP and a reference to the corresponding socket for the transmission of data. A JNIStreamEndPoint directly corresponds to an SEP at the CORBA level. When a new JNIStreamEndPoint is created, a handle (integer value) identifying the position of the SEP record within the configuration manager is propagated to the Java level.

When a bind operation to a multimedia device is performed on the same host, this handle is passed to the native code of the device (JNIDevice). After this call to bind, the StreamEndPoint and the device on one host are "connected", which means that they know how to access each other at the native code level. If, for example, an encoder device is already bound to an ATM CBR StreamEndPoint, when the start operation is performed on the device object, the device object receives the SEP record utilizing the handle from the bind call. Depending on the data in the SEP record, the native code then forwards the data to the corresponding network device. In this case, the socket data structure for the ATM CBR connection would be used.

This mechanism enables the interconnection of different devices with different SEPs. The JNI interfaces provide those operations required for the propagation of the calls in the CORBA interface (start, stop, pause, etc.).

4.4 JMF Integration

Within JMF (Java Media Framework) [10], some player objects (JMF devices) which allow the playback of registered media types as well as the registration of custom media types and players are made available. Table 1 shows the media formats and protocols supported:

Table 1 JMF 1.0 Media Formats [10]

Audio	AIFF, AU, DVI, G.723, GSM, IMA4, MIDI, MPEG-1 Layer 1 &2, PCM, RMF, WAV
Video	Apple Graphics (SMC), Apple Animation (RLE), Cinepak, H.261, H.263, Indeo 3.2, Motion-JPEG, MPEG-1
Files	AVI, QuickTime, Vivo
Protocols	File, FTP, HTTP, RTP

Future versions will add capture, transmit and transcode functionality in JMF (JMF 2.0) [10]. The access to the JMF playback functionality is provided in different abstraction layers within Java. In the highest layer, the player can be started via a *URLStreamHandler* which specifies the location and the transport protocol by means of a URL (Unified Resource Locator). Although this is the simplest way to access the media player, this method provides less flexibility. To be able to deploy our own QoS-aware transport protocol as well as our own device control, we access the media player at a lower layer. Figure 2 shows the objects involved and their dependencies. We build our *OwnDataSource* and *OwnSourceStream* objects based on the *DataSource* and *SourceStream* class that will be connected, for instance, with our *nativeATM StreamEndPoint* . The *OwnDataSource* object uses our JNI of the QoS-aware StreamEndPoint device. Registration of the *nativeATMSEP* is achieved through the *getStream* method from the *PullDataSource* class. A push model is also available. The *OwnDataSource* object is registered in the *Manager*. Using the return of the *getContentType* method, the *Manager* selects an appropriate *Player* and registers the *OwnDataSource* object within this *Player*.

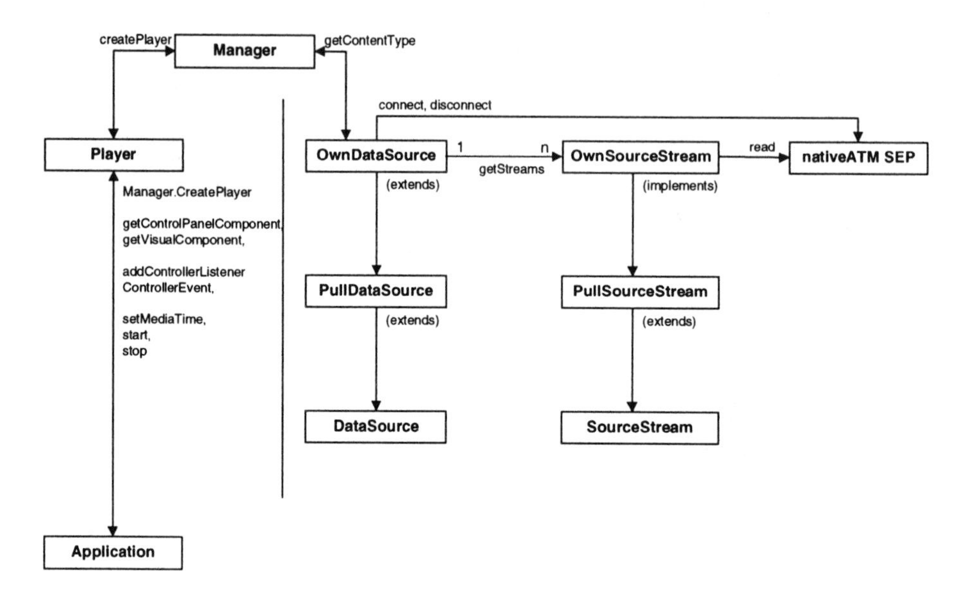

Figure 2 Integration of JMF Player Functionality

4.5 Interoperability of Java and Native Interfaces

For portability reasons, it is our objective to program the majority of the interfaces in Java. As mentioned above, there are some valid reasons for using native code. The four possibilities of connecting a device with a StreamEndPoint are shown in Figure 3. Each native object has its representation on the Java side. Therefore, the connections between Java and native objects are as simple as Java-Java connections. Connections between two native objects are more complicated, as only their references have to be passed through the Java layer and the data is directly transmitted between the native objects.

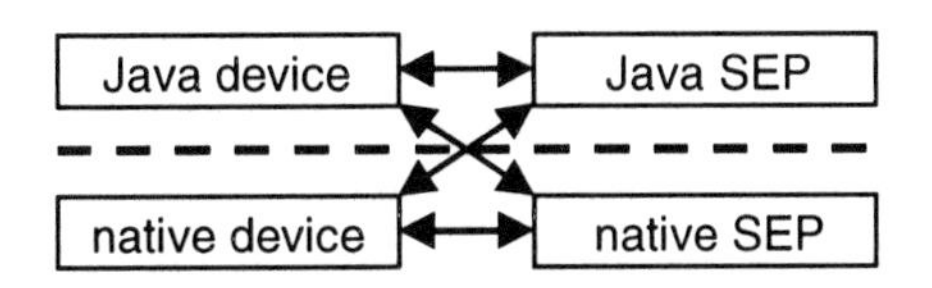

Figure 3 Interoperation of Java and Native Interfaces

5. THE PROTOTYPICAL IMPLEMENTATION AND RELATED WORK

The prototypical implementation consists of a management application that allows the remote configuration and management of the communication system and the multimedia devices with the corresponding server counterparts. Currently, the management application enables the setup of a connection between two devices. A FileSender, FileReceiver, MPEG-2 encoder and decoder are supported. As communication links, an ATM CBR connection and a best-effort TCP connection are implemented. After a setup process, a GUI allows remote access of the devices.

Within the context of CORBA, an architecture for the control and management of audio and video streams [2] has been specified. Initial implementations which partially support this standard are presently in the pipeline [5], [6], but are in an early stage of development so that adjusting them to our needs would be a fairly complex task. Therefore, our approach follows the standard as far as is suitable for our specific environment.

To support ATM within Java, a Java ATM API is under specification by the ATM Forum [7]. As stated earlier in this paper, our goal is to control the flow of multimedia data from the Java level. The flow itself is visible solely at the native level. Nevertheless, for integration with those multimedia devices already available in the Java platform, for example with the Java Media Framework (JMF), it is necessary to propagate the multimedia flow up to the Java level. In [8], we have described how this can be accomplished.

6. SUMMARY AND FUTURE WORK

The paper describes several architectural concepts for the support of a distributed multimedia environment. The architecture is based on CORBA and Java and enables the remote configuration, control and management of the multimedia devices deployed and the streams between them. Devices which are not accessible from the Java platform are integrated using the Java

Native Interface. The object-oriented character of our architecture enables the transparent integration of the various devices and StreamEndPoints.

In the future, we plan to extend the application to support additional devices, such as a video server (SGI Mediabase), as well as additional StreamEndPoints such as RSVP. In addition, we are striving to make multimedia flows available at the Java level to support a wider range of applications generically, so that either mechanism may be used as an alternative.

References

[1] "The Common Object Request Broker Architecture and Specification, 2.0 ed.," Object Management Group July 1995.

[2] D. McGrath, T. Rutt, and J. Ottensmeyer, "Control and Management of Audio/Video Streams," Object Management Group telecom/97-05-07, May 1997.

[3] D. Curtis, "Java, RMI and CORBA," OMG 1997.

[4] R. Orfali and D. Harkey, Client/Server Programming with Java and CORBA. New York: John Wiley & Sons, 1997.

[5] Iona Technologies, "OrbixMX - A Distributed Object Framework for Telecommunication Service Development and Deployment," White Paper, April 1998.

[6] N. S. Sumedh Mungee, Douglas C. Schmidt, "The Design and Performance of a CORBA Audio/Video Streaming Service," Proc. of HICSS-32 International Conference on System Sciences, minitrack on Multimedia DBMS and the WWW, Hawaii, 1999.

[7] ATM Forum SAA/API Working Group, T. Jespen and J. Shaffer, "Java ATM API Description - Proposed Outline, Revision 1," ATM Forum ATM Forum/97-1044R1, February 1998.

[8] R. Eberhardt, C. Rueß, and R. Rusnak, "Communication Application Programming Interfaces with Quality of Service Support," Proc. of IEEE International Conference on ATM (ICATM '98), Colmar, France, 1998.

[9] "The Java Native Interface Specification (JDK1.1)", Sun Microsystems, May 1997, http://java.sun.com/products/jdk/1.1/docs/guide/jni/spec/jniTOC.doc.html

[10] "The Java Media Framework API", Sun Microsystems, http://java.sun.com/products/java-media/jmf/index.html

TRAFFIC CONTROL AND RESOURCE MANAGEMENT USING A MULTI-AGENT SYSTEM

Z. Luo, J. Bigham, L.G. Cuthbert and A.L.G. Hayzelden
Dept. of Electronic Engineering, Queen Mary and Westfield College, University of London
Mile End Road, LONDON E1 4NS, U.K.

{z.luo, j.bigham, l.g.cuthbert, a.l.g.hayzelden} @elec.qmw.ac.uk

Abstract This paper describes a system that implements agent technology for traffic control and resource management in a telecommunications environment. The adoption of agent technology has allowed more flexible and negotiable resource allocation management procedures, which are relevant to the more open telecommunications service environment. The paper introduces the agent architecture developed but concentrates on the resource agent strategies and service provider negotiation mechanisms for connection admission. The architecture is being validated on a real ATM test bed.

Keywords: Multi-agent System, Resource Management, Connection Admission Control, ATM, Auction.

1. INTRODUCTION

The aim of traffic control and network resource management is to provide the customer with the required Quality of Service (QoS) yet at the same time allow the network operator to run the network in an efficient and economic manner. *Connection Admission Control* (CAC) in an ATM network is the set of actions taken by the network at connection set-up, or during call re-negotiation, in order to establish whether a Virtual Channel or Virtual Path connection can be accepted or not [1]. The purpose of the CAC software is not only to find a free route across the network from source to destination, but also to *find a route with sufficient resources* to ensure that new connections obtain sufficient resources without degrading established connections. This is of fundamental importance in ATM networks and most of the approaches use traffic descriptors (such as average cell rate, peak cell rate, burstiness, source type) as an input to some algorithm that also has knowledge of established traffic values. This is done on a link-by-link basis across the network.

This paper describes implementation of control strategies on a real ATM test bed as a society of interacting agents and use of the flexibility and brokerage capabilities of agents to provide additional functional and economic benefits. The addition of agent technology to the CAC problem brings a new dimension to traffic control and resource management in ATM networks by adding the *co-ordination between nodes* that is so far lacking in such systems [4],[5]. Benefits from using agents are expected to be as follows:

- allows different policies between user, service provider, network provider,
- faster integration of new services,
- better structure for providing open interfaces for network elements,
- applicable to ATM, IP,
- allows network optimisation,
- connection decision on entry ("one-stop shop").

The paper is organised as follows; Section 2 will introduce a multi-agent architecture to support a set of agents that manage admission and re-connection. Section 3 details the layering of agent system, including the planning and reactive layers. Section 4 discusses service provider negotiation issues. In section 5, we present the developed agent software system. Finally, section 6 describes ongoing activities and future work.

2. MULTI-AGENT ARCHITECTURE

It is assumed that network resources will be managed by exploiting dynamic bandwidth allocation to Virtual Paths (VPs). A Virtual Path (VP) is a path of specified bandwidth from a source node in the network to the destination node in the network using physical links of the network. Note that only source to destination VPs are considered in the resource management model to be described. (In general VPs can be defined for segments of a path from a source to a destination, but these are not what are being managed here.) Extensions to very large networks makes the issue more complex, but suitable partitioning is a possible means of tackling that problem. No routing is done for individual virtual connections. Rather *all new* connections are allocated to one of the relevant VPs. The bandwidth associated with any VP can change continually and is one of the controllable parameters for the Network Service Provider or negotiation commodity for Service Providers who are not the Network Provider. This is in fact a highly realistic assumption for the management of a complex network.

It is assumed that the set of VPs associated with a source-destination pair are known, fixed in terms of route (though not bandwidth), and are a small manageable subset of the set of possible VPs for that source-destination pair. Whilst this sounds limiting it is not believed to be so in practice as the set of enumerated VPs could be changed over time. Pre-enumeration of the VPs simplifies the CAC mechanism. Resource Agents (RAs) manage the VP connections. So the routing problem is not ignored but placed at a higher level (arguably where it belongs) leaving the CAC to request the appropriate Resource Agents information on the costs and feasibility of connections. Figure 1 gives a conceptual view of the problem.

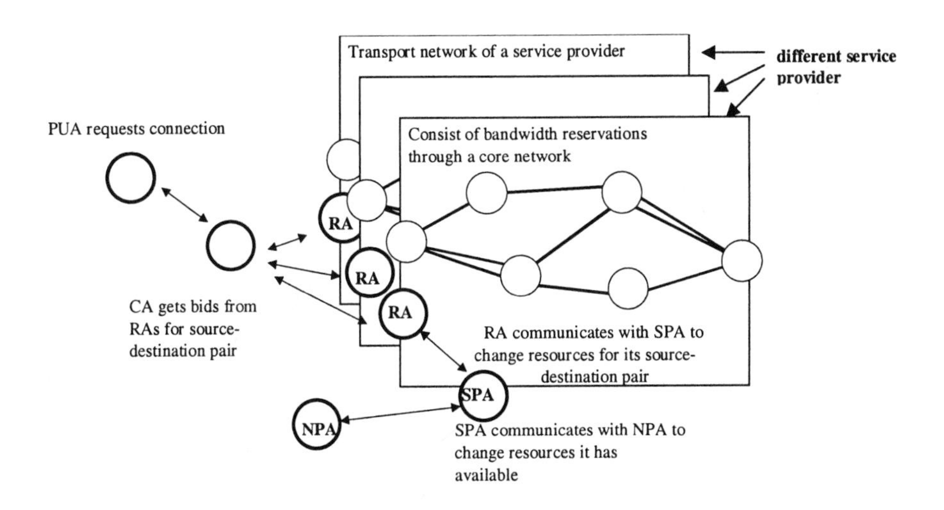

Figure 1 Basic Concepts

The main agents in the system are Connection Agents (CAs), Resource Agents (RAs), Proxy User Agents (PUAs), and Switch Wrapper Agents (SwWAs). Figure 2 shows where these agents would be located in the physical network. Note that control at network edge which is handled by PUAs, CAs and RAs can be done either in edge switches or in separate platforms.

In our model each SPA owns a RA for each source destination pair it services, and this RA manages the resources of the VPs that belong to the source destination pair. So with multiple service providers and multiple source destination pairs, there are many RAs. The user request a connection of a particular CoS along with the associated CoS parameter values via the PUA. The PUA contacts the CA placed at the entry point to the network. The CA queries the RAs for bids and gets the replies from relevant RAs then decides on the preferred service provider and the preferred offer from that service provider and instructs the chosen RA to install the connection. The chosen RA then interacts with the SwWAs in order to make the necessary connection set up on the source and destination switches. The SwWA constitutes a "virtual" software abstraction of an ATM switch and its resources, and provide a generic, vendor-independent interface for network control and management applications. Interaction between the RA and SwWAs is performed by means of some Agent Language [3], while the SwWAs have access to the switch's proprietary control method (e.g. via SNMP or some form of API).

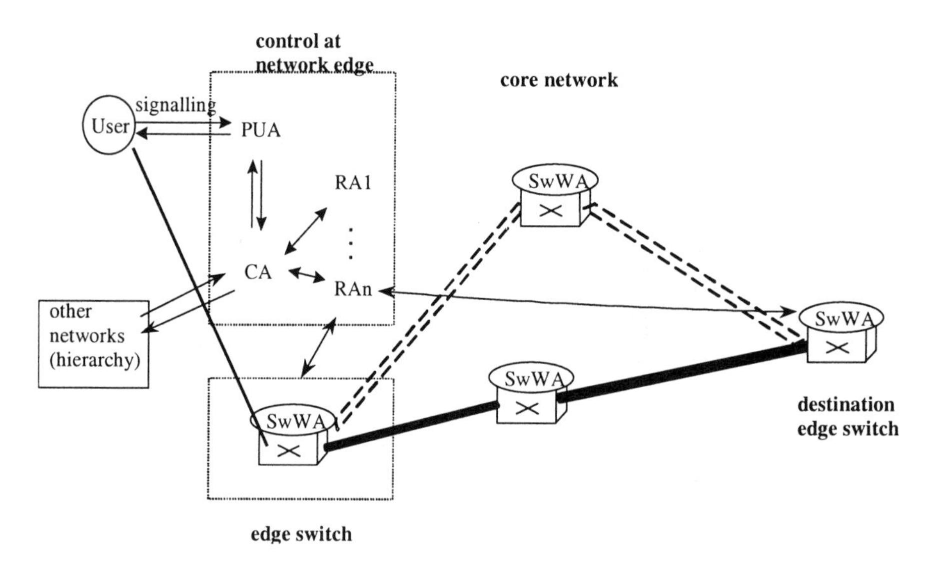

Figure 2 Multi-agent Architecture

One issue that is obviously of interest with real networks (but harder to demonstrate on the testbed) is the problem of scaleability. The architecture considered by IMPACT addresses this issue because it allows for networks to be partitioned: a modified form of PUA called a Proxy Connection Agent (PCA) could be used to manage connections between networks instead of connections at user equipment.

3. RESOURCE MANAGEMENT STRATEGIES

RA, SPA and NPA are responsible for resource management. The user of the telecommunications network tries to make the best kind of connection from A to B at the best price, and the Service Provider marshals its resources to meet customer demand and provide its contracted QoS with customers. A layering approach including a reactive and a planning layer is taken to implement resource management strategies. The RAs are reactive competences making rapid decisions regarding admission to the network. Speed and appropriate reactivity under a wide range of circumstances can be achieved by having levels of competence [8]. One of the key ideas is that each shell of competence can work with the inner shells working, but without the outer shell functioning – though not necessarily effectively over a wide range of inputs. Here the allocation competence in each RA works within a framework created by the slower planning competence hosted by the Network Service Provider (NSP) or Service Provider (SP).

For clarity we will call a SP that is not a NSP a secondary service provider (SSP). The NSP and each SSP has planning capabilities. We will only consider capacity bandwidth allocation here. The RA's can work without any input from the planning level. The planning level of a NSP or SSP monitors the state of the whole network it owns or rents. The planner changes the view of the RA's world by altering the capacities that they believe they have [5]. The relationship between the NPs and SPs and RAs is shown in Figure 3.

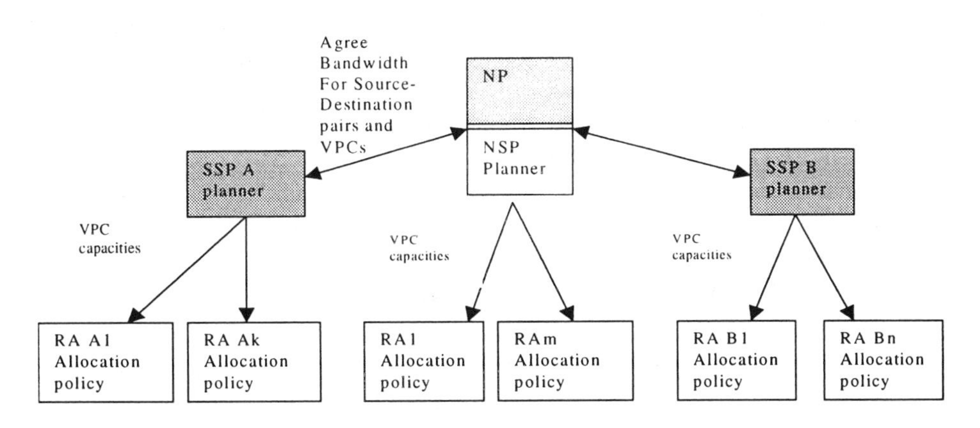

Figure 3 Competence Layers

3.1 The Planning Layer

When network flow indicates that the NP needs to optimise the capacities to allocate to each VPC of the RAs that it owns, it reads the current usage of each VPC it owns and the contracted bandwidth allocation to SSPs and computes an allocation of capacities that maximises the minimum residual in the links of the network, tie breaking with hop count [6]. This is a conservative strategy and others could be employed. However, the robustness of such a simple criterion is appropriate for this level of planning and its purpose is to ensure overall survivability of the network in a competitive environment. The planning level sets capacities for the reactive allocation policies, i.e., gives each RA *a bandwidth allocation for each VP managed by that RA*. More details of the planning approach can be found in [5]. A similar approach is adopted for SSP planning.

3.2 The Reactive Layer

The reactive layer is embodied in the RAs and has to ensure second by second control of the network. This is distributed through the network and makes decisions based on local knowledge and the most recent input from the planner. It can work without the planner periodically updating capacities, though performance would degrade over time.

A reactive strategy appropriate for a NSP is to allocate to the VPC proportional to the planned capacities and tie break by allocating to the path(s) with maximum residual capacity. A strategy appropriate for a SSP is allocate to the VPC with minimum residual capacity and tie break by allocating to path(s) with minimum hop count. For very simple measures of utilisation, proportional allocation corresponds to minimising network utilisation. So, the first rule is appropriate for a RA that belongs to a NSP as it acts to support the common good of the NSP. The second strategy attempts to minimise fragmentation of bandwidth within the constraints imposed by the higher level planner or as a result of negotiation with the NP. This strategy is also relevant for an NSP at higher levels of utilisation. Such constraints include, for example, the number of physical paths used by the RA. This strategy is appropriate for a SP that is not a NSP. It leaves as large as possible chunks of bandwidth free so that it can accept high bandwidth connections without having to restructure internally or negotiate with other SPs. This is a self-interested strategy and certainly inconsiderate in terms of a common good defined in terms of utilisation or minimum residual capacity of the network.

4. SERVICE PROVIDER NEGOTIATION

Deciding how to model the negotiation between service providers and customers is difficult as there can be various degree of lock-in, where good substitutes prior to a commitment become less good substitutes later, and different degrees of countervailing power between the interested parties when negotiating pricing. The intention here is not to presume the form of such agreements, but use such agreements as the basis of some of the bids in the auctions. We assume that the user has no user-defined procedures for managing the call for proposals but appeal to an open brokerage mechanism, currently a first price sealed bid auction. Allowing the user to upload customer specific auction code would introduce security and procedural issues, which may not be acceptable at the present time. Many would argue that such security problems can be addressed [9], but these arguments are not pursued here. Rather the user (as well as its ID and connection requirements) can provide parameters that allow specialisation of the announcer and the awarder in the auction mechanism (see Figure 4). The FIPA 97 Specification Part 2 [3] gives standards for call for proposals (a form of first-price sealed-bid), iterated call for proposals, English auction and Dutch auction. However, dynamic uploading of code by the SP is less problematic and has many attractive features, not only for the auction, but for procedures for bandwidth allocation by the RA.

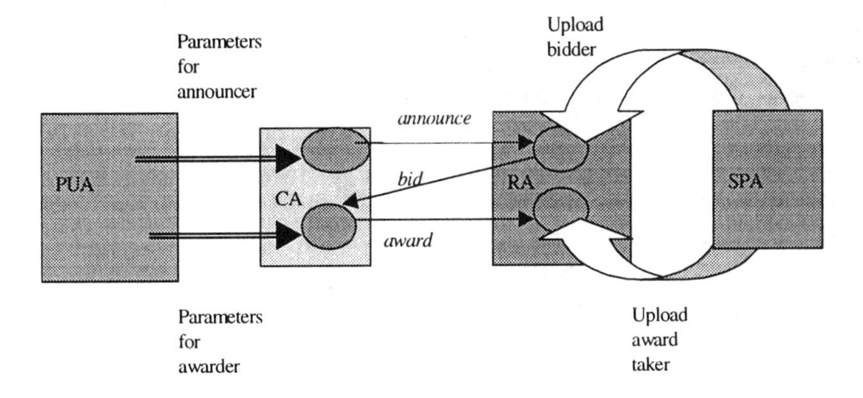

Figure 4 Auction Mechanism

In some cases the user-ID may only be used to classify the customer and bids made by the SPs would then be based only on the classification rather than individual terms. For example, the customer ID allows the most recent negotiated rates for a SP to be used so the prices quoted can be different from customer to customer as well as being history dependent for a particular customer. The awarder can be given a vector of price adjustments for named SPs, so that selection of the best price is done in a user provided context, even though the auction mechanism has not changed.

The more complex the negotiated terms the more complex the nature of the look ahead required by the PUA. Whatever the look-ahead problems, it is assumed that the calculations map into a parameter set for the awarder, such as a list of (adjustment, service provider ID) pair.

It is worth emphasising that there is no attempt here to emulate 'negotiation' in the full sense of the word. In general, negotiation between Users and Service Providers may involve rather complex packages with 'bundling' and scale reductions. It is assumed that this kind of negotiation is done off line. Connection agent 'negotiation' with service providers will use the *results* of such negotiations. The former are conducted at connection request time and done on an individual connection request basis. The *scope* of the connection request time 'negotiation' could vary from announcements to all service providers (see later) to more constrained announcements to subsets of service providers, based on a user's existing longer term contractual arrangements with SPs, if any. The procedure a User wants the connection agent to apply on its behalf depends greatly on the extent of any such prior arrangements and will be, in many cases, private to the customer.

An important assumption is that bids are taken as binding and so if awarded a connection must be made. Problems could arise if after responding with a bid that exhausts our capacity, more profitable connection request announcements are received. The capacity constraints require them to be rejected. There is then a temptation to renege on the bid. If reneging is allowed then more sophisticated protocols are needed which include penalties for such reneging [7]. It is assumed that if such a temptation arises the SP will arrange to offload the award or another less profitable award to another SP, or will resist and honour the bid. In the latter case this could be because of regulatory penalties or, more subjectively, for good customer relations it would be irrational to do otherwise. Inter SP trade and cross charging is being investigated.

5. AGENT SOFTWARE SYSTEM

To support the multi-agent system described above a purpose-designed agent software system, called the basic agent template (BAT) has been implemented. BAT is written entirely in Java and relies on Java's Remote Method Invocation (RMI) for the agent communication and Java's Reflection mechanism for the implementation of the Agent Communication Language (ACL) protocol, see Figure 5. The implementation of specific agents for special tasks (such as controlling ATM switches or computing new load schemes for a network) is very simple.

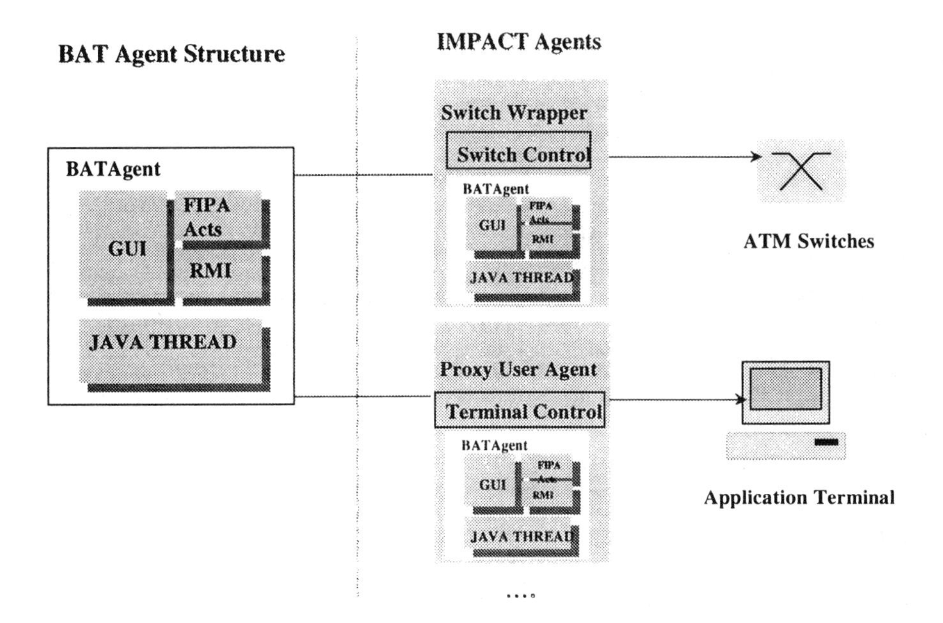

Figure 5 BAT Agent Structure

6. CONCLUSIONS

We have presented a traffic control and resource management framework using a multi-agent system. This new approach is expected to provide simple call set up procedures and better utilisation of the network through exploitation of the agents' intelligence. These expectations stem from the following features of the project's approach:

1. The CAC decision will be taken at the edge switch, thus alleviating the burden of executing the CAC procedure in every individual switching node in the network.
2. The agents will be engineered to take proper advantage of the provided resources using a more global network perspective.
3. Features of agents (such as autonomy, social ability, responsiveness and pro-activeness) will be exploited to achieve flexibility, which will be based in intelligence distributed in the network.

Although an ATM network is used as the basic infrastructure, many of the concepts are generic and the system can handle IP traffic equally well. In fact the overall concept is of a fairly dumb ATM core network with management-plane switching (cross connects) and intelligent edge switches controlling resources. Such a network represents a structure used by operators around the world.

The prototype of the agent system has been designed and implemented. The trial of a set of scenarios on an ATM test bed is in progress.

ACKNOWLEDGEMENTS

The authors gratefully acknowledge support from the European Commission under the ACTS Project AC324 "*Imp*lementation of *A*gents for *C*AC on an ATM *T*est bed". The authors also acknowledge valuable help and contributions from its partners Swisscom (the implementation of BAT system), Tele Danmark, National Technical University of Athens, Flextel, Teltec, and ASPA.

References

[1] J. L. Adams, "Asynchronous transfer mode - an overview", BT Technology Journal, Vol. 13, No. 3, pp. 9-14, 1995.
[2] M. N. Huhns, M. P. Singh, (Editors), "Readings in Agents", Morgan Kaufman, ISBN 1-55860-495-2, 1998.
[3] Foundation for Intelligent Physical Agents, http://drogo.cselt.stet.it/fipa.
[4] G. P. Kumar, P. Venkataram, "Artificial Intelligence approaches to network management: recent advances and a survey", Computer Communications, Vol. 20, pp.1313-1322, 1997.
[5] A. Hayzelden, J. Bigham. "Heterogeneous Multi-Agent Architecture For ATM Virtual Path Network Resource Configuration", Proceedings of IATA 98, Lecture Notes in Artificial Intelligence, pp. 45-59, Paris, France. 1998.
[6] J. Bigham, A. Hayzelden, "Using Survivability to Manage Uncertainty in a Telecommunications Network Management Application", Internal Report, Dept. of Electronic Engineering, Queen Mary and Westfield College, University of London, 1998.
[7] T. Sandholm, V. Lesser, "Issues in Automated Negotiation and Electronic Commerce: Extending the Contract Net Framework", Proceedings of the International Conference on Multi-agent Systems, pp.328-335, 1995.
[8] R. Brooks, "A Robust Layered Control System for A Mobile Robot" IEEE Journal of Robotics and Automation, pp.14-23, Vol. RA-2, No.1, March 1986.
[9] E. R. Harold, "Java Network Programming", Published by O'Reilly, ISBN 1-56592-227-1, 1997.

SESSION 17

Network Management

THE ADAPTABLE TMN MANAGEMENT ARCHITECTURE WITH OTHER OBJECT-BASED MANAGEMENT REQUESTS PLATFORMS

SeokHo Lee, WangDon Woo,

ATM TMN Team, Electronics and Telecommunications Research Institute,

161 Kajong-dong, Yusong-gu, Taejon, 305-350, Korea

{shsang, wdwoo}@nice.etri.re.kr

JungTae Lee

Department of Computer Engineering, Pusan National University, Korea

jtlee@hyowon.pusan.ac.kr

Abstract Over the last decade, the heterogeneous corporate network has been considered as the basic network service plan of various countries. The globalized heterogeneous corporate network is essential for network providers to provide various communication services, and for customers to acquire their various communication demands. In these cases, to manage such complicated heterogeneous corporate network, it is needed some efficient integrated management scheme.

Telecommunications management network (TMN) is a standardized network management framework for heterogeneous corporate networks. Also, asynchronous transfer mode (ATM) switches are constructed as main public backbone nodes. In this paper, we introduce implemental TMN element management layer (EML) management platform for ATM switches, which are developed by Electronics and Telecommunications Research Institute (ETRI). TMN EML manager workstation also plays a role of TMN NML agent and other agents for heterogeneous management domains, such as common object request broker architecture (CORBA) interface definition language (IDL) based management environment.

Keywords: TMN, CMIP, ATM, Management, Manager, Agent, NMS, MO, MIB, CORBA, Interoperability.

1. INTRODUCTION

Until recent times, most of distributed network elements of any involved network components from multiple vendors need to be controlled with its own proprietary management pattern. Network components and network management systems were designed specifically for each particular telecommunication equipment and service requests. Thus, it resulted a number of inter-working problems and limited integration of network management functionality. With the growth of complexity in contemporary telecommunication networks, today's trends of the worldwide telecommunication networks are integrated, distributed and heterogeneously combined. Distributed telecommunication environment needs distributed applications that include open system interconnection architecture. To control cost effectively and to make integrated network management practically, we need the one of appropriate integrated network management concepts, such as TMN, SNMP and CORBA based management. Network components are redesigned as generic object oriented information model. Telecommunication management network (TMN) [1] is a framework for the management of telecommunication networks and services of heterogeneous information management targets. The basic concept behind the TMN is to provide an organized network structure to achieve the interconnection of various types of communication infra and network elements (NE) with standardized protocols and interfaces.

On the other hand, ATM technology provides the flexibility that is needed for realizing broadband telecommunications networks. High-speed transmission systems should be expected to offer bandwidth flexibility and heterogeneous traffic accommodation. Thus, ATM architecture is predicted to become a next generation switching technology. Actually, the ATM network is being an infrastructure that will support future broadband and multimedia telecommunication service networks in Korea. Electronics and telecommunications research institute (ETRI) has developed ATM switch and TMN element layer manager/agent system for ATM switches [2]. The TMN based ATM management system, ATM TMN, will communicate with switch's agent using Q3 interface. Management aspects of the ATM switch and network are represented managed objects (MO) through guidelines for definitions of managed object (GDMO) definitions. The management operation interfaces are specified by common management information service/protocol (CMIS/P). To develop commercial TMN management system for ATM switches and network, a real implementation methodology is needed. Standards and some forum (ITU-T, ATM Forum, ETSI, GR, NMF, etc) are currently studying principles for the TMN. But, they are generally confined to information modeling and essential management application function (MAF) specifications. To realize them, we have to design implementation schemes: F interface or gateway architecture between Non-TMN environment (i.e., GUI, Manager console, Operation and Maintenance Processor of ATM switch) and TMN environment (TMN internal manager process). Each of TMN MAF is construction of MO processes as the UNIX processes or threads. CMIP handling mechanism, management information base (MIB) handling functions, application programming interfaces (API) for communication protocols and stepwise functional flow of TMN CMIP MAF scenarios are also needed [3].

Furthermore, we can expect another types of higher layered management domains, such as SNMP management domain, CORBA based domain which are located in network, service and business management layer. Thus, we have to develop agent system for those different domains of management into TMN element management system (EMS). The agent process for different domains play the role of gateway functionality which allow for the migration of information models between different domains of management technology [4]. This paper describes layered architecture of TMN management system based on EML, EL layers. The management system of TMN EML layer also plays the roles of agent system for heterogeneous management domains. In this paper they will be also described. The TMN MAF scenarios will be also described through the TMN multi-node connection management.

2. FUNCTIONAL STRUCTURE OF TMN EMS & ANOTHER AGENT FOR HETEROGENEOUS MANAGEMENT REQUESTS

The TMN EML layer management system is built in the UNIX based workstation. To support TMN management of ATM NEs, the number of workstations should be one or more. It is fully associated with workstation performance. The major factors of system performance are a number of MOs, a number of MO instances, invoked MAF concerned with process or thread, the execution size of MAF scenarios, a number of concurrently invoked MO processes, etc. The basic structure of TMN EMS system and agents are shown in Figure 1.

The EMS system plays the role of mediation device in TMN architecture; it must provide an agent role to its NMS manager and a manager role to its NE agent(s) at the same time. To satisfy more sophisticated network management requirements, the EMS must be able to handle the management requests from other management clients using different management protocols such as CORBA based management requests. As shown in Figure 1, an EMS manager consists of the manager main process, management application functions (MAF) which operate management scenarios: CN MAF (Connection), CF MAF (Configuration), FT MAF (Alarm & Fault), AC MAF (Account), SC MAF (Security), PF MAF (Performance). And, also communication protocol stacks and some of handling software blocks and graphic user interface through F interface are included. The EMS block manages several NE agents based on element management view, which represents ATM switches, and has an interface channel to a NMS agent and manager. The NML agent of network management view represents the abstraction of ATM network which is composed of the ATM switches and the network-to-network interface (NNI) links between them. – see section 3.

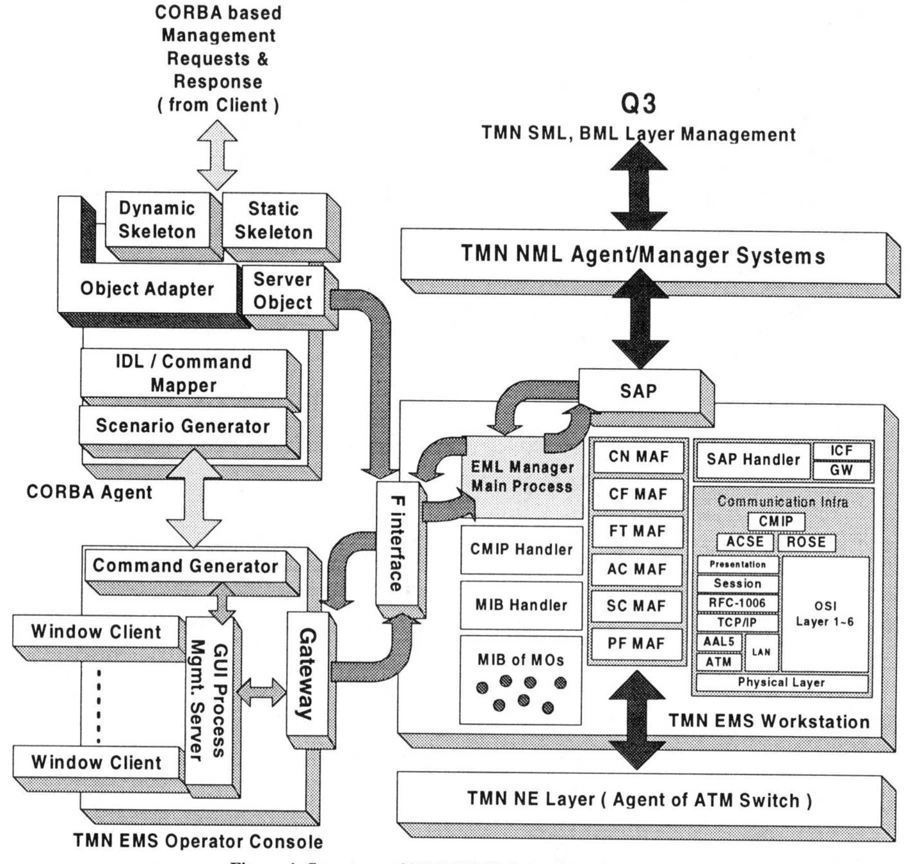

Figure 1. Structure of TMN EMS & Another Agent

An EMS management service is performed through basic and extended MAF scenarios for configuration, connection, fault, account, security and additional connection services, internally. The basic MAF operates functions associated with UNIX system, and the extended MAF provides TMN based scenario processing functions. A gateway function takes charge of interface messages between MAFs and other applications. The gateway function distributes all incoming requests (from GUI, CORBA Agent) into the proper extended MAF through basic MAF, and handles outgoing results and notifications. The NML agent receives NMS requests and dispatches messages into the EMS main process through a gateway. An Information Conversion Function (ICF) is used to bridge between NMS agent and EMS manager. It is quite different between the MO sets of the EML layer and the

NML layer. The ICF converts and propagates NMS requests to the EMS manager and EMS notifications to the NMS manager. - It performs the mediation functionality. The ICF also decouples (assembles) NMS requests into (from) EMS requests (responses) due to the difference between network view and element view. When the NMS manager makes management requests for the NMS agent, the manager checks the scope of subnetwork, which is affected by the request. To do this, the manager looks up EMS management domain, and determines which EMSs are in the proper scope.

Based on management requirements, the NMS manager may sends the same or different requests (UNI and NNI, NNI and NNI, NNI and UNI) to the selected EMS through synchronous or asynchronous binding way. The EML Manager main process inspects the requests, and dispatches the request to appropriate MO worker (process or thread). The MO workers may perform the request or forward the request to the agent in TMN NE layer. In another words, the EMS manager receives the request, it forwards the request to one of basic MAFs, which in turn create an instance of a MAF and the instance invokes management request to NE agents. The results from NE agents are collected in the EMS manager and manipulated with network view information in the NMS agent via the ICF.

TMN EML manager workstation can also plays a role of agent for another objected based management domain such as CORBA based network management. CORBA was conceived by Object Managed Group (OMG) with the purpose of making easier to exchange distributed information regardless of the system environment and location. For example, Telecommunications Information Networking Architecture (TINA) and service management systems of telecommunications network provider in Korea are also have basic management scheme with CORBA request/response architecture. There are some possible approaches to join CORBA/TMN environments. One of approaches is using F-interface through IDL/Human Machine Interface (HMI) mapping. In this case, there is no need object-to-object and service-to-service mapping between them. In this case, however, many of unwanted and non-standard additional processing steps are required. Another approach is direct method from/to CORBA to/from TMN environments. A point of excellence of this approach is that all of processing flows represented with objects and object services completely. But, in this case it is very difficult to find out the interoperable features between CORBA and TMN. A detail discussion will be described in section 4) in this paper.

3. TMN BASED MAF SCENARIO FOR MULTI-NODE CONNECTION MANAGEMENT

The information model used to implement the EMS is based on TINA, HAN/B-ISDN and ITU-T documents [6][7]. Figure 2. shows the MO relationships in the EMS which manages multiple NE agents of ATM switches and the trunk link between them. MO set that is used by the connection (CN) MAF is as follow [8]:

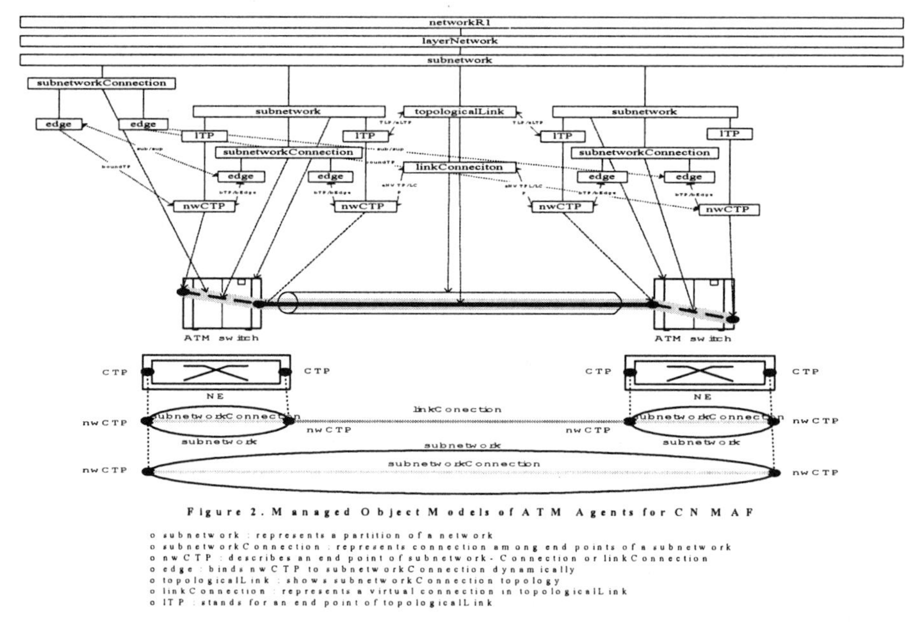

Figure 2. Managed Object Models of ATM Agents for CN MAF

- subnetwork : represents a partition of a network
- subnetworkConnection : represents connection among end points of a subnetwork
- nwCTP : describes an end point of subnetwork-Connection or linkConnection
- edge : binds nwCTP to subnetworkConnection dynamically
- topologicalLink : shows subnetworkConnection topology
- linkConnection : represents a virtual connection in topologicalLink
- ITP : stands for an end point of topologicalLink

The CN MAF includes end-to-end connection management and experimental results based on operation performed. The experimental system consists of 4 ATM NE agents, each of which is executed on a workstation, an EMS manager and a NMS agent are also executed on another workstation. The EMS and the NE agents were implemented on SPARC workstation with distributed processing environment. In order to establish an ATM end-to-end connection, the NMS manager issues M-ACTION request to the one of NMS agents with connection information such as calling party number, called party number and ATM traffic descriptors. When MO instance of a top-level subnetwork in the NMS agent receives the M-ACTION request for end-to-end connection, it starts searching the distinguished physical locations of subscribers. The NMS agent searches which NE serves calling and/or called subscriber.

It is determined by routing information which is located in each of the NMS agents. Then it issues a M-GET request to the corresponding NE agent via ICF and EMS manager. The NMS agent issues a local getting request to the ICF using proprietary IPC. Then the ICF forwards the request to the EMS manager. The EMS manager generates an M-GET request to the proper NE agent. The responses containing user location are returned to the NMS agent through the reverse steps. After the MO instance of a top-level subnetwork receives the M-GET response, it begins to search a possible and efficient path from originating NE to destination NE which are confined to single EMS boundary. The routing information is stored in MIB of the NMS agent in association with pre-configuration. Then the NMS agent sends connect M-ACTION request to each NE included in the selected path. The request is transferred with the same case of the M-GET procedure. This procedure is illustrated in Figure 3. The scenario is described with four ATM NEs. This operation could be implemented in parallel or in sequence.

4. INTEROPERATION WITH OTHER OBJECT-BASED MANAGEMENT ENVIRONMENTS

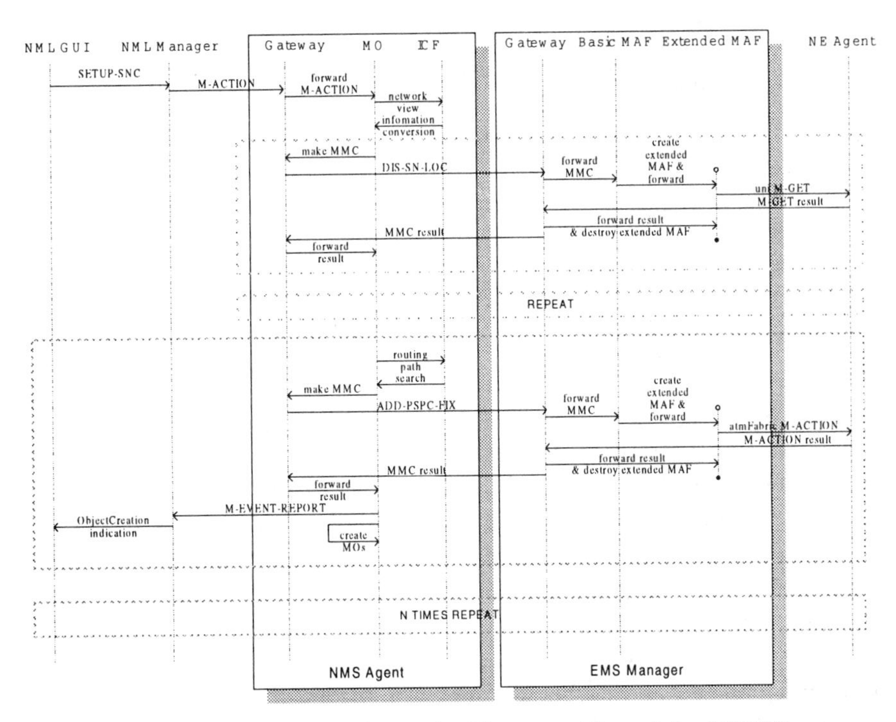

Figure 3. Multi-node Connection Management Scenario for ATM NEs

As well known, CORBA allows object-oriented communication between applications, which are distributed across networks. CORBA provides not only point-to-point object-oriented communications between management applications, but also provides the domain independent object bus for distributed communications applications. CORBA is more general-purposed than TMN CMIS/P architecture. Thus, we could expect that there are many network providers to require network management system that has the interoperability with CORBA based architecture. In fact, a few number of network providers in Korea have a plan to develop CORBA based management system in service management layer for their reliable communication services successfully. It requires a gateway system, which has the interoperability with CORBA, based management domain which are not necessary based on the TMN CMIP management domain. This is major reason that why we try to build CORBA/CMIP gateway architecture into the TMN EMS system, as described in Fig 1. The gateway process (CORBA Agent, Server Object) has to ready for the heterogeneous management requests with CORBA interface definition language (IDL) specifications [10]. In GDMO, ASN.1 and Template can be mapped into IDL specification through syntactic mapping table. And other possible mapping stacks are shown in Figure 4.

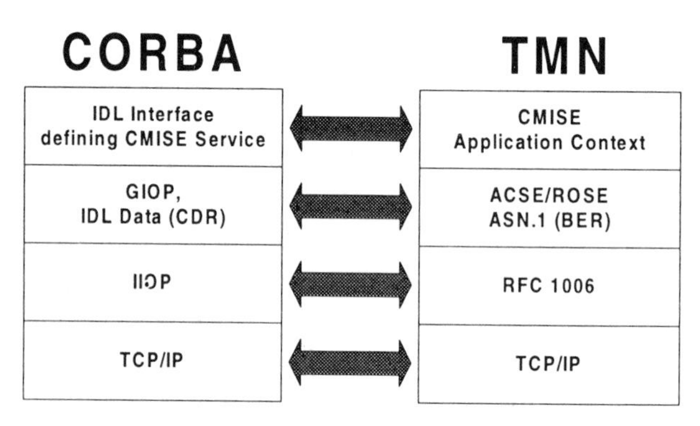

Figure 4. Possible equivalence between CORBA and TMN

In case of the CMIS/P behavior, it is need to be semantic translated. As it were, we need mapping of MO, attribute, ASN.1, and translation of actions and behaviors. During the scoped operations on the managed object, gateway gives the attribute IDs as plain string. The OIDs for given attributes are looked up in the translation information, together with the types of the attributes. Data type of abstract syntax notation (ASN.1) will be matched with CORBA IDL definitions generally. Sometimes, during the specification translation, parts of ASN.1 cannot be represented in IDL. In these cases, some omissions or mismatches are remedied during translation of the values that pass through the CORBA agent in run time. The logical position of CORBA agent is shown in Figure 1. The CORBA agent server forwards all client requests to the TMN EMS main process with form of TMN CMIS/P or operator's commands. We are specifying an algorithm for the translation of IDL specifications into GDMO definitions. The translation algorithm has flows to pick up represented GDMO managed object from IDL interfaces.

For every managed object class template, a primary IDL interface is defined which supports the operations on the managed object. CORBA based management application services will be mapped into access scheme to a managed objects in CORBA agent. The MAF behaviors of TMN are totally transparent to the IDL based CMIS services. A CORBA object service can be mapped to TMN CMIS/P management services. Table 1. shows the possible mapping interface between CMIS based management services and CORBA services.

Beyond above, there are some possible approaches for the construction of CORBA/CMIP gateway framework. Among them, we will describe internal CORBA structure.

CMIS based Management Service	**OMG CORBA Object Service**
MO FSM (Finite State Machine)	Life Cycle Service
Event Notification Service	Event + Property Service
Naming Service	Naming + Property Service
MO Filtering	Property Service
Repository	Repository Service

Table 1. A possible mapping interface between CMIS and CORBA services

All of internal interfaces are constructed with object request broker (ORB) as object bus. Its overall structure is shown in Figure 5.

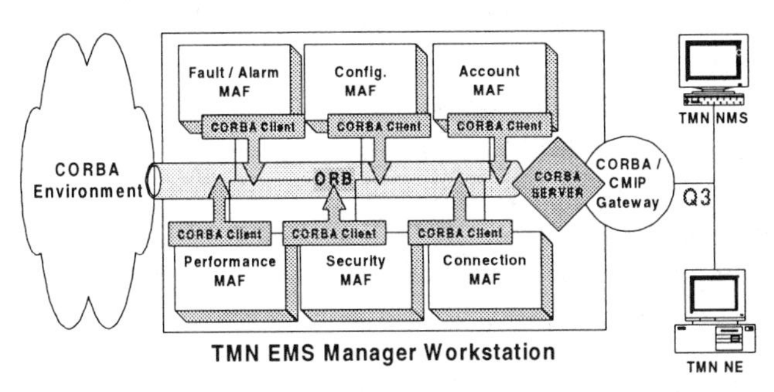

Figure 5. TMN architecture with internal CORBA

TMN EMS system distributes each of management scenarios as job-based MAF process. Each MAF process has client stub and communicates with each other through ORB bus of CORBA via server that is located in CORBA/CMIP gateway process. The CORBA server performs TMN EMS manager main process as well as the role of CORBA server object. This approach makes use of the gateway concept between domains and requires the translation of GDMO semantics and existing CMISE functionality into IDL definitions. Such an IDL interface is abstracted from the information model semantics level but offers only the required functionality by the management applications, with IDL specified object classes and operations concerned with GDMO specifications. In GDMO, ASN.1 and template can be mapped into IDL specification through syntactic mapping table. In case of the CMIS/P behavior, it is need to be semantic translated. The mapping is then made with the GDMO exhaustive model of the managed entity. This approach offers a way to encapsulate TMN CMIS/P components to be used in CORBA environment. Sometimes, however, during the specification translation, parts of ASN.1 cannot be represented in IDL. In these cases, some omissions or mismatches are remedied during translation of the values that pass through the CORBA/CMIP gateway in run time. It remains to be heavily studied.

5. CONCLUSION

In this paper, we introduced possible development structure of TMN EMS management platform for ATM NEs and network. We, mainly, described the functional framework of TMN EML management system. The MO based management application is announced through multi-node connection management of ATM network. To develop the effective TMN EMS, proposed TMN EMS architecture operates management application function (MAF)s are operated with task-oriented separated MAF processes in distributed environment. Thus, MAF's roles are also layered and categorized into F,C,A,P,S,CN based MAF processes. These are very useful to separate the manager's role into CMIP message handling and MO information handling to give flexibility and extensibility. This structure also enables that TMN EMS could be constructed to adapt other heterogeneous management platform such as CORBA based management environments. In front of EMS main manager process, we can build gateway architecture to serve CORBA based management requests, cost effectively. We are defining and developing detail gateway framework in front of TMN EMS. Designing and developing the effective mapping and translation of MOs and management applications are also topics to be heavily studied.

References

[1] ITU-T Recommendation M.3010, "Pinciples for a Telecommunication Management Network", 1996

[2] SeokHo Lee, WangDon Woo, "A Proposal on Design Scheme of TMN NEML Management Application Framework for ATM Switching Systems", IEEE ICC'97, p.1180 ~ p.1184, JUN., 1997

[3] SeokHo Lee, WangDon Woo, " An Implementation of TMN Management Application Framework for ATM Switches and Subnetwork," IEEE ICCS/ISPACS'96, Vol. 2 of 3, p.542 ~ p.546, NOV., 1996

[4] SungKee Noh, SeokHo Lee, "An Implementation of Gateway System for Heterogeneous Protocols over ATM Network", IEEE PACRIM'97, p.535 ~ p.538, 1997

[5] ATM Forum, "Customer Network Management for ATM Public Service (M3 Specification)", Rev.1.04, 1994

[6] TINA-C Deliverables, "Network Resource Information Model Specification", Doc. No. TB_LR.010_2.1_95

[7] KTSTRL, "HAN/B-ISDN Draft Specification for Network Information Model", NOV., 1996

[8] ITU-T Recommendation M.3100, "Generic Network Information Model", 1995

[9] Object Management Group, "The Common Object Request Broker: Architecture and Specification", Revision 2.0, July 1995

[10] X/Open, "Inter-Domain Management Specifications : Specification Translation", X/Open Preliminary Specification, Draft, AUG. 1995

[11] Marnix Harssema, "Integrating TMN and CORBA", Hewlett Packard Journal, OCT., 1996

[12] N.Soukouti, U.Hollberg, "Joint Inter Domain Management : CORBA, CMIP and SNMP", Integrated Network Management V, p.153 ~ 164, Chapman&Hall

LOCAL RECONFIGURATION OF ATM VIRTUAL PATH CONNECTION NETWORKS

S.A. Berezner[1], J.M. de Kock[2], A.E. Krzesinski[2] and P.G. Taylor[3] *

[1]*Dept of Maths & Stats*
University of Melbourne
Victoria 3052
Australia
berezner
@ms.unimelb.edu.au

[2]*Dept of Computer Science*
University of Stellenbosch
Stellenbosch 7600
South Africa
{jmdekock, aek1}
@cs.sun.ac.za

[3]*Dept of Applied Maths*
University of Adelaide
Adelaide 5005
Australia
ptaylor
@maths.adelaide.edu.au

Abstract Modern communication networks based on SDH/SONET or ATM can construct logical end-to-end connections between all origin-destination (O-D) pairs and thus create fully meshed logical networks upon sparse physical networks. Such logical connections are known as virtual path connections (VPCs) and a logical network formed by VPCs is known as a VPC network (VPCN). We have developed an efficient algorithm called XFG to compute optimal VPCNs for carrying multiservice traffics. The XFG algorithm needs to know the call arrival intensities between all O-D pairs. These data may not always be available. We have therefore developed a local reconfiguration algorithm to augment the XFG algorithm. The local algorithm makes use of data which are likely to be available in real networks. A call arrival at an O-D pair potentially triggers a local reconfiguration which affects only those routes that either connect the O-D pair or which use the physical link which connects the O-D pair. The information necessary to calculate the reconfiguration is kept locally at the O-D pair.

We present simulation experiments which apply local reconfiguration to a model of a small ATM network. The initial experiments indicate that the local reconfiguration algorithm performs well on the network model under investigation.

Keywords: alternative routing; ATM networks; dynamic reconfiguration; non-linear optimization; virtual path connection networks.

*The second and third authors of this work were supported by the the South African National Research Foundation Grant No. 2034100, Telkom SA Limited and Siemens Telecommunications. The fourth author was supported by the Australian Research Council Grant No. A69702505.

1. INTRODUCTION

Virtual paths (VPs) and virtual path connections (VPCs) are important concepts in ATM networks. A VP recognizes the distinct identity of a traffic stream between two communicating nodes. A VPC consists of a series of concatenated VPs and specifies a route to be traversed by a set of virtual channel connections from an originating node through a number of intermediate nodes to a destination node. Bandwidth is logically assigned to a VPC by reserving a certain part of the transmission capacity on each physical link for the exclusive use of the VPC. Using VPCs allows for faster setup of new connections along predefined routes and rapid movement of ATM cells with minimal processing at intermediate nodes.

A logical network formed by VPCs is known as a virtual path connection network (VPCN). The VPCN plays a central role in dynamic network re-configuration (see [1] and the references therein) which has been proposed as a simple and robust resource management control to manage ATM networks to respond optimally to traffic variations where these variations occur on a time scale of the order of hours. The network management control measures the offered traffic and, should a significant and persistent traffic change be detected, the control computes and implements an optimal VPCN which maximizes network profit for the new traffics. The motivation is that a dynamic VPCN will be a better method to handle slowly varying traffics, while short term traffic variations due to random mismatches between offered traffics and available bandwidth are best handled by alternative routing.

However, an excessive use of alternative routing may overload the signalling network. It is desirable to limit the number of alternatively routed calls which might arise when the traffic pattern differs from the traffic pattern which was used to design the optimal VPCN. We propose to do this by local reconfiguration: if the proportion of rejected calls for a specific origin-destination (O-D) pair exceeds the GoS requirements, the capacity of the VPC corresponding to this O-D pair is adjusted.

2. RECONFIGURATION ALGORITHMS

Anerousis and Lazar [1] survey a number of algorithms for constructing VPCs in broadband networks. They provide a taxonomy for the algorithms, distinguishing the algorithms according to a number of criteria, among them whether they are synchronous or asynchronous, whether they are centralized or decentralized, the form of the cost function being optimized, and whether signalling costs are taken into account.

Synchronous algorithms update the capacity allocated to a VPC based on observed demand for call establishment. Asynchronous algorithms maintain a capacity allocation for a fixed time. Centralized algorithms are updated at one location and require the collection of up-to-date information from all network nodes, while decentralized algorithms are executed in every switch using only local information.

Another criterion which can also be added is whether an algorithm is data-driven or data-free. A data-driven algorithm uses information about the statistics of the call arrival and service processes. A data-free algorithm does not use information about the processes which load the network: it uses information only about the network states which arise as a result of the traffic processes.

3. GLOBAL RECONFIGURATION

Consider a network which carries calls that belong to S service classes. Let b_s denote the effective bandwidth of a call of class s where $1 \leq s \leq S$. For each O-D pair (i,j) let C_{ij} denote the capacity of the link $i-j$, λ_{ij}^s the Poisson arrival rate of class s calls and $1/\mu_{ij}^s$ the mean holding time of class s calls offered to the O-D pair (i,j). C_{ij} is zero if the physical link $i-j$ does not exist. Let $\rho_{ij}^s = \lambda_{ij}^s/\mu_{ij}^s$ denote the arrival intensity of calls of class s offered to the O-D pair (i,j). Let θ_{ij}^s denote the rate at which a class s call offered to the O-D pair (i,j) generates revenue. The revenue rate does not depend on how the call is routed although this assumption can be relaxed if cross-connection costs are taken into account. The effective bandwidth b_s of a call of class s is also independent of the call's route.

Reconfiguration is performed by cross-connecting circuits using configurable switches in transit nodes to form logical direct links (VPCs) between all O-D pairs. A route is a sequence of physical links connecting an O-D pair. A route between an O-D pair (i,j) is active if circuits are reserved on each link of the route for the sole use of calls between i and j. Let $\mathcal{R}_{ij}$ denote the set of active routes between the O-D pair (i,j). We assume that the physical links are dimensioned to carry their own offered traffics so that for each O-D pair (i,j) connected by a physical link, $\mathcal{R}_{ij} = \{i-j\}$. We restrict ourselves to the design of service integrated VPCNs where each VPC carries all traffic classes.

Let $\mathcal{R} = \cup_{ij}\mathcal{R}_{ij}$ denote the set of active routes. Let $\mathbf{x} = (x_1 x_2 \ldots x_R)$ denote the transmission capacity assignments to all active routes in the network where x_r is the capacity of route r and $R = |\mathcal{R}|$. The (logical) capacity of the VPC connecting the O-D pair (i,j) is given by $\widehat{C}_{ij}(\mathbf{x}) = \sum_{r \in \mathcal{R}_{ij}} x_r$. Let $B_{ij}^s(\mathbf{x})$ denote the blocking probability experienced by a class s calls on the VPC connecting the O-D pair (i, j). The VPCN design problem can be specified in terms of the following constrained non-linear optimization problem: Maximize

$$F(\mathbf{x}) = \sum_{ijs} \theta_{ij}^{s} \rho_{ij}^{s} (1 - B_{ij}^{s}(\mathbf{x})) \tag{1}$$

Subject to:

$$\sum_{r \in \mathcal{A}_{ij}} x_r \leq C_{ij} \qquad x_r \geq 0.$$

where $\mathcal{A}_{ij}$ denotes the set of active routes that use the physical link $i - j$. Note that $\mathcal{A}_{ij}$ includes the single-link route $i - j$.

An obvious way to obtain a solution for the optimization problem 1 is to use constrained non-linear programming (NLP) methods. However, for networks of realistic size, the number of decision variables in the NLP and the computation time required to solve the NLP grow to an extent that this becomes impossible in practice.

We have therefore developed an efficient deterministic algorithm named XFG [2] to solve the optimization problem 1. The XFG algorithm computes an optimal VPCN in a sequence of steps. At each step the algorithm either adds or removes capacity to or from a route r in $\mathcal{R}_{ij}$. The route r is chosen so that the increase or decrease of capacity on route r, which is done by seizing or releasing capacity from or to the links along route r, leads to the largest increase in the rate of earning revenue. The XFG algorithm terminates when no such route r can be found. Since at each step the XFG algorithm obtains the largest possible increase in revenue rate and does not take the long-term impacts of the capacity changes performed at each step into account, the XFG algorithm is a one-step greedy algorithm. According to the classification presented above, XFG is an asynchronous, centralized algorithm which maximizes the network revenue rate and does not take signalling costs into account.

The main advantage of the XFG algorithm is its computational efficiency. For example the non-linear optimizer CFSQP [4] requires several hours to compute an optimal VPCN for the model of the NSF ATM backbone presented in section 6 The XFG algorithm solves the same model in a few seconds. Both execution times are for a Pentium II 400 MHz machine.

4. LOCAL RECONFIGURATION

The XFG algorithm constructs a fully-meshed VPCN from a (possibly sparse) physical network. Since is is likely that the traffic profiles are continuously changing, the VPCN will not remain optimal forever. The cost of installing an optimal VPCN design is a computationally expensive process for a large network. Global reconfiguration therefore cannot be performed each time the traffic offered to some O-D pair can no longer be carried to the agreed GoS.

We therefore distinguish between two VPCN design problems. A **global** re-design occurs when either the offered traffics or the physical network change significantly. In this case both the composition and the capacities of the the optimal VPCs are recalculated. Traffic initiated global re-designs should occur relatively infrequently, perhaps every hour or so. A **local** re-design occurs in response to a slow though persistent change in the offered traffic intensities. New routes are not created: instead, the capacities of the existing routes are adjusted to afford better service to the changed traffic. Local re-designs may occur every few minutes. Finally, dynamic alternate routing (DAR) [3] is used to deal with the short time-scale random mismatches between offered traffic and VPC capacity.

The effect of local reconfiguration is to make the VPCN act as an "intelligent entity" which can move logical capacity to where is it most needed: if a logical link notices that it is getting close to being overloaded, then it augments its capacity by relocating capacity from "nearby" links. Such a strategy has several desirable features: (1) The strategy is local in nature. Each logical link (i, j) needs to know the offered traffic ρ_{ij} and the current state of the physical links through which it passes. Each physical link needs to know the current state of the logical links which pass through it. An important consequence of this property is that the strategy is **scalable**: it works independently of the size of the network. (2) The strategy is dynamic in that it is able to respond to changes in traffic patterns.

Local reconfiguration is performed at an O-D pair (i, j) in response to some trigger: for example a reconfiguration can be attempted if the blocking probability B^s_{ij} for the O-D pair (i, j) violates an agreed GoS. Local reconfiguration will not be performed each time a call is blocked since this would give rise to an excessive signalling overhead. Such a call can either be queued, alternatively routed or lost.

5. DATA-FREE LOCAL RECONFIGURATION

The local reconfiguration process described above is data-driven in that it needs to know accurate values for the traffic arrival intensities. These may not be known for the following reasons: they are not measured by the network management system; they are measured but, to reduce overhead costs, they are measured infrequently; competing service providers operate separate logical networks on the same physical infra-structure – it is likely that the competitors will be denied access to each others traffic data.

We first present an abuse of notation which facilitates the description of the data-free local reconfiguration algorithm. Consider the routes that use the physical link $i - j$. One of these routes is the single-link route t which carries the traffic offered to the O-D pair (i, j). Let $\mathcal{T} \subset \mathcal{R}$ denote the set of all single-link routes. Consider a multi-link route r. With an abuse of notation let $t \in r$ denote the single-link route t where $\mathcal{R}_{ij} = \{t\}$ and $\{i - j\} \in r$, $t \in \mathcal{T}$ and $r \in \mathcal{R}$.

The data-free local reconfiguration algorithm assumes the existence of a VPCN and that the following information is available:

1. For each physical link $i - j$ the algorithm knows: the set of routes $\mathcal{A}_{ij}$ which use the link $i - j$, the allocated capacity x_r and the unused capacity y_r on each route $r \in \mathcal{A}_{ij}$ where $y_r = x_r - \sum_s n_{rs} b_s$ where n_{rs} is the number of class s calls in service on route r and b_s is the effective bandwidth of a call of class s.

2. For each O-D pair (i, j) not connected by a physical link the algorithm knows: the set of routes $\mathcal{R}_{ij}$ which connect the O-D pair (i, j), the capacity x_r allocated to each route $r \in \mathcal{R}_{ij}$, and the unused capacity y_t on each single-link route $t \in r$.

Let $X_{ij} = \sum_{r \in \mathcal{R}_{ij}} (x_r - \sum_s n_{rs} b_s)$ denote the free capacity of the VPC connecting the O-D pair (i, j). If the free capacity X_{ij} falls below a threshold D_{ij} then the local reconfiguration algorithm will attempt to add U units of capacity to X_{ij}. There are two possibilities:

1. The physical link $i - j$ exists: Select one of the longest routes $r \in \mathcal{A}_{ij}$ with unused capacity $y_r > U$. Decrease the capacity x_r of the route r by U units. This will increase the capacity x_t of each single-link route $t \in r$ by U units. Since $i - j \in r$, the free capacity X_{ij} is increased by U units.

2. The physical link $i - j$ does not exist: Select one of the shortest routes $r \in \mathcal{R}_{ij}$ such that each single-link route $t \in r$ has unused capacity $y_t > U$. Increase the free capacity X_{ij} by U units by increasing the capacity x_r of route r by U units. This will decrease the capacity x_t of each single-link route $t \in r$ by U units.

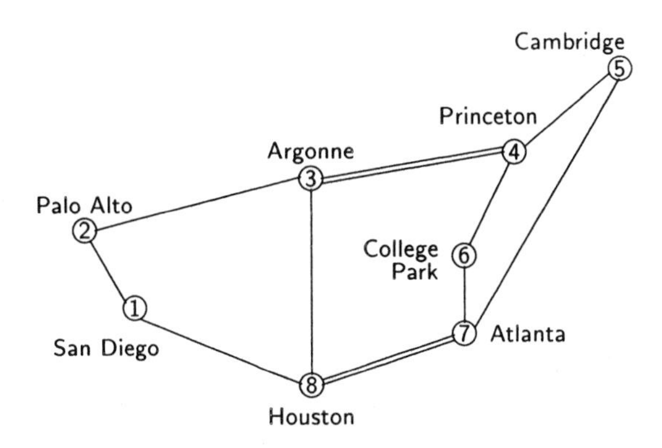

Figure 1 The core NSF network

6. RESULTS

Consider the model [5] presented in figure 1 which is a fictitious representation of the core NSF ATM backbone consisting of eight nodes each connected by one bi-directional link apart from nodes 3 and 4 and 7 and 8 which are each connected by 2 links. The network carries 6 services: the bandwidth requirement of the 1st service is 1 unit and the bandwidth requirements of services 2 through 6 are 3, 4, 6, 24 and 40 units respectively. The transmission capacity of each link is 5624 units. The traffic intensity for service 3 is given in Table 1. For all other services, the values in Table 1 are multiplied by 0.5. The revenue earned by a carried call per unit of time is assumed to be independent of how the call is routed and is equal to the bandwidth of the call.

We first perform a robust test of the local reconfiguration algorithm: we design a VPCN from the physical network using only local reconfig-

Table 1 Traffic intensity matrix

nodes	1	2	3	4	5	6	7	8
1	-	6	7	1	9	5	2	3
2	7	-	24	3	31	15	6	9
3	8	25	-	4	37	18	7	11
4	1	3	3	-	4	7	1	1
5	11	33	39	5	-	24	9	15
6	5	14	16	2	21	-	4	6
7	2	5	6	1	8	4	-	2
8	3	8	10	1	12	6	2	-

uration. This is done by initializing the simulator with the capacities of the physical links rather than with the capacities of the VPCs. We also provide the simulator with the routes which compose the VPCs as calculated by XFG: no capacities are initially assigned to the routes. In the second test we start with the fully-meshed VPCN constructed by XFG and use local reconfiguration to deal with random mismatches between offered traffics and VPC capacities. In all the simulations the threshold parameter D_{ij} is set to 12 and the reallocation unit U is set to 4. Each simulation run simulates 2 and a half million offered calls.

Table 3 presents the capacities of the VPCs resulting from these two tests. The column labeled VPCN presents the optimal VPCN constructed by XFG. Columns VPCN_0 and VPCN_1 present the VPCNs constructed by applying the local reconfiguration algorithm to the physical network and to the optimal VPCN respectively. Table 2 presents the network GoS for these two tests as well as for the optimal VPCN.

Tables 2 and 3 show that though the construction of a VPCN from a physical network is not the primary function of local reconfiguration, the near-optimal (in terms of the revenue rate) VPCN_0 constructed in this way suggests the utility of the local reconfiguration method. Perhaps the local algorithm can be used in situations where there are insufficient data to use XFG. Although the revenue rate generated by VPCN_0 is near optimal, the routes and the capacities assigned to the routes in VPCN_0 differ from those in the optimal VPCN. However, many near optimal VPCNs may exist.

The second test evaluates the local reconfiguration algorithm when it is used to deal with random mismatches between traffics and VPC capacities. Tables 2 and 3 show as expected (compare the columns labeled VPCN and VPCN_1) that the local reconfiguration algorithm makes small changes to the optimal VPCN and the revenue rate derived from VPCN_1 remains close to the optimal value.

Table 4 compares local reconfiguration and dynamic alternate routing (DAR) when the traffic intensities to all O-D pairs in the network are increased by 20%. The DAR circuit reservation parameter is set

Table 2 Global and local reconfiguration: network GoS

	VPCN	VPCN_0	VPCN_1
% blocking	1.7	2.6	1.6
% revenue lost	4.7	7.5	4.8
revenue rate	21277	20664	21265

Table 3 Global and local reconfiguration: VPC capacities

O-D	VPCN	$VPCN_0$	$VPCN_1$	O-D	VPCN	$VPCN_0$	$VPCN_1$
1,2	2130	3220	2242	3,4	2451	2892	2435
1,3	808	844	860	3,5	3353	3228	3529
1,4	148	184	120	3,6	2544	2064	2544
1,5	793	1068	909	3,7	1282	880	1282
1,6	567	556	591	3,8	1809	2620	1661
1,7	263	280	258	4,5	439	296	467
1,8	366	288	342	4,6	859	752	827
2,3	2361	2176	2433	4,7	449	220	449
2,4	366	344	350	4,8	444	264	444
2,5	2599	2420	2303	5,6	2023	1844	2023
2,6	1459	1380	1511	5,7	800	864	748
2,7	616	704	604	5,8	1241	1528	1269
2,8	903	1004	967	6,7	1296	3412	1252
				6,8	1226	800	1226
				7,8	3145	3904	3121

to 4. The results labeled LOCAL show the effect of local reconfiguration; the results labeled NONE apply when local reconfiguration and alternative routing are disabled; the results labeled DAR apply when alternative routing is enabled and local reconfiguration is disabled. Alternative routing and local reconfiguration achieve almost the same overall blocking and revenue rate. Local reconfiguration achieves a better GoS for low bandwidth calls and far fewer low bandwidth calls are lost. In

Table 4 Local reconfiguration (LOCAL) versus dynamic alternate routing (DAR)

service class s		all	1	2	3	4	5	6
bandwidth b_s			1	3	4	6	24	40
blocking	NONE	4.47	0.09	0.45	0.67	1.21	9.46	18.70
probability	LOCAL	3.77	0.01	0.02	0.06	0.24	8.54	17.42
	DAR	3.60	0.08	0.25	0.35	0.65	8.40	19.63
revenue	NONE	23567	326	976	2596	1937	7108	10624
rate	LOCAL	23848	327	980	2612	1956	7180	10793
	DAR	23324	316	946	2521	1885	7023	10633
attempts	LOCAL	9933	4834	1589	2358	827	203	122
	DAR	66163	9906	9716	19757	9786	8939	8059

particular table 4 shows that for the model under evaluation, local reconfiguration is attempted far fewer times that is alternative routing, with a corresponding reduction in signalling load. This is evident for the high bandwidth calls where alternative routing mostly fails. The revenue gain attributable to local reconfiguration is small. The rate of earning revenue for a class s call is proportional to the call's effective bandwidth requirement b_s: the improved GoS offered to low bandwidth calls therefore does not earn much revenue but will earn subscriber approval.

7. CONCLUSION

This paper is concerned with dynamic VPCN redesign as a resource management control in ATM networks where transmission capacity is reserved on the communication links in order to form dedicated logical paths for each origin-destination flow. We present a local reconfiguration algorithm designed to adapt an optimal VPCN to changes in the offered traffics. We present simulation results which indicate that, for the model under consideration, the local reconfiguration is robust, stable, and is capable of adapting the VPCN to reduce the call rejection count and thus limit the need for alternative routing.

References

[1] N. Anerousis and A.A. Lazar. Virtual Path Control for ATM Networks with Call Level Quality of Service Guarantees. *IEEE ACM Transactions on Networking* **6**:2, April 1998, pp 222–236.

[2] S.A. Berezner and A.E. Krzesinski. Call Admission and Routing in ATM Networks Based on Virtual Path Separation. IFIP TC6/WG6.2 Proceedings *4th International Conference on Broadband Communications*, Stuttgart, Germany, April 1998. Chapman & Hall (Eds PJ Kühn and R Ulrich), pp 461–472.

[3] R.J. Gibbens and F.P. Kelly. Dynamic routing in fully connected networks. IMA Journal of Mathematical Control and Information, Vol. 7 (1990) pp 77–111.

[4] C. Lawrence, J.L. Zou and A.L. Tits. User's Guide for CFSQP Version 2.5. Report Number TR-94-16r1, Electrical Engineering Department and Institute for Systems Research, University of Maryland, College Park, MD 20742 USA.

[5] D. Mitra, J.A. Morrison and K.G. Ramakrishnan. ATM Network Design and Optimization: a Multirate Loss Network Framework. Proceedings *IEEE INFOCOM '96*, pp 994–1003.

AN ALGORITHM FOR BROADBAND NETWORK DIMENSIONING

Mette Røhne, Rima Venturin, Terje Jensen, Inge Svinnset
Telenor R&D, P.O.Box 83, N-2027 Kjeller, NORWAY

Abstract Dimensioning and designing broadband networks are big challenges for a network operator. Considering the functionality accompanying ATM-based networks, there is a necessity of having network planning procedures being able to exploit the advantages provided by ATM. One of the benefits is potential use of Virtual Paths, which then could compose a set of logical networks within the physical network. This paper presents algorithms for dimensioning broadband networks and for designing VP networks. The algorithms are based on a decomposition approach, where the locations are examined one by one. Therefore, more comprehensive relationships than found in most of the design algorithms using global optimisation could be applied. Due to the modularity further enhancements could be used within the algorithms with respect to, for instance, network element functionality, traffic handling principles, cost model, network design objectives and traffic approximations.

In this paper the algorithms for dimensioning and VP design are applied on two cases; in one case the variable rate traffic is represented by Statistical Bit Rate (SBR), while in the other case the variable rate traffic is defined as best-effort type of traffic and is characterised by Available Bit Rate (ABR). At the end, comparisons of some numerical results of these two cases are presented.

Keywords: Network design, dimensioning, VP design

1. INTRODUCTION

When designing broadband networks based on Asynchronous Transfer Mode (ATM) the potential gains achieved by handling traffic flows on Virtual Channel (VC) and Virtual Path (VP) levels should be considered (e.g. [1]). VP networks design is especially interesting because the use of VPs may result in many advantages, such as reduction in the processing and delay associated with the call acceptance control functions since the VPs are cross-connected in intermediate nodes. In addition, the use of VPs may simplify procedures in case of network failure, because having defined VPs faster reconfiguration could be obtained. On the other hand, employing VP networks could increase network transmission costs and decrease network throughput, because allocation of capacity to VPs restricts sharing of resources.

The VP network design may involve consideration of many other objectives, such as flexibility to changes in demand, network processing, set-up delay, end-to-end delay, set-up cost, switching cost, transmission cost, segregation of services, assurance of quality of service, reliability and throughput.

This paper presents algorithms for the broadband network dimensioning and cost-effective VP design. The motivation for making these algorithms is described in Chapter 2. The algorithms described by main input/output data are covered in Chapter 3. The same chapter contains the proposed cost model and short overview of traffic handling mechanisms. Some numerical results are given in Chapter 4. The examples are used to illustrate the influence of the applications and service categories on the dimensioning and design of broadband networks. Applications using variable rate are characterised respectively by Statistical Bit Rate (SBR) and Available Bit Rate (ABR) and the resulting network solutions are compared. Finally, concluding remarks and plans for further work are given in Chapter 5.

2. MOTIVATION AND SCOPE

In ATM networks it is possible to have both switched VPs and VCs. The VPs are commonly handled as a part of the management plane, whereas VCs are normally handled as a part of the control plane. This categorisation implies that the VPs are changed over a longer timescale than the VCs. Therefore is the dimensioning of a VP logical network is medium-term, while each individual VC represents short-term planning.

A VP network, which consists of a set of VPs, may be seen as a logical overlay network, that may be compared to a physical network where the VPs correspond to the physical links and the switches terminating the VPs correspond to the nodes. On top of the physical infrastructure a number of VP networks can co-exist, sharing the same physical transmission and switching capacities. To illustrate how the use of VP networks affects the network performance two cases for VP networks can be looked at. In the first case the VP network is the same as the physical network, meaning that each physical link in the ATM network contains only one VP. The second case is a fully connected VP network, where every node will have one or several VPs to every other node. In the first case the network utilisation is maximised and the call blocking probability is minimised on a link. Advantage of the second case is that the processing and the cost at the intermediate nodes are at their minimal values. Therefore, the better design of VP networks will probably be somewhere between these two extreme cases depending on the cost for transmission capacity, set-up and switching.

VPs can be used for a number of reasons, like virtual networks for customers, segregating connections requesting different traffic characteristics and service quality requirements.

The challenge of designing networks can be approached in a number of ways ([1], [3], [6], [9]) During such activities, better network structures are sought where both the location of network elements and the topologies can be considered as variables.

3. PROCEDURE FOR NETWORK DESIGN

In the dimensioning procedure, presented in this paper, the locations of the network elements are given. The challenge is then to find the capacities of nodes and link sets under the constraints specified. Introducing the VP layer, the question of whether or not to cross-connect VPs should be raised. To investigate for an answer, trade-offs between possible savings in set-up/switching and the increase of other costs for having separate VPs should be compared. By introducing more VPs the link set capacities is divided into smaller units (optionally reducing the scale effect). This results typically in additional costs. Examining such trade-offs during the execution of the procedure, an objective function has to be defined. In principle, this function may differ for the different locations and could also differ from the way the network cost is calculated at the end. Compared with some other approaches (e.g. [5], [8]), the algorithm applied does not use a global optimisation formulation. Rather, the procedure has some resemblance with the decomposition approach in the sense that decisions with respect to traffic handling and capacities are made for each location in sequence, iteratively. A similar model has been applied in European projects ([2],[6]).

3.1 Overall procedure

A general network design model is illustrated in *Figure 1*.

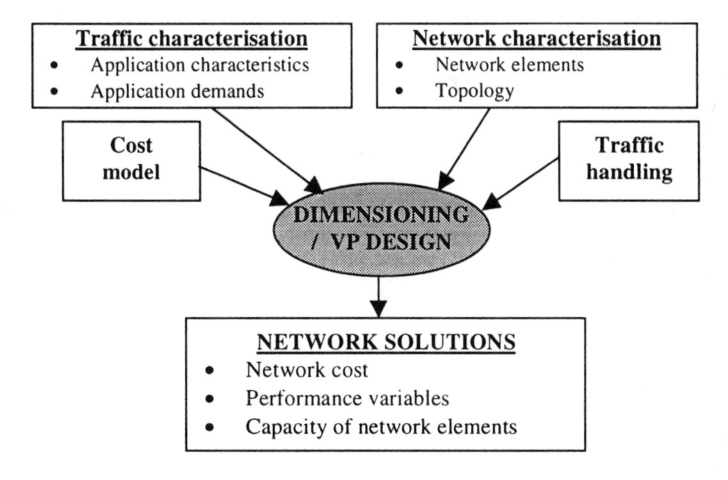

Figure 1. Input and output data for the dimensioning and VP design

The input data can be categorised as:

- Application characteristics. A number of bitstreams (connections) are associated with an application, potentially having different holding times and number of active periods during an invocation of an application. Each application is then described by a set of parameters, such as bitrates, real time requirements, acceptable cell loss ratios, and blocking probabilities, given for each bitstream associated with an application. Class of Service (CoS) indicator can be used e.g. based on these parameters. Connections belonging to the same CoS are treated corresponding to the traffic handling schemes.
- Application demands are given by the arrival intensity. This measure is commonly referring to the time periods that are considered for dimensioning (*reference periods*).
- Characteristics of potential network elements. Information about interface modules, maximal capacities, and so forth, is specified for each of the network elements.
- Topology candidates like potential locations for nodes and link sets.

- Traffic handling principles. As part of specifying traffic handling, routing rules and segregation/integration schemes are given. The routing rules are in this context given as specified alternate routes with the input data.
- Cost model. In order to find a cost-effective solution a model reflecting costs related to set-up, switching and transmission has to be described and applied in the algorithms.

3.2 Traffic handling

Traffic handling is the set of rules applied to manage the traffic in VPs and to handle establishment of connections. At the VP level, bitstreams with different characteristics can be segregated in a number of CoS. Assigning CoS could be based on various criteria, such as bandwidth demands or Quality of Service (QoS) requirements. In [4], QoS classes are associated with different cell level QoS requirements (e.g. cell delay variation, cell loss probability). Using CoS is a possible way of giving a specified quality to particular bitstreams, e.g. by separating bitstreams into different VPs. As opposed to this, bitstreams can be integrated in VPs. In this case, the VP must support the QoS of the most demanding bitstream carried by that VP. To make integration of CoS cost-effective, proper buffer management schemes with time and/or loss priorities have to be introduced in the switching nodes. Bitstreams demanding high bandwidth may experience a severe accessibility performance. Therefore, they could be separated into different CoS.

Requirements for set-up delay may have implications for routing and can be treated by introducing CoS. Traffic handling when failure occurs can differ based on a number of requirements. For instance, different survivability classes could be incorporated in CoS assignment.

3.3 Cost model

In the procedure outlined, a cost model referring to a "local view" is requested. The "local view" approach applies that decisions are made sequentially for each node (one by one).

The cost factors used when executing the algorithm are identified as:

Transmission cost, Zt_q; reflecting the cost of connections (bitstreams) between two nodes as well as termination units within those nodes, as shown in Figure 2. Considering two locations, i and j, one transmission path from i to j with capacity B_{ij}, has a cost C_{ij}. A module for terminating such transmission paths in location i has

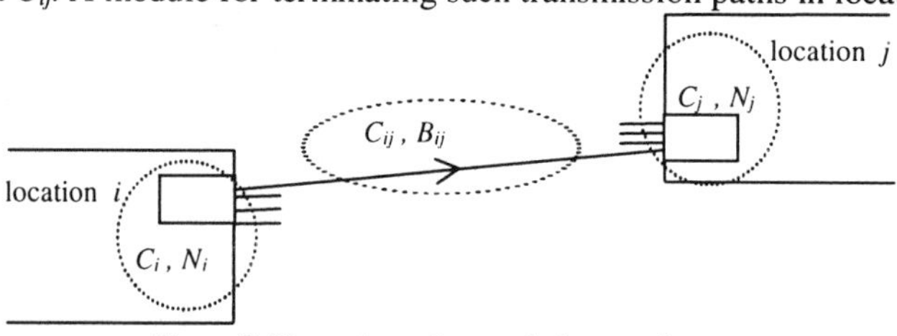

Figure 2. Illustrations of transmission cost factors

a cost C_i and may connect N_i of these transmission paths (correspondingly for location j). Assuming a VP_q having capacity demand B_q, the share of the bandwidth cost for that VP is calculated as:

$$Zt_q = \frac{B_q}{B_{ij}} C_{ij} + \frac{B_q}{N_i B_{ij}} C_i + \frac{B_q}{N_j B_{ij}} C_j$$

Switching cost, Zs_q; reflecting the effort requested for transmitting ATM cells from an input port (incoming transmission path) to an output port (outgoing transmission path). This cost component is assumed to be proportional to the traffic load, A_b, and the equivalent bandwidth, EB_b, of the bitstream b. That is,

$$Zs_q = \sum_{b \in q} Zs_b = \sum_{b \in q} A_b(\alpha \cdot EB_b + \beta)$$

where α and β are cost factors. The total cost component for a VP_q is the sum of cost components of type b carried by the corresponding VP.

Set-up cost, Zc_q; giving the processing requested for establishing (and releasing) a bitstream. Three types of processing steps could be incorporated as depicted Figure 3.

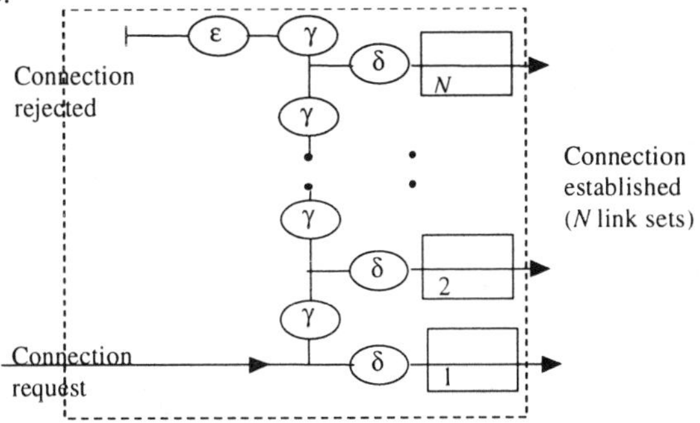

Figure 3 Illustrating the cost components for processing activities involved in connection set-up

Assuming there are N link sets that could be used for establishing the connection, where link set m as blocking probability Pb_m, this cost component could be written as:

$$Zc_q = \sum_{b \in q} Zc_b$$

$$Zc_b = \frac{A_b}{T_b}\left\{\delta(1-Pb_1) + \sum_{m=1}^{N-1}\left[(\delta + m\gamma)\prod_{k=1}^{m} Pb_k(1-Pb_{m+1})\right] + (N\gamma+\varepsilon)\prod_{k=1}^{N} Pb_k\right\}$$

where bitstreams of type b have a mean offered traffic A_b with mean holding time T_b. The cost factors δ, γ and ε express the effort related to a successful connection, an additional search for link sets and rejection of a connection request, respectively.

Considering the "local view" approach in the cost model described, different costs are compared and the solution resulting in the lowest cost is chosen. The contributions to the cost can be given weights. In this way the cost for a group of bitstreams, Z_Q, can be calculated as:

$$Z_Q = \sum_{q \in Q} \{ k_t Zt_q + k_s Zs_q + k_c Zc_q \}$$

where k_t, k_s and k_c are relative weight factors related to transmission, switching and set-up cost. The relative weighting of the cost components is a difficult issue that may have significant implications on the logical network design.

It is therefore essential to perform sensitivity analyses varying the weight factors. In principle, these weight factors can be considered as taking part of the cost or placing relative credibility, e.g. when certain components are estimated with higher accuracy.

3.4 Algorithm for network dimensioning and VP design

The objective of the dimensioning procedure is to find the number of VPs that gives the highest reduction in the total network cost, including set-up, switching and transmission costs. The main outline of the dimensioning and VP design algorithms is illustrated in *Figure 4*. A fixed routing scheme is applied.

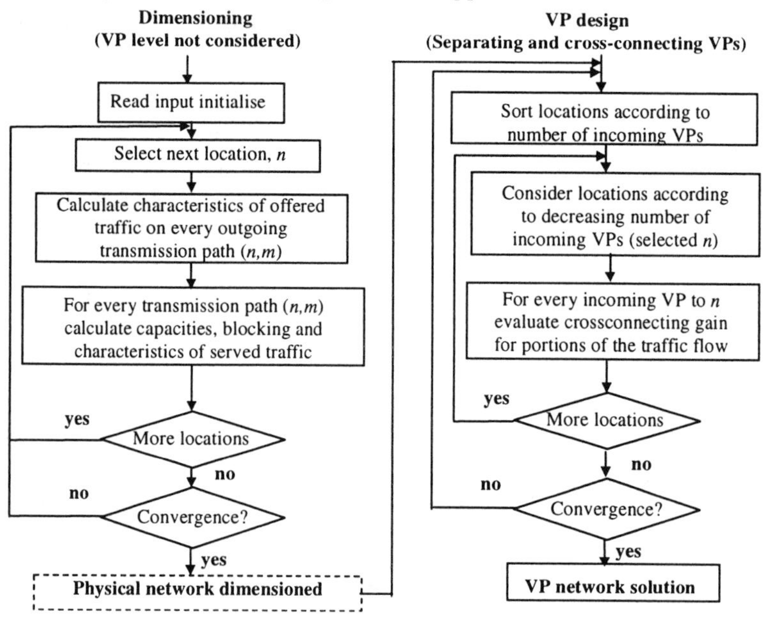

Figure 4. Outline of the broadband network dimensioning and VP design algorithm

In the first part the physical network is dimensioned without considering VPs and cross-connecting VPs. This implies switching each VC connection in every node it is crossing. In the calculations the "local view" approach has been used. The iteration is carried out until changes for the main variables (i.a. mean traffic and blocking probability) attached to the traffic streams are below the specified thresholds (convergence criteria). By separating and cross-connecting VPs the algorithm is looking for cheaper solutions. In the resulting network each physical link carries one VP with the capacity of the physical link set.

Before searching for VPs to be cross-connected the VP capacity is adjusted to the total demand whether there is segregation or not. The cost model and the relevant segregation scheme are used when deciding whether or not to cross-connect traffic streams. Traffic streams are segregated according to their CoS indicator.

Cross-connecting traffic streams means to separate the traffic streams from their previous VPs that terminate in the node and putting them into a new VP that is cross-connected in this node. This is accomplished by looking at one node at the time, starting with the node that has the highest number of incoming VPs.

VP capacity is a function of equivalent bandwidth and connection blocking probability. These are calculated based on the approximations described in [7]. Establishment/cross-connecting of VPs means a splitting of one VP into two where one of them is cross-connected through the node. The resulting sum of transmission bandwidth required for these two VPs will always be greater than or equal to the bandwidth of the VP that is split due to the scale effect. The extra bandwidth needed depends on the mixture of traffic contained in the VP to be split.

The costs, related to transmission, switching and set-up before and after the cross-connection, are calculated and compared. Splitting VPs increases the transmission cost. Cross-connecting a VP implies less switching and control activities, which leads to lower switching and set-up costs when the number of VPs increases. If the cost reduction is above a specified threshold, the traffic streams will be cross-connected in the node.

Applying these algorithms for the network dimensioning and VP design gives a cost-effective VP network solution.

4. NUMERICAL EXAMPLES

The algorithm outlined is applied on the example network depicted in *Figure 5* to give some numerical examples. The network consists of 14 ATM switches, having both VC switching and VP cross-connecting.

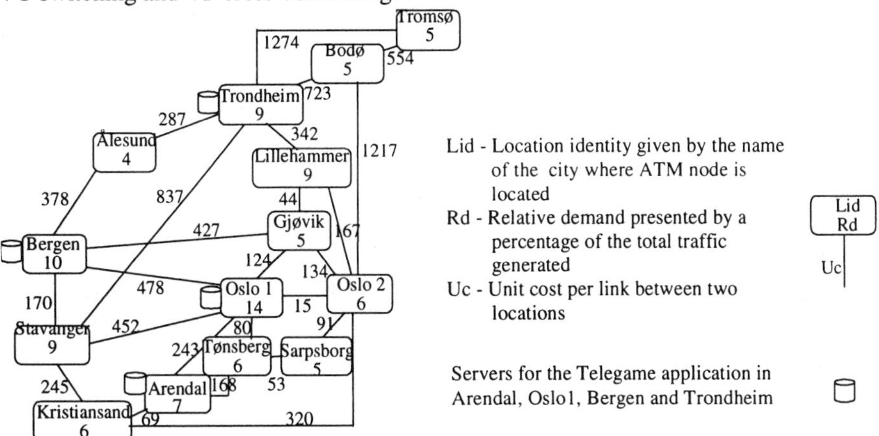

Figure 5. Network structure, location identity, relative demand and unit cost

Table 1 Applications with relevant input

Bit stream ID	Direction	ATC	MCR /SCR [kbit/s]	PCR [kbit/s]	P_{loss}	P_{block}	HT [min]	CoS
1 Telephony								
1.1	UN	DBR	64	64	10^{-6}	0.01	5	1
1.2	NU	DBR	64	64	10^{-6}	0.01	5	1
2 Video on Demand								
2.1	UN	DBR	8	8	10^{-9}	0.01	90	2
2.2	NU	SBR	1664	2064	10^{-9}	0.01	90	2
3 Videoconference								
3.1	NU	SBR	100	2000	10^{-6}	0.005	45	3
3.2	UN	SBR	8	64	10^{-6}	0.005	45	3
3.3	UN	DBR	64	64	10^{-9}	0.01	45	2
3.4	UN	DBR	384	384	10^{-9}	0.01	45	2
3.5	NU	DBR	64	64	10^{-9}	0.01	45	2
3.6	NU	DBR	384	384	10^{-9}	0.01	45	2
4 Real-time transaction								
4.1	UN	SBR	64	128	10^{-6}	0.005	2	3
4.2	NU	SBR	64	128	10^{-6}	0.005	2	3
5 Telegame								
5.1	UN	DBR	64	64	10^{-6}	0.05	20	4
5.2	NU	SBR	2000	5000	10^{-6}	0.05	20	4

Direction: UN = user → network, NU = network → user; ATC = ATM Transfer Capability; MCR = Minimum Cell Rate; SCR = Sustainable Cell Rate; PCR = Peak Cell Rate; Ploss = Cell loss ratio requirement; Pblock = Connection blocking requirement; HT = Mean holding time; CoS = Class of Service, DBR = Deterministic Bit Rate, SBR = Statistical Bit Rate

Table 2 Total demand for each application

Application name	Total demand [erlang]
1 Telephony	151 446
2 Video on Demand	2 840
3 Videoconference	97 567
4 Real-time transaction	48 502
5 Telegame	998

Table 3 Cost factors and weight factors for reference case

Cost	*Factor*	*Id*	Value
Transmission	*weight*	k_t	1.0
Switching	*weight*	k_s	k
	cost	α	200.0
	cost	β	0.0
Set-up	*weight*	k_c	k
	cost	δ	1.0
	cost	γ	1.0
	cost	ε	1.0

Five applications are considered and are characterised in *Table 1*. The total demand per application is given in *Table 2*. In order to calculate elements of the traffic matrices, the total demand is multiplied by the relative demand for the source and destination location indicated in *Figure 5*[1]. Capacity of 155 Mbit/s is considered. The unit costs, U_c, are multiplied by 100, in order to get the cost for having a single link between each pair of locations. One termination unit, able to handle one link, is assumed to have a cost equal to 10000.

The values of the cost are given in *Table 3*. The VP capacity may be higher than a single link. For the calculations presented weight factor, k, is introduced and is equal to the weight factors for set-up and switching, $k=k_c=k_s$. As a minimum 50 VPs will be established, one for each direction on every link. The maximum number of VPs that may be established with the given network structure and with four CoS values is 448.

In the reference case the configuration and the values are kept as presented in *Figure 5* and in *Table 1*, *2* and *3*. The numerical studies are made for two cases; in the first case the bitstreams with variable rate are characterised by SBR, while in the second by ABR.

[1] For example, the element in the traffic matrix for the Videoconference application as demand from Bodø to Bergen is found as: 97567 · 5/100 · 10/100.

4.1 Reference case

In this case variable rate bitstreams are characterised by SBR. For each SBR bitstream, the equivalent bandwidth has a value between the Sustainable Cell Rate (SCR) and Peak Cell Rate (PCR) dependent on the capacity of the VP.

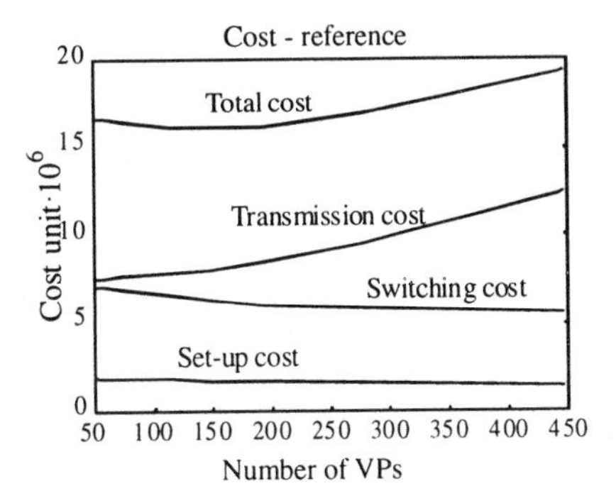

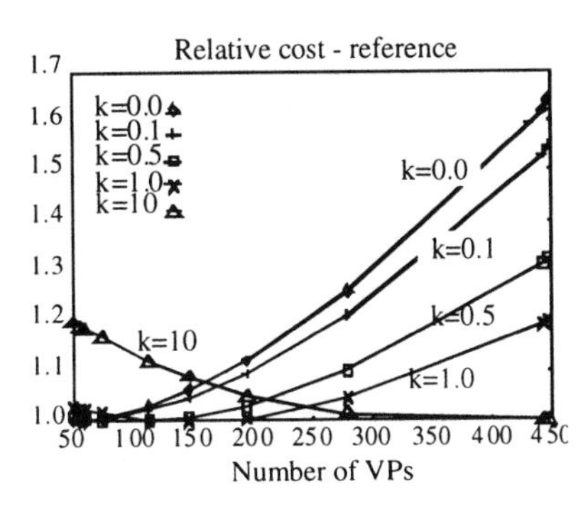

Figure 6. Results for reference case: a) Cost as function of number of VPs, b) Total cost relative minimum total cost for selected values of set-up/switching weight factors

Figure 6a) illustrates how the VP transmission cost is increasing as a function of the number of VPs.

As the costs related to set-up and switching are increasing, the establishment/cross-connection of VPs will become more profitable. A set of curves for the relative total cost for a selected number of set-up/switching weight factors, are depicted *Figure 6b).* The relative costs are found by dividing the total cost obtained by the minimum total cost for the relevant weight factor value k. The minimum total cost for the reference case, with the weight factor set to 1.0, is obtained when there are 146 VPs established.

As seen from the curves, a larger number of VPs are found as the better solutions (minimum relative cost) when greater weight is placed on set-up/switching cost. The shapes of the curves are explained by this effect. In one respect, these curves show the "goodness" of the solution found compared to alternative solutions. For instance, in case $k = 1.0$, having 50 VPs gives a total cost that is ca. 3.0 % more expensive than the better solution having 146 VPs, while the total cost having 448 VPs gives an increase of 19.9%.

4.2 ABR case

In this case, the bitstreams with variable rate is characterised by ABR. The dimensioning algorithm is using the MCR, having the same value as the SCR in the reference case, as the equivalent bandwidth.

Similar results as for the reference case is depicted in *Figure 7.* The overall network cost is reduced since the equivalent bandwidth used is significantly lower, resulting in less bandwidth needed and less switching cost since the switching cost also is influenced by the equivalent bandwidth. The equivalent bandwidth of ABR is constant and therefore independent of the VP capacity. The capacity when splitting VPs is only increased in order to satisfy the required connection blocking probability.

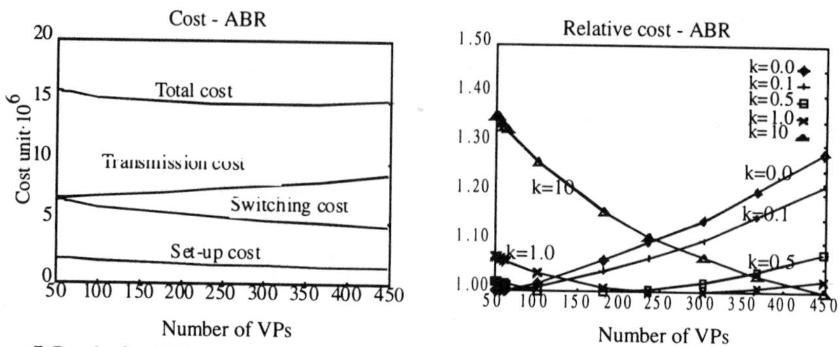

Figure 7. Results for ABR case: a) Cost as function of number of VPs, b) Total cost relative minimum total cost for selected values of set-up/switching weight factors

The minimum total cost for the ABR case with the weight factor, *k*=1.0, is obtained with 584 VPs established. Having 50 VPs gives a total cost that is. 7.3% more expensive than the better solution having 301 VPs, while the total cost is increased by 2.5% having 448 VPs. The number of VPs has less influence on the total network cost than in the reference case.

4.3 Comparing the reference and ABR cases

The network cost is considerable higher for a fully connected VP network in the reference case, since the equivalent bandwidth in case of SBR is increasing with a decreasing bandwidth per VP.

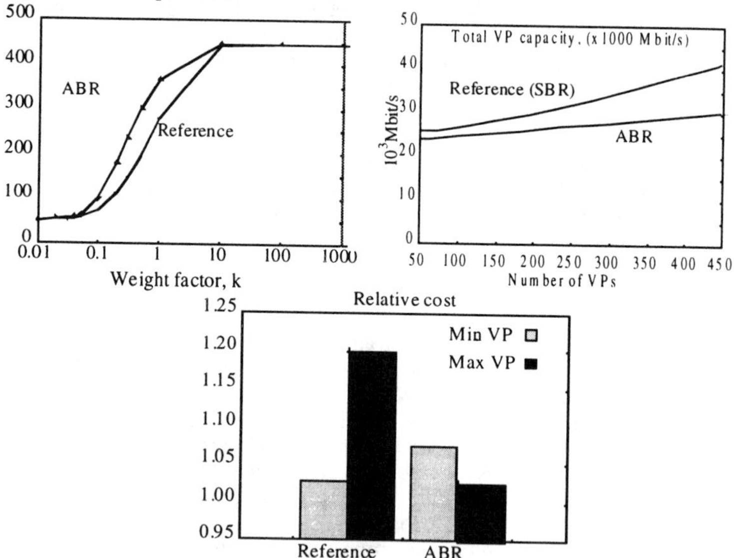

Figure 8 Comparison of reference case and ABR case: a) Number of VPs, b) Total cost relative minimum total, c) The total VP capacity in the network

In the ABR case several VPs are established for lower values of the weight factor k than in the reference case, as seen from *Figure 8a)*. As seen from *Figure 8b)*, the total capacity, when the number of VPs is 50, is quite similar for the two cases, since the VPs have capacities huge enough to make the equivalent bandwidth close to the MCR/SCR. Cross connecting the VPs will make the capacity allocated to each VP lower and the equivalent bandwidth for the variable rate bitstreams will increase. Therefore, with a fully connected network, the reference case has a total capacity that is considerably higher than for the ABR case.

The relative network cost for the two cases is illustrated in *Figure 8c)*. In the reference case the solution with the maximum number of VPs has a much higher network cost than the better solution with 146 VPs. The network cost with the maximum number of VPs is also much higher than the cost with the minimum number of VPs. In the ABR case, the relations of network costs are the opposite; the solution with a minimum number of VPs has a higher cost than the solution with the maximum number of VPs.

The various network solutions obtained, when varying the weight factor k, will provide different grade of service especially to the best-effort traffic. First, the capacity available in the reference case will be higher due to the dimensioning with equivalent bandwidth. This gives more "spare" capacity in the network solutions found for the reference case. Second, the number of VPs will reflect the degree of sharing of capacity.

5. CONCLUSIONS AND FURTHER WORK

As illustrated by the reference case and the ABR case, the characterisation of the offered traffic is essential for the network solution obtained when applying this algorithm. When the offered traffic is increased to a certain level for each CoS, the influence on the equivalent bandwidth is decreased, and the equivalent bandwidth is tending towards the mean rate. The bandwidth required has no longer a strong dependency on the number of VPs, and finding the number of VPs in the better solution is no longer the main issue.

However, the network dimensioning tool for carrying out network design presented here, may be adapted to investigate other aims, e.g. by use of other cost, equivalent bandwidth approximations, buffer allocation schemes, and capacity reservation policies.

As already discussed in Chapter 2, VPs may be used for a number of reasons. In this paper the motivation has been to segregate bitstreams having different CoS, to reduce the overall cost by cross-connecting VPs and to study the influence of application characterisation. When considering cross-connecting a VP, the switching and set-up costs in the intermediate node are compared to the cost of additional capacity necessary when splitting the VPs.

In broadband networks, the amount of offered best-effort traffic might be considerable, and the available bandwidth may be insufficient when the dimensioning is based on the MCR. The network dimensioning algorithm may be enhanced further to better incorporate best-effort type of traffic. The segregation scheme may also be changed to keep the best-effort traffic streams within VPs having a suitable traffic mixture.

References

[1] Z. Dziong, J. Zhang, L. G. Mason: "Virtual Network Design – An Economic Approach", ITC, Lund, Sweden, 1996

[2] EURESCOM P616 Deliverable D2: Congestion Control in a VC Switched ATM Network. 1998.

[3] A. Farago, S. Blaabjerg, W. Holender, B. Stavenow, T. Henk, L. Ast, S. Szekely: "Enhancing ATM Network Performance by Optimizing the Virtual Network Configuration", IFIP Data Communication and their Performance, Istanbul, Turkey, 1995.

[4] ITU-T Rec. I-356: B-ISDN ATM layer cell transfer performance, 1999.

[5] ITU-T Rec. E.737: Dimensioning methods for B-ISDN, 1997.

[6] JAMES Project Deliverable D18: Report on use of ATM network planning guidelines. 1998.

[7] K. Lindberger: "Dimensioning and Design Methods for Integrated ATM Networks", In Proceedings of 14th ITC, Vol. 1b,pg. 897-906, J. Labetouelle and J.W.Roberts(ed.), Antibes, France, June 1994

[8] J. Roberts, M. Mocci, J. Virtamo (ed.): Broadband Network Teletraffic – Final Report of Action COST 242, 1996.

[9] B. H. Ryu, H. Ohsaki, M. Murata, H. Miyahara: "Design Algorithm for Virtual Path based ATM networks", IEICE Trans. Communication, Vol. E79-B, No. 2, February 1996.

SESSION 18

Quality of Service

MOBILE RSVP

Towards Quality of Service (QoS) Guarantees in a Nomadic Internet-based Environment

Ali Mahmoodian[1], Günter Haring[2]
Institute for Applied Computer Science and Information Systems, Department for Advanced Computer Engineering, Vienna University
[1]mahmoodian@csi.com, [2]haring@ani.univie.ac.at

Abstract In this paper we introduce an extended RSVP protocol called Mobile RSVP which support multimedia and real-time traffic in a Mobile IP based environment. Our proposed protocol is appropriate for those mobile nodes that can not a-priori determine their mobility behaviour. However, we have also analysed the effect of a deterministic mobility specification. We also describe the conceptual design of the reservation mechanism of Mobile RSVP and discuss the hand-over procedure. Efforts have been made to analyse the performance of Mobile RSVP and to assess its overhead and control traffic.

Keywords: Quality of Service (QoS), RSVP, and Mobile IP

1. INTRODUCTION

Current mobile and cellular networks and protocols are only capable of providing best-effort delivery of information. Consequently, they are not suitable for real-time and multimedia communications, which often require throughput and delay guarantees from the transport system.

Recent research and development activities have been focusing on QoS issues in the wireline Internet. Protocols such as RSVP provide simplex QoS guarantees based on the requirements of heterogeneous receivers by making hop-by-hop resource reservation along the communication path.

However, QoS guarantees in a nomadic Internet-based environment remains an emerging research issue. Recognizing the need of such a protocol, we took a theoretical approach towards resource reservation in a mobile cellular network and propose the concept of an extended RSVP protocol, called Mobile RSVP, which is capable of providing QoS guarantees to mobile connections in a Mobile IP-based environment.

Mobole RSVP maintains the reserved portion of the connection until a new connection is set up, which is capable of providing QoS guarantees. Our proposed protocol also performs progressive resource reservation in advance, while the mobile host is travelling from cell to cell.

This paper is organized as follows. Sections 2 and 3 give a brief introduction to Mobile IP and RSVP, respectively. In Section 4 we discuss general service models and introduce our Mobile RSVP protocol. In Sention 5 we outline the protocol operation of Mobile RSVP and analyze various parameters. Section 6 presents various methods to enhance the performance of our proposed protocol. Section 7 discusses some open issues. We conclude in Section 8 with some future directions of this work.

2. INTRODUCING MOBILE IP

Mobile IP ([3] and [6]) is a modification to IP that allows nodes to continue receiving datagrams regardless of any movements and changes to the location and attachment to the Internet. It solves the mobility problem by maintaining two addresses for each mobile node, one for locating the mobile computer, and the other for identifying a communication endpoint on the mobile computer.

Mobile IP consists of the following three related activities:

Agent discovery - *Home agents* and *foreign agents* advertise their availability on the corresponding (sub)network they provide service. *Mobile nodes* may either listen or solicit *agent advertisement.*

Registration - Mobile node receives a so-called *care-of address* from the foreign agent corresponding to its current position and informs its home accordingly. Home agent decides whether a registration is authorised. It creates an association between the static IP address and the mobile node's current point of attachment (*mobility binding*) and subsequently sends a corresponding reply to the mobile node (Figure 1a).

Tunnelling (Figure 1b) - Datagrams received by the home agent containing the static IP address of the mobile node are encapsulated and forwarded (*tunnelled*) to the mobile node's new point of attachment. Foreign agent *decapsulates* incoming datagrams and passes them to the mobile node. Triangle routing, however, is not optimal and may cause home agent bottleneck, network congestion and an increased sensitivity to network partitions. Performance can be improved by providing the correspondent node with the mobility binding update, which enables datagrams to be sent directly to the mobile node's care-of address.

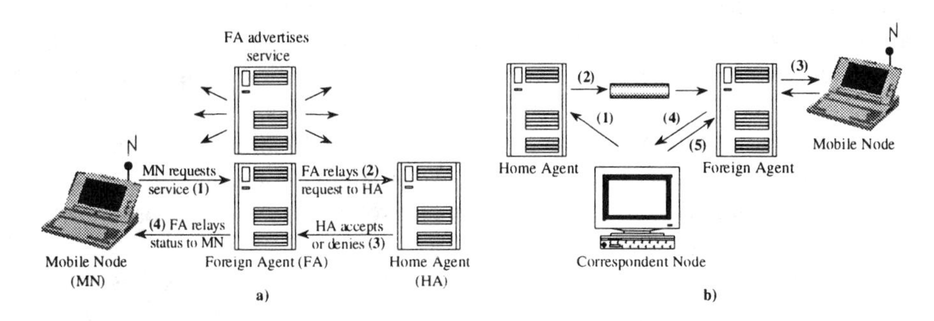

Figure 1: a) Agent advertisement and the four-step registration process of Mobile IP. b) Tunnelling process of Mobile IP using route optimisation

3. INTRODUCING RSVP

Resource Reservation Protocol (*RSVP*) is a protocol specially designed for *Integrated Services* Internet. It enables applications to set up reservations over the network for various services required. RSVP is a receiver based model and hence enables heterogeneous receivers to make reservations specifically tailored to their own needs. The operation of RSVP can be divided as follows (Figure 2):

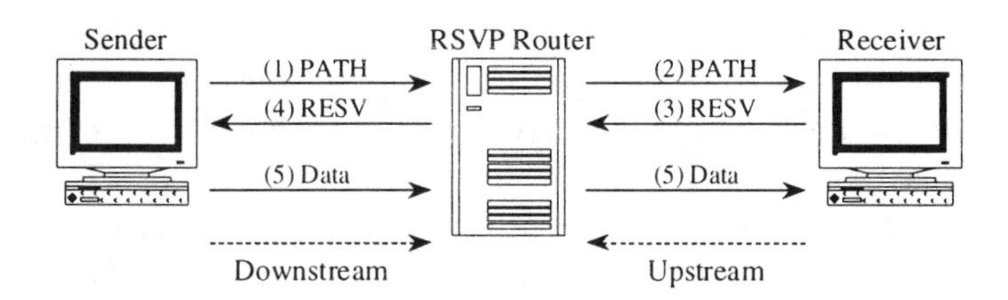

Figure 2: Operation of RSVP

(1) An RSVP sender that aims to initiate a session sends a PATH message to the corresponding receiver. The PATH message travels hop-by-hop (downstream) through the network and passes subsequent RSVP routers installing "path state" in each node along the way, which includes at least the unicast IP address of the previous upstream router. The router may also update an ADSPEC object contained in the PATH message, which summarizes the path's characteristics. The ADSPEC object is delivered to

the receiver to ensure a precise decision about how much bandwidth and delay to ask for. (2) Upon analyzing the obtained PATH message and performing necessary re-calculation of the traffic specification (ADSPEC), the PATH message is finally forwarded to the receiver. (3) The receiver evaluates the PATH message and generates a RESV message, which is eventually sent to the network. (4) The generated RESV message passes the routers in the reverse order (upstream), creates, and maintains "reservation state" in each node along the path. (5) Upon receiving the RESV message, the sender initializes appropriate traffic control parameters at the first hop (the sender) and starts transferring data.

To make best use of the reserved state in the network, RSVP supports various reservation styles and merging of similar reservations ([5] and [7]).

4. MOBILE RSVP (A PROPOSAL)

Anup K. Talukdar et al. ([4]) refers to two classes of real-time applications: *tolerant* applications, which can adapt to packet delays, and *intolerant* applications, which cannot tolerate any delay bound violations. Correspondingly, two different service models have been defined: *guaranteed* service for intolerant applications, and *predictive* service for tolerant applications. Furthermore, three service classes are proposed to which mobile nodes may subscribe.

Unfortunately, all approaches assume that the mobility of a user is predictable so that mobility can be characterized precisely by the *Mobility Specification* (*Mobility Profile*). Consequently, a mobile node is required to negotiate the desired QoS at the beginning of the data flow session. However, mobile node might not have an a-priori knowledge of its future mobility behaviour.

In our proposed reservation protocol, in particular, we have considered non-deterministic mobility behaviour. Hence, our proposed protocol will enable a higher degree of flexibility for those mobile nodes, which might not be able to adhere to a pre-defined mobility specification.

Based on the above specification, we have proposed the Mobile RSVP protocol, which is an extension of RSVP standardised by IETF for reserving resources between Internet nodes. Mobile RSVP works with Mobile IP; consequently, the assumption is made that Mobile IP is known throughout the network.

We assume that route optimisation is employed in the underlying Mobile IP protocol to facilitate the correspondent node with an up-to-date mobility binding. It is further assumed that each cell is connected to the fixed network through a *base station*. In the proposed network architecture, base stations are considered as part of Mobile IP routers with the facility to communicate with and keep track of *mobile nodes*. Preferably, there should be a wired direct connection between neighbouring base stations. Our proposed network is divided into microcellular or picocellular sub-networks capable of supporting both data and multimedia communication, whereby the bandwidth shall be divided into two portions: one for time constrained traffic on a reservation basis; and the other for connectionless non-time-constrained traffic on a contention basis, basically for control information. Each base station co-ordinates and administers channel access and bandwidth reservation in a cell.

Mobile RSVP extends the functionality of PATH and RESV message to also perform passive reservation of resources and introduces two major new messages ACTIVATE and UPDATE to activate passive reservations and to update previous reservations in case of service degradation.

5. PROTOCOL OPERATION

We assume that a mobility binding already exists to create an association between a home agent and the corresponding care-of address, along with the remaining lifetime of the association. Thus, datagrams can be tunnelled to the foreign agent and subsequently forwarded to the mobile node.

For simplicity, we consider a simplex unicast communication link between sender and receiver.

5.1 Mobile RSVP Connection Set-up

To avoid reservation of resources in the triangle route, a four-way handshake is proposed:

1. A traffic source that aims to initiate a Mobile RSVP session sends a *CONreq* message.
2. A receiver willing to obtain (multimedia) data from the traffic source responds via a *CONconf* message. The latter message contains the current care-of address of the mobile node as a parameter. Hence, the correspondent node is informed of the mobile node's current location. A *CONrej* is generated in case the receiver is not able or willing to be involved in a communication session with the traffic source.

3. The sender sends a PATH message directly to the mobile node's care-of address, which travels hop-by-hop (downstream) through subsequent RSVP routers installing (passive) QoS state in each router along the path.
4. Mobile node analyses the obtained PATH messages and generates a RESV message accordingly, which passes the routers in the reverse order (upstream). Each RSVP router has to evaluate the RESV message and make appropriate reservation.

5.2 Progressive Resource Registration

Since the mobile node is roaming, each foreign agent has to make sure that resources are available as the relevant mobile node moves to the neighbouring networks. Therefore, a modified reservation model has to be performed, which is referred to as *Progressive Resource Registration*.

Based on the latter reservation model, a foreign agent acts as a traffic source and distributes PATH messages to the neighbouring base stations. PATH messages advise the neighbouring base stations of the characteristics of the sender traffic and install path state in corresponding nodes. If R_{max} is the maximum resource contingent that can be allocated in a router (based on the *Fluid Model* and the corresponding *Leaky-Bucket* mechanism discussed in [8]) at a particular point of time and R_a is the resource requirement announced by the previous (upstream) router, then resources can be allocated only if $R_a \leq R_{max}$. We can distinguish between three different cases:

1. $R_a \leq R_{max}$: The corresponding request is forwarded to the next hop.
2. $R_a > R_{max}$: R'_a is calculated based on available resources. PATH is forwarded to the next hop using R'_a as the modified parameter.
3. The request can not be registered: Using a *REJECT* message, the PATH originating node (current foreign agent) is notified that the requested reservation can not be registered. The complete branch is then considered as a best-effort connection.

Each base station itself has finally to analyse the reservation request and either confirm or reject the request. In case of a confirmation, RESV messages are sent to the current foreign agent from each individual neighbouring base station (Figure 3a). Every RESV message makes appropriate passive reservation along its corresponding branch.

A reservation requires an entry in a database, such as $\langle SessionID, R_a\bullet\ , t_{life}\rangle$, where t_{life} indicates the lifetime of the registration, after which it will be deleted. The node, which has initiated the reservation, is automatically assigned with the highest priority for his required resources. On one hand, this prioritisation ensures the mobile node that its reserved resources are available as soon as it enters the corresponding cell. On the other hand, it enables other mobile nodes to consume reserved but still unused (passive) resources. It is obvious that registered resources have to become available as soon as the mobile node with the highest priority requests them. Thus, the resource allocation is pre-emptive. A foreign agent, who wishes to maintain the registration of its reservation request, has to refresh t_{life}, periodically. The introduction of t_{life} will prevent the blocking of other requests being registered.

The potential reservation state of each individual path to the neighbouring base station along with the corresponding branch identification is maintained by the foreign agent. This information will be of use during hand-over.

5.3 Hand-over Procedure in Mobile RSVP

The hand-over procedure of Mobile RSVP is based on the Mobile IP protocol. It assumes that a mobile node is successfully handed over to the adjacent base station and that the home agent has been notified, accordingly.

5.3.1 Activating Passive Reservation in the New Environment

After the mobile node has changed its point of attachment, the previous foreign agent has also to hand-over QoS requirements to the new foreign agent.

According to the Mobile IP protocol, the former foreign agent has to keep record of the new care-of address of the mobile node for those packets that have been sent to the old care-of address while the mobile node was changing its point of attachment. keeps this information until an appropriate link is established between the correspondent node and the mobile node. Mobile RSVP also suggests activating adequate resources between two neighbouring base stations:

To change the state of a reservation from passive to active, the former foreign agent FA_1 in the cell C_1 (Figure 3b) has to send an ACTIVATE message to the corresponding agent in the neighbouring cell C_2 (FA_2) along the same path, which has been established during the passive reservation. The correct path is identified through the relevant session identification. In case of a service degradation, which is determined by comparing the reservation state of the selected path (path to FA_2) with the current service specification, FA_1 has to inform all previous nodes involved in the session (including the traffic sender) and update their reservation states accordingly. This action is performed using an *UPDATE message* sent in the reverse path from the current foreign agent to the sender.

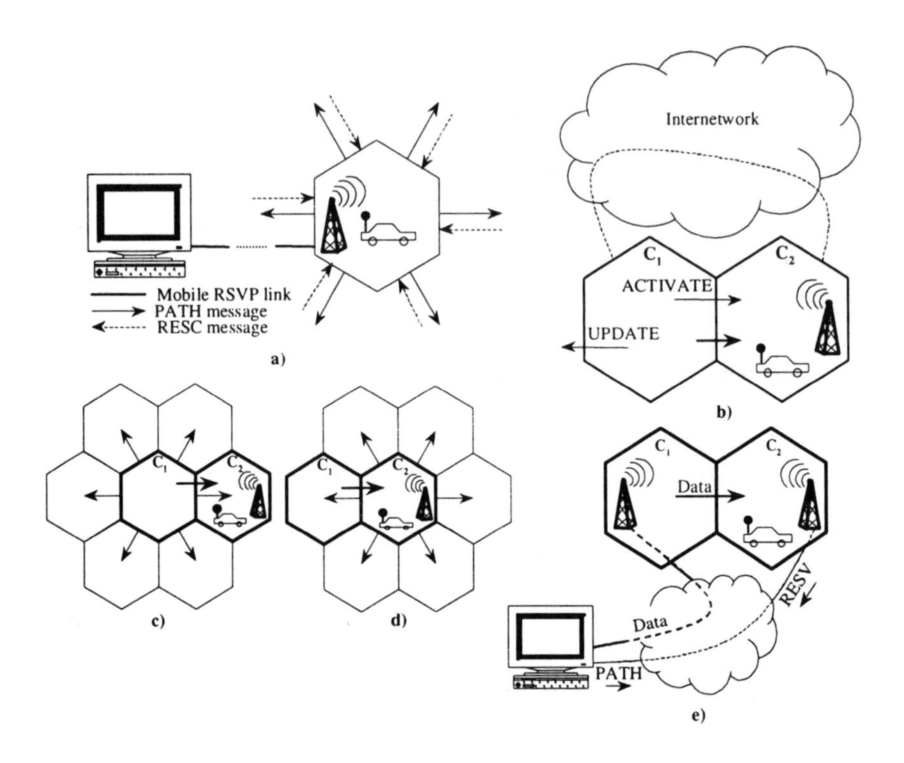

Figure 3: Operation of Mobile RSVP.

5.3.2 Release and Record Resource Reservations

When a mobile node moves to a new point of attachment, passive reservations in the old neighbourhood become obsolete. Hence, either they time out or they have to be terminated or "released". For this reason, the old foreign agent distributes a *RELEASE* message to its neighbourhood (Figure 3c) informing them that the relevant registration is not valid any longer.

On the other hand, the new foreign agent is responsible to pre-advice (PATH message) its neighbourhood with regard to the arriving mobile node (Figure 3d). This will again initialise the Progressive Resource Registration and generate reservation requests.

5.3.3 New Mobile RSVP Link to the Current Care-of Address

During the process of roaming, a mobile node has to ensure that at least two databases are updated. Firstly, a database maintained by the home agent, which contains mobility bindings of native nodes that are currently travelling (*⟨StaticIP, care_of address, lifetime⟩*) and secondly, a database maintained by correspondent nodes reflecting the mobile node's current point of attachment. When a correspondent node capable of providing Mobile RSVP services receives an updated mobility binding, it automatically initiates a new Mobile RSVP link to mobile node's new point of attachment by sending a PATH message to the mobile node (Figure 3e). However, establishing a new RSVP link is a matter of trade-off and has to be decided by the RSVP initiating entity.

At the same time, multimedia datagrams directed towards the mobile node are forwarded through the old foreign agent FA_1 to the new destination FA_2 of the mobile node along the Mobile RSVP connection between FA_1 and FA_2. Upon receiving the PATH message, FA_2 initiates a reservation by sending a correspondent RESV message. When the correspondent node receives the RESV message it automatically updates the mobility binding and starts sending datagrams to the FA_2, probably at a modified service quality. This also shows the flexibility of Mobile RSVP to adjust to variable network conditions.

The so-called "post-hand-over" procedure is realised in the Mobile RSVP layer of the corresponding protocol stack.

It is the responsibility of the Mobile RSVP instance to temporarily store updated mobility bindings until a new communication link with guaranteed QoS becomes available. Upon termination of a session, which is initiated by an *END* message, corresponding registrations are released based on the instruction of the relevant foreign agent.

5.4 Protocol Analysis

In this section, we try to calculate the response time of call set-up, connection hand-over and resource registration for a unicast mobile connection between a mobile node and a correspondent node. We also try to assess the amount of control traffic generated by Mobile RSVP.

The following parameters are used:

H_a	*Avg. number of hops along an active Mobile RSVP connection*
H_{ch}	*Avg. number of hops between home agent and correspondent node*
H_{hf}	*Avg. number of hops between home agent and foreign agent*
H_{cell}	*Avg. number of hops between two foreign agents in neighbouring cells*
$N_{neighbour}$	*Number of neighbouring cells*
τ_w	*Avg. transmission delay over a wired link*
τ_{wl}	*Avg. transmission delay over a wireless link*
τ_r	*Avg. routing delay in a router or base station*
τ_{rd}	*Avg. reservation processing delay in a router or base station*
τ_{caps}	*Avg. encapsulation or decapsulation delay of datagrams*

5.4.1 Response Time during Mobile RSVP connection set-up

Let t_s be the time it takes to establish a mobile connection, measured from the moment the mobile host makes the request until its acceptance. It is given by $t_s = t_1 + 4\tau_{wl} + H_a(3\tau_w + 2\tau_r + \tau_{rd})$, whereby $t_1 = 2\tau_{caps} + \tau_w\tau_r(H_{ch} + H_{hf})$, which includes the time for encapsulation and decapsulation of the CONreq message at both ends of the tunnel and the time to transmit and route the initial message from the correspondent node to the mobile node's current point of attachment.

5.4.2 Response Time during Progressive Resource Registration

Let t_p be the time it takes to process a Progressive Resource Registration measured from the moment the current foreign agent initiates the event until its eventual establishment. It is given by $t_p = N_{neighbour} H_{cell}(2\tau_w + \tau_r + \tau_{rd})$.

5.4.3 Response Time during Mobile RSVP hand-over[1]

Let t_h express the hand-over latency. If the mobile node is transmitting or receiving when it crosses cell boundary, then t_h is measured from the moment the mobile node greets the new foreign agent until it can resume transmission or the switchover occurs at the crossover node. It is given by $t_h = \tau_w(\tau_{rd}H_a + H_{cell}) + t_n$, where t_n refers to the time it takes to establish a new Mobile RSVP link to the mobile node's new point of attachment, to release registered resources and to perform appropriate Progressive Resource Registration given by $t_n = t_p + H_{cell}\tau_w(N_{neighbour} - 1) + t_{new}$, whereby $t_{new} = H_a(2\tau_w + \tau_r + \tau_{rd}) + 2t_{wl}$ and refers to the time to establish a new Mobile RSVP link from the correspondent node to mobile node's new point of attachment.

5.4.4 Protocol Overhead and Control Traffic

Classified in Table 1 are the protocol messages (including acknowledgements) generated in the events of connection establishment, Mobile RSVP hand-over and connection termination. In general, the amount

[1] It should be noted that the response time of the Mobile IP handoff procedure is not included in our calculation.

of control traffic is linearly proportional to the number of neighbouring cells $N_{neighbour}$.

Our proposed protocol generates even less control traffic when the reservation is initiated by the mobile node, i.e. the mobile node sends a PATH message to the corespondent node, provided the correspondent node is connected to a fixed ("wired") network. The a-priori knowledge of the correspondent node's address makes initial set-up messages CONreq and CONconf superfluous, since the mobile node does not need to go through the home agent to reach a correspondent node. In this case, the total number of control messages will reduce by four.

	MN↔FA	**FA↔fixed network**	**FA↔FA**	**Total Number**
Connection set-up	1 × CONreq 1 × CONconf 1 × PATH 1 × RESV	1 × CONreq 1 × CONconf 1 × PATH 1 × RESV	$N_{neighbor}$ × (PATH + RESV)	$2N_{neighbor} + 8$
Mobile RSVP hand-over		1 × UPDATE	1 × ACTIVATE $N_{neighbor}$ × RELEASE $N_{neighbor}$ × (PATH + RESV)	$3N_{neighbor} + 2$
Post hand-over	1 × PATH 1 × RESV	1 × PATH 1 × RESV	$N_{neighbor}$ × (PATH + RESV)	$2N_{neighbor} + 4$
Connection termination	1 × END	1 × END	$N_{neighbor}$ × RELEASE	$N_{neighbor} + 2$

Table 1: Control messages in the Mobile RSVP

5.5 Performance Optimisation of Mobile RSVP

The main goal of performance optimisation in Mobile RSVP is to reduce control traffic and protocol overhead. As we have seen, the amount of control traffic increases linearly proportional to the number of neighbouring cells $N_{neighbour}$. Hence, one major source of improvement is to reduce the number of communicating neighbours. One may take the mobility behaviour and the trajectories of mobile node into account. For example, it is quite unlikely that a mobile node return to the cell that it just left. A foreign agent can also determine the direction of motion of mobile nodes from the sequence PATH messages associated with it (Figure 4a).

This information can be used by the foreign agent to decide whether to proceed Progressive Reservation Registrations, or it can be used by the network to adapt algorithms such as the *Adaptive Prioritised Link Partitioning* proposed by [2]. The result would be less control overhead and better network efficiency.

Another approach would be (and this applies also to the policy control modul of RSVP) the implementation of a priority mechanism that allows users to send reservation requests with higher priority than others. The *Integrated Services (IS)* may be coupled with a billing system that charges the user according to priority level of his reservation request.

An a-priori knowledge of the mobile node's mobility behaviour, where the corresponding foreign agents are pre-determined might help to reduce control traffic and improve service availability. In fact, for this kind of applications Mobile RSVP does not represent an appropriate reservation model. In such cases, it is more convenient to perform resource reservation in a timely manner, where QoS can be guaranteed starting at a particular point of time and lasting for a certain period.

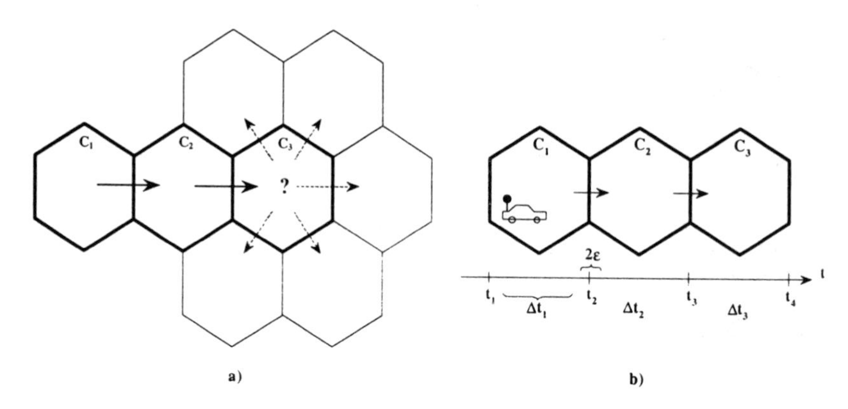

Figure 4: a) Assessing the mobility behaviour of the mobile node based on the previous motion path. b) *Resource reservation model for mobile nodes with deterministic behaviour.*

Hence, reservation has to be made in advance when the mobile node decides to leave its home network, so each foreign agent knows approximately when a mobile node will arrive in the respective cell (network) and how long it plans to stay. Foreign agents might be notified of any potential delay that might occur. This reservation model is best applicable if the mobile node is travelling for example by car or train, where the mobility route is somehow pre-determined. However, it does not provide the flexibility a mobile node might require, but is exemplifies an equitable approach for a certain class of mobility behaviour.

Figure 4b illustrates such a reservation model. In a deterministic mobility behaviour model the equation $t_{total,i} = \Delta t_i + 2\varepsilon$ should hold, where $t_{total,i}$ is total time a reservation should be maintained in the corresponding cell i for a specified mobile node and $\Delta t_i = t_{i+1} - t_i$, which is the time the mobile node will occupy cell i. ε refers to the average time required for hand-over. 2ε is required to enable a smooth hand-over and depends upon the corresponding hand-over algorithm.

5.6 Open Issues

A major problem of Mobile RSVP will be its applicability in the current network infrastructure. It is required that relevant routers of the underlying network are aware of Mobile RSVP, otherwise the control traffic will be very high and will even affect the best-effort performance. As a first approach, it is suggested that Mobile RSVP be realised in corporate Intranets to provide multimedia or other real-time data to mobile nodes.

Another pitfall of Mobile RSVP is its control traffic and protocol overhead, which increases rapidly when both communication counterparts are mobile hosts. In case the correspondent node is mobile himself, a huge amount of additional control messages is necessary, which would make Mobile RSVP quite inefficient, since the underlying network structure might get overloaded with required protocol overhead.

In this case, both involved parties (mobile hosts) have to employ progressive resource registration. This will extremely increase the complexity of a potential mobile resource reservation protocol.

In a multicast scenario, the control traffic will be even higher. Appropriate measures have to be considered to overcome this pitfall. The concept of *Differentiated Services* ([8] and [9]) might help in this respect to reduce control traffic. This approach will remain for further investigation and research.

5.7 Concluding Remarks

Motivated by the challenge of providing QoS guarantees to mobile users, an extended RSVP protocol (Mobile RSVP) has been proposed, which could support multimedia and real-time traffic in a Mobile IP based environment. Mobile RSVP is appropriate for those mobile nodes, which can not a-priori determine their mobility behaviour. In addition, the effect of a deterministic mobility specification has been analysed and proposals for more appropriate protocols have been outlined.

In addition, we have described the conceptual design of the reservation mechanism of Mobile RSVP and discussed the hand-over procedure. Efforts have been made to analyse the performance of Mobile RSVP and to assess its overhead and control traffic.

As far as future work is concerned, the focus will be at simulating the behaviour of Mobile RSVP under various network conditions based on a performance model for Mobile IP networks. It is also necessary to evaluate, experiment and propose mechanisms for assessing mobility behaviour, which would eventually lead to the definition of a mobility profile.

We have also started to investigate and research a liaison between Mobile IPv6 and RSVP, since IPv6 provides more sophisticated flow control and QoS provision mechanisms. Furthermore, mobility support concepts have already been drafted and proposed in [1].

References

[1] David B. Johnson and Charles E. Perkins, "Mobility Support in IPv6", *Internet Draft,* draft-ietf-mobileip-ipv6-03.txt, July 30, 1997.

[2] Kam Lee, "Supporting mobile multimedia in integrated services networks", *Wireless Networks,* Vol. 2, pp. 205-217, 1996.

[3] Charles E. Perkins, "Mobile IP, Adding Mobility to the Internet", *MobiCom'97, The Third Annual ACM/IEEE International Conference on Mobile Computing and Networking,* September 26, 1997.

[4] Anup K. Talukdar, B. R. Badrinath and Arup Acharya, "On Accomodating Mobile Hosts in an Integrated Services Packet Network", *IEEE Proceedings of the Infocom 97,* April 1997, pp. 1064-1053.

[5] Paul P. White, "RSVP and Integrated Services in the Internet: A Tutorial", *IEEE Communications Magazine,* May 1997, pp. 100-106.

[6] Charles E. Perkins and David B. Johnson, "Route Optimization in Mobile IP", *Internet Draft,* available at draft-ietf-mobileip-optim-06.txt, July 29, 1997.

[7] R. Braden, L. Zhang, S. Berson, S. Herzog and S. Jamin, "Resource ReSerVation Protocol (RSVP), Version 1 Functional Specification", *RFC 2205* , available at ftp://ds.internic.net/rfc/rfc2205.txt.

[8] C. Metz, "IP QoS: Travelling in First Class on the Internet", *Internet Computing,* Vol. 3, Nr. 2, pp. 84-88, March/April 1999.

[9] S. Blake et al., "An Architecture for Differentiated Services", *RFC 2475,* available at ftp://ds.internic.net/rfc/rfc2475.txt.

THE EFFECT OF ARQ BLOCK SIZE ON THE EFFICIENCY OF A WIRELESS ACCESS LINK USING ADAPTIVE MODULATION

Fraser Cameron and Moshe Zukerman
Department of Electrical and Electronic Engineering
The University of Melbourne, Parkville, Victoria 3052, Australia
Email: fkc@ee.mu.oz.au

Abstract This paper optimizes a comprehensive set of parameters with the aim to minimize the bandwidth required for wireless transmission on a single link investigating the effect of ARQ blocksize on the adaptive link. The paper takes into consideration a comprehensive set of coding, modulation and teletraffic issues including the effects of retransmissions of erroneous packets using ARQ; modulation efficiency; FEC redundancy; and effect of traffic burstiness on loss and delay. Simulations are used to derive BER values for a range of modulation efficiency and FEC Code Rate parameters. These parameters are optimized to obtain maximal efficiency when looking at TCP over wireless multi-service media. We show that an adaptive system provides a relatively insensitive response to variation in transmitted block size with advantages.

Keywords: Wireless networks, FEC, ARQ, Adaptive modulation

1. INTRODUCTION

Personal and terminal mobility requires better area coverage, improved signal quality and reliability as well as the ability to interface with wired multimedia networks based on TCP/IP and/or ATM. According to current research and developments trends, the future Internet will support provision of specified Quality of Service (QoS) to customers which can be measured and guaranteed [6]. Unlike wired optical fibre networks where QoS requirements are mainly related to cell or packet loss probability due to congestion, in wireless broadband networks we must also consider access delay as an important QoS requirement affecting network dimensioning, and traffic control mechanisms.

To overcome error in a transparent mode of transmission, it is possible to use Forward Error Correction (FEC) and/or to reduce the number of symbols in the transmitted constellation in order to improve BER. Both of these solutions result in decreased channel capacity. In [2] and [3] it was shown that adaptive modulation is a powerful tool for performance optimisation – the size of the constellation is an essential part of data services in wireless networks and the choice of that constellation has a high impact on the BER. In this paper it will be shown that adaptive modulation is a powerful tool in reducing wireless link sensitivity to ARQ block size which is an important result that leads to increased utilisation of the wireless link.

The basic idea behind an adaptive modulation scheme is to adjust the method of transmission to ensure that the maximum efficiency is achieved over the link at all time. For comments on implementability of adaptive modulation and the use of this research to GSM and other mobile networks see [3]. In general, a modulator accepts M binary digits at a time and transmits a single symbol. This provides Modulation Gain. We will henceforth use the term Modulation Gain, denoted G, to represent the increase in capacity (bits/sec) using given modulation constellation relative to that of binary modulation. That is, G is given by $G\ p= \log_2 (M)$ where M is the number of symbols in the modulation constellation. However, the penalty for such gain is an increase in BER and with it, an increase in the required number of retransmissions by the higher layer protocols (link layer ARQ or end-to-end ARQ). There is a clear trade-off between the block size of ARQ retransmissions will have a clear impact on the amount of redundant data that must be retransmitted on error, in which the block size of the ARQ plays an important role [7].

2. THE MODEL

The basic idea of the analytical model relies on assumptions of separation between Queuing Based Efficiency (QBE) and Error Based Efficiency (EBE) (see particularly [3] for a fuller explanation). The overall efficiency, or simply efficiency, is the ratio between the mean user bit rate generated by the customer and the required channel capacity: it is the product of the QBE and the EBE. In this paper we focus on the effect of block size on EBE. For a discussion of QBE, the reader is referred to [2], which shows how the values of QBE were obtained, and for a discussion of EBE optimisation not taking into account the effect of block size the reader is referred to [3].

Briefly, the EBE is the ratio between the mean user data rate and the user data rate plus all overheads associated with error corrections and retransmission of erroneous blocks of data, namely, FEC, and ARQ and adjusted for the Modulation Gain. Accordingly the EBE is a product of (1) the FEC code rate R defined by the ratio: R = (user data) / (user data + FEC redundancy); (2) the ARQ efficiency denoted A and defined by the ratio: A = (user data + FEC redundancy) / (user data + FEC redundancy + additional capacity for ARQ retransmissions); and (3) the above defined Modulation Gain (G).

In [3] an analytical formula for the *Optimal EBE* denoted EBE^* and defined as the maximal value of EBE subject to meeting QoS requirements was derived for an adaptive wireless multi-service system. We give the formula here for use in this paper but we refer the reader to [3] for derivation. The equation optimises values of G and R for a given ARQ block size and a particular FEC scheme. If N is the ARQ block size in bits, then the probability that an ARQ block will be transmitted successfully is denoted by α and is given by:

$$\alpha = (1 - e(G,R))^N \tag{1}$$

Given that α is a function of G and R the EBE is given by and and the utilization of the adaptive system is optimised by:

$$EBE^* = \max_{G,R}[G \times R \times \alpha] \tag{2}$$

We must optimize G and R, considering the trade-off between the FEC redundancy and Modulation Gain versus the volume of retransmission of erroneous blocks of data (which will be affected by the blocksize, N). We conservatively assume independent bit errors.

Notice that the above does not consider packet delay due to ARQ retransmission. Nevertheless, in most wireless or mobile systems the propagation delay over the radio link is insignificant and hence a large number of retransmissions are possible without violating delay QoS requirements. However, when the radio link is long, eg for a satellite, or when ARQ is performed end to end at the TCP level over large distances, a constraint on the delay must be added to the optimisation of Eq. (2).

2.1 Considering Block size, *N*

From Eq.(1), it is clear as that BER increases, there will be a rapid fall in α and thus a fall in efficiency for given G and R as $e(G,R)$ for the fixed G and R becomes significant (see Eq. (2)). A larger N will make this decline occur at a higher SNR (or $e(G,R)$) on the link, and increase the rate of decline of EBE against SNR than for a smaller N. In the non-adaptive system, this decline is not correctable and leads to the much discussed TCP 'slow start' problem [5]. That problem identified by authors concerning ATM over wireless links is that the window protocol mechanisms built into TCP are designed to avoid network congestion and do not take account of characteristics of the wireless environment (which was not a consideration in the design of TCP). Hence cell loss on any link of the network – which TCP assumes is wired, having low BER – must be as the result of network congestion rather than an errored packet.

One solution, considered in [9] suggests the implementation of an additional small block ARQ in a specific Wireless Data Link Control Layer (rather than end-to-end ARQ) to deliver a better quality channel over the wireless link to the higher layer protocols. This implementation of small blocks at the data link layer provides error free transmission to higher layers thus avoiding inefficiencies in retransmission of TCP larger blocks. In this paper, we find that in our adaptive modulation model, such a scheme may not be necessary.

In the case where N is variable, as for TCP over the wireless link, by Eq. (2), it is possible to obtain a condition for when G and R need to be changed from their origional values, G_1 and R_1 to new optimal values as a result of a change of N from $N(1)$ to $N(2)$.

$$G_1 \times R_1 \times (1 - e(G_1, R_1))^{N(1)} < \max_{G,R}[G \times R \times \alpha_{N(2)}] \quad (3)$$

Eq. (3) gives this condition for a change of G or R in the adaptive system. Where the value of N is variable, an increase in the value of N will lead to a reassessment of optimal values of G and R. In the case of a fixed modulation scheme, the optimization is only available over R or not at all meaning that EBE will decline rapidly as SNR increases, leading to a smaller N imposed by TCP/IP and a large loss in user utilisation. The adaptive scheme on the other hand is able to alter R and G to maintain service levels.

Since there is no analytical solution for the BER for a given set of G and R, we have to rely on simulations where we commence with the optimisation of Modulation Gain and FEC and then consider blocksize N for the optimized adaptive system. The details of the simulations are discussed in [3].

3. THE SIMULATION

Simulations used in this paper were developed using the Matlab Communications Toolbox [6]. The wireless environment was simulated using a Varying Rayleigh Fading Channel (VRFC) in which the parameters of the Rayleigh Fading Channel were varied randomly with uniform distribution. The simulation uses Minimum Phase Shift Keying (MPSK) which was preferred for simplicity, however the analysis presented is equally relevant for High Level Modulation including Enhanced Data rates for GSM Evolution (EDGE) [4] which uses Quaternary Offset QAM (Q-O-QAM). This approximation is justified since the analysis is independent of modulation scheme and the preferred modulation scheme for future wireless (or WATM) networks has not yet emerged.

4. RESULTS

In this section we present numerical results displaying the relative insensitivity of our scheme to block size of the transmission, effectively providing a solution to concerns regarding the performance of TCP over WATM and other wireless multi-media systems.

Since Go Back N requires end-to-end (not just wireless link) retransmission of all subsequent packets to the errored packet, and there is a sliding window, larger window size is equivalent to the retransmission of larger ARQ block sizes. Figure 1 demonstrates this argument. In this case we consider EBE with varying window size. As *SNR* decreases and $e(R,G)$ increases, the protocol reduces TCP window size, and hence throughput.

Figure 1 shows firstly the EBE gain in lower SNRs for the fixed modulation case achieved as a result of reduction of window size from 53000 bytes to 530 bytes. It is clear that the utilisation curve for the non-adaptive case with smaller window goes rapidly to zero at a lower SNR, showing that a reduction in window size gives an increased EBE in cases of higher $e(R,G)$. Thus, although TCP reduces the window size assuming network congestion, it is also carrying out the correct EBE action for wireless cell loss. We note that our maximum window size is large, however, for an inter-continental link, this is a realistic size window for a bit rate of 20Mbs.

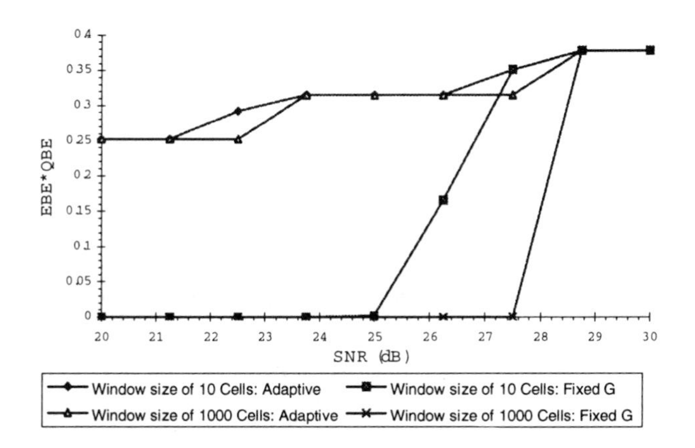

Figure 1. Adaptive and fixed modulation for uncoded MPSK system with ARQ. TCP window size equal to 530 and 53000 bytes. The fixed case is $G = 6$. Overall Transmission Efficiency (EBE and QBE) is shown.

Secondly, Figure 1 shows that, for the case of adaptive modulation, the action of the transmitter is to move from one constellation to another when error becomes significant using the rule in Eq. (3). In this case, delay increases as the system comes closer to a transition point, since error and hence retransmission increases at this point. TCP decreases window size until the transmitter chooses a modulation constellation with greater redundancy, lower efficiency and consequential lower packet error.

The TCP window can then increase in response to the new conditions. Hence, TCP acts to 'cushion' the transfer between MEs. Note that as the TCP window size becomes small, there is a larger delay penalty in the overall utilisation that is not captured by the EBE (especially over long end-to-end distances), however when this becomes significant, the EBE in the larger window size is likely to be so poor as to render the smaller window size nonetheless more efficient.

In the case, as suggested in [7] that a data link layer small block ARQ is used exclusively on the wireless link in order to make the wireless link appear to the higher TCP/IP layer as if it is error free, Figure 2 compared with Figure 3 show that the adaptive system is relatively insensitive to blocksize (compared with the fixed system in Figure 1. Hence another advantage of the adaptive scheme is that any block size ARQ can be used in the wireless ARQ to provide good efficiency.

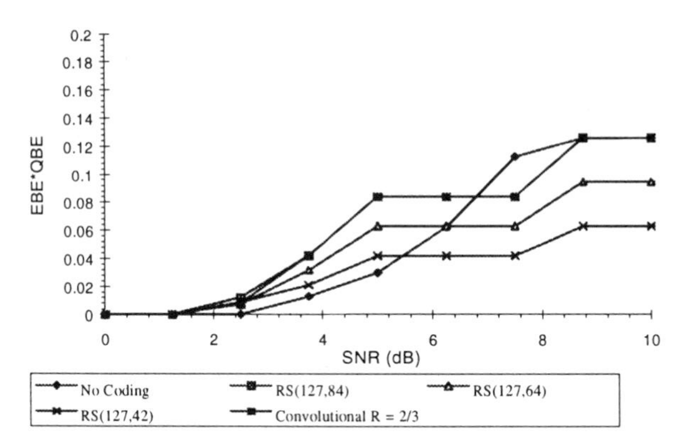

Figure 2. Efficiency comparison amongst several coding schemes; ARQ block size = one ATM cell (53 bytes) at low SNRs.

In [3] we find that optimal EBE is given by the adaptive modulation system in the absence of coding for most SNRs. In Figure 1 this is the case – changes to the system are made exclusively to G and the maximum efficiency is given in the uncoded case.

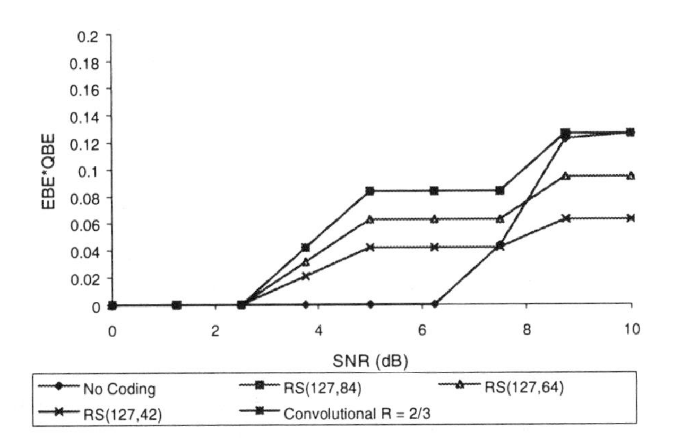

Figure 3. Efficiency comparison amongst coding schemes – ARQ block size=100 cells (5300 bytes) at low SNRs.

In Figures 2 and 3 we show that for very low SNR values, changes in coding are required to achieve optimal performance, particularly for larger block sizes. We note however that, even in this situation, the higher rate codes still give better performance than the more powerful codes with greater redundancy.

5. CONCLUSION

This paper considers the tradeoff of providing QoS for WATM at both the physical (modulation) and data link level (ARQ and FEC). It also considers the effect of the introduction of lower level ARQ (AAL). We show that the use of FEC with associated delay and complexity constraints can be avoided without violating the QoS bound. Our results show that that adaptive modulation offers significant efficiency and coverage gains over the fixed system when ARQ is considered and that in many instances an adaptive modulation system renders FEC redundant. We leave the implementation of the adaptive change of modulation parameters for another paper, however our findings suggest that preference should be given to systems that adaptively change their parameters according to the instantaneous level of interference and noise. We further observe that (1) the TCP assumption of a wired network, and the 'slow start' problem in particular, may not be a difficulty in wireless networks (2) that reduction in TCP window size has a beneficial effect in wireless transmission and (3) that even for certain realistic SNR conditions, the wastage due to errors can be very high because of retransmission.

References

[1] H. ArmbrÜster, "The Flexibility of ATM: Supporting Future Multimedia and Mobile Communications," *IEEE Personal Comm.*, pp.76-84, Feb. 1995.

[2] F. Cameron, M. Zukerman and M. Gitlits, "Wireless Link Dimensioning and Transmission Parameters Optimization," in K. Leung and B. Vojcic (eds.) *Multiaccess, Mobility and Teletraffic for Wireless Communications: Volume 3*, pp. 295-308, Kluwer Academic Publishers, Boston,1999.

[3] F. Cameron, M. Zukerman, and M. Gitlits, "Adaptive Transmission Parameters Optimisation in Wireless Multi-access Communication", *ICON '99 (Forthcoming)*

[4] Ericsson, "EDGE Feasibility Study Work Item 184; Improved Data Rates through Optimised Modulation (Preliminary version 0.1)", SMG #22, Munich, Germany, 12-16 May 1997.

[5] J. Cain and D. McGregor, "A recommended error control architecture for ATM networks with wireless links", (1997) 15(1) *IEEE J Select Areas Comm* 16.

[6] The Math Works Inc., Communications Tool Box for use with MATLAB and SIMULINK, April 1996.

[7] D. Raychaudhuri et al., "WATMnet: a prototype wireless ATM system for multimedia personal communication," *IEEE J. Select. Areas Commun.* vol. 15, no. 1, pp. 83-94, January 1997.

[8] M. Schwartz, "Network Management and Control Issues in Multimedia Wireless Networks," *IEEE Personal Comm.*, pp. 8-16, June 1995.

[9] D. Turina, "Performance Evaluation of a Single-Slot Packet Data Channel in GSM", *IEEE VTC*, pp544-548, 1997

[10] M. Zukerman, P. L. Hiew and M. Gitlits, "FEC code rate and bandwidth optimisation in WATM networks", in D. Everitt and M. Rumsewicz (eds.), Multiaccess, Mobility and Teletraffic: Advances in Wireless Networks, Kluwer, Boston, 1998.

[11] M. Zukerman, P. L. Hiew and M. Gitlits, "Teletraffic Implications of a Generic ATM Wireless Access Protocol", IEEE Globecom '97.

"SUPER-FAST" ESTIMATION OF CELL LOSS RATE AND CELL DELAY PROBABILITY OF ATM SWITCHES

Junjie Wang, K. Ben Letaief, and M. Hamdi *

The Hong Kong University of Science and Technology
Clear Water Bay, Hong Kong

Abstract In this paper, we consider the evaluation of the cell loss rate (CLR) and cell delay probability (CDP) in nonblocking ATM switches using computer simulations. Specifically, we investigate the application of *importance sampling* techniques as a "super-fast" alternative to conventional Monte Carlo simulation in finding the CLR and CDP in nonblocking ATM switches. A novel "*split switch*" model is developed to decouple the input and output queueing behavior so as to reduce the simulation complexity. Numerical results demonstrate that considerable computation cost can be saved using importance sampling techniques while achieving a high degree of accuracy.

Keywords: ATM Switch, Performance Evaluation, Importance Sampling, Cell Loss Rate, Cell Delay Probability.

1. INTRODUCTION

In this paper, we consider the performance of ATM switches with respect to cell loss rate (CLR) and cell delay probability (CDP, i.e., the probability distribution of the cell delay). These are clearly some of the important issues in the switch design and a sizable amount of work has been done on the performance evaluation of ATM switches with regard to these QoS parameters [1]-[3]. Due to the complex traffic model, it turns out that close-form solution is difficult to achieve, if not impossible. Alternatively, researchers resort to simulation-based methodologies. Unfortunately, the required CLR for a typical ATM switch is smaller than 10^{-6} for most practical applications. Likewise, some real-time traffic requires a delay threshold probability (i.e., the probability

*This research work was supported in part by the Hong Kong Research Grant Council under the Grant RGC/HKUST 100/92E.

that a cell experiences a delay in excess of a particular threshold) is also around 10^{-6} or even smaller [7]. Hence, a prohibitive amount of time is needed in conventional Monte Carlo (MC) simulations to obtain an estimate of these rare events for a particular accuracy requirement.

Importance Sampling (IS), as a promising technique, can significantly reduce the simulation time required to obtain accurate estimates [4]-[6]. In this paper, we consider the application of IS to the estimation of the CLR and CDP in nonblocking ATM switches. Note that the application of IS to the estimation of the CLR in ATM switches has been considered in [3; 6; 8]. However, in these studies, the switch model was a space-division ATM switch with *only* output queues. In this paper, we consider the more practical case of ATM switches with both input and output queues, which makes the analysis more complicated.

The rest of the paper is organized as follows. Section 2 gives a brief introduction of the concept of IS for the estimation of rare events. In Section 3, we propose a notion of "*split switch*" model, where we divide the performance estimation into two sub-problems which deal with the input and output queueing respectively, some IS biasing schemes are developed. Section 4 includes some simulation results illustrating the accuracy and efficiency of the proposed IS schemes. Finally, we conclude in Section 5.

2. IMPORTANCE SAMPLING OF RARE EVENTS

Using conventional MC simulation, the estimator for α is simply the sample mean estimator based on a sequence of *i.i.d.* samples $X^{(1)}, ..., X^{(L)}$ from the density $f(.)$. That is,

$$\hat{\alpha} = \frac{1}{L} \sum_{\ell=1}^{L} I_E(X^{(\ell)}). \tag{1}$$

where I_E is an indicator function of event E. Practically, $100/\alpha$ samples is required to to obtain a reliable estimate within a 10% accuracy. Hence, the estimation of probabilities of the order of $10^{-6} - 10^{-9}$ (which are quite common in communication networks) would be difficult to achieve because of the prohibitive simulation time.

Importance sampling (IS) involves choosing $f^*(.)$ as the simulation density such that $f^*(x) > 0$ whenever $f(x) > 0$ [4]. The IS estimator is then given by

$$\hat{\alpha}^* = \frac{1}{L} \sum_{\ell=1}^{L} I_E(X^{(\ell)}) w(X^{(\ell)}) \tag{2}$$

where $X^{(1)}, ..., X^{(L)}$ are now *i.i.d* samples from the IS simulation density $f^*(.)$ and $w(.)$ is called the *importance sampling weight*, which is defined as the ratio of the true density $f(.)$ to $f^*(.)$.

A close observation of (2) indicates that IS is completely dependent on the selection of $f^*(.)$. Then, it is easily shown that $\texttt{Var}^*[\hat{\alpha}^*] = \texttt{Var}^*[I_E(X)\,w(X)]/L$. Let $L_\zeta \triangleq \min\{L : \frac{\sqrt{\texttt{Var}^*[\hat{\alpha}^*]}}{\alpha} \leq \zeta\}$. Then, L_ζ is the minimum required number of IS runs to obtain a $100 \times \zeta\%$ accuracy. To minimize N_α or equivalently maximize the computational efficiency, we need to minimize $\texttt{var}^*[I_E(X)\,w(X)]$. The optimal solution $f_o^*(.)$ is given by $f_o^*(z) = I_E(x)\,f(x)/\alpha$. This optimal solution is quite general and results in an IS estimator with a zero variance [8]! Unfortunately, it is not practical since it involves the unknown estimate. However, it is useful since it indicates the features of good IS densities. For example, it implies that good IS densities which achieve high efficiencies should be biased in a way that "favors" the "important" or rare events of interest to occur more frequently. Thus, the fundamental problem in any IS scheme is to find a suitable IS method which can reduce the variance of the IS estimator and hence the number of runs to achieve a given accuracy.

3. FAST ESTIMATION OF CLR AND CDP

In this section, we consider the application of IS to the estimation of CLR and CDP in nonblocking ATM switches as shown in Fig. 1. The dimension of the switch is N, and the capacity of the input and output buffers is K and L, respectively. The switch has a speed-up factor m. Hence, up to m HOL cells can be selected simultaneously in an output conflict. ATM cell arrivals on the N input ports are governed by *i.i.d.* Bernoulli processes with an intensity of λ.

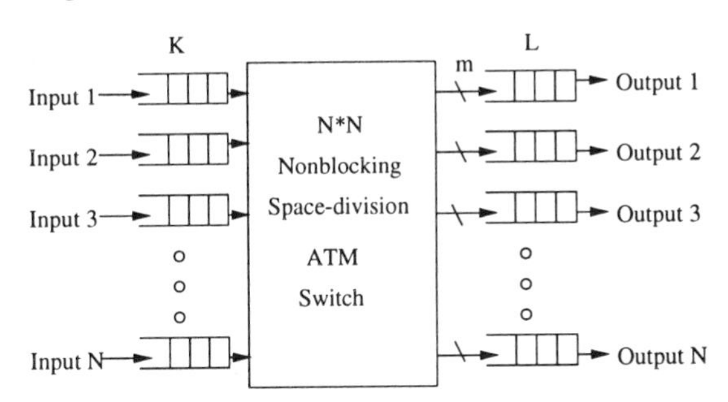

Figure 1 A Nonblocking Space-division ATM switch.

The application of IS to CLR estimation in output-queueing switches was investigated in [8], some efficient schemes were proposed. In this paper, we will extend the methods in [8] to ATM switches with both output and input queueing. Note that for ATM switches with I/O queueing, incoming cells may be lost at both the input and output queues when the cells coming to the queues find no space for them. We denote the total number of arriving cells at the switch as N_a and the number of lost cells due to input and output buffer overflow as N_i and N_o, respectively. Let γ_i and γ_o be the CLR in input and output queues, respectively. It then follows that $\gamma_i = \frac{N_i}{N_a}$ and $\gamma_o = \frac{N_o}{N_a - N_i}$. Since N_i is quite small compared with N_a, we can obtain the CLR for the whole switch γ as follows without loss of accuracy.

$$\gamma = \frac{N_o + N_i}{N_a} \approx \gamma_i + \gamma_o. \tag{3}$$

For CDP, if we denote the cell delay incurred at the input queue and output queue as d_i and d_o (measured in slot) respectively, we can then easily compose the probability distribution of the overall delay d by convolution. That is,

$$\mathtt{Pr}(d = n) = \sum_{j=0}^{n} \mathtt{Pr}(d_i = j)\,\mathtt{Pr}(d_o = n - j). \tag{4}$$

One of the key contributions in this paper is the proposal of a "*split switch*" model, which decouples the queueing behavior of the input and output buffers according to (3) and (4). Therefore, we can estimate γ_i and γ_o (or d_i and d_o) separately and then combine these results to yield the overall result. In our "*split switch*" model, two variants of input/output queuing schemes, named VIQ (*Variant of Input Queuing*) and VOQ (*Variant of Output Queuing*) as shown in Fig. 2, are developed in order to deal with the input and output queueing, respectively.

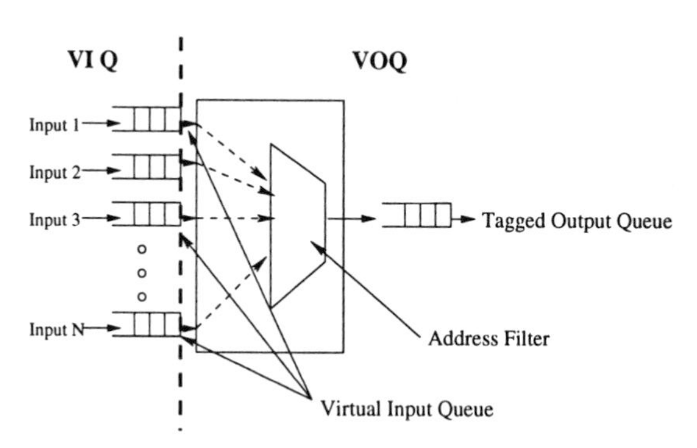

Figure 2 VIQ and VOQ schemes.

In the traditional input queuing scheme, the speed-up factor m is usually equal to 1 since the HOL blocking forms a bottleneck for the throughput of the output line. However, in order to be equivalent to the original switch model, the original speed-up factor m ($m > 1$) is kept in the VIQ scheme. On the other hand, in the traditional output queuing schemes, some cells may be lost if not selected in the output contention. In our VOQ scheme, they can stay in the HOL of the "*virtual input queue*". Specifically, the VOQ can be viewed as a combination of: (1) a group of *virtual input queues*; (2) an address filter to solve output contention; and (3) a tagged output queue. It can be seen that, the "split switch" model can precisely describe the operations of the original switch. The key advantage of this model is that CLR and CDP estimation in the original ATM switches can be divided into 2 sub-problems which are considered in VIQ and VOQ parts separately.

3.1 IS SCHEMES FOR VIQ

In the VIQ scheme, we can focus on a single input queue to obtain the CLR and CDP estimate. Suppose that I_n is the length of the tagged queue in the n-th slot, and let the number of arriving and departing cells in the n-th slot be H_n and G_n, respectively. It can be written

$$I_n = \max\{\min\{K, I_{n-1} + H_n - G_n\}, 0\} \tag{5}$$

where

$$\begin{aligned} \texttt{Pr}(H_n = 1) &= \lambda & (6) \\ \texttt{Pr}(G_n = 1) &= \min\{1, m/D_n\}. & (7) \end{aligned}$$

In the above equations, D_n is the number of HOL cells which have the same destination as the head of the tagged queue. Note that although we focus on a single input queue, there exists strong correlations between the HOL cells in parallel queues, thereby, forming an N-dimensional queueing process. It is hence difficult to get a closed form solution to such a problem without some independence approximations [1; 2].

The basic idea behind the IS scheme for VIQ is to bias the probability that the cell is selected in the output contention. That is, we make the cell in the tagged queue less likely to be selected in the output contention. Thus, it is more likely to stay in the queue to hold back the arriving cells, thereby, incurring longer delay and more loss. This is done as follows: Suppose that the head of the tagged queue is destined for output j, then all the HOL cells which are also destined for output j except the tagged one, i.e., $D_n - 1$ cells, contend for $(m-1)$ winners. After that, all the cells that failed in the first round of selection plus the tagged one contend for the last chance. As a result,

$$\texttt{Pr}^*(G_n = 1) = \begin{cases} 1, & \text{when } D_n \le m \\ \frac{1}{D_n-(m-1)}, & \text{when } D_n > m \end{cases} \tag{8}$$

3.2 IS SCHEMES FOR VOQ

Suppose A_n^j is the number of cells destined for output port j which come to the HOL of input queues in the n-th slot. Next, let C_n^j be the number of cells destined for output port j in the n-th slot. It follows

$$C_n^j = \max\left\{C_{n-1}^j - m, 0\right\} + A_n^j. \tag{9}$$

Now consider the tagged output queue and suppose that O_n^j is the length of the output queue j during the n-th slot. Likewise, let S_n^j denote the number of cells which arrive at output j during the n-th slot. Thus we have

$$O_n^j = \max\{O_{n-1}^j - 1, 0\} + S_n^j, \tag{10}$$

$$\mathtt{Pr}(S_n^j = k) = \begin{cases} \mathtt{Pr}(C_n^j = k), & \text{when } k < m \\ \sum_{l=m}^{N} \mathtt{Pr}(C_n^j = l), & \text{when } k = m \\ 0 & \text{otherwise} \end{cases} \tag{11}$$

A close observation of (9)-(11) indicates that the only randomness in the system is A_n^j, which is binomially distributed:

$$\mathtt{Pr}(A_n^j = k) = \binom{F_n}{k} (\frac{1}{N})^k (1 - \frac{1}{N})^{F_n - k} \tag{12}$$

where F_n is the total number of cells coming to the HOL of all input queues in the n-th slot, i.e., $F_n = \sum_{j=1}^{N} A_n^j$. Intuitively, we can derive the probability mass function of A_n^j in (12), then we can use some IS schemes to bias it to improve the estimation efficiency. However, note that F_n is not constant but depends on all cell sources. This makes it difficult to derive an explicit probability mass function of A_n^j, which will in turn makes the application of IS not straightforward.

The *"virtual input queue"* in the VOQ scheme is next introduced to generate traffic subject to Eqns. (9)-(12). Two biasing schemes are developed for the VOQ scheme as follows.

Accurate Biasing Scheme

In the original switch model, each incoming cell has an equal probability to be destined for any output port. That is,

$$\mathtt{Pr}(Destination = j) = \frac{1}{N}, \qquad j = 1, 2, ..., N. \tag{13}$$

To apply IS, we bias the routing probability such that the incoming cells are more likely to be destined for the tagged output queue, i.e.

$$\mathtt{Pr}^*(Destination = j) = \frac{M}{N}, \qquad j = 1, 2, ..., \frac{N}{M} \tag{14}$$

where M is defined as the *routing weight*. Hence we have

$$\mathtt{Pr}^*(A_n^j = k) = \binom{F_n}{k} (\frac{M}{N})^k (1 - \frac{M}{N})^{F_n - k}. \tag{15}$$

The scheme is called "accurate" since no approximation is made here (in contrast to the other biasing scheme we formulate below).

Approximate biasing scheme

In [1], it has been demonstrated that when the size of ATM switches, N, goes to infinity, A_n^j is subject to a Poisson distribution with intensity $\rho = \bar{F}_n/N$ ($\bar{F}_n$ is the mean of F_n). That is, we have

$$\mathtt{P_k} = \mathtt{Pr}(A_n^j = k) = \frac{\rho^k e^{-\rho}}{k!}. \tag{16}$$

Such an approximation is reasonable when a large-scale ATM switch is considered and if only a rough estimate is required. Therefore, we can apply some IS schemes developed in [8] to directly bias the arrival process A_n^j in (12). For example, we can use an exponential biasing scheme. As a result, we get:

$$\mathtt{P_k}^* = \frac{\xi^k \mathtt{P_k}}{\sum_{i=0}^{\infty} \xi^i \mathtt{P_i}} = \frac{(\xi\rho)^k e^{-\xi\rho}}{k!}, \qquad k > 0 \tag{17}$$

where ξ is the bias parameter and $\xi > 1$. The parameter ξ should then be chosen in such a way that the sample variance is as small as possible.

4. NUMERICAL RESULTS

The CLR estimates under VIQ and VOQ schemes is shown in Fig. 3 as a function of the buffer size. For comparison, we also list the MC estimates which are obtained with the original switch model.

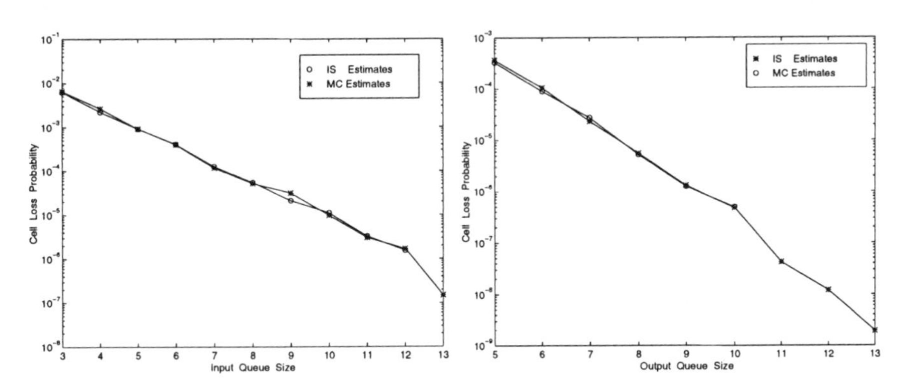

Figure 3 CLR of VIQ for $\lambda = 0.8$, $N = 16$, $m = 2$ and VOQ for $\lambda = 0.5$, $N = 16$, $K = 3$ and $m = 2$.

After obtaining the CLR in VIQ and VOQ schemes respectively, we can combine these estimates to yield the overall CLR of the ATM switches according to Eqn. (3). The simulation is run under the assumption that the total buffer size of the input and output queues is fixed at 32 so that different CLRs are observed in different allocation approaches. In Fig. 4, the overall CLR is plotted as a function of the input queue size. Thus, it is clear that more buffers should be allocated to the output queue than to the input queue in order to achieve the lowest CLR.

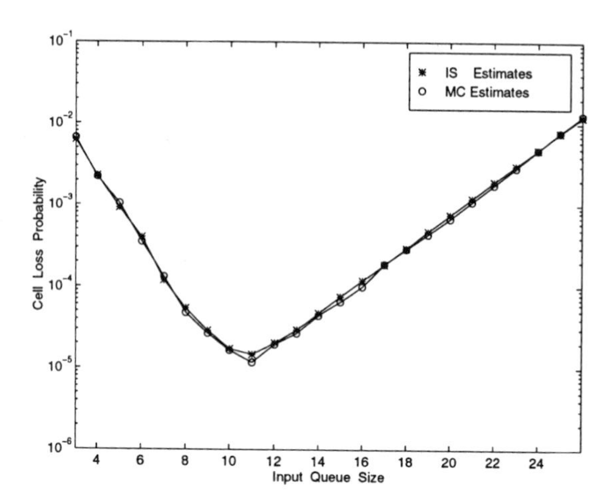

Figure 4 CLR estimation for $\lambda = 0.8$, $N = 16$ and $m = 2$.

In the CDP estimation, we use IS based on *"split switch"* model, and then combine the CDP estimates from the VIQ and VOQ schemes according to (4). The results are compared with MC simulation, as shown in Fig. 5. We denote the delay threshold as t (measured in slot) and the delay threshold probability as η. A close observation of this figure indicates an excellent agreement between the two approaches. However, the IS scheme highly relieves the computation burden. Such computational saving is illustrated in Table 1.

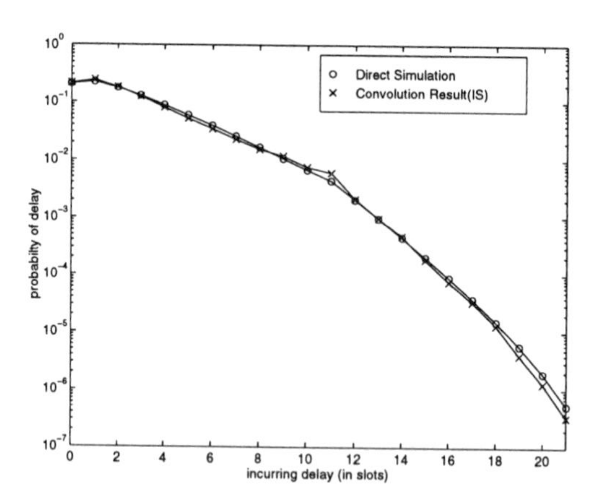

Figure 5 CDP estimates for λ=0.8, $N = 16$, $K = 8$, $L = 12$, m=2.

Table 1 The computation gains with IS in CDP estimation($N = 16, K = 8, L = 12$ and $m = 2$)

λ	t	η	CPU Time (IS)	CPU Time (MC)
0.6	13	1.12×10^{-7}	6.9 minutes	6.31 days
0.7	15	3.14×10^{-7}	4.4 minutes	2.05 days
0.8	21	2.51×10^{-7}	5.8 minutes	2.51 days

5. CONCLUSION

In this paper, we considered the application of IS to the estimation of the CLR and CDP of non-blocking ATM switches. We proposed the *"split switch"* model as an analytical tool in the performance evaluation of ATM switches with I/O queues. The IS estimates obtained using the proposed methodologies were shown to be in excellent agreement with MC simulations, which are indicative of exact system performance. In addition, it has been demonstrated that a considerable computation burden can be saved using our IS schemes which is an indication of the good potential that these IS techniques have in being used in conjunction with real-time admission control algorithms. Finally, we plan to extend these results in the future to include more realistic traffic models, and to investigate the potential of using IS techniques as a real-time method for estimating *Quality of Service* parameters in ATM networks and their integration with real-time admission control algorithms.

References

[1] M. J. Karol, M. G. Hluchuj and S. P. Morgan, "Input and output queuing on a space-division packet switch," *IEEE Trans. on Commun.* , vol. 35, pp. 1347-1356, Dec. 1987.

[2] M. J. Lee and David S. Ahn, "Cell loss analysis and design trade-offs of nonblocking ATM switches with nonuniform traffic," *IEEE/ACM Trans. on Networking* , vol. 3, No.2, pp. 199-209, Apr. 1995.

[3] Q. L. Wang and V. S. Frost, "Efficient estimate of cell loss blocking probability for ATM systems," *IEEE/ACM Trans. on Networking* , vol. 1, No. 2, pp. 230-235, Apr. 1993.

[4] P. Heidelberger, "Fast simulation of rare events in queuing and reliability models," *ACM Trans. on Modeling and Computer Simulation*, vol. 5, No.

INDEX OF CONTRIBUTORS

KEYWORD INDEX